Lecture Notes in Mathematics

Edited by A. Dold and B. Eckmann

Series: Mathematisches Institut der Universität Bonn
Adviser: F. Hirzebruch

676

T0184578

Differential Geometrical Methods in Mathematical Physics II

Proceedings, University of Bonn,
July 13–16, 1977

Edited by
K. Bleuler, H. R. Petry and A. Reetz

Springer-Verlag
Berlin Heidelberg New York 1978

Editors:
Konrad Bleuler
Axel Reetz
Herbert Rainer Petry
Institut für theoretische Kernphysik
der Universität Bonn
Nussallee 14–16
D-5300 Bonn

Library of Congress Cataloging in Publication Data
Main entry under title:

Differential geometrical methods in mathematical physics.

(Lecture notes in mathematics ; 676)
Bibliography: p.
Includes index.
1. Geometry, Differential--Congresses. 2. Mathematical physics--Congresses. I. Bleuler, Konrad, 1912- II. Reetz, Axel, 1937- III. Series: Lecture notes in mathematics (Berlin) ; 676.
QA3.L28 no. 676 [QC20.7.D52] 510'.8s [530.1'5'636]
78-12108

AMS Subject Classifications (1970): 17A30, 53AXX, 53BXX, 53CXX, 53C50, 55BXX, 55FXX, 55F25, 57D15, 57D20, 57D30, 58A10, 58F05, 70H15, 81-02, 83CXX

ISBN 3-540-08935-7 Springer-Verlag Berlin Heidelberg New York
ISBN 0-387-08935-7 Springer-Verlag New York Heidelberg Berlin

This work is subject to copyright. All rights are reserved, whether the whole or part of the material is concerned, specifically those of translation, reprinting, re-use of illustrations, broadcasting, reproduction by photocopying machine or similar means, and storage in data banks. Under § 54 of the German Copyright Law where copies are made for other than private use, a fee is payable to the publisher, the amount of the fee to be determined by agreement with the publisher.

© by Springer-Verlag Berlin Heidelberg 1978

Printing and binding: Beltz Offsetdruck, Hemsbach/Bergstr.
2141/3140-543210

P R E F A C E

Applications of modern differential geometry to theoretical physics became of greatest importance during recent years. In particular, the new gauge theories, which yield the fundamental coupling schemes between elementary particles, exhibit most interesting geometrical and topological aspects. Differential geometry plays, however, also an important role in various other branches of physics: a characteristic example constitutes geometric quantization which is not only of basic physical interest but became, at the same time, a fruitful field of pure mathematical research. In addition, geometrical viewpoints remain decisive tools in our understanding of Einstein's theory of gravitation, both in its classical form and in the more recent developments connected with the transition to quantum theory. These three topics were discussed during the last Bonn conference, and the various contributions are collected in the three corresponding chapters of this volume (comp. the index overleaf). In addition, there is an introductory chapter by S. Sternberg in which the basic geometrical structures of physical field theories are worked out explicitely.

Chapter I on gauge theories starts with a lecture of M.E. Mayer who emphasizes the role of characteristic classes in the classification of the underlying principal bundles of the physical gauge fields. The contributions of M. Carmeli and R.N. Sen deal with the same subject from different viewpoints, whereas various physical applications are discussed by T.T. Wu. On the other hand, the role of differential forms and cohomology classes in electrodynamics and gravity is investigated by W. Thirring and A.O. Barut. These arguments are enlarged in an essential way by Y. Ne'eman who introduces a gauge--theoretical reformulation of general relativity. Liftings of principal bundles to coverings of the structure group are discussed by W. Greub, whereas the non-uniqueness of such liftings is related by H.R. Petry to well-known physical phenomena in superconductivity Finally, V. de Alfaro, S. Fubini and G. Furlan present their results on conformal invariance in field theory in connection with symmetry breaking and solutions of classical Yang-Mills-theories.

Chapter II collects papers dealing with topics in geometric quantisation and symplectic structures. An important new development in the original quantization scheme, as given by B. Kostant and J.M. Souriau, is the incorporation of the Maslow correction; this amounts physically to a better determination of quantum mechanical ground-state energies and leads, in addition, to a deeper geometrical understanding of the so-called WKB-method. These improvements of the general theory are discussed by N.Woodhouse, J.H.Rawnsley and J.Czyz. In contrast to these papers J.A.Wolf goes back to the mathematical origin of geometric quantization, namely to the representation theory of Lie groups. In addition, D.J.Simms shows how to include the time variable properly into the geometrical formalism, and J.Sniatycki describes the quantum mechanics of charged particles in external electromagnetic and gravitational fields within this framework. On the other hand, J.M.Souriau presents a new formulation of thermodynamics in terms of symplectic geometry, and S.Sternberg emphasizes the usefulness of similar geometrical methods in the calculus of variation. The important relations between gauge theories and symplectic structures are investigated on the classical level by P.L.Garcia, and M.Moshinsky discusses unitary representations of canonical transformations, i.e. general symplectic diffeomorphisms.

Chapter III is devoted to the connection between quantum field theory and general relativity. A first step towards the great goal of a final unification of gravitation and quantum theory consists in the construction of quantized fields on the background of a given curved classical space-time continuum. In this connection C.J.Isham investigates covariant field quantization by considering the properties of Wightman functions under such generalized geometrical conditions, whereas G.W.Gibbons introduces Feynman's functional integration method for this purpose. In addition, P.Hajicek discusses a description of physical observables with the help of C^*-algebras and P.Spindel presents a special example. On the other hand, S.Deser describes a new approach to the quantization of the basic gravitational field itself which uses within the framework of so-called supersymmetry the mathematical concept of graded Lie algebras; the corresponding representations are,in turn,analysed and surveyed by V.Kac.

The organizers wish to thank the "Stiftung Volkswagenwerk" again for a most generous financial support of this conference which constitutes a sequel to a similar meeting held in 1975 (comp.Springer Lecture Notes in Mathematics 57o). It was thus possible to unite scientists from a large number of foreign countries in order to intensify a fruitful exchange between mathematics and physics.

Bonn, June 1978 K.Bleuler H.R.Petry A.Reetz

TABLE OF CONTENTS

ON THE ROLE OF FIELD THEORIES IN OUR PHYSICAL CONCEPTION OF GEOMETRY

by

Shlomo Sternberg

The study of our conception of space has been central to natural philosophy from the very beginning. The purpose of the present essay in taking up this ancient question is to stress certain aspects of the role of field theories which have not been emphasized in the standard treatments. Let us begin, as is traditional in essays of this kind, with Descartes. According to Descartes the world is a plenum filled with three types of matter: luminescent matter which emits light; transparent matter which transmits light; and opaque matter which reflects light. By a complicated and intricate scheme of vortices and pressures, forces are transmitted from one material body to another. In particular, the transmission of light depends upon the nature of the transmitting body and the laws of refraction were deduced from the hypotheses concerning the nature of the transmitting medium. I do not want to get into the great debate about the impact of Cartesian philosophy on the subsequent development of scientific thought. The two points that I want to emphasize here are, (1) that from the grandiose scheme of the structure of the media of space, the correct laws of optics, at least those of geometrical optics, were deduced; and, (2) that the gradiose scheme, to whatever extent it was carried out in detail, was based on a very mechanistic philosophy of the interaction of one portion of the plenum on another. By mechanistic I mean in the real sense of a machine, one gear turing on another, one vortex influencing another, one particle pressing on another. The case has been made that the development of industrial machinery in the Renaissance is what ultimately influenced Descartes' conception of the universe as one gigantic machine. (Compare with the impact of the development of computers on our view of "life" or of the brain as an intricate computer.)

The conflict between Cartesian and Newtonian philosophy is well known. Newton himself takes a firm position against action at a distance. We recall a famous quotation from his Principia,

> "that one body may act upon another at a distance through a vacuum, without a mediation of anything else ... is to me so great an absurdity that I believe no man has in philosophical matters a competent faculty for thinking, can ever fall into it."

Nevertheless, Newtonian theory was a successful theory of action at a distance and the Newtonian school ultimately rejected contiguity as a fundamental principle of physics. The conflict between Cartesian and Newtonian philosophy reached its high point in the early part of the eighteenth century and was pithily summarized by Voltaire:

> "A Frenchman who arrives in London will find philosophy, like everything else, very much changed there. He had left the world of plenum and now he finds it a vacuum." (Lettres philosophiques, quatorzieme lettre.)

Cartesian physics on the continent was overthrown, due in no small part to the efforts of Voltaire and his girlfriend, Madame du Châtelet.

By the end of the nineteenth century, the relative positions of England and the continent had been reversed. The great, French, mathematical physicists had brought the theory of analytical dynamics to a state of perfection. In the meanwhile, however, in England the concept of a theory of fields was being developed, principally due to the geometrical intuition and physical experiments of Faraday, and of the conceptual genius of Maxwell. It is interesting and instructive to see the psychological difficulties which such field theories had in gaining acceptance amonst the French mathematical physicists. We quote extensively from Poincaré's introduction to his lectures on electricity and optics of 1888 and 1889.

> "The first time that a French reader opens Maxwell's book, a feeling of unease and often even of defiance is mingled, at least at first, with a sense of admiration. It is only after a prolonged interaction, and at the cost of much effort, that this feeling disappears...
>
> "Why is it that the ideas of the English scientist have such difficulty in becoming accepted amongst us? It is undoubtedly because the education received by the majority of educated Frenchmen disposes them to prefer precision and logic above any other quality. The old theories of mathematical physics gave us complete satisfaction in this respect. All of our masters from Laplace to Cauchy proceeded in the same manner. Starting from clearly announced hypotheses, they deduced all of the consequences with mathematical rigor and then compared them with experiment. They seemed to want to give to each of the branches of physics the same precision possessed by celestial mechanics.
>
> "For a spirit accustomed to admiring such models, it is difficult for a theory to be satisfying. Not only can it not tolerate the slightest appearance of contradiction, but it requires that the diverse parts be logically related to one another and that the number of distinct hypotheses be reduced to a minimum.
>
> "This is not all; there are other requirements which seem to me less reasonable. Behind the matter which reaches our senses and with which we are acquainted by experiment, he wishes to see another matter, the true one to his eyes, which has only purely geometrical qualities and whose atoms are only mathematical points subjected only to the laws of dynamics. Yet he seeks, (with an unrecognized contradiction), to have these indivisible and colorless atoms represent and consequentially approximate as closely as possible ordinary matter...

"Thus, in opening Maxwell, a Frenchman expects to find there a theoretical and logical collection, as precise as the theory of physical optics founded upon the hypothesis of the ether. He thus prepares himself for a disappointment that I wish to have the reader avoid by announcing immediately what he should expect to find in Maxwell and what he will be unable to find.

"Maxwell does not give a mechanical explanation of electricity and magnetism; he limits himself to showing that such an explanation is possible...

Assuming that the reader will thus limit his aspirations, he will still have more difficulties to overcome: The English scientist does not attempt to construct a definitive, well ordered and unique edifice. Rather, it seems that he raises a great numer of provisional and independent constructions between which connections are difficult and sometimes impossible.

"Let us take, for example, the chapter where he explains electrostatic attractions by pressures and tensions which are present in the dielectric medium. This chapter could be omitted without making the rest of the volume less clear or less complete and, on the other hand, it contains a theory which is sufficient in and of itself and which one could understand without having read a single line of what proceeded or followed it. But it is not only that it is independent of the rest of the work; it is difficult to reconcile with the fundamental ideas of the book, as is shown later by a thorough discussion. Maxwell doesn't even attempt such a reconciliation. He limits himself to saying, 'I have not been able to make the next step, namely, to account by mechanical considerations for these stresses in the dielectric.'...

"One need not, thus, avoid all contradiction... Two contradictory theories could, in effect, provided one does not mix them and one does not seek the ultimate sources of things, both be useful instruments of research; and, perhaps, the reading of Maxwell would be less suggestive if he had not opened up so many new and divergent courses for us.

"But the fundamental idea thus becomes a bit obscured. In fact, it is so well obscured that in most popularizations it is the one point that is completely left aside.

"I, therefore, wish to explain in this introduction what this fundamental idea consists of.

"In all physical phenomena there exists a certain number of parameters which can be obtained and measured directly from experiment. I will call them

$$q_1, q_2, \cdots, q_n \quad .$$

"Observation then teaches us the laws of variation of these parameters and these laws can usually be put in the form of differential equations relating these parameters q to one another and to time.

"What must one do to give a geometrical interpretation to such a phenomenon?

"One seeks to explain it by movements of ordinary matter, whether by one or several hypothetical fluids. These fluids will be considered as formed by a great number of isolated molecules; let $m_1, m_2, \cdots, m_p$ be the masses of these molecules and let x_i, y_i, z_i be coordinates of the i-th molecule.

"One must also suppose that there is conservation of energy and, therefore, that there exists a certain function U of the $3p$ coordinates x_i, y_i, z_i which plays the role of potential energy. The $3p$ equations of motion can then be written as:

$$(1)\qquad \begin{aligned} m_i \frac{d^2x_i}{dt^2} &= -\frac{dU}{dx_i} \\ m_i \frac{d^2y_i}{dt^2} &= -\frac{dU}{dy_i} \\ m_i \frac{d^2z_i}{dt^2} &= -\frac{dU}{dz_i} \end{aligned}$$

"The kinetic energy of the system is equal to:

$$T = \frac{1}{2}\, m_i(x'^2_i + y'^2_i + z'^2_i) \quad .$$

"The potential energy is equal to U and the equation which expresses the conservation of energy is written:

$$T + U = \text{constant}$$

"One would thus have a complete mechanical explanation of the phenomenon when one knew, on the one hand, the potential energy function U and, on the other hand, one knew how to express the $3p$ coordinates x_i, y_i, z_i by means of the n parameters, q .

"If we replace these coordinates by their expressions in terms of q's , the equations (1) take a different form. The potential energy U becomes a function of the q's and as far as the kinetic energy T , it depends not only on the q's , but also on the first derivatives and will be homogeneous of the second order with respect to these derivatives. The laws of motion will then be expressed by Lagrange's equations:

(2) $$\frac{d}{dt}\frac{dT}{dq'_k} - \frac{dT}{dq_k} + \frac{dU}{dq_k} = 0$$

"If the theory is good, these equations will be identical to the experimental laws observed directly. Thus, in order that a mechanical explanation of a phenomenon be possible, it is necessary that one be able to find two functions U and T depending the first on the q's alone and the second on these parameters and their derivatives; that T be homogeneous of the second order with respect to these derivatives and that these differential equations deduced from experiment can be put in the above form.

"The converse is also true. Whenever one can find two such functions T and U one is sure that the pehnomenon is susceptible to mechanical explanation...

Poincaré then proceeds to show that if we are given the functions $U = U(q)$ and $T = T(q, q')$ of the above form, we can always find constants $m_1, \cdots, m_p$ and 3p functions $x_i(q)$, $y_i(q)$ and $z_i(q)$ such that $T = \frac{1}{2}\Sigma\, m(x_i'^2 + y_i'^2 + z_i'^2)$ and a potential function $V(x,y,z)$ so that $U(q) = V(x(q), y(q), z(q))$. But he points out, these m's and x, y, z, etc. are not unique. He writes,

"Thus, as soon as the functions U and T exist, one can find an infinity of mechanical explanations of the phenomena. If, thus, a phenomenon has a complete mechanical explanation, it will have an infinity of others, which will render equally good account of all of the particularities revealed by experiment.

"The proceeding is confirmed by the history of all branches of physics. In optics, for example, Fresnel believed the vibration to be perpendicular to the plane of polarization. Neumann regarded it as parallel to this plane. People searched for a long time for an experimentum crucis which allowed one to decide between these two theories, but couldn't find one. Similarly, without going out of the domain of electricity, we can state that the two fluid theory, or the single fluid theory, both gave an equally satisfactory explanation of all the observations of electrostatics. All of these facts are easily explained by the properties of Lagrange's equations which I have just recalled. It is now easy to understand Maxwell's fundamental idea. In order to demonstrate the possibility of a mechanical explanation of electricity we don't have to occupy ourselves with finding this explanation itself. It is enough for us to know the expression of the two functions T and U which are both parts of the energy, and to construct, with these two functions, the Lagrange equations and then compare these equations with the experimental law."

Poincaré has thus clearly explained the need for a mechanical explanation of the properties of the so called vacuum. But notice that for Poincaré the notion of "mechanical explanation" has already become quite a bit more abstract than the meaning that this notion had for Descartes. For Poincaré the notion of a mechanical explanation is Lagrange's equations with a specified form of the Lagrangian, namely that it be dependent only on the generalized coordinates and their first derivatives and that it be homogeneous and quadratic in these derivatives. Poincare's requirement, modified by the current dogmas of quantum mechanics, has been accepted as a necessary desideratum for a physical explanation to this very day.

The triumph of Maxwell's theory, its overthrow of Newtonian physics and the development of the special and general theories of relativity are stories so well known that they bear no repeating. One of the legacies of general relativity has been the development of modern differential geometry with its emphasis on the invariant formulation of its constructs. A present day geometer is not satisfied with a geometrical object when presented in a coordinate system, even if he is at home with its law of transformation under changes of variable. He requires a genuinely invariant definition in terms of reasonably familiar concepts which are accepted in the mathematical community. It is only then that he recognizes or admits the "reality" of the concept. It is from this point of view that I wish to examine some of the fundamental objects of classical and modern field theories and see what their impact is on our hypotheses concerning the geometry of space and time.

There has also been a recent, and as yet not generally accepted, change in our concept of a mechanical system. It has been recognized since the time of Lagrange and Hamilton that the mathematical analysis of a dynamical system requires us to admit, in the phase space of the dynamical system, transformations which mix the coordinate and momenta variables. These transformations are subjected only to the requirement that they be "canonical", that is that they preserve the fundamental symplectic form:

$$\omega = \Sigma \, dp_i \wedge dq^i \quad .$$

In admitting such local transformations we destroy the interpretation of our space as a phase space, retaining only its symplectic structure. However, it has only been in recent years, due principally to the efforts of Souriau, that we have recognized that mechanical systems should consist of symplectic manifolds which do not necessarily admit any global interpretation as the phase space of some configuration space. Thus the stage setting for dynamics will be a general symplectic manifold X or other (generally presymplectic) manifolds associated with it, but X itself will not necessarily be the cotangent bundle of any manifold. It is only with the introduction of such spaces into mechanics that one can find the classical formulations of such notions as spin. In admitting these types of mechanical systems, one must reject the Lagrangian and, therefore, the variational formulation of mechanics, but substitute for it a formulation which is more in character with symplectic geometry. We shall assume this perspective in our analysis.

Most of the discussion will be in the framework of classical as opposed to quantum theories, although, from time to time, we shall point out applications to quantum mechanics. For example, we shall spend a good bit of what follows in showing how various fields are related to symplectic structures. A symplectic structure determines canonical coordinates which are intimately related to the problem of quantization.

We shall begin with relatively modest mathematical tools, mainly the calculus of differential forms, then (around Section 5) make use of more sophisticated ideas of differential geometry. In order to make the text read smoothly, I have deferred all bibliographical references, including basic references to the mathematical ideas that we use, to a final section. I have also deferred various comments, not essential to the main ideas of argument to this final section. Although I have deferred acknowledgements to the final section, I cannot close this introduction without expressing my indebtedness to Prof. J. M. Souriau whose discoveries have greatly influenced the views presented here.

TABLE OF CONTENTS

This essay builds from a reformulation of familiar facts to the more abstract. The key presentation of principles is in Section 8, and the more logically inclined reader might prefer to start there and then proceed to details.

1. Electrostatics: The Dielectric Properties of the Vacuum Determine Euclidean Geometry

We begin by formulating in geometrical terms the fundamental objects of electrostatics. The electric field strength E is a linear differential form which, when integrated along any path, gives the voltage drop across that path; thus the units of E will be voltage/length. Since voltage has units energy/charge and force has units energy/length, we can also write the units of E as force/charge. We emphasize that E, as a geometrical object, is a linear differential form on three dimensional space, because, by its physical definition, it is something which assigns voltage differences to paths by integration. The second fundamental object in electrostatics is the dielectric displacement. It is a two-form, D, which, when integrated over the boundary surface of any region, gives the total charge contained in that region. Thus D is a two-form which satisfies the equation

$$dD = 4\pi\rho\, dx \wedge dy \wedge dz$$

where the three form $\rho dx \wedge dy \wedge dz$ represents the density of charge in any three dimensional region. By its definition D assigns charge to surfaces by a process of integration. Therefore, by its defining properties, it is given as two form on three dimensional space (and not as a vector field as prescribed in the standard treatments). There are two fundamental equations in electrostatics. The first of these postulates a relationship between E and D determined by the medium. The second asserts that $dE = 0$. Now E is a one form and D is a two form. We cannot have functional relationship between a one form and a two form on a three dimensional space without imposing severe geometrical restrictions on the space. More specifically, the relationship between E and D in what is known as a homogeneous isotropic medium is as follows. One postulates that there is a preferred rectangular (i.e. Euclidean) coordinate system $x, y,$ and z and, therefore, a well defined $*$ operator given by:

$$*dx = dy \wedge dz$$
$$*dy = -dx \wedge dz$$
$$*dz = dx \wedge dy$$

which relates one forms to two forms and that

$$D = \varepsilon * E$$

where ε is a function of the medium known as its dielectric constant. In vacuo the function ε is a constant, ε_0, known as the dielectric constant of the vacuum. Now in three dimensions the $*$ operator associated to a Riemann metric completely determines the metric. Giving the $*$ operator is the same as giving the metric. Therefore, the law of electrostatics $D = \epsilon_0 * E$ completely determines the Riemannian geometry of space which is asserted by the experimental laws of electrostatics to be Euclidean.

The statement that it is the dielectric properties of the vacuum that determine Euclidean geometry is not merely a mathematical sophistry. In fact, the forces between charged bodies in any medium are determined by the dielectric properties of that medium. Since the forces that bind together macroscopic bodies as we know them are principally electrostatic in nature, it is the dielectric property of the vacuum which fixes our rigid bodies. We use rigid bodies as measuring rods to determine the geometry of space. It is in this very real sense that the dielectric properties of the vacuum determine Euclidean geometry.

We should emphasize once again, in view of what is going to follow, that it is not the field itself, the E or the D, that determines the Euclidean geometry, but rather the response of the vacuum to the presence of a field in potentio; the fixing of the relationship between the E and the D which determines Euclidean geometry. Euclidean geometry, in turn, determines the equations of motion of a free, uncharged particle in space. Giving a Riemannian geometry determines a scalar product on the tangent space at any point and, therefore, on the cotangent space as well. The equations of motion of a free particle of mass m are determined by the Hamiltonian function H_m on $T^*\mathbb{R}^3$ where

$$H_m(q, p) = \frac{1}{2m} \| p \|^2$$

where $q = (x, y, z)$ is the position and $p = (p_x, p_y, p_z)$ is the momentum of the free particle.

The effect of an actual electrostatic field is to modify the equations of motion of a charged particle. We consider a particle whose charge e is sufficiently small so that its effect on the electromagnetic field can be ignored. So we are dealing with the passive equations of a test particle of small charge in the presence of a given electrostatic field. The field equation $dE = 0$ is locally equivalent to the existence of a function ϕ such that $E = -d\phi$. Let us assume that we are dealing with a region of space which is simply connected so that we will take ϕ to be globally defined in our region. Then the equations of motion of a charged particle are given by a modified Hamiltonian $H_{m,e,\phi}$ where

$$H_{m,e,\phi}(q,p) = \frac{1}{2m} \| p \|^2 + e\phi(q) \quad .$$

In all of the above discussion we have been regarding charge as an independent unit. Therefore, strictly speaking, we should consider the electric field strength E not as a numerical valued linear differential form, but rather as a vector valued linear differential form with values in a one dimensional space dual to the charges. Then the choice of units of charge would amount to the choice of a basis in this one dimensional vector space. Actually, as we shall see later on, the correct formulation will be to consider the field strength as a vector bundle valued differential form and the precise geometrical character of this vector bundle will be elucidated.

2. Magnetostatics: The Magnetic Field Determines the Symplectic Structure on Phase Space

In the proceeding section we wrote down the Hamiltonian for a charged test particle in the presence of an electric field. In writing such a Hamiltonian, we implicity took for granted that one would derive Hamilton's equations from this Hamiltonian by the standard procedure. That is, we took for granted that there existed a symplectic form ω on the phase space $T^*\mathbb{R}^3$ and that the vector field describing the differential equations of motion of the particle was derived from the Hamiltonian by the standard procedure

$$\xi_H \lrcorner\, \omega = -dH \quad .$$

We also took for granted that the symplectic form ω was the canonical symplectic form carried by $T^*\mathbb{R}^3$ in virtue of its being a cotangent bundle. That is we took

$$\omega = dq_x \wedge dp_x + dq_y \wedge dp_y + dq_z \wedge dp_z \quad .$$

In magnetostatics, there are also two fundamental quantities, the "magnetic flux" B and the "magnetic loop tension". Faraday's law of induction says that associated with any system of magnets there is a certain "flux". If γ is a closed circuit and S is a surface whose boundary is γ, then the change of flux through S (by, for example, moving the magnets) induces an electromotive force around γ. Put more mathematically, let

$$B = B_x dy \wedge dz - B_y dx \wedge dz + B_z dx \wedge dy \quad .$$

Let S be a surface with boundary γ, then Faraday's law of induction says that

$$\frac{d}{dt}\int_S B = -\int_\gamma E \quad .$$

From its definition as a flux we see that we must regard B as a two form on three dimensional space.

Now, the presence of a magnetic flux changes the motion of a small test particle in its presence. The change in the equations of motion can be described as follows. We are given B as a two-form on $\mathbb{R}^3$. By the standard projection, π, of $T^*\mathbb{R}^3$ on to $\mathbb{R}^3$, assigning to each point in phase space the corresponding point in configuration space, we can regard B equally well as being defined on $T^*\mathbb{R}^3$. With this new identification we would write

$$B = B_x dq_y \wedge dq_z - B_y dq_x \wedge dq_z + B_z dq_x \wedge dq_y \quad .$$

Here we write q_x for $x \circ \pi$ et cetera.

It is a fundamental law of electromagnetism that the integral of B around any closed surface in space vanishes ("There is no true magnetism," to quote Hertz), thus

$$dB = 0 \quad .$$

This means that the form

$$\omega_{e,B} = \omega + eB$$

is a closed two form defined on $T^*\mathbb{R}^3$. We claim that $\omega_{e,B}$ is also non-degenerate. Indeed, let us examine the equations

$$\xi_H \lrcorner\, \omega_{e,B} = -\, dH \quad .$$

Let us write

$$\begin{aligned} \xi_H &= a\frac{\partial}{\partial q} + b\frac{\partial}{\partial p} \\ &= a_x \frac{\partial}{\partial q_x} + a_y \frac{\partial}{\partial q_y} + a_z \frac{\partial}{\partial q_z} + b_x \frac{\partial}{\partial p_x} + b_y \frac{\partial}{\partial p_y} + b_z \frac{\partial}{\partial p_z} \quad . \end{aligned}$$

Then

$$\xi_H \lrcorner\, \omega_{e,B} = a_x(-\, dp_x - eB_y dq_z + eB_z dq_y) + b_x dq_x + \text{two similar terms} \quad .$$

On the other hand,

$$\begin{aligned} dH &= \frac{\partial H}{\partial q} dq + \frac{\partial H}{\partial p} dp \\ &= \frac{\partial H}{\partial q_x} dq_x + \frac{\partial H}{\partial p_x} dp_x + \text{two similar terms} \quad . \end{aligned}$$

Comparing the coefficient of dp_x shows that

$$a_x = \frac{\partial H}{\partial p_x}$$

and comparing the coefficient of dq_x shows that

$$b_x = -\frac{\partial H}{\partial q_x} + e\left(\frac{\partial H}{\partial p_y} B_z - \frac{\partial H}{\partial p_z} B_y\right)$$

plus similar equations for a_y, a_x, b_y and b_z. In particular, we see that the a's and b's are completely determined and, hence, that the form $\omega_{e,B}$ is non-degenerate. If we take the Hamiltonian H to be $H_{m,e,\phi}$ as described in the proceeding section, then

$$\frac{\partial H}{\partial p_x} = \frac{1}{m} p_x \qquad \text{so} \qquad \frac{dq_x}{dt} = \frac{1}{m} p_x \qquad \text{etc.}$$

$$\frac{\partial H}{\partial q_x} = e \frac{\partial \phi}{\partial x} = - eE_x \qquad \text{so} \qquad \frac{dp_x}{dt} = e(E_x + \frac{p_y}{m} B_z - \frac{p_z}{m} B_y)$$

and the differential equations that we obtain are precisely the classical differential equations for a charged particle in the presence of a given external electric and magnetic field. Thus, the presence of the magnetic field B modifies the equations of motion of a charged paricle by modifying the symplectic structure on the cotangent bundle.

The magnetic flux also makes its presence felt by affecting the equations of motion of a magnet considered as a spinning electrical particle. To describe these equations of motion, we must replace the six dimensional phase space $T^*\mathbb{R}^3$ by the eight dimensional space given as the direct product $T^*\mathbb{R}^3 \times S^2$ where S^2 is the standard two dimensional sphere. Let Ω denote the standard value form on the unit sphere S^2. On the product space $T^*\mathbb{R}^3 \times S^2$ we can put the symplectic structure given by

$$\omega_{e,B} + s\Omega$$

where the parameter s is called the spin of the particle. Now, we can think of a vector $u \in S^2$ as determining a vector $su \in \mathbb{R}^3$ of length s. In turn, we can regard, in view of a fixed orientation on space, the vector su as determining a bi-vector in $\mathbb{R}^3$:

$$S = *(su) \in \wedge^2 \mathbb{R}^3$$

and we can then, at each point of $\mathbb{R}^3$, take the scalar product of S with the magnetic flux at that point, also regarded as an exterior two vector at the point in question. Thus, $S \cdot B$ is a well defined function on $T^*\mathbb{R}^3 \times S^2$. Let us now introduce the modified Hamiltonian $H_{m,e,\phi,\mu,B}$ given by

$$H_{m,e,\phi,\mu,B}(q,p,u) = \frac{1}{2m} \|p\|^2 + \phi(q) + \mu S \cdot B \quad .$$

Here the parameter μ is called the magnetic moment of the particle and we will leave it as an exercise to the reader to verify that the equations of motion associated to this Hamiltonian relative to the symplectic structure described above gives precisely what we want for the equations of a charged spinning particle with magnetic moment μ in the presence of an external electric and magnetic field.

We will discuss the magnetic field strength in conjunction with Maxwell's equations in the next section.

3. Maxwell's Equation: The Constituitive Properties of the Vacuum Determine the Conformal Geometry of Space Time

Electrostatics and magnetostatics are only approximately correct. They must be replaced by Maxwell's theory which we now quickly review.

We begin by rewriting Faraday's law of induction $\frac{d}{dt}\int_S B = -\int_\gamma E$ in a form which is more congenial from the space time approach. Consider an interval $[a, b]$ in time and the three dimensional cylinder $S \times [a, b]$ whose boundary is the two dimensional cylinder $\gamma \times [a, b]$ together with the top and the bottom of this two dimensional cylinder. See the accompanying figure.

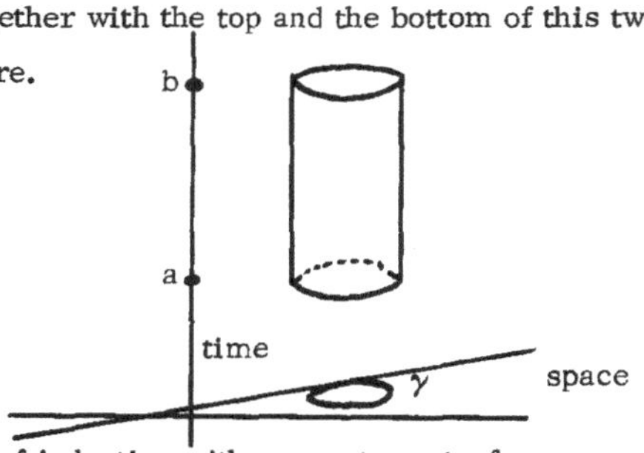

Integrating Faraday's law of induction with respect to t from a to b gives the equation

$$\int_{S \times \{b\}} B - \int_{S \times \{a\}} B + \int_{\gamma \times [a, b]} E \wedge dt = 0 .$$

Let us set

$$F = B + E \wedge dt$$

so that F is a two form defined on four dimensional space. Let C denote the three dimensional cylinder, $C = S \times [a, b]$ so that

$$\partial C = S \times \{b\} - S \times \{a\} + \gamma \times [a, b] .$$

Now, B is a two-form involving just the spatial differentials and, therefore, must vanish when restricted to the side $\gamma \times [a, b]$ of the cylinder, while dt and, hence, $E \wedge dt$ must vanish on the top and the bottom. Thus, we can write Faraday's law of induction as

$$\int_{\partial C} F = 0 .$$

We can also consider a three dimensional region C lying entirely in space at one fixed constant time. In this case, ∂C will be a surface on which $dt = 0$ so that $\int_{\partial C} F = \int_{\partial C} B$ and by the absence of true magnetism, this surface integral must vanish.

Thus, $\int_{\partial C} F = 0$ for all three dimensional cubes whose sides are parallel to any three of the four coordinate axes. This is enough to imply that $dF = 0$ where now, of course, d stands for the exterior derivative in four space. Since $F = B_x dy \wedge dz - B_y dx \wedge dz + B_z dz \wedge dy + E_x dy \wedge dt + E_y dy \wedge dt + E_z dz \wedge dt$, the equation $dF = 0$ is equivalent to the four equations

$$\frac{\partial B_x}{\partial x} + \frac{\partial B_y}{\partial y} + \frac{\partial B_z}{\partial x} = 0 \qquad \frac{\partial B_y}{\partial t} - \frac{\partial E_z}{\partial x} + \frac{\partial E_x}{\partial z} = 0$$

$$\frac{\partial B_x}{\partial t} - \frac{\partial E_y}{\partial z} + \frac{\partial E_z}{\partial y} = 0 \qquad \frac{\partial B_z}{\partial t} - \frac{\partial E_x}{\partial y} + \frac{\partial E_y}{\partial x} = 0 \quad .$$

We will use Faraday's law to define B so that the units of B are

$$\frac{\text{voltage} \cdot \text{time}}{\text{area}} = \frac{\text{energy} \cdot \text{time}}{\text{charge} \cdot (\text{length})^2}$$

Ampère's Law relates current to magnetism. It says that the "electric current flux" through a surface S whose boundary is γ equals the "magnetic loop tension" around γ. According to Maxwell's great discovery we must write the "electric current flux" as the sum of two terms $\frac{\partial D}{\partial t} + 4\pi j$ where D is the dielectric displacement and j is the current density of moving charges. (For slowly varying fields the first term is negligible in comparison with the second and did not appear in Ampère's original formulation.) The "magnetic loop tension" is obtained by integrating a linear differential form, H, called the "magnetic field strength" around γ. Thus Ampère's law says

$$\int_S \left(\frac{\partial D}{\partial t} + 4\pi j\right) = \int_\gamma H \quad .$$

Consider the three dimensional cylinder $C = S \times [a, b]$ as before and set

$$G = D - H \wedge dt \quad .$$

We integrate Ampère's law from a to b with respect to t and get

$$\int_{\partial C} G = -4\pi \int_C j \quad .$$

If we consider a three dimensional region R at constant time, then dt vanishes on ∂R and the integral of G over ∂R is the same as the integral of D. This equals $4\pi \times$ (the total charge in R) which is $4\pi\rho\, dx \wedge dy \wedge dz$. Thus for regions which lie in constant time we have

$$\int_{\partial R} G = 4\pi\rho\, dx \wedge dy \wedge dz \quad .$$

Let us set

$$J = \rho\, dx \wedge dy \wedge dz - j \wedge dt$$

and we see that

$$\int_{\partial C} G = 4\pi \int_C J$$

for any three dimensional cube whose sides are parallel to the coordinate axes. Thus

$$dG = 4\pi J$$

from which it follows that $dJ = 0$.

We summarize

Set	then Maxwell's equations say
$F = B + E \wedge dt$	$dF = 0$
$G = D - H \wedge dt$	$dG = 4\pi J$.
$J = \rho\, dx \wedge dy \wedge dz - j \wedge dt$	

Notice that Maxwell's equations are invariant under smooth orientation preserving charges of coordinates.

We will use Ampère's law to define H. D has units $\frac{\text{charge}}{\text{area}}$ so $\frac{\partial D}{\partial t}$ has units $\frac{\text{charge}}{\text{area} \cdot \text{time}}$ and j has units $\frac{\text{current}}{\text{area}} = \frac{\text{charge}}{\text{area} \cdot \text{time}}$. Thus

$$H \text{ has units } \frac{\text{charge}}{\text{time} \cdot \text{length}} \quad .$$

In vacuo we have the constituitive relations

$$D = \epsilon_0 * E \qquad \text{and} \qquad B = \mu_0 * H$$

ϵ_0 has units

$$\frac{\text{charge}}{\text{area}} \times \frac{\text{length}}{\text{voltage}} = \frac{(\text{charge})^2}{\text{energy} \cdot \text{length}}$$

μ_0 has units

$$\frac{\text{energy} \cdot \text{time}}{\text{charge} \cdot (\text{length})^2} \times \frac{\text{time} \cdot \text{length}}{\text{charge}} = \frac{\text{energy} \cdot (\text{time})^2}{(\text{charge})^2 \cdot \text{length}} \quad .$$

Thus $\frac{1}{\epsilon_0 \mu_0}$ has units $\frac{(\text{length})^2}{(\text{time})^2} = (\text{velocity})^2$. Thus the theory of electromagnetism has a fundamental velocity built into it. It was Maxwell's great discovery that this velocity is exactly c - the velocity of light ! So introduce cdt instead of dt and the four dimensional star operator

$$* (dx \wedge dy) = - cdz \wedge dt$$

$$* (dy \wedge dz) = cdy \wedge dt$$

$$* (dy \wedge dz) = - cdx \wedge dt$$

$$* (dx \wedge cdt) = dy \wedge dz$$

$$* (dy \wedge cdt) = - dx \wedge dz$$

$$* (dz \wedge cdt) = dx \wedge dy$$

Then

$$* F = - c(B_z dz \wedge dt + B_y dy \wedge dt + B_z dx \wedge dt) + \frac{1}{c}(E_x dy \wedge dz - E_y dx \wedge dz + E_z dx \wedge dy) .$$

The constituitive equations can be written as

$$G = \sqrt{\frac{\epsilon_0}{\mu_0}} \times * F \quad .$$

Now in four dimensions, the $*$ operator from $\wedge^2 \to \wedge^2$ does <u>not</u> determine the metric. It does determine it up to scalar multiple at each point. Thus the constituitive properties of the vacuum determine the conformal Lorentzian structure of space-time.

From now on we shall use coordinates in which $c = 1$. Thus x_0, x_1, x_2, x_3 will be coordinates on space-time with

$$ds^2 = dx_0^2 - dx_1^2 - dx_2^2 - dx_3^2$$

the Lorentz metric, and $* dx_1 \wedge dx_2 = dx_0 \wedge dx_3$, etc.

4. The Lorentz Force: The Electromagnetic Field Determines the Symplectic Structure on the Phase Space of Space-Time

We denote our four dimensional space time by M so that T^*M denotes its cotangent bundle, and ω the canonical two form on T^*M. Suppose we are given an electromagnetic field F, which is a two form on M. Via the projection $\pi : T^*M \to M$ we can pull F back to T^*M to obtain a two form on T^*M which we shall continue to denote by F. We define

$$\omega_{e,F} = \omega + eF \quad ,$$

where e is an electric charge. One half of Maxwell's equations assert that $dF = 0$ so that $d\omega_{e,F} = 0$. Since F involves only dq's, i.e. differentials coming from M, it is easy to check, using the same argument as in the preceeding section, that $\omega_{e,F}$ is non-degenerate. Thus e and F determine a symplectic structure on T^*M. Let H be a function on T^*M and suppose that the Hamiltonian equations corresponding to H and to the canonical symplectic form, ω, describe the equations of motion of a "free uncharged particle". Then the Hamiltonian equations corresponding to H relative to the symplectic form $\omega_{q,F}$ describe the equations of motion of a charged particle of charge q in the presence of the external electromagnetic field, F. As before, we are assuming that the charge q is sufficiently small that we can neglect the influence of the particle on the electromagnetic field. If we take $M = \mathbb{R}^{1,3}$ to be Minkowski space with the standard Lorentz metric, and we take

$$H(p, q) = \tfrac{1}{2} \| p \|^2 = \tfrac{1}{2}(p_0{}^2 - p_1{}^2 - p_2{}^2 - p_3{}^2)$$

then it is easy to check, using the same methods as in Section 3, that this gives the standard Lorentz equations. We shall redo this calculation in a more general setting later on. We have introduced the effect of the electromagnetic field by keeping the same Hamiltonian H but modifying the symplectic structure. In the standard physics literature, there is another procedure, called "minimal coupling" for obtaining the Lorentz equation which keeps the original symplectic structure but modifies the Hamiltonian. Let us pause to show that the two procedures are formally equivalent. The minimal coupling prescription is as follows: We can, at least locally, find a one form A on M called a four-potential, satisfying

$$dA = F \quad .$$

We can think of A as a section of T^*M and introduce a modified Hamiltonian $H_{e,A}$ where

$$H_{e,A}(q, p) = H(q, p - eA(\pi(q))) \quad .$$

The second order differential equations determined by the Hamiltonian equations of $H_{e,A}$ relative to the canonical form ω are again the Lorentz equations. To see why the two procedures are the same, let us introduce the map $\varphi_{e,A}$ of T^*M into itself defined by

$$\varphi_{e,A}(q,p) = (q, p + eA(q)) \quad .$$

Let $\theta = p \cdot dq$ be the canonical (action) form on T^*M so that $\omega = d\theta$. Then

$$\varphi_{e,A}{}^*\theta = \theta + e\pi^*A \quad .$$

Here A is a one form on M so π^*A is a one form on T^*M and $d\pi^*A = \pi^*dA = \pi^*F$. We have been writing F for π^*F so, applying d to the preceding equation, we get

$$\varphi_{e,A}{}^*\omega = \omega + eF = \omega_{e,F} \quad .$$

On the other hand, $\varphi^{-1}_{e,A}(q,p) = (q \cdot p - eA(q))$ so

$$(\varphi^{-1}_{e,A})^*H = H_{e,A} \quad .$$

It is now clear that the solution curves of $H = \varphi_{e,A}{}^*H_{e,A}$ relative to $\omega_{e,A} = \varphi_{e,A}{}^*\omega$ are the images, under $\varphi_{e,A}$ of the solution curves of $H_{e,A}$ relative to ω. Since $\pi \circ \varphi_{e,A} = \pi$, we obtain the same trajectories on M from one system as from the other.

There are some important differences between the space-time treatment described in this section, and the discussion in Sections 1 and 2. First of all, in contrast to the electrostatic case, the Riemannian (Lorentzian) metric of space-time is not completely determined by the electromagnetic constituitive properties of the vacuum, and hence the Hamiltonian H must be specified. The special theory of relativity assumes that $M = \mathbb{R}^{1,3}$ with its flat metric, and $H = \frac{1}{2}\|p\|^2$ as above. (The constituitive properties of the vacuum would allow any metric conformally equivalent to the flat metric, and hence any Hamiltonian of the form $H(q,p) = \frac{1}{2}\lambda(q)\|p\|^2$ where λ is any function on $\mathbb{R}^{1,3}$.) The general theory of relativity supposes the existence of a Lorentzian metric g determined from Einstein's equation which provides a theory of the gravitational force in addition to electromagnetism. Then Maxwell's equations are replaced by the equations $dF = 0$ and $d * F = J$ where now the $*$ operator is that determined by g. (To the extent that g is not conformally equivalent to a flat metric, this implies a modification in Maxwell's equations.) Then H is taken to be $H(q,p) = \frac{1}{2}\|p\|^2$ where now $\|p\|^2$ now means the square length of the covector p relative to g. The trajectories of the Hamiltonian vector field corresponding to H relative to the canonical form ω - of the "free" Hamiltonian equations - project onto the geodesics of g. The presence of an electromagnetic field F modifies these equations for a charged particle as indicated above.

A second difference between the relativistic formulation of this section and the non-relativistic formulation in Sections 1 and 2 lies in the concept of mass. In Section 1 we introduced the mass as a parameter in the Hamiltonian on a six dimensional phase space. Here T^*M is eight dimensional and the mass is introduced by considering the seven dimensional manifold $H = \frac{1}{2}m^2$. The vector field ξ_H is tangent to this submanifold, where ξ_H denotes the Hamiltonian vector field associated to H by the form $\omega_{e,F}$. Let $\omega_{e,F,m}$ denote the restriction of $\omega_{e,F}$ to the submanifold $H = \frac{1}{2}m^2$. Then $\omega_{e,F,m}$ is a closed two form of rank six on a seven dimensional manifold and so has a one dimensional null foliation which is spanned by ξ_H. The projection onto M of the integral curves of this null foliation are the world lines of the particles of mass m.

Let us now examine the relativistic equations for a spinning particle. In three dimensions, we identified the spin of a particle at a point in space as an anti-symmetric tensor, i.e. as an element of $\wedge^2 \mathbb{R}^3$. Under the identification of $\wedge^2 \mathbb{R}^3$ with $\mathbb{R}^3$ we could regard the set of all spin vectors of constant length as a sphere, and hence were able to use the symplectic structure of the sphere. In four dimensions, we should regard the spin as an element of $\wedge^2 (\mathbb{R}^{1,3})$ at each point. We must describe the analogue of the sphere and the corresponding symplectic structure. To this purpose we first observe that for any vector space V carrying a non-degenerate scalar product we may identify $o(V)^*$ with $\wedge^2(V)$. Here $O(V)$ denotes the orthogonal group of V and $o(V)$ the Lie algebra of $O(V)$ and $o(V)^*$ the dual space of $o(V)$. The identification of $o(V)$ with $\wedge^2(V)$ is just the usual identification of infinitesimal orthogonal transformations with anti-symmetric tensors of degree two. The identification of $o(V)$ with $o(V)^*$ comes from the scalar product on V (or if you like, from the Killing form of $o(V)$). Now if G is any Lie group and g its Lie algebra, then G acts on g via the adjoint representation and on g^*, the dual space of g by the contragredient to the adjoint representation, known as the coadjoint representation. It is known that the orbits of G acting on g^* are symplectic manifolds; i.e. if $\ell \in g^*$ and $\mathfrak{G} = G \cdot \ell$ then $\mathfrak{G}$ carries a canonical symplectic structure which is invariant under the action of G. Let us denote the symplectic form on $\mathfrak{G}$ by $\Omega_{\mathfrak{G}}$. If M is a (pseudo)Riemannian manifold, let B_M denote the bundle of orthonormal frames of M. If $O(V)$ is the orthogonal group of the corresponding metric, then we can form the associated bundle $B_M \times \wedge^2 V/O(V)$ which may be identified with the bundle of anti-symmetric two tensors $\wedge^2 TM$. With each orbit, $\mathfrak{G}$, we may form the associated bundle $B_M \times \mathfrak{G}/O(V)$ which will be a submanifold of $\wedge^2 TM$ which we shall denote by $\wedge^2 TM_{\mathfrak{G}}$. In the case of special relativity, where we may choose a global flat frame, we may use this frame to identify $\wedge^2 TM_{\mathfrak{G}}$ with $M \times \mathfrak{G}$.

A different choice of frame modifies this identification by the action of an element of $O(V)$ on $\mathfrak{G}$. Since $\Omega_{\mathfrak{G}}$ is invariant under the action of $O(V)$, we see that $\Omega_{\mathfrak{G}}$ is well defined as a two form on $\wedge^2 TM_{\mathfrak{G}}$.

We can use the projection $\pi : T^*M \to M$ to pull the bundle $\wedge^2 TM_{\mathfrak{G}}$ back to T^*M, i.e. we can consider it as a bundle over T^*M. Again, in the case of special relativity where we can choose a globally defined flat frame on M, we may, using this frame, identify the pulled back bundle with $T^*M \times \mathfrak{G}$, and observe that the form $\Omega_{\mathfrak{G}}$ is well defined on it. Finally, we may define the form

$$\omega_{e,F,\mathfrak{G}} = \omega_{e,F} + \Omega_{\mathfrak{G}}$$

on $T^*M \times \mathfrak{G}$. Since $\omega_{e,F}$ and $\Omega_{\mathfrak{G}}$ are each closed and non-degenerate, and since they involve different variables, it is clear that $\omega_{e,F,\mathfrak{G}}$ is a well defined symplectic structure on the pulled back bundle $\pi^{\#}\wedge^2 TM_{\mathfrak{G}}$. (Our construction made use of the global trivialization available in special relativity. Later on we shall see how to define the symplectic structure in general, and shall see that the Riemannian curvature will enter.) In case $M = \mathbb{R}^3$, the orbits $\mathfrak{G}$ are the spheres of radius s (and the origin corresponding to $s = 0$) and the construction reduces to the construction in Section 2. For $M = \mathbb{R}^{1,3}$ we must examine the orbits of the Lorentz group acting on $\wedge^2(\mathbb{R}^{1,3})$. If $S \in \wedge^2(\mathbb{R}^{1,3})$ then $S \wedge S \in \wedge^4(\mathbb{R}^{1,3})$, which, up to choice of orientation, we may identify with $\mathbb{R}$. Thus $|S \wedge S|$ is a function on $\wedge^2(\mathbb{R}^{1,3})$ invariant under the action of the Lorentz group. Also, the scalar product on $\mathbb{R}^{1,3}$ induces a scalar product on $\wedge^2(\mathbb{R}^{1,3})$ and so $\|S\|^2$ is a second invariant function. It is easy to see that these invariant functions are independent and that, in fact all the orbits are four dimensional (except for $\{0\}$ which is zero dimensional). Thus, for non-zero orbits, the symplectic manifold $\pi^{\#}\wedge^2 TM_{\mathfrak{G}}$ is twelve dimensional. Since, in the non-relativistic limit, the symplectic manifold corresponding to a spinning particle was eight dimensional, we must describe a procedure for cutting down four dimensions. Actually, as in the case of a particle without spin, we shall describe a presymplectic manifold whose dimension is one more than the dimension of the non-relativistic limiting manifold, and whose null foliations project onto the world lines of the spinning particle. We thus want to get to a nine dimensional manifold, i.e. cut down three dimensions. This is done as follows: For any $p \in \mathbb{R}^{1,3}$ consider the constraint

$$*S \wedge p = 0 \quad .$$

For example, if $p = re_0$ this implies that S is a linear combination of $e_1 \wedge e_2$, $e_1 \wedge e_3$ and $e_2 \wedge e_3$. (Here e_0, e_1, e_2, e_3 is an orthonormal basis of $\mathbb{R}^{1,3}$.) In other words, "S is spinning in space in the rest frame of p". By readjusting the basis e_1, e_2, e_3 of space we can arrange that

$$S = se_2 \wedge e_3$$

so that $S \wedge S = 0$ and $\|S\|^2 = s^2$. Thus, for $\|p\|^2 > 0$, the condition $*S \wedge p = 0$ requires us to restrict attention to orbits satisfying $S \wedge S = 0$. We assume that $\|S\|^2 = s^2 > 0$.

Now S is an anti-symmetric two tensor as is F. We can therefore form their scalar product $S \cdot F$ which is a scalar valued function. Let $f > 0$ be any differentiable function of a real variable and consider the submanifold $\mathcal{V}_9$ defined by the equations

$$*S \wedge p = 0$$

and

$$H(q, p) = f(eS \cdot F) \quad .$$

We claim that $\mathcal{V}_9$ is nine dimensional. Indeed, since, for fixed q, these equations are invariant under the action of the Lorentz group acting on p and S it is enough to show that the fiber of $\mathcal{V}_9$ over q is five dimensional, and for this it suffices to show that the isotropy group of some point (q, p, S) of $\mathcal{V}_9$ is a one dimensional subgroup of the Lorentz group. Since $f > 0$ we can choose $p = re_0$ and hence $S = se_2 \wedge e_3$. It is clear that the isotropy group consists of those orthogonal transformations of space which fix the e_1 axis and so is one dimensional. For generic f the restriction of $\omega_{e,F,\mathfrak{G}}$ to $\mathcal{V}_9$ has rank eight, and so its null foliation is one dimensional. The corresponding integral curves describe the motion of a spinning particle. We shall return to the relation of these equations to spinors later on. Arguments based on a deeper study of the Dirac equation, and which I do not understand, show that we should take f to be a linear function of the form $f(x) = m_0 + gx$ where m_0 and g are constants.

Notice that the notion of a Lagrangian has been entirely eliminated and the notion of "force" has also been almost completely removed from the theory. It only makes its appearance, in the form of "inertial forces" in writing down H. As we shall see at the end of Section 6 this last vestige can also be removed, and the entire classical theory of particle motion reduced to the construction of presymplectic manifolds.

5. Gauge Theories: The Internalization of Geometrical Constructs

In the very early days of the theory of general relativity, Hermann Weyl brought out the point that Einstein's theory assumed a definite Lorentzian metric whereas Maxwell's equations, i.e. the constituitive properties of the vacuum, only determine the metric up to scalar multiple. This suggests that the bundle of orthonormal frames be replaced by the bundle of orthogonal frames, and hence that the Lorentz group $O(1, 3)$ be replaced by $O(1, 3) \times \mathbb{R}^+$. Then we could no longer insist on a connection which preserved a Lorentzian metric, but only on one which preserved it conformally. In particular, parallel transport around a closed path might result in a change of scale. In passing from one system of local frames (i.e. section of the frame bundle) to another, one must not only apply a Lorentz transformation at each point, but also a gauge transformation which has the effect changing our measuring rods and clock rates by a scalar factor at each point. A connection form on the bundle of orthogonal, as opposed to orthonormal, frames would have one additional component, which could be thought of as a linear differential form. Weyl identified this form with the four potential of the electromagnetic field, and thus the field F itself as a component of the curvature. In this way, he was able to propose a purely geometric theory which unifies gravitation and electromagnetism.

Einstein raised some objections to Weyl's theory. It is instructive to read Einstein's comments, Weyl's response at the time and his own comments on his theory, written almost forty years later.

Einstein's Comments

> "If light rays were the only means to determine metrical relationships in the neighborhood of a world point empirically, then, of course there would be an undermined factor in the distance ds (as well as in the g_{ik}). But this indeterminacy is not present in the definition of ds if one brings in results of measurements made with (infinitely small) rigid bodies (yardsticks) and clocks. One can then measure a time-like ds directly by means of a standard clock whose world line contains ds.
>
> Such a definition of the distance element ds would only be illusory if the notions of standard yardstick and standard clock were based on a fundamentally false assumption; this would be the case if the length of a standard yardstick (resp. the rate of a standard clock) depended on the previous history. If this were really so in nature, then there could be no chemical elements with a well determined frequency of spectral lines, but rather the relative frequency of two (specially close) atoms of the same type would, in general, be different. Since this is not the case, it seems to me that the basic hypothesis of the theory is unfortunately unacceptable; even though the

theory cannot fail to impress the reader with its depth and boldness."

Weyl's reply:

"Author's response. I thank Mr. Einstein for giving me the opportunity to reply immediately to the objection that he has raised. In fact I do not believe it to be justified. According to the theory of special relativity, a rigid yardstick always has the same rest length when it comes to rest in an appropriate frame of reference; and, under the same conditions a correctly running clock will always have the same period when measured in proper time (Michelson experiment, Doppler effect). No one suggests that a clock subjected to an arbitrarily violent motion will measure the proper time $\int ds$ ... all the more so if the clock (the atom) is subjected to the influence of a strongly varying electromagnetic field. The most that one can claim in the theory of general relativity, therefore, is that a stationary clock in a static gravitational field and in the absence of an electromagnetic field measures the integral $\int ds$. How a clock moving in an arbitrary way, and under the common influence of an arbitrary electromagnetic and gravitational field behaves, can only be learned after we have developed a dynamics based on the physical laws...

According to the theory developed here, the quadratic form ds^2 behaves with great approximation as in the theory of special relativity, at least outside if the interiors of atoms and with an appropriate choice of coordinates and of the undetermined proportionality factors, ...

From the mathematical point of view, we should emphasize that the geometry developed here is the true infinitesimal geometry. It would be very strange if a partial and inconsequential infinitesimal geometry, with an electromagnetic field stuck on afterwards, were to be realized in nature instead of this one. But naturally, I could be on the wrong track with my whole approach. What we have here is pure speculation, which, it goes without saying, must be compared with experiment. To do this, however, one must first draw the consequences of the theory. I hope for the help of my colleagues in this demanding task."

Postscript June 1955

"This work stands at the beginning of a series of attempts to build a "unified field theory" which was later continued by many others without very convincing success, it seems to me. In particular, as is well known, Einstein himself was occupied without interruption with this problem until the very end. The strongest argument for my theory seemed to be the following: that gauge invariance corresponds to the principle of conservation of electrical charge in the same way as coordinate invariance corresponds to the theorem of conservation of energy-momentum. Later, quantum theory introduced the Schrödinger Dirac potential Ψ of the electron-positron field in which a principle of gauge invariance, obtained from experiment and guaranteeing conservation of charge appeared which related the Ψ's with the electromagnetic potentials ϕ_i in a way similar to that which my speculative theory related the gravitational potentials g_{ik} to the ϕ_i ... Furthermore the ϕ_i were measured in known atomic units rather than unknown cosmological ones. There seems to me to be no doubt, that the principle of gauge invariance has its right place here, and not, as I believed in 1918 in the interplay of gravity and electricity. See, in this connection, my essay 'Geometry and Physics'. "

Let us explain Weyl's last comments in somewhat more modern language. In atomic or subatomic physics, one neglects gravitational effects entirely, so that one deals with special relativity and the Riemannian curvature of the metric vanishes. This means that in Weyl's theory, the only curvature component that enters is that associated to the change of gauge. More precisely, this means that instead of considering the bundle of orthogonal frames, we can, by a choice of a global flat orthonormal basis, reduce the bundle of orthogonal frames to an $\mathbb{R}^+$ bundle consisting of all orthogonal frames which are multiples of some fixed orthonormal frame. Then a connection on this bundle can be regarded as a linear differential form on M which we identify with the four-potential and its curvature with the electromagnetic field, and Weyl's theory reduces to Maxwell's theory with a particular geometric interpretation.

In the currently accepted theories of electromagnetism, the gauge transformation, instead of being regarded as a <u>dilitation</u> factor multiplying our units of length by $\lambda(x) = e^{\psi(x)}$ at each point x of M, is reagarded as a <u>phase</u> factor, multiplying the "electron positron field" by the complex number $e^{i\psi(x)}$ at each point. Thus, instead of the multiplicative group, $\mathbb{R}^+$, introduced by Weyl, one deals with the compact group $U(1)$.

As a consequence of the changed viewpoint, we must consider a principal $U(1)$ bundle. In contrast to Weyl's theory, we are unable to interpret this bundle as a sub-bundle of the bundle of frames, but must consider it as an abstractly given principal bundle.

In the 1920's Cartan introduced a variant of the theory of general relativity in which parallel translation around a closed curve involved a translation of the origin in addition to a Lorentz transformation of the tangent space - the so called "affine connections." This meant that the principal Lorentz group bundle of general relativity had to be replaced by a Poincaré group bundle. While it is possible to regard this bundle as the bundle of affine bases of the tangent space at each point, if we do that, we lose the geometric interpretation of the tangent space at x as the space of tangent vectors to curves passing through x. Thus, once again, we are led to the study of a principal bundle over M, this time with the Poincaré group as structure group. But the main impetus to this more abstract point of view came from the discovery of the various "internal" symmetry groups such as $SU(2)$ and $SU(3)$ relating various types of elementary particles.

This led to the idea, proposed by Yang and Mills that we consider a principal bundle P with general structure group G. The fundamental geometrical object, replacing the electromagnetic field, will be a connection on P. Such a connection, Θ, is called a Yang-Mills field in the physics literature. There are various mathematically equivalent ways of defining a connection, and to fix our sign conventions we shall explicitly recall one of them. The group G acts on P by right multiplication. We shall let R_a denote the map of P into itself defined by

$$R_a(p) = pa^{-1} \quad , \quad a \in G \quad .$$

Let g denote the Lie algebra of G. Then any $\xi \in g$ gives rise to a vector field, ξ_P on P which is vertical and satisfies $R_a{}^*\xi_P = (\mathrm{Ad}\,a\,\xi)_P$ for all $a \in G$. A connection form on P is a g-valued linear differential form, Θ, which satisfies

$$\left.\begin{aligned} R_a{}^*\Theta &= \mathrm{Ad}\,a(\Theta) \\ \Theta(\xi_P) &= \xi \end{aligned}\right\} \text{and} \qquad \text{for all } a \in G \quad .$$

Here $\mathrm{Ad}\,a$ on the right-hand side of the first equation means that we apply $\mathrm{Ad}\,a$ to the image of Θ. The $\Theta(\xi_P)$ in the second equation denotes the value of the one form Θ on the vector field ξ. Evaluating Θ on a general vector field would give us a g valued function on P. The second equation asserts that $\Theta(\xi_P)$ is the constant function assigning to each point of P the element ξ of g. At each point $p \in P$ the linear map $\Theta_p : TP_p \to g$ has a kernel which is denoted by hor_p and called the horizontal subspace at p. It is clear that $dR_a(\mathrm{hor}_p) = \mathrm{hor}_{R_a(p)}$ for all $p \in P$ and $a \in G$ and that giving the family of horizontal subspaces $\{\mathrm{hor}_p\}$ is the same as giving Θ.

We can write any vector field (or any tangent vector at a point) as

$$\zeta = \zeta_{ver} + \zeta_{hor}$$

where ζ_{ver} and ζ_{hor} are the vertical and horizontal components. The curvature form, $\widetilde{F}$, of the connection is the g-valued two form on P defined by

$$\widetilde{F}(\zeta_1, \zeta_2) = d\Theta(\zeta_{1hor}, \zeta_{2hor}) \quad .$$

It is clear that $\widetilde{F}$ is a horizontal form in the sense that $\zeta_{ver} \rfloor \widetilde{F} = 0$ for any vertical vector ζ_{ver} and that

$$R_a^* \widetilde{F} = Ad_a \widetilde{F} \qquad \text{for all} \qquad a \in G \quad .$$

Let $g(P)$ denote the bundle associated to P by the adjoint representation of G on g ; so that $g(P) = P \times g/G$ where $a \in G$ acts on $P \times g$ by sending (p, ξ) into $(pa^{-1}, Ada\, \xi)$. Then the preceeding two properties of $\widetilde{F}$ show that $\widetilde{F}$ determines and is determined by a $g(P)$ valued two form on M which we shall denote by F . With some ambiguity, we shall also call F the curvature of Θ .

A section $s : U \to P$, where U is some open subset of M is called, in the physics literature, a choice of local gauge. Then $A_s = s^*\Theta$ is a g - valued one form defined on U known as the expression of the Yang-Mills field in the local gauge, s . The curvature, F is called the field strength, and $F_s = s^*\widetilde{F}$ is a g - valued two form on U called the field strength in local gauge.

Going back to the special cases $G = \mathbb{R}^+$ or $G = U(1)$, we see that F is now regarded as a $g(P)$ valued two form where g is the (common) Lie algebra of $\mathbb{R}^+$ and $U(1)$ and the adjoint representation is trivial since the groups are abelian. We may thus regard $g(P)$ as the trivial bundle $M \times g$ and F as a g - valued two form in this case. But now, in order to think of eF as a scalar valued two form, we must regard e as lying in g^* , the dual space of g . In the next section, we discuss the generalization of the discussion of Sections 2 and 4 when G is an arbitrary Lie group.

6. The Symplectic Mechanics of a Classical Particle in the Presence of a Yang-Mills Field

Let $P \to M$ be a principal bundle with structure group G. Let $P^{\#} \to T^*M$ denote the pullback of P to T^*M via the projection of $T^*M \to M$ so $P^{\#} = \pi^{\#}P$. Let Θ be a connection on P and $\Theta^{\#}$ the induced connection on $P^{\#}$. Let Q be a Hamiltonian G - space ; this means that

i) Q is a symplectic manifold with symplectic form, Ω, that

ii) G acts on Q as a group of symplectic diffeomorphisms, so that there is a homomorphism of the Lie algebra g of G into the algebra of Hamiltonian vector fields, and that

iii) we are given a lifting of this homomorphism to a homomorphism of g into the Lie algebra of functions on Q (where the Lie algebra structure is given by Poisson bracket).

Thus to each $\xi \in g$ we get a function f_ξ on Q and a Hamiltonian vector field ξ_Q on Q so that

$$\xi_Q \lrcorner \Omega = f_\xi \quad .$$

We can thus form the moment map $\Phi : Q \to g^*$ where

$$\langle \xi , \Phi(z) \rangle = f_\xi(z) \qquad z \in Q \quad .$$

Here g^* denotes the dual space of g and $\langle \ , \ \rangle$ denotes the pairing between g and g^*. The group G acts on g^* via the contragredient to the adjoint representation and the map Φ commutes with the action of G. Since $P^{\#}$ is a principal G bundle and G acts on Q, we can form the associated bundle $Q(P^{\#}) = P^{\#} \times Q/G$. The point of this section is to show how the choice of a connection on P determines a symplectic structure on $Q(P^{\#})$, generalizing the constructions of Sections 2 and 4. We must define a symplectic form σ on $Q(P^{\#})$. We shall give two equivalent definitions. From the first it will be clear that $d\sigma = 0$ and from the second it will be clear that σ is non-degenerate. Let $\rho : P^{\#} \times Q \to Q(P^{\#})$ be the natural projection and let ω denote the symplectic form of T^*M pulled back to $P^{\#}$. Then we define a form $\omega_{Q,\Theta}$ by

$$\omega_{Q,\Theta} = \omega + \langle \Theta^{\#} , \Phi \rangle + \Omega \quad .$$

Since Φ is a g^* valued function on Q (considered as defined on $P^{\#} \times Q$) and $\Theta^{\#}$ is a g valued one-form on $P^{\#}$ (considered as defined on $P^{\#} \times Q$) we see that $\langle \Theta^{\#}, \Phi \rangle$ is a scalar valued one-form and thus $d\langle \Theta^{\#} , \Phi \rangle$ is a scalar valued two-form on $P^{\#} \times Q$. It is clear that

$$d\omega_{Q,\Theta} = 0$$

since all three summands are closed. We claim that

$$\omega_{Q,\Theta} = \rho^*\sigma$$

where σ is a two-form on $Q(P^{\#})$. To prove this we must show

a) that $\omega_{Q,\Theta}$ is invariant under the action of G

and

b) that $\xi^{\#} \lrcorner \omega_{Q,\Theta} = 0$ for any $\xi^{\#}$ tangent to the fiber of the projection ρ.

To prove a) observe that ω and Ω are invariant under G and that Φ and Θ transform in contragredient manner so that $\langle \Theta^{\#}, \Phi \rangle$ is invariant.

To prove b) we may choose $\xi^{\#} = \xi_{P^{\#}} + \xi_Q$.

Now

$$R_a{}^*\Theta^{\#} = Ad_a\Theta^{\#}$$

and

$$\Theta^{\#}(\xi_{P^{\#}}) = \xi \qquad \text{for all} \qquad \xi \in g \quad .$$

If we let $D_{\xi_{P^{\#}}}$ denote the Lie derivative with respect to the vector field $\xi_{P^{\#}}$ then the infinitesimal version of the first condition is

$$D_{\xi_{P^{\#}}}\Theta = ad_\xi\Theta \quad .$$

Since $D_{\xi_{P^{\#}}}\Theta = \xi_{P^{\#}} \lrcorner d\Theta + d(\xi_{P^{\#}} \lrcorner \Theta)$ and, by the second condition $\xi_{P^{\#}} \lrcorner \Theta = \xi$ is a constant so $d(\xi_{P^{\#}} \lrcorner \Theta) = 0$, we have

$$\xi_{P^{\#}} \lrcorner d\Theta = ad_\xi\Theta \quad .$$

Now

$$d\langle \Theta^{\#}, \Phi \rangle = \langle d\Theta^{\#}, \Phi \rangle - \langle \Theta^{\#} \wedge d\Phi \rangle$$

where $\langle \wedge \rangle$ denotes the exterior multiplication of a g valued form with a g^* valued form derived from the pairing of g with g^*. Then

$$\begin{aligned}\xi^{\#} \lrcorner d\langle \Theta^{\#}, \Phi \rangle &= \langle \xi_{P^{\#}} \lrcorner d\Theta^{\#}, \Phi \rangle - \langle \xi_{P^{\#}} \lrcorner \Theta^{\#}, d\Phi \rangle + \langle \Theta^{\#}, \xi_Q \lrcorner d\Phi \rangle \\ &= \langle ad_\xi\Theta^{\#}, \Phi \rangle - \langle \xi, d\Phi \rangle + \langle \Theta^{\#}, ad^{\#}{}_\xi\Theta^{\#} \rangle \quad .\end{aligned}$$

The first and third terms cancel so that

$$\begin{aligned}\xi^{\#} \lrcorner d\langle \Theta^{\#}, \Phi \rangle &= -\langle \xi, d\Phi \rangle \\ &= -\xi_Q \lrcorner \Omega \quad .\end{aligned}$$

Since $\xi^{\#} \lrcorner \omega = 0$ as ω comes from T^*M, and since $\xi_{P^{\#}} \lrcorner \Omega = 0$ so

$\xi^{\#} \rfloor \Omega = \xi_Q \rfloor \Omega$ we see that

$$\xi^{\#} \rfloor \omega_{Q,\Theta} = \xi^{\#} \rfloor (\langle d\Theta^{\#}, \Phi \rangle + \Omega) = 0$$

proving b).

We have thus defined a form σ on $Q(P^{\#})$ and since $d\rho^*(\sigma) = 0$ it follows that $d\sigma = 0$. We must show that σ is non-degenerate. There is an alternative way of describing the symplectic structure on $Q(P^{\#})$ which is useful for this purpose and for other computations. Given the connection, Θ, we can write every tangent vector ζ_z at $z \in Q(P^{\#})$ as the sum of a horizontal vector $\widetilde{V}_{Qz}$ and a vertical vector w_z. Here V is a vector field on T^*M and $\widetilde{V}_Q$ the corresponding horizontal vector field on $Q(P^{\#})$. if we let $\widetilde{V}$ denote the horizontal vector field on $P^{\#}$ corresponding to V, and $\rho : P^{\#} \times Q \to Q(P^{\#})$ the natural projection, then, for any $(n, y) \in P^{\#} \times Q$ with $\rho(n, y) = z$ we have

$$d\rho_{(n,y)} \widetilde{V} = \widetilde{V}_{Qz} .$$

If w_z is a vertical tangent vector at z then we can write

$$w_z = d\rho_{(n,y)} v_{(n,y)}$$

where $v_{(n,y)}$ is a tangent vector to Q at y. If we choose some different $v'_{(na,a^{-1}y)}$ with

$$w_z = d\rho_{(na,a^{-1}y)} v'_{(na,a^{-1}y)}$$

then we must have

$$v = dL(a)_{a^{-1}} v'$$

where $L(a) : Q \to Q$ denotes the action of $a \in G$ on Q. In particular, let w_1 and w_2 be two vertical vectors at $z \in Q(P^{\#})$. Let us choose tangent vectors v_1 and v_2 to Q at y, corresponding to w_1 and w_2. Then $\Omega(v_1 v_2)$ is independent of the choice of v_1 and v_2. We let $\widetilde{\Omega}$ denote the form on $Q(P^{\#})$ given by

$$\widetilde{\Omega}(\zeta_1 \wedge \zeta_2) = \Omega(v_1 \wedge v_2)$$

where

$$\zeta_1 = \widetilde{V}_1 + w_1$$

$$\zeta_2 = \widetilde{V}_2 + w_2 .$$

The map $\Phi : Q \to g^*$ induces a map $\widetilde{\Phi} : Q(P^{\#}) \to g^*(P^{\#})$ given by

$$\widetilde{\Phi}([n, y]) = [n, \Phi(y)]$$

where $[n, y] = \rho(n, y)$ denotes the equivalence class of (n, y) in $Q(P^{\#})$ with similar notation for $g^*(P^\#)$. We recall that curvature form, F, is a $g(P^\#)$ valued two-form on T^*M corresponding to the "basic" g valued two-form $\widetilde{F}$ on $P^\#$ as described in the preceding section. We can pull F back to $Q(P^\#)$ via the projection $Q(P^\#) \to T^*M$ and consider it as $g(P^\#)$ valued two-form defined on $Q(P^\#)$ which we continue to denote by F. Then $\widetilde{\Phi} \cdot F$ is a scalar two-form on $Q(P^\#)$. Finally, we denote the pullback of the symplectic form on T^*M to $\widetilde{P} \times_G Q$ by ω. Then we claim that

$$(*) \qquad \sigma = \omega + \widetilde{\Phi} \cdot F + \widetilde{\Omega} \quad .$$

To verify this formula, it is enough to check it when σ is evaluated on $\widetilde{V}_{Q1} \wedge \widetilde{V}_{Q2}$, on $\widetilde{V}_{Q1} \wedge w_2$ and on $w_1 \wedge w_2$. In the first case we have

$$\begin{aligned} \sigma(\widetilde{V}_{Q1} \wedge \widetilde{V}_{Q2}) &= \rho^*\sigma(\widetilde{V}_1 \wedge \widetilde{V}_2) \\ &= (\omega + \Phi \cdot d\Theta + d\Phi \wedge \Theta + \Omega)(V_1 \wedge V_2) \quad . \end{aligned}$$

But $\Theta(\widetilde{V}) = 0$ for any horizontal field $\widetilde{V}$ and $\Omega(\widetilde{V}_1 \wedge \widetilde{V}_2) = 0$ so the last two terms vanish. Also $\widetilde{\Omega}(\widetilde{V}_{Q1} \wedge \widetilde{V}_{Q2}) = 0$. Thus

$$\sigma(\widetilde{V}_{Q1} \wedge \widetilde{V}_{Q2}) = (\omega + \Phi \cdot d\Theta)(\widetilde{V}_1 \wedge \widetilde{V}_2) = (\omega + \widetilde{\Phi} \cdot F)(\widetilde{V}_{Q1} \wedge \widetilde{V}_{Q2}) \quad .$$

In the second case, the value of both sides on $\widetilde{V}_1 \wedge w_2$ is seen to be zero. In the third case $\sigma(w_1 \wedge w_2) = \Omega(v_1 \wedge v_2) = \widetilde{\Omega}(w_1 \wedge w_2)$ and the first two terms on the right of $(*)$ vanish, completing the proof of $(*)$.

Notice that from $(*)$ it is clear that σ is non-singular. Indeed, if $(f + w) \rfloor \sigma = 0$, then we must have $w \rfloor \widetilde{\Omega} = 0$ implying that $w = 0$. Then $V \rfloor \sigma = \widetilde{V} \rfloor (w + \Theta \cdot F)$. Now F is really a form on M (since the connection Θ comes from a connection on M). Thus if $V = a\frac{\partial}{\partial p} + b\frac{\partial}{\partial q}$ then the coefficient of dp in $\widetilde{V} \rfloor (\omega + \Theta \cdot F)$ is b so $\widetilde{V} \rfloor (\omega + \Theta \cdot F) = 0$ implies $B = 0$. But then $\widetilde{V} \rfloor (\omega + \Phi \cdot F) = adq$ so we also conclude that $a = 0$.

Suppose that H is a function defined on T^*M and let us consider the Hamiltonian vector field ξ_H on $Q(P^\#)$ determined by

$$\xi_H \rfloor \sigma = -dH \quad .$$

Let us write

$$\xi_H = \widetilde{V}_q + \widetilde{V}_p + w$$

where $V_q = a\frac{\partial}{\partial q}$ and $V_p = b\frac{\partial}{\partial p}$ in terms of local coordinates on T^*M. Then

$$w \rfloor \sigma = w \rfloor \widetilde{\Omega} = 0$$

implying

$$w = 0 \quad .$$

The coefficient of dp in $\xi_H \lrcorner \sigma$ is $-a$ so we see that

$$a = \frac{\partial H}{\partial p} \quad .$$

The dq term in $\xi_H \lrcorner \sigma$ is

$$bdq + \widetilde{\Phi} \cdot (a \frac{\partial}{\partial q} \lrcorner F)$$

which must equal $-\frac{\partial H}{\partial q} dq$.

Now $\widetilde{\Phi} \cdot (a \frac{\partial}{\partial q} \lrcorner F)$ is a covector which we can consider as a point of T^*M . Thus we obtain the equations

$$\frac{dq}{dt} = \frac{\partial H}{\partial p}$$

$$\frac{dp}{dt} + \widetilde{\Phi} \cdot \left(\frac{dq}{dt} \lrcorner F\right) = \frac{\partial H}{\partial q} \quad .$$

These equations, together with the equation $w = 0$ are the Hamiltonian equations corresponding to an H that comes from T^*M . Notice that $w = 0$ says that the solution curves are all horizontal, i. e. obtained by parallel transport from their projections onto T^*M .

Let $s : U \to P_{|U}$ be a section of P defined on some open set $U \subset M$, and let $A = s^*\Theta$. Then s induces an identification of $P^\#$ with $T^*U \times G$ and of $P^\# \times Q$ with $T^*U \times Q$. It follows that the local expression for σ is

$$\sigma_{|T^*U} = \omega + d(\Phi \cdot A) + \Omega \quad .$$

We can give a slightly different interpretation to this local expression for σ and for the corresponding Hamiltonian equations for a Hamiltonian H coming from T^*M . On the space $T^*U \times Q$ we have the product symplectic structure with form $\omega + \Omega$. We can think of $\Phi \cdot A$ as a section of $T^*U \times Q$ over $U \times Q$ and hence as giving a transformation ψ_A of $T^*U \times Q$ into itself defined by

$$\psi_A(q, p, r) = (q, p + \Phi(r) \cdot A(q), r) \qquad (p, q) \in T^*U$$
$$r \in Q \quad .$$

Then we can rewrite $(*)$ as

$$\sigma_{|U} = \psi_{A*}(\omega + \Omega) \quad .$$

Let H be a Hamiltonian coming from T^*M . If C is a solution curve of H relative to σ , then (locally) $\psi_A^{-1} \circ C$ is a solution curve of

$$H_A = H \circ \psi_A^{-1}$$

relative to the product symplectic form $\omega + \Omega$.

In what follows we will be particularly interested in the following choices for Q :

i) $Q = \mathfrak{O}$ is an orbit of G acting on g^* in which case Φ is the injection of $\mathfrak{O}$ into g^* , or

ii) $Q = T^*G$ where G acts on itself by right multiplication, and we consider the corresponding induced action of G on T^*G . We make the left invariant identification of T^*G with $G \times g^*$ so that $b \in G$ acts on (c, α) by $b(c, \alpha) = (cb^{-1}, b\alpha)$. The moment map in this case is easily seen to be given as

$$\Phi(c, \alpha) = \alpha \quad .$$

For example, we may take $G = U(1)$. Since $U(1)$ is abelian, its action on $u(1)^*$ is trivial, and so its orbits consist of points e , where we may identify e with the electric charge. If we set $H(q, p) = \frac{1}{2} p^2$ then equations of motion become the Lorentz equations, supplemented by the equation $\frac{de}{dt} = 0$.

The equation $\frac{de}{dt} = 0$ asserts that charge is conserved. For a more general group, the analogous assertion is that the lifted curve lies on the bundle associated to a fixed orbit. (For instance, in case we take $Q = T^*G$ then, in local coordinates, the assertion is that $\alpha(t)$ must lie on a fixed orbit, or, more generally, that $(\alpha(t))$ lies in a fixed orbit.) Now, the program of geometric quantization associates (at least in many cases) unitary representations to certain orbits. Thus the condition of "staying on a fixed orbit" is the classical version of conservation of "internal quantum number." For example, for $SU(2)$ the orbits are spheres, and the spheres of half integer radius $s = k/2$ correspond to representations of $SU(2)$ of spin $k/2$. Thus the condition of lying on a fixed sphere corresponds to conservation of isotopic spin.

Another case of importance is where P is an extension of the bundle of frames of some (pseudo) Riemannian manifold, M , of signature (k, ℓ) . If we let P_M denote the bundle of orthonormal frames, then P_M is a principal bundle with structure group $O(k, \ell)$. Let P_G be some other principal bundle with structure group G and we suppose that P is the product bundle $P = P_M \times P_G$ with structure group $O(k, \ell) \times G$.

The metric induces a connnection (the Levi-Civita connection) on P_M and we assume that we are given a connection on P_G . These together determine a connection on P whose curvature we shall write as $R + F$. (Here R is the Riemannian curvature which is a two-form with values in the Lie algebra of $O(k, \ell)$ while F is the curvature of P_G with values in g .)

An orbit of $O(k, \ell) \times G$ acting on the dual of its Lie algebra can be written as $\mathcal{O}_M \times \mathcal{O}$ where $\mathcal{O}_M$ is an orbit of $O(k, \ell)$ and $\mathcal{O}$ is an orbit of G, each group acting on the dual of its Lie algebra.

Let $o(k, \ell)^*$ denote the dual of the Lie algebra of $O(k, \ell)$. As indicated in Section 4, we have a natural identification of $o(k, \ell)^*(P)$ with $\wedge^2 T(M)$, the anti-symmetric two tensors on M. Now we can pair an anti-symmetric two tensor, S, at $x \in M$ with F_x to obtain a point of $g(P^{\#})$ which we can then pair with $\widetilde{\Phi} \in g^*(P^{\#})$ to obtain real number $\widetilde{\Phi} \cdot F \cdot S$. For any function f of a real variable, we can consider the constraints $*S \wedge p = 0$ and $H = f(\Phi \cdot F \cdot S)$ and proceed as in Section 4 to get the equations of a spinning particle.

There is an alternative method for deriving the equations of motion which avoids the cotangent bundle altogether and is more in the spirit of Cartan's approach. Let L be a closed subgroup of G and let P_L be a principal L bundle over M and P_G and extension of P_L to a principal G bundle. Let Q be a symplectic (or presymplectic) manifold with a Hamiltonian G action, and let U be a submanifold of Q invariant under L. We can identify $U(P_L)$ as a submanifold of $Q(P_G)$. Now let Θ be a connection on P_G. We can construct the closed two form $d(\Phi \cdot \Theta) + \Omega$ on $P_G \times Q$. If ρ now denotes the projection of $P_G \times Q$ onto $Q(P_G)$, the proof we gave above shows that there is a form σ on $Q(P_G)$ so that

$$d(\Phi \cdot \Theta) + \Omega = \rho^* \sigma \quad .$$

Of course σ will be degenerate (unless M reduces to a point). We can form the restriction of σ to $U(P_L)$ which we denote by υ. The form υ is closed, and still has some degeneracy. However, it will sometimes be the case that the degeneracy of υ (i.e. the dimension of its null space at any point) is smaller than that of σ, and, in fact, υ might have a one dimensional null space at each point. The differential equations giving this null foliation can then be considered as giving the equations for trajectories on M. For example, let us take L to be the Lorentz group and G to be the Poincaré group. Suppose, to fix the ideas, we look at the six dimensional orbits of G acting on the dual of its Lie algebra which correspond to positive mass and zero spin. Let Q be one such orbit with mass m and let U be the submanifold of Q which projects into 0 under the projection $g^* \to \ell^*$ dual to the injection of ℓ into g. Then U is three dimensional. If we take P_L to be the bundle of orthonormal frames on a Lorentzian manifold, we may take P_G to be the bundle of affine frames. The Lorentzian metric gives rise to an affine connection, i.e. a connection on P_G. The bundle $U(P_L)$ is seven dimensional and the null trajectories of the induced presymplectic structure give the geodesics on M. The point of this construction is that it proceeds directly from the connection and eliminates the Hamiltonian altogether.

Suppose that L is a closed subgroup of two different groups, G_1 and G_2. This induces homomorphisms of $\ell \to g_1$ and $\ell \to g_2$, together with a linear representation of L on g_1 and g_2. Suppose that we are given a linear map: $\lambda : g_1 \to g_2$ which is equivarient for the action of L and such that the diagram

$$\begin{array}{ccc} & \nearrow & g_1 \\ \ell & & \downarrow \lambda \\ & \searrow & g_2 \end{array}$$

commutes. Let Q_1 and Q_2 be presymplectic manifolds with forms Ω_1 and Ω_2 with Hamiltonian actions of G_1 and G_2 respectively and let f_1 and f_2 be L equivalent maps of manifolds U_i on which ℓ acts into Q_1 and Q_2. We say that f_1 and f_2 are consistent with an H map $f : U_1 \to U_2$ if

$$f_1{}^*\Omega_1 = f^*f_2{}^*\Omega_2$$

and the diagram

$$\begin{array}{ccccccc} U_1 & \xrightarrow{f_1} & Q_1 & \xrightarrow{\Phi_1} & g_1{}^* & & \\ & & & & & \searrow & \\ f \downarrow & & & & \uparrow \lambda^* & & \ell^* \\ & & & & & \nearrow & \\ U_2 & \xrightarrow[f_2]{} & Q_2 & \xrightarrow[\Phi_2]{} & g_2{}^* & & \end{array}$$

commutes. (Notice that the right-hand triangle in this diagram automatically commutes, and that if Q_1 and Q_2 are orbits of G_1 and G_2 acting on $g_1{}^*$ and $g_2{}^*$ so that the Φ_i determine the Ω_i then this second condition implies the first.) Let P_L be a principal L bundle and P_{G_1} and P_{G_2} extensions of this bundle to G_1 and G_2. We say that the connections Θ_1 and Θ_2 on P_{G_1} and P_{G_2} are consistent if

$$\Theta_2 = \lambda\Theta_1 \quad \text{when restricted to } P_L \quad .$$

Notice that on $P_L \times U$ the above consistency conditions imply that $f_1{}^*\Omega_1 = f^*f_2{}^*\Omega_2$ and

$$(\Phi_1 \circ f_1) \cdot \Theta_1 = (\lambda_1{}^*\Phi_2 \circ f_2 \circ f) \cdot \Theta_1 = (\Phi_2 \circ f_2 \circ f) \cdot \Theta_2 \quad .$$

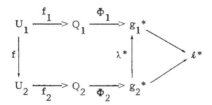

From these two equations it clearly follows that the induced map $f : U_1(P_L) \to U_2(P_L)$ satisfies $f^*U_2 = U_1$ where U_i denote the induced presymplectic forms on $U_i(P_L)$.

Let us examine this construction in the case that the Lie algebra, ℓ, has a complement n in g which is invariant under the adjoint action of ℓ in g, so that

$$g = \ell \oplus n \quad \text{and} \quad [\ell, n] \subset n \quad .$$

Then we can write $\Theta = \Theta_\ell + \Theta_n$ where Θ_ℓ denotes the ℓ component of Θ and Θ_n the n component. Both of these forms are defined on P_G but notice that the restriction of Θ_ℓ to P_L induces a connection on P_L. Let $\mathcal{H}$ denote projection onto the horizontal component of this connection. Since the tangent space to the fiber of P_L is spanned by vectors coming from ℓ we see that Θ maps vertical tangent vectors into ℓ and hence that Θ_n vanishes on vectors tangent to the fiber of P_L. Thus Θ_n, when restricted to P_H, determines a one form $\hat{\Theta}_n$ on M with values in the vector bundle $n(P_H)$. Let F^ℓ denote the curvature of Θ_ℓ so that F^ℓ is a two form on M with values on $\ell(P_L)$ which corresponds to the ℓ - valued two form $d\Theta_\ell \circ \mathcal{H}$ on P_L. The connection Θ_ℓ induces a covariant (exterior) derivative on any vector bundle associated to P_H which we shall denote by d_{Θ_ℓ}. Thus, for example, we can form $d_{\Theta_\ell} \hat{\Theta}_n$ which will be a two form on M with values in $n(P_L)$ which corresponds to the n - valued two from $d\Theta_n \circ \mathcal{H}$ on P_L. We can also decompose the function Φ into components, $\Phi = \Phi_\ell + \Phi_n$ and thus $\tilde{\Phi} = \tilde{\Phi}_\ell + \tilde{\Phi}_n$. It is easy to modify the preceeding arguments to show that the following formula for ν holds:

$$\nu = \tilde{\Phi}_\ell F^\ell + d\tilde{\Phi}_n \wedge \hat{\Theta}_n + \tilde{\Phi}_n d_{\Theta_h} \hat{\Theta}_n + \tilde{\Omega}_\ell \quad .$$

In this formula F^ℓ, Θ_n and $d_{\Theta_\ell} \hat{\Theta}_n$ are vector bundle valued forms on M, pulled back to $U(P_L)$. The form $\tilde{\Omega}_\ell$ is obtained from Ω via the projection, as described above, but this time relative to the connection Θ_ℓ.

We will illustrate the use of this construction in Section 10 in the case where g is the Poincaré algebra so that n corresponds to translations, and also in the case that g is the conformal algebra (so that n will be nine dimensional).

7. The Energy Momentum Tensor and the Current

In the preceeding section we wrote down equations of motion for point particles in the presence of an external Yang-Mills field. In the present section we set up some formalism which allows us to discuss the behavior of continuous media. In particular we shall write down the analogues of the equations expressing the conservation of charge and the Einstein identity involving the covariant divergence of the energy momentum tensor when the electromagnetic field is replaced by an arbitrary Yang-Mills field. The justification for these equations will be given in the next section, and the reader might prefer to look at that section first, before delving into the formalism developed here.

We need to introduce some differential complexes associated to the bundle P and the connection Θ. We will be considering differential forms with values in the associated vector bundles $g(P)$ and $g^*(P)$.

Let E_1, E_2 and E_3 be vector bundles over the same manifold M and suppose that we are given a bilinear map $\beta : E_1 \times E_2 \to E_3$. Then the tensor product of β with exterior multiplication induces a map $E_1 \otimes \wedge^k T^*M \times E_2 \otimes \wedge^\ell T^*M \to E_3 \otimes \wedge^{k+\ell} T^*M$. Taking sections, we see that β induces a "exterior multiplication" of E_1 valued forms with E_2 valued forms to obtain E_3 valued forms. If B_1, B_2 and B_3 are vector spaces on which we are given linear representations of G, then each of them gives rise to associated bundles $E_1 = B_1(P)$ etc. A G - equivariant map from $B_1 \times B_2 \to B_3$ gives a $\beta : E_1 \times E_2 \to E_3$ and hence a corresponding exterior multiplication. We wish to use this construction for the following choices:

$B_1 = B_2 = B_3 = g$ and the map of $g \times g \to g$ is given by Lie bracket

$B_1 = g$ and $B_2 = B_3 = g^*$ and $g \times g^* \to g^*$ is given by the coadjoint representation of g acting on g^*

$B_1 = g$, $B_2 = g^*$, $B_3 = \mathbb{R}$ and the map is evaluation.

Let

$\wedge^k(M)$ denote the space of smooth scalar valued exterior differential forms on M of degree k

$\mathfrak{a}^k(M)$ the space of smooth $g(P)$ valued k forms, i.e. of smooth sections of $g(P) \otimes \wedge^k T^*M$

and

$\mathfrak{a}^{*k}(M)$ the space of smooth $g^*(P)$ valued k forms.

Since g acts on itself via the adjoint representation and on g^* by the co-adjoint representation, we get pairings

$$\mathcal{A}^k \times \mathcal{A}^\ell \to \mathcal{A}^{k+\ell} \quad \text{denoted by} \quad \omega^k \otimes \omega^\ell \to [\omega^k, \omega^\ell]$$

$$\mathcal{A}^k \times \mathcal{A}^{*\ell} \to \mathcal{A}^{*(k+\ell)} \quad \text{denoted by} \quad \omega^k \otimes \sigma^\ell \to \omega^k \# \sigma^\ell$$

and since we have the evaluation map $g^* \times g \to \mathbb{R}$

$$\mathcal{A}^k \times \mathcal{A}^{*\ell} \to \Lambda^{(k+\ell)} \quad \text{denoted by} \quad \omega^k \otimes \sigma^\ell \to \omega^k \wedge \sigma^\ell \quad .$$

The direct sum $\mathcal{A} = \oplus \mathcal{A}^k$ is a Lie superalgebra under the bracket and it is represented on the space $\mathcal{A}^* = \oplus \mathcal{A}^{*k}$. Further one has the identity

$$[\omega^p, \omega^q] \wedge \sigma^r = \omega^p \wedge (\omega^q \# \sigma^r) \quad .$$

Let $\mathrm{Aut}(P)$ denote the group of all automorphisms of the principal bundle, P. Thus $\varphi \in \mathrm{Aut}(P)$ is a diffeomorphism of P onto itself which satisfies

$$R_a \varphi = \varphi R_a \qquad \text{for all} \qquad a \in G \quad .$$

Any such φ carries fibers into fibers and hence induces a tranformation of M into itself which we denote by $\bar{\varphi}$. The subgroup of $\mathrm{Aut}(P)$ consisting of those φ for which $\bar{\varphi} = \mathrm{id}$ is known as the gauge group and will be denoted by $\mathrm{Gau}(P)$.

The "Lie algebra" of $\mathrm{Aut}(P)$, denoted by $\mathrm{aut}(P)$ consists of all smooth vector fields, ζ, on P which satisfy

$$R_a{}^* \zeta = \zeta \qquad \text{for all} \qquad a \in G \quad .$$

We can think of such a vector field as being a section of a vector bundle defined on M: that is, for each $x \in M$, define the vector space $E(P)_x$ by

$$E(P)_x = \{\Gamma TP|_{\pi^{-1}x}\}^G = \{\text{invariant sections of } TP \text{ along } \pi^{-1}x\} \quad .$$

These fit together to form a vector bundle $E(P)$ and we have the exact sequence of vector bundles

$$0 \longrightarrow g(P) \xrightarrow{\ i\ } E(P) \xrightarrow{\ \dot{\pi}\ } T(M) \longrightarrow 0$$

Taking smooth sections of these bundles gives the sequence

$$0 \longrightarrow \mathcal{A}^0(M) \longrightarrow \mathrm{aut}(P) \longrightarrow \mathcal{V}(M)$$

where $\mathcal{V}(M)$ denotes the algebra of all smooth vector fields on M. This shows that we can regard $\mathcal{A}^0(M)$ as $\mathrm{gau}(P)$, the "Lie algebra" of the gauge group $\mathrm{Gau}(P)$.

Giving a connection, Θ, on P is equivalent to giving a splitting of the sequence

$0 \to g(P) \to E(P) \to T(M) \to 0$. That is, giving Θ is the same as giving two vector bundle maps $\ell : E(P) \to g(P)$ and $\lambda : T(M) \to E(P)$ which satisfy

$$\ell i = \mathrm{id} \quad \text{and} \quad \lambda \dot{\pi} = \mathrm{id}$$

so we have the diagram

$$0 \longrightarrow g(P) \xrightarrow[\ell]{i} E(P) \xrightarrow[\lambda]{\dot{\pi}} T(M) \longrightarrow 0 \quad .$$

Two connections, Θ and Θ' give us two maps, λ and λ', which satisfy $\dot{\pi}(\lambda - \lambda') = 0$ so

$$\lambda - \lambda' = i\tau$$

where $\tau : T(M) \to g(P)$ so $\tau \in \mathfrak{a}^1(M)$. Conversely, given Θ and $\tau \in \mathfrak{a}^1(M)$ we get a new connection Θ'. Thus the set of all connections is an affine space associated to the linear space $\mathfrak{a}^1(M)^1$. In particular we may identify the space of "infinitesimal connections" i.e. the "tangent space to the space of connections at Θ" with $\mathfrak{a}^1(M)$.

An element, J, of $\mathfrak{a}^{n-1*}(M)$ is called a current, where $n = \dim M$. Let $\mathfrak{a}_0^1(M)$ denote the subspace of $\mathfrak{a}^1(M)$ consisting of forms with compact support. Then, if M is oriented (as we shall assume, for simplicity from now on) we can regard $J \in \mathfrak{a}^{n-1*}(M)$ as defining a linear function on $\mathfrak{a}_0^1(M)$ by integration:

The value of J on $\tau \in \mathfrak{a}^1(M)$ is given by

$$\langle J, \tau \rangle = \int_M J \wedge \tau \quad .$$

A connection Θ defines a covariant differential, $d_\Theta : \mathfrak{a}^k \to \mathfrak{a}^{k+1}$ and $\mathfrak{a}^{*k} \to \mathfrak{a}^{*k+1}$.

It is defined as follows: Let E be any vector space with a given linear representation of G on E. We can form the associated bundle $E(P)$ and consider the space $\wedge^k(M, E(P))$ consisting of k forms on M with values in the vector bundle $E(P)$. We can identify the space $\wedge^k(M, E(P))$ with the set of E valued k - forms Ω on P with the properties

$$v \lrcorner \Omega = 0 \qquad \text{for any vertical vector field}$$

and

$$R_a^* \Omega = a \cdot \Omega \qquad \text{for all} \quad a \in G \quad .$$

Let $\mathcal{H}$ denote projection onto the horizontal space (determined by Θ) and, similarly, for any E - valued form, σ, define $\mathcal{H}\sigma$ by

$$\mathcal{H}\sigma(\xi_1, \cdots, \xi_k) = \sigma(\mathcal{H}\xi_1, \cdots, \mathcal{H}\xi_k) \quad .$$

Then for any $\Omega \in \wedge^k(M, E)$ (thought of as a form on P) define

$$d_\Theta \Omega = \mathcal{H}(d\Omega) \quad .$$

Taking $E = g$ or g^* we get $d_\Theta : \mathfrak{a}^k \to \mathfrak{a}^{k+1}$ and $d_\Theta : \mathfrak{a}^{*k} \to \mathfrak{a}^{*k+1}$. It is easy to check that

$$d(\omega^k \wedge \sigma) = d_\Theta \omega^k \wedge \sigma + (-1)^k \omega^k \wedge d_\Theta \sigma$$

for $\omega^k \in \mathfrak{a}^k$ and $\sigma \in \mathfrak{a}^k$.

Now let us consider an $s \in \mathfrak{a}^0$. We can think of s as a vertical vector field on P which satisfies

$$R_a^* s = s \quad .$$

Equally well, we can think of s as giving a g - valued function, call it $\tilde{s} : P \to g$ satisfying

$$R_a^* \tilde{s} = (Ad_a)\check{S} \quad .$$

The relation between s and $\tilde{s}$ is clearly given by

$$\tilde{s} = s \,\lrcorner\, \Theta$$

(for any connection form Θ).

Let D_s denote Lie derivative with respect to the vector field s. Then

$$\begin{aligned} D_s \Theta &= d(s \,\lrcorner\, \Theta) + s \,\lrcorner\, d\Theta \\ &= d\tilde{s} + s \,\lrcorner\, d\Theta \quad . \end{aligned}$$

We claim that for any vertical vector field s we have

$$s \,\lrcorner\, d\Theta = [\tilde{s}, \Theta]_g$$

(where the bracket on the right is the bracket in g; i.e. $[\tilde{s}, \Theta](\xi) = [\tilde{s}, \Theta(\xi)]$ for any bector field ξ). To prove this equality, it is enough to prove it at every point and since both sides depend purely algebraically on s, it is enough to prove it in the special case where $s = \hat{\xi}$, the vector field on P coming from the right action of $\xi \in g$ on P. Now

$$\hat{\xi} \,\lrcorner\, \Theta = \xi \quad , \quad \text{a constant element of } g$$

so

$$d(\hat{\xi} \,\lrcorner\, \Theta) = 0$$

while

$$D_{\hat{\xi}}\Theta = [\xi, \Theta] \quad .$$

But $D_{\hat{\xi}}\Theta = d(\hat{\xi} \lrcorner \Theta) + \hat{\xi} \lrcorner d\Theta$ proving that $\hat{\xi} \lrcorner d\Theta = [\xi, \Theta]$. Thus

$$D_s\Theta = d\tilde{s} + [\tilde{s}, \Theta] \quad .$$

But for any element of $\mathfrak{G}$ it is easy to check that $d_\Theta \omega = d\omega + [\omega, \Theta]$. Thus we have proved that

$$D_s\Theta = d_\Theta s \quad .$$

Now suppose that s has a compact support. Then

$$J \wedge D_s\Theta = J \wedge d_\Theta s$$

$$= d(J \wedge s) - d_\Theta J \wedge s$$

and so

$$\int J \wedge D_s\Theta = -\int d_\Theta J \wedge s \quad .$$

In the next section we shall present a geometrical argument which suggests that a reasonable condition to impose on a current J is that it be orthogonal to all $\tau \in \mathfrak{a}_0^1(M)$ of the form $\tau = D_s\Theta$ for s of compact support. From the above equation we see that this is equivalent to the condition

$$d_\Theta J = 0 \quad .$$

This equation is the generalization to the Yang-Mills case of the equation $dJ = 0$ which asserts the conservation of charge in electromagnetic theory.

Recall that a pseudo-Riemannian metric g induces a connection on the bundle $T(M)$ and all associated tensor bundles. It therefore also induces a covariant differentiation operator $g : T(M) \to T(M) \otimes T^*(M)$ and, more generally, form $R(M) \to R(M) \otimes T^*(M)$ where $R(M)$ denotes any tensor bundle.

We now examine the Einstein identity involving the covariant divergence of the energy momentum tensor, T. In local coordinates this is usually written as $T^{ik}{}_{|k} = 0$ where $T^{ij}{}_{|k}$ denote the components of the covariant differential of T and the summation convention is used. We can write this equation in coordinate free notation as follows: The metric g allows us to identify $T^*(M) \otimes T^*(M)$ with $T^*(M) \otimes T(M)$. Let us denote the section of $T^*M \otimes TM$ corresponding to T by $\overline{\overline{T}}$. The volume form on M determined by g gives an identification of TM with $\wedge^{n-1}T^*M$ where $n = \dim M$. Thus a section of $T^*M \otimes TM$ corresponds to a section of $T^*M \otimes \wedge^{n-1}T^*M$; the section corresponding to $\overline{\overline{T}}$ will be denoted by $\overline{T}$. Thus the symmetric tensor field T corresponds to a section of $T^*M \otimes \wedge^{n-1}T^*M$. Now covariant differentiation induces a covariant exterior derivative, $\hat{d}$ from sections of $T^*M \otimes^k T^*M$ to sections of $T^*M \otimes \wedge^{k+1}T^*M$. It is easy to check that the Einstein identity is equivalent to

$$\hat{d}\,\overline{T} = 0$$

as a section of $T^*M \otimes \wedge^n T^*M$.

Let V be a vector field on M. We can form the Lie derivative, $D_V g$ of the metric g with respect to the vector field V, so that $D_V g$ is a symmetric tensor field. In particular, we can take the scalar product (pointwise) of T with $D_V g$ to obtain a function $T \cdot D_V g$. On the other hand, if S is a section of $T^*M \otimes \wedge^n T^*M$ we can contract the T^*M component with the vector field V so as to obtain $S \cdot V$ which is a section of $\wedge^n T^*M$. We wish to prove the following fact: for any symmetric tensor field T and any vector field V of compact support we have

$$\int \tfrac{1}{2} T \cdot D_V g(\mathrm{vol}) = -\int (\hat{d}\,\overline{T}) \cdot V$$

where (vol) denotes the volume form associated to the metric.

Let V be any vector field on M and $\check{V}$ the linear differential form associated to V by the metric, thus, in local coordinates, if

$$V = V^i \frac{\partial}{\partial x^i} \quad \text{then} \quad \check{V} = V_j\, dx^j$$

where

$$V_i = g_{ij} V^i \quad .$$

Let $\mathscr{S}$ denote the symmetrization operator on $T^* \otimes T^*$ so $\mathscr{S}(u \otimes v) = u \otimes v + v \otimes u$. We claim that the Lie derivative of g with respect to V, $D_V g$ is given by

$$D_V g = \mathscr{S}(g\,\check{V}) \quad .$$

Indeed, in local coordinates we have

$$D_V g = \frac{\partial g_{ij}}{\partial x^k} V^k\, dx^i\, dx^j + g_{ij} \frac{\partial V^i}{\partial x^k} dx^k\, dx^j$$

$$+ g_{ij} \frac{\partial V^j}{\partial x^k} dx^i\, dx^k \quad .$$

Let us choose normal coordinates so that at a given point we have

$$\frac{\partial g_{ij}}{\partial x^k} = 0$$

and

$$\hat{d}(V) = \hat{d}(g_{ij} V^i dx^j)$$

$$= g_{ij} \frac{\partial V}{\partial x^k} dx^k\, dx^j \quad .$$

Comparing the two preceeding equations establishes our formula.

Let T be a symmetric tensor field, i. e. a section of $S^2(T(M))$. We can write

$$\frac{1}{2} T \cdot D_V = T \cdot \delta V \quad .$$

Let $\overline{\overline{T}}$ denote the section of $T(M) \otimes T^*(M)$ equivalent to T under the isomorphism of $T(M)$ with $T^*(M)$ determined by the metric. Then we can write the preceeding equation as

$$\frac{1}{2} T \cdot D_V = \overline{\overline{T}} \cdot \delta V$$

(since δV as section of $T(M) \otimes T^*M$ corresponds to δV). Finally let $\overline{T}$ denote the section of $T^*(M) \otimes \wedge^{n-1} T^*(M)$ defined by

$$\overline{T} = \overline{\overline{T}}(\mathrm{vol}) \quad .$$

Thus

$$\frac{1}{2} \overline{\overline{T}} \cdot D_V g(\mathrm{vol}) = \overline{T} \wedge \delta V \quad .$$

Now $\hat{d}T$ is a section of $T^*M \otimes \wedge^n T^*M$ and hence $\hat{d}T \cdot V$ is an n - form, and we have

$$\hat{d}\overline{T} \cdot V + \overline{T} \wedge \hat{d}V = d(\overline{T} \cdot V) \quad .$$

Thus, if V has compact support we can write

$$\int (\frac{1}{2} \overline{T} \cdot D_V g)(\mathrm{vol}) = - \int \hat{d}\overline{T} \cdot V \quad .$$

In general relativity, in the absence of an electromagnetic field, the Einstein identity concerning the vanishing of the covariant divergence of the energy momentum can be formulated, as we have seen, as

$$\hat{d}\overline{T} = 0 \quad .$$

We can think of an "infinitesimal variation", ν, in the psuedo-Riemannian metric g as being a section of $S^2 T^*M$ and we can think of the symmetric tensor T as defining a linear function on the space of symmetric tensor fields of compact support by setting

$$\langle T, \nu \rangle = \int \tfrac{1}{2} T \cdot \nu(\mathrm{vol}) \quad .$$

We see from the above discussion that the Einstein identity $\hat{d}\overline{T} = 0$ is equivalent to the assertion that

$$\langle T, D_V g \rangle = 0$$

for all vector fields V of compact support. In the presence of an electromagnetic field the Einstein identity is modified so as to read

$$\hat{d}\overline{T} + \hat{J} \lrcorner F(\text{vol}) = 0 \quad .$$

Here J is an $n-1$ form and $\hat{J}$ denotes the vector field corresponding to J by the relation $\hat{J} \lrcorner (\text{vol}) = J$. Since F is a two form, the interior product $\hat{J} \lrcorner F$ is a one form and thus $\hat{J} \lrcorner F(\text{vol})$ is a section of $T^*M \otimes \wedge^n T^*M$ as is $\hat{d}\overline{T}$. When we replace the electromagnetic field by an arbitrary Yang-Mills field, the same equation makes sense where now F is a section of $g(P) \otimes \wedge^2 T^*M$ and $J \in \mathfrak{a}^{n-1*}(M)$ so that $\hat{J}$ is a section of $g^*(P) \otimes TM$. Thus the contraction or "interior product" $\hat{J} \lrcorner F$ makes sense, and we can write the preceeding equation with this new interpretation.

We can also describe the equation $\hat{d}\overline{T} + \hat{J} \lrcorner F(\text{vol}) = 0$ as an orthogonality condition. Indeed, for any vector field V of compact support on M, let $\widetilde{V}$ denote the horizontal lifting of V to P. Then $\widetilde{V}$ is an element of $\text{aut}(P)$ and thus $D_{\widetilde{V}}\Theta = \tau_V \in \mathfrak{a}^1(M)$ is an "infinitesimal condition". We claim that

$$J \wedge \tau_V = -((\hat{J} \lrcorner F) \cdot V)(\text{vol}) \quad .$$

Indeed, $D_{\widetilde{V}}\Theta = \widetilde{V} \lrcorner d\Theta + d(\widetilde{V} \lrcorner \Theta) = \widetilde{V} \lrcorner d\Theta$ since $\widetilde{V} \lrcorner \Theta = 0$ because $\widetilde{V}$ is horizontal. But for any vector fields V and W on M, $F(V, W)$ is that section of $g(P)$ corresponding to the g valued function $d\Theta(\widetilde{V}, \widetilde{W})$. From this it is easy to see that the preceeding formula holds.

Now let us consider the pair (T, J) as defining a linear function on the space of all pairs (ν, τ) where ν is a compactly supported section of S^2T^*M and $\tau \in \mathfrak{a}^2(M)$ by

$$\langle (T, J), (\nu, \tau) \rangle = \langle T, \nu \rangle + \langle J, \nu \rangle = \int \tfrac{1}{2} T \cdot \nu(\text{vol}) + \int J \wedge \tau \quad .$$

For any vector field, V, on M let us set

$$D_{\widetilde{V}}(g, \Theta) = (D_V g, D_{\widetilde{V}}\Theta) \quad .$$

Then the preceeding computation show that for V of compact support we have

$$\langle (T, J), D_{\widetilde{V}}(g, \Theta) \rangle = -\int (\hat{d}\overline{T} + \hat{J} \lrcorner F(\text{vol})) \cdot V \quad .$$

Thus the condition that (T, J) be orthogonal to all $D_{\widetilde{V}}(g, \Theta)$ is equivalent to the condition $d\overline{T} + J \lrcorner F(\text{vol}) = 0$.

In the case of general relativity, it was pointed out by Einstein, Hoffman and Infeld that the equations of motion of a particle can be derived as a limiting form of the equation $d\overline{T} = 0$ if we let T approach a generalized tensor field concentrated along a curve. In other words we can get the equations for a point particle if the energy momentum tensor becomes concentrated into a point mass (which in space time will be represented by its world line which is a curve). The same holds true in the Yang-Mills case. But, following Souriau, it is more instructive to show how the equations

$$d_{\Theta} J = 0$$

and

$$\hat{d}\overline{T} + J \lrcorner F(\mathrm{vol}) = 0$$

of the present section, and the equations of motion described in the preceeding section both derive from a common geometrical principle, which we shall explain in the next section.

8. The Principle of General Covariance: The Structure of Unified Field Theories.

In the preceeding sections, we wrote down equations describing the motion of particles, or of continuously distributed matter in the presence of an "external" field. The matter moved under the influence of the field, but we assumed that we could ignore any effect that the motion of the matter has on the field itself. We know that this view of matter as "passive" is at best only an approximation. A moving electron produces it own electromagnetic field. One of the goals of this section is to elucidate the nature of the "passive" approximation. A general theory should include not only equations specifying the motion of matter in the presence of a field, but also equations determining the field itself, such as Maxwell's equation for the electromagnetic field and Einstein's equation for the gravitational field. In this section we shall present Souriau's analysis of the structure of such theories, and show how this analysis applies not only to the cases of electromagnetism and gravity but also to more general field theories.

Let $\mathfrak{G}$ be a group acting on a space $\mathfrak{X}$. In what follows we shall pretend that $\mathfrak{G}$ is a Lie group acting on a manifold $\mathfrak{X}$; but in the applications that we have in mind, the group $\mathfrak{G}$ will be infinite dimensional, for example $\mathfrak{G} = \mathrm{Aut}(P)$ and so will $\mathfrak{X}$, for example $\mathfrak{X}$ might be the space of all connections on P. By a <u>covariant theory</u> we shall mean a (smooth) function, $\mathfrak{F}$, defined on $\mathfrak{X}$ which is invariant under the action of $\mathfrak{G}$, so

$$\mathfrak{F}(a\mathfrak{x}) = \mathfrak{F}(\mathfrak{x}) \quad \text{for all} \quad a \in \mathfrak{G} \quad \text{and} \quad \mathfrak{x} \in \mathfrak{X} \quad .$$

Let $\mathfrak{x}$ be a point of $\mathfrak{X}$ and let $\mathfrak{B} = \mathfrak{G} \cdot \mathfrak{x}$ be the orbit through $\mathfrak{x}$. Let μ be an element of $T^*\mathfrak{X}_{\mathfrak{x}}$ so that μ is a linear function on the tangent space, $T\mathfrak{X}_{\mathfrak{x}}$, to $\mathfrak{X}$ at the point $\mathfrak{x}$.

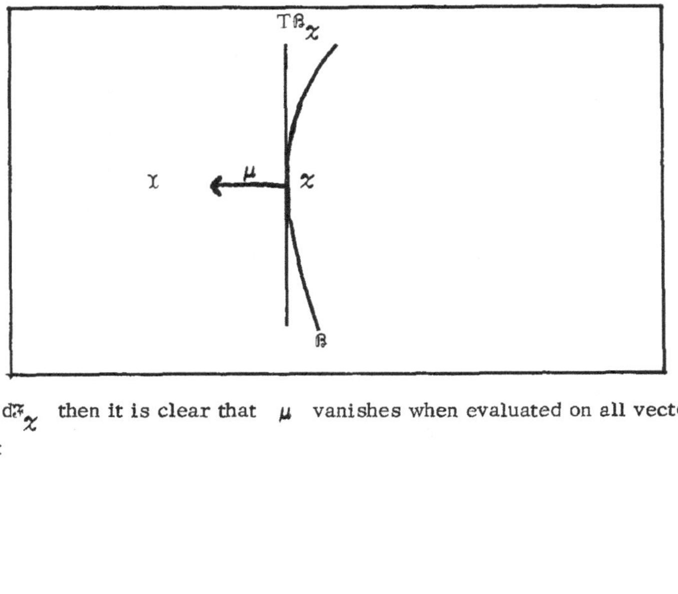

If $\mu = d\mathfrak{F}_{\mathfrak{x}}$ then it is clear that μ vanishes when evaluated on all vectors tangent to $\mathfrak{B}$, i.e. that

(I) $$\mu \in T\mathcal{B}_{\mathfrak{X}}^{\perp} \quad .$$

This equation is, as we shall see, the prototype of generalized Einstein identities written at the end of the preceeding section and the various equations of motion derived in the preceeding paragraphs.

Now let us suppose that the tangent bundle $T\mathfrak{X}$ has a given global trivialization. Thus we have a definite identification of all the various tangent spaces $T\mathfrak{X}_{\mathfrak{x}}$ with some fixed vector space Z. Then given $\mu \in Z^*$, we can try to find a $\mathfrak{x}$ such that

(S) $$d\mathfrak{F}_{\mathfrak{x}} = \mu \quad .$$

This equation for $\mathfrak{x}$ is, as we shall see, the prototype of the various "source equations" or "field equations". Naturally, (I) is a necessary condition for (S) so the "identities" are always a consequence of the "field equations". Since we will be dealing with infinite dimensional spaces, it will sometimes be necessary to consider situations where μ will not be defined on all of Z but only on some subspace, Z_0 which contains all the $T\mathcal{B}_{\mathfrak{x}}$ so that (I) still makes sense. Similarly, it may be that $\mathfrak{F}$ is not defined on all of $\mathfrak{X}$, and yet (S) makes good sense where both sides are regarded as linear functions on Z_0. Let us illustrate what we have in mind for the case of general relativity.

Let M be a (four dimensional) manifold. Let $\mathfrak{X}$ be the space of all pseudo-Riemannian metrics on M (of signature $+---$). Let $\mathfrak{G} = \mathrm{Diff}_0(M)$ be the group of all diffeomorphisms of M of compact support. (A diffeomorphism has compact support if it equals the identity outside some compact subset.) The action of $\mathfrak{G}$ on $\mathfrak{X}$ is the obvious one: a diffeomorphism φ sends the metric $\mathfrak{g}$ into the metric $\varphi^{-1*}\mathfrak{g}$. As we saw in the preceeding section, the "tangent space" to $\mathfrak{X}$ at $\mathfrak{g}$ can be identified with the space of all symmetric tensor fields on M, i.e. with $\Gamma S^2 T^*M$, the space of all smooth sections of S^2T^*M. Notice that this space $Z = \Gamma S^2 T^*M$ is a well defined vector space, and does not depend on the choice of $\mathfrak{g}$. We have thus identified all $T\mathfrak{X}_{\mathfrak{g}}$ with Z and so have trivialized $T\mathfrak{X}$. We shall let $Z_0 = \Gamma_0 S^2 T^*M$ denote the space of symmetric tensor fields of compact support. The "Lie algebra" of $\mathfrak{G}$ consists of all smooth vector fields of compact support on M. The orbit, $\mathcal{B}$, through $\mathfrak{g}$ consists of those metrics that can be obtained from $\mathfrak{g}$ by a "global change of variables" (of compact support). The "tangent space to the orbit", $T\mathcal{B}_{\mathfrak{g}}$ consists of those symmetric tensor fields ν of the form $\nu = D_V\mathfrak{g}$ for some smooth vector field V of compact support. Notice that $T\mathfrak{X}_{\mathfrak{g}} \subset Z_0$ for all $\mathfrak{g}$. A (continuous) linear function, μ, on Z_0 is called a "generalized" or "distributional" tensor field. We say that $\mu = \mu_T$ is smooth if there is a smooth symmetric tensor field T such that

$$\langle \mu_T, \nu \rangle = \int \tfrac{1}{2} T \cdot \nu(\mathrm{vol}) \quad .$$

We saw in the preceeding section that if $\mu = \mu_T$ then condition (I) (for $\mathfrak{X} = g$) is equivalent to the Einstein identity $\hat{\partial}\overline{T} = 0$ when we identify T with the energy momentum tensor. Einstein chooses, as his function, $\mathfrak{F}$, the function given, up to a constant by,

$$\mathfrak{F}(g) = -\int_M g^{ik} R_{ik}(g)\,(\mathrm{vol})_g$$

where $R_{ik}(g)$ is the Ricci curvature of g. There is no guarantee that this integral will converge, and so, strictly speaking, $\mathfrak{F}$ is not well defined. On the other hand, since $\mathfrak{F}$ is given by the integral of a local expression, the formal expression for $d\mathfrak{F}_g(\nu)$ will also be given by the integral of a local expression; in fact, a bit of computation will show that

$$d\mathfrak{F}_g(\nu) = \int_M \tfrac{1}{2} T \cdot \nu(\mathrm{vol})_g$$

where

$$T_{ik}(g) = R_{ik}(g) - \tfrac{1}{2} R(g) g_{ik}$$

where R is the scalar curvature. Now if $\nu \in Z_0$ then the integral in the definition of $d\mathfrak{F}_g(\nu)$ makes sense since the integrand has compact support. Thus, in interpreting (S) we can, for each fixed $\nu \in Z_0$ replace, in the definition of $\mathfrak{F}$, integration over M by integration over a sufficiently large compact region. With this interpretation, it is clear that (S) is equivalent to the Einstein gravitational field equations. The "passive approximation" to a theory consists in retaining (I) but dropping (S). For the case of general relativity, we have seen that if $\mu = \mu_T$ is smooth, then condition (I) is equivalent to the Einstein identity $\hat{\partial}\overline{T} = 0$. Suppose that μ is a generalized tensor field concertrated along a curve, that is, suppose that

$$\langle \mu, \nu \rangle = \int_\gamma \tfrac{1}{2} T \cdot \nu\, ds$$

where γ is a smooth curve and T is a smooth section of S^2T^*M over γ. Then it is not hard to prove that γ is a geodesic and that T is a decomposable tensor, where in suitable parametrization, $T(s) = p(s) \otimes p(s)$ where $(q(s), p(s))$ is a solution of the equations of geodesic flow. Thus the equations of motion of a "particle" are included in the identity equations (I).

We now describe the data associated to a Yang-Mills theory. Let $\pi : P \to M$ be a principal bundle with structure group G. We take $\mathcal{G} = \mathrm{Aut}_0(P)$, the group of all automorphisms of P which equal the identity outside $\pi^{-1}C$ where C is a compact subset of M. We have a homomorphism $\mathrm{Aut}_0(P) \to \mathrm{Diff}_0(M)$ sending any $\varphi \in \mathcal{G} = \mathrm{Aut}_0(P)$ into the transformation, $\overline{\varphi}$, it induces on M. We let $\mathfrak{X}$ consist of all pairs $\mathfrak{X} = (g, \Theta)$

where g is a pseudo-Riemannian metric on M (of suitable signature) and Θ is a connection on P. The group $\mathcal{G}$ acts on $\mathfrak{X}$ by

$$\varphi \cdot (g, \Theta) = (\bar{\varphi}^{-1}{}_{*} g, \varphi^{-1}{}_{*}\Theta) \quad .$$

The tangent space $T\mathfrak{X}_{\chi}$ at any point χ can be identified with $Z = \Gamma S^2 T^*M \oplus \mathfrak{a}^1(M)$ as we have seen, and we shall take $Z_0 = \Gamma_0 S^2 T^*M \oplus \mathfrak{a}_0^1(M)$. A linear function, μ, on this space will thus be a pair, $\mu = (T, J)$ where T is a "generalized tensor field" and J is a "generalized element of $\mathfrak{a}^{n-1}{}_*(M)$". In case T is a smooth tensor field and J is an actual element of $\mathfrak{a}^{n-1}{}_*(M)$, i.e. J is a smooth section of $g^*(P) \otimes \Lambda^{n-1} T^*M$, then we say that $\mu = \mu_{T,J}$ is smooth and we have

$$\langle \mu_{T,J}, (\nu, \tau) \rangle = \int_M (\tfrac{1}{2} T \cdot \nu (\mathrm{vol})_g + J \wedge \tau) \quad .$$

The "Lie algebra" of $\mathcal{G}$ is $\mathrm{aut}_0(P)$, the set of all vector fields, ξ, on P such that $R_a{}^*\xi = \xi$ and $\xi = 0$ outside $\pi^{-1}(C)$ for some compact subset C of M.

Any vector field ξ in $\mathrm{aut}(P)$ induces a vector field $\bar{\xi}$ on M (by projection). The "tangent space to the orbit" $T\mathcal{B}_{(g,\Theta)}$ consists of all (ν, τ) satisfying

$$\nu = D_{\bar{\xi}}\, g \quad , \quad \tau = D_{\xi}\Theta \quad \text{for some} \quad \xi \in \mathrm{aut}(P) \quad .$$

We can write every $\xi \in \mathrm{aut}_0(P)$ as a sum of its vertical and horizontal components,

$$\xi = \xi_{\mathrm{vert}} + \xi_{\mathrm{hor}}$$

both belonging to $\mathrm{aut}_0(P)$ with

$$\xi_{\mathrm{hor}} = \tilde{\bar{\xi}} \quad ,$$

the horizontal lift of the vector field $\bar{\xi}$ on M. As we have seen in the preceeding section, if $\mu = \mu_{(T,J)}$ is smooth, then the condition that μ be orthogonal to all $D_{\xi_{\mathrm{hor}}}(g, \Theta) = (D_{\bar{\xi}}\, g, D_{\tilde{\bar{\xi}}}\Theta)$ is the same as $\hat{d}\bar{T} + \tilde{J} \lrcorner F (\mathrm{vol})_g = 0$. Now $D_{\xi_{\mathrm{ver}}}(T, J) = (0, D_{\xi_{\mathrm{ver}}}\Theta)$ and we have also seen in the last section that the condition that J be orthogonal to all $D_{\xi_{\mathrm{ver}}}\Theta$, which is the as that μ be orthogonal to all $D_{\xi_{\mathrm{ver}}}(g, \Theta)$ is equivalent to the condition $d_{\Theta} J = 0$. Thus for smooth μ the condition (I) is equivalent to the Einstein-Yang-Mills identities

$$d_{\Theta} J = 0 \quad \text{and} \quad \hat{d}\bar{T} + \hat{J} \lrcorner F (\mathrm{vol})_g = 0 \quad .$$

Now suppose that μ is given by a smooth section, T, of $S^2 T^*M$ and a smooth section $\tilde{J}$ of $g^*(P) \otimes T(M)$ concentrated along a curve, γ, so that

$$\langle \mu, (\nu, \tau) \rangle = \int_\gamma (\tfrac{1}{2} T \cdot \nu + \hat{j} \cdot \tau) ds$$

where $\hat{j} \cdot \tau$ denotes the obvious contraction of a section of $g^*(P) \otimes T(M)$ with a section of $g(P) \otimes T^*(M)$. It is not hard to show that for such a μ, condition (I) implies the following: a) that $J(s) = \ell(s) \otimes \gamma'(s)$ where γ' denotes the tangent vector to γ at $\gamma(s)$ and ℓ is a section of $g^*(P)_\gamma$; b) that there is one fixed orbit, $\mathfrak{O}$, in g^* so that ℓ is a section (over γ) of the sub-bundle $\mathfrak{O}(P)$; c) that $T(s) = p(s) \otimes P(s)$; and d) that with suitable parametrization, $q(s)$ of γ the curve $(q(s), p(s), \ell(s))$ is a solution curve of the equations of Section 6 with the Hamiltonian $H = \frac{1}{2} \| p \|^2$ determined by the metric, g. Thus the law of "conservation of internal quantum number" together with the equations of Section 6 are consequences of (I). For the electromagnetic case Souriau has shown that if we consider μ which are first order distributions concentrated on a curve (i. e. are linear functions given by integrating appropriate expressions times τ, ν and their first derivatives, the integral being taken over γ) then, with appropriate supplemental physically reasonable hypotheses, (I) implies the equation of a spinning particle as described in Section 5. It is reasonable to suppose that a straightforward generalization will yield the corresponding result in the Yang-Mills case.

As a third example of our abstract set up we consider the following situation. Let X be a symplectic manifold, with symplectic form ω. Let $\mathcal{G}$ be the group of symplectic diffeomorphisms with compact support. Thus $\mathcal{G}$ consists of diffeomorphisms φ of X which satisfy $\varphi^*\omega = \omega$ and which equal the identity outside some compact subset. The "Lie algebra" of $\mathcal{G}$ consists of Hamiltonian vector fields of compact support. We shall assume, for simplicity, that the topology of X is such that every Hamiltonian vector field is of the form ξ_f where $\xi_f \omega = -df$ where f is a smooth function on X. Then, for any function, h, on X, we have $D_{\xi_f} h = \{f, h\}$ where $\{\ ,\ \}$ denotes Poisson bracket. We can identify the "Lie algebra" of $\mathcal{G}$ with the space of smooth functions of compact support under Poisson bracket. We will take $\mathfrak{X}$ to be the vector space consisting of all smooth functions on X. A point, χ, of $\mathfrak{X}$ is just a smooth function, H, and since $\mathfrak{X}$ is a vector space, we may identify all the tangent spaces $T\mathfrak{X}_H$ with $\mathfrak{X}$ itself. Thus we take $Z = \mathfrak{X}$ and take Z_0 to consist of smooth functions of compact support. We can write the most general continuous linear function, μ, on Z_0 as

$$\langle \mu, f \rangle = \int u f \omega^n$$

where $\dim X = 2n$ so that ω^n denotes Liouville measure on X and where u is a generalized function on X. The "tangent space to the orbit" $T\mathcal{B}_H$ consists of all

$D_{\xi_f} H = \{f, H\}$ where f has compact support. Condition (I) becomes

$$\int u\{f, H\}\, \omega^n = 0$$

for all f of compact support. Since $D_{\xi_H} \omega^n = 0$, this last expression is the same as $-\int \{u, H\} f \omega^n$. As this is to vanish for all f of compact support, we see that (I) is equivalent to

$$\{u, H\} = 0 \quad ,$$

i.e. that u be invariant under the flow generated by H. For example, if u is a generalized function concentrated on a curve, this curve must be a trajectory of H.

We now make some cursory remarks about the nature of the equation (S). It is, of course, a non-linear partial differential equation in most of the cases of interest and the condition (I), while necessary may be far from being sufficient. (There are, on occasions, μ which satisfy (I) and represent convenient physically idealized situations for which the corresponding (S) has no solutions.)

Nevertheless, to get a feeling for the structure of the equation, let us first look at the simplest model of the equation (S); suppose that $\mathfrak{X}$ is a finite dimensional vector space and $\mathfrak{F}$ a quadratic function on $\mathfrak{X}$. The linear structure of $\mathfrak{X}$ gives a trivialization of its tangent bundle and hence for any function $\mathfrak{F}$ the map sending $\mathfrak{X}$ into $d\mathfrak{F}_{\mathfrak{X}}$ can be thought of as a map from $\mathfrak{X}$ to $\mathfrak{X}^*$. In the case that $\mathfrak{F}$ is a quadratic function, this is the standard linear map of a vector space into its dual induced by a quadratic form. Solving (S) then reduces to solving a linear equation. For more general $\mathfrak{F}$ on a vector space $\mathfrak{X}$ the map from $\mathfrak{X}$ to $\mathfrak{X}^*$ determined by $\mathfrak{F}$ is called a Legendre transformation. Thus solving (S) becomes inverting a Legendre transformation. There is a more suggestive way of writing a Legendre transformation which makes use of symplectic geometry. Consider $T^*\mathfrak{X} = \mathfrak{X} \oplus \mathfrak{X}^*$ with its natural symplectic structure. Then the graph of the map sending $\mathfrak{X}$ into $d\mathfrak{F}_{\mathfrak{X}}$, i.e. the set of all pairs $(\mathfrak{X}, d\mathfrak{F}_{\mathfrak{X}})$ is a Lagrangian submanifold, Λ, of $T^*\mathfrak{X}$. Thus the problem of solving (S) amounts to the following: we have projections, $\pi_{\mathfrak{X}}$ and $\pi_{\mathfrak{X}^*}$ of Λ and $\mathfrak{X}$ and $\mathfrak{X}^*$ respectively. Given $\mu \in \mathfrak{X}^*$ our problem is to find $\lambda \in \Lambda$ such that $\pi_{\mathfrak{X}^*}(\lambda) = \mu$ and then take $\mathfrak{X} = \pi_{\mathfrak{X}}(\lambda)$. This suggests that in the general case one might want to modify (S) by considering a more general Lagrangian submanifold of $T^*\mathfrak{X}$ than one arising from a function on $\mathfrak{X}$.

In any event, for the case of a Yang-Mills theory, to get (S) we must specify a function $\mathfrak{F}$. In the case that the Lie algebra, g of the structure group G has a g invariant scalar product (for example if g were semi-simple) the bundle g(P) gets a scalar product, and using the metric g on M we obtain a scalar product on

$g(P) \otimes \wedge^2 T^*M$. Since F is a section of this bundle, it makes sense to talk of $\|F\|^2$ as a function on M. Thus the "simplest" function one could write down would be of the form

$$\mathfrak{F}(g, \Theta) = \int_M (a \|F\|^2 + bg^{ik} R_{ik}(g)) (\mathrm{vol})_g$$

where a and b are constants.

In the present section we have treated the gravitational force as being different from the other forces in that we have taken $\mathcal{I}$ to connect all pairs (g, Θ). As indicated at the end of Section 6, we could take an approach more in the spirit of Cartan's ideas, and put everything in terms of a connection (coming from a larger group). We will not expound on this point here.

9. Conserved Quantities

In the preceeding sections, we identified the (generalized) symmetric tensor field, T, and the (generalized) element, J, of $\mathfrak{C}^{n-1}*(M)$ with the "energy momentum tensor" and the "current density" respectively. The only real justification we gave for these identifications was that the generalized Einstein identities hold, and, in particular, if T and J are concentrated along a curve, we could identify this curve with the world line of a point particle. From the point of view of physical interpretation, we would like to be able to associate conserved quantities to T and J. For the case of point particles, we pointed out in the preceeding section that (I) implies conservation of "total internal quantum number", i. e. that we are dealing with the Hamiltonian equations of motion associated to a fixed orbit in g^*. For Hamiltonian systems we know how to associate conservation laws to symmetries of the system. We would like to have a similar procedure for continuous T and J. Since we have not derived the equations of our theory from a Lagrangian, we do not have Noether's theorem at our disposal to relate infinitesimal symmetries to conservation laws. We will instead, following Souriau, use the identity (I) directly to relate infinitesimal symmetries to conserved quantities. (It would be interesting and useful to analyze the relation between Souriau's procedure and Noether's.)

We will suppose that our space time manifold M is so constructed as to admit some global time variable, i. e. some function $t : M \to \mathbb{R}$ so that it makes sense to talk of a "cosmic future" consisting of all points in M for which $t > b$ and a "cosmic past" consisting of all points where $t < a$. We will let $Z_{a,b}$ denote the set of (ν, τ) such that $\operatorname{supp} \nu$ and $\operatorname{supp} \tau$ are both contained in $t^{-1}[a, b]$ where $a < b$. Thus $Z_{a,b}$ consists of those "infinitesimal variations" of (g, Θ) which are temporally limited to the interval $a \leq t \leq b$. We have $Z_0 \subset Z_{a,b} \subset Z$. We shall assume that our linear function μ is "spatially compactly supported" that is, we assume that $\operatorname{supp} \mu \cap t^{-1}[a, b]$ is compact for any interval $[a, b]$ of time. Then μ, which was initially defined as a linear functional on Z_0, can be unambiguously extended so as to be a linear functional on $Z_{a,b}$ for any $[a, b]$.

Now let ξ be an element of aut(P) such that $D_\xi(g, \Theta) = 0$. Thus ξ is an infinitesimal symmetry of (g, Θ) (and, in particular the vector field $\bar{\xi}$ on M is a Killing vector field for the metric g). Let $\lambda : M \to \mathbb{R}$ such that $\lambda = 0$ for $t < a$ and $\lambda = 1$ for $t > b$. Then we can form the vector field $\lambda\xi$ which also belongs to aut(P) and consider the variation $D_{\lambda\xi}(g, \Theta)$. Since $\lambda\xi = 0$ for $t < a$ we see that $D_{\lambda\xi}(g, \ ) = 0$ for $t < a$ and since $\lambda\xi = \xi$ for $t > b$ we see that $D_{\lambda\xi}(g, \Theta) = 0$ for $t > b$. Thus $D_{\lambda\xi}(g, \Theta) \in Z_{a,b}$ and we can evaluate μ on $D_{\lambda\xi}(g, \Theta)$, i. e. form the quantity

$$\langle \mu, D_{\lambda\xi}(g, \Theta) \rangle \quad .$$

Notice that since $\lambda\xi$ does not necessarily belong to $\text{aut}_0(P)$, it does not follow from (I) that this expression vanishes. However we claim that it does follow from (I) that $\langle \mu, D_{\lambda\xi}(g, \Theta) \rangle$ is independent of λ, a and b. Indeed, by definition $\langle \mu, D_{\lambda\xi}(g, \Theta) \rangle = \langle \mu, D_{\varphi\lambda\xi}(g, \Theta) \rangle$ where φ is any smooth function which is identically one on the support of μ, and which is spatially compact. If we made a second choice, φ' and λ', then $\lambda\varphi - \lambda'\varphi'$ has compact support and hence $\lambda\varphi\xi - \lambda'\varphi'\xi$ belongs to $\text{aut}_0(P)$. Hence it follows from (I) that $\langle \mu, D_{\lambda\xi}(g, \Theta) \rangle = \langle \mu, D_{\lambda'\xi}(g, \Theta) \rangle$. Let us denote $\langle \mu, D_{\lambda\xi}(g, \Theta) \rangle$ by $\mu(\xi)$ is independent of λ can be construed as a conservation law. In fact, we may choose λ to pass from zero to one in an arbitrarily small neighborhood of some space like hypersurface S_0 (for example we might choose S_0 as given by $t = t_0$). If we replace S_0 by some other space like surface S_1 (say $t = t_1$) we get the same value for $\mu(\xi)$. We thus get conservation under arbitrary time displacements.

For example, let ζ be an element of the center of the Lie algebra g and take $\xi = \zeta_P$, the infinitesimal generator of the one parameter group consisting of right multiplication by the group generated by ζ. Then $\bar{\xi} = 0$ and $D_\xi\Theta = 0$ for any connection Θ. Thus ζ gives rise to a conserved quantity for any (g, Θ). In the case of electromagnetism, where $G = U(1)$ is abelian its Lie algebra is its own center. The conservation law corresponding to a generator of u(1) is charge conservation. In case P is a trivial bundle and Θ the trivial connection, every Killing vector field of g will give rise to conserved quantitites. For the case of special relativity, the Lie algebra of Killing vector fields is the Poincaré algebra. In this case the invariants associated by T to the elements of the Poincaré algebra by T are precisely the invariants given by the energy momentum tensor in special relativity.

10. Conformal Models

In the preceeding four sections we have studied general properties of Yang-Mills fields. To proceed further, we must specify the model. That is, we must choose the correct group G and function $\mathfrak{F}$. The search for the correct group G has, for the past twenty years, been one of the central topics of research in physics. The remarkable successes of the theory of unitary symmetry have shown that the group $SU(3)$ must be a part, in some sense, of the internal symmetry group. On the other hand, classical electromagnetic theory shows that somehow the group of conformal symmetries of (flat) space time should be brought into the picture. In the present section we shall study a class of groups (and other algebraic structures) related to the conformal group and see that some of these groups can also be related to $SU(3)$. Needless to say, the results of this section are preliminary, and, from the physical point of view highly speculative.

We begin with a rapid survey of some basic facts concerning the group of conformal transformations on a real vector space of arbitrary dimension, and then to some of the special properties inherent in the four dimensional case. Let V be a finite dimensional real vector space with non-degenerate scalar product, $(\, , \,)$. We let $O(V)$ denote the orthogonal group of V and $o(V)$ the Lie algebra of $O(V)$.

We can identify vector fields on V with V - valued functions, using the linear structure on V. If ξ is any V - valued function, its differential $d\xi$ can be thought of as a $V \otimes V^*$ - valued function, i.e. as a $\mathrm{Hom}(V, V)$ - valued function. The condition that a vector field ξ be conformal can then be written as

$$d\xi \in o(V) \oplus Z \qquad \text{at all points,}$$

where $o(V)$ denotes the orthogonal algebra of V and Z the scalar multiples of the identity on V. It is well known that, if $\dim V \geq 3$, the only solutions to the above equations must be polynomial vector fields of degree at most two. Furthermore, if we break up any conformal vector field into its homogeneous components, this corresponds to a decomposition of the conformal algebra (if $\dim V \geq 3$) into a vector space sum

$$V + (o(V) + Z) + V' \quad ,$$

where V consists of the constant vector fields, $(o(V) + Z)$ consists of the linear vector fields, and V' consists of the quadratic vector fields. The subspaces V and V' are nonsingularly paired under the Lie bracket into $o(V) + Z$, and in fact into the Z component if we project onto the centre. In fact the structure of the conformal algebra can be most succinctly summarized as follows. We can, using the scalar product on V, identify $o(V)$ with $\wedge^2(V)$. Here the element $u \wedge v$ is identified with the linear transformation sending w into $(v, w)u - (u, w)v$. Let us construct a new vector space W two dimensions greater, obtained by adjoining two isotropic vectors f_{-1} and f_4. The scalar product on W is defined by keeping the old scalar product between elements of V and setting

$$(f_{-1}, v) = (f_4, v) = (f_{-1}, f_{-1}) = (f_4, f_4) = 0 \quad \text{and} \quad (f_{-1}, f_4) = 1 \quad .$$

Then the conformal algebra is isomorphic to the orthogonal algebra $o(W)$, with the constant vector fields being identified with elements of the form $f_{-1} \wedge v$, the quadratic fields with elements of the form $f_4 \wedge v$, and the infinitesimal scale transformation identified with the element $f_{-1} \wedge f_4$ (and $o(V)$ with $\wedge^2(V) \subset \wedge^2 W)$. In particular, if $V = \mathbb{R}^{1,3}$ then $W = \mathbb{R}^{2,4}$ so that the Lie algebra of conformal vector fields on Minkowski space is isomorphic to the algebra of infinitesimal orthogonal transformations of six dimensional space with a scalar product whose signature is $++----$. We denote this algebra by $o_{2,4}$. It is fifteen dimensional.

Special relativity asserts that the group of symmetries of nature is the subgroup of the conformal group which consists of those transformations which also preserve the class of "inertial frames". (We shall elaborate on this point below.) This group is then the eleven-dimensional group, which we shall denote by G_{11}, consisting of the (globally defined) Lorentz transformations, the scale transformations (i. e. dilatations of M) and the translations. Let g_{11} denote the Lie algebra of G_{11} and let g_{15} denote the Lie algebra of conformal vector fields on M. Thus each element of g_{15} is a globally defined vector field on M, but not every element of g_{15} can be exponentiated to a global transformation on M. The Lie algebra g_{11} is a subalgebra of g_{15} (and consists precisely of those vector fields which can be exponentiated). Let G_{15} be a Lie group whose Lie algebra is g_{15}, so that G_{11} is a closed subgroup of G_{15}. (We could choose G_{15} to be the simply connected group. However, we might prefer to make a different choice; for instance, we might want to take G_{15} to be the group isomorphic to $SO_{2,4}$.) Having chosen G_{15} it becomes natural to "complete" the Minkowski space M to obtain a manifold $\widetilde{M}$ on which the entire group G_{15} acts. Let us examine what is involved in this procedure of "conformal completion".

We are looking for a manifold $\widetilde{M}$ which is a homogeneous space for the group G_{15} together with a smooth map, $f : M \to \widetilde{M}$, such that

$$f(am) = \widetilde{a}f(m) \qquad \text{for all} \qquad a \in G_{11}$$

and

$$df_m(\xi_m) = \widetilde{\xi}_{f(m)} \qquad \text{for all} \qquad \xi \in G_{15} \quad ,$$

where ξ_m denotes the value of the vector field ξ at $m \in M$, where $\widetilde{\xi}$ is the vector field on $\widetilde{M}$ given by the action of G_{15} on $\widetilde{M}$, and where we denote the action of $a \in G$ on $x \in \widetilde{M}$ by $\widetilde{a}x$. Now this problem, as it stands, does not admit a unique solution.

Indeed, suppose that $(\widetilde{M}, f)$ is such a solution and let k be any element of G_{11} and let us define a new action of G_{15} on $\widetilde{M}$, by setting

$$\widetilde{a'x} = (kak^{-1})\widetilde{x} \quad ,$$

and a new map f' by

$$f'(m) = f(km) \quad .$$

Then it is easy to see that the new action and the new map so defined are again a solution to our problem.

On the other hand, a solution to our problem does exist. Indeed, let us pick some point $x_0 \in M$. Let $\overset{0}{g}_{15}$ consist of the vector fields in g_{15} which vanish at x_0, and let $\overset{0}{g}_{11} = g_{11} \cap \overset{0}{g}_{15}$. As we shall see below, the algebra $\overset{0}{g}_{15}$ generates a closed subgroup, call it H, of G_{15}. The $H \cap G_{11}$ is the subgroup generated by $g_{11} \cap \overset{0}{g}_{15}$ and coincides with the isotropy group of x_o in G_{11}. We now define $\widetilde{M} = G_{15}/H$ and map $M = G/(H \cap G_{11})$ into $\widetilde{M}$ by setting

$$f(a(H \cap G_{11})) = aH \quad .$$

If we take x_0 to be the origin, then $\overset{0}{g}_{15}$ consists of the linear and quadratic vector fields. By the above description, this is the eleven-dimensional subalgebra of g_{15} isomorphic to g_{11}, and, indeed, conjugate to g_{11} by an element of the adjoint group of g_{15}. Thus $\overset{0}{g}_{15}$ does indeed generate a closed subgroup of G_{15} and we obtain a conformal completion $\widetilde{M}$ as indicated above. It is not difficult to see that $\widetilde{M}$ admits a conformal structure invariant under G_{15} and that our embedding of M into $\widetilde{M}$ is conformal. Notice that, once we have picked a point x_0 and specified its image point $f(x_0)$ in $\widetilde{M}$, then the question of conformal completion has a unique solution. This is because the algebra $\overset{0}{g}_{15} \cap g_{11}$ is the reductive subalgebra $o_{1,3} + Z$ which acts completely reducibly on g_{15}. It breaks g_{15} up into three inequivalent subspaces, and hence there is only one way of enlarging the subalgebra $o_{1,3} + Z$ to a subalgebra, $\overset{0}{g}_{15}$ of g_{15} with $\overset{0}{g}_{15} \cap g_{11} = o_{1,3} + Z$. There is thus only one candidate for H. Hence, if we specify the image of x_o, we can identify $\widetilde{M}$ with G_{15}/H.

It is clear from the above discussion that g_{11} can be characterized as the normalizer, in g_{15}, of the subalgebra consisting of the constant vector fields. We can think of the concept of a "family of inertial frames" as being the same as some subalgebra of g_{15} acting as constant vector fields, i. e. as "infinitesimal translations". It is in this sense that we can regard the group G_{11} as the group preserving both Maxwell's equations and the notion of inertial frame.

To proceed further, it will be convenient for us to have an explicit model for $\widetilde{M}$.

Let us take G_{15} to be the group $SO_{2,4}$, the identity component of the group of orthogonal transformations of $R^{2,1}$, where $R^{2,4}$ is the six-dimensional real space endowed with a metric of signature $++----$. Let $\overline{M}$ denote the projective null quadric, i.e. a point x in $\overline{M}$ is a null line in $R^{2,4}$. Let us choose some point $x_\infty \in \overline{M}$, and let P_∞ denote the isotropy group of x_∞. In view of the discussion above, we know that P_∞ is an eleven-dimensional group of $SO_{2,4}$ which is isomorphic to G_{11}. Let us set

$$M = \{x \in \overline{M},\ x \text{ not orthogonal to } x_\infty\}$$

and

$$\Xi = \{x \in \overline{M},\ x \text{ orthogonal to } x_\infty \text{ but } x \neq x_\infty\} \quad .$$

It is clear that the four-dimensional submanifold M, the three-dimensional submanifold Ξ and the zero-dimensional submanifold $\{x_\infty\}$ are all stable under P_∞. We claim that P_∞ acts transitively on each of these, so that they provide the orbit decomposition of $\overline{M}$. Indeed, let us choose some null vector $f_{-1} \in x_\infty$. Then, if x_0 is some point in M, we can choose some $f_4 \in x_0$ with $(f_{-1}, f_4) = 1$. If we choose some other $x \in M$, then we can find a vector f, lying in x, with $(f_{-1}, f) = 1$. By standard linear algebra (Witt's theorem) we can find an element of $SO_{2,4}$ which carries the pair f_{-1}, f_4 into the pair f_{-1}, f. In particular, it carries x_∞ into x_∞ (and so lies in P_∞) and maps x_0 into x. Thus P_∞ acts transitively on M. A similar argument shows that P_∞ acts transitively on Ξ. We have thus proved

The isotropy group of a point $x_\infty \in \overline{M}$ is an eleven-dimensional subgroup P_∞ of $SO_{2,4}$ isomorphic to Poin $\times \mathbb{R}^+$ where Poin is the Poincaré group and where $\mathbb{R}^+$ acts as scale transformations. Under P_∞ the manifold $\overline{M}$ decomposes into three orbits: the open (four-dimensional) orbit M, the three-dimensional orbit Ξ and the zero-dimensional orbit $\{x_\infty\}$.

In this precise sense the conformal completion of Minkowski space has been achieved by "adjoining a light cone at infinity".

Let N denote the nilradical of P_∞, so that N is a four-dimensional commutative (vector) group. The group P_∞ is the semi-direct product of P_∞/N with N, and the group P_∞/N acts on N as Lorentz transformations followed by dilatations. Thus N has a Minkowski metric determined only up to scale, i.e. the Minkowski "angle" is well defined. If we choose an "origin" $x_0 \in M$, then we get a map $f_0 : N \to M$ given by $f_0(v) = v \cdot x_0$ for $v \in N$. The subgroup preserving both the "antipode" x_∞ and the origin x is a seven-dimensional group G_7, isomorphic to $R^+ \times SO_{1,3}$. We let $S = S(x_0, x_\infty)$ denote the one-parameter subgroup of dilatations in G_7. Thus S is the

centre of G_7 and consists of the dilatations in P_∞. It follows that the set of Minkowski metrics on N, and hence on M, is a homogeneous space of S.

Let us choose x_∞ and x_0 as above, and let U be the two-dimensional space that they span. Since x_∞ and x_0 are nonorthogonal null lines in U, it follows that the restriction of the metric of $R^{2,4}$ to U is nondegenerate and, in fact, has signature $+-$. Let V be the orthogonal complement of U in $R^{2,4}$, so that V is a four-dimensional subspace carrying an induced metric of signature $+---$. An element of S acts by multiplying a vector on the line x_∞ by some positive number r and on a vector in the line x_0 by multiplication by r^{-1}, and so as a hyperbolic transformation on U, and acts trivially on V. (Indeed, if we choose some vector $f_{-1} \in x_\infty$ and a vector $f_4 \in x_0$, the element $f_{-1} \wedge f_4$ is an infinitesimal generator of S.) The semi-simple part of G_7 acts as $SO_{1,3}$ on V and trivially on U (and is generated by linear combinations of $v_1 \wedge v_2$ with $v_1 \in V$). Each choice of $f_{-1} \in x_\infty$ determines a linear map of V into N, by sending $v \in V$ into $f_{-1} \wedge v \in N \subset o_{2,4}$ and this map is equivariant with respect to the action of $SO_{1,3}$. Replacing f_{-1} by rf_{-1} means that we now identify v with $rf_{-1} \wedge v$, which means that the same element of N is identified with a vector in V which is r^{-1} as large. Thus replacing the vector f_{-1} by rf_{-1} has the effect of multiplying the Lorentz metric by r^{-2}.

The group S also acts transitively on the space of positive-definite lines in U (and also on the space of negative-definite lines). If $f_{-1} \in x_\infty$ and $f_4 \in x_0$ are chosen so that $(f_{-1}, f_4) = 1$, then we can parametrize the set of all positive-definite lines in U by the unit vectors

$$e_+ = af_{-1} + bf_4 , \qquad a > 0 , \; 2ab = 1$$

and this vector is transformed into

$$raf_{-1} + r^{-1}bf_4 ,$$

with the negative-definite lines parametrized similarly, except that $2ab = -1$.

The preceeding results (despite our notation) did not depend on the four-dimensional character of space-time. They all would go through with $o_{1,3}$ replaced by $o_{1,n}$ and $o_{2,4}$ by $o_{2,n+1}$. (With suitable modifications they would apply to other signatures as well.) We now develop some results which are "four-dimensional accidents". Ultimately, they stem from the fact that the Dynkin diagrams  and 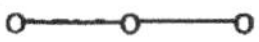 are the same, so that the algebras A_3 and D_3 are isomorphic. More specifically, they will follow from the fact, to be proved below, that $SU(2, 2)$ is locally isomorphic to $SO(2, 4)$.

Let $\mathbb{C}^{2,2}$ denote complex four-dimensional space endowed with a pseudo-Hermitian metric of signature $++--$. The group $U(2,2)$ denotes the group of complex linear transformations which preserve the metric and $SU(2,2)$ the subgroup of those elements of $U(2,2)$ with determinant one. A choice of an orthonormal basis, b, of (the one-dimensional) space $\wedge^4(\mathbb{C}^{2,2})$ determines a $*$ operator which is an anti-linear map $*: \wedge^k(\mathbb{C}^{2,2}) \to \wedge^{4-k}(\mathbb{C}^{2,2})$ determined by $u \wedge *v = (u,v)b$. (Here we have continued to denote by $(\ ,\ )$ the induced scalar produce on $\wedge^k(\mathbb{C}^{2,2})$.) In particular, $*$ maps $\wedge^2(\mathbb{C}^{2,2})$ into itself, and it is easy to check that $*^2 = \mathrm{id}$. We can thus think of $*$ as defining a complex conjugation on the six-dimensional complex vector space $\wedge^2(\mathbb{C}^{2,2})$. We let $\wedge^2(\mathbb{C}^{2,2})_{\mathbb{R}}$ denote the real points of $\wedge^2(\mathbb{C}^{2,2})_{\mathbb{R}}$ under this complex conjugation, i.e. those w which satisfy $*w = w$. Since $*$ is invariantly defined under $SU(2,2)$, we see that $SU(2,2)$ acts on $\wedge^2(\mathbb{C}^{2,2})_{\mathbb{R}}$ as real linear transformations. Let B denote the complex bilinear form on $\wedge^2(\mathbb{C}^{2,2})$ determined by $B(w,w)b = w \wedge w$ where b is the basis vector we have chosen in $\wedge^4(\mathbb{C}^{2,2})$. Notice that B is invariant under $Sl(\mathbb{C}^{2,2})$ and hence under $SU(2,2)$. The complex quadratic variety defined by $B(w,w) = 0$ $(w \neq 0)$ consists of the decomposable elements and the corresponding projective variety consists of all complex planes in $\mathbb{C}^{2,2}$ (or, in projective language, it is $G(1,3)$, the variety of all lines in projective three-space). For $w \in \wedge^2(\mathbb{C}^{2,2})_{\mathbb{R}}$, we have

$$Q(w,w)b = w \wedge w = w \wedge *w = (w,w)b$$

so that $Q(w,w) = (w,w)$. Thus Q restricts to a real quadratic form on $\wedge^2(\mathbb{C}^{2,2})_{\mathbb{R}}$ which is invariant under $SU(2,2)$. If h_1, h_2, h_3, h_4 is an orthogonal basis of $\mathbb{C}^{2,2}$ with $\|h_1\|^2 = \|h_2\|^2 = 1$ and $\|h_3\|^2 = -1$ then $\{h_i \wedge h_j\}_{1 \leq i \leq j \leq 4}$ is and orthogonal basis for $\wedge^2(\mathbb{C}^{2,2})$ under the pseudo-Hermitian form $(\ ,\ )$. Furthermore, $\|h_1 \wedge h_2\|^2 = \|h_3 \wedge h_4\|^2 = 1$, while $\|h_i \wedge h_j\|^2 = -1$ for all other choices of i and j. Thus $(\ ,\ )$ is a pseudo-Hermitian scalar product of signature $++----$ on $\wedge^2(\mathbb{C}^{2,2})$. The real scalar product on $\wedge^2(\mathbb{C}^{2,2})_{\mathbb{R}}$ obtained by restriction of $(\ ,\ )$ will thus also have the same signature. We may thus identify $\wedge^2(\mathbb{C}^{2,2})_{\mathbb{R}}$ with $\mathbb{R}^{2,4}$ and have defined a homomorphism of $SU(2,2) \to SO(2,4)$ whose kernel is easily seen to consist of $\{\pm I\}$ where I is the four-dimensional identity matrix. Now $Sl(\mathbb{C}^{2,2})$ has dimension fifteen over the complex numbers and hence $SU(2,2)$ is fifteen-dimensional over the real numbers. Thus $SU(2,2)$ and $SO(2,4)$ are locally isomorphic. Notice that the preceeding identification allows us to regard $\overline{M}$ as the set of real points of the complex projective variety, $G(1,3)$. If we think of $G(1,3)$ as being the set of all planes in $\mathbb{C}^{2,2}$, it is easy to see that $\overline{M}$ consists precisely of the null planes, i.e.

those planes which are isotropic under $(\ ,\ )$.

We now describe the gradation determined by a choice of x_0 and x_∞ in terms of the Lie algebra $su(2,2)$ of $SU(2,2)$. For this purpose it will be convenient to choose coordinates so that the pseudo-Hermitian form is given by the matrix $\frac{1}{2}\begin{pmatrix} 0 & I \\ I & 0 \end{pmatrix}$ i.e. so that

$$2(z,z) = z_1\bar{z}_3 + z_2\bar{z}_4 + z_3\bar{z}_1 + z_4\bar{z}_2 \quad \text{where} \quad z = \begin{pmatrix} z_1 \\ z_2 \\ z_3 \\ z_4 \end{pmatrix} .$$

Relative to this basis it is easy to see that $su(2,2)$ consists of all four by four complex matrices of the form

$$\begin{pmatrix} A & B \\ C & -A^* \end{pmatrix} \quad \text{where} \quad \operatorname{im} \operatorname{tr} A = 0 \quad \text{and} \quad B = -B^* , \quad C = -C^* .$$

We can take x_∞ to be the null plane determined by $z_3 = z_4 = 0$, in which case P_∞ is the (parabolic) subalgebra determined by the condition $C = 0$. We may take x_0 to consist of vectors satisfying $z_1 = z_2 = 0$ so that the stabilizer subalgebra of x_0 consists of matrices with $C = 0$. The Lie algebra g_7 then has $B = C = 0$ while the subalgebra $o(1,3)$ consists of those A with $\operatorname{tr} A = 0$, i.e. is isomorphic to $sl(2,\mathbb{C})$. It is, of course, well known that the group $Sl(2,\mathbb{C})$ is the universal covering group of $SO(1,3)$. Indeed, let us set $B = iX$ where $X = X^* = \begin{pmatrix} x_0 + x_3 & x_1 + ix_2 \\ x_1 - ix_2 & x_0 - x_3 \end{pmatrix}$ so that $\det X = x_0^2 - x_1^2 - x_2^2 - x_3^2$. The matrix $\begin{pmatrix} 0 & iX \\ 0 & 0 \end{pmatrix}$ corresponds to infinitesimal translation through the vector X. The Lie bracket of $\begin{pmatrix} A & 0 \\ 0 & -A^* \end{pmatrix}$ with $\begin{pmatrix} 0 & iX \\ 0 & 0 \end{pmatrix}$ is $\begin{pmatrix} 0 & iY \\ 0 & 0 \end{pmatrix}$ where $Y = AX + XA^*$ which corresponds, infinitesimally to the action of $Sl(2,\mathbb{C})$ where $a \in Sl(2,\mathbb{C})$ sends X into aXa^*. Since $\det aXa^* = |\det a|^2 \det X = \det X$ because $\det a = 1$, we see that this gives the action of $Sl(2,\mathbb{C})$ as Lorentz transformations.

The group $Sl(2,\mathbb{C})$ has two inequivalent irreducible representations on a complex two-dimensional vector space known as the spin representations of type $(1/2, 0)$ and $(0, 1/2)$. They are given by $u \rightsquigarrow au$ and $u \rightsquigarrow a^{*-1}$ for $u \in \mathbb{C}^2$ and $a \in Sl(2,\mathbb{C})$. At the Lie algebra level these are $u \rightsquigarrow Au$ and $U \rightsquigarrow -A^*u$. Notice that these are precisely the actions of $\begin{pmatrix} A & 0 \\ 0 & -A^* \end{pmatrix}$ on the invariant subspaces x_∞ and x_0

For any two-by-two matrix $D = \begin{pmatrix} a & b \\ c & d \end{pmatrix}$ set $D^a = \begin{pmatrix} d & -b \\ -c & a \end{pmatrix}$ so that $DD^a = (\det D) I$. For the self-adjoint two-by-two matrix X define the four-by-four matrix $\gamma(X) \in su(2, 2)$ by

$$\gamma(X) = \begin{pmatrix} 0 & iX \\ iX^a & 0 \end{pmatrix} .$$

It follows immediately from the definition that

$$\gamma(X)^2 = \det X \begin{pmatrix} I & 0 \\ 0 & I \end{pmatrix} = - \| X \|^2 I_4$$

where $\| X \|^2 = \det X$ and I_4 denotes the four-by-four identity matrix. (Thus the map $X \to \gamma(X)$ gives the spin representation of the Clifford algebra determined by the opposite Lorentz metric on $\mathbb{R}^{1,3}$.) We can use these γ matrices to provide an alternative description of the symplectic manifolds associated to spinning particles.

There is a real quadratic map ψ of $\mathbb{C}^{2,2}$ into $su(2, 2)^*$ sending $z \in \mathbb{C}^{2,2}$ into the linear function $\psi(z)$ on $su(2, 2)$ given by

$$\psi(z)(A) = i(Az, z) .$$

The orbit of a $z \in \mathbb{C}^{2,2}$ under the action of $SU(2, 2)$ is the seven dimensional "pseudo-sphere" consisting of all y with $(y, y) = (z, z)$. The map ψ will carry this pseudo-sphere into a six dimensional orbit of $SU(2, 2)$ acting on $su(2, 2)^*$, these being the orbits of minimal dimension. All points of the form $e^{i\theta} z$ are carried into the same point by ψ and it is easy to check that indeed the fibers of the map ψ are circles.

Let us choose a pair of isotropic planes, x_0 and x as before. This gives an identification of $\mathbb{C}^{2,2}$ with $\mathbb{C}^2 \oplus \mathbb{C}^2$ and we write $z = \binom{u}{v}$ so that $(z,z) = 2 \operatorname{Re} u \cdot v$ where $u \cdot v$ denotes the standard Hermitian scalar product in $\mathbb{C}^2$.

Notice that elements of the subgroup $Sl(2, \mathbb{C}) \subset SU(2, 2)$ preserve $u \cdot v$ and not merely $\operatorname{Re} u \cdot v$. Thus $\operatorname{Im} u \cdot v$ is an additional invariant. It is not difficult to see that so long as $\langle z, z \rangle = \operatorname{Re} u \cdot v \neq 0$ this is the only additional invariant. Indeed, since $u \neq 0$ we can find some element of $Sl(2, \mathbb{C})$ taking u into $\binom{1}{0}$. The isotropy group of $\binom{1}{0}$ in $Sl(2, \mathbb{C})$ consists of all matrices $\begin{pmatrix} 1 & b \\ 0 & 1 \end{pmatrix}$. The transpose inverse is $\begin{pmatrix} 1 & 0 \\ -b & 1 \end{pmatrix}$ and, if $v = \binom{x}{y}$, we have

$$\begin{pmatrix} 1 & 0 \\ -b & 1 \end{pmatrix} \begin{pmatrix} x \\ y \end{pmatrix} = \begin{pmatrix} x \\ y - bx \end{pmatrix} .$$

Thus, so long as $x \neq 0$, we can modify y as we like. Thus, the pseudo-spheres of non-zero radius decompose into six dimensional orbits under the action of $Sl(2, \mathbb{C})$.

The null pseudo-sphere will have six dimensional orbits (when $\operatorname{Im} u \cdot v \neq 0$) and four dimensional orbits (when $\operatorname{Im} u \cdot v = 0$).

We can now use the γ matrices and the map ψ above to construct a quadratic map of $\mathbb{C}^{2,2}$ into the dual space to the Poincaré algebra. For notational convenience let us set g_p = Poincaré algebra and g_c = conformal algebra. We let ℓ denote the Lorentz algebra which is a subalgebra of both g_p and g_c. We write $g_p = \ell + n_p$ where n_p denotes the translations in the Poincaré algebra. The γ matrices define a linear map, $\lambda : g_p \to g_c$ given by

$$\lambda(A, X) = \begin{pmatrix} A & 0 \\ 0 & -A^* \end{pmatrix} + \begin{pmatrix} 0 & iX \\ iX^a & 0 \end{pmatrix}$$

$$= \begin{pmatrix} A & 0 \\ 0 & -A^* \end{pmatrix} + \gamma(X)$$

where $A \in \ell$ and $X \in n$.

We thus have the commutative diagram

$$\begin{array}{ccc} & \nearrow & g_p \\ \ell & & \downarrow \lambda \\ & \searrow & g_c \end{array} ,$$

and the dual diagram

$$\begin{array}{ccc} & \swarrow & g_p^* \\ \ell^* & & \uparrow \lambda^* \\ & \nwarrow & g_c^* \end{array} .$$

We can compose $\psi : \mathbb{C}^{2,2} \to g_c^*$ and $\lambda^* : g_c^* \to g_c^*$ to get a map $\lambda^* \circ \psi : \mathbb{C}^{2,2} \to g_p^*$. Let us write an element of g_p^* as $S + \vec{p}$ where $S \in \ell^*$ and $\vec{p} \in n_p^*$. In g_p^* there are two invariant functions, m^2 of degree two given by

$$m^2(S, \vec{p}) = \|\vec{p}\|^2$$

and w^2, of degree four given by

$$w^2(S, \vec{p}) = \|S \wedge \vec{p}\|^2$$

(where, in this last equation, we think of S as an element of $\wedge^2(\mathbb{R}^{1,3})$ and hence $S \wedge \vec{p}$ is an element of $\wedge^3(\mathbb{R}^{1,3})$).

We can form the functions $m^2 \circ \lambda^* \circ \psi$ and $w^2 \circ \lambda^* \circ \psi$ both mapping $\mathbb{C}^{2,2} \to \mathbb{R}$, which, by abuse of language, we shall also denote by m^2 and w^2. We claim that, for $z = \binom{u}{v}$, we have the formulas

$$m^2(z) = |u \cdot v|^2$$

and

$$w^2(z) = \tfrac{1}{4}(\mathrm{Re}\, u \cdot v)^2 \, |u \cdot v|^2 = \tfrac{1}{4}(z, z)\, |u \cdot v|^2 \quad .$$

To verify the first formula, let us compute the n_p component of $\lambda^* \circ \psi$. For any $X \in n_p$ we have

$$\begin{aligned}
\lambda^* \circ \psi(z)(X) &= -i(\gamma(X)z, z) \\
&= -i\left(\begin{bmatrix} 0 & iX \\ iX^a & 0 \end{bmatrix}\begin{pmatrix} u \\ v \end{pmatrix}, \begin{pmatrix} u \\ v \end{pmatrix}\right) \\
&= \left(\begin{pmatrix} X & v \\ X^a & u \end{pmatrix}, \begin{pmatrix} u \\ v \end{pmatrix}\right) \\
&= Xv \cdot v + X^a u \cdot u
\end{aligned}$$

where in this last expression $Xv \cdot v$ denotes the Hermitian scalar product of Xv with v in $\mathbb{C}^2$. Now

$$Xv \cdot v = \operatorname{tr} X \times (v \otimes v^*)$$

and

$$X^a u \cdot u = \operatorname{tr} X^a \times (u \otimes u^*) = \operatorname{tr} X \times (u \otimes u^*)^a \quad .$$

So, writing $\vec{p}(z)$ for the n_p component of $\lambda^* \circ \psi(z)$, we have

$$\vec{p}(z)(X) = \operatorname{tr} X \times [v \otimes v^* + (u \otimes u^*)^a]$$

so that

$$\vec{p}(z) = v \otimes v^* + (u \otimes u^*)^a$$

and

$$\begin{aligned}
m(z) &= \det[v \otimes v^* + (u \otimes u^*)^a] \\
&= \det \begin{pmatrix} |v_1|^2 + |u_2|^2 & v_1 \bar{v}_2 - u_1 \bar{u}_2 \\ v_2 \bar{v}_1 - u_2 \bar{u}_1 & |v_2|^2 + |u_1|^2 \end{pmatrix} \\
&= |v_1|^2 \, |u_1|^2 + |u_2|^2 \, |u_1|^2 + u_1 \bar{u}_2 \bar{v}_1 v_2 + \bar{u}_1 u_2 v_1 \bar{v}_2 \\
&= (u_1 \bar{v}_1 + u_2 \bar{v}_2)(\bar{u}_1 v_1 + \bar{u}_2 v_2) \\
&= |u \cdot v|^2 \quad .
\end{aligned}$$

Now let us compute the ℓ^* component of $\lambda^* \circ \psi$. We have

$$\lambda^* \quad \psi(z)(A) = \psi(z)\begin{pmatrix} A & 0 \\ 0 & -A^* \end{pmatrix} = -i\left(\begin{pmatrix} A & u \\ -A^* & v \end{pmatrix}, \begin{pmatrix} u \\ v \end{pmatrix}\right)$$

$$= -\frac{i}{2}[Au \cdot v - A^*v \cdot u]$$

$$= -\frac{i}{2}\operatorname{tr} A \times (u \otimes v^*) - \frac{i}{2}\operatorname{tr} A^* \times (v \otimes u^*)$$

$$= -\frac{i}{2}[\operatorname{tr} A \times u \otimes v^* - \overline{\operatorname{tr} A \times u \otimes v^*}] \quad .$$

So, writing $S(z)$ for the ℓ component of $\lambda^* \circ \psi(z)$ we see that

$$S(z)(A) = \operatorname{Im} \operatorname{tr} A \times (u \otimes v^*) \quad .$$

Now let us suppose, for the moment, that $u \cdot v \neq 0$, so, in particular, that $u \neq 0$. Then we can transform z by some element of $Sl(2, \mathbb{C})$ into the form

$$z = \begin{pmatrix} 1 \\ 0 \\ x \\ 0 \end{pmatrix} ,$$

so that

$$p(z) = \begin{pmatrix} |x|^2 & 0 \\ 0 & 1 \end{pmatrix} = \frac{1 + |x|^2}{2} e_0 + \frac{1 - |x|^2}{2} e_3$$

and, for $A = \begin{pmatrix} a & b \\ c & -a \end{pmatrix}$

$$S(z)A = \operatorname{Im} a\bar{x} = \operatorname{Im} a \operatorname{Re} x + \operatorname{Re} a \operatorname{Im} x \quad .$$

Now the matrix $\begin{pmatrix} \frac{i}{2} & 0 \\ 0 & -\frac{i}{2} \end{pmatrix}$ corresponds to the generator of rotations in the e_1, e_2 plane and $\begin{pmatrix} \frac{1}{2} & 0 \\ 0 & -\frac{1}{2} \end{pmatrix}$ corresponds to hyperbolic rotation in the e_0, e_3 plane. Thus

$$S(z) = \frac{1}{2} \operatorname{Re} x \, e_1 \wedge e_2 + \frac{1}{2} \operatorname{Im} x \, e_0 \wedge e_3$$

and

$$S(z) \wedge \vec{p}(z) = \frac{1}{4}[1 + |x|) \operatorname{Re} x] e_0 \wedge e_1 \wedge e_2 + [(1 - |x|) \operatorname{Re} x] e_1 \wedge e_2 \wedge e_3 \quad .$$

So

$$\| S(z) \wedge \vec{p}(z) \|^2 = \frac{1}{4} |x|^2 (\operatorname{Re} x)^2$$

$$= \frac{1}{4} |u \cdot v|^2 (\operatorname{Re} u \cdot v)^2$$

$$= \frac{1}{4} |u \cdot v|^2 (z, z)^2 \quad .$$

Thus

$$w^2(z) = \frac{1}{4} |u \cdot v| (z, z)^2$$

$$= \frac{1}{4} m^2(z) (z, z)^2 \quad .$$

We have verified this equality under the assumption that $u \cdot v \neq 0$, but of course this implies that the equality holds for all z. In particular if we define the spin s by $s^2 = w^2/m^2$ for $m \neq 0$ we see that

$$s(z) = \frac{1}{2} |(z, z)| \quad .$$

We have thus defined a map $\lambda \circ \psi : \mathbb{C}^{2,2} \to g_p^*$ which is equivariant under the action of the Lorentz group and which takes the six dimensional Lorentz orbits given by specified values of $\langle z, z \rangle$ and $u \cdot v$ on the Lorentz suborbits of the Poincaré orbits given by specified values of the mass and spin as described by the preceeding equations. We are thus in the situation described at the end of Section 6:

Let M be a Lorentzian manifold M. We assume that M admits a spin structure. This means that we have a principal P_L bundle $(L = Sl(2, \mathbb{C}))$ which is a double cover of the bundle of orthonormal frames. Let P_{G_p} denote the extension of P_L to a principal G_p bundle, where G_p is the semidirect product of L with $\mathbb{R}^{1,3}$, as above. We may identify P_{G_p} with the double cover of the bundle of affine frames and the Levi-Civita connection gives rise to a connection Θ_{g_p} on P_{G_p}. Let P_{G_c} denote the extension of P_L to a principal G_c bundle where $G_c = SU(2, 2)$ is the conformal group.

We can view the map $\lambda : n_p \to n_c$ as defining an $n_c(P_L)$ valued form $\hat{\Theta}_{n_c}$ on M. (A more intrinsic defintion of $\hat{\Theta}_{n_c}$ is that it is the symbol of the Dirac operator on M.) Then, by our discussion in Section 6, this gives rise to a connection, Θ_{g_c} on P_{G_c}. Let U_c be a Lorentz orbit in $\mathbb{C}^{2,2}$ and let $U_p = \lambda^* \psi(U_c)$ its image in g_p^*. Our construction at the end of Section 6 gives rise to closed two forms

$$\upsilon_p \text{ on } U_p(P_L)$$

and

$$\upsilon_c \text{ on } U_c(P_L)$$

together with an induced map, $\tilde{\lambda} : U_c(P_L) \to U_p(P_L)$ satisfying

$$\tilde{\lambda}^* \upsilon_p = \upsilon_c \quad .$$

In the general situation U_c will be six dimensional and U_p will be five dimensional so that $U_c(P_L)$ will be ten dimensional and $U_p(P_L)$ will be nine dimensional and the fibers of $\tilde{\lambda}$ will be circles. Now the point of this whole discussion is that there is a one form α defined on $U_c(P_L)$ such that (for $m > 0$)

$$d\alpha = U$$

and, for any $y \in U_p(P_L)$,

$$\alpha_{|\tilde{\lambda}^{-1}(y)} = 2 s i d\theta$$

where $s = \frac{1}{2}(z, z)$ is the spin of U_p and $d\theta$ is the natural angular coordinate on $\tilde{\lambda}^{-1}(y)$.

To construct α we proceed as follows: let β be the one form on $\mathbb{C}^{2,2}$ given by

$$\beta = \operatorname{im}(z, dz) \quad .$$

If ξ is any element of $SU(2,2)$ and $\xi_{\mathbb{C}^{2,2}}$ denotes the corresponding vector field on $\mathbb{C}^{2,2}$, that $\xi_{\mathbb{C}^{2,2}} \lrcorner dz = \xi z$. Thus

$$\xi_{\mathbb{C}^{2,2}} \lrcorner \beta = \operatorname{im}(z, Az) \quad .$$

Now we claim that the linear differential form

$$\Phi \cdot \Theta + \beta$$

defined on $P_{G_c} \times \mathbb{C}^{2,2}$ really comes from a form on $\mathbb{C}^{2,2}(P_{G_c})$ which we shall call α. Since the above form is clearly invariant under G_c, all that remains is to show that the above vector vanishes on $\xi_{P_{G_c}} + \xi_{\mathbb{C}^{2,2}}$. Now $\xi_{P_c} \lrcorner \Theta = \xi$ and $\Phi(z) = \psi(z)$ so

$$\begin{aligned} \xi_{P_{G_c}} \lrcorner \Phi \cdot \Theta &= \psi(z)(A) \\ &= -i(Az, z) \\ &= \operatorname{im}(Az, z) \end{aligned}$$

since $(Az, z) = -(z, Az) = -\overline{(Az, z)}$ is purely imaginary. Comparing this with the above formula for $\xi_{\mathbb{C}^{2,2}} \lrcorner \beta$, proves the desired result. Now it is easy to check that $d\beta_{|U_c}$ is precisely the presymplectic structure on U_c pulled back from the symplectic form on U_c by $\psi : U_c \to g_c^*$. Comparing this with the construction of υ described in Section 6 shows that

$$d\alpha = \upsilon_c \quad .$$

Finally, the kernel of $\tilde{\lambda}$ really comes from the fact that $\psi(e^{i\theta}z) = \psi(z)$. Using θ as a parameter on this circle, we see that

$$\begin{aligned} \beta &= \operatorname{im}(z, iz\,d\theta) \\ &= -(z, z)\,d\theta \quad . \end{aligned}$$

Thus our construction is immediately applicable to the problem of prequantization. We thus have the diagrams:

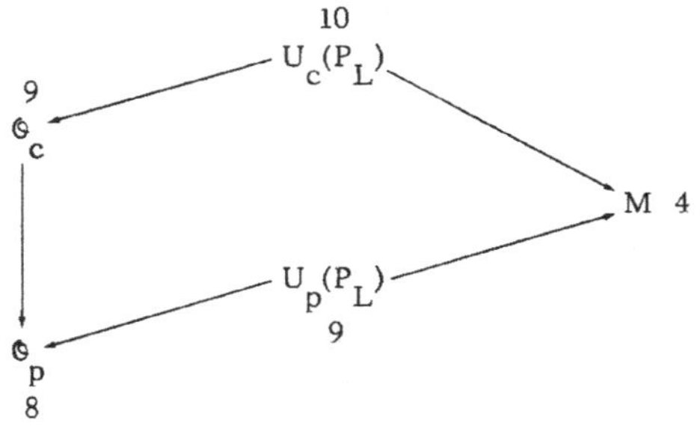

(provided that the quotient manifolds $\mathfrak{G}_c$ and $\mathfrak{G}_p$ exist):

The main algebraic device in the preceeding equation was the quadratic map ψ. Now we can think of $\mathbb{C}^{2,2}$ as the odd piece of a superalgebra whose even piece is SU(2, 2), and then (under the identification of SU(2, 2) with su(2, 2)* given by the Killing form) we can regard ψ as simply the square mapping $\mathbb{C}^{2,2}$ into u(2, 2). We can enlarge u(2, 2) to u(2, 2) × u(k) where k is any integer and $\mathbb{C}^{2,2}$ to $\mathbb{C}^{2,2} \times \mathbb{C}^k$. This then allows an "internal" U(k) group and a larger superalgebra with its corresponding ψ. By appropriate choice of connection forms one can then introduce associated line bundles and connections relevant to specific physical models.

The interesting fact in this construction is that it unites the complex structure of $\mathbb{C}^{2,2}$ with the real structure of the group SU(2, 2). This seems also to be connected with commutativity versus anti-commutativity and Lie algebras versus Lie superalgebras (graded Lie algebras). In fact we may introduce the notion of a Hermitian Lie algebra as follows: Let h be a real Lie algebra, represented by linear transformations of a complex vector space V and suppose that $H : V \times V \to h \otimes \mathbb{C} = h$ is an equivariant hermitian form. Here "hermitian" means that H(u, v) is linear in u and conjugate-linear in v with

$$H(v,u) = H(u,v)^{-} \;, \quad {}^{-} = \text{conjugation of } \mathbb{C} \text{ over}$$

and "equivariant" means that

$$[\xi, H(u,v)] = H(\xi u, v) + H(u, \xi v) \quad \text{for} \quad \xi \in \quad \text{and} \quad u, v \in V$$

where $[\ ,\ ]$ is extended as usual from h to $h_{\mathbb{C}}$.

The map $2\,\mathrm{Im}\,H : V \times V \to h$ is antisymmetric and $\mathbb{R}$ - bilinear, so one tries to use it to make $h + V$ into a Lie algebra by: the usual bracket $h \times h \to h$, the representation $h \times V \to V$ (i.e., $[\xi, u] = \xi u = -[u, \xi]$), and $V \times V \to h$ given by

$$[u, v] = 2\,\mathrm{Im}\,H(u,v) = \frac{1}{i}\{H(u,v) - H(v,u)\} \quad \text{for} \quad u, v \in V \quad .$$

This defines a Lie algebra if and only if the Jacobi identity holds, and that is the case just when it holds for any three elements of V:

$$[[u,v],w] + [[v,w],u] + [[w,u],v] = 0 \quad \text{for} \quad u,v,w \in V \quad .$$

In other words, the above definition of $[\ ,\ ]$ gives a Lie algebra structure on $h + V$ just when

(JI) $$\{H(u,v)w + H(v,w)u + H(w,u)v\} - \{H(v,u)w + H(w,v)u + H(u,w)v\} = 0 \quad .$$

Similarly $2\,\mathrm{Re}\,H : V \times V \to h$ is symmetric and $\mathbb{R}$ - bilinear, so one tries to use it to make $h + V$ into (a $\mathbb{Z}_2$-graded Lie algebra) a Lie superalgebra.

We recall that a Lie superalgebra is a $\mathbb{Z}_2$-graded vector space

$$g = g_{\mathrm{even}} + g_{\mathrm{odd}}$$

with a graded bilinear map $[\ ,\ ] : g \times g \to g$ i.e.

$$[g_{\mathrm{even}}, g_{\mathrm{even}}] \subset g_{\mathrm{even}}$$

$$[g_{\mathrm{even}}, g_{\mathrm{odd}}] \subset g_{\mathrm{odd}}$$

$$[g_{\mathrm{odd}}, g_{\mathrm{odd}}] \subset g_{\mathrm{even}} \quad .$$

It is assumed that g_{even} is an ordinary Lie algebra, that the map $g_{\mathrm{even}} \times g_{\mathrm{odd}} \to g_{\mathrm{odd}}$ gives a representation of g_{even} on g_{odd}, that the map of $g_{\mathrm{odd}} \times g_{\mathrm{odd}} \to g_{\mathrm{even}}$ is symmetric and that the graded version of Jacobi's identity holds. Letting $\deg x = 0$ for $x \in g_{\mathrm{even}}$ and $\deg x = 1$ for $x \in g_{\mathrm{odd}}$, the graded (super) version of Jacobi's identity says that left multiplication is a graded derivation, i.e. that

$$[x,[y,z]] = [[x,y],z] + (-1)^{\deg x \deg y}[y,[x,z]] \quad .$$

Let us define $g_{\mathrm{even}} = h$ and $g_{\mathrm{odd}} = V$ and define the symmetric map

$[\ \ ]_G : V \times V \to h$ by

$$[u, v]_G = 2\,\mathrm{Re}\,H(u, v) = H(u, v) + H(v, u) \quad \text{for} \quad u, v \in V$$

the graded version. Again, $[u, v]_G$ defines a Lie superalgebra if and only if the graded version of the Jacobi Identity holds, that is the case just when it holds for any three elements of V, and the latter is equivalent to

(GJI) $$\{H(u,v)w + H(v,w)u + H(w,u)v\} + \{H(v,u)w + H(w,v)u + H(u,w)v\} = 0 \quad .$$

Notice that we obtain both a Lie algebra and a graded Lie algebra, i.e., that both Jacobi Identities are satisfied, just when

(HJI) $$H(u,v)w + H(v,w)u + H(w,u)v = 0 \quad \text{for} \quad u, v, w \in V \quad .$$

The unitary algebras,

$$\{u(k,\ell) \oplus u(1)\} \oplus \mathbb{C}^{k,\ell} \quad \text{and} \quad \{(u(k,\ell)/u(1)) \oplus u(1)\} \oplus \mathbb{C}^{k,\ell}$$

to be described below, have the interesting property that both Jacobi identities hold where $\mathbb{C}^{k,\ell}$ is complex $(k + \ell)$-space with hermitian scalar product $\langle z, w\rangle = -\sum_1^k z_j \overline{w}_j + \sum_{k+1}^{k+\ell} z_j \overline{w}_j$, and where

$$u(k,\ell) = \{\xi : \mathbb{C}^{k,\ell} \to \mathbb{C}^{k,\ell} \text{ linear: } \langle \xi z, w\rangle + \langle z, \xi w\rangle = 0\}$$

is the Lie algebra of its unitary group.

$u(k, \ell)$ has complexification $g\ell(k + \ell\,;\mathbb{C})$, the Lie algebra of all complex $(k + \ell) \times (k + \ell)$ matrices. Let $*$ denote adjoint relative to $\langle\ ,\ \rangle$, that is $\langle \xi z, w\rangle = \langle z, \xi^* w\rangle$. Then $u(k,\ell) = \{\xi \in g\ell(k + \ell\,;\mathbb{C}) : \xi^* = -\xi\}$, and $\xi^- = -\xi^*$ is complex conjugation of $g\ell(k + \ell\,;\mathbb{C})$ over $u(k,\ell)$. Now the map H_0 ·

$$H_0 : \mathbb{C}^{k,\ell} \times \mathbb{C}^{k,\ell} \to g\ell(k + \ell\,;\mathbb{C}) \quad \text{defined by} \quad H_0(u, v)w = i\langle w, v\rangle u$$

is hermitian and $u(k,\ell)$ - equivariant, for

$$\langle H_0(u,v)w, z\rangle = i\langle w,v\rangle\langle u,z\rangle = i\langle w,v\rangle\langle z,u\rangle = -\langle w, H_0(v,u)z\rangle$$

and, for $\xi \in u(k,\ell)$,

$$[\xi, H_0(u,v)]w = i\{\langle w,v\rangle \xi u - \langle \xi w, v\rangle u\} = H_0(\xi u, v)w + H_0(u, \xi v)w \quad .$$

As to the Jacobi Identities, note

$$H_0(u,v)w + H_0(v,w)u + H_0(w,u)v = i\{\langle w,v\rangle u + \langle u,w\rangle v + \langle v,u\rangle w\} \quad .$$

The trick here is to give $u(1)_{\mathbb{C}} = g\ell(1\,;\mathbb{C})$ the conjugate of its usual complex structure, so that

$$H_1 : \mathbb{C}^{k,\ell} \times \mathbb{C}^{k,\ell} \to g\ell(1;\mathbb{C}) \quad \text{given by} \quad H_1(u,v) = i\langle v, u\rangle$$

is also hermitian. This done,

$$H : \mathbb{C}^{k,\ell} \times \mathbb{C}^{k,\ell} \to u(k,\ell)_{\mathbb{C}} \oplus u(1)_{\mathbb{C}} \quad \text{by} \quad H = H_0 \oplus H_1$$

is hermitian and $u(k,\ell)$ - equivariant, and satisfies both Jacobi Identities. Thus we have both ordinary and Lie superalgebra structures on our unitary algebras $\{u(k,\ell) \oplus u(1)\} + \mathbb{C}^{k,\ell}$.

More generally, fix non-negative integers $a + b = c$ and $k + \ell = m$, and let $\mathbb{C}^{k,\ell} \oplus \mathbb{C}^{a,b}$ denote complex $(m+c)$ - space with the hermitian scalar product

$$\langle z, w\rangle = -\sum_{1}^{k} z_j \overline{w}_j + \sum_{k+1}^{m} z_j \overline{w}_j - \sum_{m+1}^{m+a} z_j \overline{w}_j + \sum_{m+a+1}^{m+c} z_j \overline{w}_j \quad .$$

Let $u(k,\ell;a,b)$ denote the Lie algebra of the unitary group of $\mathbb{C}^{k,\ell} \oplus \mathbb{C}^{a,b}$. It consists of all linear transformations ξ such that $\langle \xi z, w\rangle + \langle z, \xi w\rangle = 0$. In matrices, it consists of all

$$\underbrace{\begin{pmatrix} A_{11} & A_{12} & Z_{11} & Z_{12} \\ A_{12}^* & A_{22} & Z_{21} & Z_{22} \\ -Z_{11}^* & Z_{21}^* & B_{11} & B_{12} \\ Z_{12}^* & -Z_{22}^* & B_{12}^* & B_{22} \end{pmatrix}}_{k \quad \ell \quad a \quad b} \begin{matrix} \}k \\ \}\ell \\ \}a \\ \}b \end{matrix} \quad \text{with} \quad A_{jj}^* = -A_{jj} \quad \text{and} \quad B_{jj}^* = -B_{jj} \quad .$$

To simply this denote $\begin{pmatrix} U_{11} & U_{12} \\ U_{21} & U_{22} \end{pmatrix}^{\#} = \begin{pmatrix} -U_{11}^* & U_{21}^* \\ U_{12}^* & -U_{22}^* \end{pmatrix}$. Then

$$u(k,\ell;a,b) = \left\{ \underbrace{\begin{pmatrix} A & Z \\ Z^{\#} & B \end{pmatrix}}_{m \quad c} \begin{matrix} \}m \\ \}c \end{matrix} : \begin{matrix} A = A^{\#}, & \text{i.e.,} & A \in u(k,\ell) \\ B = B^{\#}, & \text{i.e.,} & B \in u(a,b) \end{matrix} \right\} .$$

Now $u(k,\ell;a,b)$ evidently is of the form $h + V$ where

$$h = \left\{ \begin{pmatrix} A & 0 \\ 0 & B \end{pmatrix} : A \in u(k,\ell),\ B \in u(a,b) \right\} \cong u(k,\ell) \oplus u(a,b) \quad ,$$

$$V = \left\{ \begin{pmatrix} 0 & Z \\ Z^{\#} & 0 \end{pmatrix} : Z \text{ is } m \times c \right\} \cong \mathbb{C}^{k,\ell} \otimes \mathbb{C}^{a,b}$$

and the Lie algebra h acts on the complex vector space V by

$$\left[\begin{pmatrix} A & 0 \\ 0 & B \end{pmatrix}, \begin{pmatrix} 0 & Z \\ Z^{\#} & 0 \end{pmatrix}\right] = \begin{pmatrix} 0 & AZ - ZB \\ (AZ - ZB)^{\#} & 0 \end{pmatrix} .$$

Now $h_{\mathbb{C}} = \{\begin{pmatrix} A & 0 \\ 0 & B \end{pmatrix} : A$ is $m \times m$, B is $c \times c\}$. Give $h_{\mathbb{C}}$ the complex structure that is linear in A and conjugate linear in B. Then the map $H : V \times V \to h_{\mathbb{C}}$ defined by

$$H(z, w) = izw, \text{ that is } H\left(\begin{pmatrix} 0 & Z \\ Z^{\#} & 0 \end{pmatrix}, \begin{pmatrix} 0 & W \\ W^{\#} & 0 \end{pmatrix}\right) = \begin{pmatrix} iZW^{\#} & 0 \\ 0 & iZ^{\#}W \end{pmatrix}$$

is hermitian, is ℓ - equivariant, and defines both Lie algebra and Lie superalgebra structures because it satisfies (HJI). Further, the Lie algebra structure is the original one on $u(k, \ell; a, b)$.

Proof. The statement on $h_{\mathbb{C}}$ is standard, and the complex structure there is specified so that $H(z, w)$ is linear in z, conjugate linear in w. An elementary calculation shows $(iZW^{\#})^{\#} = iWZ^{\#}$, and this proves H hermitian. ℓ - equivariance is seen by direct calculation, using

$$\begin{aligned} i[A, ZW^{\#}] &= i\{(AZ - ZB)W^{\#} - Z(W^{\#}A - BW^{\#})\} \\ &= i\{(AZ - ZB)W^{\#} + Z(AW - WB)^{\#}\} \quad \text{for } A = A^{\#}, \ B = B^{\#} \end{aligned}$$

and it counterpart

$$i[B, Z^{\#}W] = i\{(AZ - ZB)^{\#}W + Z^{\#}(AW - WB)\} \quad \text{for } A = A^{\#}, \ B = B^{\#} .$$

To check (HJI) we calculate

$$[H(z, w), u] = i\left[\begin{pmatrix} ZW^{\#} & 0 \\ 0 & Z^{\#}W \end{pmatrix}, \begin{pmatrix} 0 & U \\ U^{\#} & 0 \end{pmatrix}\right] = i\begin{bmatrix} 0 & ZW^{\#}U - UZ^{\#}W \\ Z^{\#}WU^{\#} - U^{\#}ZW^{\#} & 0 \end{bmatrix},$$

so

$$[H(z, w), u] + [H(w, u), z] + [H(u, z), w] = \begin{pmatrix} 0 & T \\ T^{\#} & 0 \end{pmatrix}$$

where

$$\frac{1}{i}T = ZW^{\#}U - UZ^{\#}W + WU^{\#}Z - ZW^{\#}U + UZ^{\#}W - WU^{\#}Z = 0 .$$

And finally $2 \operatorname{Im} H(z, w) = [z, w]$ direct.

The unitary algebras mentioned above are the case $a = 1$ and $b = 0$. For there V is identified with $\mathbb{C}^{k,\ell}$ and the preceeding definition specializes to

$$H(z, w) : u \mapsto i\{ZW^{\#}U - UZ^{\#}W\} = i\langle u, w\rangle z - i\langle w, z\rangle u .$$

Here note the necessity of mapping to $u(k, \ell)_{\mathbb{C}} \oplus u(1)_{\mathbb{C}}$ to ensure that H be hermitian, even though that algebra acts on $\mathbb{C}^{k,\ell}$ as $u(k, \ell)_{\mathbb{C}}$. Taking $b = 3$ gives an "internal U(3)".

When $k = \ell$ our Lie superalgebra structure has a more refined graduation. Split $\mathbb{C}^{k,k} = W_1 + W_{-1}$, sum of two complementary k - dimensional isotropic subspaces, and use bases of $W_{\pm 1}$ that are dual to each other. So we have a basis of $\mathbb{C}^{k,k}$ in which its hermitian scalar product has matrix $\begin{pmatrix} 0 & I \\ I & 0 \end{pmatrix}$. In that basis,

$$u(k, k) = \left\{ \begin{pmatrix} M & E \\ F & -M^* \end{pmatrix} : M \text{ is } k \times k \text{ and } E, F \in u(k) \right\} ,$$

which gives the vector space direct sum splitting $u(k, k) = g_{-2} + g_0 + g_2$ where

$$g_{-2} = \left\{ \begin{pmatrix} 0 & 0 \\ F & 0 \end{pmatrix} : F \in u(k) \right\}, \quad g_0 = \left\{ \begin{pmatrix} M & 0 \\ 0 & -M^* \end{pmatrix} : M \text{ is } k \times k \right\}, \quad g_2 = \left\{ \begin{pmatrix} 0 & E \\ 0 & 0 \end{pmatrix} : E \in u(k) \right\} .$$

Similarly V, which maps $\mathbb{C}^{a,b}$ to $\mathbb{C}^{k,k}$, splits as $g_{-1} + g_1$ where

$$g_{-1} = \left\{ z \in V : \mathbb{C}^{a,b} \to W_{-1} \right\} \quad \text{and} \quad g_1 = \left\{ z \in V : z : \mathbb{C}^{a,b} \to W_1 \right\} .$$

Now we have a vector space direct sum decomposition

$$u(k, k; a, b) = \sum g_i \quad \text{with} \quad g_i = 0 \quad \text{for} \quad |i| > 2$$

such that

$$[g_i, g_j] \subset g_{i+j} \quad \text{in general,} \quad H : g_i \times g_j \to (g_{i+j})_{\mathbb{C}} \quad \text{for} \quad i, j \quad \text{odd} .$$

So the graded Lie algebra product also satisfies $[g_i, g_j]_G \subset g_{i+j}$, refining the $\mathbb{Z}_2$ - grading to a $\mathbb{Z}$ - grading.

11. Additional Comments

0. To the Introduction. For a description of DesCartes' theory see Whittaker [41] and Mach [21]. The quotations from Newton and Voltaire are both taken from Whittaker. The long quotation from Poincaré is from the introduction to [23]. I would like to thank Melissa Mills and Don Zagier for improving and correcting my original translation. It is interesting to remark that Poincaré wrote the comments about the impossibility, in principle, of deciding between the one fluid and two fluid theories of electricity in 1889. In 1897 J. J. Thompson published his results on the discovery of the electron, which, in a sense, settled the question. The moral is that while there is a danger in trying to extend our naive intuition into realms far from everyday experience, there is also a danger in being overly abstract and not admitting the "reality" of one's models. I have tried to make the point elsewhere, cf. [34], that a similar phenomenon took place in the history of astronomy: the Ptolemaic theory of the motions of the moon were such that one had to consider this theory as abstract; that the heavenly bodies are different, in principle from the terrestial ones. One of Kepler's great psychological breakthroughs was to put the planets, the earth and the moon on the same footing.

1. To Section 1. For the basic facts about elementary differential geometry and the exterior calculus we refer to [20] whose notation we follow. We have defined E to be a linear differential form rather than a vector field. This reflects a point of view that we wish to develop. If one considers the electric field strength as a force field, measured by its effect on a small charged test particle then we should take E to be a vector field. On the other hand, in actual laboratory measurements we would use a potentiometer, and measure potential drop along paths. From this point of view it is natural to think of E as a linear differential form.

2. For the role of the cotangent bundle and its symplectic structure see Chapter 13 of [20]. For more detailed information about symplectic geometry, see Chapter 4 of [12]. The idea of using the magnetic field to modify the symplectic structure appears in [32]. The sphere is an orbit of $O(3)$ in $o(3)^*$. The orbits of a Lie group G acting on g^*, the dual of its Lie algebra are always symplectic manifolds, cf. [19] or [12] Chapter 4.

3. Our treatment of Maxwell's equations follows the spirit of Sommerfeld's presentation [30] but in modern geometric notation using the exterior calculus instead of vector calculus, and there is a slight difference in interpretation of the relative roles of the

magnetic flux and the magnetic field strength. The point about the star operator determining the metric only up to scalar multiple is a case of the following well known observation in multilinear algebra, cf. [37].

Let V be a vector space endowed with a nondegenerate scalar product of arbitrary signature. The $*$ - operator, cf. [20] maps $\wedge^k V^* \to \wedge^{n-k} V^*$ where $n = \dim V$, and depends on the choice of scalar product. If we modify the scalar product by multiplying it by a nonzero scalar, the various $*$ - operators get multiplied by powers of that scalar. It is easy to check that if $n = 2m$ and we take $k = m$, the map $* : \Lambda^m V^* \to \Lambda^m V^*$ is unchanged. The equations $dF = 0$, $d * F = 0$ are thus conformally invariant, if F is an exterior form of degree m on a $2m$ - dimensional pseudo-Riemannian manifold. If we take $m = 1$ and $F = u\,dx + v\,dy$ on the Euclidean plane, we get the Cauchy-Riemann equations. If we take F to be the two-form giving the electromagnetic field on Minkowski space, we get the Maxwell equations.

4. The idea that we can obtain the Lorentz force by regarding the electromagnetic field as modifying or determining the symplectic structure is found in Souriau [32]. See also Sniatetski [29]. The general relation between this procedure and the more standard "minimal coupling" method is explained in [35]. The proposed equations for a particle with spin are due to Souriau [31]. These differ by very small terms from the Bargman-Michel-Telegdi equations [3]. Under normal circumstances these small corrections would be masked by quantum effects, but presumably suitable experimental conditions these differences can be detected. It should be remarked that it is the classical equation that it used for the experimental determination of the magnetic moment. A specific model for the function f is derived by Duval in [6].

5. The exchange between Einstein and Weyl is from [40]. I would like to thank Prof. Don Zagier for help in the translation. In case the reader suspects that I have mistranslated Weyl's audacious statement about his being the "true infinitesimal geometry" here is the original passage:

> Die hier entwickelte Geometrie ist, das muss vom mathematischen Standpunkt aus betont werden, die wahre Nahegeometrie. Es wäre merkwürdig, wenn in der Natur statt deser wahren eine halbe und inkonsequente Nahegeometrie mit einem angeklebten elektromagnetischen Felde realisiert wäre.

For a modern presentation of the theory of connections see [11]. Let us give a brief description of Cartan's viewpoint, as we shall amplify on this approach in the following sections. In the classical theory of connections there were two approaches. One, due to

Levi-Civita stresses the role of the tangent bundle as a vector bundle, and a connection as providing a means of parallel transport of vectors along curves. Thus the principal bundle involved will be a $Gl(n)$ bundle, where n is the dimension of the base manifold, or more generally, an H bundle where H is some subgroup of $Gl(n)$ such as the orthogonal group for some quadratic form. The principal bundle P_H is thus a subbundle of the bundle of frames. From Cartan's point of view, the tangent space at any point is just some homogeneous space having the same dimensionality as M which we are using to approximate M, and the role of a connection is to allow us to "develop" M onto this space along a curve, much as a surface is developed onto the plane by rolling it along a curve. Thus we must have some larger group, G, which contains H as a subgroup. For example we might take G to be the group of affine motions or Euclidean motions of $\mathbb{R}^n$. In this case $G/H = \mathbb{R}^n$. If $Z = G/H$ has the same dimension as M then we might think of G/H as a model space with which we wnat to compare M. For this purpose we consider a principal bundle P_G over M with structure group G together with a reduction of P_G to a principal H bundle P_H. We can form the associated bundle $Z(P_H)$ and the fact that we are given a reduction of P_G to P_H is the same as saying that we are given a section s of $Z(P_G)$ over M which we can think of as a choice of origin in each fiber of $Z(P_G)$. We can consider $T(M)$ and the bundle $V(Z)_s$ of vertical tangent vectors to $Z(P_G)$ along s as n - dimensional vector bundles over M. They both admit H as structure group. Suppose we are given an identification of these two bundles. In other words, we have not only identified each point of M with a "base point" of the fiber over m but also have made this identification up to first Then, a connection of P_G (which has no null vectors when restricted to P_H) is called a Cartan connection, and can be used to formulate the intuitive notion of development along curves. The precise modern definition of Cartan connection was first given by Ehresmann [8]. For further developments of the theory, see the papers by Kobayashi [16] and [17]. In what follows we will be making use of two principal bundles, P_H and P_G with H a closed subgroup of G and P_H a restriction of P_G and a connection on P_H. We shall drop, however, the requirement that G/H have the same dimension as M.

6. The construction of the symplectic structure associated to a connection and a Hamiltonian G space comes from [35]. For details about Hamiltonian group actions see [15]. Equations of motion in a Yang-Mills field have been written down by Wong [42]. For spinning particles see also [2] and [26]. The derivation of the equations of motion follows [13]. It has been pointed out by Duval [7] that if we take $G = SU(3)$ then there is a quadratic map from $\mathbb{C}^3 \to SU(3)^*$ and the image of $\langle z, z \rangle = 1$ gives

precisely the orbit associated with quarks. More generally there is a quadratic map of $\mathbb{C}^{p,q} \to SU(p, q)^*$ and the pseudo-spheres go over into the minimal orbits, cf. [38], and the discussion at the end of Section 10. The direct construction of the presymplectic structure without using the cotangent bundle follows [36].

7. The discussion here is motivated by Souriau's ideas in [31] and follows [13]. For the case of compact groups G see [1], as well, for the differential form formalism.

8. The main principal, as enunciated in this section, is a straightforward generalization of the principal enunciated by Souriau in [31]. The fact that it yields the equations of particle motion of Section 6 and the continuum equations of Section 7 was proved in [13]. Of course, the particle equations can be obtained as a limiting case of the continuum equations, as was pointed out by Einstein, Infeld and Hoffmann [9]. On the other hand, the continuum equations can be derived from a statistical mechanics viewpoint from the particle equations, cf. [39] where this is done in special relativity. The Einstein condition $T^{ij}{}_{|j} = 0$ has been the subject of some controversy, cf. for example [25].

9. The discussion here follows Souriau [31]. As mentioned in the text, it would be important to understand the relation between the derivation of conservation laws as presented here, and those coming from a Lagrangian via Noether's theorem, cf. for example [10].

10. The discussion here is sketchy and incomplete. The description of conformal geometry follows [37]. The use of the γ matrices to prequantize (in the sense of Kostant [19]) is a modification of the method of Souriau [33]. It is interesting to observe that the element in $SU(2, 2)$ associated by γ to time translation $\frac{\partial}{\partial x_0}$ is precisely the compact generator used by Segal in [27] in his theory of the red shift. For an introduction to Lie superalgebras (graded Lie algebras) see [5]. For the general theory of supermanifolds see [18]. For the classification of the simple Lie superalgebras see Kac [14]. The concept of Hermitian Lie algebras was introduced in [38]. The method used here to introduce supergravity seems similar to the construction in [24].

Our space $\mathbb{C}^{2,2}$ is "twistor space" in the sense of Penrose. The relation is as follows: Consider a point z on the pseudo-sphere of radius 0. By applying a suitable element of $SU(2, 2)$ we may assume that $z = \binom{u}{v}$ with $u = \binom{0}{0}$ and $v = \binom{0}{1}$. Let us examine which matrices $\begin{pmatrix} A & iX \\ 0 & -A^* \end{pmatrix}$ in the Poincaré algebra stabilizer z. If we write $A = \begin{pmatrix} a & b \\ c & -a \end{pmatrix}$ and $X = \begin{pmatrix} x_0+x_3 & x_1+ix_2 \\ x_1-ix_2 & x_0-x_3 \end{pmatrix}$ it is easy to check that the condition is

$a = c = 0 = x_1 = x_2$ and $x_0 = x_3$. This subalgebra is precisely the subalgebra preserving the null line through the origin in the direction of the null vector with components $x_1 = x_2 = 0$ and $x_0 = x_3$. The vectors z_0 and λz_0 determine the same null line. The Poincaré group acts transitively on the space of null lines and the group SU(2, 2) acts transitively on the pseudo-sphere, and hence on the projective variety determined by the null pseudo-sphere. We can thus identify this projective variety with the (conformal completion of) the set of null lines in Minkowski space. This is the point of view espoused by Penrose in his theory of twistors [22].

Bibliography

[1] Atiyah, M. F. and Bott, R., On the Yang-Mills theory over Riemann surfaces, to appear.

[2] Balachandaran, A. P., Salomonson, P., Skagerstam, B. -S. and Winnberg, J. -O., Classical description of a particle interacting with a non-Abelian gauge field, Phys. Rev. Lett. D, vol. 15 (April, 1977).

[3] Bargmann, V., Michel, L., Telegdl, V. L., Procession of the polarization of particles moving in a homogeneous electromagnetic field, Phys. Rev. Lett. 2, (1959), p. 435.

[4] Birkhoff, George David, Dynamical systems, New York, A. M. S., 1927.

[5] Corwin, L., Ne'eman, Y., and Sternberg, S., Graded Lie algebras in mathematics and physics, Reviews of Modern Physics, vol. 47 (1975), pp. 573-603.

[6] Duval, Ch., The general relativistic Dirac-Pauli particle: an underlying classical model, American Institute Henri Poincaré, vol. XXV (1976), pp. 345-362.

[7] ______, Mouvements classiques dans un champ de Yang-Mills, preprint 77/P 987 CNRS, Marseilles.

[8] Ehresmann, C., Les connexions infinitesimales dan un espace fibré differentiable, Colloque de topologie (Bruxelles, 1950).

[9] Einstein, A., Infeld, L., and Hoffman, B., The gravitational equations and the problems of motion, Annals of Mathematics, vol. 39, Princeton University Press, 1938, pp. 65.

[10] Goldschmidt, H., and Sternberg, S., The Hamilton-Cartan formalism in the calculus of variations, Ann. Inst. Fourier, 23 (1973), pp. 203-267.

[11] Greub, Warner, Halperin, Stephen, VanStone, Ray, Connections, curvature, and cohomology Vol. II - Lie groups, principal bundles and characteristic classes: [Pure and Applied Mathematics: A Series of Monographs and Textbooks, Vol. 47.] New York, Academic Press, 1973.

[12] Guillemin, V., and Sternberg, S., Geometric asymptotics, A. M. S., Providence, 1977.

[13] ______, and ______, On the equations of motion of a classical particle in a Yang-Mills field and the principle of general covariance, Hadronic Journal, vol. I, Nonantum, Mass., 1978.

[14] Kac, V., Lie superalgebras, Advances in Mathematics, vol. 26 (1977), pp. 8-96.

[15] Kazhdan, D., Kostant, B., and Sternberg, S., Hamiltonian group actions and dynamical systems of Calogero type, to appear in Communications in Pure and Applied Mathematics.

[16] Kobàyashi, Shôshichi, On connections of Cartan, Canadian Journal of Mathematics, 8 (1956), pp. 145-156.

[17] ______, Theory of connections, Ann. Mat. Pura Appl. (4) 43 (1957), pp. 119-194.

[18] Kostant, B., Graded manifolds, graded Lie theory and prequantization, Differential Geometric Methods in Mathematical Physics, Lecture notes in Mathematics, Springer-Verlag, vol. 570 (1977), pp. 177-306.

[19] Kostant, B., Quantization and unitary representations, Lectures on modern analysis and application III, Springer-Verlag, vol. 170 (1970), pp. 87-207.

[20] Loomis, L., and Sternberg, S., Advanced calculus, Addison-Wesley, Reading, MA., 1968.

[21] Mach, Ernst, The principles of physical optics; an historical and philosophical treatment, New York, Dover Publications (1953).

[22] Penrose, R., Twistor algebra, Journal of Mathematical Physics, vol. 8, no. 2, February, 1967, p. 347.

[23] Poincaré, Henri, Electricité et optique. La lumière et les theories electrodynamiques; Pari G. Carré et C. Wand, Paris, 1901.

[24] Regge, T., and Ne'eman, Y., Gauge theory of gravity and supergravity on a group manifold, to appear.

[25] Sacks, R. K., and Wu, H., General relativity for mathematicians, Springer-Verlag, 1977, esp. pp. 97-98.

[26] Salomonson, P., Skagerstam, B. -S., and Winnberg, J. -O., Equations of motion of a Yang-Mills particle, Phy. Rev. D. Vol. 16 (1977), pp. 2581-2585.

[27] Segal, I. E., Mathematical cosmology and extragalactic astronomy, New York: Academic Press, 1975.

[28] Simms, D. J., and Woodhouse, N. M. J., Lectures on geometric quantization, Lecture notes in Physics, vol. 53, Springer-Verlag, (1976).

[29] Sniatetski, N., Geometric quantization, to appear.

[30] Sommerfeld, Arnold Johannes Wilhelm, Lectures on theoretical physics, Vol. 3, Electrodynamics. New York, Academic Press, 1950-1956.

[31] Souriau, Jean-Marie, Modele de particule à spin dan le champ électromagnétique at gravitationel, Annales de l'Institut Henri Poincaré, Sec. A, Vol. XX, no. 4, 1974.

[32] ______, Structure des systemes dynamiques, [Maitrises de Mathematiques]. Paris, Dunod, 1970 xi, pp. 414.

[33] ______, Structure of dynamical systems, to appear.

[34] Sternberg, Shlomo, Celestial mechanics, Vol. I, W. A. Benjamin, Inc., New York, 1969.

[35] ______, Minimal coupling and the symplectic mechanics of a classical particle in the presence of a Yang-Mills field, Pro. of National Academy of Sciences, Vol. 74, No. 12, pp. 5253-5254, December, 1977.

[36] ______, and Ungar, T., Classical and prequantized mechanics without Lagrangians or Hamiltonians, Hadronic Journal, Nonantum, Mass., 1978.

[37] ______, and Wolf, J. A., Charge conjugation and Segal's cosmology, 11 Nuovo Cimento, Vol. 28A, N. 2, July, 1975.

[38] ______, and ______, Hermitian Lie algebras and metaplectic representations, to appear in Transactions of the American Mathematical Society.

[39] Synge, John Lighton, Relativity: the special theory, Amsterdam, North-Holland Publishing Co., New York, Interscience Publishers, 1956.

[40] Weyl, Hermann, Gravitation und Elektrizität, Sitzungs-berichte der Königlich PreuBischen Akademie der Wissenschaften zu Berlin, pp. 465-480, (1918).

[41] Whittaker, Edmund Taylor, A history of the theories of aether and electricity from the age of Descartes to the close of the nineteenth century. London, New York, Longmans, Green and Co., 1910.

[42] Wong, S. K., Field and particle equations for the classical Yang-Mills field and particles with isotropic spin, Nuovo Cimento, Vol. A65 (1970), p. 689.

CHARACTERISTIC CLASSES AND SOLUTIONS OF GAUGE THEORIES

Meinhard E. Mayer

Theoretische Physik, ETH Hönggerberg
8093 Zürich/Switzerland

and

University of California*
Irvine, CA 92717/USA

Introduction

Since this is the first talk of this Conference, I was asked by the organizers to start with a general introduction to the subject of gauge theories and their differential-geometric apsects. Since a number of reviews (among them a set of lecture notes of my own, Mayer (1977), to be quoted as LN) have appeared in the literature I will be brief in the written version, referring the reader to LN for details. I will also use the bibliography of my lecture notes, referring to papers quoted there, e. g., as LN-61 for Hermann Weyl's pioneering papers. References in this paper will be given in the form: Name of author (year).

The second half of my contribution to this Conference will review the geometric interpretation of those solutions of the classical Yang-Mills equations which have gained popularity under the names "instantons" or "pseudoparticles" and their surprising relation to the second Chern class of bundles with self-dual curvatures over compact 4-dimensional manifolds. Since to many of us this is new music, I must beg the indulgence of the experts in the audience, asking them to pay less attention to the "interpretation" of this music and more to the new "melody". In particular, since Professor Singer was unable to attend this conference, I will take the liberty of telling you what I learned from him about the recent results he has obtained in collaboration with Atiyah and Hitchin (Atiyah (1977)) on the number of instantons, and on the Atiyah-Ward Ansätze for the construction of these solutions. I must apologize in advance if my "interpretation" will lack the virtuosity proper to I. M. Singer. I also refer the reader to a recent review by Stora (1977) with a good bibliography, and to Römer (1977).

---

*Permanent address.

Two years ago, at the 1975 Bonn Symposium (LN-37) I quoted Beethoven's "Der schwer gefasste Entschluss" from his String Quartet Op. 135: "Muss es sein?" (i. e., must we use bundles in dealing with gauge theories?). Now the answer is a resounding yes: "Es muss sein, es muss sein!".

The recent rapprochement between physicists and mathematicians brought about by the discovery that instantons are self-dual curvatures and that even algebraic geometry is useful, is a very exciting phenomenon toward which these conferences have aimed since 1971 . I am certain that I am expressing our unanimous appreciation to the organizers for making it again possible for all of us to be here together.

If there was one person who was singularly responsible for the development of gauge theories and their differential geometry, it was certainly Hermann Weyl, who would be happy to know that physics is finally accepting many of the ideas he has put forward (the idea of a manifold appears in his "Idee der Riemannschen Fläche", the concept and term "Eichinvarianz" = gauge invariance come from his 1918 attempt to introduce electromagnetism by a local change in the scale of length in general relativity, gauge theory, as we now know it was developed in his 1929 paper, which implicitly also contains the concept of a principal bundle; that paper also contains his famous two-component equation for massless fermions and the idea of parity nonconservation; his theory of asymptotic distribution of eigenvalues of elliptic operators was a precursor of many of the recent developments in the theory of elliptic operators on manifolds; and let us not forget that together with Elie Cartan he taught us all we know today about the theory of compact Lie groups and much of differential geometry).

The more mathematical work on gauge theories during the past two years (I will not have time to speak on the numerous "practical" successes, such as the almost complete unification of weak and electromagnetic interactions made possible by the discovery of the "Higgs mechanism", the various attempts to understand the quark-gluon model of hadrons in "quantum chromodynamics", etc.) has concentrated mainly on properties of "instantons", and on the lattice version of gauge theories, where some important results have been obtained (Glimm-Jaffe (1976), Osterwalder (1976)).Due to a lack of time and competence, I will not discuss the latter. Neither will it be possible to discuss here problems related to quantized gauge theories.

1. A Brief Introduction to Gauge Theories

For a short historical overview of the development of gauge theory, from the Schwarzschild action principle for electrodynamics of 1903 and Hermann Weyl's 1918 paper, in which he "gauges" the scale of length, through the Yang-Mills generalization to nonabelian gauge groups of 1954, to the recent developments which will be discussed in more detail in the second part of this lecture, we refer the reader to LN. The only omission in the historical account there is to the attempt by London (1927) to extend Weyl's gauge principle of 1918 to de Broglie waves, thus anticipating some of the results of Weyl's 1929 paper (LN-61). It is not clear whether Weyl was aware of London's work, he refers briefly to Wigner and others. Although London's paper is quasiclassical in nature, it contains several interesting remarks and is well worth reading.

As an introduction to gauge theory for the mathematicians in the audience I will paraphrase in more modern notation (differential forms) the contents of the relevant parts of the Weyl (LN-61) and Yang and Mills (LN-63) papers. For more details, cf., e. g., LN. For typographical reasons I will deviate somewhat from the standard notations, using wherever possible latin letters and subscripts.

We will operate throughout on a 4-dimensional real manifold, with a pseudoriemannian metric. Most of the time the manifold will be Minkowski space M_4 with the quadratic form $(x^0)^2 - (x^1)^2 - (x^2)^2 - (x^3)^2 = g_{ik}x^i x^k$ determining the metric. Later on we will pass to R^4 by means of the substitution $x^0 \to ix^4$ and an overall sign-change in the metric, and a compactification of R^4 into S^4 by means of a stereographic projection.

Following Weyl we consider two-component (Weyl) spinors on M_4, corresponding to massless charged particles of spin 1/2 (although the spin of the particle is unimportant for our purpose, and we could just as well have discussed a complex scalar field, describing spin 0 charged particles, cf. LN). Since we will not be interested in the "gravitational" aspects of Weyl's work, M_4 will be considered as Minkowski space, but we will use arbitrary (moving) frames at different points. In other words, a moving frame is a section of the trivial bundle $M_4 \times SO(1,3)$. A two-component spinor is a section u(x) in a complex vector bundle of rank 2, associated to the frame bundle by the representation SL(2,C) of the Lorentz group SO(1,3) (we consider only proper orthochronous Lorentz transformations, since the improper transformations require doubling the number of spinors).

The spinor u can be thought of as a complex two-component column $u = \binom{u_1}{u_2}$ and its adjoint $u^* = (\bar{u}_1, \bar{u}_2)$, as a row-vector. The relation between SL(2,C) and SO(3,1) is established by the correspondence between 4-vectors and spinors:

$$x^k = u^* S^k u, \qquad (1.1)$$

where S^0 is the two-dimensional unit matrix and S^1, S^2, S^3 are the three Pauli matrices, $S^a = \sigma_a$, a = 1, 2, 3 (we use Weyl's notation). If M_4 is a Riemannian manifold, one can define a covariant differentiation of spinors, by using the homomorphism between SO(1,3) and SL(2,C) to carry over the connection from frames to spinors (this is done explicitly in Weyl's paper).

The Weyl equation for the two-component spinor can be considered a consequence of an action principle with a Lagrangian density proportional to

$$u^*(x) S^k \partial_k u(x) \quad + \text{c.c.} \qquad (1.2)$$

where c.c. denotes a "complex conjugate" term required to make the action real (one can see that this can be achieved by adding an exact 4-form to the action density).

In addition to the obvious Lorentz invariance of (1.2), it is also invariant under the action of an additional U(1) group:

$$u(x) \to \exp(ig)u(x), \quad u^*(x) \to u^*(x)\exp(-ig), \qquad (1.3)$$

where g is a real number (in fact, an element of the Lie algebra iR of U(1)).

It was Hermann Weyl's and Fritz London's observation that if one is free to choose frames arbitrarily (but smoothly) at different points of M_4 one should also be free to choose the phase g arbitrarily at each space-time point, i. e., replace exp(ig) by exp(ig(x)), a section of the (trivial) principal bundle with structure group U(1). But then the action density is no longer invariant (since the ordinary partial derivative ∂_k does not commute with the action of the local gauge transformation), but this can be repaired by introducing the gauge-covariant differential

$$Du(x) = du(x) + Au , \qquad (1.4)$$

or the gauge-covariant derivative

$$D_k u(x) = \partial_k u(x) + A_k(x)u(x), \qquad (1.5)$$

where the 1-form

$$A = A_k(x)dx^k \qquad (1.6)$$

(summation convention) is subject to the gauge transformation:

$$A \to A - dg(x) = A + iG(x)^{-1}dG(x) \qquad (1.7)$$

where $G(x) = \exp(ig(x))$ is the section of the U(1) bundle which determines the gauge transformation ($g(x)$ is usually called a gauge function in physics). A physicist sees in (1.7), written in terms of coordinates A_k, the gauge transformation for the vector potential of classical electrodynamics, whereas a mathematician will recognize the characteristic transformation property of a connection 1-form (with values in the abelian Lie algebra of U(1) - the factor i takes care of the reality of A). The mathematician Hermann Weyl immediately introduced the curvature 2-form

$$F = dA = \tfrac{1}{2}(\partial_i A_k - \partial_k A_i)dx^i \wedge dx^k, \qquad (1.8)$$

which the physicist Hermann Weyl recognized as the components of the gauge-invariant electromagnetic field-strength tensor.

The Bianchi identity

$$dF = d^2A = 0 \qquad (1.9)$$

automatically yields the homogeneous set of Maxwell equations. In order to obtain the inhomogeneous Maxwell equation, linking F to the current density $eu^*S^k u$, Weyl proposes to add to the action density

$$u^*S^k D_k u + \text{c. c.} \qquad (1.10)$$

of the spinor field, the only reasonable term proportional to the norm of the curvature F:

$$(1/4)\ F\wedge{}^*F = (1/4)\ F_{jk}F^{jk}d^4x \qquad (1.10)$$

where the dual two-form $^*F$ has as coefficients the components of the dual field strength tensor (the other 4-form $F\wedge F = F\wedge dA$ is exact and can only be used to modify boundary conditions; cf. LN for details).

The inhomogeneous Maxwell equation is then of the form

$$d{}^*F = -{}^*J, \qquad (1.11)$$

where $^*J$ is the three-form dual to the current density.

The whole discussion can be generalized to a nonflat space-time by replacing gauge-covariant derivatives with gauge and space covariant ones, and also to other fields, such as Dirac spinors, scalars, etc. Current conservation, and thus charge conservation are automatic, since $^*J$ is exact and hence closed:

$$d{}^*J = 0. \qquad (1.12)$$

In 1954 Yang and Mills (LN-63) extended Weyl's gauge principle to a theory with a nonabelian internal symmetry group. In particular, they considered a doublet of spinors (describing, e. g., the proton and the neutron, or, in today's language, two massless quarks) u_α, $\alpha = 1, 2$, with a Lagrangian density

$$\sum_\alpha u_\alpha^* S^k \partial_k u_\alpha + \text{c.c.}, \qquad (1.13)$$

obviously invariant under the group SU(2):$u \to Gu$ where u is the "two-spinor" , or spinor-isospinor, transforming under SL(2,C)×SU(2) and G is a matrix belonging to SU(2) (tensored with the SL(2,C) unit matrix, which we will suppress in the subsequent discussion).

By Noether's theorem there are three conserved currents

$$j_a^k = u^* S^k T_a u, \; a = 1, 2, 3, \; k = 0,\ldots,3, \qquad (1.14)$$

transforming according to the adjoint representation of SU(2) (T_a are the Hermitean generators of the Lie algebra of SU(2) -- the isospin Pauli matrices). Following Weyl's argument, Yang and Mills proposed to replace the global invariance under SU(2) by a local invariance, i. e., the three parameters of the group SU(2) become (smooth) functions of the space-time coordinate x. In other words, the matrix G becomes a smooth function G(x) of the point x in M_4. As before, differentiation then no longer commutes with the "gauge transformation", and the three currents (1.14) have to be coupled to an "isotriplet of compensating vector fields" $Y_{ka}(x)$ which have the following transformation properties when u is replaced by G(x)u. It is convenient to combine the vector fields Y_{ka} with the generators T_a and the basis one-forms dx^k to obtain the Lie-algebra (matrix) valued one-form

$$Y = i\sum Y_{ka}(x)\, T_a dx^k \qquad (1.15)$$

(the factor i has been introduced to make Y anti-Hermitean). The one-form Y transforms under a gauge transformation exactly as a connection one-form does:

$$Y \to G(x)^{-1} A G(x) + G(x)^{-1} dG(x), \qquad (1.16)$$

i. e., in addition to the adjoint transformation of the isovector Y_a there appears the Maurer-Cartan form $G^{-1}dG$ which compensates for the noninvariance of (1.13). Just as for the Abelian case one achieves this by replacing the ordinary differential of spinors, du, by the gauge-covariant differential

$$Du = du + Yu = (\partial_k + i\sum Y_{ka} T_a) u dx^k. \qquad (1.17)$$

The Yang-Mills field strength is obtained by taking the covariant exterior differential of the matrix one-form Y, yielding the matrix two-form (the curvature of the connection Y):

$$M = DY = dY + Y \wedge Y$$

$$= i\sum T_a M^a_{jk}(x) dx^j \wedge dx^k, \tag{1.18}$$

where the field strengths M^a_{jk} transform according to the adjoint representation of SU(2) when u is subjected to a gauge transformation:

$$M \to G^{-1}MG, \tag{1.19}$$

and is therefore an object from which one can form invariants.

The first set of Yang-Mills equations (not involving u) is the Bianchi identity

$$DM = dM + Y \wedge M = 0. \tag{1.20}$$

In distinction from the Maxwell equation, this equation is nonlinear since it involves the Yang-Mills potential Y and field strength M.

In order to obtain the second set of Yang-Mills equations, we copy the electromagnetic case, replacing in (1.13) the partial derivative by the gauge-covariant derivative (1.17) and adding to the Lagrangian density the invariant four-form

$$(1/4)\mathrm{Tr}(M \wedge {*M}). \tag{1.21}$$

If we denote as before the three-form matrix dual to the Noether currents (1.14) by $*J$, the action principle yields as Euler equations for the fields u, Y, M:

$$\langle Du, S^k \rangle = 0, \tag{1.22}$$

$$D{*M} = {*J}, \tag{1.23}$$

where $\langle\ ,\ \rangle$ denotes evaluation of the one-form on the four-vector S^k (this is short-hand for the gauge-covariant Weyl equations for the massless spinor fields u, with the Yang-Mills potential Y). The second equation, (1.23), shows us that the three-form $*J$ is not closed, i. e., the Noether current of the isospinor field is not conserved by itself. However, if we rewrite the equation in the form

$$d{*M} = {*J} - Y \wedge {*M}, \tag{1.24}$$

we see that the total "Yang-Mills current"

$$J - {*(Y \wedge {*M})} = J_{YM} \tag{1.25}$$

is the dual of a closed three-form and hence is a conserved current. The physical interpretation of this fact is that the Yang-Mills fields themselves "carry SU(2)-charge", and that only the total current is conserved. This situation is analogous to the conservation of energy and momentum in general relativity, where in addition to the energy-momentum tensor of matter one must consider the energy-momentum tensor of the gravitational field in order to obtain a conserved quantity.

Strocchi (1977) has emphasized the fact the the exactness of the three-form J_{YM} is important for a consistent formulation of a local quantum gauge field theory and the discussion of the Higgs phenomenon.

Although it contradicts the original idea of Weyl's, one often considers uncoupled Yang-Mills fields, i. e., a pair of matrix-valued forms, Y, M, satisfying the equations:

$$M = DY, \quad DM = 0, \quad D{*}M = 0. \tag{1.26}$$

It is clear that any solution of these equations can be written in the form

$$M = M^{+} + M^{-}, \tag{1.27}$$

where

$$M^{+} = {*}M^{+}, \quad M^{-} = -{*}M^{-}. \tag{1.28}$$

We will discuss such self-dual (or anti-self-dual) curvatures in detail in the next sections.

The Yang-Mills construction can be generalized by replacing SU(2) by an arbitrary compact Lie group; to a certain extent one may consider general relativity as a gauge theory with SO(1,3) as gauge group, but, as already remarked by Weyl, Einstein's equations do not follow as analogues of (1.23).

To summarize, we have the following formulation of classical gauge theories:

a. A gauge group is a (possibly trivial) principal bundle over space-time. A gauge is the choice of a (local) section and a gauge transformation is locally a change of chart, corresponding to a different choice of trivializing section. The structure group G is compact.

b. A classical field is a section of a (complex) vector bundle E, associated to the principal bundle by a representation R(G) of the structure group G.

c. A gauge potential is a connection one-form Y, with the transformation property (1.16) under a change of chart. It defines covariant differentiation of sections (via the representation R) in any associated vector bundle.

d. The curvature form M = DY is of the adjoint type and defines the field strength of the gauge field.

e. The Yang-Mills equations are: i) the Bianchi identity $DM = 0$, ii) the equation $D{*}M = {*}J$, which follows from an action principle in which the norm of the curvature M is required to be minimized simultaneously with the action of the spinor field (one can choose the appropriate coupling constants so that none appears explicitly in the equations).

It is clear that the Yang-Mills fields, being a connection and curvature in a bundle, are of a different geometric nature than the fields (which are sections of a vector bundle). Therefore, in a discussion of a quantum theory for such fields it is not appropriate to treat both objects on the same footing. The fields can be described, e. g., as operator-valued distributions on sections of the associated vector bundle E, but the connection and curvature lose their meaning if one replaces them by operators (they are Lie-algebra-valued differential forms in the classical theory and the quantum analog is much more complicated, cf. Mayer (1975)). One possible approach, which is related to the lattice gauge theories, is to replace the differential formalism in terms of Y and M by an "integral" formalism, in terms of elements of the holonomy group of the connection associated to (spacelike) loops in space-time. This approach is related to the quantization of electrodynamics proposed by Mandelstam (1962), where the loops are replaced by lines from the point at infinity and will be discussed in detail elsewhere.

Returning to the classical theory, it is convenient (because of the existence of mathematical results) to consider principal bundles over compact manifolds. The compactification of the SU(2) gauge theory proceeds in two steps: i) passage from M_4 to the Euclidean space R^4 (there is some physical justification for this when one uses a path-integral formalism for quantization, as well as the experience with Euclidean quantum field theory in lower dimensions); ii) since there is no scale of length in the theory the Yang-Mills equations are conformally invariant (one says that the Y-M field is "massless") hence one can compactify R^4 into S^4 by means of a stereographic projection. This leads to the "instanton" solutions which we will discuss in more detail in the following sections.

We now turn to a discussion of the "Chern form"

$$c_2(M) = (8\pi^2)^{-1} M \wedge M, \qquad (1.29)$$

which characterizes the topological properties of the gauge bundle.

2. The BPST Instanton and its Chern Number

Without going into the physical motivation and interpretation, which at best is ambiguous, I want to show how the original instanton solution (also called the pseudoparticle solution, since one possible interpretation is as a "particle" in Euclidean spacetime = "pseudo-particle" in Minkowski space) was introduced by Belavin, Polyakov, Schwartz and Tyupkin (LN-4) and 't Hooft (LN-54), in the notations introduced in the preceding section.

The instanton is an outgrowth of the search for solutions of classical field theories which are localized in spacetime and for which the stability properties stem from a "topological" conservation law, i. e., a homotopy invariant which characterizes a class of solutions. These properties were discovered by Polyakov, 't Hooft and others and a clear description of the situation as of 1975 can be found in Sidney Coleman's Cargese Lectures (LN-11). Following a suggestion of Polyakov, Belavin, Polyakov, Schwartz and Tyupkin (to be abbreviated as BPST henceforth) posed the problem of finding a connection Y in an SU(2)-bundle over R^4 which is asymptotically (i.e., for large $r^2 = x_1^2 + x_2^2 + x_3^2 + x_4^2$) a "pure gauge", i. e., a Maurer-Cartan form $G(x)^{-1}dG(x)$, hence flat, $M \to 0$, minimizes the classical action with density (1.21), and is such that the curvature M falls off rapidly enough at infinity so that the action integral is finite:

$$\int \mathrm{Tr}(M \wedge {*M}) < \infty. \tag{2.1}$$

Mathematically this integral is the norm squared of the curvature. One obvious solution is $Y = 0$, which corresponds to the "classical vacuum", but BPST noticed that there exist solutions to the uncoupled Yang-Mills equations (1.26) which are not identically zero and lead to local minima of the integral (2.1) (in physical papers the integral is multiplied by $(1/4g^2)$, where g^2 is the Yang-Mills coupling constant measuring the self-coupling of the fields; for simplicity we shall set $g = 1$). Since each solution of the Yang-Mills equations which is asymptotically Maurer-Cartan can be characterized by a function $G(x)$ on R^4 with values in the structure group SU(2) of the bundle (i. e., by a section of the principal bundle defined outside a sufficiently large ball, or on its boundary S^3 of points "at infinity"), BPST have argued that equivalence classes of solutions (the equivalence being given by a gauge transformation, or a change of chart, or by a continouous deformation of the functions G(x) into each other) can be labelled by the homotopy classes of mappings of the three-dimensional sphere S^3 into the group SU(2) (which is homeomorphic to S^3), i. e., by elements of the third homotopy group $\pi_3(SU(2))$ which is the additive group of integers. They also noticed that since the 4-form (1.29) is closed (for the special case of SU(2) this follows from an elementary calculation, using the definition (1.20) and the fact that the trace of four Pauli matrices vanishes; in the general case it can be proved using the structure equations) it can be written, at least asymptotically, as the exterior differential (divergence) of a 3-form

$$\begin{aligned} Tr(M \wedge M) &= dTr(Y \wedge dY + (2/3)Y \wedge Y \wedge Y) \\ &= d{*}J_5(8\pi^2), \end{aligned} \qquad (2.2)$$

where J_5 is an "axial vector current", and 't Hooft recognized that it is the term which occurs in the Adler-Bell-Jackiw anomaly (which is thus reduced to a purely geometrical effect in a vector bundle). One can thus use Stokes' theorem to reduce the integral of $c_2(M)$ over R^4, which is finite on account of the finiteness of the action and the self-dual or anti-selfdual character of the curvature M ((1.27), (1. 28)), to an integral of $*J_5$ over the "boundary" S^3, where Y has the Maurer-Cartan form $G^{-1}dG$. In the special gase of SU(2) it turns out that the wedge-product of the three Y-forms is just the Haar measure on SU(2) (normalized by the factor $(8\pi^2)$) and thus the integral

$$q = (8\pi^2)^{-1}\int(M \wedge M), \qquad (2.3)$$

where the parentheses stand for the Killing inner product in the Lie algebra, or the trace in a matrix representation, is an integer which counts the number of times the map G(x) (the asymptotic gauge) applies the sphere S^3 on the group manifold SU(2), i. e. the <u>degree</u> of G(x). The trivial vacuum is obtained by choosing G(x) homotopic to the unit matrix e in SU(2) and solutions with nonvanishing q have been obtained by BPST for $q = \pm 1$ (the instanton and anti-instanton) and by 't Hooft and others for arbitrary q. These solutions were obtained by considering SU(2) as one of the two such factors to which SO(4) is isomorphic, and using a rigid coupling of the SO(4) transformations of R^4 which leave S^3 invariant and the isospin transformations. This leads to a particularly simple expression of the self-duality condition for M, and without going into the details of the calculations, yields the following solution for the connection form Y:

$$Y^{\pm} = (r^2/(r^2 + \lambda^2))\ G(x)^{-1}dG(x), \qquad (2.4)$$

where

$$G(x) = r^{-1}(x_4 \pm i \sum_1^3 T_a x_a), \qquad (2.5)$$

λ is a parameter characterizing the size of the instanton, and T_a are the isospin Pauli matrices (notice that the isospin index a is identified with the 3-space index of x_a). There are other equivalent forms of the solution (2.4), given by 't Hooft. I **will** only list the solution for q = n, which is written in terms of a four-dimensional harmonic function:

$$Y^n = \sum \eta_k^{aj}(\partial_j \ln\rho)(iT_a)dx^k, \quad a = 1, 2, 3, \qquad (2.6)$$

where the 't Hooft matrix is

$$\eta_k^{aj} = (1/2)\mathrm{Tr}(S_k S^j T^a), \quad j, k = 1,\ldots,4, \qquad (2.7)$$

and we have used the notation S^k as introduced in the previous section

$$S^j = (S_1, S_2, S_3, 1), \quad S_k = (-S_1, -S_2, -S_3, 1),$$

S_a, T_a are the spin and isospin Pauli matrices, which are multiplied due to the rigid coupling of the internal and spacetime groups, and

$$\rho(x) = 1 + \sum_1^q \lambda_i^2/(x - x_i)^2. \qquad (2.8)$$

It was noticed by Jackiw, Nohl and Rebbi (1977) that the solutions (2.4), (2.8) are invariant under an SO(5) subgroup of the conformal group SO(5,1) of R^4, and that by means of a stereographic projection the connections Y of an SU(2)-bundle over R^4 (which is necessarily trivializable, since there always exists a section) is pulled back into a connection (which we continue to denote by Y) of an SU(2) bundle over S^4 (which is nontrivial, since one needs two charts for local trivializations, the two charts which are most convenient being the mapping of S^4 minus the north or south pole onto a ball of large radius in R^4 - we leave it as an exercise to the reader to draw the appropriate pictures and derive the pull-back formulas in terms of bases in R^4 and the tangent space to S^4). SO(5) rotations of S^4 correspond to conformal mappings of R^4, in particular, the stereographic projection associates to "big S^3-spheres" on S^4 3-dimensional hyperplanes in R^4, and to the S^3-sphere at infinity in R^4 used to calculate the Chern number q, a "parallel" S^3-sphere near the north or south pole of S^4.

In their original paper BPST call the integer q defined by (2.3) a "Pontryagin class", which is justified since they consider an SO(4) bundle (in fact, the 4-form which is the integrand of (2.3) is a characteristic class, the integer resulting from integration over the fundamental cell has been called in the literature "Pontryagin index" or "Pontryagin number"). Since we will concentrate on complex vector bundles it is more appropriate to consider the Chern characteristic classes (of which the Pontryagin classes are special cases) and that is why I called (1.29) the <u>second Chern class</u> and the integer (2.3) the <u>Chern number</u> or <u>Chern index</u>. Most early papers on instantons did not discuss the characteristic class aspect of q, but concentrated on its interpretation in terms of the third homotopy group or degree of the map G(x). Since I have been emphasizing the Chern class aspect since the summer of 1976 (Mayer (1976)), let me quickly remind you how Chern classes are defined for complex vector bundles and what their basic properties are. Cf. also Wu and Yang, LN-62.

Let E be a complex vector bundle of rank n, associated by an n-dimensional representation to the principal bundle P with structure group G. Without loss of generality we can think of G as either SU(n) or as $GL(n, \mathbb{C})$ (the frame bundle of E is the principal bundle with the latter as structure group). Let D be a connection (covariant differentiation) with connection 1-form ω and curvature 2-form Ω. Since the curvature form is of the "adjoint" type and is of even degree one can form invariant polynomials (with respect to the adjoint action of G on its Lie algebra), in particular the determinant

$$c(E, s) = \det(I + i\Omega(s,\omega)/2\pi), \qquad (2.9)$$

where s is a (local) frame, in terms of which the connection and curvature are given as matrices of differential forms. The factor multiplying the curvature is chosen so as to yield integers in the end. It can be shown easily (cf. the lectures of Chern and the Bott-Chern paper quoted in LN) that the inhomogeneous differential form (2.9) does not depend on the choice of frame s and that moreover it is a <u>closed form</u>. Consequently it defines a de Rham cohomology class, which can be shown not to depend on the choice of connection, and is thus a <u>characteristic class</u> of E, called the <u>Chern class</u> c(E). Expanding c(E) into forms of various degrees (necessarily even and labelled by half of the degree)

$$c(E) = 1 + \sum c_i(E) \ , \ c_i \in H^{2i}(B, R) \qquad (2.10)$$

where R can be replaced by Z, and the base space B is <u>compact</u>.

For the sequel we shall assume E to be associated to a principal bundle with structure group U(n), and we introduce the formal factorization (which is possible since the $c_i(E)$ are forms of even degree)

$$\sum_1^n c_i(E) = \prod_1^n (1 + \gamma_i). \tag{2.11}$$

(If E is a Whitney sum of line bundles the γ_i are in fact the first Chern classes of these line bundles.)

The factorization (2.11) allows us to define the Chern character

$$\mathrm{ch}(E) = \sum_1^n \exp(\gamma_i), \tag{2.12}$$

which is additive and multiplicative under Whitney sum and tensor product of vector bundles:

$$\mathrm{ch}(E \oplus E') = \mathrm{ch}(E) + \mathrm{ch}(E'),\ \mathrm{ch}(E \otimes E') = \mathrm{ch}(E).\mathrm{ch}(E').$$

The Todd class of the vector bundle E can also be defined in terms of the factorization (2.11):

$$\mathrm{td}(E) = \prod_1^n \gamma_i/(1 - \exp(\gamma_i)). \tag{2.13}$$

These two characteristic classes enter into the computation of the topological index of a differential operator or an elliptic complex.

Finally, if the structure group of the bundle is O(n) (or SO(n)) and the base space has dimension multiple of 4 (in particular, space-time), the appropriate characteristic classes are the Pontryagin classes

$$p_i(E) = (-1)^i c_{2i}(C(E)) \in H^{4i}(B, Z), \tag{2.14}$$

where C(E) denotes the complexification of the real vector bundle E. The Chern class of highest rank defines the Euler class e(E'), where E' is the 2n-dimensional real oriented vector bundle induced by the complex structure of E.

3. Instanton Counting and the Atiyah-Singer Index Theorem

The solution (2.6) - (2.8) of the Yang-Mills equations was generalized by Jackiw, Nohl, and Rebbi (by replacing the 1 in Eq. (2.8) by $1/(x - x_0)$) to a family of solutions depending on 5q + 4 parameters. By using qualitative arguments they made it plausible that there should exist 8q - 3 independent instanton solutions for Chern number q.

The first to prove that this is indeed so was A. S. Schwartz (1977), who used the index theorem for the linearized equations and gave arguments for stability of the number of solutions under deformations. Schwartz noticed that the index is a linear combination of the Chern number q, the Hirzebruch signature of the base space and the Euler characteristic of the base space, and thus could be calculated by considering three special cases.

Another proof, which I will essentially reproduce below, was found at about the same time by Atiyah, Hitchin, and Singer (1977). The Atiyah, Hitchin and Singer paper also gives in footnotes the result for an SU(n)-bundle over S^4 ($h^1 = 4nq - n^2 + 1$), the number of zero-eigenvalue harmonic spinors (d = q), as well as the modifications required when B is an arbitrary compact riemannian manifold. Explicit calculations of these numbers can be found in a recent preprint by Römer (1977).

Let us consider the Yang-Mills equations (1.26) with the curvature M satisfying the self-duality condition (1.28). We denote by A^0, A^1, and A^2_- the spaces of smooth sections of the bundles of 0-forms (gauge functions), 1-forms (Y-M potentials), and 2-forms (Y-M fields) which are anti-selfdual, with values in the Lie algebra of G. We denote by D_0 the linear part of the exterior covariant differential (1.17), and by D_1 the projection of that operator on the anti-selfdual space A^2_-. Since we consider only self-dual solutions we have

$$D_1 D_0 = 0. \tag{3.1}$$

We thus arrive at an <u>elliptic complex</u>:

$$0 \longrightarrow A^0 \xrightarrow{D_0} A^1 \xrightarrow{D_1} A^2_- \longrightarrow 0, \tag{3.2}$$

the operators D_0 and D_1 being elliptic of first order (as can be easily seen by computing their symbols). The analytical index of an elliptic complex over a compact manifold is given by its <u>Euler characeristic</u> (the alternating sum of dimensions of its "cohomology spaces")

$$\mathrm{ind}_a(D_0, D_1) = h^0 - h^1 + h^2 = \chi(C), \tag{3.3}$$

where,

$$\begin{aligned} h^0 &= \dim\ker D_0, \\ h^1 &= \dim\ker D_1 - \dim\operatorname{im} D_0, \\ h^2 &= \dim A^2_- - \dim\operatorname{im} D_1, \end{aligned} \tag{3.3}$$

and $\chi(C)$ denotes the Euler characteristic of the complex.

If we are dealing with an SU(n)-bundle which does not reduce to a lower-dimensional holonomy group the kernel of D_0 is trivial (i. e. the Yang-Mills connection is irreducible) and $h^0 = 0$. The number h^1 is the number we are interested in (the number of <u>moduli</u>). h^2 is also equal to the dimension of the kernel of D_1^*, the formal adjoint of D_1, and can be shown to vanish for the case under consideration (e. g., by noticing that $D_1D_1^*$ is strongly elliptic, or invoking a Bochner type vanishing theorem). Thus, the number of independent instantons (or the dimension of the space of moduli, in geometric terms) is the number of linear variations of the gauge potential and equals

$$h^1 = - \mathrm{ind}_a(D_0, D_1). \qquad (3.4)$$

The Atiyah-Singer index theorem (cf., e. g., Atiyah-Singer(1968), Palais (1965), Booss (1977)) tells us that the analytical index of an elliptic complex coincides with the topological index, which can be expressed in terms of characteristic classes of the vector bundles involved (in particular, the Chern characters of the vector bundles of which the A^i are spaces of sections and the Todd class of the complexified tangent bundle of the base-manifold. The explicit calculation is a bit lengthy and I will omit it here, but ultimately it expresses the topological index of the complex (3.2) in terms of the simpler invariants of S^4, its Euler - Poincaré characteristic $\chi(S^4) = 2$, and the Hirzebruch siganture $L = (1/3)\int p_1 = 0$, and the Chern characters of the sympbols of D_0, D_1, which in turn can be expressed in terms of the second Chern class $c_2(E)$, the dimension of the adjoint representation of the structure group, and the Casimir operator of the group (for an explicit calculation, cf. the preprint of Römer (1977)). The final expression for the topological index is ($G = SU(n)$)

$$\mathrm{ind}_t(C) = (n^2-1)(\chi(B) - L(B))/2 - 4nq \qquad (3.5)$$

which for $B = S^4$ yields the value quoted on the preceding page and for $n = 2$ reduces to $8q - 3$.

So far we have discussed the linear complex (3.2). In order to show that the number of independent moduli remains the same, Atiyah, Hitchin and Singer appeal to a theorem of Kuranishi (1964) which guarantees that infinitesimal variations of the connection form, δY, really integrate to give genuine local variations, when $h^2 = o$. In addition, the theorem asserts that the family of solutions thus obtained is locally complete and non-redundant. The space of moduli is however not necessarily connected.

The case of massless spinors (harmonic spinors) has also been treated on the basis of the index theorem. The number of harmonic spinors of positive and negative helicity in the field of an instanton connection can be calculated in terms of the "heat equation" method of computing the index (Atiyah, Bott and Patodi (1973), Schwartz (1977)).

4. Instantons and Algebraic Geometry

Under this title Atiyah and Ward (1977), cf. also Ward (1977), have announced a series of Ansätze which allow one to construct solutions to the Yang-Mills equations (for SU(2) principal bundles over S^4) for all values of q. Since complete proofs of the Atiyah-Ward results are not yet available, and even the formulation requires a substantial amount of algebraic geometry, I will limit myself here to a very brief summary of their announcement, and an incomplete interpretetation of the complexifications of the bundles based on an explanation given to me by I. M. Singer.

I will start out with the statement of the Atiyah-Ward theorem and will then try, as best as I can, to explain what is involved. For details we will have to await the appearance of the detailed paper.

Theorem. There is a natural one-one correspondence between

a) self-dual solutions Y of the SU(2) Yang-Mills equations on S^4 up to gauge equivalence, and

b) isomorphism classes of 2-dimensional algebraic vector bundles E over three-dimensional projective space (over the complex numbers) $P_3(\mathbb{C})$ satisfying the conditions:

1) E has a symplectic structure;

2) the restriction of E to every real line of $P_3(\mathbb{C})$ is algebraically trivial.

The restriction 2) can be understood if one takes into account the fact that algebraic vector bundles of rank 2 over a projective line are isomorphic to a direct sum of line bundles, and are characterized by the first Chern classes. The assumption 2) amounts to setting the appropriate integers equal to zero for all real lines of $P_3(\mathbb{C})$.

In order to understand how one gets from an SU(2) principal bunlde with self-dual connection Y to an isomorphism class of (algebraic) vector bundles over the projective space P_3 (we omit the $\mathbb{C}$ henceforth, since we will be dealing exclusively with the complex projective 3-space) one has to make use of the twistor theory of Penrose (1972). Since S^4 has no global complex structure, one can complexify S^4 into a manifold Q_4 by considering the quadric

$$z_1^2 + \dots + z_5^2 = z_6^2 \quad \text{in } P_5 \tag{4.1}$$

where the z_i are homogeneous coordinates, which can be identified with the Grassmann manifold of lines in P_3. Any line x,y in P_3 can be described in terms of its Plücker coordinates $\pi_{\alpha\beta} = x_\alpha y_\beta - x_\beta y_\alpha$ satisfying the identity:

$$\pi_{01}\pi_{23} + \pi_{02}\pi_{31} + \pi_{03}\pi_{12} = 0 \qquad (4.2)$$

Thus, one complexifies S^4 into Q_4 and then uses the Klein correspondence of Plücker coordinates and homogeneous coordinates, to transform the bundle into a bundle over P_3. The bundle so obtained (Ward(1977)) is a 2-dimensional complex holomorphic vector bundle E over P_3, with the connection of the original bundle transformed into a complex structure (in the process the structure group SU(2) gets complexified into SL(2,$\mathbb{C}$) and the Newlander-Nirenberg condition for the integrability of vector fields in holomorphic bundles is the holomorphic translation of the existence of a connection with self-dual curvature in the real bundle (note that connections are easily pulled back, but cannot in general be "pushed forward")). The real form SU(2) of the structure group of the vector bundle then imposes the restrictions 1) and 2) mentioned above (to see this one extends the complex numbers to quaternions and uses the symplectic structure on E induced by multiplication by j). Finally, it is important to note that the vector bundles E are algebraic (i. e., their transition functions are algebraic).

There has been a lot of purely mathematical activity devoted to the classification of algebraic vector bundles of rank 2 over P_3, a subject which is actively pursued by leading algebraic geometers, and the correspondence with Yang-Mills theory, which may be more than an accident, has led to intensive interactions between mathematicians and physicists.

Atiyah and Ward also describe a series of Ansätze, which allow, in principle, to construct instantons for all values of q, and thus to find the missing solutions, not given by the 't Hooft Ansatz (2.8).

The Atiyah-Ward Ansätze involve so-called jumping lines, i. e., lines on which triviality of the vector bundles breaks down, and which can be characterized by their degree and genus, subject to restrictions involving q. The simplest ones correspond to the solutions of 't Hooft and Jackiw, Nohl, and Rebbi, and the jumping lines are obtained by intersecting Q_4 with a projective line. For q = 1 this results in the parameter space being the hyperbolic 5-space, a fact which has also been noticed by Yang. The higher Ansätze involve invariant wave equations (e. g., Maxwell equations for l = 2).

5. CONCLUDING REMARKS

This brief (and unfortunately incomplete) survey of recent points of contact between Euclidean space gauge theories and developments in differential topology and algebraic geometry shows us that we may be on the threshold of important new theoretical developments in physics. Without going into the multiple attempts at physical intepretation of instantons, such as "vacuum periodicity" and "tunneling", the role instantons might play in quark confinement, or in "steepest descent" calculations of path integrals, many of which will probably not survive the test of time, let me point out several things which may not have received their due attention.

First, it is obvious that as soon as the Yang-Mills connection is coupled to its sources, i. e., the second Yang-Mills equation is $D{*}M = {*}J$, the curvature M ceases to be self-dual and the results described above will have to be modified. But this is exactly the situation required by Weyl's gauge principle: a globally defined nonzero curvature without sources is not the result of a gauge principle. The Dirac (or Weyl) spinor fields as sections of a spin-bundle with Riemannian and Yang-Mills connection can be treated by the methods discussed here, and have in fact led to the surprizing recognition that the Adler-Bell-Jackiw anomaly for the axial vector current is of a purely geometric nature, and follows easily from the heat-equation version of the index theorem (cf. the discussion of spin-index in Atiyah-Bott-Patodi (1973) and the brief discussion in Stora (1977). On the other hand, if $D{*}M = {*}J \neq 0$ the whole discussion of Sections 3 and 4 breaks down, and so far only a "perturbation" approach was used, considering expansions of the connection around the instanton solutions.

A second problem which will have to be tackled is the passage to the noncompact Minkowski space as base space. Here the approach of Atiyah and Ward, involving complexifications and the use of SL(2,C) bundles may lead to a satisfactory solution.

A third problem, which has preoccupied me for some years, but to which I can offer no solution, is the problem of a consistent quantum field theory of Yang-Mills fields. There is no problem of formulating Wightman-Garding type axioms for a spinor field with a classical Yang-Mills connection. The fields are treated as operator-valued linear functionals (distributions) on spaces of appropriately chosen sections of a vector bundle associated to the gauge and space-time groups (in the latter the bundle is usually trivial). However, it is not at all clear how to interpret the connection and its curvature from the operator point of view. A straightforward quantization seems meaningless, since Y and M are of quite a different nature than the spinors. Some time ago (Mayer (1975)) I have proposed to treat the quantized connection and curvature as "morphisms" of the operator-valued distributions implemented by representations of the Lie algebra of the gauge group. My current thinking (and I was gratified to hear that to some extent this is shared by I. M. Singer) is that the most promising approach is to think of the quantized gauge field as an operator-automorphism associated to the holonomy group of the connection (for an irreducible connection the holonomy group and the curvature determine each other completely, if the base space is simply connected). In other words, to each loop in the base space (in Minkowski space one must consider only spacelike loops and the problem becomes more involved) one associates a "morphism" of operator-valued distributions on smooth sections of the spinor bundle, and it remains to be seen whether one can associate field operators (e. g., in the sense of generators) to the curvature or the transverse components of the connection. The approach of Mandelstam (1962) is of this nature (his loops are based on a point at spacelike infinity, and he represents the transformations by integrals of the potentials along the lines to infinity). Details of this approach need to be worked out, and some constructive ideas might come from lattice gauge theories, which are formulated ab initio in terms of elements of the holonomy group.

In this connection I would like to remark on a formal analogy of the present day situation with the Bohr-Sommerfeld quantization rule. There the quantum condition $\int pdq = n$ can be interpreted (Arnol'd) as the evaluation of a characteristic class on a fundamental cycle. It might be worthwhile considering the integrality of the Chern number q as characterizing a quasiclassical approximation of a yet to be discovered quantum theory of connections. At any rate, this integrality has to be reckoned with in any "quantization" scheme.

There has been some related "fallout" from the physicists preoccupation with the index theorem: the zeta function for elliptic operators (cf., e. g., Singer (1974)) yields a natural technique for computing determinants of Laplacians or Dirac operators, and hence for regularizing propagators and their products. This technique has already been used successfully by Hawking in quantum gravity and is slowly making its way into gauge theory. I forecast a much wider use of this technique in the future.

In conclusion let me say that the fact that many theoretical physicists were forced to learn techniques from topology, differential geometry and algebraic geometry, whether or not instantons will play an important role in physics, is bound to lead to a revolution in the way we teach mathematical physics to our students. One can already see a new trend (the use of differential forms in mechanics and electromagnetism , cf. e. g., Thirring's talk and his new text on mathematical physics, and lectures of R. Jost at the ETH on mechanics and electrodynamics several years ago). Just as the resistance to vectors, tensors and operators was slowly overcome 50 years ago or longer, the resistance to differential forms, bundles and cohomology on the part of today's physicist is bound to crumble, and we shall see a new style in mathematical physics emerging over the next decade.

ACKNOWLEDGEMENTS

First of all, I would like to thank Isadore M. Singer for illuminating discussions on the topics of this talk, which took place at Irvine, Berkeley, Oxford, Warwick, and Ellmau in 1977. I wish to thank Lisl and Daniel Kastler for an invitation to Ellmau (Tirol), providing a pleasant setting for some of these discussions).

I am indebted to the members of the Institut für theoretische Physik der ETH Zürich, particularly to Klaus Hepp and Res Jost, for their kind hospitality, financial support, and patient listening to my lectures during the winter semester 1977/78.

Last but not least I wish to express my gratitude to Konrad Bleuler and the other organizers of this Conference for their generous hospitality and for making my participation possible.

REFERENCES

NOTE. This list of references is a supplement to the references in my lecture notes (Mayer(1977)) quoted as LN-number. Due to circumstances beyond my control I was unable to produce a complete list of recent references, and I would like to apologize to those authors who have not been cited due to my oversight or ignorance of their work. A rather exhaustive list of references up to July 1977 can also be found in Stora (1977).

ATIYAH, M. F., R. BOTT, and V. K. PATODI (1973): On the Heat Equation and the Index Theorem, Inventiones Math. 19, 279-330 (1973). Errata: 28, 277-280 (1975).

--- and G. B. SEGAL: The Index of Elliptic Operators, II, Ann. of Math. 87, 531-545 (1968).

--- and I. M. SINGER (1968): The Index of Elliptic Operators, I, III, Ann. of Math. 87, 484-530 & 546-604 (1968).

---, N. J. HITCHIN, and I. M. SINGER (1977):Deformations of Instantons, Proc. Nat. Acad. Sci (USA) 74, 2662-2663 (1977).

--- and R. S. WARD (1977): Instantons and Algebraic Geometry, Commun. Math. Phys. 55, 117-124 (1977).

BOOSS,B. (1977): Topologie und Analysis (Eine Einführung in die Atiyah-Singer-Indexformel), Springer-Verlag,Berlin,Heidelberg, New York,1977.

GLIMM, J and A. JAFFE (1976): Cargese Lectures, 1976 (to appear); also: Instantons in a U(1) Lattice Gauge Theory, Commun. math. Phys. 56, 195-212 (1977).

JACKIW, R., C. NOHL, and C. REBBI (1977): Phys. Rev. D15, 1642 (1977).

---,and C.REBBI (1977): Degrees of Freedom in Pseudoparticle Systems, Phys. Lett. 67B, 189-192 (1977). Spinor Analysis of Yang-Mills Theory, Phys. Rev. D16, 1052-1060 (1977).

KURANISHI, M. (1965): Proc. of the Conference on Complex Analysis, p. 142 (H. Röhrl, Editor), Springer Verlag, Berlin-Heidelberg-New York, 1965.

LONDON, F. (1927): Quantenmechanische Deutung der Theorie von Weyl, Z. Physik 42, 375-389 (1927).

MAYER, M. E. (1975): Gauge Fields as Quantized Connection Forms, in: Conference on Differential-Geometrical Methods in Mathematical Physics (eds. K. Bleuler & A. Reetz), Lecture Notes in Math. vol. 570, Springer Verlag, Berlin-Heidelberg-New York, 1977.

--- (1976): Gauge Theories and Characteristic Classes, Preprint, UCI, September 1976. (Submitted to LMP, fate unknown).

MANDELSTAM, S.(1962): Quantum Electrodynamics without Potentials, Ann. Phys. (N. Y.) 19, 1-24 (1962).

OSTERWALDER, K. (1976): Gauge Theory on the Lattice, Cargese Lectures 1976 (to appear); also with E. Seiler, Gauge Theory on the Lattice, Ann. Phys. (N. Y.), to appear.

PALAIS, R. S. (1965): Seminar on the Atiyah-Singer Index Theorem, Ann. of Math. Studies, No. 57, Princeton University Press, 1965.

PENROSE, R. and M. A. H. MAC CALLUM (1972): Twistor Theory, Phys. Reports 6, 241 (1972).

H. RÖMER (1977): Number of Parameters for Instanton Solutions on Arbitrary Four-Dimensional Compact Manifolds, Preprint, CERN, TH. 2409, 28 October 1977.

SCHWARZ, A. S. (1977): On Regular Solutions of Euclidean Yang-Mills Equations, Phys. Lett. 67B, 172-174 (1977).

SINGER, I. M.(1974): Eigenvalues of the Laplacian and Invariants of Manifolds, Proc. Intern. Congr. of Mathematicians, Vancouver, B.C. 1974, pp. 187-200. Canad. Math. Congr., 1975.

STORA, R. (1977): Yang-Mills Instantons, Geometrical Aspects, Erice Lectures, 27 June- 9 July, 1977. CNRS/CPT Preprint, Marseille, Septem ber 1977.

F. STROCCHI (1977): Spontaneous Symmetry Breaking in Local Gauge Quantum Field Theory; the Higgs Mechanism, Commun. math. Phys. 56,57 -78 (1977).

't HOOFT, G. (1976): Symmetry Breaking through Bell-Jackiw Anomalies, Phys. Rev. Lett. 37, 8-11 (1976); Cal Tech Seminar, 1977.

WARD, R. S. (1977): On Self-Dual Gauge Fields, Phys. Lett. 61A, 81 - 82 (1977).

ADDENDUM: After this manuscript was completed the following reference became available in Zürich:

BERNARD, C.W., N. H. CHRIST, A. H. GUTH, and E. J. WEINBERG (1977): Pseudoparticle Parameters for Arbitrary Gauge Groups, Phys. Rev. D16, 2967-2977 (1977).

In this paper the authors embed the SU(2) connection in an arbitrary compact simple Lie group connection, and use the index theorem to determine the number of independent parameters. Their base space is S^4. The paper contains a detailed discussion of h^0 for the case in which the connection is reducible to the holonomy group.

CLASSIFICATION OF CLASSICAL YANG-MILLS FIELDS

by

M. Carmeli*

Institute for Theoretical Physics
State University of New York at Stony Brook
Stony Brook, New York 11794

and

Department of Physics
Ben Gurion University, Beer Sheva 84120, Israel

ABSTRACT

Classification of classical SU(2) gauge fields is described. Invariants of the fields are isolated and expressed in terms of gauge fields spinors. A review is given to the classification of the electromagnetic and gravitational fields and their invariants. Comparison between the three fields is made.

CONTENT:

Based on a lecture delivered at the Conference on Differential Geometric Methods in Mathematical Physics, Bonn, 1977.

---

*Work supported in part by the National Science Foundation under Grant No. PHY-76-15328.

1. INTRODUCTION

In the following the classification of the classical Yang-Mills field is reviewed. This subject is new as compared to that of the Petrov classification of the Weyl tensor in general realtivity theory [1]. It is also different from the Petrov classification since more than one group is involved here, namely, the spacetime Lorentz group and the internal symmetry group which will be assumed here to be SU(2). Of course we may classify gauge fields that are associated with larger groups such as SU(3), SU(4), etc. The degree of complexity will, of course, be higher. We can also classify gauge fields, moreover, in the Euclidean space instead of in the Minkowskian space. The problem of classification of a field is deeply related to the physical meaning of the field and to the exact solutions of the field equations. This is the situation in general relativity theory where a great deal of insight was obtained through the Petrov classification of the free-space gravitational field.

It is sometimes argued that problems of exact solutions and classification are highly mathematical topics that are hardly needed or at least one can manage without them. It is clear now, however, that this is not the case. Moreover, there seems to be little hope to obtain a deep and accurate insight into the physics of gauge fields without understanding exactly their classification. This is in fact the situation in general relativity theory.

Before we discuss the classification of gauge fields, we review the subject of classification of the electromagnetic field and the gravitational field. The comparison between the three fields is useful since, even though the physics is different, there is a lot of similarity between them.

2. PRELIMINARIES: THE ELECTROMAGNETIC FIELD

In the following an elementary review of the classification of the electromagnetic field is given. This could be used as an introduction to both the gravitational and gauge fields for readers who are unfamiliar with the problem of classification.

Let $F_{\mu\nu}$ be the electromagnetic field tensor. We may then form the two invariants

$$F_{\mu\nu}F^{\mu\nu} = 4\ K_1 \text{ (scalar)}, \tag{2.1}$$

$$F_{\mu\nu}\,{}^*F^{\mu\nu} = 4\ K_2 \text{ (pseudoscalar)}. \tag{2.2}$$

Here ${}^*F^{\mu\nu}$ is the dual to the tensor $F_{\mu\nu}$ defined as usual by

$${}^*F^{\mu\nu} = \frac{1}{2}\,\varepsilon^{\mu\nu\rho\sigma}F_{\rho\sigma}\ , \tag{2.3}$$

where $\varepsilon^{\mu\nu\rho\sigma}$ is the totally skew-symmetric Levi-Civita contravariant tensor density of weight +1, whose values are +1 or -1, depending upon whether $\mu\nu\rho\sigma$ is an even or an odd permutation of 0123, and zero otherwise.

In terms of the electromagnetic potentials we can also write the second invariant in the form

$$F_{\mu\nu} \, {}^{*}F^{\mu\nu} = 4 K_2 = 2 \partial_\alpha (\varepsilon^{\alpha\beta\gamma\delta} A_\beta \partial_\gamma A_\delta), \tag{2.4}$$

where

$$F_{\mu\nu} = \partial_\nu A_\mu - \partial_\mu A_\nu \, . \tag{2.5}$$

The following identification between the electric field $\vec{E}$ and the magnetic field $\vec{H}$, and the electromagnetic field tensor $F_{\mu\nu}$ will be made:

$$\vec{E} = (E_x, E_y, E_z) = (E_1, E_2, E_3), \tag{2.6}$$

$$\vec{H} = (H_x, H_y, H_z) = (H_1, H_2, H_3), \tag{2.7}$$

where

$$E_i = F_{i0}, \qquad H_i = \frac{1}{2} \varepsilon_{ijk} F_{jk} \, . \tag{2.8}$$

We then have for the electromagnetic field tensor and its dual the following:

$$F_{\mu\nu} = \begin{pmatrix} 0 & -E_x & -E_y & -E_z \\ E_x & 0 & H_z & -H_y \\ E_y & -H_z & 0 & H_x \\ E_z & H_y & -H_x & 0 \end{pmatrix}, \tag{2.9a}$$

$$^{*}F^{\mu\nu} = \begin{pmatrix} 0 & H_x & H_y & H_z \\ -H_x & 0 & -E_z & E_y \\ -H_y & E_z & 0 & -E_x \\ -H_z & -E_y & E_x & 0 \end{pmatrix}. \tag{2.9b}$$

In terms of the electric and magnetic fields the two invariants are then given by:

$$\frac{1}{2}(H^2 - E^2) = K_1 , \quad -\vec{E}\cdot\vec{H} = K_2. \tag{2.10}$$

One can see that K_2 is a pseudoscalar since the electric field is a polar vector whereas the magnetic field is an axial vector. Of course, $K_2^{\ 2}$ is a true scalar.

We may present the two invariants K_1 and K_2 in a somewhat different form. Define the complex field

$$F_i = F^+_{i0} = E_i + i H_i, \tag{2.11}$$

where $F^+_{\mu\nu}$ is defined by

$$F^+_{\mu\nu} = F_{\mu\nu} + i \ {}^*F_{\mu\nu}. \tag{2.12}$$

Then under a Lorentz transformation the complex vector $\vec{F}$ will undergo a complex three-dimensional "rotation".

A Lorentz transformation along the x-axis, for instance, will give the following transformation rule for the components of the complex field:

$$\begin{aligned} F_x &= F'_x \\ F_y &= F'_y \cosh\psi - i F'_z \sinh\psi \\ F_z &= iF'_y \sinh\psi + F'_z \cosh\psi. \end{aligned} \tag{2.13}$$

The "angle of rotation" ψ is related to the velocity v of the Lorentz frame by

$$\tanh \psi = v \tag{2.14}$$

where the speed of light is taken as unity. Now the only invariant one can construct out of a vector under rotation is its square,

$$\vec{F}^2 = (\vec{E} + i\vec{H})^2 = (\vec{E}^2 - \vec{H}^2) + 2i\vec{E}\cdot\vec{H} = -2 \ (K_1 + iK_2). \tag{2.15}$$

The classification of the electromagnetic field can be seen as follows. We have three possibilities:

(1) General Field: Not both of the two invariants K_1 and K_2 are zero. In this case we may make the following statements [2].

(a) There exists a Lorentz frame such that in the new system the electric field and the magnetic field are parallel to each other. The velocity of the Lorentz frame that characterizes such a system is then given by

$$(1 + v^2)^{-1} \vec{v} = (\vec{E}^2 + \vec{H}^2)^{-1} \vec{E} \times \vec{H} . \tag{2.16}$$

(b) If $K_2 = 0$, but $K_1 < 0$, then there exists a Lorentz transformation such that in the new frame the magnetic field vanishes. If $K_2 = 0$, but $K_1 > 0$, then there exists a Lorentz transformation such that in the new frame the electric field vanishes, instead.

(2) Null Field: Both K_1 and K_2 vanish. We may then make the following statements:

(a) There exists a Lorentz transformation such that in the new frame one has:

$$E_x = H_x = 0, \quad E_y = H_z , \quad E_z = - H_y . \tag{2.17}$$

(b) There exists a Lorentz transformation such that in the new frame the electric field and the magnetic field can be made arbitrarily small or arbitrarily large.

(3) Zero Field: The electromagnetic field vanishes.

The two invariants of the electromagnetic field may be related to the eigenvalues of the electromagnetic field tensor. The eigenvalue-eigenvector equation has the form

$$(F_{\mu\nu} - \lambda\eta_{\mu\nu}) V^{\nu} = 0 , \tag{2.18a}$$

thus yielding

$$|F_{\mu\nu} - \lambda\eta_{\mu\nu}| = 0 \tag{2.18b}$$

for the eigenvalues. Here $\eta_{\mu\nu}$ is the Minkowskian metric tensor defined by $\eta_{00} = -\eta_{11} = -\eta_{22} = -\eta_{33} = 1$, and zero otherwise. When written explicitly, Eq. (2.18b) gives

$$\lambda^4 + (H^2 - E^2)\lambda^2 - (\vec{E}\cdot\vec{H})^2 = 0 , \tag{2.19}$$

or

$$\lambda^4 + 2K_1\lambda^2 - K_2{}^2 = 0 \tag{2.20}$$

when written in terms of the two invariants K_1 and K_2.

Equation (2.20) may now be solved, giving

$$\lambda^2 = -K_1 \mp (K_1^{\,2} + K_2^{\,2})^{\frac{1}{2}} . \qquad (2.21)$$

Corresponding to the vanishing or nonvanishing of the two invariants K_1 and K_2, we can evaluate the eigenvalues and eigenvectors of Eq.(2.18a).

According to the three different cases of the field, we have

$$\begin{aligned} &K_1 \neq 0, \quad K_2 \neq 0; \quad \lambda_1^2 \neq \lambda_2^2 \neq 0 \\ &K_1 \neq 0, \quad K_2 = 0; \quad \lambda_1^2 = 0, \ \lambda_2^2 = -2K_1 \\ &K_1 = 0, \quad K_2 \neq 0; \quad \lambda_1^2 = K_2, \ \lambda_2^2 = -K_2, \end{aligned} \qquad (2.22)$$

$$K_1 = K_2 = 0; \quad \lambda = 0 \qquad (2.23)$$

The eigenvectors can also be found from Eq. (2.18a). We obtain

$$\vec{E}\cdot\vec{V} + \lambda\phi = 0, \quad \vec{H}\times\vec{V} - \lambda\vec{V} - \phi\vec{E} = 0, \qquad (2.24)$$

where use has been made of the three-dimensional notation $V^\mu = (V^0, V^m) = (\phi, \vec{V})$.

Finally it is worthwhile mentioning that we can classify the electromagnetic field by means of the tensor $F^+_{\mu\nu}$ defined by Eq. (2.12) and satisfies

$$^*F^+_{\mu\nu} = -i\, F^+_{\mu\nu} . \qquad (2.25)$$

One then finds that $F^+_{\mu\nu}$ has two null directions determined by the vectors $k^\mu \neq 0$. They may or may not coincide, and satisfy

$$F^+_{\mu[\nu}\, k_{\rho]}\, k^\mu = 0 . \qquad (2.26)$$

If the two directions coincide, the electromagnetic field is null, otherwise it is general. In the null case we have

$$F^+_{\mu\nu}\, k^\nu = 0 . \qquad (2.27)$$

Of course

$$F^+_{\mu\nu} = 0 \qquad (2.28)$$

for the zero field case.

3. PRELIMINARIES: THE GRAVITATIONAL FIELD

We recall that the classification of the gravitational field is made by means of classifying the Weyl conformal tensor. The Riemann curvature tensor is defined by

$$(\nabla_\gamma \nabla_\beta - \nabla_\beta \nabla_\gamma) V_\alpha = R^\delta{}_{\alpha\beta\gamma} V_\delta , \tag{3.1}$$

where

$$R^\delta_{\alpha\beta\gamma} = \partial_\beta \Gamma^\delta_{\alpha\gamma} - \partial_\gamma \Gamma^\delta_{\alpha\beta} + \Gamma^\mu_{\alpha\gamma} \Gamma^\delta_{\mu\beta} - \Gamma^\mu_{\alpha\beta} \Gamma^\delta_{\mu\gamma} . \tag{3.2}$$

Here ∇_α denotes covariant differentiation and Γ are the affine connections.

The Riemann curvature tensor has the following symmetry properties:

$$R_{\alpha\beta\gamma\delta} = - R_{\beta\alpha\gamma\delta} = - R_{\alpha x\delta\gamma} \tag{3.3}$$

$$R_{\alpha\beta\gamma\delta} = R_{\gamma\delta\alpha\beta} \tag{3.4}$$

$$R_{\mu\alpha\beta\gamma} + R_{\mu\beta\gamma\alpha} + R_{\mu\gamma\alpha\beta} = 0 . \tag{3.5}$$

From the Riemann curvature tensor one constructs the Ricci tensor,

$$R_{\alpha\beta} = R^\rho{}_{\alpha\rho\beta} \tag{3.6}$$

and the Ricci scalar curvature $R = g^{\alpha\beta} R_{\alpha\beta}$.

The Einstein field equations are then given by

$$R_{\mu\nu} = \kappa (T_{\mu\nu} - \frac{1}{2} g_{\mu\nu} T), \tag{3.7}$$

where $T_{\mu\nu}$ is the energy-momentum tensor and T is its trace. The tensor $g_{\mu\nu}$ is the metric tensor, and κ is Einstein's gravitational constant.

Because of the symmetry properties (3.3) - (3.5) the Riemann curvature tensor in four dimensions has only twenty independent components. The tensor, however, is not irreducible, and can be decomposed as follows:

$$R_{\rho\sigma\mu\nu} = C_{\rho\sigma\mu\nu} + \frac{1}{2} (g_{\rho\mu} R_{\sigma\nu} + g_{\sigma\nu} R_{\rho\mu} - g_{\rho\nu} R_{\sigma\mu} - g_{\sigma\mu} R_{\rho\nu})$$

$$+ \frac{1}{6} (g_{\rho\nu} g_{\sigma\mu} - g_{\rho\mu} g_{\sigma\nu}) R \tag{3.8}$$

or in the form

$$R_{\rho\sigma\mu\nu} = C_{\rho\sigma\mu\nu} + \frac{1}{2} (g_{\rho\mu} S_{\sigma\nu} + g_{\sigma\nu} S_{\rho\mu} - g_{\rho\nu} S_{\sigma\mu} - g_{\sigma\mu} S_{\rho\nu})$$

$$- \frac{1}{12} (g_{\rho\nu} g_{\sigma\mu} - g_{\rho\mu} g_{\sigma\nu}) R. \qquad (3.9)$$

In the above formulas $C_{\alpha\beta\gamma\delta}$ is the Weyl conformal tensor, and

$$S_{\alpha\beta} = R_{\alpha\beta} - \frac{1}{4} g_{\alpha\beta} R \qquad (3.10)$$

is the treefree Ricci tensor.

The Weyl conformal tensor satisfies the same symmetries that the Riemann tensor satisfies,

$$C_{\alpha\beta\gamma\delta} = - C_{\beta\alpha\gamma\delta} = - C_{\alpha\beta\delta\gamma} \qquad (3.11)$$

$$C_{\alpha\beta\gamma\delta} = C_{\gamma\delta\alpha\beta} \qquad (3.12)$$

$$C_{\mu\alpha\beta\gamma} + C_{\mu\beta\gamma\alpha} + C_{\mu\gamma\alpha\beta} = 0 . \qquad (3.13)$$

In addition, it is traceless,

$$C^{\rho}{}_{\alpha\rho\beta} = 0. \qquad (3.14)$$

Hence it has only ten independent components.

The decomposition given above for the Riemann tensor may be written symbolically as

$$R_{\alpha\beta\gamma\delta} = C_{\alpha\beta\gamma\delta} \oplus S_{\alpha\beta} \oplus R , \qquad (3.15)$$

namely, the Riemann curvature tensor is decomposed into the Weyl conformal tensor, the tracefree Ricci tensor, and the Ricci scalar curvature.

When the Einstein field equations are satisfied, the Ricci tensor is replaced by the energy-momentum tensor. Hence the only components of the Riemann tensor that describe gravitation are those of the Weyl conformal tensor. It is for this reason that the Weyl conformal tensor is sometimes said to describe the gravitational field.

4. PROPERTIES OF THE WEYL TENSOR

We now classify the Weyl conformal tensor thus, in effect, classifying the gravitational field. This will be done in a very brief way since the spinor method will be used later on.

We introduce the following two 3 x 3 real matrices E and H whose matrix elements are defined by

$$E_{ij} = C_{0i0j}, \qquad H_{ij} = {}^{*}C_{0i0j} . \tag{4.1}$$

Here $C_{\alpha\beta\gamma\delta}$ is the Weyl conformal tensor and ${}^{*}C_{\alpha\beta\gamma\delta}$ is its dual,

$$ {}^{*}C_{\alpha\beta\gamma\delta} = \frac{1}{2}(-g)^{\frac{1}{2}}\, \varepsilon_{\alpha\beta\mu\nu}\, C^{\mu\nu}{}_{\gamma\delta} . \tag{4.2}$$

Since the classification scheme is made in a local Lorentz frame, the matrix H can also be written in the form

$$H_{ij} = \frac{1}{2}\, C_{0imn}\, \varepsilon_{jmn} \tag{4.3}$$

where low case Latin indices take the values 1, 2, 3.

By definition, the matrix E is symmetric,

$$E_{ij} = E_{ji}. \tag{4.4}$$

When written explicitly, the above matrices have the forms

$$E_{ij} = \begin{pmatrix} 0101 & C_{0102} & C_{0103} \\ C_{0201} & C_{0202} & C_{0203} \\ C_{0301} & C_{0302} & C_{0303} \end{pmatrix}, \tag{4.5}$$

$$H_{ij} = \begin{pmatrix} C_{0123} & C_{0131} & C_{0112} \\ C_{0223} & C_{0231} & C_{0212} \\ C_{0323} & C_{0331} & C_{0312} \end{pmatrix}. \tag{4.6}$$

If we now calculate the trace of the matrix H we find that it vanishes,

$$\mathrm{Tr}\, H = C_{0123} + C_{0231} + C_{0312} = 0, \tag{4.7}$$

by virtue of Eq. (3.5). Moreover, the Weyl tensor is traceless,

$$C^{\rho}{}_{\alpha\rho\beta} = \eta^{\rho\sigma} C_{\rho\alpha\sigma\beta} = C_{0\alpha 0\beta} - C_{1\alpha 1\beta} - C_{2\alpha 2\beta} - C_{3\alpha 3\beta} = 0. \tag{4.8}$$

Taking $\alpha = \beta = 0$, Eq. (4.8) yields

$$\mathrm{Tr}E = C_{0101} + C_{0202} + C_{0303} = 0. \tag{4.9}$$

Taking $\alpha\beta = 01, 02, 03$, Eq. (4.8) yields

$$C_{0212} = C_{0331}, \; C_{0112} = C_{0323}, \; C_{0131} = C_{0223} \tag{4.10}$$

thus showing that the matrix H is also symmetric

$$H_{ij} = H_{ji}. \tag{4.11}$$

Likewise, taking $\alpha\beta = 23, 31, 12, 11, 22, 33$, we obtain from Eq. (4.8)

$$C_{1213} = C_{0203}, \; C_{2321} = C_{0301}, \; C_{3132} = C_{0102}, \tag{4.12a}$$

$$C_{1212} = -C_{0303}, \; C_{1313} = -C_{0202}, \; C_{2323} = -C_{0101}. \tag{4.12b}$$

Hence the ten components of the Weyl tensor are presented by the two 3x3 symmetric and traceless matrices E and H, each of which has only five independent components.

We now define the 3×3 symmetric and traceless complex matrix

$$C_{ij} = E_{ij} + i H_{ij} = \overset{+}{C}_{0i0j} \tag{4.13}$$

where

$$\overset{+}{C}_{\mu\nu\rho\sigma} = C_{\mu\nu\rho\sigma} + i \, {}^{*}C_{\mu\nu\rho\sigma} \,. \tag{4.14}$$

We will examine the eigenvalue-eigenvector equation

$$C_{ij} V_j = \lambda V_i \tag{4.15}$$

where V_j is a complex vector and λ are complex eigenvalues that are related to the invariants of the Weyl conformal tensor.

The Weyl tensor can then be classified according to the possible numbers of eigenvalues and eigenvectors of the complex matrix C. The maximum number of eigenvalues for the matrix C is three. Corresponding to each eigenvalue, there exists at least one eigenvector.

The invariants of the field can also easily be found. Since the matrix C is traceless, we consider the invariants

$$\mathrm{Tr}\, C^2 = I, \quad \mathrm{Tr} C^3 = J. \tag{4.16}$$

The eigenvalue equation gives a cubic equation for λ,

$$\lambda^3 - \frac{1}{2} I \lambda - \frac{1}{3} J = 0, \tag{4.17}$$

where use has been made of the fact that $J = 3 \det C$ and of the Cayley-Hamilton theorem according to which

$$C^3 - \frac{1}{2} IC - \frac{1}{3} JI = 0. \tag{4.18}$$

(Notice that the first I in the above equation is an invariant of the field whereas the second one is the 3×3 unit matrix.) One can show that there are no further invariants since $\mathrm{Tr}\, C^n$, with $n = 4,5,\ldots$, can be expressed in terms of the two invariants I and J. In terms of the conformal tensor we can write

$$I = \frac{1}{8} C^{\alpha\beta\gamma\delta} C^{+}_{\alpha\beta\gamma\delta} = \frac{1}{16} C^{+\alpha\beta\gamma\delta} C^{+}_{\alpha\beta\gamma\delta} \tag{4.19}$$

$$I = \frac{1}{16} C^{\alpha\beta}{}_{\mu\nu} C^{\mu\nu}{}_{\rho\sigma} C^{+\rho\sigma}{}_{\alpha\beta}. \tag{4.20}$$

Let λ_1, λ_2 and λ_3 be the three eigenvalues of the matrix C. They may or may not be distinct. From the eigenvalue equation we obtain

$$\begin{aligned} &\lambda_1 + \lambda_2 + \lambda_3 = 0 \\ &\lambda_1\lambda_2 + \lambda_2\lambda_3 + \lambda_3\lambda_1 = -I/2 \\ &\lambda_1\lambda_2\lambda_3 = J/3. \end{aligned} \tag{4.21}$$

If the three eigenvalues are equal, then they all vanish. Thus the two invariants $I = J = 0$. This is the case of gravitational fields of types III, N, and O.

If, however, two of the eigenvalues, let us say λ_1 and λ_2, are equal and $\lambda_3 \neq \lambda_1 = \lambda_2 \neq 0$, Eqs. (4.21) give

$$I = 6\lambda_1^2 = 6 \lambda_2^2 = 3 \lambda_3^2/2, \tag{4.22}$$

$$J = -6\lambda_1^3 = -6\lambda_2^3 = 3\lambda_3^3/4. \tag{4.23}$$

The last equations then eimply that

$$I^3 = 6J^2 \neq 0. \tag{4.24}$$

This is the case of gravitational fields of types II and D.

Finally, if the three eigenvalues are different from each other, then $I^3 \neq 6J^2$. The gravitational field is then of type I, namely general.

It is worthwhile pointing out that the classification of the gravitational field given above is invariant under a change of Lorentz frame. This can easily be seen since under a Lorentz transformation the matrix C transforms into

$$C' = PCP^t, \tag{4.25}$$

where P is a 3×3 complex orthogonal matrix, $P^{-1} = P^t$, with determinant unity. It is given by

$$P = \begin{pmatrix} ad+bc & i(ac+bd) & ac-bd \\ -i(ab+cd) & \frac{1}{2}(a^2+b^2+c^2+d^2) & -\frac{i}{2}(a^2-b^2+c^2-d^2) \\ ab-cd & \frac{i}{2}(a^2+b^2-c^2-d^2) & \frac{1}{2}(a^2-b^2-c^2+d^2) \end{pmatrix}. \tag{4.26}$$

Here a,b,c,d are four complex numbers given by

$$g = \begin{pmatrix} a & b \\ c & d \end{pmatrix}; \quad ad - bc = 1, \tag{4.27}$$

where g is an element of the group SL(2,C).

The matrix P gives a three-dimensional representation for the proper, orthochronous, homogeneous, Lorentz group. For Lorentz transformations (boosts) along the x-, y- and z-axis, for instance, we obtain for P:

$$P_{IL}(\psi) \quad \begin{pmatrix} \cosh\psi & i\sinh\psi & 0 \\ -i\sinh\psi & \cosh\psi & 0 \\ 0 & 0 & 1 \end{pmatrix}, \tag{4.28a}$$

$$P_{2L}(\psi) = \begin{pmatrix} \cosh\psi & 0 & -i\sinh\psi \\ 0 & 1 & 0 \\ i\sinh\psi & 0 & \cosh\psi \end{pmatrix}, \quad (4.28b)$$

$$P_{3L}(\psi) = \begin{pmatrix} 1 & 0 & 0 \\ 0 & \cosh\psi & -i\sinh\psi \\ 0 & i\sinh\psi & \cosh\psi \end{pmatrix}. \quad (4.28c)$$

We also obtain

$$P_{1R}(\psi) = \begin{pmatrix} \cos\psi & -\sin\psi & 0 \\ \sin\psi & \cos\psi & 0 \\ 0 & 0 & 1 \end{pmatrix}, \quad (4.29a)$$

$$P_{2R}(\psi) = \begin{pmatrix} \cos\psi & 0 & \sin\psi \\ 0 & 1 & 0 \\ -\sin\psi & 0 & \cos\psi \end{pmatrix}, \quad (4.29b)$$

$$P_{3R}(\psi) = \begin{pmatrix} 1 & 0 & 0 \\ 0 & \cos\psi & \sin\psi \\ 0 & -\sin\psi & \cos\psi \end{pmatrix}, \quad (4.29c)$$

for the rotations around the x-, y- and z-axis.

Suppose now that V_i is an eigenvector of the matrix C, with eigenvalue λ. Then, because of Eq. (4.25), it follows that $P\vec{V}$ is an eigenvector of the transformed matrix C' with the same eigenvalue,

$$C'(P\vec{V}) = \lambda(P\vec{V}). \quad (4.30)$$

The opposite is also correct.

Finally, it is worthwhile pointing out that one can also use the equation

$$k_{[\alpha}\overset{+}{C}_{\beta]\gamma\delta[\rho}\; k_{\sigma]}\; k^{\gamma}\; k^{\delta} = 0 \quad (4.31)$$

to classify the Weyl tensor. The vector k^{α} is called the principal null vector and Eq. (4.31) is known as the Debever-Penrose equation. It has four solutions k^α in general that determine four directions.

The Weyl tensor is of type I if the four directions are different; of type II if two of them coincide with the remaining two being distinct; of type D if the directions coincide in pairs; of type III if three directions coincide; and of type N is all four directions coincide.

When the gravitational field is of types II and D, then the Weyl tensor satisfies

$$C^+_{\beta\gamma\delta[\rho} \; k_{\sigma]} \; k^\gamma k^\delta = 0. \tag{4.32}$$

If it is of type III, it satisfies

$$C^+_{\beta\gamma\delta[\rho} \; k_{\sigma]} \; k^\delta = 0 . \tag{4.33}$$

When it is of type N it satisfies

$$C^+_{\beta\gamma\delta\rho} \; k^\rho = 0 . \tag{4.34}$$

And, of course, one has

$$C^+_{\beta\gamma\delta\rho} = 0 \tag{4.35}$$

for the zero field.

5. TWO-COMPONENT SPINORS

In the following,use will be made of 2-component spinors. These quantities appear in the theory of representations of the group SL(2,C) [3] .

Let P_{mn} denote the aggregate of all polynomials $p(z,\bar{z})$ of degree smaller or equal to m in the variable z and of degree smaller or equal to n in the variable $\bar{z}$. Here m and n are fixed non-negative integers. P_{mn} is then a linear vector space of dimension (m+1) (n+1). Let

$$g = \begin{pmatrix} a & b \\ c & d \end{pmatrix}; \qquad ad - bc = 1, \tag{5.1}$$

be an element of the group SL(2,C), and define the operator D(g) by

$$D(g)\; p(z,\bar{z}) = (bz+d)^m (\bar{b}\bar{z}+\bar{d})^n \; p\left(\frac{az+c}{bz+d}, \; \frac{\bar{a}\bar{z}+\bar{c}}{\bar{b}\bar{z}+\bar{d}} \right) . \tag{5.2}$$

The mapping $g \to D(g)$ is then a linear representation of the group SL(2,C):

$$D(g_1)\; D(g_2) = D(g_1 g_2); \qquad D(1) = 1. \tag{5.3}$$

The above representation is known as the spinor representation of the group SL(2,C) and is of dimensions (m+1) (n+1).

To obtain the usual 2-component spinors, we consider all systems of numbers

$$\phi_{A_1 \cdots A_m X'_1 \cdots X'_n} \tag{5.4}$$

that are symmetrical in its indices $A_1 \cdots A_m$ and in $X'_1 \cdots X'_n$ where these indices take the values 0,1 and 0', 1'. The system of such numbers provides a linear space, denoted by $\tilde{P}_{mn}$, also of dimensions (m+1)(n+1).

A one-to-one correspondence between the spaces P_{mn} and $\tilde{P}_{mn}$ can be obtained by the substitution

$$p(z\,,\,\bar{z}\,) = \sum \phi_{A_1 \cdots A_m X'_1 \cdots X'_n}\, z^{A_1 + \cdots + A_m}\, \bar{z}^{X'_1 + \cdots + X'_n} \quad . \tag{5.5}$$

The polynomial (5.5) is of degree $\leqslant$ m in z and $\leqslant$ n in $\bar{z}$. It therefore belongs to the space P_{mn}. On the other hand, every polynomial

$$p(z,\bar{z}) = \Sigma\, p_{rs} z^r z^s \tag{5.6}$$

in the space P_{mn} can be written in the form of Eq. (5.5), if we make the substitution

$$\begin{pmatrix} m \\ r \end{pmatrix} \begin{pmatrix} n \\ s \end{pmatrix} \phi_{A_1 \cdots A_m X'_1 \cdots X'_n} = p_{rs} \tag{5.7}$$

with $A_1 + \cdots + A_m = r$ and $X'_1 + \cdots + X'_n = s$.

The spinor representation can now be realized in the space $\tilde{P}_{mn}$. We obtain

$$D(g)p(z,\bar{z}) = \sum \phi'_{A_1 \cdots A_m X'_1 \cdots X'_n}\, z^{A_1 + \cdots + A_m}\, \bar{z}^{X'_1 + \cdots + X'_n} \tag{5.8}$$

where

$$\phi'_{A_1 \cdots A_m X'_1 \cdots X'_n} = \sum g_{A_1 B_1} \cdots g_{A_m B_m}\, \bar{g}_{X'_1 Y'_1} \cdots \bar{g}_{X'_n Y'_n}\, \phi_{B_1 \cdots B_m Y'_1 \cdots Y'_n} \tag{5.9}$$

In the following we outline the application of spinors in general relativity theory in a very brief way [4]. The connection between tensors and spinors is obtained by means of mixed quantities $\sigma^{\mu}_{AB'}$ where A and B' are spinor indices taking the values 0,1 and 0',1'. They provide four Hermitian 2×2 matrices

$$\sigma^{\mu}_{AB'} = \bar{\sigma}^{\mu}_{B'A} . \tag{5.10}$$

When a locally Cartesian coordinate system is used, and the space is flat, the above matrices are then the unit matrix and the three Pauli matrices. In general relativity one usually does not have to know these matrices explicitly.

The relationship between the above matrices and the geometrical metric tensor $g_{\mu\nu}$ is then given by

$$g_{\mu\nu}\sigma^{\mu}_{AB'}\sigma^{\nu}_{CD'} = \varepsilon_{AC}\,\varepsilon_{B'D'} , \quad \sigma_{\mu AB'}\,\sigma^{AB'}_{\nu} = g_{\mu\nu} . \tag{5.11}$$

Here ε_{AB}, ε^{AB}, $\varepsilon_{A'B'}$, and $\varepsilon^{A'B'}$ are the Levi-Civita skew-symmetric spinors defined by

$$\varepsilon = \begin{pmatrix} 0 & 1 \\ -1 & 0 \end{pmatrix} . \tag{5.12}$$

Raising and lowering indices are made by means of these spinors according to the convention

$$\xi^{A} = \varepsilon^{AC}\,\xi_{C} , \qquad \xi_{A} = \xi^{C}\,\varepsilon_{CA} \tag{5.13a}$$

$$\eta^{A'} = \varepsilon^{A'C'}\,\eta_{C'} , \qquad \eta_{A'} = \eta^{C'}\,\varepsilon_{C'A'} \tag{5.13b}$$

The spinor equivalent of a tensor $T_{\alpha\beta}$, for instance, is given by

$$T_{AB'CD'} = \sigma^{\alpha}_{AB'}\,\sigma^{\beta}_{CD'}\,T_{\alpha\beta} . \tag{5.14}$$

The condition for a vector V_{α} to be real is that its spinor equivalent be Hermitian,

$$V_{AB'} = \bar{V}_{B'A} . \tag{5.15}$$

The electromagnetic field tensor $F_{\mu\nu}$, for instance, is equivalent to the spinor that can be decomposed as follows:

$$F_{AB'CD'} = \varepsilon_{AC}\,\bar{\phi}_{B'D'} + \phi_{AC}\,\varepsilon_{B'D'} \tag{5.16}$$

Here use has been made of the notation

$$\phi_{AB} = \frac{1}{2}\,F_{AC'\,BD'}\,\varepsilon^{C'D'} \tag{5.17}$$

where ϕ_{AB} is a symmetric spinor, $\phi_{AB} = \phi_{BA}$. The spinor ϕ_{AB} is the electromagnetic field spinor and is completely equivalent to the electromagnetic field tensor $F_{\mu\nu}$.

In the same fashion we decompose the Riemann tensor $R_{\alpha\beta\gamma\delta}$(see Section 3). It is more appropriate with the spinor spirit, however, to obtain what might be called the curvature spinor. The latter spinor is obtained by applying the commutator of covariant differentiations on a spinor instead of on a vector when one obtains the usual Riemann tensor. We obtain:

$$(\nabla_\nu\nabla_\mu - \nabla_\mu\nabla_\nu)\,\xi^Q = F^Q{}_{P\mu\nu}\,\xi^P , \tag{5.18}$$

where

$$F^Q{}_{P\mu\nu} = \Gamma^Q_{P\mu,\nu} - \Gamma^Q_{P\nu,\mu} + \Gamma^B_{P\mu}\,\Gamma^Q_{B\nu} - \Gamma^B_{P\nu}\,\Gamma^P_{B\mu} . \tag{5.19}$$

Here $\Gamma^Q_{P\mu}$ are the spinor affine connections. Notice that the curvature spinor is a complex quantity.

The spinor equivalent of the Riemann tensor is then related to the curvature spinor $F^Q{}_{P\mu\nu}$ by

$$R_{AB'CD'EF'GH'} = F_{ACEF'GH'}\,\varepsilon_{B'D'} + \varepsilon_{AC}\,F_{B'D'EF'GH'} . \tag{5.20}$$

The curvature spinor $F_{ACEF'GH'}$ may now be decomposed as follows:

$$F_{PQAB'CD'} = -\,(\chi_{PQAC}\,\varepsilon_{B'D'} + \phi_{PQB'D'}\,\varepsilon_{AC}) \tag{5.21}$$

where

$$\chi_{ABCD} = \psi_{ABCD} + \frac{1}{6}\,(\varepsilon_{AD}\varepsilon_{BC} + \varepsilon_{AC}\,\varepsilon_{BD})\,\lambda. \tag{5.22}$$

In the above formulas ψ_{ABCD} is the totally symmetric Weyl conformal spinor, $\phi_{PQB'D'}$ is the tracefree Ricci spinor, and $\lambda = -R/4$, where R is the Ricci scalar curvature.

It is the Weyl spinor ψ_{ABCD} that is classified in the theory of general relativity, just as the Maxwell spinor ϕ_{AB} that is being classified in electrodynamics.

6. CLASSIFICATION OF THE ELECTROMAGNETIC AND GRAVITATIONAL FIELDS

We now briefly review the spinor method in classifying the electromagnetic and the gravitational fields [5]. The method of classifying the Yang-Mills field will be partially similar to the presentation of this section.

A. The Electromagnetic Field

Classification of the electromagnetic field is made by means of the electromagnetic field spinor ϕ_{AB}. We write the eigenspinor and eigenvalue equation

$$\phi^A{}_B\, \alpha^B = \lambda\, \alpha^A. \tag{6.1}$$

To study this equation we introduce a basis of two spinors ℓ_A and n_A in our space, with the condition $\ell_A n^A = 1$.

The null tetrad induced by these two spinors are given by

$$\ell^\mu = \sigma^\mu_{AB'}\, \ell^A \bar{\ell}^{B'}, \tag{6.2a}$$

$$m^\mu = \sigma^\mu_{AB'}\, \ell^A \bar{n}^{B'}, \tag{6.2b}$$

$$\bar{m}^\mu = \sigma^\mu_{AB'}\, n^A \bar{\ell}^{B'}, \tag{6.2c}$$

$$n^\mu = \sigma^\mu_{AB'}\, n^A \bar{n}^{B'}. \tag{6.2d}$$

Using the notation $\xi_0{}^A = \ell^A$ and $\xi_1{}^A = n^A$, the above null vectors can be collectively denoted by

$$\sigma^\mu_{ab'} = \begin{pmatrix} \ell^\mu & m^\mu \\ \bar{m}^\mu & n^\mu \end{pmatrix}, \tag{6.3}$$

and satisfy the orthogonality relation

$$\sigma^\mu_{ab'}\, \sigma_{\mu cd'} = \epsilon_{ac}\, \epsilon_{b'd'}, \quad \sigma^\mu_{ab'}\, \sigma^{\nu ab'} = g^{\mu\nu}. \tag{6.4}$$

Define now the three scalars

$$\phi_0 = F_{\mu\nu}\ell^\mu m^\nu \tag{6.5a}$$

$$\phi_1 = \frac{1}{2} F_{\mu\nu}(\ell^\mu n^\nu + \bar{m}^\mu m^\nu) \tag{6.5b}$$

$$\phi_2 = F_{\mu\nu} \bar{m}^\mu n^\nu \quad , \tag{6.5c}$$

then the eigenspinor equation (6.1) can be written in the form

$$\Phi\alpha = \lambda\alpha \tag{6.6}$$

where Φ is the 2×2 complex matrix given by

$$\Phi = \begin{pmatrix} \phi_1 & \phi_2 \\ -\phi_0 & -\phi_1 \end{pmatrix} \tag{6.7}$$

and α is the column matrix given by

$$\alpha = \begin{pmatrix} \alpha^0 \\ \alpha^1 \end{pmatrix}, \tag{6.8}$$

with $\alpha^a = \zeta^a{}_A \alpha^A$.

Equation(6.6) has two eigenvalues that are given by

$$\lambda_{1,2} = \pm(-\det \Phi)^{\frac{1}{2}} = \pm(\phi_1{}^2 - \phi_0\phi_2)^{\frac{1}{2}}. \tag{6.9}$$

Hence we have the following three cases:

(1) $\phi_1^2 \neq \phi_0\phi_2$. Here we have two distinct eigenvalues and two distinct eigenspinors. This is a general type field.

(2) $\phi_1^2 = \phi_0\phi_2$. The eigenvalues vanish and there is only one eigenspinor. This is the null field case.

(3) $\phi_0 = \phi_1 = \phi_2 = 0$. Here we have one eigenvalue and two eigenspinors. This is the zero field case.

It will be noted that the eigenvalues described above can be given in terms of the electromagnetic spinor ϕ_{AB} by

$$\phi_{AB}\phi^{AB} = 2(\phi_0\phi_2 - \phi_1{}^2) = -\,\mathrm{Tr}\,\Phi^2 = K, \tag{6.10}$$

where

$$K = K_1 + iK_2 = \tfrac{1}{4}(F_{\mu\nu}F^{\mu\nu} + iF_{\mu\nu}{}^{*}F^{\mu\nu}) \tag{6.11}$$

(see Section 2). The detail of the above discussion is summarized in Table 1 and in Figure 1.

TABLE 1

Classification of the electromagnetic field. The types of fields are as follows. Type I is general, type N is null, and type 0 is zero field.

DISTINCT EIGENSPINORS	2		1
DISTINCT EIGENVALUES	2	1	1
TYPE OF ELECTROMAGNETIC FIELD	I	0	N

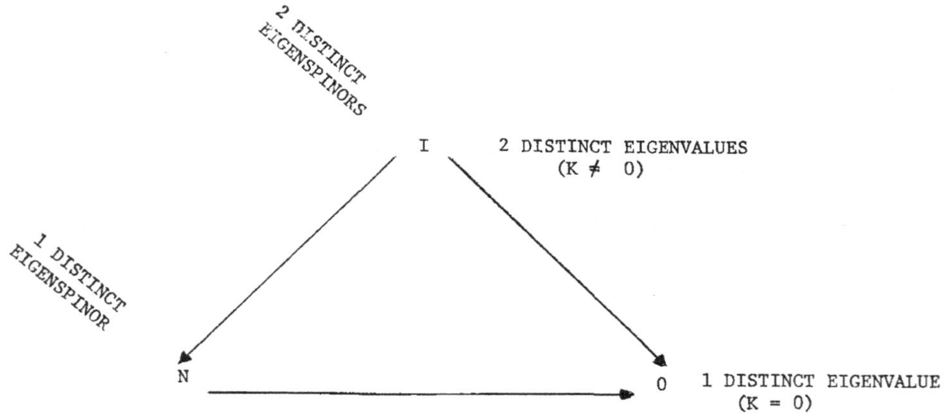

Figure 1

Classification of the electromagnetic field in terms of the field invariant K, and the eigenvalues and eigenspinors.

One more way to see the above classification is by means of decomposing the electromagnetic field spinor ϕ_{AB}. Since ϕ_{AB} is symmetric, one can factorize it by use of the fundamental theorem of algebra. For consider the invariant

$$\phi_{AB}\zeta^A\zeta^B = \phi_{00}\zeta^0\zeta^0 + 2\phi_{01}\zeta^0\zeta^1 + \phi_{11}\zeta^1\zeta^1$$

$$= (\zeta^1)^2 (\phi_{00} z^2 + 2\phi_{01} z + \phi_{11}) , \qquad (6.12)$$

where $z = \zeta^0/\zeta^1$, and the spinor ζ^A is arbitrary. The polynomial in Eq.(6.12) can be factorized and we obtain

$$\phi_{AB} = \alpha_{(A} \beta_{B)} \quad . \qquad (6.13)$$

Here α_A and β_A are arbitrary one-index spinors. The invariant of the field K is subsequently given by

$$K = - \frac{1}{2}(\alpha_A\beta^A)^2. \qquad (6.14)$$

The classification of the electromagnetic field can now be made in terms of the decomposition (6.13). If the two spinors α_A and β_A are distinct from each other, then $K \neq 0$ and we have a field of a general type. If, on the other hand, they are equal or proportional, then $K = 0$ and we have the null field and the zero field cases (see Figure 1).

B. The Gravitational Field

To classify the gravitational field using the spinor method, we classify the Weyl spinor ψ_{ABCD}. The eigenspinor-eigenvalue equation is now given by

$$\psi^{AB}{}_{CD}\ \phi^{CD} = \lambda\phi^{AB} \qquad (6.15)$$

Equation (6.15) is then solved in terms of the Weyl spinor components $\psi_0, \cdots, \psi_4$, a set of five scalars defined by

$$\psi_0 = -C_{\mu\nu\rho\sigma}\ell^\mu m^\nu\ \ell^\rho m^\sigma \qquad (6.16a)$$

$$\psi_1 = -C_{\mu\nu\rho\sigma}\ell^\mu\ n^\nu\ \ell^\rho\ m^\sigma \qquad (6.16b)$$

$$\psi_2 = -\frac{1}{2} C_{\mu\nu\rho\sigma} \ell^\mu n^\nu (\ell^\rho n^\sigma - m^\rho \bar{m}^\sigma) \tag{6.16c}$$

$$\psi_3 = -C_{\mu\nu\rho\sigma} \bar{m}^\mu m^\nu \ell^\rho n^\sigma \tag{6.16d}$$

$$\psi_4 = -C_{\mu\nu\rho\sigma} \bar{m}^\mu n^\nu \bar{m}^\rho n^\sigma . \tag{6.16e}$$

The eigenvalue equation (6.15) can then be written as

$$\Psi\chi = \lambda\chi \tag{6.17}$$

where Ψ is the 3×3 complex matrix given by

$$\Psi = \begin{pmatrix} -2\psi_2 & i(\psi_1 + \psi_3) & \psi_3 - \psi_1 \\ i(\psi_1 + \psi_3) & \frac{1}{2}(\psi_0 + 2\psi_2 + \psi_4) & \frac{i}{2}(\psi_0 - \psi_4) \\ \psi_3 - \psi_1 & \frac{i}{2}(\psi_0 - \psi_4) & \frac{1}{2}(-\psi_0 + 2\psi_2 - \psi_4) \end{pmatrix} . \tag{6.18}$$

The matrix Ψ can be compared to the matrix C used in Section 4 to classify the Weyl conformal tensor. The relationship between them is easily found if we write the spinor equivalent of the Weyl tensor and its dual,

$$C^+_{\alpha\beta\gamma\delta} = C_{\alpha\beta\gamma\delta} + i\, {}^*C_{\alpha\beta\gamma\delta} . \tag{6.19}$$

One then easily finds that the spinor equivalent to the tensor $C^+_{\alpha\beta\gamma\delta}$ is given by

$$C^+_{AB'CD'EF'GH'} = -2\, \psi_{ACEG}\, \varepsilon_{B'D'}\, \varepsilon_{F'G'} \tag{6.20}$$

As a consequence we obtain for the matrix C the following expression

$$C_{mn} = C^+_{0m0n} = -2\, \psi_{ACEG}\, \sigma^{AC}_{0m}\, \sigma^{EG}_{0n} \tag{6.21}$$

in terms of the Weyl spinor ψ_{ABCD}. In Eq. (6.21) use has been made of the notation according to which

$$\sigma_{0m}^{\ \ AB} = \sigma_0^{\ A}{}_{C'}\, \sigma_m^{\ BC'} . \tag{6.22}$$

A simple calculation then gives

$$\sigma^{AB}_{01} = \frac{1}{2}\begin{pmatrix} 1 & 0 \\ 0 & -1 \end{pmatrix}, \quad \sigma^{AB}_{02} = \frac{1}{2}\begin{pmatrix} -i & 0 \\ 0 & -i \end{pmatrix}, \quad \sigma^{AB}_{03} = \frac{1}{2}\begin{pmatrix} 0 & -1 \\ -1 & 0 \end{pmatrix}. \tag{6.23}$$

The matrix obtained is then given by

$$C_{mn} = \begin{pmatrix} \frac{1}{2}(-\psi_0 + 2\psi_2 - \psi_4) & \frac{i}{2}(\psi_0 - \psi_4) & (\psi_1 - \psi_3) \\ \frac{i}{2}(\psi_0 - \psi_4) & \frac{1}{2}(\psi_0 + 2\psi_2 + \psi_4) & -i(\psi_1 + \psi_3) \\ (\psi_1 - \psi_3) & -i(\psi_1 + \psi_3) & -2\psi_2 \end{pmatrix}. \tag{6.24}$$

Comparing this matrix with the matrix Ψ of Eq. (6.18) obtained from the spinor method we see that the two matrices are identical if one reverse the counting of the columns and rows of the matrices and change ψ_1 and ψ_3 into $-\psi_1$ and $-\psi_3$, respectively.

Our problem, using the spinor formalism, is then reduced to that of using the usual tensor method. The two invariants I and J of the gravitational field, defined in Section 4, can now be written in terms of the Weyl spinor by

$$I = \psi_{ABCD}\,\psi^{ABCD} = \mathrm{Tr}\ \Psi^2$$

$$= 2(\psi_0\psi_4 - 4\psi_1\psi_3 + 3\psi_2^2), \tag{6.25}$$

$$J = \psi_{AB}{}^{CD}\,\psi_{CD}{}^{EF}\,\psi_{EF}{}^{AB} = \mathrm{Tr}\ \Psi^3$$

$$= 6(\psi_0\psi_2\psi_4 - \psi_0\psi_3{}^2 - \psi_1{}^2\psi_4 + 2\psi_1\psi_2\psi_3 - \psi_2{}^3). \tag{6.26}$$

The eigenvalues and eigenspinors of Eq. (6.15) are summarized in Table 2 and in Figure 2.

Table 2

Classification of the gravitational field in terms of eigenspinors and eigenvalues.

DISTINCT EIGENSPINORS	3			2		1
DISTINCT EIGENVALUES	3	2	1	2	1	1
PETROV TYPE OF FIELD	I	D	0	II	N	III

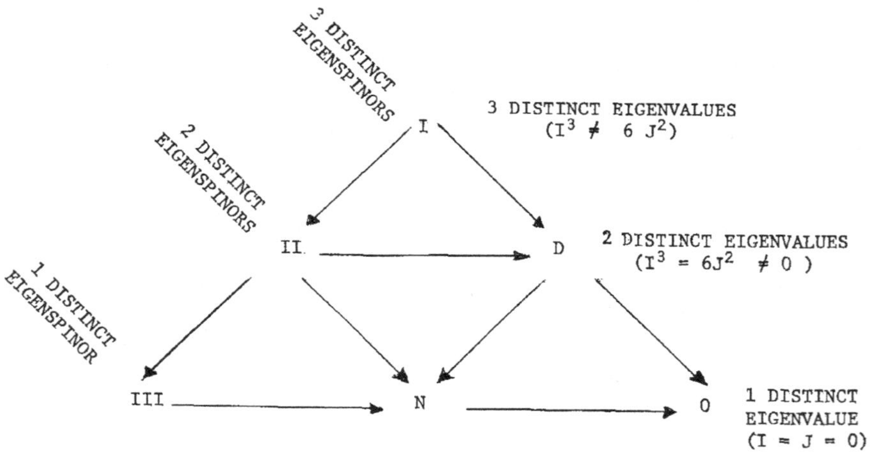

Figure 2

Classification of the gravitational field in terms of the field invariants I and J, the eigenvalues and eigenvectors. The symbols in the diagram stand for Petrov fields of types I, II, D, III, N, and 0.

Finally, here again we may decompose the Weyl spinor into products of one-index spinors. We then obtain:

$$\psi_{ABCD} = \alpha_{(A}\beta_B\gamma_C\delta_{D)} . \tag{6.27}$$

Again, according to the identification of the spinors α_A, β_A, γ_A and δ_A we obtain all the different types of fields. If the four spinors are distinct from each other, we obtain type I; if two of them coincide, then we obtain type II; if they coincide in pairs, we then obtain type D; if three of the four spinors coincide, we then obtain type III; if all of them coincide, we obtain type N; and finally, if the Weyl spinor vanishes, we have type 0.

If the gravitational field is of types III, N, or 0, for instance, we then have for the invariants I and J

$$I = J = 0. \tag{6.28}$$

For type D, on the other hand, we can present the gravitational field spinor in the form

$$\psi_{ABCD} = \alpha_{(A}\alpha_B\delta_C\delta_{D)} . \tag{6.29}$$

We then find for the two invariants of the gravitational field the following expressions:

$$I = \frac{1}{6}(\alpha_A\delta^A)^4, \qquad J = -\frac{1}{36}(\alpha_A\delta^A)^6. \tag{6.30}$$

Hence we have $I^3 = 6J^2 \neq 0$ in this case.

The latter result can also be seen in a different way. For type D we can choose all five components of the Weyl spinor ψ_n to be zero except for ψ_2. Hence we can write for the Weyl spinor in this case

$$\psi_{ABCD} = 6\psi_2\, \ell_{(A}\ell_B n_C n_{D)} , \tag{6.31}$$

with $\ell_A n^A = 1$. Equations (6.25) and (6.26) then yield

$$I = 6\psi_2^2 , \qquad J = -6\psi_2^3 . \tag{6.32}$$

Hence $I^3 = 6J^2 = 6^3\psi_2{}^6 \neq 0$, where

$$\psi_2 = \frac{1}{6}(\alpha_A\delta^A)^2 . \tag{6.33}$$

A similar calculation can be made for the type II gravitational field.

7. GAUGE FIELDS: METHOD OF INFINITESIMAL HOLONOMY GROUP

The classification of unquantized gauge fields was first discussed by Eguchi, using the method of infinitesimal holonomy group [6]. He found ten classes of fields that were classified by him into nonradiative, radiative, and mixtures of the two. The method of the infinitesimal holonomy group uses the fact that one can make a roundtrip by means of successive Lorentz transformations, starting from the laboratory, going into a canonical frame, and then coming back to the laboratory.

If $F_{\mu\nu}$ describes the electromagnetic field tensor, for instance, then the net result of the trip described above is given by the expression $G_{\mu\nu}{}^{\rho\sigma}F_{\rho\sigma}$, which is again equal to $F_{\mu\nu}$. Hence we may write

$$G_{\mu\nu}{}^{\rho\sigma} F_{\rho\sigma} = F_{\mu\nu} \tag{7.1}$$

where $G_{\mu\nu\rho\sigma}$ is a certain expression of the Lorentz transformations. Equation (7.1) has the simple meaning, namely, the electromagnetic field tensor is the eigenbivector of $G_{\mu\nu\rho\sigma}$ with eigenvalue unity. This is the situation for both nonradiative and radiative fields.

The quantity $G_{\mu\nu\rho\sigma}$ satisfies a closure property:

$$G_{\mu\nu}{}^{\rho\sigma} G_{\rho\sigma\alpha\beta} = G_{\mu\nu\alpha\beta}. \tag{7.2}$$

In addition, it satisfies the following relationships:

$$G_{\mu\nu\rho\sigma} = -G_{\nu\mu\rho\sigma} = -G_{\mu\nu\sigma\rho} , \tag{7.3}$$

$$G_{\mu\nu\rho\sigma} = G_{\rho\sigma\mu\nu}, \tag{7.4}$$

$$G_{\mu\alpha\beta\gamma} + G_{\mu\beta\gamma\alpha} + G_{\mu\gamma\alpha\beta} = 0, \tag{7.5}$$

$$G^{\alpha}{}_{\mu\alpha\nu} = \frac{1}{2} g_{\mu\nu}. \tag{7.6}$$

Equations (7.3) to (7.6) show that the symmetry properties of $G_{\mu\nu\rho\sigma}$ are the same as those of the curvature tensor in vacuum with a cosmological constant equals to $\frac{1}{2}$. Hence the classification of $G_{\mu\nu\rho\sigma}$ becomes the same as that of the Weyl tensor. Because of the additional constraint given by Eq. (7.2), however, the analogy between $G_{\mu\nu\rho\sigma}$ and the Weyl conformal tensor is not complete and some fields are excluded from the classification scheme.

Using a six-dimensional notation $G_{AB} = G_{\mu\nu\rho\sigma}$ where A, B run over 01, 02, 03, 23, 31 and 12, we then have for the nonradiative case,

$$G_A{}^B = \frac{1}{2}\begin{pmatrix} L & 0 \\ 0 & L \end{pmatrix}, \tag{7.7}$$

where 0 is the 3×3 zero matrix and L is given by

$$L = \begin{pmatrix} 0 & 0 & 0 \\ 0 & 0 & 0 \\ 0 & 0 & 1 \end{pmatrix}.$$

Likewise, we obtain

$$H_A{}^B = \frac{1}{4}\begin{pmatrix} M & N \\ P & Q \end{pmatrix}, \tag{7.8}$$

where M, N, P, Q are four 3×3 matrices given by

$$M = \begin{pmatrix} 1 & 0 & 0 \\ 0 & 0 & 0 \\ 0 & 0 & 0 \end{pmatrix}, \qquad N = \begin{pmatrix} 0 & -1 & 0 \\ 0 & 0 & 0 \\ 0 & 0 & 0 \end{pmatrix},$$

$$P = \begin{pmatrix} 0 & 0 & 0 \\ -1 & 0 & 0 \\ 0 & 0 & 0 \end{pmatrix}, \qquad Q = \begin{pmatrix} 0 & 0 & 0 \\ 0 & 1 & 0 \\ 0 & 0 & 0 \end{pmatrix},$$

for the radiative case.

Let us now define the matrices

$$K_{ij} = F_{i\mu\nu}\, F_j{}^{\mu\nu}, \quad J_{ij} = F_{i\mu\nu}\, {}^*F_j{}^{\mu\nu}. \tag{7.9}$$

Here $F_{k\mu\nu}$ is the Yang-Mills field strength and ${}^*F_{k\mu\nu}$ is its dual,

$${}^*F_k{}^{\mu\nu} = \frac{1}{2}\, \varepsilon^{\mu\nu\rho\sigma}\, F_{k\rho\sigma}. \tag{7.10}$$

The matrices K and J are not gauge invariant by themselves and the field strengths may in general change types under SU(2) rotations. If one requires the field strengths to remain of the same type under SU(2) rotations, one then obtains a strong constraint on the matrices K and J possible forms.

The classification given by Eguchi is as follows:

(1) All field strengths remain nonradiative in all isospin gauges. We then have

$$F^i_{\mu\nu} = G^i_{\mu\nu\rho\sigma}\, F^{i\rho\sigma}, \tag{7.11}$$

where i = 1, 2, 3, and there is no summation over i. Here the G's are determined by the round trip to the canonical frame of $F^i_{\mu\nu}$.

It follows that the G's are actually isoscalars, namely, independent of the isospin index i. By means of a Lorentz transformation the F's are simultaneously brought to their canonical forms. We then obtain:

$$\vec{F}_{\mu\nu} = \vec{\rho} A_{\mu\nu} + \vec{\zeta} A^*_{\mu\nu} \tag{7.12}$$

where

$$A_{\mu\nu} = \xi_{[\mu}\, \eta_{\nu]}\ , \qquad A^*_{\mu\nu} = 2p_{[\mu}\, q_{\nu]} \tag{7.13}$$

and $\vec{\rho} = \vec{E}_z$ and $\vec{\zeta} = \vec{H}_z$ in the canonical frame. In Eq. (7.13) use has been made of the notation

$$\xi_\mu = (1,0,0,-1) \qquad \eta_\mu = (1,0,0,1)\ , \tag{7.14}$$

$$p_\mu = (0,\ 1,0,0), \qquad q_\mu = (0,0,1,0). \tag{7.15}$$

In arbitrary frames $A_{\mu\nu}$ satisfies $A_{\mu\nu}A^{\mu\nu} = -2$ and $A_{\mu\nu}A^{*\mu\nu} = 0$.

(2) Only one of $F^i_{\mu\nu}$ can be made radiative by a particular choice of SU(2) gauge. Then the remaining two nonradiative fields have a common canonical frame. The form of the field strengths is now given by

$$\vec{F}_{\mu\nu} = \vec{\rho}\, A_{\mu\nu} + \vec{\zeta}\, A^*_{\mu\nu} + \vec{\sigma}\, B_{\mu\nu}\ , \tag{7.16}$$

where $\vec{\rho}\cdot\vec{\sigma} = \vec{\zeta}\cdot\vec{\sigma} = 0$, and $B_{\mu\nu}$ satisfies $B_{\mu\nu}B^{\mu\nu} = B_{\mu\nu}B^{*\mu\nu} = 0$.

(3) Two of the $F^i_{\mu\nu}$ become radiative under a particular gauge choice. In this case we obtain:

$$\vec{F}_{\mu\nu} = \vec{\rho}\, A_{\mu\nu} + \vec{\zeta}\, A^*_{\mu\nu} + \vec{\sigma}_1 B^1_{\mu\nu} + \vec{\sigma}_2\, B^2_{\mu\nu}\ , \tag{7.17}$$

where

$$\vec{\rho}\cdot\vec{\sigma}_1 = \vec{\rho}\cdot\vec{\sigma}_2 = \vec{\zeta}\cdot\vec{\sigma}_1 = \vec{\zeta}\cdot\vec{\sigma}_2 = \vec{\sigma}_1\cdot\vec{\sigma}_2 = 0. \tag{7.18}$$

The B's satisfy $B^i_{\mu\nu}\, B^{i\mu\nu} = B^i_{\mu\nu}\, B^{*i\mu\nu} = 0$, and they may or may not be equal to each other.

(4) Three of the $F^i_{\mu\nu}$ become radiative under a particular choice of gauge. We then have:

$$\vec{F}_{\mu\nu} = \vec{\sigma}_1 B^1_{\mu\nu} + \vec{\sigma}_2 B^2_{\mu\nu} + \vec{\sigma}_3 B^3_{\mu\nu}\ , \tag{7.19}$$

with

$$\vec{\sigma}_1 \cdot \vec{\sigma}_2 = \vec{\sigma}_2 \cdot \vec{\sigma}_3 = \vec{\sigma}_3 \cdot \vec{\sigma}_1 = 0 . \qquad (7.20)$$

In the extreme case when $B^1 = B^2 = B^3$, all components $F^i_{\mu\nu}$ become radiative in all SU(2) gauges. This special class is obtained directly if we assume that

$$F^i_{\mu\nu} = H^i_{\mu\nu\rho\sigma} F^{i\rho\sigma} . \qquad (7.21)$$

In this case $\vec{F}_{\mu\nu}$ has a particularly simple form,

$$\vec{F}_{\mu\nu} = \vec{\sigma} F_{\mu\nu} . \qquad (7.22)$$

8. CLASSIFICATION OF SU(2) GAUGE FIELDS

In the last section we briefly reviewed the problem of classifying the SU(2) gauge fields using the method of infinitesimal holonomy group. As has been pointed out before, the classification obtained in this method is not completely gauge invariant. The physical meaning of such a classification is therefore not clear. Other methods were therefore used in order to obtain a satisfactory classification of the SU(2) gauge fields.

This problem was investigated by Roskies, who constructed the invariants of the SU(2) gauge fields [7]. He also classified the field strengths, according to their asymptotic behavior at large spatial distances, into three classes. The eigenvector structure of matrices, constructed from the field strengths, was also studied. A total of nine real invariants were found that describe a complete set of independent invariants.

The method proved to be useless, however, for the classification problem. The eigenvalue-eigenvector calculation becomes so cumbersome that computer use was needed without achieving the desired classification. The problem of classification was thus left unsolved. It was pointed out, however, that the three types of fields obtained may be associated with different values of the invariants. These are those fields for which (1) all invariants are different from each other; (2) all invariants are zero; and (3) the invariants satisfy a certain algebraic relation between themselves.

We have seen in previous sections, however, that when the invariants of the field satisfy a certain relation between themselves, it is not necessary that one obtains only one kind of field. Both the electromagnetic field and the gravitational field are of such nature. When all invariants of the gravitational field vanish, for instance, we obtain three different types of fields rather than just one field. These are types III, N and O (see Figure 2). Another example is that when the two invariants of gravitation I and J satisfy the relationship of $I^3 = 6J^2$. We then obtain the two types of fields II and D.

In gauge fields, such as the Yang-Mills field, the situation is even more complicated because of the double group structure. Even so, the situation will be seen to be similar in SU(2) gauge fields to that of gravitation. In the following we will see that we have more independent relations between the SU(2) gauge fields than just three. We will also see that associated with these relations between the field invariants there are ten independent and physically different types of SU(2) gauge fields.

The invariants of the SU(2) gauge fields can be expressed in terms of the matrices defined in the last section by Eqs. (7.9),

$$K_{ij} = F_{i\mu\nu}F_j^{\mu\nu} , \quad J_{ij} = F_{i\mu\nu}{}^*F_j^{\mu\nu} . \tag{8.1}$$

A possible set of invariants of the field can be taken as follows: Tr K, Tr J, t, t', det K, det J, $\mathrm{Tr}K^2$, $\mathrm{Tr}J^2$, and TR(JK), where the two invariants t and t' are given by

$$t = \frac{1}{6} \varepsilon_{ijk} F_{i\mu}{}^{\nu} F_{i\nu}{}^{\rho} F_{k\rho}{}^{\mu} , \tag{8.2}$$

$$t' = -\frac{1}{6} \varepsilon_{ijk} {}^*F_{i\mu}{}^{\nu}\, {}^*F_{j\nu}{}^{\delta *}F_{k\rho}{}^{\mu} . \tag{8.3}$$

The classification given by Roskies in terms of three types of fields is as follows:

(1) Fields for which all the invariants vanish.

(2) Fields for which the invariants satisfy

$$\det J = \det K = t = t' = 0 , \tag{8.4a}$$

$$\mathrm{Tr}(JK) = \mathrm{Tr}J \ \ \mathrm{Tr}K, \tag{8.4b}$$

$$\mathrm{Tr}K^2 - (\mathrm{Tr}K)^2 = \mathrm{Tr}J^2 - (\mathrm{Tr}J)^2. \tag{8.4c}$$

(3) Anything else.

A more detailed classification of SU(2) gauge fields was subsequently given by Carmeli [8]. Ten classes of fields were isolated using an extended version of the spinor methods. In the following we review these results. A different approach to the classification of SU(2) gauge fields was also given by Wang and Yang [9]. The invariants of the SU(2) gauge field are given and discussed in detail in the next section.

The spinor equivalent of an SU(2) gauge field strength $F_{k\mu\nu}$ is a complex function

χ_{kAB} that is obtained from $F_{k\mu\nu}$ in the same way that the electromagnetic field spinor ϕ_{AB} is obtained from the electromagnetic field tensor $F_{\mu\nu}$.

Accordingly, and in analogy to Eq. (5.16) for the electromagnetic field, we can write for gauge fields the following:

$$F_{kAB'CD'} = \varepsilon_{AC} \bar{\chi}_{kB'D'} + \chi_{kAC} \varepsilon_{B'D'} \tag{8.5}$$

where the gauge field spinor $F_{kAB'CD'}$ is given by

$$F_{kAB'CD'} = \sigma^{\mu}{}_{AB'} \sigma^{\nu}{}_{CD'} F_{k\mu\nu} \tag{8.6a}$$

and

$$\chi_{kAC} = \frac{1}{2}\varepsilon^{B'D'} F_{kAB'CD'} \quad . \tag{8.6b}$$

Here the indices A and B are ordinary SL(2,C) spinor indices taking the values 0 and 1, whereas k = 1,2,3 describe the isospin vector components in the internal SU(2) space.

The spinor χ_{kAB} is symmetric in its two spinor indices A and B: $\chi_{kAB} = \chi_{kBA}$.

Hence it has 3×3 complex components, χ_{k00}, $\chi_{k01} = \chi_{k10}$, and χ_{k11}. These are equivalent to the eighteen real components of the gauge field strengths $F_{k\mu\nu}$. Using the null tetrad of Section 6, we can define the three isospin vectors (but scalars in spacetime) as follows:

$$\chi_{k0} = F_{k\mu\nu} \ell^{\mu} \ell^{\nu} \tag{8.7a}$$

$$\chi_{k1} = \frac{1}{2} F_{k\mu\nu} (\ell^{\mu} \ell^{\nu} + \bar{m}^{\mu} m^{\nu}) \tag{8.7b}$$

$$\chi_{k2} = F_{k\mu\nu} \bar{m}^{\mu} m^{\nu} \quad . \tag{8.7c}$$

The gauge field spinor χ_{kAB} will be referred to as the Yang-Mills spinor in analogy to the Maxwell spinor ϕ_{AB}.

The Yang-Mills spinor can also be described as a quantity having two SL(2,C) spinor indices and two SU(2) spinor indices, χ_{MNAB}, where M and N take the values 1 and 2. The quantity χ_{AB} may thus be regarded as a matrix whose rows and columns are fixed by the indices M and N, namely $(\chi_{AB})_{MN}$.

The matrix χ_{AB} is Hermitian and traceless. The relationship between χ_{MNAB} and χ_{kAB} is given by

$$\chi_{MNAB} = 2^{-\frac{1}{2}} \sigma^{k}_{MN} \chi_{kAB} \tag{8.8a}$$

$$\chi_{kAB} = 2^{-\frac{1}{2}} \chi_{MNAB} \sigma^{k}_{NM} \quad . \tag{8.8b}$$

Using matrix notation, the above relations can be written as

$$\chi_{AB} = 2^{-\frac{1}{2}} \sigma^k \chi_{AB}, \tag{8.9a}$$

$$\chi_{kAB} = 2^{-\frac{1}{2}} \mathrm{Tr}(\chi_{AB}\sigma^k). \tag{8.9b}$$

Here σ^k are the usual Pauli matrices. It should be noticed, however, that χ_{MNAB} is not symmetric with respect to its SU(2) indices M and N.

A general Yang-Mills spinor χ_{MNAB} may or may not be decomposed into products of irreducible components of one or both types of SL(2,C) and SU(2) spinors. Hence we may have Yang-Mills spinors with an isospin index having the form of a vector, χ_{kAB}, or having the form of products of SU(2) spinors, $\chi_{MNAB} = \alpha_{M(A}\beta_{NB)}$, for example. Here brackets indicate on symmetrization in the SL(2,C) spinor indices, namely,

$$\alpha_{M(A}\beta_{NB)} = \frac{1}{2} (\alpha_{MA}\beta_{NB} + \alpha_{MB}\beta_{NA}). \tag{8.10}$$

The decomposition of the Yang-Mills spinor into irreducible products does not seem to follow the pattern of decomposition of the Maxwell spinor or the Weyl spinor (see Section 6). In the previous case, the decomposition was possible because of the fundamental theorem of algebra. The fact that a Weyl spinor, for instance, can be written as a product of four one-index spinor (with an appropriate symmetrization) is a general property of symmetric spinors in the complex plane. Because we here deal with mixed indices spinors, however, the method cannot be used with the same generality. Hence the decomposition process itself here can be considered as part of the classification scheme itself.

We may call a Yang-Mills spinor with an isospin index χ_{kAB} a vector-type gauge field whereas a Yang-Mills spinor with SU(2) indices a spinor-type gauge field. Vector-type gauge fields may always be written as spinor-type fields. The contrary, however, is not correct.

The vector-type and the spinor-type gauge fields can be further decomposed into irreducible products. The field χ_{kAB} can be decomposed into $\alpha_{(A}\beta_{kB)}$ and $\alpha_{(A}\beta_{B)}\gamma_k$. Of course $\alpha_{(A}\beta_{kB)}$ can be decomposed into $\alpha_{(A}\beta_{B)}\gamma_k$ or $\alpha_{(A}\alpha_{kB)}$ or $\alpha_{(A}\alpha_{B)}\gamma_k$, where α_{kA} is defined by $\alpha_{kA}\alpha_{kB} = \alpha_{(A}\delta_{B)}$. All of these fields can naturally go over into the zero field. A spinor-type gauge field, likewise, can be decomposed into $\alpha_{M(A}\beta_{B)}\delta_N$, $\alpha_{(A}\beta_{B)}\gamma_M\delta_N$, $\alpha_{M(A}\alpha_{B)}\delta_N$, $\alpha_A\alpha_B\gamma_M\delta_N$ and, of course, the zero field.

9. INVARIANTS OF SU(2) GAUGE FIELDS

From the Yang-Mills mixed indices spinor χ_{kAB}, which is an isospin vector and a symmetric SL(2,C) spinor, we can construct all the invariants of the SU(2) gauge field. We have seen in previous sections how efficiently and elegantly a similar procedure can be used in order to construct the invariants of the electromagnetic and the gravitational fields. To this end a few more spinors, which are extremely interesting from the geometrical point of view, will be defined in the following. The invariants of the SU(2) gauge field will be associated with these newly defined spinors.

We first define the spinor

$$\xi_{ABCD} = \chi_{kAB}\,\dot{\chi}_{kCD} \quad . \tag{9.1}$$

From its definition the spinor ξ_{ABCD} satisfies the following symmetries:

$$\xi_{ABCD} = \xi_{BACD} = \xi_{ABDC} , \tag{9.2}$$

$$\xi_{ABCD} = \xi_{CDAB}. \tag{9.3}$$

Hence it can be decomposed into a totally symmetric spinor and a scalar as follows:

$$\xi_{ABCD} = \eta_{ABCD} + \frac{P}{6}\,(\varepsilon_{AC}\varepsilon_{BD} + \varepsilon_{AD}\varepsilon_{BC}) \tag{9.4}$$

where the scalar P is the trace of ξ_{ABCD}:

$$P = \xi_{AB}{}^{AB} = \varepsilon^{AC}\,\varepsilon^{BD}\,\xi_{ABCD} \quad . \tag{9.5}$$

A simple calculation, moreover, shows that

$$\xi_{AC}{}^{C}{}_{B} = \varepsilon^{CD}\xi_{ACDB} = \frac{P}{2}\,\varepsilon_{AB}. \tag{9.6}$$

The spinor ξ_{ABCD} resembles in its properties the gravitational field spinor χ_{ABCD} (see Section 5) that combines the Weyl spinor and the Ricci scalar curvature. The difference between the two spinors being in their trace structure: the trace of the gravitational field spinor is $\chi_{AB}{}^{AB} = \lambda = -R/4$, where R is the Ricci scalar curvature, and thus it is real. Here, however, P is a complex function. The spinor η_{ABCD} in Eq. (9.4), on the other hand, is a totally symmetric spinor in all of its four indices. It is completely analogous to the Weyl conformal spinor, and has only five independent complex components.

A third four-index spinor can subsequently be defined as follows:

$$\zeta_{ABC'D'} = \chi_{kAB}\ \bar{\chi}_{kC'D'} \ . \qquad (9.7)$$

It satisfies the symmetries that the tracefree Ricci spinor $\phi_{ABC'D'}$ satisfies (see Section 5):

$$\zeta_{ABC'D'} = \zeta_{BAC'D'} = \zeta_{ABD'C'} , \qquad (9.8)$$

$$\zeta_{ABC'D'} = \bar{\zeta}_{C'D'AB} = \overline{\zeta_{CDA'B'}} \ . \qquad (9.9)$$

It therefore has nine real independent components. The spinor $\zeta_{ABC'D'}$ is, moreover, irreducible.

A fourth spinor that can be constructed out of the Yang-Mills spinor is given by

$$\chi_{ABCDEF} = \varepsilon_{ijk}\ \chi_{iAB}\ \chi_{jCD}\ \chi_{kEF} . \qquad (9.10)$$

It satisfies the following symmetry:

$$\chi_{ABCDEF} = \chi_{BACDEF} = \chi_{ABDCEF} = \chi_{ABCDFE} \ . \qquad (9.11)$$

In addition, the spinor χ_{ABCDEF} keeps or changes its sign, depending upon whether the pairs of indices AB, CD, EF are an even or an odd permutation of the pairs of numbers 00, 01(= 10), 11, and zero otherwise. Hence it can be decomposed as follows:

$$\chi_{ABCDEF} = \frac{Q}{24} (\varepsilon_{AC}\ \varepsilon_{BE}\ \varepsilon_{DF} + \varepsilon_{AF}\ \varepsilon_{BC}\ \varepsilon_{DE}$$

$$+ \varepsilon_{AC}\ \varepsilon_{BF}\ \varepsilon_{DE} + \varepsilon_{AE}\ \varepsilon_{BC}\ \varepsilon_{DF} + \varepsilon_{AD}\ \varepsilon_{BF}\ \varepsilon_{CE}$$

$$+ \varepsilon_{AD}\ \varepsilon_{BE}\ \varepsilon_{CF} + \varepsilon_{AF}\ \varepsilon_{BD}\ \varepsilon_{CE} + \varepsilon_{AE}\ \varepsilon_{BD}\ \varepsilon_{CF}), \qquad (9.12)$$

where Q is a complex quantity, the trace of the spinor χ_{ABCDEF}:

$$Q = \chi_{A\ C\ E}^{\ \ C\ E\ A} = \varepsilon^{CB}\ \varepsilon^{ED}\ \varepsilon^{AF}\ \chi_{ABCDEF} \ . \qquad (9.13)$$

Finally, two more mixed indices spinors, with unprimed and primed indices, can be defined as follows:

$$\phi_{ABCDE'F'} = \varepsilon_{ijk}\ \chi_{iAB}\ \chi_{jCD}\ \bar{\chi}_{kE'F'} , \qquad (9.14)$$

$$\phi_{ABC'D'E'F'} = \epsilon_{ijk}\, \chi_{iAB}\, \bar{\chi}_{jC'D'}\bar{\chi}_{kE'F'} ,\tag{9.15}$$

The relationship between them can easily be shown to be given by

$$\phi_{ABCDE'F'} = \bar{\phi}_{E'F'ABCD} ,\tag{9.16}$$

$$\phi_{ABC'D'E'F'} = \bar{\phi}_{C'D'E'F'AB} .\tag{9.17}$$

The invariants of the SU(2) gauge field can now be constructed from the above five spinors. We already have two complex invariants P and Q defined above by Eqs. (9.5) and (9.13). More invariants may be defined as follows:

$$R = \zeta_{ABC'D'}\, \zeta^{ABC'D'} ,\tag{9.18}$$

$$S = \xi_{ABCD}\, \xi^{ABCD} = G + \frac{P^2}{3}\tag{9.19}$$

$$T = \phi_{ABCDE'F'}\, \phi^{ABCDE'F'},\tag{9.20}$$

where the invariant G is given by

$$G = \eta_{ABCD}\, \eta^{ABCD} .\tag{9.21}$$

Finally, we may define two more invariants F and H by means of

$$F = \xi_{AB}{}^{CD}\xi_{CD}{}^{EF}\, \xi_{EF}{}^{AB} = H + PG + \frac{P^3}{9}\tag{9.22}$$

and

$$H = \eta_{AB}{}^{CD}\, \eta_{CD}{}^{EF}\, \eta_{EF}{}^{AB} .\tag{9.23}$$

It will be noted that the seven invariants P, Q, S, T, F, G, and H are complex functions, whereas R is real. The reality of the invariant R can easily be seen if we calculate its complex conjugate:

$$\bar{R} = \overline{\zeta_{ABC'D'}}\ \overline{\zeta^{ABC'D'}} = \bar{\zeta}_{A'B'CD}\ \bar{\zeta}^{A'B'CD}$$

$$= \zeta_{CDA'B'}\, \zeta^{CDA'B'} = \zeta_{ABC'D'}\, \zeta^{ABC'D'} = R,\tag{9.24}$$

We recall that the number of invariants of the SU(2) gauge field in terms of real functions is nine. Hence we obviously have interdependence relations between the above-defined seven complex and one real invariants. A selection should be made here that is based on physical grounds just as in the gravitational and the electro-magnetic cases. It will also be noted that the two invariants G and H are constructed from the totally symmetric spinor η_{ABCD} in precisely the same way as the gravitational field invariants I and J are constructed from the totally symmetric Weyl conformal spinor ψ_{ABCD} [see Section 6, Eqs. (6.25) and (6.26)]. The following three sets of invariants can be taken, for instance, as the invariants of SU(2) gauge field:

$$P,\ Q,\ R,\ S,\ T\ ; \tag{9.25a}$$

$$P,\ T,\ R,\ G,\ H\ ; \tag{9.25b}$$

$$P,\ F,\ R,\ S,\ T\ . \tag{9.25c}$$

We may calculate the invariants for a simple exact solution to the classical Yang-Mills field equations. For a monopole, for instance, that has both electric and magnetic charges, we obtain [10]:

$$F_{k0j} = -\frac{e}{g}\frac{x^k x^j}{r^4}, \qquad F_{kij} = \frac{1}{g}\,\varepsilon_{ijm}\frac{x^k x^m}{r^4}. \tag{9.26}$$

The Yang-Mills spinor χ_{kAB} is then given by

$$\chi_{kAB} = \gamma_k\ \ell_{(A}n_{B)}, \qquad \gamma_k = -\frac{e+i}{g}\ \frac{x^k}{r^3}, \tag{9.27}$$

where ℓ_A and n_A are two arbitrary SL(2,C) spinors satisfying $\ell_A n^A = 1$. We obtain:

$$\left.\begin{aligned} &P = -\frac{1}{2}\left(\frac{e+i}{gr^2}\right)^2, \quad Q = 0, \quad R = P\bar{P}, \quad S = P^2, \\ &T = 0, \quad F = P^3, \quad G = \frac{2}{3}P^2, \quad H = \frac{2}{9}P^3. \end{aligned}\right\} \tag{9.28}$$

It is probably remarkable that the two invariants G and H, which are analogous to the two gravitational invariants I and J, satisfy the relation

$$G^3 = 6H^2 \neq 0, \tag{9.29}$$

for the monopole solution. We recall that the relation $I^3 = 6J^2$ is satisfied by all type II and D gravitational fields. In this sense the monopole solution of the Yang-Mills field equations belongs to type Dv gauge fields, just as the Schwarzschild metric belongs to type D gravitational fields in the Petrov classification. Hence we get Figure 3 for the vector-type SU(2) gauge fields.

Finally, it is worthwhile giving the expressions for the above invariants in terms of the matrices J and K defined by Eq. (8.1). We obtain, for instance,

$$P = \frac{1}{4}(\mathrm{Tr}K + i\mathrm{Tr}J) = \frac{1}{4} F_{k\mu\nu}(F_k^{\mu\nu} + i\, {}^*F_k^{\ \mu\nu}),$$

$$Q = \frac{3}{2}(t + it'), \qquad R = \frac{1}{16}(\mathrm{Tr}K^2 + \mathrm{Tr}J^2), \tag{9.30}$$

$$S = \frac{1}{16}(\mathrm{Tr}K^2 - \mathrm{Tr}J^2) + \frac{i}{8}\mathrm{Tr}(JK), \text{ etc.}$$

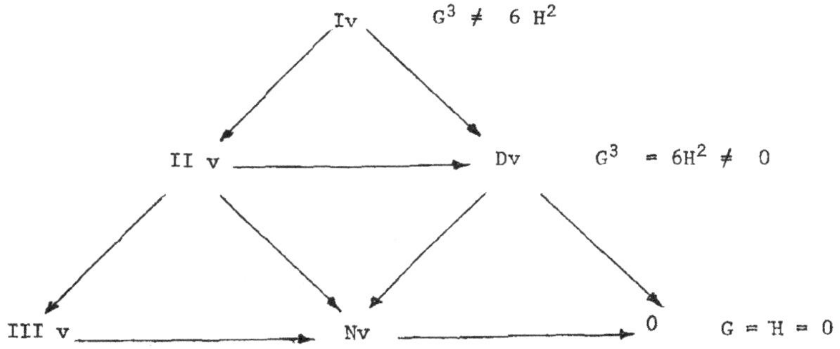

Figure 3.

Isovector diagram of classification of SU(2) gauge fields. The symbols in the diagram are as follows: Iv = χ_{kAB}, IIv = $\alpha_{(A}\beta_{kB)}$, Dv = $\alpha_{(A}\beta_{B)}\gamma_k$, IIIv = $\alpha_{(A}\alpha_{kB)}$, where α_{kB} is defined by $\alpha_{kB}\alpha_{kC} = \alpha_{(B}\delta_{C)}$, Nv = $\alpha_A\alpha_B\gamma_k$ and 0 is the zero field. The independent set of invariants chosen for this diagram is given by Eq. (9.25b), namely P, Q, R, G and H. The completely symmetrical spinor η_{ABCD} has identical decomposition to the Weyl conformal spinor ϕ_{ABCD} for each one of the six classes of fields in the diagram. For type Iv we have $\eta_{ABCD} = \alpha_{(A}\beta_B\gamma_C\delta_{D)}$; for type IIv: $\eta_{ABCD} = \alpha_{(A}\alpha_B\gamma_C\delta_{D)}$; for type Dv: $\eta_{ABCD} = \alpha_{(A}\alpha_B\delta_C\delta_{D)}$; for type IIIv: $\eta_{ABCD} = \alpha_{(A}\alpha_B\alpha_C\delta_{D)}$; for type Nv: $\eta_{ABCD} = \alpha_A\alpha_B\alpha_C\alpha_D$; and for type 0: $\eta_{ABCD} = 0$.

ACKNOWLEDGMENT

I have benefited from discussions with Y. Aharonov, Y. Ne'eman, A. Salam, P. van Nieuwenhuizen and especially C. N. Yang.

APPENDIX A: TRANSFORMATION RULES FOR YANG-MILLS SPINORS

In Section 8 the Yang-Mills spinor χ_{kAB} was defined and from it we defined the quantities χ_{km}, where m = 0,1,2, with $\chi_{k0} = \chi_{k00}$, $\chi_{k1} = \chi_{k01} = \chi_{k10}$ and $\chi_{k2} = \chi_{k11}$. Here k = 1,2,3 denotes an SU(2) isospin vector index, whereas A,B, = 0,1 denote SL(2,C) spinor indices. In this appendix the transformation rules for these quantities under the proper, orthochronous, homogeneous, Lorentz transformations are given. We will not be concerned, however, with the transformations of the SU(2) isospin index k, since it transforms under 3×3 real orthogonal rotations.

A-1 GENERAL TRANSFORMATION PROPERTIES

The transformation law for the spinor χ_{kAB} is given by the usual law of transformation for spinors. Accordingly under a proper, orthochronous, homogeneous Lorentz transformation we obtain for the transformed components the following:

$$\chi'_{kAB} = g_A{}^C g_B{}^D \chi_{kCD} = g_A{}^C \chi_{kCD} (g^t)^D{}_B \tag{A.1}$$

where $g_A{}^B$ are the matrix elements of g, and g is an element of the group SL(2,C). Using matrix notation the above formula can then be written in the form

$$\chi'_k = g \chi_k g^t \tag{A.2}$$

where χ_k denotes the 2×2 matrix

$$\chi_k = \begin{pmatrix} \chi_{k00} & \chi_{k01} \\ \chi_{k10} & \chi_{k11} \end{pmatrix}. \tag{A.3}$$

Let us denote a matrix g of the group SL(2,C) by

$$g = \begin{pmatrix} a & b \\ c & d \end{pmatrix}; \qquad ad - bc - 1. \tag{A.4}$$

Using Eq. (A.2), we then obtain for the transformed components of the spinor χ_k the following:

$$\begin{pmatrix} \chi'_{k0} & \chi'_{k1} \\ \chi'_{k1} & \chi'_{k2} \end{pmatrix} = \begin{pmatrix} a^2\chi_{k0} + 2ab\chi_{k1} + b^2\chi_{k2} & ac\chi_{k0} + (ad+bc)\chi_{k1} + bd\chi_{k2} \\ ac\chi_{k0} + (ad+bd)\chi_{k1} + bd\chi_{k2} & c^2\chi_{k0} + 2cd\chi_{k1} + d^2\chi_{k2} \end{pmatrix}. \qquad (A.5)$$

It is sometimes more convenient to work with three-dimensional transformations and Eq. (A.5) can then be written in the form

$$\chi'_k = Q\,\chi_k \qquad (A.6)$$

where χ_k now denotes the three-by-one column matrix

$$\chi_k = \begin{pmatrix} \chi_{k0} \\ \chi_{k1} \\ \chi_{k2} \end{pmatrix} \qquad (A.7)$$

and Q is given by the 3×3 complex matrix

$$Q = \begin{pmatrix} a^2 & 2ab & b^2 \\ ac & ad + bd & bd \\ c^2 & 2cd & d^2 \end{pmatrix}. \qquad (A.8)$$

The matrix Q gives a complex 3×3 representation for the proper, orthochronous, homogeneous Lorentz group [5]. Matrices of higher orders can be obtained by use of spinors having more indices.

A-2 TRANSFORMATIONS UNDER ROTATIONS AND BOOSTS

According to the choice of the parameters a,b,c, and d of the matrix g of the group SL(2,C), we obtain the desired one-parameter subgroups of transformations for both rotations and boosts. For a rotation around the x-axis, $a = d = \cos\frac{\psi}{2}$ and $b = c = i\sin\frac{\psi}{2}$. The corresponding matrix is then given by

$$Q_{1R}(\psi) = \begin{pmatrix} \cos^2\frac{\psi}{2} & 2i\sin\frac{\psi}{2}\cos\frac{\psi}{2} & -\sin^2\frac{\psi}{2} \\ i\sin\frac{\psi}{2}\cos\frac{\psi}{2} & 1 - 2\sin^2\frac{\psi}{2} & i\sin\frac{\psi}{2}\cos\frac{\psi}{2} \\ -\sin^2\frac{\psi}{2} & 2i\sin\frac{\psi}{2}\cos\frac{\psi}{2} & \cos^2\frac{\psi}{2} \end{pmatrix}. \qquad (A.9a)$$

For a rotation around the y-axis, a=d=cos $\frac{\psi}{2}$, and b = -c = -sin $\frac{\psi}{2}$:

$$Q_{2R}(\psi) = \begin{pmatrix} \cos^2\frac{\psi}{2} & -2\sin\frac{\psi}{2}\cos\frac{\psi}{2} & \sin^2\frac{\psi}{2} \\ \sin\frac{\psi}{2}\cos\frac{\psi}{2} & 1-\sin^2\frac{\psi}{2} & -\sin\frac{\psi}{2}\cos\frac{\psi}{2} \\ \sin^2\frac{\psi}{2} & 2\sin\frac{\psi}{2}\cos\frac{\psi}{2} & \cos^2\frac{\psi}{2} \end{pmatrix}, \quad \text{(A.9b)}$$

and for a rotation around the z-axis, a = exp(iψ/2), b = c = 0, and d = exp(-iψ/2):

$$Q_{3R}(\psi) = \begin{pmatrix} e^{i\psi} & 0 & 0 \\ 0 & 1 & 0 \\ 0 & 0 & e^{-i\psi} \end{pmatrix}. \quad \text{(A.9c)}$$

For a boost along the x-axis, a = d = cosh $\frac{\psi}{2}$ and b = c = sinh $\frac{\psi}{2}$:

$$Q_{1L}(\psi) = \begin{pmatrix} \cosh^2\frac{\psi}{2} & 2\sinh\frac{\psi}{2}\cosh\frac{\psi}{2} & \sinh^2\frac{\psi}{2} \\ \sinh\frac{\psi}{2}\cosh\frac{\psi}{2} & 1+2\sinh^2\frac{\psi}{2} & \sinh\frac{\psi}{2}\cosh\frac{\psi}{2} \\ \sinh^2\frac{\psi}{2} & 2\sinh\frac{\psi}{2}\cosh\frac{\psi}{2} & \cosh^2\frac{\psi}{2} \end{pmatrix}. \quad \text{(A.10a)}$$

For a boost along the y-axis, a = d = cosh $\frac{\psi}{2}$ and b = -c = i sinh $\frac{\psi}{2}$:

$$Q_{2L}(\psi) = \begin{pmatrix} \cosh^2\frac{\psi}{2} & 2i\sinh\frac{\psi}{2}\cosh\frac{\psi}{2} & -\sinh^2\frac{\psi}{2} \\ -i\sinh\frac{\psi}{2}\cosh\frac{\psi}{2} & 1+2\sinh^2\frac{\psi}{2} & i\sinh\frac{\psi}{2}\cosh\frac{\psi}{2} \\ -\sinh^2\frac{\psi}{2} & -2i\sinh\frac{\psi}{2}\cosh\frac{\psi}{2} & \cosh^2\frac{\psi}{2} \end{pmatrix}, \quad \text{(A.10b)}$$

and for a boost along the z-axis, a = exp(ψ/2), b = c = 0, and d = exp (-ψ/2):

$$Q_{3L}(\psi) = \begin{pmatrix} e^{\psi} & 0 & 0 \\ 0 & 1 & 0 \\ 0 & 0 & e^{-\psi} \end{pmatrix} , \qquad \text{(A.10c)}$$

A-3 ROTATIONS AROUND NULL VECTORS

The above parametrization of the matrix Q can be extended to other parameterizations known in the theory of general relativity. One such parameterization is done by noticing that the matrix g of the group SL(2,C) can be factorized as a product of three matrices of the form

$$g_1(z) = \begin{pmatrix} 1 & 0 \\ z & 1 \end{pmatrix}, \quad g_2(z) = \begin{pmatrix} z & 0 \\ 0 & z^{-1} \end{pmatrix}, \quad g_3(z) = \begin{pmatrix} 1 & z \\ 0 & 1 \end{pmatrix} , \qquad \text{(A.11)}$$

where z is a complex variable.

The transformations $g_1(z)$ and $g_3(z)$ describe complex one-parameter null rotations about the null vectors ℓ_μ and n_μ , respectively. The transformation $g_2(z)$, on the other hand, corresponds to an ordinary Lorentz transformation (boost) in the $\ell_\mu - n_\mu$ plane, along with a spatial rotation in the $m_\mu - \bar{m}_\mu$ plane. As in the text, use is being made of the notation according to which $\zeta_0^{\ A} = \ell^A$ and $\zeta_1^{\ A} = n^A$.

We may also factorize the group SL(2,C) in a different fashion. The three alternative basis matrices for the group SL(2,C) can then be given by

$$g_0 = \begin{pmatrix} 0 & 1 \\ -1 & 0 \end{pmatrix}, \quad g_1(z) = \begin{pmatrix} 1 & 0 \\ z & 1 \end{pmatrix}, \quad g_2(z) = \begin{pmatrix} z & 0 \\ 0 & z^{-1} \end{pmatrix} . \qquad \text{(A.12)}$$

The two sets of matrices given by Eqs. (A.11) and (A.12) differ in the last and first matrices, respectively. The one-parameter subgroup $g_3(z)$ can then be obtained from the set of the three matrices g_0, $g_1(z)$ and $g_2(z)$ of Eq. (A.12). We find that

$$g_3(z) = -g_0 g_1(-z) g_0 . \qquad \text{(A.13)}$$

We may now use the matrix Q of Eq. (A.8) in order to find the one-parameter matrices corresponding to the matrices g_0, $g_1(z)$, $g_2(z)$ and $g_3(z)$. A straightforward calculation then gives:

$$Q_0(z) = \begin{pmatrix} 0 & 0 & 1 \\ 0 & -1 & 0 \\ 1 & 0 & 0 \end{pmatrix} , \qquad \text{(A.14a)}$$

$$Q_1(z) = \begin{pmatrix} 1 & 0 & 0 \\ z & 1 & 0 \\ z^2 & 2z & 1 \end{pmatrix}, \qquad (A.14b)$$

$$Q_2(z) = \begin{pmatrix} z^2 & 0 & 0 \\ 0 & 1 & 0 \\ 0 & 0 & z^{-2} \end{pmatrix}, \qquad (A.14c)$$

$$Q_3(z) = \begin{pmatrix} 1 & 2z & z^2 \\ 0 & 1 & z \\ 0 & 0 & 1 \end{pmatrix}. \qquad (A.14d)$$

A-4 CHANGE OF BASIS FOR SPINORS

The spinor χ_{km} can be presented differently so that it transforms under 3×3 complex, orthogonal matrices, with determinant unity. This is done by defining the new components

$$\begin{aligned} \tilde{\chi}_{k0} &= \sqrt{2}\; i \; \chi_{k1} \\ \tilde{\chi}_{k1} &= \frac{1}{\sqrt{2}} (\chi_{k0} + \chi_{k2}) \\ \tilde{\chi}_{k2} &= \frac{i}{\sqrt{2}} (\chi_{k0} - \chi_{k2}) . \end{aligned} \qquad (A.15)$$

Let us now denote the transformation law of $\tilde{\chi}_{km}$ by

$$\tilde{\chi}'_k = P\, \tilde{\chi}_k \qquad (A.16)$$

where $\tilde{\chi}_k$ denotes the 3×1 column matrix

$$\tilde{\chi}_k = \begin{pmatrix} \tilde{\chi}_{k0} \\ \tilde{\chi}_{k1} \\ \tilde{\chi}_{k2} \end{pmatrix}, \qquad (A.17)$$

and P is given by the 3×3 complex orthogonal matrix with determinant unity given by Eq. (4.26) of the text.

The one-parameter subgroups corresponding to Lorentz transformations and rotations along and around the x-, y-, and z-axis are given in the text by Eqs.(4.28) and (4.29), respectively. Here we give the four one-parameter subgroups corresponding to g_0, $g_1(z)$, $g_2(z)$ and $g_3(z)$. We obtain

$$P_0(z) = \begin{pmatrix} -1 & 0 & 0 \\ 0 & 1 & 0 \\ 0 & 0 & -1 \end{pmatrix}, \qquad (A.18a)$$

$$P_1(z) = \begin{pmatrix} 1 & iz & z \\ -iz & 1+z^2/2 & -iz^2/2 \\ -z & -iz^2/2 & 1-z^2/2 \end{pmatrix}, \qquad (A.18b)$$

$$P_2(z) = \begin{pmatrix} 1 & 0 & 0 \\ 0 & \frac{1}{2}(z^2+z^{-2}) & -\frac{i}{2}(z^2-z^{-2}) \\ 0 & \frac{i}{2}(z^2-z^{-2}) & \frac{1}{2}(z^2+z^{-2}) \end{pmatrix}, \qquad (A.18c)$$

$$P_3(z) = \begin{pmatrix} 1 & iz & -z \\ -iz & 1+z^2/2 & iz^2/2 \\ z & iz^2/2 & 1-z^2/2 \end{pmatrix}. \qquad (A.18d)$$

APPENDIX B: VECTOR AND SPINOR PRESENTATION OF THE ELECTROMAGNETIC FIELD

Let $F_k = \overset{+}{F}_{k0} = E_k + iH_k$, where $\overset{+}{F}_{\mu\nu}$ is equal to $F_{\mu\nu} + i\,{}^*F_{\mu\nu}$ (see Section 2). A simple calculation then shows that the spinor equivalent $\overset{+}{F}_{AB'CD'}$ to the tensor $\overset{+}{F}_{\mu\nu}$ is given by

$$\overset{+}{F}_{AB'CD'} = 2\,\phi_{AC}\,\varepsilon_{B'D'} \quad . \tag{B.1}$$

Hence we obtain:

$$F_k = \overset{+}{F}_{k0} = -\overset{+}{F}_{0k} = -\overset{+}{F}_{AB'CD'}\,\sigma_0^{AB'}\,\sigma_k^{\ CD'} \quad , \tag{B.2}$$

or, using Eq. (B.1),

$$F_k = -2\phi_{AC}\,\sigma_0{}^{A}{}_{D'}\,\sigma_k^{\ CD'} = -2\,\phi_{AC}\sigma_{0k}^{AC} \quad . \tag{B.3}$$

Here σ_{0k}^{AC}, with k = 1,2,3, are three matrices given by Eqs. (6.22) and (6.23) in the text.

A straightforward calculation then gives the following results:

$$F_1 = \phi_2 - \phi_0, \quad F_2 = i(\phi_2 + \phi_0), \quad F_3 = 2\,\phi_1, \tag{B.4}$$

where use has been made of the notation $\phi_0 = \phi_{00}$, $\phi_1 = \phi_{01} = \phi_{10}$, and $\phi_2 = \phi_{11}$. The inverse transformation is given by

$$\phi_0 = -\frac{1}{2}(F_1 + iF_2)$$

$$\phi_1 = \frac{1}{2}F_3 \tag{B.5}$$

$$\phi_2 = +\frac{1}{2}(F_1 - iF_2)$$

It will be noted that the components ϕ_0, ϕ_1, ϕ_2 transform under the proper orthochronous, homogeneous Lorent transformations by means of the matrices Q of Appendix A.

Just as for the gravitational field, we may define a new set of spinor components $\tilde{\phi}_k$, with k = 0,1,2, that transform under the 3×3 complex orthogonal matrices with determinant unity P of Appendix A. These components are defined by

$$\tilde{\phi}_0 = \sqrt{2}\, i\phi_1$$

$$\tilde{\phi}_1 = \frac{1}{\sqrt{2}}(\phi_0 + \phi_2) \qquad (B.6)$$

$$\tilde{\phi}_2 = \frac{i}{\sqrt{2}}(\phi_0 - \phi_2)$$

In terms of the Cartesian components F_k we obtain:

$$\tilde{\phi}_0 = \frac{i}{\sqrt{2}} F_3 , \quad \tilde{\phi}_1 = -\frac{i}{\sqrt{2}} F_2, \quad \tilde{\phi}_2 = -\frac{i}{\sqrt{2}} F_1 . \qquad (B.7)$$

REFERENCES

1. A. Z. Petrov, Einstein Spaces, Pergamon Press, New York, 1969.
2. J. L. Anderson, Principles of Relativity Physics, Academic Press, New York, 1967.
3. M. A. Naimark, Linear Representations of the Lorentz Group, Pergamon Press, New York, 1964.
4. R. Penrose, A spinor approach to general relativity, Ann. Phys. (N.Y.) 10, 171 (1970).
5. M. Carmeli, Group Theory and General Relativity, McGraw-Hill, New York, 1977.
6. T. Eguchi, Classification of unquantized Yang-Mills fields, Phys.Rev. D13, 1561 (1976).
7. R. Roskies, Invariants and classification of Yang-Mills fields, Phys. Rev. D15, 1722 (1977).
8. M. Carmeli, Classification of Yang-Mills fields, Phys. Rev. Letters 39, 523 (1977). The number of classes of fields mentioned in this reference was twelve, rather than ten; the zero field was counted twice and it follows that the two classes of fields Ds and IIs can actually be identified. I am grateful to J. Anandan and M. Kugler for pointing out this identification to me.
9. L.-L. Wang and C. N. Yang, Classification of SU(2) gauge fields, ITP-SB-78-1, to be published.
10. M. Carmeli, Monopole solution of Yang-Mills equations, Phys. Letters 68B, 463 (1977).

BUNDLE REPRESENTATIONS AND THEIR APPLICATIONS

R.N. Sen
Department of Mathematics
Ben Gurion University of the Negev
Beersheba 84120, Israel

INTRODUCTION

In this talk I shall define *bundle representations* and try to show that they are of interest both mathematically and physically. I shall devote much of my time to mathematical questions. I had set out to do exactly the opposite, but what happened was the following. Bundle representations were first defined in ref. [1], and the original physical motivations were adequately discussed there. One might imagine that some knowledge of bundles is required to define bundle representations, but this turns out to be untrue. In fact, we had noticed at the very beginning that the actual construction of *canonical* bundle representations was just the first part of the Mackey inducing construction: we hold ourselves back from taking the last step. Now most physicists know this construction but might be less familiar with fibre bundles, and therefore I thought to economise on the required mathematical background by taking this approach. If one does not carry out the inducing construction in full, one might not be using all the hypotheses which the latter demands, a possibility which raises some very interesting questions which are impossible to ignore.

In order to be able to induce sufficiently many representations by the Mackey construction, we need the topological group G to be *locally compact*, because Haar measure is an indispensable tool. The point at which we stop, so as to be left with bundle representations, is just *before* we make full explicit use of the local compactness of G. This does not yet mean that bundle representations can be defined when G is not locally compact, because one crucial sub-construction might fail. The interesting unsolved mathematical problems lie here, and I shall make an effort to give a clear picture of the situation.

Turning to physics, the reason why bundle representations are significant is that they provide the natural framework for studying the quantum-mechanical behaviour of large systems. One expects, from the Bohr correspondence principle, that the larger a system is the less quantum-mechanical it becomes, and it is perfectly legitimate to ask how much quantum-mechanical behaviour, if any, is left to a system which is composed of infinitely many massive particles. The answer appears to be: just as much as is allowed by bundle representations!

In other words, what we are suggesting is the following. If we have a physical system composed of a large number of particles, each of which is small enough to be effectively subject to the laws of quantum mechanics, then this system *as a whole* has a residuum of quantum-mechanical behaviour which is determined *exactly* by the fundamental relativity group (Galilei or Poincaré) and its bundle representations.

BUNDLE REPRESENTATIONS

We shall deal with a topological group G, a closed subgroup H of G, a continuous unitary representation U of H, and shall trace the steps involved in constructing the unitary representation of G which is *induced* by the representation U of H. Now the category of topological groups, without further restrictions, is too large; essential tools such as Haar measures do not always exist. We could restrict ourselves to Lie groups, which might suffice for physics. There everything works smoothly - too smoothly, in fact, to leave scope for interesting unanswered mathematical questions. Therefore, we shall try to explore the land between Lie and topological groups. The basic mathematical problem, so far unsolved, is to determine just what this *Lebensraum* is, that is, to characterize the subclass of topological groups which admit enough bundle representations.

We have, first of all, to ensure that H has a continuous unitary representation U. Actually we ought to ask for a little more, namely that the family of continuous unitary representations be separating for H. This is assured if H is locally compact [2], and we therefore assume that the closed subgroup H of G is locally compact (recall, however, that if G is T_0 and H is locally compact in the relative topology, then H is necessarily closed [3]). In that case G/H is (is not) locally compact depending on whether G is (is not) locally compact. For definiteness, we take G/H to be the space of *left*-cosets, with the quotient topology, and write

$$G/H = M .$$

We shall denote by $\mathcal{H}_2$ the carrier space of the representation U of H. To avoid getting bogged down in too many details, we shall assume that $\mathcal{H}_2$ is actually a subspace of $L_2(H,d\mu)$, where $d\mu$ is a Haar measure on H. In physics one often works with $L_2(H/K,d\mu)$ where K is a closed subgroup of H and $d\mu$ an invariant or quasi-invariant measure on H/K; if H/K is an abelian group, it might be preferable to work with its Pontrjagin dual. As these choices appear to simplify rather than complicate matters, we shall not dwell upon them.

The first step of the inducing construction is to form the topological product $M \times \mathcal{H}_2$: all our Hilbert spaces will be furnished with the strong topology. The second step is to construct an action of G on $M \times \mathcal{H}_2$. The third step is to replace $M \times \mathcal{H}_2$ by $\mathcal{H}_1 \times \mathcal{H}_2$, where $\mathcal{H}_1 = L_2(M,d\mu)$, and $d\mu$ is an invariant or quasi-invariant measure on M. At this stage one needs the hypothesis of local compactness of M, and therefore of G. The fourth step is to transfer the action of G on $M \times \mathcal{H}_2$ to $\mathcal{H}_1 \times \mathcal{H}_2$. If the former action had been satisfactory, the latter action would be continuous and unitary, and would solve the problem.

That is the gist of the inducing construction. The two crucial steps are (i) defining an appropriate action of G on $M \times H_2$, and (ii) lifting this action to $H = H_1 \times H_2$. As already explained, (ii) cannot be carried out if M is not locally compact.

A *bundle representation* is obtained if one stops after (i). Thus the carrier space of this "representation" is $M \times H_2$, of which only the second factor is a linear space. Such an object is a special kind of a vector bundle, and thus the name *bundle representations*. In ref. [1] we called these *canonical* bundle representations, to distinguish them from somewhat more general ones.

At first sight it seems that if step (ii) is not carried out, then M no longer has to be locally compact. However, it could well happen that if M were *not* locally compact, then one would be unable to define a continuous action of G on $M \times H_2$. We now turn to the question of this action.

Observe that we already have a natural action of G on $M = G/H$, as well as an action U of H on H_2. The remainder is synthesized as follows. First let us fix some notation.

Let $x \in M$, $g \in G$ and $\pi: G \to M$ be the natural projection of G onto M; π sends an element $g \in G$ to the left-coset $gH = x \in M$. We shall denote the point H itself by x_0. The natural action

$$\bar{h}: G \times M \to M$$

of G on M is given by

$$\bar{h}(g,x) = \pi\{g\pi^{-1}(x)\}$$

where $\pi^{-1}(x) = \{a;\ a \in G,\ \pi(a) = x\}$. Since only one $\bar{h}$ is under consideration, we shall abbreviate $\bar{h}(g,x)$ to gx, and release the letter h for other use. It is a standard result that when G/H has the quotient topology, the natural action of G upon it is continuous and open.

Defining an action of G on $M \times H_2$ means the following. Let $\phi \in H_2$, $(x,\phi) \in M \times H_2$. Denote by $g(x,\phi)$ the *image*, in $M \times H_2$, of the point (x,ϕ). Then we should demand that

$$g(x,\phi) = (gx,\ u(g,x)\phi) \tag{1}$$

where $u(g,x)$ is an unitary operator, depending on g and x, which acts upon H_2. We want the right-hand side of (1) to be of this form for the following reason. Assume that U is the identity representation of H, so that H_2 is one-dimensional and $u(g,x) = 1$ $\forall\ g \in G,\ x \in M$. Then the action of G on $M \times H_2$ becomes essentially the action of G on M itself, which we would like to be the natural action. This explains the term gx on the right of (1). Also we would like to demand, for obvious reasons, that

$$u(e,x)\phi = \phi \qquad \forall\, x \in M, \quad \phi \in H_2 \ ,$$

i.e.

$$u(e,x) = 1 \ . \tag{2a}$$

The only real information comes from the associative law

$$g_1[g_2(x,\phi)] = [g_1g_2](x,\phi)$$

which gives

$$u(g_1,g_2x)\ u(g_2,x) = u(g_1g_2,x) \ , \tag{2b}$$

and finally we must require u to be continuous; $\bar{h}$, as remarked earlier, is already continuous.

The process of finding a continuous $u(g,x)$ satisfying (2a,b) is again broken down into two steps. In the first step we define a *Wigner rotation* of G into H, i.e. a *surjective* map

$$k\colon G \times M \to H \tag{3}$$

satisfying

$$k(g_1,g_2x)\ k(g_2,x) = k(g_1g_2,x) \qquad \forall\, g_1,\ g_2 \in G,\ x \in M \tag{4a}$$

and, in particular,

$$k(e,x) = e \qquad \forall\, x \in M \ . \tag{4b}$$

Surjectivity is assured if we impose on k the very natural condition

$$k(h,x_0) = h \ \forall\, h \in H \ . \tag{4c}$$

Such a Wigner rotation will be called *standard*.

The existence of a *continuous* Wigner rotation is an exceedingly restrictive condition on G and H. For our purpose, a weaker condition, which will be called *sufficient continuity*, is enough. This is defined as follows. We can ask whether or not the map (3) is continuous at any given point. Let the set of discontinuities of k be denoted by Δ. We say that k is *sufficiently continuous* if $k(\Delta)$ is of *zero Haar measure* on H.

If a sufficiently continuous Wigner rotation k exists, then, transferring k to H_2 by the substitution

$$u(g,x) = U(k(g,x)) \tag{5}$$

we obtain a $u(g,x)$ which is continuous, since $\mathcal{H}_2 \subset L_2(H,d\mu)$. Thus the only remaining problem is the *construction* of a sufficiently continuous Wigner rotation. For this one generally introduces a new notion.

We shall define an *algebraic cross-section* of H in G to be a map

$$\eta: M \to G \tag{6}$$

satisfying the condition

$$\pi\eta(x) = x \qquad \forall\, x \in M . \tag{7}$$

If, additionally

$$\eta(x_0) = e \tag{8}$$

we shall call η *standard* (recall that $x_0 = H$). If an algebraic cross-section is *globally continuous*, we simply drop the adjective "algebraic". If M can be covered by a family of open sets such that the restriction of η to each member of the family is piecewise-continuous, we call η *locally continuous*. If η is defined not over all M but only over a proper open subset *and is continuous thereon*, we call it a *local cross-section*. (The interested reader is referred to Varadarajan's book [4] for Chevalley's theorem on the existence of local cross-sections in Lie groups, and to Steenrod's book [5] which is still a basic reference on the whole subject. Our terminology is identical with Steenrod's on the intersection, which differs somewhat from Varadarajan's.)

Given an algebraic cross-section, we can always define a Wigner rotation: for if η is an algebraic cross-section then

$$k(g,x) = \eta(gx)^{-1}\, g\eta(x) \tag{9}$$

is a Wigner rotation; the verification is routine. Moreover, if η is standard then so is k. These facts are very well known.

What is less well known is that the converse is also true; first observe that equation (9) may be solved for η in terms of k as follows:

$$\eta(x) = ak(a,x_0)^{-1} \tag{10}$$

where, if $x \neq x_0$, then a is any element of G such that

$$ax_0 = x . \tag{11}$$

One simply substitutes (10) into the right of (9), uses (4a) and obtains an identity.

If $\eta(x_0) = e$ and $g = h \in H$, (9) immediately yields (4c). This may be interpreted as follows.

We assume that we are given a standard Wigner rotation $k(g,x)$, independently of any algebraic cross-section. We may then use it to *define* a standard cross-section by formulae (10) and (11). That the function η so defined is an algebraic cross-section, i.e. satisfies (7), is self-evident; that it is standard follows from (4c).

The fact that a standard Wigner rotation defines standard algebraic cross-sections was explained to me by M. Magidor.

We have, essentially, solved all algebraic problems, and only the questions of continuity remain. If G is Lie then there are no problems. If G is not Lie but nevertheless locally-compact, there should not be too many problems, in view of Mostert's result: if G is a locally-compact and finite-dimensional group, and H is a closed subgroup of G, then H has a local cross-section in G ([6], as quoted in [5], p. 218). When G is *not* locally compact, I do not know of any results.

I would like to close this section with the following remark. Our definition of bundle representations is not coordinate-free, and therefore has to be supplemented by a discussion of equivalence classes. A complete discussion of the latter was not given in [1] and will be provided elsewhere. Some new features appear to emerge from this discussion.

MATHEMATICAL PROBLEMS

In the following, all algebraic cross-sections and Wigner rotations are assumed to be standard, which ensures nontriviality and does not imply any loss of generality. The first two problems are purely technical; the third is not.

1. Let G be a topological group and H a locally compact subgroup of G such that H has a local cross-section in G. Is the resulting Wigner rotation sufficiently continuous?

2. Let G and H be as above, and such that G admits a sufficiently continuous Wigner rotation into H. Is there an associated algebraic cross-section which is locally continuous, or does H have a local cross-section in G?

3. Let G and H be as above. What are the necessary and sufficient conditions on G and H which ensure the existence of (a) local cross-sections of H in G? (b) sufficiently continuous Wigner rotations of G into H?

The fundamental problem is whether or not there exists a "global harmonic analysis" which picks up precisely at the point where ordinary harmonic analysis ceases to function, namely when the topological group G is not longer locally compact.

Additionally, interesting "arithmetical" problems might arise when the subgroup H is discrete. We cannot discuss these questions here.

A PHYSICAL APPLICATION

Turning to the question of physical applications, it seems preferable to give a nontrivial example in detail, rather than several in an indigestible concentration. More examples may be found in [1], or in a series of forthcoming works under the general title: "Theory of Symmetry in the Quantum Mechanics of Infinite Systems".

Consider a system of infinitely many identical (massive, spinless, Bose) non-relativistic particles, with a reasonable two-body interaction. The relativity group for a single particle is the inhomogeneous Galilei group, and its state space is a Hilbert space; we shall take this space to be $L_2(R^3, d^3p)$, the square-integrable function over *momentum* space. When we try to describe the *entire* system, i.e. find for it a state space and an action of the Galilei group upon it, we encounter a difficulty.

At first glance it might appear that the state space for the whole system ought also to be a Hilbert space; but then we would not be able to implement the boosts upon it, because to change the velocity of an infinite system we would have to impart to it an infinite amount of energy and momentum. No unitary operator upon a Hilbert space can do this, which is intuitively plausible and easy enough to justify rigorously.

Physicists, by and large, have accepted this situation; it is called a *broken symmetry*. The idea is that, in the limit of infinitely many particles, *the state space does not change character but the relativity group shrinks:* no attempt is made to implement the boosts. A "dual" situation is encountered in high-energy physics: just as, for very large systems, the relativity group shrinks, so, for very small systems, it "expands".

The idea that I would like to put forward is the exact opposite: instead of modifying the relativity group, let us try to modify the state space in such a way that the whole relativity group can act upon it. For small systems, i.e. high energies, this remains a formidable challenge, which may require the development of much new mathematics. For large systems, however, it can be shown to work. Let us return to our concrete example.

Assume that our infinite system does indeed possess some residual quantum-mechanical behaviour, that is, there *is* a Hilbert space associated with it. The boosts cannot be implemented upon it, but time translations ought surely to be implementable, and also possibly the Euclidean transformations. Therefore we look at bundle representations of G, with $M = G/H$, G being the connected inhomogeneous Galilei group and $H = E_3 \times T$, E_3 being the Euclidean group (rigid motions) in three dimensions and T

the time translations. M is then the space of the boosts, and the Hilbert space in $M \times H_2$ carries only a representation of $E_3 \times T$. The boosts act on M, and the difficulty which we had encountered earlier disappears. Observe that $E_3 \times T$ is locally compact and therefore closed.

We parametrize an element $g \in G$ as follows:

$$g = (b, \vec{a}, \vec{v}, R)$$

where b is a time translation, $\vec{a}$ is a space translation, $\vec{v}$ a velocity boost and R a proper three-dimensional rotation (the symbol R is also used to denote the real line; this should not cause any confusion). The multiplication law in G is

$$(b', \vec{a}', \vec{v}', R')(b, \vec{a}, \vec{v}, R) = (b'+b, \vec{a}' + R'\vec{a} + \vec{v}'b, \vec{v}' + R'\vec{v}, R'R)$$

and the identity element is

$$e = (0, 0, 0, 1) .$$

The subgroup $H = E_3 \times T$ consists of all elements of the form

$$(b, \vec{a}, 0, R) ,$$

and the space of the boosts M too is a subgroup, consisting of elements of the form

$$(0, 0, \vec{w}, 1) .$$

We shall denote a point on M by $\vec{w}$. Then the natural projection is

$$\pi(b, \vec{a}, \vec{v}, R) = \vec{v}$$

and the natural action of G on M is

$$(b, \vec{a}, \vec{v}, R)\vec{w} = \vec{v} + R\vec{w} .$$

Moreover, there exists a globally-continuous natural standard cross-section

$$\eta(\vec{w}) = (0, 0, \vec{w}, 1)$$

and the Wigner rotation $k(g,\vec{w})$, obtained by inserting this η into (9), namely

$$k(g,\vec{w}) = (b, \vec{a}, -b(\vec{v} + R\vec{w}), R) ,$$

is both strictly continuous and standard. We only have to choose a representation U of H.

We take $H_2 = L_2(R^3, d^3p)$, represent the translations by characters, and choose the following representation of $E_3 \times T$:

$$(U(b, \vec{a}, 0, R)\phi)\vec{p} = \{\exp i(bE(\vec{p}^2) - \vec{a}\cdot p\}(\phi \circ R^{-1})\vec{p} \,. \tag{12}$$

This representation is not irreducible. An irreducible representation of E_3 corresponds to $\vec{p}^2$ const., and one of T is simply a character $\exp iEb$. The functional dependence of E on $\vec{p}^2$ in (12) means that we have chosen one specific irreducible of T corresponding to a specific irreducible of E_3. The representation (12) is a direct integral of all such irreducibles, each occurring with same "multiplicity".

Now we can write down an explicit formula for the bundle representation corresponding to the choice (12). It is:

$$(b, \vec{a}, \vec{v}, R)(\vec{w}, \phi) = (\vec{v} + R\vec{w}, e^{i\Lambda}(\phi \circ R^{-1})p) \tag{13}$$

where

$$\Lambda = -\vec{a}\cdot\vec{p} + b\{E + (\vec{v} + R\vec{w})\cdot\vec{p}\} \,, \tag{14}$$

and we have not displayed explicitly that E depends on $\vec{p}^2$.

Let us consider, in order to facilitate physical interpretation, the following special cases of (13) and (14).

<u>(a) $\vec{v} = 0, R = 1, \vec{w} = 0$.</u> Then

$$(b, \vec{a}, 0, 1)(0, \phi) = (0, e^{i(bE-\vec{a}\cdot\vec{p})}\phi) \,. \tag{15}$$

<u>(b) $\vec{v} \neq 0, R = 1, \vec{w} = 0$.</u> We use

$$(b, \vec{a}, \vec{v}, 1) = (b, \vec{a}, 0, 1)(0, 0, \vec{v}, 1)$$

to write

$$\begin{aligned}(b, \vec{a}, 0, 1)(0, 0, \vec{v}, 1)(0, \phi) &= (b, \vec{a}, 0, 1)(\vec{v}, \phi) \\ &= (\vec{v}, e^{i\{b(E+\vec{v}\cdot\vec{p})-\vec{a}\cdot\vec{p}\}}\phi) \,.\end{aligned} \tag{16}$$

Comparing (15) with (16) we find that the function $E(\vec{p}^2)$ in the rest-fibre $\vec{w} = 0$ transforms to the function

$$E' = E(\vec{p}^2) + \vec{p}\cdot\vec{v}$$

in the fibre with $\vec{w} = \vec{v}$, whereas $\vec{p}$ remains unchanged:

$$\vec{p}\,' = \vec{p}\ .$$

These are well-known transformation properties of the energy and momenta of low-lying excitations in superfluid Helium, which are crucial to Landau's criterion of superfluidity.

That is, if we choose for $E(\vec{p}^{\,2})$ the Landau spectrum, the mathematical object which we have defined corresponds precisely to the low-lying excitations in superfluid Helium. However, it is worth noting that we have obtained it as a representation - a *bundle* representation - of the *full* Galilei group. We have found that the mathematical arbitrariness in the choice of the function $E(\vec{p}^{\,2})$ arises from the reducibility of the representation (12). These reducible representations correspond to physically elementary systems - *the elementary excitations*. In the quantum mechanics of infinite systems, *unlike* the situation in high-energy physics, *there is no correspondence between mathematical irreducibility and physical elementarity*. Additionally, the bundle representation (13), (14) describes a nonrelativistic system of *zero mass*. Finally, in the pair $(\vec{w}, \phi)$, $\vec{w}$ labels an *inertial frame* in the sense of Einstein, and ϕ a state in the Hilbert space belonging to that inertial frame.

Bundle representations tell us, among other things, how to take quantum mechanics from one inertial frame to another.

ACKNOWLEDGEMENTS

I would like to thank my colleagues M. Magidor and H. Gauchman for discussions, Professor K. Bleuler and Dr. A. Reetz for their hospitality in Bonn, and Mrs. Yael Ahuvia of Ben Gurion University for typing this article.

REFERENCES

1. H.J. Borchers and R.N. Sen, *Relativity Groups in the Presence of Matter*, Comm. math. Phys. 42, 101-126 (1975).
2. S.A. Gaal, *Linear Analysis and Representation Theory*, Vol. 198 in the *Grundlehren* series, Springer-Verlag, 1973, p. 306 *et seq*.
3. E. Hewitt and K.A. Ross, *Abstract Harmonic Analysis*, Vol. I, Vol. 115 in the *Grundlehren* series, Springer-Verlag, 1963, Theorem 5.11, p. 35.
4. V.S. Varadarajan, *Lie Groups, Lie Algebras and their Representations*, Prentice-Hall, 1974, p. 79 *et seq*.
5. N. Steenrod, *The Topology of Fibre Bundles*, Princeton University Press, 1951 (with an appendix added in 1956).
6. P.S. Mostert, *Local Cross-Sections in Locally Compact Groups*, Proc. Amer. Math. Soc. 4, 645-649 (1953).

INTRODUCTION TO GAUGE THEORY

Tai Tsun Wu+
CERN, Geneva

1. Introduction

Professor Bleuler has kindly asked me to give an elementary overall view of gauge theory from a physicists's point of view. Gauge theory consists of two types:

a) Abelian gauge theory, which is essentially electromagnetism of Faraday and Maxwell, and

b) Non-Abelian gauge theory, or Yang-Mills theory (20).

Both types can be studied from three points of view:

i) Differential formelism,

ii) Integral formelism, and

iii) Global formalism.

It is in the global formalism where fiber bundles play a central role. This classification is summarized in Fig. 1.

2. Differential Formalism for Abelian Gauge Theory

The concept of a gauge transformation was first introduced by Weyl (15) in his attempt to unify gravitation and electromagnetism. By gauge transformation he means a scale change that depends on the space-time point x^μ. Consider two neighbouring space-time points x^μ and $x^\mu+dx^\mu$. In going from x^μ to $x^\mu+dx^\mu$ the gauge transformation gives a factor $1+s_\mu dx^\mu$ say. Since a function f becomes $f+\frac{\partial f}{\partial x^\mu}dx^\mu$, combination with the scale change gives

$$f + dx^\mu\left(\frac{\partial}{\partial x^\mu} + s_\mu\right) f \tag{2.1}$$

Weyl attempted to identify s_μ with the electromagnetic vector potential A_μ, but he was not successful. It is remarkable that this attempt of Weyl was made before the advent of quantum mechanics.

As we know now, there is a missing factor of i. As studied by Fock (8), London (10) and Weyl (16) himself, what is needed is a phase change instead of a scale change. Instead of identifying s_μ with A_μ the paper identification is

---

+) On sabbatical leave from Harvard University. Work supported in part by U.S. Department of Energy Contract No. EY-76-S-02-3227.

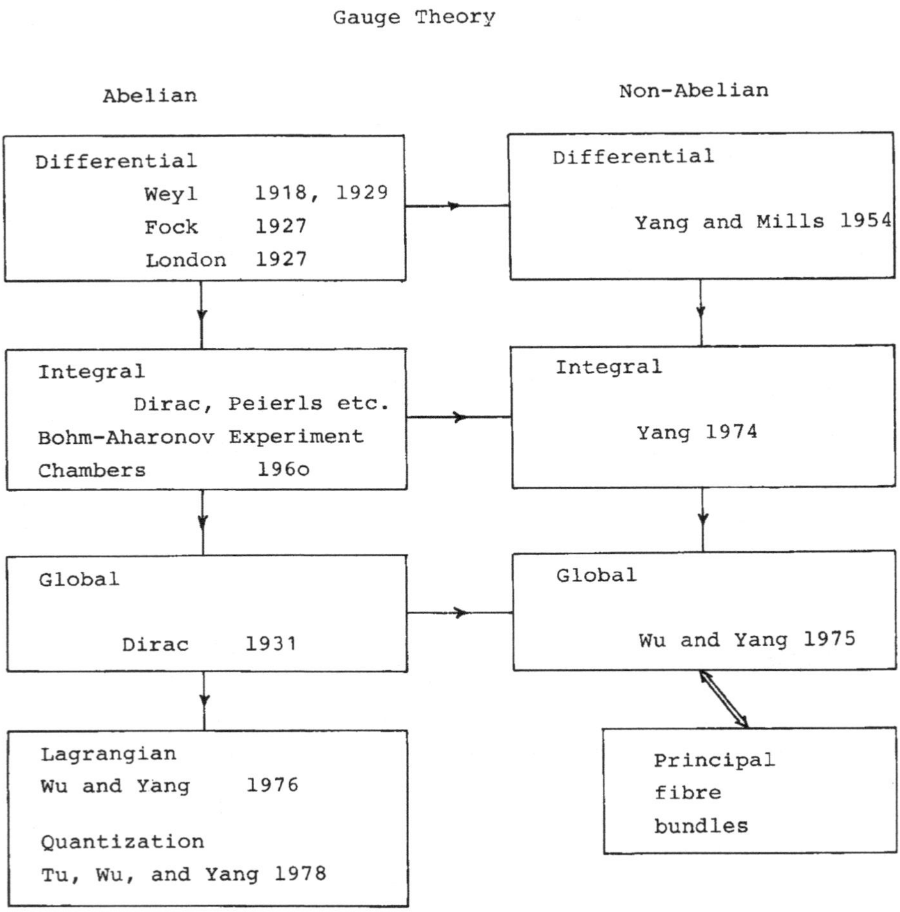

Fig. 1 Schematic outline of gauge theory

$$s_\mu = -i \frac{e}{\hbar c} A_\mu \quad . \tag{2.2}$$

Strictly speaking, the term gauge transformation is no longer appropriate but the term sticks.

3. Differential Formalism for Non-Abelian Gauge Theory

The next conceptual advance came quarter of a century later, when Yang and Mills (20) generalized space-time-dependent gauge transformation to the non-Abelian case SU(2) . It was known at that time that, in the absence of electromagnetic interaction, isotopic spin is conserved. In other words, strong interactions are invariant under isospin rotation. This means that there is freedom of convention to choose, for example, what we call a proton rather than a neutron or a linear combination. Yang and Mills pointed out that there is an independent freedom for each space-time point, and generalized

$$\frac{\partial}{\partial x^{\mu}} - \frac{ie}{\hbar c} A_{\mu} \tag{3.1}$$

of electromagnetism to

$$\frac{\partial}{\partial x^{\mu}} - \frac{i\varepsilon}{\hbar c} B_{\mu} , \tag{3.2}$$

which operators on, for example, the two-component wave function

$$\psi = \binom{\psi_p}{\psi_n} \tag{3.3}$$

of the proton and neutron. Here B_{μ} has isospin indices and is the Yang-Mills field.

4. Integral Formalism for Abelian Gauge Theory

Maxwell's equations are partial differential equations for $f_{\mu\nu}$, which consists of $\vec{E}$ and $\vec{H}$. In classical mechanics, the motion of an electron in an electromagnetic field is governed by the Lorentz force, which is expressable in terms of $f_{\mu\nu}$ and the electron world line. Therefore, classically the introduction of the vector potential A_{μ} is entirely for mathematical convenience. The situation is quite different in quantum mechanics, where the Dirac equation for the electron in an electromagnetic field involves A_{μ} explicitly. Therefore, here A_{μ} plays an essential role.

If $f_{\mu\nu}=0$ in the entire spac-time continuum, then the obvious choice for A_{μ} is $A_{\mu}=0$. More generally, if $f_{\mu\nu}=0$ in a simply connected space-time region, then we can choose $A_{\mu}=0$ in this region. The important point made by Aharonov and Bohm (1), also previously by Ehrenberg and Siday (7), is that if $f_{\mu\nu}=0$ in a doubly connected (or more generally a multiply connected) space--time region, then it may not be consistant to choose $A_{\mu}=0$.

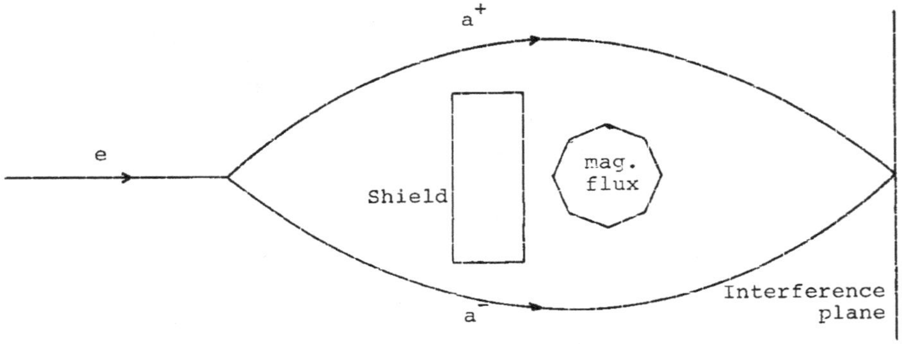

Fig.2 Bohm-Aharonov experiment

The Bohm-Aharonov experiment (1) first carried out by Chambers (3), is shown schematically in Fig. 2. Suppose the static magnetic field is confined in a cylindrical region in space, with the flux return path far away. Scattering experiment with an incident electron beam is then carried out such that the electron does not enter the region of magnetic flux. Interference pattern is then observed on the screen. In the absence of the magnetic flux, let $a^+(a^-)$ be the electron wave function at the screen when a shielding block is introduced to prevent the electron from reaching the screen via the path below (above) the region of magnetic flux. Then, when the magnetic flux is turned off, the amplitude at the screen is

$$a^+ + a^- . \tag{4.1}$$

If the magnetic flux is turned on, the interference pattern is determined by

$$a^+ + a^- \exp\left(\frac{ie}{\hbar c}\,\Omega\right) \tag{4.2}$$

where Ω is the total magnetic flux in the cylinder. Since

$$\Omega = \oint A_\mu dx^\mu \tag{4.3}$$

with the line integral taken outside the cylinder, interference pattern depends on the phase factor

$$\exp \quad \frac{ie}{\hbar c} \oint A_\mu dx^\mu \tag{4.4}$$

A more detailed consideration (17) shows that $f_{\mu\nu}$ underdescribes electromagnetism (ie. , different physical situations in a space--time region may have the same $f_{\mu\nu}$), and $\oint A_\mu dx^\mu$ overdescribes electromagnetism (i.e., different $\oint A_\mu dx^\mu$ in a space-time region may describe the same physical situation). The phase factor $e^{\frac{ie}{\hbar c} \oint A_\mu dx^\mu}$ is just right to describe electromagnetism. Note that this phase factor gives $f_{\mu\nu}$ in the limit where the line of integration is an infinitesimal loop.

For theoretical treatments, this phase factor (4.4), which is invariant under the gauge transformation

$$A_\mu \rightarrow A_\mu + \frac{\hbar c}{e} \frac{\partial\alpha}{\partial x^\mu} , \tag{4.5}$$

is less easy to use than the concept of a path-dependent phase factor

$$e^{\frac{ie}{\hbar c} \int_P^Q A_\mu dx^\mu} . \tag{4.6}$$

Such a path-dependent (or non-integrable) phase factor is not gauge invariant. Instead under (4.5),

$$e^{\frac{ie}{\hbar c} \int_P^Q A_\mu dx^\mu} \rightarrow e^{i\alpha(Q)} \, e^{\frac{ie}{\hbar c} \int_P^Q A_\mu dx^\mu} \, e^{-i\alpha(P)} \tag{4.7}$$

Electromagnetism is thus the gauge-invariant manifestation of a non-integrable phase factor (2, 11, 17).

5. Integral Formalism for Non-Abelian Gauge Theory

The generalization of integral formalism to Yang-Mills theory was carried out by Yang (19) in 1974. The starting point of this consideration is the non-integrable phase factor

$$(e^{\int_P^Q \frac{-\varepsilon}{\hbar c} b_\mu^k X_k dx^\mu})\text{ ordered} \tag{5.1}$$

where b_μ^k is the Yang-Mills field (20) and X_k are the generators of the underlying Lie group. If the group is U_1, (5.1) reduces to (4.6). The Yang-Mills field is the gauge-invariant manifestation of this non-integrable phase factor.
For the properties of (5.1), see Yang (19).

6. Global Formalism for Abelian Gauge Theory

The early work of Dirac (6) in 1931 leads to the recognition that in general A_μ and the phase factor (4.4) can only be properly defined in each of several overlapping space-time regions. These overlapping space-time regions are like coordinate neighbourhoods and their union is the entire space-time region of interest. Analogous to the static electric field

$$\vec{E} = \frac{e}{r^2}\hat{r} \tag{6.1}$$

of a point charge, Dirac (6) considered the static magnetic field

$$\vec{H} = \frac{g}{r^2}\hat{r} \tag{6.2}$$

of a magnetic monopole. The important point is that this magnetic field (6.2) cannot be obtained from a vector potential A_μ which is regular for $r>0$. For example, the choices in spherical coordinate

$$\vec{A}_a = \frac{g}{r\sin\theta}(1-\cos\theta)\hat{\phi} \tag{6.3}$$

$$\vec{A}_b = -\frac{g}{r\sin\theta}(1+\cos\theta)\hat{\phi} \tag{6.4}$$

are respectively singular at $\theta=\pi$ and $\theta=0$. However, if we define two regions R_a and R_b by

$$R_a:\quad 0\leqslant\theta<\frac{\pi}{2}+\delta\ ,\quad r>0\qquad 0\leqslant\phi<2\pi\ ,$$

and

$$R_b: \ \frac{\pi}{2} - \delta<\theta\leqslant 2\pi, r>0 \qquad 0\leqslant\phi<2\pi \tag{6.5}$$

with $0<\delta\leqslant\frac{\pi}{2}$, then we get a global description of the electromagnetic field of a magnetic monopole by using $\vec{A}_a$ in R_a and $\vec{A}_b$ in R_b (17). In the overlap $R_a \cap R_b$, the gauge transformation, according to (4.7), is given by

$$e^{i\alpha} = e^{\frac{2ieg}{\hbar c}\phi} \tag{6.6}$$

This $e^{i\alpha}$ is single-valued, and hence well-defined, if and only if

$$\frac{2eg}{hc} = \text{integer} . \tag{6.7}$$

This is the Dirac quantization condition (6).

7. Global Formalism for Non-Abelian Gauge Theory

The necessity of dealing with several overlapping space-time regions has nothing to do with the nature of the underlying group for the gauge theory. For the non-Abelian as well as the Abelian case, the Yang-Mills field (20) b_μ^k and the non-integrable phase factor (5.1) are defined in each of the overlapping space-time regions. In the overlap of any two such regions, there is a gauge transformation relating the phase factors defined for the two regions. The phase factor takes on a slightly more complicated form if the points P and Q are in different space-time regions. With this setup, there is clearly a very close relation with fibre bundles (9,13). A short dictionary (17) between the physical and mathematical terminologies is given in Table 1.

8. Discussions

We have described briefly the development of the language appropriate for gauge theory. It is my belief that this basic language needs no further major revision.

The next problem is to develop the dynamics of gauge theory on the basis of the global formalism. Some steps have been taken in this direction in the context of electromagnetism. The Lagrangian with positrons, monopoles and electromagnetic field has been written down (18). To avoid the Rosenbaum paradox (12), the condition needs to be imposed that the positron world lines and the monopole world lines do not intersect. This condition causes no difficulty because it excludes a set of probability zero. The quantum field theory for this system has also been carried out recently (14) within the global formalism.

There is the following most chellenging and important problem that is as yet unsolved. Because of the Dirac (6) quantization condition (6.7) which plays a central role in the global Lagrangian formalism it is not possible for the two coupling constants $\frac{e^2}{\hbar c}$ and $\frac{g^2}{\hbar c}$ to be both small. How can we develop a systematic approximate scheme for calculating the interaction of positrons, magnetic monopoles, and photons in the context of quantum field theory?

I thank CERN and the University of Bonn for their kind hospitality. I am greatly indebted to Professor Chen Ning Yang from whom I learn gauge theory.

References

1. Y. Aharonov and D. Bohm, Phys. Rev. 115, 485 (1959).

2. I. Biaĺynicki-Birula, Bull. Acad. Pol. Sci., Ser. Sci. Math. Astron. Phys. 11, 135 (1963).

3. R.G. Chambers, Phys. Rev. Letters 5, 3 (1960).

4. Shiing-shen Chern, Ann. Math. 45, 747 (1944).

5. Shiing-shen Chern, Geometry of Characteristic Classes, Proc. 13th Biennial Seminar, Canadian Math. Congress (1972).

6. P.A.M. Dirac, Proc. Roy. Soc. London A 133, 60 (1931).

7. W. Ehrenberg and R.E. Siday, Proc. Phys. Soc. London B 62, 8 (1949).

8. V. Fock, Zeit. f. Physik 39, 226 (1927).

9. S. Kobayashi and K. Nomizu, Foundations of Differential Geometry (Interscience, New York, Vol. I -1963, Vol. II-1969).

10. F. London, Zeit. f. Physik 42, 375 (1927).

11. S. Mandelstam, Ann. Phys. (N.Y.) 19, 1 and 25 (1962).

12. D. Rosenbaum, Phys. Rev. 147, 891 (1966)

13. N. Steenrod, The Topology of Fibre Bundles (Princeton University Press, Princeton, N.J., 1951).

14. Tung Sheng Tu, Tai Tsun Wu, and Chen Ning Yang. Acad. Sinica (to be published)

15. H. Weyl, Stzber. Preuss. Akad. Wiss. 465 (1918)

16. H. Weyl, Zeit. f. Physik 56, 330 (1929)

17. Tai Tsun Wu and Chen Ning Yang, Phys. Rev. D 12, 3845 (1975)

18. Tai Tsun Wu and Chen Ning Yang, Phys. Rev. D 14, 437 (1976)

19. Chen Ning Yang, Phys. Rev. Letters 33, 445 (1974)

20. Chen Ning Yang and Robert L. Mills, Phys. Rev. 96, 191 (1954).

Table I. Translation of Terminology

Gauge Theory	Bundle
gauge (or global gauge)	principal coordinate bundle
gauge type	principal fibre bundle
gauge potential b_μ^k	connection on principal fibre bundle
field strength $f_{\mu\nu}^k$	curvature
phase factor (5.1)	parallel displacement
source J_μ^k	?
electromagnetism	connection on U_1 bundle (i.e. principal fibre bundle with group U_1).
Yang-Mills field for isospin	connection on SU_2 bundle
Dirac's monopole quantization	classification of U_1 bundle according to first chern class (4,5)
electromagnetism without monopole	connection on trivial U_1 bundle
electromagnetism with monopole	connection on non-trivial U_1 bundle

THE USE OF EXTERIOR FORMS IN FIELD THEORY

W. Thirring and R. Wallner

Institut für Theoretische Physik

Universität Wien

Cartan's calculus formalizes the quantities appearing in Maxwell's and Einstein's equations and is therefore particularly suited for expressing them. Unfortunately, it is not yet as commonly used as it deserves because the practitioner frequently don't know it and the mathematicians seldomly do explicit calculations. We would like to summarize here the main points of this formalism. For details see [1].

1. Basis

The central objects in field theory are differential forms ($\equiv$ antisymmetric covariant tensor fields). The ones of rank p, $p = 0,1,\ldots,$ m = dimension of the manifold M (= p-forms) are a module E_p over E_o, the real-valued functions. The exterior product $\wedge : E_p \times E_q \to E_{p+q}$ makes

$$\bigoplus_{p=0}^{m} E_p$$

to a graded algebra. If e^α is a local basis in E_1 one can construct a ($\binom{m}{p}$-dimensional) basis in E_p with :

$$e^{\alpha_1 \ldots \alpha_p} := e^{\alpha_1} \wedge e^{\alpha_2} \wedge \ldots \wedge e^{\alpha_p} \quad , \quad (\alpha_k = 0,\ldots,m-1) \ .$$

Thus every p-form ω can be written

$$\omega = \omega_{|\alpha_1 \ldots \alpha_p|} \, e^{\alpha_1 \ldots \alpha_p} \ , \qquad \omega_{(\alpha)} \in E_o \ , \tag{1.1}$$

where vertical bars mean summation over $\alpha_1 < \alpha_2 < \ldots < \alpha_p$.

2. Exterior Derivative

The differential processes of vector calculus are synthesized in the operation $d : E_p \to E_{p+1}$ with the properties

(i) $d(\omega_1+\omega_2) = d\omega_1 + d\omega_2, \quad \omega_{1,2} \in E_p$

(ii) $d(\omega_1 \wedge \omega_2) = (d\omega_1)\wedge\omega_2 + (-)^p\, \omega_1 \wedge d\omega_2, \quad \omega_1 \in E_p, \quad \omega_2 \in E_q$ (2.1)

(iii) $d\, d\omega = 0, \quad \omega \in E_p.$

Functions x^α over a neighbourhood are called independent if the $dx^\alpha \in E_1$ are independent. Using them as a basis $e^\alpha = dx^\alpha$ the exterior derivative becomes simply

$$d\omega = d(\omega_{|\alpha_1\ldots\alpha_p|}\, e^{\alpha_1\ldots\alpha_p}) = \frac{\partial\omega_{|\alpha_1\ldots\alpha_p|}}{\partial x^\alpha}\, e^{\alpha\,\alpha_1\ldots\alpha_p} .$$

3. Pseudo-Riemannian Structure

A scalar product $<|>$ in E_1

$$<e^\alpha|e^\beta> = g^{\alpha\beta} = g^{\beta\alpha} , \qquad \mathrm{Det}\, g^{\alpha\beta}(x) \neq 0 \quad \forall\, x \in M ,$$

also gives an isomorphism $* : E_p \leftrightarrow E_{m-p}$. Because of linearity one has to specify it only for a basis:

$$* e^{\alpha_1\ldots\alpha_p} := \varepsilon^{\alpha_1\ldots\alpha_p|\alpha_{p+1}\ldots\alpha_m|}\, e_{\alpha_{p+1}\ldots\alpha_m} \tag{3.1}$$

$$\varepsilon := \sqrt{(-)^s g}\; e^0 \wedge e^1 \wedge \ldots \wedge e^{m-1} = \varepsilon_{|\alpha_1\ldots\alpha_m|}\, e^{\alpha_1\ldots\alpha_m}$$

$g := \mathrm{Det}\,(g_{\alpha\beta})$, $\qquad (-)^s$ = signature of g ,

ε = generalized Levi-Civita tensor.

It allows to define the co-derivative $\delta : E_p \to E_{p-1}$:

$$\delta := * \circ d \circ * \; (-)^{m(p+1)+s} . \tag{3.2}$$

4. Integration

A p-form ω defines a measure over p-dimensional submanifolds N_p. The corresponding integral is simply denoted by $\int_{N_p} \omega$. It is the inverse operation to d in the sense that partial integration generalizes to Stokes' theorem

$$\int_{N_{p+1}} d\omega = \int_{\partial N_{p+1}} \omega , \tag{4.1}$$

∂N = boundary of N, ω with compact support.

$\nu \in E_{m-p}$ defines

$$\nu(\omega) = \int_M \nu \wedge \omega , \qquad \omega \in E_p ,$$

a linear functional over E_p. Correspondingly we shall also admit distribution-type p-forms (currents) [2], understanding that due care has to be exercised when multiplying them. Of greater importance will be linear maps $E_p \rightarrow E_q|_{\bar{x}}$ where the latter denotes the antisymmetric covariant tensors at the point $\bar{x} \in M$ [3]. In particular one can define a "δ-function" $\delta_{\bar{x}} \in E_p|_{\bar{x}} \otimes E_{m-p}$ which reproduces the value at a point $\bar{x}$ of a p-form

$$\int \delta_{\bar{x}} \wedge \omega = \omega|_{\bar{x}} . \tag{4.2}$$

$\delta_{\bar{x}}$ has its support in $\bar{x}$ and there it can be written in a natural basis $\bar{e}$ for $E_p|_{\bar{x}}$ and e of E_{m-p}

$$\delta_{\bar{x}} = \bar{e}^{|\alpha_1 \ldots \alpha_p|} \otimes {}^*e_{\alpha_1 \ldots \alpha_p} \; (-)^{p(m-p)} \; \delta(x^0 - \bar{x}^0) \; \delta(x^1 - \bar{x}^1) \ldots \; \delta(x^{m-1} - \bar{x}^{m-1}) . \tag{4.3}$$

5. Differential Equations

The differential equations encountered in classical field theory are of such a structure that exterior and co-derivative of p-forms are specified The appropriate tool for solving such an equation is the "Green-function" $G_{\bar{x}}$. Like $\delta_{\bar{x}}$ it is actually a tensor-valued current $\in E_p|_{\bar{x}} \otimes E_{m-p}$ and satisfies the equation

$$(\delta d + d\delta) G_{\bar{x}} = - \delta_{\bar{x}} . \tag{5.1}$$

As a consequence of Stokes' theorem and the rules for derivatives one obtains for $\bar{x} \in N \subset M$, Dim N = Dim M

$$\omega|_{\bar{x}} = (-)^{p+m} \int_N [dG_{\bar{x}} \wedge \delta\omega - \delta G_{\bar{x}} \wedge d\omega] - \int_{\partial N} [\delta G_{\bar{x}} \wedge \omega + (-)^{m+p+s} {}^*dG \wedge {}^*\omega] . \tag{5.2}$$

Since we have not yet demonstrated the existence of $G_{\bar{x}}$ so far this result is only formal. By explicit construction of $G_{\bar{x}}$ in certain cases one sees that (5.2) is valid provided ∂N is nowhere tangent to the light cone of $\bar{x}$. If this is the case, (5.2) solves the problem as it expresses ω at $\bar{x}$ by $d\omega$ and $\delta\omega$ in N and ω and ${}^*\omega$ at ∂N. For some manifolds like $\mathbb{R}^m$, $g = -dx^0 \otimes dx^0 + dx^1 \otimes dx^1 + \dots + dx^{m-1} \otimes dx^{m-1}$, there is a $G_{\bar{x}}$ with the additional property that its support is contained in the past light cone of $\bar{x}$. In this case (5.2) solves the initial-value problem: If $N = \{x^\alpha \in \mathbb{R}^m : t_0 \le x^0 \le t_1\}$ only the part of ∂N with $x^0 = t_0$ contributes and ω is determined by $d\omega$, $\delta\omega$ and the values of ω and ${}^*\omega$ for $x^0 = t_0$.

6. Electrodynamics

Maxwell's Equations

They refer to the case m = 4, p = 2, and for the field form $F \in E_2$ they read

$$dF = 0 , \qquad \delta F = J = \text{charge-current} \in E_1 , \tag{6.1}$$

or in integral form

$$\int_{\partial N_3} F = 0 , \qquad \int_{N_3} {}^*J = - \int_{\partial N_3} {}^*F . \tag{6.2}$$

Since $\delta \circ \delta = 0$, they imply current conservation: $\delta J = 0$, in integral form

$$\int_{\partial N_4} {}^*J = 0 .$$

In some manifolds like $\mathbb{R}^4$, $\delta J = 0$ implies the existence of $F \in E_2$ such that $\delta F = J$. In general the latter equation is stronger than $\delta J = 0$. In particular, if $N_3 = \{x^\alpha \in \mathbb{R}^4 : x^0 = t\}$ is compact and without boundary (like in $\mathbb{R} \times T^3$) then the total charge

$$Q(t) = \int_{N_3} {}^*J$$

is not only constant in t but zero:

$$\int_{N_3} {}^{*}J = \int_{\partial N_3} {}^{*}F = \int_{\emptyset} {}^{*}F = 0 \; . \tag{6.3}$$

The usual procedure for solving (6.1) is to conclude from $dF = 0$ that $F = dA$, go to the Lorentz-gauge $\delta A = 0$ and then solve $(\delta d + d\delta)A = J$. This is not so good because one wants to express F in terms of the initial values of F and ${}^{*}F$ and not in terms of A and its derivatives. (5.2) gives in $\mathbb{R}^4$ directly the solution of the Cauchy problem if one uses

$$G_{\bar{x}}^{ret} = -\frac{1}{4\pi}\,\bar{e}_{\alpha\beta} \otimes {}^{*}e^{\alpha\beta}\;\delta((\bar{x}-x)^2)\;\Theta(\bar{x}^0-x^0) \;: \tag{6.4}$$

$$F_{|\bar{x}} = \int_N dG_{\bar{x}}^{ret} \wedge J - \int_{\partial N} [\delta G_{\bar{x}}^{ret} \wedge F + {}^{*}dG_{\bar{x}}^{ret} \wedge {}^{*}F] \; . \tag{6.5}$$

If $N = \{x \in \mathbb{R}^4 : x^0 > t\}$ and the limit $t \to -\infty$ exists, these equations express F in terms of an incoming field and the usual Lienard-Wiechert potentials. In some situations other Green-functions are called for. For instance on a metallic surface B the boundary condition is simply $F_{|B}$:= restriction of F to B = 0. In such a situation $\partial N = B \cup S$ in (5.2) will have a space-like part S corresponding to an initial surface and B. To have only S contribute to $\int_{\partial N}$ we need a Green-function such that

$$ {}^{*}dG_{\bar{x}}{}_{|B} = 0 \; . \tag{6.6}$$

With this Green-function (5.2) solves the initial-value problem

$$F_{\bar{x}} = \int_N dG_{\bar{x}} \wedge J - \int_S [\delta G_{\bar{x}} \wedge F + {}^{*}dG_{\bar{x}} \wedge {}^{*}F] \quad . \tag{6.7}$$

If $B = \{x \in \mathbb{R}^4 : x^1 = 0\}$ (metallic mirror) a Green-function satisfying (6.6) is simply $(1 + R)G_{\bar{x}}^{ret}$ where R is the reflection $x^1 \to -x^1$. Similarly the $G_{\bar{x}}$ for a rectangular wave-guide is constructed by a sequence of reflections. In this way the causal structure of G becomes obvious, whereas in usual procedure [4] after piecing plane waves together remain the questions

(a) Has one found all solutions?

(b) Does nothing propagate faster than light?

7. Gravitation

Here the e^α play the role of potentials and the connection-forms $\omega^\alpha{}_\beta \in E_1$:

$$de^\alpha = -\omega^\alpha{}_\beta \wedge e^\beta , \qquad dg_{\alpha\beta} = g_{\alpha\mu}\omega^\mu{}_\beta + g_{\beta\mu}\omega^\mu{}_\alpha =: \omega_{\alpha\beta} + \omega_{\beta\alpha} \tag{7.1}$$

correspond to the field strength. However, by a change of the basis:

$$e^\alpha \to A^\alpha{}_\beta e^\beta , \qquad A^\alpha{}_\beta \in E_o , \tag{7.2}$$

the $\omega^\alpha{}_\beta$ transform inhomogeneously. Only in the curvature-forms $R_{\alpha\beta} \in E_2$:

$$R_{\alpha\beta} = d\omega_{\alpha\beta} - \omega_{\mu\alpha} \wedge \omega^\mu{}_\beta \tag{7.3}$$

these inhomogeneous terms drop out. Thus, if one wants to construct a Lagrange-function $L \in E_4$ which is invariant under (7.2) the simplest possibility after $*1\!\!1$ is (a constant apart)

$$L = R_{|\alpha\beta|} \wedge *e^{\alpha\beta} + L_{\text{matter}} . \tag{7.4}$$

Variation of e^α gives Einstein's equations in a form advocated by Trautman [5]

$$-R_{|\alpha\beta|} \wedge *e^{\alpha\beta\mu} = \frac{\delta L_{\text{matter}}}{\delta e_\mu} =: *T^\mu \in E_3 . \tag{7.5}$$

However, it seems desirable to have an equation like the inhomogeneous Maxwell equation where the co-derivative of a 2-form is a current. This can easily be done by writing in the left side of (7.5) the contribution $d\omega_{\alpha\beta} \wedge *e^{\alpha\beta\mu}$ as $d(\omega_{\alpha\beta} \wedge *e^{\alpha\beta\mu}) + \omega_{\alpha\beta} \wedge d{*e^{\alpha\beta\mu}}$. Retaining only the first term on the left hand side we obtain (in units $8\pi\kappa = 1$)

$$d(\omega_{|\alpha\beta|} \wedge *e^{\alpha\beta\mu}) = -(*T^\mu + *t^\mu)$$

$$*t^\mu = -\frac{1}{2}\omega_{\alpha\beta} \wedge (\omega^\mu{}_\nu \wedge *e^{\alpha\beta\nu} + \omega^\beta{}_\nu \wedge *e^{\alpha\nu\mu}) . \tag{7.6}$$

The interpretation of (7.6) is obviously that the currents of the energy ($\mu = 0$) and momentum ($\mu = 1,2,3$) have a contribution T from matter and one t from gravitation. (7.6) has as a consequence not only the conservation law

$$\delta(T^\mu + t^\mu) = 0 \tag{7.7}$$

but the stronger statement

$$\int_{N_3} ({}^*T^\mu + {}^*t^\mu) = \int_{\partial N_3} \omega_{|\alpha\beta|} \wedge {}^*e^{\alpha\beta\mu} \quad . \tag{7.8}$$

In particular, if we have a space-like compact N_3 without boundary then

$$\int_{N_3} ({}^*T^\mu + {}^*t^\mu) = 0 \quad ,$$

that is, the total energy and momentum in a closed universe are zero.

In order to define a suitable angular momentum, we construct a 3-form corresponding to a symmetric energy-momentum-tensor $\langle e^\nu, t^\mu\rangle = t^{\mu\nu}$. This can be done by extracting $1/\sqrt{-g}$ out of $d(\omega_{|\alpha\beta|} \wedge {}^*e^{\alpha\beta\mu})$

$$d(\omega_{|\alpha\beta|} \wedge {}^*e^{\alpha\beta\mu}) = \frac{1}{\sqrt{-g}}\, d(\sqrt{-g}\,\omega_{|\alpha\beta|} \wedge {}^*e^{\alpha\beta\mu}) - \omega^\nu{}_\nu \wedge \omega_{|\alpha\beta|} \wedge {}^*e^{\alpha\beta\mu}$$

from which we get

$$d(\sqrt{-g}\,\omega_{|\alpha\beta|} \wedge {}^*e^{\alpha\beta\mu}) = -\sqrt{-g}\,({}^*T^\mu + {}^*t^\mu_{LL}) \tag{7.9}$$

$${}^*t^\mu_{LL} = {}^*t^\mu - \omega^\nu{}_\nu \wedge \omega_{|\alpha\beta|} \wedge {}^*e^{\alpha\beta\mu} \quad . \tag{7.10}$$

We call ${}^*t_{LL}$ Landau-Lifschitz-form because in a natural basis, i.e. $e^\mu = dx^\mu$, its corresponding energy-momentum-tensor can be shown to be exactly the symmetric Landau-Lifschitz-expression [6] for the pseudo-energy-tensor [7], ${}^*t^\mu_{LL}$ and ${}^*t^\mu$ coincide in a basis of constant $\langle e^\mu, e^\nu\rangle \; \forall\, \mu, \nu$, since there $\omega^\rho{}_\rho = 0$. Therefore in the natural basis we have a local conservation law

$$\delta[\delta^{\alpha\beta}_{\mu\nu}\, x^\mu (T^\nu + t^\nu_{LL})] = 0 \;, \qquad \delta^{\alpha\beta}_{\mu\nu} := \delta^\alpha_\mu\,\delta^\beta_\nu - \delta^\alpha_\nu\,\delta^\beta_\mu \;. \tag{7.11}$$

The argument concerning total energy and momentum runs as before if we define the total 4-momentum by

$$P^\mu = \int_{N_3} \sqrt{-g}\,({}^*T^\mu + {}^*t^\mu_{LL}) \quad . \tag{7.12}$$

However, for the total angular momentum

$$J^{\alpha\beta} = \int_{N_3} \sqrt{-g}\,\delta^{\alpha\beta}_{\mu\nu}\, x^\mu ({}^*T^\nu + {}^*t^\nu_{LL}) \tag{7.13}$$

a similar argument fails because

(a) The coordinates will not be globally defined

(b) The integrand is not an exact 3-form.

To verify the last claim we rewrite

$$- \delta^{\alpha\beta}_{\mu\nu}\, x^{\mu}\, d(\sqrt{-g}\;\omega_{|\rho\sigma|} \wedge {}^{*}e^{\rho\sigma\nu}) =$$

$$= -d(\sqrt{-g}\;\delta^{\alpha\beta}_{\mu\nu}\, x^{\mu}\, \omega_{|\rho\sigma|} \wedge {}^{*}e^{\rho\sigma\nu}) - \frac{1}{2}\,\delta^{\alpha\beta}_{\mu\nu}\,\sqrt{-g}\; g^{\mu\rho}\, d{}^{*}e^{\nu}{}_{\rho}$$

($dx^{\nu} = e^{\nu}$). The last term is exact if we can absorb $\sqrt{-g}\; g^{\mu\rho}$ within $d{}^{*}e^{\nu}{}_{\rho}$. This is trivially possible if $\sqrt{-g}\;\langle dx^{\mu}, dx^{\rho}\rangle$ is constant on N_3, that means if N_3 is flat. But this is also possible within the linearized theory

$$g^{\mu\nu} = \eta^{\mu\nu} + h^{\mu\nu}\,, \qquad |h^{\mu\nu}| << 1$$

because $\sqrt{-g}\; g^{\mu\rho} \sim \eta^{\mu\rho} + O(h)$ and $d{}^{*}e^{\nu}{}_{\rho} \sim O(h)$.

Therefore we have a global statement about the angular momentum only in special circumstances although the local conservation law is always valid if we use a natural basis.

References

[1] Y. Choquet-Bruhat, C. DeWitt-Morette, M. Dillard-Bleick: Analysis, Manifolds and Physics, North Holland, 1977

H. Flanders, Differential Forms, Academic Press 1963

E. Hlawka, Acta Phys. Austr. Suppl. 7, Springer 1970

H. Holman and H. Rummler, Alternierende Differentialformen, Bibliographisches Institut 1972

M. Spivak, Calculus on Manifolds, Benjamin 1965

W. Thirring, Klassische dynamische Systeme, Springer 1977

[2] G. de Rham, Variêtés Diffêrentiables, Hermann, Paris, 1955

[3] W. Thirring, Klassische Feldtheorie, Springer 1978

[4] J.D. Jackson, Classical Electrodynamics, J. Wiley & Sons, 1967

[5] A. Trautman, Theory of Gravitation, in: The Physicist's Conception of Nature, J. Mehra ed., Reidel 1973

[6] L.D. Landau, E.M. Lifschitz, Klassische Feldtheorie, Akademie-Verlag, Berlin 1966

[7] W. Thirring, R. Wallner, to appear in the Revista Brasileira de Fisica

ELECTROMAGNETIC FIELDS ON MANIFOLDS: BETTI NUMBERS, MONOPOLES AND STRINGS, MINIMAL COUPLING

A. O. Barut
Department of Physics
The University of Colorado
Boulder, Colorado 80309

ABSTRACT. The differential geometric study of electromagnetic fields on manifolds is extended to nonexact and even to nonclosed forms. Duality and variational principles are discussed and minimal coupling is derived.

I. INTRODUCTION

The main motivation for considering the electromagnetic field on manifolds is to formulate a unified theory of matter and field: Matter would be singularities of the field, or, alternatively, "matter" would be interpreted as due to the cuts and holes in the topological structure of space-time. This description would remove the duality between fields and matter and their (largely arbitrary) couplings; there would be just one kind of object: "field," not on a flat Minkowski space, but on a complicated manifold. It is a program akin to general relativity, but we have in mind the space-time structure in the small, and aim at a formulation suitable for the electron and other fundamental particles.

The results reported here are continuations of several recent papers[1)-3)] to which we refer for other references and for historical background.

II. NOTATIONS AND MATHEMATICAL FORMULAS

In the following calculations we make use of the following notions and rules.[4),5)]

M is the (compactified) manifold of space-time, of dimension $n=4$, oriented. A p-chain C^p is a cycle, if $\partial C^p = 0$; it is exact, if $C^p = \partial C^{p+1}$. We shall mainly use a 2-form F representing the electromagnetic field, 1-forms A, α, β, representing various potentials, and J, k representing currents. A p-form ω^p is closed if $d\omega^p = 0$; it is exact if $\omega^p = d\alpha$. In M, for each p, there are finitely many, B_p, fundamental p-cycles C_i such that every p-cycle C is homolog to a linear combination: $C = k_1 C_1 + \cdots + k_{B_p} C_{B_p}$, and C is exact only if all $k_i = 0$. [In spaces with torsion with suitable identification, a C is said to be <u>exact</u> even if $C \neq \partial C'$ but only $kC = \partial C'$.] Then B_p

is the pth Betti number of M with Euler - Poincaré characteristic

$$\chi_M = \sum_{i=1}^{n} (-1)^p B_p$$

which is a topological invariant. For a closed form ω, the integrals

$$\pi_i = \int_{C_i} \omega, \qquad i = 1 \ldots B_p$$

are called the fundamental p-periods of ω in M^n.

All periods of exact forms vanish and, conversely, if all periods π_i of ω in M vanish, ω is exact (the first Cartan-de Rham theorem). If π_i are not zero, there exists in M an everywhere regular and closed form ω whose periods are the prescribed π_i (second Cartan-de Rham theorem).

The rules for multiplication and differentials of forms and cycles are

$$\omega_1^{p_1}\omega_2^{p_2} = (-1)^{p_1 p_2}\omega_2^{p_1}\omega_1^{p_2}$$

$$d(\omega_1\omega_2) = \omega_2 d\omega_1 + (-1)^{p_1}\omega_1 d\omega_2$$

$$\partial(C_2C_1) = C_2\partial C_1 + (-1)^{n-p}\partial C_2 C_1 .$$

A 0-cycle is a finite linear combination of points P_i: $C^o = k_1P_1 + \cdots + k_\nu P_\nu$. The index of C^o is $I(C^o) = \sum_i k_i$. The necessary and sufficient condition for C^o to be exact is $I(C^o) = 0$. Similarly, the index of an n-form ω^n in M^n is

$$I(\omega^n) = \int_{M^n} \omega^n$$

and vanishes if ω^n is exact.

A p-form ω^p and an (n-p)-chain C^{n-p} are associated if the fundamental periods of ω^p are equal to the fundamental indices of C^{n-1}:

$$\pi_i^p = \int_{C_i^p} \omega^p = I\left(C^{n-p}C_i^p\right).$$

(More generally, there is an isomorphism between the p-homology group $H_p(M)$ of p - cycles modulo p-exact cycles and the (n-p)-cohomology group $H^{n-p}\Omega(M)$, of closed (n-p)-form moduls (n-p)-exact forms.) In physics, this implies that physical quantities are represented sometimes as p-chains, sometimes as forms ω^{n-p}. [For examples, see Reference 1b.)

An elementary <u>p-current</u> (or flow) of dimension p is the set (C^{p+k}, ω^k), $k = 0, 1, \ldots n-p$, where ω^k is defined at least on C^{p+k}. There are (n-p+1) types of p-currents. In M^4, the 2-currents are (C^2, ω^0), (C^3, ω^1) and (C^4, ω^2). Or, the following three currents contain a 2-form ω^2: (C^2, ω^2), (C^3, ω^2) and (C^4, ω^2).

The <u>adjoint</u> ω^* of a form ω^p is a form of degree (n-p) which in local coordinates is given by (M Riemannian)

$$\omega^* = A^*_{j_1 \ldots j_{n-p}} dx^{j_1} \ldots dx^{j_{n-p}},$$

where

$$A^*_{i_{p+1} \ldots i_n} = \epsilon_{i_1 \ldots i_n} A^{i_1 \ldots i_p}$$

$$\epsilon_{i_1 \ldots i_n} = \pm\sqrt{g}\, \delta^{1 \ldots n}_{i_1 \ldots i_n}$$

$$g = \det(g_{ij}), \quad \pm = \text{orientation of } M, \quad \delta^{1 \ldots n}_{i_1 \ldots i_n} = (-1)^p, \quad p = \text{permutation}$$

$$A_{i_1 \ldots i_p} = g_{i_1 k_1} \ldots g_{i_p k_p} A^{k_1 \ldots k_p}, \quad g^{ik} g_{kj} = \delta^i_j .$$

<u>Rules</u>:

$$1^* = \epsilon_{1 \ldots n} dx^1 \ldots dx^n = \sqrt{g}\, dx^1 \ldots dx^n = dv^n$$

$$(\lambda_1 \omega_1 + \lambda_2 \omega)^* = \lambda_1 \omega_1^* + \lambda_2 \omega_2^*$$

$$(\omega^*)^* = (-1)^{pn+1} \omega$$

$$\omega_1 \omega_2^* = \omega_2 \omega_1^*$$

$$\omega \omega^* = F\, 1^*,$$

where F is a positive definite quadratic form of the coefficients of ω.

The <u>scalar product</u> of two forms ω_1^p, ω_2^p,

$$(\omega_1, \omega_2) = \int_{M^n} \omega_1^p \omega_2^{p*} = (\omega_2, \omega_1)$$

is commutative and distributive. $(\omega, \omega) = 0 \Rightarrow \omega = 0$, and $(\omega_1^{p_1}, \omega_2^{p_2}) = 0$, $p_1 \neq p_2$, by definition.

<u>Definition</u>:

$$\delta\omega = (d\omega^*)^* = (*d^*)\omega .$$

The degree of $\delta\omega^p$ is (p-1). A form ω^p is <u>co-closed</u> if $\delta\omega^p = 0$; it is <u>coexact</u> if $\omega^p = \delta\omega^{p+1}$.

Rules:

$$\delta\delta\omega = 0$$

$$\int_{M^n} d(\omega_1\omega_2^*) = 0, \qquad \text{for closed form } \omega_1^p,\ \omega_2^{p+1}.$$

$$(d\omega_1, \omega_2) = (-1)^{pn+1}(\omega_1, \delta\omega_2).$$

Definition: Laplace-Beltrami operator Δ

$$\Delta\omega^p \equiv (-1)^{np}\delta d\omega^p + (-1)^{np+n} d\delta\omega^p.$$

A form ω^p is harmonic if it is of class C^2 and $\Delta\omega^p = 0$.

Every p-form ω^p of class C^2 can be uniquely decomposed into the sum $\omega^p = \omega_1 + \omega_2 + \omega_3$ of exact ($\omega_1 = d\alpha$), co-exact ($\omega_2 = \delta\beta$) and harmonic forms ($d\omega_3 = 0$, $\delta\omega_3 = 0$), respectively, which are mutually orthogonal.

There exists one and only one harmonic p-form having arbitrarily given fundamental periods (Theorem of Hodge). Hence, the number of linearly independent harmonic forms of degree p is equal to the pth Betti number B_p. Furthermore, if $\alpha = d\omega$, then among all such ω there exists a unique one for which ω^* is exact ($\omega^* = d^*\beta$) and which makes $(\omega, \omega) = \int_M \omega\omega^*$ a minimum.

Remark: When applying certain theorems valid for closed manifolds to the space-time manifold M^4, we assume that physical quantities vanish at infinity sufficiently rapidly that compactification at infinity has no effect to physics at small distances (in the distribution sense). In fact, it is sometimes necessary to compactify space in order that physics makes sense. For example, the Hamiltonian of a monopole with a singular potential going to infinity is gauge invariant and well-defined only in a one-point compactified space, the point at infinity being the opposite monopole.

III. MAXWELL-LORENTZ AND MAXWELL-DIRAC FIELDS

The standard Maxwell-Lorentz field F^M on a manifold M is a closed 2-form,

$$dF^M = 0, \tag{1}$$

satisfying

$$\delta F^M = J, \tag{2}$$

where J is the 1-form representing the electric current. (Magnetic current will be introduced later.)

If the current is zero ("free field"), $F \equiv F^o$ is both closed and co-closed, hence harmonic. According to the Hodge's theorem, the number of linearly independent free Maxwell fields on M is equal to the second Betti number B_2 of M, and each independent electromagnetic field is uniquely determined if the B_2 periods,

$$\pi_i = \int_{C_i^2} F^o, \qquad i = 1, 2, \ldots B_2, \tag{3}$$

are prescribed. Here C_i^2 are the B_2 fundamental linearly independent 2-cycles.

Thus, on a manifold M, even a "free field" has new fundamental numbers π_i associated with it. In contrast to the Minkowski space for which all π_i vanish, these fundamental period π_i will have physical significance. So far there are no restrictions on the spectra of the periods π_i.

If all π_i of F^M in M vanish, then according to de Rham's first theorem, F^M is exact. (The converse is, of course, also true.) Then the potential A can be introduced by

$$F = dA. \tag{4}$$

Here A is not unique; an exact 1-form can be added to it without changing F. If all periods of F^* vanish, F^* is exact, hence a co-potential β can be introduced such that $F^* = d^*\beta$, or

$$F = \delta\beta. \tag{5}$$

The second more general field introduced in physics is the so-called Maxwell-Dirac field F^D satisfying

$$\delta F^D = J \quad \text{and} \quad dF^D = k^*, \tag{6}$$

where J is the electric current 1-form as before, but k^* is the 3-form dual to the magnetic 1-current k. [The second equation can also be written as $\delta(F^{D*}) = 2k$.] The field F^D is neither closed nor exact.

The quantities J and k are phenomenological quantities introduced to represent matter. The equations (1), (2) or (6) have to be supplemented by additional equations of motion of matter.

IV. DUALITY

We propose now to express J and k in terms of the topological quantities π_i given in Eq. (3). This will have the fundamental merit that everything will be of electromagnetic or geometrical nature: We do not have to supplement our equations by phenomenological matter equations and their coupling to the field.

A duality principle was introduced[1b)] to eliminate k. We now give a general version of this principle including J.

Let us first reduce the Maxwell-Dirac field F_D, Eq.(6) to a standard Maxwell Lorentz field F^M, Eqs. (1)-(2), and then the field F^M to a harmonic field F^o.

Maxwell-Dirac Field

Formulation 1: F is closed, $dF = 0$, but Manifold M is such that the cycles are not exact, hence periods do not vanish:

$$\pi_i = \int_{C_i^2} F \neq 0.$$

Example: $M = M^4 - \{\vec{x} = 0\}, \quad C_i^2 = \{t = 0, \vec{x} = R\} = S^2$

$$\pi = \int_{S^2} F = \int_{S^2} \vec{B}\cdot df = 4\pi g = \text{total magnetic charge inside } S^2.$$

Formulation 2: F is not closed but almost closed, i.e., deviates from a closed form by a de Rham-"current"

$$F^D = F^M + \Lambda, \quad dF^M = 0,$$

but M is such that every cycle C_i^2 is exact. Then

$$\pi_i = \int_{C_i^2} F^D = \int_{\partial C_i^3} F^D = \int_{C_i^3} dF^D = \int_{C_i^3} d\Lambda \neq 0.$$

Example: $M = M^4, \quad C_i^2 = S^2 = \partial v^3, \quad d\Lambda = k^*$

$$\pi_i = \int_{v^3} k^* = \int_{v^3} k^o d^3x = 4\pi g.$$

The duality consists in the fact that the magnetic flux is calculated in the first formulation purely geometrically as a surface integral over a non-exact cycle, and in the second formulation as a volume integral of an assumed physical matter (magnetic charge) density.

The remaining equations are: $\delta F^M = J_{tot}$, in the first formulation, and, in the second formulation,

$$dF^D = d\Lambda \equiv k^*$$

$$\delta F^D = \delta F^M + \delta\Lambda = J_{tot} + \delta\Lambda = J$$

with $J_{tot} = J - \delta\Lambda$. The boundary $d\Lambda$ of the "matter-term" represents the magnetic current, and the co-boundary $-\delta\Lambda$ represents a new current J_Λ associated with de Rham form Λ.

Using the de Rham-homomorphism between cycles C^p and forms ω^{n-p}, the two-formulations can be mapped into each other: C_i^2 in the first formulation and the form Λ in the second formulation, which are special instances of de Rham currents. The physical interpretation of the various forms of Λ has been discussed in References 2 and 3.

Maxwell-Lorentz Field (electric currents only)

Formulation 1: F is harmonic $= F^{O}$, $dF^{O} = 0$, $\delta F^{O} = 0$, but cycles are not all exact. For example:

$$\pi_i = \int_{C_i^2} {}^*F^{O} = 4\pi Q \neq 0.$$

Formulation 2: F is closed, but not exact: $F^{M} = F^{O} + \Lambda_2$, such that $d\Lambda_2 = 0$, $\delta\Lambda_2 = J$. And all cycles are exact. Then

$$\pi_i = \int_{C_i^2} {}^*F^{M} = \int_{\partial C_i^3} {}^*F^{M} = \int_{C_i^3} d{}^*F^{M} = \int_{C_i^3} d{}^*\Lambda_2$$

$$= \int_{C_i^3} {}^*J = 4\pi Q.$$

Again, the period (total charge) is once calculated purely geometrically as a surface integral, and once as a volume integral over the physical matter (charge) density.

V. VARIATIONAL PRINCIPLES AND MINIMAL COUPLING

Consider first the equation $\delta F = J$, or $d^*F = {}^*J$, only, i.e., J^* is closed and all periods of J^* vanish. Then, according to the minimum principle on harmonic forms, there exists among all such ${}^*F$'s a unique one F^{M} for which F^{M} is exact, i.e., all periods of F^{M} vanish, which makes $(F^*,F^*) = \int_M {}^*FF$ a minimum. Nothing is said about the periods of F^*; they are in general different from zero.

In order to construct the minimum, we introduce the difference $F^{O} = F^{M} - dA$ from exactness such that $\delta F^{O} = 0$, or $\delta F^{M} = \delta dA = J$. Then

$$(F^{O},F^{O}) = (F^{M},F^{M}) - 2(F^{M},dA) + (dA,dA)$$

$$= (F^{M},F^{M}) + 2(\delta F^{M},A) - (\delta dA,A)$$

$$= (F^{M},F^{M}) + (J,A).$$

The equation

$$(F^{M},F^{M}) + (J,A) = \text{minimum}$$

is the standard variational principle of electrodynamics with so-called minimal coupling. Indeed, starting from this equation, $F^{M} = dA$, $\delta F^{M} = J$ makes the above minimum zero. In fact, the minimum of (F^{O},F^{O}) is zero, which means $F^{O} = 0$. Also, because $dF^{o} = dF^{M}$, at the minimum, F^{o} is harmonic and, since all its periods vanish, it must itself vanish.

There is, of course, the symmetric case, if we start from $dF = k^{*}$. Then among all such F's there is a unique one for which F^{*} is exact and which makes (F,F) a minimum. Minimizing ${}^{*}F^{o} = {}^{*}F - d^{*}\beta$ or $F^{o} = F - \delta\beta$, we have the minimal coupling with a co-potential

$$(F,F) + (\beta, k^{*}) = \text{minimum}.$$

Next we are looking at a particular ${}^{*}F$, namely, F^{D}, such that F^{D} is not exact but of the form $F^{D} = F^{M} + \Lambda$. Then

$$dF^{D} = d\Lambda, \quad \delta F^{D} = \delta F^{M} + \delta\Lambda = J_{tot} + \delta\Lambda.$$

Hence we can interpret $d\Lambda = k^{*}$, $\delta\Lambda = -J_{\Lambda}$, $J_{tot} = J + J_{\Lambda}$ and obtain the Maxwell-Dirac equations.

The variational principle becomes

$$(F^{D}, F^{D}) - 2(F^{D}, \Lambda) + (\Lambda, \Lambda) + (A, J) + (A, J_{\Lambda}) = \text{minimum}$$

or

$$(F^{M}, F^{M}) + (J_{tot}, A) = \text{minimum}.$$

For the second form with potentials and co-potentials, we set

$$F^{D} = d\alpha + \delta\beta$$

$$dF^{D} = d\delta\beta - k^{*}, \quad \delta F^{D} = \delta d\alpha = J.$$

We now minimize again the difference $F^{o} = F^{D} - d\alpha - \delta\beta$ and obtain

$$(F^{D}, F^{D}) + (\alpha, J) + (\beta, k^{*}) = \text{minimum}$$

which is again a minimal coupling principle with two potentials.

Variational Principle for Potentials

Let $F = dA$. Then among all such A's there exists a unique one, $\overset{o}{A}$, for which ${}^{*}A$ is exact: ${}^{*}A = d^{*}\omega$, or $\overset{o}{A} = -\delta\omega$ (true only for $n=4$), and which makes

$$(A, A) = \int_{M} A^{*}A = \text{minimum}.$$

Period of A need not vanish: $\int_{C^{1}} A \neq 0$.

Generalized "Gauge" Transformations

In a general manifold, the equation $F = dA$ is invariant under $A' = A + B$ with $dB = 0$. But B need not be exact, $B \neq d\phi$. Thus, the periods of A will be different:

$$\oint A' = \oint A + \oint B.$$

(The second term would be zero if B is exact.)

REFERENCES

1. A. O. Barut, in Quantum Theory and Structure of Time and Space (edited by L. Castell et al.), C. Hanser Verlag, Vol. I (1975), Vol. II (1977).
2. A. O. Barut, De Rham currents, extended singularities of fields and magnetic monopoles, Reports Math. Phys. 11, 415-422 (1977).
3. A. O. Barut, Charge quantization condition with N strings. A new quantum number of charge-monopole systems, Lett. Math. Phys. 1, 367-370 (1977).
4. P. Bidal and G. de Rham, Les formes différentielles harmoniques, Comm. math. Helvetici, 19, 1-49 (1946).
5. W. V. D. Hodge, The Theory and Applications of Harmonic Integrals (Cambridge Univ. Press, 1946).

GRAVITY IS THE GAUGE THEORY OF THE PARALLEL - TRANSPORT MODIFICATION OF THE POINCARE GROUP

Yuval Ne'eman *
Tel-Aviv University, Tel-Aviv, Israel

Abstract

We prove that only the Dynamically - Restricted Anholonomized General Coordinate Transformation Group reproduces Einstein's theory of Gravitation directly when gauged. This amounts to a Modified Poincaré group where translations are replaced by Parallel transport. We also explain the role of GL(4R) and explore the Modified Affine Group. Using the Ogievetsky theorem, we present several No-Go theorems restricting the joint application of Conformal and Affine Symmetries.

1. Introduction: Gauge Theories

The first local gauge invariance principle (LGIP, or just "gauge") to be suggested [Weyl, 1919] dealt with dilations, and was introduced as an addition to Einstein's Gravity. H. Weyl was looking for a geometrical derivation of Electromagratism, which would thereby also "unify" it with Gravitation. His first theory invoked dilation invariance, and failed at the time since macroscopic evidence appeared to be clearly in disagreement with such a postulate. This particular theory has recently been revived at the quantum level as a gauge invariance with "spontaneous breakdown" [Englert et. al., 1975]. The geometrical derivation itself was revived after the advent of quantum mechanics as a U(1) gauge [Weyl, 1929] i.e. a locally dependent phase for complex charged matter fields instead of scale invariance. We would now render it as a Principal Bundle B $(\mathcal{M}^4$, G) with Minkowski Space-Time $\mathcal{M}^4$ as base space, and G = U(1) as structure group. The gauge transformations are given by the set of Bundle automorphisms whose action on $\mathcal{M}^4$ is the identity, i.e. leaving a point $x \in \mathcal{M}^4$ invariant. They thus act only in the fiber above that point, and can be written as $g(x)$, $g \in G$. They belong to the "stability group" of translations in $\mathcal{M}^4$.

This abstract "internal" gauge invariance was H. Weyl's second definition, and it won wide acceptance. Three decades later, it served as a model for the (G = SU(2)) local Non-Abelian internal gauge of C.N. Yang and R.L. Mills [1954; see also Shaw 1954]. The method was further generalized [Ne'eman 1961, Gell-Mann 1962, Salam and

* Partially supported by the U.S. - Israel Binational Science Foundation.

Ward 1961] to SU(3) and in principle to any Semi-simple group [Gell-Mann and Glashow 1961, Ionides 1961]. In recent years, this SU(3) universal (and therefore gauge-like) coupling which is indeed observed in the coupling of hadrons to massive vector-mesons (the ρ, ω, ϕ, ψ, Υ, with $J^{PC} = 1^{--}$) has to be regarded as a pole-dominance approximation for phenomenological vector fields [Gell-Mann, 1962]. On the other hand, an $SU(2)_{Left} \times U(1)$ LGIP involving a subgroup of that SU(3) but acting on leptons and on SU(3) invariant quarks as well is favored as a Weak and Electromagnetic Unified Gauge [Weinberg 1967, Salam 1968] (though other groups are still possible), and an $SU(3)_{color}$ LGIP is believed to represent the quark-glueing [Nambu 1965, Fritzsch and Gell-Mann 1972, Weinberg 1973] (and confining?) part of the Strong Interactions. Those applications have become serious candidate dynamical theories since the achievement of G. 't Hooft and M. Veltman ['t Hooft 1971, 't Hooft and Veltman 1972, Lee and Zinn-Justin 1972] in completing the renormalization of the Yang-Mills interaction [Feynmann 1963; De Witt 1964, 1967; Faddeev and Popov 1967; Fradkin and Tyutin 1970; Veltman 1970], including the case of "spontaneous breakdown" [Higgs 1964a, 1964b, 1966; Englert and Brout 1964; Guralnik et al 1964; Kibble 1967] of the local gauge coupled with a Goldstone- Nambu realization of the global symmetry [Goldstone 1961, Nambu and Jona - Lasinio 1961]. For the Strong Interactions, renormalization has also led to the discovery of Asymptotic Freedom [Politzer 1973, Gross and Wilczek 1973] which seems particularly fitting for short range quark interactions, and appears to support the $SU(3)_{color}$ gauge idea. We review the highlights of a Yang-Mills type gauge.

The dynamical variables in a $B(\mathcal{M}^4, G)$ gauge theory may include matter fields (quarks) $q^a(x)$ which are generally represented as sections of a vector bundle E associated to B,

$$E = B \times_G \Lambda(G)$$

where $\Lambda(G)$ is the (3 x 3 for quarks) appropriate representation of G on q^a:

$$(b(g_o(x)),\ q(x) \circ g = (b(g_o(x)g(x)),\quad \Lambda(g^{-1})q(x))$$

The covariant derivative in E involves matrix connections (potentials)

$$\rho(x) = \rho_\mu{}^A(x)\, X_A\, dx^\mu \tag{1.1}$$

where X_A is the Lie-algebra of G in the $\Lambda(g)$ representation. The covariant derivative in E is then

$$(D\, q)^a = d\, q^a - (\rho)^a{}_b\, q^b \tag{1.2}$$

and the dynamical theory is derived by the replacement

$$(\delta_b{}^a\ \partial_\mu) \rightarrow (D_\mu)_b{}^a \tag{1.3}$$

known as a "minimal" or "universal" coupling. Indeed, with a free Lagrangian

$$\mathcal{L}_o = - \bar{q}_a\ (\gamma^M\ \partial_\mu + m)\ q^a \tag{1.4}$$

the unwanted contribution due to $\partial_\mu\ g \neq 0$

$$- \bar{q}_a\ \gamma^\mu\ (g^{-1}\ \partial_\mu\ g)_b{}^a\ q^b$$

is cancelled by

$$\rho \rightarrow (g\ \rho\ g^{-1} + g^{-1}\ d\ g) \tag{1.5}$$

For an infinitesimal transformation $(\Lambda(g))_b{}^a = \delta^a_b + (i\alpha^A\ X_A)_b{}^a$, the unwanted $\partial_\mu \alpha^A$ term arises in

$$\partial_\mu \alpha^A \frac{\partial}{\partial(\partial_\mu q^a)} (X_A)_b{}^a\ q^b = \partial_\mu \alpha^A\ J_A{}^\mu$$

where $J_A{}^\mu$ is the Noether current, satisfying a covariant conservation law

$$D\ J_A = 0 \tag{1.6}$$

The curvature

$$R = (d\ \rho - \rho \wedge \rho) = (d\rho - \frac{1}{2}\ [\rho,\ \rho]) = (R^A\ X_A) \tag{1.7}$$

similarly satisfies the Bianchi identity

$$(DR) = 0 \tag{1.8}$$

The equations of motion are

$$(D\ {*R}\ ) = {*J}^A \tag{1.9}$$

where * stands for the duals

$$*R_{\mu\nu}{}^{A} = \frac{1}{2}\,\varepsilon_{\sigma\tau\mu\nu}\,R_{\sigma\tau}{}^{A} \tag{1.10a}$$

$$*J_{\mu\sigma\tau}{}^{A} = \frac{1}{6}\,\varepsilon_{\nu\mu\sigma\tau}\,J_{\nu}{}^{A} \tag{1.10b}$$

The equations of motion can be used to turn (1.6) into a non-G-covariant conservation law for a new current

$$d\,J'_A = 0 \tag{1.11}$$

where J'_A will include contributions from the ρ^A potentials themselves. This will be more problematic in Gravitation.

Connections, covariant derivatives and curvatures can also be introduced in B itself, where they will regulate their own gauge invariance (no "sources"). The matrices X_A will now belong to the adjoint representations,

$$(X_A)_B{}^C = -\,C_{AB}{}^{C}$$

The definitions are

$$R^A = d\,\rho^A - \tfrac{1}{2}\,\rho^B \wedge \rho^C\, C_{BC}{}^{A} = \tfrac{1}{2}\rho^B \wedge \rho^C\, R_{BC}{}^{A} \tag{1.12}$$

$$(D\rho)^A = d\,\rho^A - \rho^B \wedge \rho^C\, C_{BC}{}^{A} \tag{1.13}$$

$$(DR)^A = 0 \tag{1.14}$$

using contractions with vector-fields D_A,

$$D_A = \Delta_A{}^{\mu}\,\partial_\mu \qquad\qquad \rho^A\,(\Delta_B) = \delta^A{}_B \tag{1.15}$$

and with the resulting commutator (from double contraction of (1.12)),

$$[D_A,\, D_B] = (C^C{}_{AB} + R^C{}_{AB})\, D_C \tag{1.16}$$

where $C^C{}_{AB} = 0$ in $\mathcal{M}^4$ but not in "Superspace" $\mathfrak{R}^{4/4N}$ as we shall later see.

Notice that in the adjoint representation, (1.13) can also be written as $D\rho = d\rho - [\rho,\, \rho]$ and is not equivalent to R. This is due to the antisymmetry of $C_{BC}{}^{A}$ or $(-X_B)_C{}^{A}$ in the (B,C) indices, as against $(\Lambda_B)_b{}^a$ in (1.2) for $(\rho^B\,\lambda_B)_b{}^a\,q^b$ where there is no such link between B and b. The antisymmetry implies a factor 2 in contracting with $(\tfrac{1}{2}\,dx^\mu \wedge dx^\mu)$ as against the curl $d\rho$.

2. The First Step: Gauging the (intrinsic) Lorentz Group

We first return to Gravity when R. Utiyama [1956] attempts to derive that theory from a Gauge Principle. Since not much was known at the time about the renormalizability of Yang-Mills LGIP theories, this was in the main an aesthetic urge. Utiyama gauged the (homogeneous) Lorentz group G = SL(2,C) = :L using the equivalent of connection one-forms

$$\rho^{ij} = \rho_\mu{}^{ij}\, dx^\mu \quad (i,j = 0,1\ldots 3 \text{ in a local frame; } \mu = 0,1\ldots 3 \text{ holonomic} \tag{2.1}$$

However, to reproduce Einstein's theory it appeared that he had to introduce a-priori curvilinear coordinates, and a set of 16 "parameters" $\Delta_k{}^\mu(x)$. These were initially treated as given functions of x and later became field variables, to be identified with orthonormal vector fields Δ_k reciprocal to a vierbein frame,

$$\rho^i\,(\Delta_k) = \delta^i_k \tag{2.2a}$$

the $\rho_\mu{}^i$ thus arising as vierbein fields, with (η_{ij} in the Minkowski metric)

$$\eta_{ij}\,\rho_\mu{}^i\,\rho_\nu{}^j = g_{\mu\nu}(x) \tag{2.2b}$$

Still, the relationship of $\rho_\mu{}^{ij}$, to the Christoffel connection $\Gamma_{\mu\nu}{}^\lambda$ was incomplete, since the formula he derived was forced by an arbitrary assumption to select only ($\mu\ \nu$) symmetric contributions to $\Gamma_{\mu\nu}{}^\lambda$. As we shall see, this role of the Connection ("Affinity" in holonomic - "world tensor" - language) as a Gauge Potential has since been perfected. However, it contrasted sharply with the physical intuition of workers in Gravitation [e.g. Thirring 1977] who regard the metric (or vierbein) as the Gravitational Potential, and consider the Connection as the analogue to the Field Strength in Electrodynamics.

Sciama [1962] and Kibble [1961] continued Utiyama's project. Although they were aiming at a full Poincaré gauge (G = ISL(2,C):=P), their main achievement consisted in clarifying the Lorentz gauge. They showed that this consisted only in the stability group over $\mathcal{M}^4$, i.e. the "internal" action of H = SL(2,C) = :L, which we generally describe as the Spin of the Matter fields (though it does not include contributions to physical spin due to the holonomic - "Greek" - indices of gauge fields, curvatures etc., i.e. in particular, the photon or the Yang-Mills' fields own spins). This "Latin" or "anholonomic" spin $s_{ij}{}^\mu$ gives rise to a new interaction term, in which it is minimally coupled to the connection $\rho_\mu{}^{ij}$.

$$A_S = \int d^\mu x \quad e \; \rho_\mu{}^{ij} \; S_{ij}{}^\mu \;=\; \int \rho^{\,ij} \wedge {}^*S_{ij} \tag{2.3}$$

where ${}^*S_{ij}$ is a dual three-form

$${}^*S_{ij} = \frac{1}{6}\,\varepsilon_{\tau\nu\rho\sigma} \quad S_{ij}{}^{\tau} \; dx^\nu \wedge dx^\rho \wedge dx^\sigma \tag{2.4}$$

Indeed, this arises when we perform the replacement

$$\delta_i{}^\mu \, \partial_\mu \longmapsto \Delta_i{}^\mu \, D_\mu^{(H)} \quad , \qquad D_\mu^{(H)} = \partial_\mu + \tfrac{1}{2}\rho_\mu{}^{ij} \; f_{ij} \tag{2.5}$$

where $\frac{1}{2}f_{ij}$ is a representation of the Lorentz generators, appropriate for action on the ψ matter field in $\mathcal{L}_M(\psi,\, \partial_\mu\, \psi)$

$$\frac{\partial \mathcal{L}_M}{\partial \rho_\mu{}^{ij}} = - \frac{\partial \mathcal{L}}{\partial(\partial_\mu \psi)}\, f_{ij} \;\; \psi =: e \;\; s_{ij}{}^\mu \tag{2.6}$$

The factor $e = \det \rho_\mu{}^i$ arises in the replacement $d^u x \longmapsto e \; d^u x$ of the matter action measure. However, the variation of the action by $\delta \rho^{ij}$ also receives a contribution from the Einstein free action ($\lambda := 8\pi c^{-3} G$, G being Newton's constant; $[K] = [L^2]$ in "natural" units)

$$A = \frac{1}{8} \int \frac{1}{k}\, R^{ij} \wedge \rho^k \wedge \rho^\ell \;\; \varepsilon_{ijk\ell} = \frac{1}{8k} \int R^{ij} \wedge \zeta_{ij} \tag{2.7}$$

so that one has a new equation of motion (besides Einstein's) involving three-forms

$$R^k \wedge \rho^t \;\; \varepsilon_{ijk\ell} \;=\; - \frac{1}{2}\lambda \;\, {}^*s_{ij} \tag{2.8}$$

In these equations, R^{ij} and R^i are the curvature two-forms, with

$$R^{ij} = dx^\mu \wedge dx^\nu \; R_{\mu\nu}{}^{\cdot\cdot ij} = d\rho^{ij} + \rho^i{}_k \wedge \rho^k{}_\ell = \tfrac{1}{2}\rho^k \wedge \rho^\iota \; R_{k\iota}{}^{\cdot\cdot \; ij} \tag{2.9}$$

$$R^i = dx^\mu \wedge dx^\nu \; R_{\mu\nu}{}^{\cdot\cdot i} = d\rho^i + \rho^i{}_k \wedge \rho^k = D^{(L)}\rho^i = \tfrac{1}{2}\rho^k \wedge \rho^\iota \; R_{k\iota}{}^{\;i} \tag{2.10}$$

In conventional nomenclature, R^{ij} is the Riemannian curvature and R^i the Cartan "torsion". For empty space, (2.8) becomes $R^i = 0$ and solving (2.10) for $\rho^i{}_k$ (in $D^{(L)}\,\rho^i$) then produces the Christoffel symbol formula. However, when (Latin) spinning matter is present, solving for $\rho^i{}_k$ will produce in addition an antisymmetric contribution to the Christoffel connection.

Contracting R^k with two vector-fields we find,

$$R_{ij}^{::\ k} = \Delta_i^{\ \mu} \Delta_j^{\ \nu} R_{\mu\nu}^{\ \ k} = (\Delta_i, \Delta_j, R^k) \tag{2.11}$$

$$= (\Delta_i, \Delta_j, (d\rho^k + \rho^k_{\ \iota} \wedge \rho^\iota))$$

$$= \Omega_{ij}^{::} + \tfrac{1}{2}(\rho_i^{\ k}{}_\iota \delta_j^{\ \iota} - \rho_j^{\ k}{}_\iota \delta_i^{\ \iota})$$

Thus,

$$R_{ij}^{\ \ k} = \Omega_{ij}^{::\ k} + \tfrac{1}{2}(\rho_i^{\ k}{}_j - \rho_j^{\ k}{}_i) \tag{2.12}$$

The doubly contracted exterior derivative $\Omega_{ij}^{::\ k}$ has been called "the object of Anholonomity", [Schouten 1954; Hehl et. al 1976a]. Using the Minkowski metric (in the tangent space) we can lower the ^{k} index, and remembering that the antisymmetry of the Lorentz generators imposes

$$\rho_{i\,k\,j} = -\rho_{i\,j\,k} \tag{2.13}$$

we can extract $\rho_{i\,j\,k}$,

$$\rho_{i\,j\,k} =: \Omega_{i\,j\,k} - \Omega_{j\,k\,i} + \Omega_{k\,i\,j} + R_{j\,k\,i} - R_{k\,i\,j} - R_{i\,j\,k} \tag{2.14}$$

The last three terms, making up together the "Contortion tensor" $K_{i\,j\,k}$, vanish for $R^k = 0$ and represent the contribution of ("Latin") spinning matter when present.

$$K_{i\,j\,k} = R_{j\,k\,i} - R_{k\,i\,j} - R_{i\,j\,k}$$

Inserting the last expression for R^i in (2.10) into (2.8), and replacing the holonomic index in $S_{ij}^{\ \ \mu}$

$$S_{ij}^{::\ k} = \rho_\mu^{\bullet\ k} S_{ij}^{\ \ \mu} \tag{2.15}$$

we get the equation of motion,

$$R_{i\,j}^{\ \ k} - \delta_i^k R_{\iota\,j}^{\ \ \iota} - \delta_j^k R_{i\,\iota}^{\ \ \iota} =: T_{i\,j}^{\ \ k} = k\, S_{i\,j}^{\ \ k} \tag{2.16}$$

$T_{ij}^{::\ k}$ is sometimes named the "Modified Torsion". We can also contract the upper (naming) index of the torsion tensor $R_{\mu\nu}^{\ \ k}$ in (2.11)

$$R_{\mu\nu}{}^{\rho} = \Delta_k{}^{\rho} R_{\mu\nu}{}^{k} = \Delta_k{}^{\rho} \rho_\mu{}^{i} \rho_\nu{}^{j} R_{ij}{}^{k}$$

If we now insert (2.12) we find

$$R_{\mu\nu}{}^{\rho} = - \tfrac{1}{2} (\rho_{\mu\nu}{}^{\rho} - \rho_{\nu\mu}{}^{\rho}) = - \tfrac{1}{2} (K_{\mu\nu}{}^{\rho} - K_{\nu\mu}{}^{\rho}) \qquad (2.17)$$

Holonomically, torsion thus corresponds to the antisymmetric part of the connection. Note that these are not the indices which are antisymmetric in the anholonomic connection due to the Lorentz gauge. Returning to the equation of motion we derived, we note that Eq. (2.16) being algebraic (due to 2.17) rather than differential (due to the particular choice of the Einstein Lagrangian which is linear in the canonical momenta), the connection potential $\rho_\mu{}^{ij}$ does not propagate. Instead, like a gauge connection in a current-field identity [Lee et al 1967] it is replaced by the spin-current itself, so that (2.3) becomes a spin-spin term with very weak coupling k^2, a contact term. Sciama and Kibble thus rediscovered Cartan's modification [1922-25] of Einstein's Relativity. At the same time, this can be regarded as a "first order" or Palatini [1969] formalism for that theory (independent variations for ρ^i and ρ^{ij}). It then differs from it by that $k^2 s^i s_i$ term only [Weyl 1950]. This theory, further analyzed by Hehl [1970] and by Trautman [1972] is known as the Einstein - Cartan - Sciama - Kibble theory (or U_4 theory), and is thus indeed derivable <u>in its spin-torsion parts</u> from a Lorentz gauge.

3. Difficulties in Gauging Translations; Pseudo-Invariance

The attempt to reproduce Gravity had of course to come to grips with the main part of the theory - the universal coupling of the Energy-Momentum tensor-current to the gravitational potential (i.e. to the metric $g_{\mu\nu}$ or in a vierbein formalism, to the $\rho_\mu{}^i$ of (2.2)). Indeed, varying ρ^i in (2.7) yields Einstein's equation for empty space,

$$R^{ij} \wedge \rho^k \varepsilon_{ijk\ell} = 0 \tag{3.1}$$

which becomes, in holonomic language, after some manipulations,

$$R^\mu{}_\nu - \tfrac{1}{2} R\, \delta^\mu{}_\nu = 0 \tag{3.2}$$

In the presence of matter we have

$$R^{ij} \wedge \rho^k \varepsilon_{ijk\ell} = k\, {*t_\ell} \tag{3.3}$$

where $*t_\ell$ is the energy-momentum current 3-form

$$*t_\ell = \frac{1}{6} \varepsilon_{ijkm}\, \rho^j \wedge \rho^k \wedge \rho^m\, t^i{}_\ell \tag{3.4}$$

for the density $t^i{}_\ell$.

The Sciama-Kibble approach fell short of attaining this goal by a gauge principle. Kibble noted that the Lorentz-gauge invariance having been ensured by the covariant derivative (2.5), the remaining unwanted gradient term corresponding to translations is a homogeneous term, in contradistinction to the Yang-Mills case,

$$\delta(D_\mu{}^{(L)}\psi) = \tfrac{1}{2}\varepsilon^{ij} f_{ij} (D_\mu{}^{(L)}\psi) - \partial_\mu \xi^\nu D_\nu{}^{(L)}\psi$$

its removal is achieved by a multiplicative application rather than by the usual additive construction. Indeed, taking

$$D_k = \Delta_k{}^\mu D_\mu{}^{(L)} \quad , \qquad \delta\Delta_k{}^\mu = - \varepsilon^i{}_k \Delta_i{}^\mu + \partial_\nu \xi^\mu \Delta_k{}^\nu \tag{3.5}$$

yields

$$\delta D_k\psi = \tfrac{1}{2} \varepsilon^{ij} f_{ij} D_k\psi - \varepsilon^i{}_k D_i\psi \tag{3.6}$$

Kibble thus attributed to (the vector field) $\Delta_k{}^\mu$ the role of a translation gauge field, with ξ^μ as the translation parameter. This fitted an analysis of the action of the Poincaré group on fields, in which the intrinsic Lorentz action was given anholomic indices, but where all the rest (both orbital angular momentum action and translations) was incorporated in the General Coordinate Transformation and represented holonomically,

$$\delta\psi = \tfrac{1}{2}\,\varepsilon^{ij}\, f_{ij}\,\psi \qquad\qquad \delta x^\mu = \xi^\mu = \varepsilon^\mu{}_\nu x^\nu + \varepsilon^\mu \tag{3.7}$$

with in addition

$$\delta_o\,\psi = -\xi^\mu\,\partial_\mu\psi + \tfrac{1}{2}\,\varepsilon^{ij}\,f_{ij}\,\psi \tag{3.8}$$

The separation (3.7) in which the "orbital" action of ε^{ij} appeared as $\varepsilon^{\mu\nu}$ and was incorporated in ξ^μ corresponds indeed to the Fiber Bundle picture, in which the gauged group is the stability subgroup of P, the Poincaré group. However, the assignment of ξ^μ to coordinate transformations precluded any form of gauging for trans lations. A variation $\delta_o\psi$ had been introduced so as to reproduce

$$\delta_o\psi = \psi'(x) - \psi(x) = S\psi - \xi^\mu\,\partial_\mu\psi - \psi(x) \tag{3.9a}$$

since

$$S\psi(x) = \psi'(x') \tag{3.9b}$$

i.e. a resetting of the value of the argument to its original value x, after the action of a Lorentz transformation, in view of the latter's simultaneous effect on the coordinates (its orbital action).

Sometime using $\delta_o\psi$ in his interpretation, Kibble remarked that one could also regard (3.5) as involving a translation-gauge field $(\Delta_k{}^\mu - \delta_k{}^\mu)$

$$D_k = \delta_k{}^\mu\,\partial_\mu + (\Delta_k{}^\mu - \delta_k{}^\mu)\partial_\mu + \delta_k^\mu\,\tfrac{1}{2}\,\rho_\mu{}^{ij}\,f_{ij} \tag{3.10}$$

where the second term could correspond to ∂_μ as the $\delta_o\psi$ algebraic generator in (3.8) multiplied by its gauge field. A recent attempt to pursue this idea [Cho 1976] appears to have failed due to the difficulty of expressing $e = \det \rho^k$ in that interpretation.

A tetrad field is defined by erecting at every point x a frame of vectors $r_x{}^i(x)$

$$\rho_\mu{}^i\,(X) := \left(\frac{\partial r_x{}^i(x)}{\partial x^\mu}\right)_{x = X} \tag{3.11}$$

and its variation under a coordinate transformation $x^\mu \to x^\mu + \xi^\mu$ is given by

$$\delta \rho_\mu{}^i (x) = - \partial_\mu \xi^\nu \rho_\nu{}^i \tag{3.12}$$

which is indeed the inverse of the ξ^μ variation of $\Delta_k{}^\mu$ in (3.5). However, one can replace the linear connection ρ^{ij} corresponding to gauging the Lorentz group, by a Cartan connection [Kobayashi 1956, 1972, Trautman 1973], in which the Bundle Structure Group G is the Affine (or Poincaré) Group. For the Poincaré group this means having in the Bundle "intrinsic" translations and altogether 10 connections. By choosing the origin in that fiber, one can make the translation-connection coincide with the frame, except that we now have an anholonomic translation gauge, with variation,

$$\delta \rho_\mu{}^i (x) = \partial_\mu \varepsilon^i + \rho_\mu{}^i{}_j \varepsilon^j - \varepsilon^i{}_j \rho_\mu{}^j = D_\mu^{(P)} \varepsilon^i \tag{3.13a}$$

or for the forms

$$\delta \rho^i = d\varepsilon^i + \rho^{ij} \wedge \varepsilon^j - \rho^j \wedge \varepsilon^{ij} = D^{(P)} \varepsilon^i \tag{3.13b}$$

Such a translation-gauge was indeed suggested by Trautman [1973] and by Petti [1976]. It yields the universal coupling of eq. (3.3) through Noether's theorem or the Bianchi identities. The difficulty is that the Einstein Lagrangian itself is not Poincaré-gauge invariant [e.g. Ne'eman and Regge 1978a]. Under the translation gauge (3.13) we find terms in $D^{(P)} \varepsilon^i$ arising from ζ_{ij} in (2.7). Integration by parts then makes the action produce a variation proportional to the torsion R^i. (The Action is of course trivially translation-invariant.)

One way out of this dilemma is to abandon the concept of invariance for a weaker "pseudo-invariance", holding only after the application of the equations of motion. This was done somewhat half-heartedly in Supergravity [Freedman et al 1976; Deser and Zumimo 1976; Freedman and van Nieuwenhuizen 1976] so emphasized by C. Teitelboim [1977], and generalized to gravity by J. Thierry-Mieg [1978]. Indeed, applying (2.8) for empty space (i.e. $R^i = 0$) after the variation makes Einstein's free Lagrangian invariant under (3.13). However this interpretation does not guarantee the possibility of exponentiation to a finite gauge, i.e. group action. In addition, $\delta \rho^{ij} = 0$ under the translation gauge, which has to be modified so as to fit $R^i = 0$, with no gauge mechanism to provide for the new $\delta \rho^{ij}$. Moreover, the interpretation fails when spinning matter is present.

4. The Parallel Transport Gauge (AGCT)

The next step in solving the mystery of the translation gauge is due to von der Heyde [1976; see also Hehl et al 1976a]. Returning to Kibble's $\delta_o\psi$ concept (eq. 3.9a - 3.9b) he noticed that with space being already "curved" due to the Lorentz gauge, the transport term had to involve parallel transport, i.e. the covariant derivative $D_\mu^{(L)}\psi$ rather than $\partial_\mu\psi$. Moreover, to preserve the Poincaré group appartenance of the translation generators, the operator D_k of (3.5) should be used, with the anholonomic ("Latin") indices covering P rather than just L. This solution thus combines the idea of 10 connections, including the vierbeins ρ^i, with that of parallel trànsport.

$$\overline{\delta}_o\ \psi^m(x) = \tfrac{1}{2}\,(\varepsilon^{ij}\ f_{ij})^m_n\ \psi^n - (\varepsilon^k\ D_k)^m_n\ \psi^n \tag{4.1}$$

We have seen that $\varepsilon^{ij} = \delta^i_\mu\ \delta^j_\nu\ \varepsilon^{\mu\nu}$, but for ε^k

$$\varepsilon^k = \varepsilon^\mu\ \rho_\mu{}^k \tag{4.2}$$

This is due to the "flatness" of the fiber (parameters $\varepsilon^{ij} = \varepsilon^{\mu\nu}$) as against the curvature induced by the Lorentz gauge in $\mathcal{M}^4$. Equation (4.1) can be interpreted as an active Lorentz transformation followed by a passive resetting of the coordinate frame to the original value of x. We can convince ourselves of the role of $-\ \varepsilon^k\ D_k$ as a translation-gauge by noting that the entire $\overline{\delta}_o\psi$ transformation amounts to a trivial action on the base space, a conclusion which would still be true in the principal bundle when taking the $\rho^A(x)$ for $\psi(x)$, except for the gauge term. Thus as the homogeneous part of an infinitesimal Poincaré transformation in the extended bundle with G = P, it should be considered as a gauge transformation. The interaction Lagrangian is still produced by the replacement

$$\delta_k{}^\mu\ \partial_\mu \longmapsto D_k \tag{4.3}$$

and the Equivalence Principle to maintained, independently of the existence of microscopic torsion. Indeed, Special-relativistic matter in a non-inertial frame is always locally equivalent to the same matter in a gravitational field [von der Heyde 1975]. Note also that in this derivation, the appearance of curvature is natural, due to our improved understanding of geometry: Utiyama and Kibble had to make a jump to curvilinear coordinates, whereas the Fiber Bundle picture tells us that curvature is nothing but the base-space effect of gauging a group in the Fiber. Indeed, even the electromagnetic U(1) or the modern SU(3) gauges induce curvature in space time (the $F_{\mu\nu}{}^a$).

We have recently generalized this approach [Ne'eman and Regge 1978a,b] , showing that the Supersymmetric ("local") transformations of Supergravity correspond to a similar parallel-transport action in Superspace with a further restriction to $\mathcal{M}^4$.

We now analyze the geometric and algebraic structure of the parallel-transport gauges.

To understand these gauges and indeed to analyze the entire problem of gauging a "non-Internal" group, i.e. a group with some action on space-time, we revert to a new manifold. Noting that in gauging P, the Fiber was L with 16 dimensions, and the base-space $\mathcal{M}^4$ had 4 dimensions, we observe that the Bundle dimensionality was 10, the same as that of the Poincaré group. In Supergravity, with a 14-dimensional group, workers in Superspace $\mathfrak{R}^{4/4}$, an 8-dimensional manifold, found that they had to restrict the pure gauge group to L = SO(3.1). Adding, we find again 8 + 6 = 14, the group manifold dimensionality.

In the formalism we recently developed with T. Regge [1978a,b] for the gauging of non-internal groups, we work in the Group Manifold. Generalized curvatures R^A (A = i, [ij] in P) appear as the non-vanishing right-hand side of the Cartan-Maurer equations for Left invariant forms ω^A, when such forms are replaced by a "perturbed" set ρ^A (a ten-bein) (see (1.12))

$$d\rho^A - \tfrac{1}{2}\rho^B \wedge \rho^E C_{BE}{}^A = R^A = \tfrac{1}{2}\rho^B \wedge \rho^E R_{BE}{}^A \tag{4.4a}$$

or

$$d\rho^A - \tfrac{1}{2}\rho^B \wedge \rho^E (C_{BE}{}^A + R_{BE}{}^A) = 0 \tag{4.4b}$$

For an orthonormal basis of vector fields D_B orthogonal to the ρ^A,

$$\rho^A (D_B) = \delta^A_B \qquad \omega^A (D_B^{L.I.}) = \delta^A_B \qquad R^A \xrightarrow[\rho\to\omega]{} 0 \tag{4.5}$$

$$[D_B, D_E] = (C_{BE}{}^A + R_{BE}{}^A)\, D_A \tag{4.6}$$

$$[D_B^{L.I.}, D_E^{L.I.}] = C_{BE}{}^A\, D_A \tag{4.7}$$

(4.7) is the Left-invariant generator algebra. In (4.8) we have an algebra with "structure functions" instead of constants.

We can now also calculate the variation of D_E:

$$(\delta D)_E = [\varepsilon^B D_B, D_E] = \varepsilon^B (C_{BE}{}^A + R_{BE}{}^A)\, D_A \tag{4.8a}$$

$$\frac{(\delta D)_E}{\delta\varepsilon^B} = (C_{BE}{}^A + R_{BE}{}^A)\, D_A = (\delta D)_{EB} \tag{4.8b}$$

and since the product $\rho^E D_E$ is invariant, we can derive the variations of the adjoint representation ρ^E from those of the co-adjoint D_E (the difference is important when the group is not semi-simple, which is the case for P, GP, Extended GP, GA(4R) etc. but not for the Conformal SU(2,2), Graded-Conformal SU(2,2/1) or Extended G. Conformal SU(2,2/N). The factor $(-1)^{be}$ takes care of the grading in case of a Graded (or Super)Group.

$$\delta(\rho^E D_E)_B = (\delta\rho)_B{}^E D_E + (-1)^{be} \rho^E (\delta D)_{EB} = 0$$

$$(\delta\rho)_B{}^A D_A = -(-1)^{be} \rho^E (C_{BE}{}^A + R_{BE}{}^A) D_A$$

$$\delta_B \rho^A = -(-1)^{be} \rho^E (C_{BE}{}^A + R_{BE}{}^A) \tag{4.8c}$$

If we treat $\varepsilon^B(Z)$ as a local gauge (Z is 10-dimensional for P), we have to add the necessary gradient term. Summing over the B index we get, (we leave out the gradings for simplicity, $D^{(G)}$ is the covariant derivative defined over the group G)

$$\delta\rho^A = d\varepsilon^A - \rho^E \wedge \varepsilon^B (C_{EB}{}^A + R_{EB}{}^A) = D^{(G)}\varepsilon^A - \rho^E \varepsilon^B R_{EB}{}^A \tag{4.9}$$

or also (see definition following (4.17))

$$\delta\rho^A = D^{(G)} \varepsilon^A - 2 (\varepsilon, R^A) \tag{4.10}$$

$$(D^{(G)}\eta)^A = d\eta^A - \rho^B \wedge \eta^E C_{BE} \tag{4.11}$$

We have shown that all this is unchanged when a subgroup H (the Lorentz group L for both P and GP) is factorized out in the group manifold (so that we are left with $\mathcal{M}^4$ for P and $\mathcal{R}^{4/4}$, i.e. "Superspace", for GP, as base spaces M). This also corresponds to H being gauged, as in section 2. In that case, denoting by E, F the indices in the range of G/H , and by A, B $\in$ H , ρ^A contains only dx differentials ($x \in M = G/H$) and ω^A itself, ρ^F only dx differentials,

$$D_A = D_A{}^{L.I.} \tag{4.12a}$$

since

$$\rho^A(D_B{}^{L.I.}) = \omega^A (D_B{}^{L.I.}) = \delta^A{}_B, \quad \rho^F(D_B{}^{L.I.}) = 0 \tag{4.12b}$$

which also implies

$$R_{BJ}{}^K = 0 \; ; \quad I,J,K \in G \; ; \; A,B \in H \; ; \; E,F \in G/H \tag{4.13}$$

Similarly, for holonomic indices, since the only "perturbed" forms are constructed of M differentials,

$$R_{QU}{}^{K} = 0 \quad ; \quad Q,R \in H \quad ; \quad V,U \in G \quad ; \quad Y,Z \in G/H \tag{4.14}$$

To further our understanding of these parallel transport gauges, we analyze the effect of a general coordinate transformation (either in the G-manifold, or after factorization in G/H) on our one forms ρ^K

$$\delta x^U = \varepsilon^U \tag{4.15}$$

$$\delta\rho^K = \delta(dx^U \rho_U{}^K) = Dx^V \frac{\partial\varepsilon^U}{\partial x^V} \rho_U{}^K + dx^U \varepsilon^V \frac{\partial}{\partial x^V} \rho^U$$

$$= dx^V \{\frac{\partial\varepsilon^K}{\partial x^V} + \varepsilon^U \frac{\partial}{\partial x^U} \rho_V{}^K - \varepsilon^U \frac{\partial}{\partial x^V} \rho_U{}^K\}$$

where we have defined (see (4.2))

$$\varepsilon^K = \varepsilon^U \rho_U{}^K \tag{4.16}$$

Since

$$d\rho^K = -\tfrac{1}{2} (dx^V \wedge dx^U) (\frac{\partial}{\partial x^U} \rho_V{}^K - \frac{\partial}{\partial x^V} \rho_U{}^K)$$

we can regroup the terms in $\delta\rho^K$,

$$\delta\rho^K = d\varepsilon^K - 2 (\varepsilon, d\rho^K) \tag{4.17}$$

where the scalar product parenthesis represents contraction with the second factor in the two-form. Also,'

$$\delta\rho^K = D^{(G)} \varepsilon^K + \rho^I \wedge \varepsilon^U \rho_U{}^J C_{IJ}{}^K - 2 (\varepsilon, d\rho^K) =$$

$$= D^{(G)} \varepsilon^K + (\varepsilon, -2 d\rho^K + \rho^I \wedge \rho^J C_{IJ}{}^K) =$$

$$= D^{(G)} \varepsilon^K - 2(\varepsilon, R^K) \tag{4.18}$$

$$= D^{(G)} \varepsilon^K - \rho^I \varepsilon^J R_{IJ}{}^K \tag{4.19}$$

The algebra of parallel transport operators in G is thus in fact an algebra generating "anholonomized" (see (4.16)) General Coordinate Transformations (AGCT) on the G manifold. That gauge invariance is thus guaranteed by the General Covariance of the Lagrangian. Indeed, it is this gauge which reproduces the General Covariance Group, rather than GL(4R)-gauging, as commonly believed.

5. Dynamically Restricted AGCT gauges

We can construct the D_E for the factorized case. These correspond to translations in the quotient space G/H. From (4.5), (4.12b)

$$\rho^B\ (D_E) = 0 \quad , \qquad \rho^F\ (D_E) = \delta^F_E \tag{5.1}$$

we find (still using the indices as in (4.13) - (4.14))

$$D_E = \Delta_E{}^Y\ D_Y{}^{(H)} \tag{5.2}$$

$$D_Y{}^{(H)} = \frac{\partial}{\partial x^Y} - \tfrac{1}{2} \sum_{A \in \{H\}} \rho_Y{}^A\ S_A \tag{5.3}$$

where $D^{(H)}$ is the H-covariant derivative, ρ^A is the post-factorization form on M itself and S_A is the Right-Invariant algebra (of left-translation), which commutes with the $D_A^{L.I.}$ and has structure constants $- C_{BE}{}^A$. For the Poincare group with Ξ^{ab} the SO(3.1) variable to be factorized,

$$\tilde\rho^{ij}\ (\Xi,x) = (\Xi^{-1}\ d\Xi)^{ij} + \rho^{kl}(x)\ \ \Xi^{lj}\ \ \Xi^{ki} \tag{5.4}$$

$$\tilde\rho^{i}\ (\Xi,x) = \Xi^{ki}\ \rho^k\ (x) \tag{5.5}$$

The $\rho^{ij}(x)$ are the connection potentials we introduced in (2.1) and used in Section 2, while we have used $\tilde\rho^{ij}$ and $\tilde\rho^{i}$ in these last equations to denote the pre-factorization one-forms. Our previous discussion of the parallel-transport or AGCT gauges holds for either set.

For the parallel-transport modified translation gauge we thus get the variations,

$$\delta\rho^{ij} = D^{(P)}\ \varepsilon^{ij} - \rho^k\ \varepsilon^l\ \ R_{kl}{}^{ij} \tag{5.6}$$

$$\delta\rho^{i} = D^{(P)}\ \varepsilon^{i} - \rho^k\ \varepsilon^l\ \ R_{kl}{}^{i} \tag{5.7}$$

where

$$D^{(P)}\ \eta^{ij} = d\eta^{ij} + \rho^{ik} \wedge \eta^{kj} - \rho^{kj} \wedge \eta^{ik} \tag{5.8}$$

$$D^{(P)}\,\eta^{i} = d\eta^{i} + \rho^{ik} \wedge \eta^{k} - \rho^{k} \wedge \eta^{ik} \tag{5.8b}$$

Compare (5.7) with (3.13b) and with (3.12)!

The parallel-transport gauges((5.6)-(5.7)) introduced by Von der Heyde [1976; see also Hehl et al 1976a] (in space-time; here generalized to the Group manifold), are still "semi-trivial", since they only reproduce General Covariance. Note that one very important point is guaranteed: we realize that AGCT form a group and can be exponentiated, since they are just a subset of the Group of Diffeomorphisms.

Now once a Lagrangian is introduced, it will yield equations of motion. These equations will restrict the values of the $R_{IJ}{}^{K}$ components in (4.19), (5.6-5.7). For instance, we have seen in (4.13) and (4.14) the results of the Lagrangian being gauge-invariant under a subgroup H (the Lorentz group in P and GP). First, the parallel transport generators in the H direction coincide with the Lie-Algebra L.I. generators, so that H-gauging is "conventional". Secondly, applying the equations of motion produces the cancellations (notation as in (4.13))

$$R_{AB}{}^{J} = 0 \qquad\qquad R_{IJ}{}^{E} = 0 \tag{5.9}$$

which makes the <u>Dynamically Restricted</u> AGCT gauge for translations <u>coincide</u> for ρ^{E} itself (the vierbein ρ^{i} in Gravity), with an ordinary translation gauge (but not for ρ^{A}, the connection ρ^{ij}). In Supergravity, where H is also the Lorentz group, D.R. A.G.C.T. translations thus also look like an ordinary gauge for ρ^{i} itself, but not when acting on ρ^{ij} or ρ^{α}, the spinor potential. Supersymmetry D.R. A.G.C.T. also produce a variation involving R^{ij} for both vector and spinor variations [Ne'eman and Regge, 1978a,b]:

$$\delta\rho^{ij} = D^{(GP)}\,\varepsilon^{ij} - \rho^{c}\,\varepsilon^{d}\,R_{cd}{}^{ij} - \rho^{c}\,\varepsilon^{\dot\alpha}\,R_{c\dot\alpha}{}^{ij} \tag{5.10}$$

$$\delta\rho^{i} = D^{(GP)}\,\varepsilon^{i} \tag{5.11}$$

$$\delta\rho^{\alpha} = D^{(GP)}\,\varepsilon^{\alpha} - \rho^{c}\,\varepsilon^{d}\,R_{cd}{}^{\alpha} \tag{5.12}$$

The components $R_{c\dot\alpha}{}^{ij}$ in (5.10) are essential to the "local supersymmetry" transformations of Supergravity [Freedman et al 1976; Deser and Zumino 1976; Freedman and van Nieuwenhuizen 1976]

$$D^{(GP)}\,\eta^{ij} = D^{(P)}\,\eta^{ij} \tag{5.13}$$

$$D^{(GP)}\,\eta^{i} = D^{(P)}\,\eta^{i} + \bar{\rho}\gamma^{i}\eta \tag{5.14}$$

$$D^{(GP)}\,\eta^{\alpha} = d\eta^{\alpha} + \tfrac{1}{2}\,(\rho^{ij}\,\sigma^{ij}) \wedge \eta^{\alpha} - \tfrac{1}{2}\,(\sigma^{ij}\rho)^{\alpha} \wedge \eta^{ij} \tag{5.15}$$

The action in supergravity is given by

$$A = \frac{1}{8} \int_{\mathfrak{M}^4} (R^{ij} \wedge \zeta_{ij} + R^{\alpha} \wedge \bar{\zeta}_{\alpha}) \tag{5.16}$$

with, on a generic $\mathfrak{M}^4$, the equations (anholonomic spinor indices are not explicited)

$$R^i = 0 \tag{5 .17}$$

$$R^{ij} \rho^k \varepsilon_{ijk\ell} - 2i \bar{R}\gamma_5 \gamma_\ell \rho = 0 \tag{5.18}$$

$$\gamma^i \rho^i R = 0 \tag{5.19}$$

from which one derives

$$R_{IJ}{}^i \equiv 0 \quad \forall I,J; \quad R_{\alpha\beta} = 0; \quad R_{i\alpha} = 0; \quad R_{[ij]K} = 0 \quad \forall K \tag{5.20a}$$

$$R_{\alpha m}{}^{ij} \varepsilon_{ijk\ell} - R_{\alpha k}{}^{ij} \varepsilon_{ijm\ell} = 4i R_{mk} (\gamma_5 \gamma_\ell)_\alpha \tag{5.20b}$$

Equation (5.20b) and the $\bar{\varepsilon}^{\dot{\alpha}}$ variation in (5.10) are essential to supergravity. Indeed, the Supersymmetry transformations $\bar{\varepsilon}^{\dot{\alpha}}$ of Supergravity, which were derived directly, posed the problem of what we now know is a Dynamically Restricted AGCT, before it had ever been raised in Gravity, although the survival of the ε^ℓ transformation in (5.6) is completely analogous. In both theories, ρ^{ij} does not propagate and is extracted from $R^i = 0$ in terms of the other potentials, which tended to hide the physical importance of either (5.6) or (5.10).

We still have to discuss one more aspect of these theories. Working in the Group Manifold, how come we only use $\mathfrak{M}^4$ for the integration in either (2.7) or (5.16)?

First, the reduction of the base space to the quotient of G (P or GP) by its subgroup H (L in both cases, although other such subgroups exist for GP): if $\mathcal{G}$, $\mathcal{H}$ and $\mathcal{F}$ are the Lie algebras of G, H and G/H, the conditions may be

(a) weak reducibility: $[\mathcal{H}, \mathcal{F}] \subset \mathcal{F}$

(b) a symmetric manifold: $[\mathcal{F}, \mathcal{F}] \subset \mathcal{H}$

(c) an ideal: $[\mathcal{F}, \mathcal{F}] \subset \mathcal{F}$

For G = P and H = L, all three hold, but for G = GP and H = L, we have (a) and (c); for $H = L \otimes A^2$ (left or right handed supersymmetry) we have (a) and (b); for $H = GP_1$ (supersymmetry with only nilpotent elements in the ring of parameters) we have (a), (b) and (c). Each case induces a different theory, with ordinary Supergravity corresponding to H = L. The MacDowell-Mansouri [1977] version of de Sitter

Gravity follows (a) and (b).

The homogeneous spaces $P/L = \mathcal{M}^4$, $GP/L = \mathcal{R}^{4/4}$ correspond to the "factorized" theories. We conjecture that if a Lagrangian is H gauge-invariant, then it is H-factorizable as a consequence of the equations of motion. A heuristic proof of this hypothesis exists for solutions infinitesimally close to a factorized one. All such solutions can be reduced to factorized ones by an infinitesimal coordinate transformation on G. However, discrete families of factorized solutions with the same boundary conditions but topologically distinct may exist in the large.

This explains restricting the action integral to $\mathcal{M}^4$ in Gravity. In Supergravity, factorization reduces us to $\mathcal{R}^{4/4}$. However, physics is seen to be completely determined by what happens on a simple $\mathcal{M}^4$. The transfer of information from any $\mathcal{M}^4$ to any other $\mathcal{M}^4$ in $\mathcal{R}^{4/4}$ corresponds to our AGCT gauges. Partial $\mathcal{M}^4$ slices correspond to all possible supersymmetry-related conventional Supergravity theories.

6. GL(4R) and Affine Gauges

In trying to reproduce Gravity as a gauge theory, several authors[43-46] gauged GL(4R). This group seemed to fit that role, judging from the fact that in holonomic ("world tensor") coordinates, the covariant derivative is

$$D_{\mu}\,\phi = \partial_{\mu} + \Gamma_{\mu\nu}{}^{\rho}\; G^{\nu}{}_{\rho}\,\phi \tag{6.1}$$

where $G^{\nu}{}_{\rho}$ is the GL(4R) matrix representation corresponding to the world-tensor field ϕ. Indeed, world tensors are classified by the finite irreducible (non-unitary) representations of GL(4R).

However, as proved by DeWitt [1964a] this has nothing to do with a Yang-Mills type gauge. We are dealing with the General Coordinate Transformation Group, and its structure constants do not correspond to a GL(4R) gauge. Indeed, as we have shown, they correspond to an AGCT translation gauge. However, as a group, the G.C.T.G. is represented over its linear subgroup, which happens to be GL(4R). This is true of any such non-linear group, owing to the role played by the Jacobian determinant. We refer the reader to De Witt's text[47] and to the work of A. Joseph and A. I. Solomon [1970], who, in working out the theory of Global and Infinitesimal Nonlinear Chiral transformations, explained the construction of representations and covariant derivatives for such non-linear (and non gauge-factorizable) groups. (In Chiral symmetry, Isospin is the linear subgroup).

One more general point about GL(4R). It had always been assumed in the folklore of general relativity (and often written in texts) that GL(nR) has no double-valued or spinorial representations. E. Cartan [1938] is referred to for this prevalent belief, in two of his theorems. As can be seen in the text, one theorem refers explicitly to spinors with a finite number of components. The other theorem is an overstatement: "the three Unimodular Groups in two dimensions have no multi-valued representations". SU(2) is of course compact and simply connected; it is the covering group $SU(2) = \overline{SO(3)}$ of SO(3), where spinors are bivalued. SL(2C) has SU(2) as compact subgroup, and thus has the same topology. Indeed, $SL(2C) = \overline{SO(1.3)}$ is the covering group of the Lorentz group, and Lorentz bivalued representations become single valued here. Now it is true that $SL(2R) = \overline{SO(1.2)}$ and the bivalued representations of SO(1.2) become single-valued in SL(2R), which may explain the error in Cartan's theorem. However, SL(2R) has like SO(2) an infinite covering, and we can find in Bergmann's analysis [1947] double-valued representations of SL(2R), which become single valued in $\overline{\overline{SL(2R)}} = \overline{\overline{SO(1.3)}}$, etc.

In a recent study [Ne'eman 1977, 1978], we have proved the existence of double-valued representations of SL(nR), GL(nR) and the G.C.T.G. in <u>n</u>. These reduce to infinite direct sums of SO(n) or O(n) spinors. They are single valued in $\overline{SL(nR)}$, $\overline{GL(nR)}$ and $\overline{GCTG(n)}$. For such "polyfields", (6.1) can be used, provided the $G^{\nu}{}_{\rho}$ are infinite-dimensional.

These band-spinors or bandors are all known for SL(2R) [Bargmann 1947] and SL(3R) [Joseph 1970; Sigacki 1975]. They have now also been listed for SL(4R) [Sigacki 1978]. Note that $\overline{SL(4R)} = \overline{\overline{SO(3,3)}}$ and some of these representations had been included in a study of SO(3.3) by A. Kihlberg [1966].

Gauging GL(4R) [Yang 1974] prior to the introduction of bandors implied that spinor matter fields would not be minimally coupled therein. Note that most of these theories did not really exploit GL(4R) anyhow, and added metric restrictions [Mansouri and Chang 1976] which reduced GL(4R) to SO(1.3) or alternatively reduced GA(4R) - the Affine group in 4 dimensions, i.e. GL(4R) x T_4 - to Poincare SO(1.3) x T_4. However, we shall further discuss one consequence of starting with a larger group which is generally disregarded: the representation structure.

We now study the result of a GL(4R) gauge, in the context of a GA(4R) mixed-gauge (ordinary for GL(4R), D.R.AGCT for the translations).

It is [Hehl et al 1976b, 1977a] in the Metric-Affine theory and in its Spinor version [Hehl et al 1977b, 1978] and gauge [Lord, 1978] that the actual enlargement of the sets of connections, curvatures and currents are used, rather than an immediate restriction to Einstein's theory. The spinor matter field is now a polyfield, i.e. an infinite representation of GL(4R), with physical states given by GL(3R) bandors (this is the little group). One such bandor is $\mathfrak{D}$ (½,0) which reduces under the spin to the sum $\frac{1}{2} \oplus \frac{5}{2} \oplus \frac{9}{2} \oplus \frac{13}{2} \oplus \ldots$

The connections now include in addition to those of P, ten β^{ij} symmetric in (i,j). The $D_{(ij)}$ generators in the (flat) group space generate shear (for traceless $D_{(ij)}$) and scaling (for the trace). We thus enlarge the angular momentum current tensor into the <u>hypermomentum</u> tensor, with shear, scale and spin currents in its intrinsic part:

$$h_{ab}{}^{\mu} = s_{ab}{}^{\mu} + \tfrac{1}{4}\,\eta_{ab}\,h^{\mu} + \bar{h}_{ab}{}^{\mu} \qquad (6.2)$$

where η_{ab} is the Minkowski metric, h^{μ} is the scale (or dilation) current and $\bar{h}_{ab}{}^{\mu}$ is the shear current. Note that the "orbital" part of hypermomentum can be reduced to the set of time-derivatives of gravitational quadrupole moments [Dothan et al 1965; Hehl et al 1977b].

The Noether currents of the theory are given by

$$t_a{}^{\mu} = e^{-1}\ \frac{\delta \mathcal{L}}{\delta \rho_{\mu}{}^{a}} \qquad (6.3)$$

$$h_{ab}{}^{\mu} = - e^{-1} \frac{\delta \mathcal{L}}{\delta \rho_{\mu}{}^{ab}} \tag{6.4}$$

The field equations are ($\mathcal{L}_o$ is the gravitational field Lagrangian)

$$\frac{\delta \mathcal{L}_o}{\delta \rho_{\mu}{}^{a}} = - 2\kappa \; e \; t_a{}^{\mu} \tag{6.5}$$

$$\frac{\delta \mathcal{L}_o}{\delta \rho_{\mu}{}^{ab}} = 2 \; e \; h_{ab}{}^{\mu} \tag{6.6}$$

Choosing the free action

$$A_o = \frac{1}{8K} \int (R^{[ab]} \wedge \zeta_{[ab]} + \beta \, Q^2) \tag{6.7}$$

where

$$Q_{\mu} = \frac{1}{4} \rho_{\mu\sigma}{}^{\sigma} \tag{6.8}$$

we find that equation (6.6) becomes again algebraic in relating connections to hypermomenta, and the $\rho_{\mu}{}^{(ab)}$ does not propagate. Note that the holonomic $\rho_{\mu(\sigma\tau)}$ corresponds to the Non-metricity tensor,

$$\rho_{\mu(\sigma\tau)} = - D_{\mu} \, g_{\sigma\tau} \tag{6.9}$$

which appears in the identity,

$$\rho_{\mu\nu}{}^{\sigma} \equiv g^{\sigma\tau} \; \Delta^{\sigma\beta\gamma}_{\nu\mu\tau} \; (\tfrac{1}{2} \, \partial_{\alpha} \, g_{\beta\gamma} - g_{\gamma\varepsilon} \, R^{\cdot\cdot\varepsilon}_{\alpha\beta} - \tfrac{1}{2} \, D_{\alpha} \, g_{\beta\gamma}) \tag{6.10}$$

$$\Delta^{\alpha\beta\gamma}_{\nu\mu\tau} := \delta^{\alpha}_{\nu} \; \delta^{\beta}_{\mu} \; \delta^{\gamma}_{\tau} + \delta^{\alpha}_{\mu} \; \delta^{\beta}_{\tau} \; \delta^{\gamma}_{\nu} - \delta^{\alpha}_{\tau} \; \delta^{\beta}_{\nu} \; \delta^{\gamma}_{\mu} \tag{6.11}$$

When no polyfields are present, there is no non-metricity. In the presence of intrinsic hypermomentum, non-metricity exists but is confined to the region where that matter exists, without propagating over intermediary regions. Again, the linearity of the Einstein Lagrangian in derivatives preserves the Riemannian properties of space-time. Macroscopically, one can always define a local Minkowski metric.

7. Extending the Poincaré Group: No-go Theorems

There are three main ways in which one has extended the Poincaré group:

- Conformally, into the simple group
  $\mathrm{Con}(4R) = SU(2.2) = \overline{SO(4.2)}$
- Linearly into the Affine group GA(4R)
- Spinorially, into GP

In fact, the latter extension can also be performed for SU(2.2), extending it thus further into SU(2.2/1) or SU(2.2/N). We have recently shown [Ne'eman and Sherry, 1978] that GA(nR) can be similarly extended into infinite-dimensional graded Lie Groups g GL(nR). Although we have constructed these graded-Affine groups for n = 2,3 only as yet, it appears plausible that g GL(4R) should also exist.

There is one important point we should note when gauging a group G larger than P. Although we may afterwards introduce constraints which will reduce the theory to Einstein's General Relativity, there are still traces of the larger group $G \supset P$. For example, the matter fields physical states have to fit in unitary representations of G. In our case, these would be Polyfields (with either integer or half-integer spins). In Conformal Relativity resulting from gauging the Conformal group [Englert et al 1975; Harnad and Pettitt 1976, 1977; Kaku et al 1977], these would be Mack's [1977] Unitary representations of the Conformal group.

Ogievetsky [1973] has proved that in a holonomic representation of Con(4R) U GL(4R) generator algebras, closure occurs only over the entire analytical General Coordinate Transformations Group A. This is due to the commutators of the Special Conformal Transformation generators K_μ and the Shears $S_{(\mu\nu)}$, which keep generating operators

$$x_1^m \, x_2^n \, x_3^r \, x_4^s \, \partial_\mu$$

with ever-increasing powers (m, n, r, s). In more recent work [Borisov and Ogievetsky, 1974; Cho and Freund, 1975] this theorem has been applied to Gravitational theories. We would like to note the following theorems that can be drawn from Ogievetsky's:

(1) Assuming a theory to be (globally) invariant under Con(4R):= C and GL(4R):=G reduces it to a trivial S-matrix. Indeed, we find that if the Lagrangian $\mathcal{L}$ obeys

$$[\mathcal{L}, C] = 0 \quad , \quad [\mathcal{L}, G] = 0 \tag{7.1}$$

then

$$[\mathcal{L}, [C,G]] = [\mathcal{L}, A] = 0 \tag{7.2}$$

so that we have an infinite number of active-Symmetry Noether theorems.

(2) Gauging both C and G imposes a trivial S-matrix. This results from (7.2) because a local gauge includes the case of a constant (global) gauge.

These theorems are not modified by spontaneous breakdown via a Goldstone mechanism, since this still yields all global Noether currents.

A Higgs-Kibble mechanism breaks the local gauge group but preserves the global conservation laws. Thus, only a Higgs mechanism breaking the A gauge down to global (or local) P invariance can release the S-matrix from triviality.

It is important to remember that the Ogievetsky algebra is a representation of the Diffeomorphisms, but as such is purely a holonomic construct with no (active) Symmetry connotation. Symmetries and their local extension as Gauges are entirely anholonomic.

Acknowledgements

We would like to thank Dr. J. Thierry-Mieg for an enlightening discussion.

References

Bargmann, V. 1947, Ann. Math. 48, 568-640.

Borisov, A. B. and V. I. Ogievetsky, 1974, Teoret. i. Mat. Fiz. 21, 329-342.

Cartan, E. 1922, Com. R. Ac. Sci. (Paris) 174, 593;
1923, 24, 25, Ann. Ec. Nor. Sup. 40, 325; 41, 1; 42, 17;
1938, Lecons sur la Theorie des Spineurs II (Hermann Editeurs, Paris) articles 85-86, pp. 87-91, 177, pp. 89-91.

Cho, Y. M. 1976, Phys. Rev. D14, 2521-2525.

Cho, Y. M. and P. G. O. Freund, 1975, Phys. Rev. D12, 1711-1720.

Deser, S. and B. Zumino, 1976, Phys. Lett. 62B, 335.

De Witt, B. S., 1964a "Dynamical Theory of Groups and Fields" in Relativity, Groups and Topology (Les Houches 1963 Seminar), ed. C. and B. De Witt, Gordon and Breach (N.Y), pp. 587-826. See in particular article 13, pp. 688-689.
- 1964b, Phys. Rev. Lett. 12, 742.
- 1967, Phys. Rev. 162, 1195; 162, 1239.

Dothan, Y., M. Gell-Mann and Y. Ne'eman 1965, Phys. Lett. 17, 148-151.

Englert, F. and R. Brout, 1964, Phys. Rev. Lett. 13, 321.

Englert, F., E. Gunzig, C. Truffin and P. Windey 1975, Phys. Lett. 57B, 73-77.

Feymman, R. P., 1963, Acta. Phys. Polon. 24, 697.

Faddeev, L. D., and V. N. Popov, 1967, Phys. Lett. 25B, 29.

Fradkin, E. S., and I. V. Tyutin, 1970, Phys. Rev. D2, 2841.

Freedman, D. Z., P. van Nieuwenhuizen and S. Ferrara, 1976, Phys. Rev. D13, 3214.

Freedman, D. Z. and P. van Nieuwenhuizen, 1976, Phys. Rev. D14, 912-916.

Fritzsch, H., and M. Gell-Mann, 1972, in Proc. 16th Intern. Conf. H.E.P., J.D.Jackson and A. Roberts,eds., NAL pub. (Batavia), 2, pp. 135-165.

Gell-Mann, M., 1962, Phys. Rev. 125, 1067-1084.

Gell-Mann, M. and S. L. Glashow, 1961, Ann. Phys. 15, 437.

Goldstone, J., 1961, Nuovo Cim. 19, 154.

Gross, D. and F. Wilczck, 1973, Phys. Rev. Lett. 30, 1343.

Guralnik, G. S., C. R. Hagen and T. W. B. Kibble, 1964, Phys. Rev. Lett. 13, 585.

Harnad, J. P., and R. B. Pettitt, 1976, J. Math. Phys. 17, 1827-1837;
- 1977, in Group Theoretical Methods in Physics (5th Int. Coll.) J. Patera and P. Winternitz, eds., Academic Press (N.Y.), pp. 277-301.

Hehl, F. W., 1970, "Spin und Torsion in der allgemeinen Relativitatstheorie, oder die Riemann-Cartansche Geometrie der Welt", Clausthal Tech. Univ. Thesis.

References - continued:

Hehl, F. W., P. v. d. Heyde, G. D. Kerlick and J. M. Nester, 1976a, Rev. Mod.Phys. 48, 393-416.

Hehl, F. W., G. D. Kerlick and P. v. d. Heyde, 1976b, Phys. Lett. 63B, 446-448.

Hehl, F. W., G. D. Kerlick, E. A. Lord and L. L. Smalley, 1977a, Phys. Lett. 70B, 70-72.

Hehl, F. W., E. A. Lord, and Y. Ne'eman, 1977b, Phys. Lett. 71B, 432-434.

Hehl, F. W., E. A. Lord and Y. Ne'eman, 1978, Phys. Rev. D17, 428-433.

von der Heyde, P., 1975, Lett. al Nuovo Cim. 14, 250-252.

von der Heyde, P., 1976, Phys. Lett. 58A, 141-143.

Higgs, P. W., 1964a, Phys. Lett. 12, 132;
- 1964, Phys. Rev. Lett. 13, 508;
- 1966, Phys. Rev. 145, 1156.

't Hooft, G., 1971, Nucl.Phys. B33, 173 and B35, 167.

't Hooft, G., and M. Veltman, 1972, Nucl. Phys. B50, 318.

Ionides, P. 1962, London University Thesis.

Joseph, A. and A. I. Solomon, 1970, J. Math. Phys., 11, 748-761.

Joseph, D. W. 1970, "Representations of the Algebra of SL(3R) with $\Delta j = 2$", University of Nebraska preprint, unpublished.

Kaku, M., P. K. Townsend and P. van Nieuwenhuizen 1977, Phys. Lett. 69B, 304-308.

Kibble, T. W. B. 1961, J. Math. Phys. 2, 212-221;
1967, Phys. Rev. 155, 1554.

Kihlberg, A. 1966, Arkiv f. Fysik, 32, 241-261.

Kobayashi, S. 1956, Canad. J. of Math. 8, 145-156.

Kobayashi, S. 1972, Transformation Groups in Differential Geometry, Springer Pr. (Berlin).

Lee, B. W. and J. Zinn-Justin, 1972, Phys. Rev. D5, 3121, 3137 and 3155.

Lee, T. D., S. Weinberg and B. Zumino 1967, Phys. Rev. Lett. 18, 1029.

Lord, E. A. 1978, Phys. Lett. 65A, 1-4.

MacDowell, S. W. and F. Mansouri 1977, Phys. Rev. Lett. 38, 739.

Mansouri, F. and L. N. Chang 1976, Phys. Rev. D13, 3192-3200.

Mack, G. 1975, DESY 75/50 report.

Nambu, Y. 1965, in Symmetry Principles at High Energy (Coral Gables 1965), B. Kursunoglu, P. Perlmutter and I. Sakmar, eds., Freeman Pub. (San Francisco) pp. 274-283.

Nambu, Y. and G. Iona-Lasinio, 1961, Phys. Rev. 122, 345.

Ne'eman, Y. 1961, Nucl. Phys. 26, 222-229.

Ne'eman, Y. 1977, Proc. Natl. Acad. Sci. U.S.A. 74, 4157-4159.

Ne'eman, Y. 1978, Ann. Inst. Henri Poincare 28.

Ne'eman, Y. and T. Regge 1978a, Rivista del Nuovo Cim.

References - continued:

Ne'eman, Y. and T. Regge 1978b, Phys. Lett.

Ne'eman, Y. and T. Sherry 1978

Ogievetsky, V. I. 1973, Lett. al Nuovo Cim. 8, 988-990.

Palatini, A. 1919, Rend. Circ. Mat. Palermo 43, 203.

Petti, R. J. 1976, Gen. Rel and Grav. 7, 869-883.

Politzer, H. D. 1973, Phys. Rev. Lett. 30, 1346.

Salam, A. 1968, in Proc. Eighth Nobel Symposium, N. Svartholm ed., Almquist and Wiksell pub. (Stockholm) 367-378.

Salam, A. and J. C. Ward 1961, Nuovo Cim. 20, 419.

Schouten, J. A. 1954,Ricci Calculus, Springer Pub. (Berlin).

Sciama, D. W. 1962, in Recent Developments in General Relativity, Pergamon Press (N.Y.-Warsaw) pp. 415-440.

Shaw, T. 1954, Thesis.

Sijacki, Dj. 1975, J. Math. Phys. 16, 298.

Sijacki, Dj. 1978, to be pub.

Teitelboim, C. 1977, Phys. Rev. Lett. 38, 1106-1110.

Thierry-Mieg, J. 1978, to be pub.

Thirring, W., 1978, in this volume.

Trautman, A. 1972, Bull. Acad. Polon. Sci., Ser. Sci. Mat. Ast. et Phys. 20, 185 and 583.

Trautman, A. 1973, Symposia Mathematica 12, 139-162.

Utiyama, R. 1956, Phys. Rev. 101, 1597.

Veltman, M. 1970, Nucl. Phys. B21, 288.

Weinberg, S. 1967, Phys. Rev. Lett. 19, 1264.

Weinberg, S. 1973, Phys. Rev. Lett. 31, 494.

Weyl, H. 1918, Sitzungsberichte d. Preuss. Akad. d. Wissensch.

Weyl, H. 1929, Zeit. f. Phys. 56, 330.

Weyl, H. 1950, Phys. Rev. 77, 699.

Yang, C. N. 1974, Phys. Rev. Lett. 33, 445.

Yang, C. N. and R. L. Mills 1954, Phys. Rev. 96, 191.

ON THE LIFTING OF STRUCTURE GROUPS

Werner Greub
Department of Mathematics, University of Toronto
and
Herbert-Rainer Petry
Institut für Theoretische Kernphysik, Universität Bonn

§ 1. General principal bundles

1. Γ-structures

Let $P=(P,\pi,B,G)$ be a principal bundle where P and B are topological spaces and G is a topological group. Let $\rho: \Gamma\to G$ be a continuous homomorphism from a topological group Γ onto G with kernel K. ρ will be called central, if

(i) K is discrete

(ii) K is contained in the center of Γ.

Thus, in particular, K is abelian.

It is easy to check that (ii) follows from (i) if Γ is connected. Thus every covering projection $\rho: \Gamma\to G$ (Γ connected) satisfies the conditions above.

Throughout this paper we shall assume that ρ is central.

A Γ-structure on P is a Γ-principal bundle $\tilde{P}=(\tilde{P},\tilde{\pi},B,\Gamma)$ together with a strong bundle map $\eta: \tilde{P}\to P$ which is equivariant under the right actions of the structure groups; that is

$$\eta(\tilde{z}\cdot\gamma) = \eta(\tilde{z})\cdot\rho(\gamma) \quad , \tilde{z}\in\tilde{P},\ \gamma\in\Gamma .$$

Example: Let A be a discrete abelian group and set $\Gamma=G\times A$. Let $\rho: G\times A\to G$ be the obvious projection. Then every principal G-bundle admits a Γ-structure.

In fact, set $\tilde{P}=P\times A$, $\tilde{\pi}(z,a)=\pi(z)$, $(z\varepsilon P,\ a\varepsilon A)$, and define η by $\eta(z,a) = z$.

A principal bundle need not admit a Γ-structure. It is the purpose of this paper to define an element $\kappa(P,\rho)$ of $H^2(B,K)$ (the second Cech cohomology group of B with coefficients in K) with the following property: P admits a Γ-structure if and only if $\kappa(P,\rho)=e$.

Remark: If K is written as an additive group this means, of course, that $\kappa(P,\rho)=0$.

Two Γ-structures $\tilde{P}$ and $\tilde{P}'$ are called isomorphic, $\tilde{P}\cong\tilde{P}'$, if there is a strong bundle isomorphism $\phi\colon \tilde{P}\to\tilde{P}'$ satisfying the conditions

$$\phi(\tilde{z}\cdot\gamma) = \phi(\tilde{z})\cdot\gamma\ ,\quad \tilde{z}\varepsilon\tilde{P}\ ,\quad \gamma\varepsilon\Gamma\ ,\qquad \eta'\ \phi=\eta\ .$$

2. Local cross-sections

Recall that an open covering $\{U_i\}$ of a topological space is called simple if all the non-empty intersections $U_{i_1}\cap\dots\cap U_{i_p}$ are contractible. A space is called an L-space, if every open covering has a simple refinement. In particular, every smooth manifold is an L-space. From now on we shall assume that the base space B is an L-space.

Let $\{U_i\}$ be a simple covering of B by open sets. Then P is trivial overy every U_i. Choose a system of local cross-sections $\sigma_i\colon U_i\to P$. Then the corresponding transition functions $g_{ij}\colon U_{ij}\to G$ are defined by

$$\sigma_j(x) = \sigma_i(x)\cdot g_{ij}(x),\quad x\varepsilon U_{ij}\ .$$

They satisfy the consistency relations

$$g_{ij}(x)g_{jk}(x) = g_{jk}(x),\quad x\varepsilon U_{ijk}\ .$$

Lemma I: A G-principal bundle admits a Γ-structure if and only if there are continuous maps $\gamma_{ij}: U_{ij} \to \Gamma$ such that

$$(a)\ \gamma_{ij}(x)\gamma_{jk}(x) = \gamma_{ik}(x) \quad , \quad x\varepsilon U_{ijk} ,$$

and

$$(b)\ \rho\circ\gamma_{ij} = g_{ij} .$$

Proof: 1) If P admits a Γ-structure, let $\tilde{\sigma}_i$ be a system of local cross-sections for $\tilde{P}$, denote the transition map by γ_{ij} and set $\sigma_i=\eta\circ\tilde{\sigma}_i$. Then the σ_i form a system of local cross-sections for P . Thus we have

$$\sigma_j(x) = \eta\tilde{\sigma}_j(x) = \eta(\tilde{\sigma}_i(x)\gamma_{ij}(x)) =$$

$$= \eta\tilde{\sigma}_i(x)(\rho\gamma_{ij}(x)) = \sigma_i(x)(\rho\gamma_{ij}(x)) \quad , \quad x\varepsilon U_{ij} .$$

It follows that $\rho\gamma_{ij}=g_{ij}$.

2) Conversely, assume that these conditions are satisfied. Then there exists a Γ-bundle $\tilde{P}$ over P and a system of local cross-sections $\tilde{\sigma}_i$ in $\tilde{P}$ with the γ_{ij} as transition functions[1]. Moreover, by a standard result on principal bundles, there is a strong equivariant bundle map $\eta: \tilde{P}\to P$. Thus, $(\tilde{P},\eta)$ is a Γ-structure on P .

3. Γ-obstructions

Let $U=\{U_i\}$ be an open covering of B such that P is trivial over U_i . Since B is an L-space we may assume that the covering U is simple. Then the transition maps $g_{ij}:U_{ij}\to G$

lift to continuous maps $\gamma_{ij}: U_{ij} \to \Gamma$.
Now consider a non-empty triple intersection U_{ijk} and set for $x \in U_{ijk}$

$$p_{ijk}(x) = \gamma_{jk}(x)\gamma_{ik}(x)^{-1}\gamma_{ij}(x) .$$

Then

$$\rho p_{ijk}(x) = g_{jk}(x)g_{ik}(x)^{-1}g_{ij}(x) = e$$

and so

$$p_{ijk}(x) \in K , x \in U_{ijk} .$$

Since U_{ijk} is connected and K is discrete, the p_{ijk} must be constant. Thus they define a 2-cochain in the nerve $N(U)$ with values in the abelian group K . This cochain will be denoted by p ,

$$p(i,j,k) = p_{ijk}(x) , x \in U_{ijk} .$$

Lemma II: p is a cocycle.

Proof: In fact, the coboundary of p is given by

$$(Dp)(i,j,k,l) = p(j,k,l)p(i,k,l)^{-1}p(i,j,l)p(i,j,k)^{-1} .$$

Now

$$p(j,k,l)p(i,k,l)^{-1} = p(i,k,l)^{-1}p(j,k,l)$$

because K is abelian. One finds that for $x \in U_{ijkl}$

$$p(j,k,l)p(i,k,l)^{-1} = \gamma_{ik}(x)^{-1}\gamma_{il}(x)\gamma_{jl}^{-1}(x)\gamma_{jk}(x) .$$

Inserting this into the first equation and observing that K is in the center of Γ we obtain

$$(Dp)(i,j,k,l) = \gamma_{ij}(x)^{-1}\gamma_{il}(x)\gamma_{jl}(x)^{-1}p(i,j,l)p(i,j,k)^{-1}\gamma_{jk}(x) .$$

Finally,

$$p(i,j,l)p(i,j,k)^{-1} = \gamma_{jl}(x)\gamma_{il}(x)^{-1}\gamma_{ij}(x)\gamma_{ij}(x)^{-1}\gamma_{ik}(x)\gamma_{jk}(x)^{-1} =$$
$$= \gamma_{jl}(x)\gamma_{il}(x)^{-1}\gamma_{ik}(x)\gamma_{jk}(x)^{-1} .$$

Inserting this into the equation above yields $(Dp)(i,j,k,l)=e$.

Lemma III: The cohomology class represented by the cocycle p is indendent of the choice of the local sections σ_i and the liftings γ_{ij} .

Proof: Let σ_i^1 and σ_i^2 be two systems of local cross-sections defined over the same simple covering. Denote the corresponding transition functions by g_{ij}^1 and g_{ij}^2 and let γ_{ij}^1 and γ_{ij}^2 be liftings of g_{ij}^1 and g_{ij}^2 respectively. Denote by p^1 and p^2 the cocycles formed via the two liftings. Define continuous functions $\lambda_i\colon U_i \to G$ by the equation

$$\sigma_i^2(x) = \sigma_i^1(x)\cdot\lambda_i(x) \qquad (x\in U_i) .$$

Then

$$g_{ij}^2(x) = \lambda_i(x)^{-1}g_{ij}^1(x)\lambda_j(x) .$$

Since the U_i are contractible the λ_i lift to continuous functions $\tilde{\lambda}_i\colon U_i \to \Gamma$ such that $\rho\cdot\tilde{\lambda}_i=\lambda_i$. Now define maps $f_{ij}\colon U_{ij} \to \Gamma$ by setting

$$f_{ij}(x) = \gamma_{ij}^1(x)[\tilde{\lambda}_i(x)\gamma_{ij}^2(x)\tilde{\lambda}_j(x)^{-1}]^{-1} .$$

It is easy to check that

$$\rho \cdot f_{ij}(x) = e \qquad (x \varepsilon U_{ij}) ,$$

and so the f_{ij} are constant. Thus they define a 1-cochain in $N(U)$ with values in K. We now show that

$$Df = p^1 \cdot (p^2)^{-1} .$$

First observe that, since $p^2(i,j,k)$ is in the center of Γ,

$$p^2(i,j,k) = \lambda_j(x)p^2(i,j,k)\lambda_j(x)^{-1} \qquad (x \varepsilon U_{ijk}) .$$

Using this and the fact that $f_{ij}(x)$ is in the center of Γ, as well, we obtain

$$p^1(i,j,k)p^2(i,j,k)^{-1} = f_{jk}(x)f_{ik}(x)^{-1}f_{ij}(x) \quad (x \varepsilon U_{ijk}) .$$

4. The obstruction class

In view of lemmas II and III, p determines an element of $\check{H}^2(N(U),K)$. Passing to the direct limit (over all simple coverings) we obtain an element $\kappa(P,\rho)\varepsilon\check{H}^2(B,K)$ which we call the Γ-obstruction.

Theorem I: A principal bundle admits a Γ-structure if and only if the class $\kappa(P,\rho)$ vanishes.

Proof: 1) Assume that P admits a Γ-structure. Then, by lemma 1 there are liftings γ_{ij} of g_{ij} such that

$$p(i,j,k) = e ,$$

whence

$$\kappa(P,\rho) = 0 .$$

2) Conversely assume that $\kappa(P,\rho)=0$. Then there is a 1-cochain f_{ij} in $N(U)$ with values in K such that

$$f_{jk}\cdot f_{ik}^{-1}f_{ij} = p(i,j,k) .$$

Define maps $h_{ij}:U_{ij} \to \Gamma$ by setting

$$h_{ij}(x) = \gamma_{ij}(x)f_{ij}^{-1} \quad (x\in U_{ij})$$

where the γ_{ij} are liftings of the g_{ij} again. Then a straight forward calculation shows that

$$h_{ik}(x) = h_{ij}(x)\cdot h_{jk}(x) \quad (x\in U_{ijk}) .$$

Now apply lemma 1.

5. <u>The naturality of $\kappa(P,\rho)$</u>:

Let $\hat{P}=(\hat{P},\hat{\pi},\hat{B},G)$ and $P=(P,\pi,B,G)$ be principal G-bundles and let

$$\begin{array}{ccc} \hat{P} & \overset{\Phi}{\longrightarrow} & P \\ {\scriptstyle\hat{\pi}}\downarrow & & \downarrow{\scriptstyle\pi} \\ \hat{B} & \overset{\Psi}{\longrightarrow} & B \end{array}$$

be a bundle map which restricts to homeomorphisms on the fibres B .

<u>Lemma 4</u>: Denote by $\psi^{\#}:\check{H}(\hat{B},K)\leftarrow H(B,K)$ the homomorphism induced by ψ .

Then:

$$\kappa(\hat{P},\rho) = \psi^{\#}\cdot\kappa(P,\rho) .$$

Proof: In fact choose simple coverings $\{\hat{U}_\alpha\}_{\alpha\varepsilon A}$ and $\{U_i\}_{i\varepsilon I}$ of $\hat{B}$ and B such that $\psi(\hat{U}_\alpha) \subset U_i$ for some i. Then there is a selection function $\tau: A \to I$ such that $\psi(\hat{U}_\alpha) \subset U_{\tau(\alpha)}$.

Now choose a system of local cross-sections $\sigma_i: U_i \to P$ and set

$$\hat{\sigma}_\alpha(\hat{x}) = \phi_{\hat{x}}^{-1}\cdot\sigma_{\tau(\alpha)}(\psi(\hat{x})) \qquad (\hat{x}\varepsilon\hat{U}_\alpha)$$

where $\phi_{\hat{x}}$ denotes the restriction of ϕ to the fibre over $\hat{x}$.

Then the $\hat{\sigma}_\alpha$ form a system of local cross-sections for $\hat{P}$. The corresponding transition functions are connected by the relation

$$\hat{g}_{\alpha\beta}(\hat{x}) = g_{\tau(\alpha),\tau(\beta)}(\psi(\hat{x})) , \quad (\hat{x}\varepsilon\hat{U}_{\alpha\beta}) .$$

Now let $\gamma_{ij}: U_{ij} \to \Gamma$ be liftings of the g_{ij} and set

$$\hat{\gamma}_{\alpha\beta}(\hat{x}) = \gamma_{\tau(\alpha),\tau(\beta)}(\psi(\hat{x})) \qquad (\hat{x}\varepsilon\hat{U}_{\alpha\beta}) .$$

Then

$$\rho\cdot\hat{\gamma}_{\alpha\beta}(\hat{x}) = \rho\cdot\gamma_{\tau(\alpha),\tau(\beta)}(\psi(\hat{x})) = g_{\tau(\alpha),\tau(\beta)}(\psi(\hat{x})) = \hat{g}_{\alpha\beta}(\hat{x})$$

and so the $\hat{\gamma}_{\alpha\beta}$ are liftings of the $\hat{g}_{\alpha\beta}$. Thus the obstruction cocycle for $\hat{P}$ is given by $(x=\psi(\hat{x}))$:

$$\hat{p}(\alpha,\beta,\gamma) = \hat{\gamma}_{\beta\gamma}(\hat{x})\hat{\gamma}_{\alpha\gamma}(\hat{x})^{-1}\gamma_{\alpha\beta}(\hat{x}) =$$

$$= \gamma_{\tau(\beta)\tau(\gamma)}(x)\gamma_{\tau(\alpha)\tau(\gamma)}(x)^{-1}\gamma_{\tau(\alpha)\tau(\beta)}(x) =$$

$$= p(\tau(\alpha),\tau(\beta),\tau(\gamma))(\psi(\hat{x})) \quad .$$

Hence we have indeed

$$\kappa(\hat{P},\rho) = \psi^{\#}\,\kappa(P,\rho) \ .$$

6. The bundle $P_1 \dot{\times} P_2$

Let $P_1=(P_1,\pi_1,B\ ,G_1)$ and $P_2=(P_2,\pi_2,B\ ,G_2)$ be principal bundles over the same base. Choose local systems of cross--sections for P_1 and P_2 over the same simple covering of B . Then the maps

$$g_{ij}(x) = g^1_{ij}(x) \times g^2_{ij}(x) \quad , x\varepsilon U_{ij} \ ,$$

satisfy

$$g_{ij}(x)g_{jk}(x) = g_{ik}(x) \quad , x\varepsilon U_{ijk} \ ,$$

and so they define a principal bundle

$$P_1 \dot{\times} P_2 = (P,\pi,B,\ G_1{\times}G_2) \quad .$$

Next, let $\rho_1\colon \Gamma_1 \to G_1$ and $\rho_2\colon \Gamma_2 \to G_2$ be central homomorphisms with kernels K_1 and K_2 .
Then

$$\rho_1{\times}\rho_2\colon \Gamma_1{\times}\Gamma_2 \to G_1{\times}G_2$$

is a central homomorphism with kernel $K_1 \times K_2$. We shall express the class $\kappa(P_1 \dot{\times} P_2)$ in terms of the classes $\kappa(P_1)$ and $\kappa(P_2)$. First observe that if $f \varepsilon C^p(U;K_1)$ and $g \varepsilon C^p(U,K_2)$ are cochains then a cochain $f \times g \varepsilon C^p(U, K_1 \times K_2)$ is defined by

$$(f \times g)(i_o \ldots i_p) = f(i_o \ldots i_p) \times g(i_o \ldots i_p) .$$

This pairing of cochains induces a pairing of cohomology classes which is also denoted by $\dot{\times}$.

Proposition I: The obstruction class $\kappa(P_1 \dot{\times} P_2)$ is given by

$$\kappa(P_1 \dot{\times} P_2) = \kappa(P_1) \dot{\times} \kappa(P_2) .$$

Proof: Choose liftings $\gamma^1_{ij}: U_{ij} \to \Gamma_1$ and $\gamma^2_{ij}: U_{ij} \to \Gamma_2$ of the g^1_{ij} and the g^2_{ij} and define maps $\gamma_{ij}: U_{ij} \to \Gamma_1 \times \Gamma_2$ by setting

$$\gamma_{ij}(x) = \gamma^1_{ij}(x) \times \gamma^2_{ij}(x) , \quad x \varepsilon U_{ij} .$$

Then

$$\rho \gamma_{ij} = g^1_{ij} \times g^2_{ij} = g_{ij}$$

and so the γ_{ij} are liftings of the g_{ij}. Denote by p_1 p_2 and p the cocycles for P_1, P_2 and $P_1 \dot{\times} P_2$ obtained via these liftings. Then

$$p(i,j,k) = p_1(i,j,k) \times p_2(i,j,k)$$

and so

$$p = p_1 \dot{\times} p_2$$

Passing to cohomology yields the proposition.

7. λ-extensions:

Let $P=(P,\pi,B,G)$ be a principal G-bundle and let $\lambda\colon G\to G'$ be a homomorphism. Then λ determines a principal bundle $P_\lambda=(P_\lambda,\pi_\lambda,B,G)$ over the same base in the following way: Choose a covering U_i of B with a system of local sections σ_i and transition functions g_{ij}. Define maps

$$g'_{ij} = \lambda \circ g_{ij} \quad .$$

Then the g'_{ij} satisfy the relation

$$g'_{ij}(x)g'_{jk}(x)g'_{ki}(x) = e$$

and consequently, there is a principal bundle P_λ with a system of local sections such that the g'_{ij} are the corresponding transition functions. P_λ is called the <u>λ-extension</u> of P .

Next assume that $\rho\colon \Gamma\to G$ and $\rho'\colon \Gamma'\to G'$ are central homomorphisms with kernels K and K' resp. Let $\lambda\colon G\to G'$ be a homomorphism. We shall say that ρ and ρ' are <u>related</u> if there is a continuous map $\tilde\lambda$ (not necessarily a homomorphism) such that

(1) the diagram

$$\begin{array}{ccc} \Gamma & \xrightarrow{\tilde\lambda} & \Gamma' \\ \rho\downarrow & & \downarrow\rho' \\ G & \xrightarrow{\lambda} & G' \end{array}$$

commutes and

(2) $\quad \tilde{\lambda}(k\cdot g) = \tilde{\lambda}(k)\tilde{\lambda}(g) \quad$ for all $k \varepsilon K$, $g \varepsilon \Gamma$.

$\tilde{\lambda}$ restricts to a homomorphism of K into K'. In fact, if $k\varepsilon K$ then $\rho'\tilde{\lambda}(k) = \lambda\cdot\rho(k)=\lambda(e)=e'$ whence $\tilde{\lambda}(k)\varepsilon K'$.
Next let P_λ be a λ-extension of P and assume that the homomorphisms $\rho: \Gamma\to G$ and $\phi: \Gamma'\to G'$ are related. Choose a simple covering $\{U_i\}$ of B and let γ_{ij} be a lifting of g_{ij} . Set

$$\gamma'_{ij} = \tilde{\lambda}\cdot\gamma_{ij} .$$

Then

$$\rho'\gamma'_{ij} = g'\tilde{\lambda}\gamma_{ij} = \lambda\rho\gamma_{ij} = \lambda g_{ij} = g'_{ij}$$

and so the γ'_{ij} are liftings of the g'_{ij} .
Now set

$$\theta_{ijk}(x) = \tilde{\lambda}(\gamma_{ij}(x))\tilde{\lambda}(\gamma_{jk}(x))\tilde{\lambda}(\gamma_{ij}(x)\gamma_{jk}(x))^{-1}, \ x\varepsilon U_{ijk} .$$

Then we have

$$\begin{aligned}\rho'\theta_{ijk}(x) &= \lambda\rho\{\gamma_{ij}(x)\}\lambda\rho\{\gamma_{jk}(x)\}\cdot\lambda\rho\{\gamma_{ij}(x)\}\gamma_{jk}(x)\}^{-1} = \\ &= \lambda\{g_{ij}(x)\}\lambda\{g_{ik}(x)\}\lambda\{g_{ij}(x)g_{ik}(x)\}^{-1} = \\ &= \lambda[g_{ij}g_{ik}(g_{ij}g_{ik})^{-1}] = \lambda(e) = e' \quad .\end{aligned}$$

It follows that

$$\theta_{ijk}(x) \ \varepsilon \ K' , \ x\varepsilon U_{ijk} ,$$

and so these functions are constant. Hence they define a 2-cochain θ in the nerve $N(U)$ with values in K'.

Proposition II: Let p and p' be denote the 2-cocycles for P and P_λ obtained via the liftings γ_{ij} and γ'_{ij}. Then

$$p'(i,j,k) = \theta(i,j,k)\tilde{\lambda}p(i,j,k) \quad .$$

Proof: Applying $\tilde{\lambda}$ to the equation

$$\gamma_{ik} = p_{ijk}^{-1}\gamma_{ij}\gamma_{jk}$$

and using (2) we obtain

$$\gamma'_{ik} = \tilde{\lambda}(p_{ijk}^{-1})\tilde{\lambda}(\gamma_{ij}\gamma_{jk}) = \tilde{\lambda}(p_{ijk})^{-1}\tilde{\lambda}(\gamma_{ij}\gamma_{jk}) \quad .$$

On the other hand,

$$\gamma'_{ik} = p'^{-1}_{ijk}\,\gamma'_{ij}\gamma'_{jk} = p'^{-1}_{ijk}\,\tilde{\lambda}(\gamma_{ij})\tilde{\lambda}(\gamma_{jk}) \quad .$$

These equations yield

$$p'^{-1}_{ijk}\,\tilde{\lambda}(\gamma_{ij})\tilde{\lambda}(\gamma_{jk}) = \tilde{\lambda}(p_{ijk})^{-1}\tilde{\lambda}(\gamma_{ij}\gamma_{jk})$$

whence

$$p'^{-1}_{ijk} = \tilde{\lambda}(p_{ijk})^{-1}\tilde{\lambda}(\gamma_{ij}\gamma_{jk})\tilde{\lambda}(\gamma_{jk})^{-1}\tilde{\lambda}(\gamma_{ij})^{-1} \quad .$$

It follows that

$$p'_{ijk} = \tilde{\lambda}(\gamma_{ij})\tilde{\lambda}(\gamma_{jk})\tilde{\lambda}(\gamma_{ij}\gamma_{jk})^{-1}\tilde{\lambda}(p_{ijk}) =$$

$$= \theta(i,j,k)\tilde{\lambda}(p_{ijk}) \; .$$

Since p and p' are cocycles and since the restriction of $\tilde{\lambda}$ to K is a homomorphism, it follows from the lemma that θ is a cocycle. Thus it represents an element $\theta \varepsilon \check{H}^2(B,K')$.

Now we have as an immediate consequence of proposition II the following

Corollary: The obstruction classes of P and P_λ are connected by the relation

$$\kappa(P_\lambda) = \theta \cdot \tilde{\lambda}_* \kappa(P)$$

where

$$\tilde{\lambda}_*: \check{H}(B,K) \to \check{H}(B,K') \quad .$$

denotes the homomorphism induced by the homomorphism $\tilde{\lambda}: K \to K'$.

In particular, if $\tilde{\lambda}$ is a homomorphism of Γ, then

$$\kappa(P_\lambda) = \tilde{\lambda}_* \kappa(P) \quad .$$

8. The difference class

Let $\rho: \Gamma \to G$ be a fixed central homomorphism and assume that a G-principal bundle P admits two Γ-structures $(\tilde{P},\eta)$ and $(\tilde{P}',\eta)$. Let σ_i be a system of local cross-sections for P defined over a simple covering $\{U_i\}$. Then there are local cross-sections $\tilde{\sigma}_i$ and $\tilde{\sigma}_i'$ in $\tilde{P}'$, respectively such that

$$\eta \circ \tilde{\sigma}_i = \sigma_i \quad \text{and} \quad \eta' \circ \tilde{\sigma}_i' = \sigma_i \quad .$$

Let $\gamma_{ij}: U_{ij} \to \Gamma$ and $\gamma'_{ij}: U_{ij} \to \Gamma$ be the transition functions and define maps $\delta_{ij}: U_{ij} \to \Gamma$ by

$$\delta_{ij}(x) = \gamma_{ij}(x)\gamma'_{ij}(x)^{-1} , \quad x \varepsilon U_{ij} .$$

Then

$$\rho\delta_{ij}(x) = e$$

and so the δ_{ij} are constant. Thus they define a 1-cochain δ in $N(U)$ with values in K . A straightforward computation shows that δ is a cocycle and so it determines an element of $H^1(B,K)$. This element is denoted by $\delta(\tilde{P},\tilde{P}')$ and is called the <u>difference class</u> for $\tilde{P}$ and $\tilde{P}'$.
It follows directly from the definition that

(a) $$\delta(\tilde{P},\tilde{P}')\cdot\delta(\tilde{P}',\tilde{P}'') = \delta(\tilde{P},\tilde{P}'')$$

and

$$\delta(\tilde{P},\tilde{P}') = \delta(\tilde{P}',\tilde{P})^{-1} .$$

<u>Theorem II</u>: Let $\tilde{P}$, $\tilde{P}_1$ and $\tilde{P}_2$ be Γ-structures on P . Then $\tilde{P}_1 \cong \tilde{P}_2$ if and only if

$$\delta(\tilde{P},\tilde{P}_1) = \delta(\tilde{P},\tilde{P}_2) .$$

<u>Proof</u>: In view of relation (a) it is sufficient to show that $\tilde{P}_1 \cong \tilde{P}_2$ if and only if $\delta(\tilde{P}_1,\tilde{P}_2)=0$.
1) Assume that $\phi: \tilde{P}_1 \to \tilde{P}_2$ is an isomorphism of Γ-structures. Let $\sigma^1_i: U_i \to \tilde{P}_1$ be a local system of cross-sections for $\tilde{P}_1$

and set $\sigma_i^2 = \phi\sigma_i^1$. Then the σ_i^2 form a local system of cross-sections for $\tilde{P}_2$. The corresponding transition maps satisfy

$$\gamma_{ij}^2 = \gamma_{ij}^1$$

and so we have $\delta(i,j)=e$ whence $\delta(\tilde{P}_1,\tilde{P}_2)=e$.

2) Assume that $\delta(\tilde{P}_1,\tilde{P}_2)=e$. Then there is a 1-cochain λ in $N(U)$ such that

$$\lambda(i)\lambda(j)^{-1} = \delta(i,j) \quad .$$

Thus,

$$\gamma_{ij}^2(x) = \lambda(i)^{-1}\gamma_{ij}^1(x)\lambda(j) \quad , \quad x\varepsilon U_{ij} \quad .$$

Now set $W_i = \tilde{\pi}^{-1}(U_i)$ and define maps $\phi_i \colon W_i \to \tilde{P}_2$ by

$$\phi_i(\tilde{z}) = \sigma_i^2(\tilde{\pi}_1(\tilde{z})) \cdot \lambda(i)\gamma_i(\tilde{z}), \quad \tilde{z}\varepsilon W_i \ ,$$

where $\gamma_i(\tilde{z})\varepsilon\Gamma$ is the unique element satisfying

$$\tilde{z} = \tilde{\sigma}_i^1\{\tilde{\pi}_1(\tilde{z})\}\gamma_i(\tilde{z}) \quad , \quad \tilde{z}\varepsilon W_i \, .$$

It is easy to check that

$$\phi_i(\tilde{z}) = \phi_j(\tilde{z}) \quad , \quad \tilde{z}\varepsilon W_i \cap W_j \quad ,$$

and so the ϕ_i define a global map $\phi \colon \tilde{P}_1 \to \tilde{P}_2$.

A straightforward calculation shows that

$$\phi(\tilde{z}\cdot\gamma) = \phi(\tilde{z})\cdot\gamma$$

and

$$\eta_2 \circ \phi = \eta_1 \quad .$$

Thus, ϕ is an isomorphism of Γ-structures.

<u>Theorem III</u>: Assume that $(\tilde{P},\eta)$ is a Γ-structure on P and let $\alpha \varepsilon \check{H}^1(B,K)$ be given. Then there exists precisely one Γ-structure $(\tilde{P}',\eta')$ such that the difference class is α.

<u>Proof</u>: The uniqueness follows directly from theorem II. To prove existence, let a be a cocycle in $N(U)$ with values in K representing α. Let γ_{ij} be the transition maps for $\tilde{P}$ and set

$$\text{(b)} \qquad \gamma'_{ij}(x) = a_{ij}^{-1} \cdot \gamma_{ij}(x) \; , \; x \varepsilon U_{ij} \; .$$

Then we have

$$\gamma'_{jk}\gamma'^{-1}_{ik}\gamma'_{ij} = a_{jk}^{-1}\gamma_{jk}\gamma_{ik}^{-1}a_{ik}a_{ij}^{-1}\gamma_{ij} =$$

$$= (a_{ij}a_{ik}^{-1}a_{ij})^{-1}\gamma_{jk}\gamma_{ik}^{-1}\gamma_{ij} \quad .$$

Since a is a cocycle, it follows that

$$\gamma'_{jk}\gamma'^{-1}_{ik}\gamma'_{ij} = \gamma_{jk}\gamma_{ik}^{-1}\gamma_{ij} = e \; .$$

Thus there is a principal bundle $\tilde{P}' = (\tilde{P}',\tilde{\pi}',B,\Gamma)$ with the γ'_{ij} as transition maps.
Next observe that, in view of (a)

$$\rho\gamma'_{ij} = \rho\gamma_{ij} = g_{ij} \quad .$$

Thus, by a standard theorem[1], there exists a strong equivariant bundle map $\eta': \tilde{P}' \to P$. This shows that $(\tilde{P}', \eta')$ is a Γ-structure on P. Finally it follows from the construction that

$$\delta(\tilde{P}', \tilde{P}) = \alpha \quad .$$

This completes the proof.

§2. Bundles with structure group $O(n)$ and $U(n)$

9. The class $\kappa_1(P)$

Let $P = (P, \pi, B, O(n))$ be an $O(n)$-bundle. Choose a system of local cross-sections σ_i over a simple covering of B. Then the transition functions are denoted by a_{ij},

$$a_{ij}(x) \in O(n) \ , \ x \in U_{ij} \ ,$$

and so

$$\det a_{ij}(x) = \pm 1 \ , \ x \in U_{ij} \ .$$

Since the U_{ij} are connected, the functions $\det a_{ij}$ must be constant. Thus they define a 1-cochain q in $N(U)$ with values in the abelian group S^0: $\{1,-1\}$. Its coboundary is given by

$$(Dq)(i,j,k) = q(j,k)\, q(i,k)^{-1} q(i,j)$$

$$\det(a_{jk} a_{ik}^{-1} a_{ij}) = \det 1 = 1$$

and so q is a cocycle.

Next let $\Phi: S^{o} \rightarrow Z_2$ be the isomorphism defined by

(a) $$\Phi(1) = 0 \text{ , } \Phi(-1) = 1$$

and set

$$\Phi(q) = A \quad .$$

Then A is a 1-cocycle with values in Z_2 . Thus it determines an element

$$\kappa_1(P) \varepsilon \check{H}^1(B; Z_2) \text{ .}$$

10. The group Pin(n)

Let $C(n)$ denote the Clifford algebra over an n-dimensional Euclidean space R^n . Recall[2)] that there is a canonical imbedding $R^n \rightarrow C(n)$ and that $C(n)$ is a Z_2-graded algebra. The Z_2-degree of a homogeneous element a will be denoted by deg a . Thus every element $a \varepsilon C(n)$ can be uniquely decomposed in the form

$$a = a^+ + a^- \text{ , } \deg a^+ = 1 \text{ , } \deg a^- = -1 \quad .$$

The <u>degree invoultion</u> ω of $C(n)$ is defined by

$$\omega(a) = a^+ - a^- \text{ .}$$

Let $C^+(n)$ denote the group of invertible elements in $C(n)$. Then a representation of $C^+(n)$ in $C(n)$ is defined by

$$ad(a)u = \omega(a)ua^{-1} \text{ , } a \varepsilon C^+(n) \text{ , } u \varepsilon C(n) \text{ ,}$$

Finally let T denote the antiqutomorphism of $C(n)$ which fixes every vector $x \varepsilon R^n$.
The group $Pin(n)$ is the subgroup of $C^+(n)$ consisting of the elements a which satisfy

(1) $ad(a)(R^n) \varepsilon R^n$

(2) $T(a) \cdot a = e$ (the unit element of $C(n)$).

The representation ad restricts to a representation ρ of $Pin(n)$ in R^n by orthogonal transformations with kernel $\{1,-1\}$ and every orthogonal transformation is obtained in this way. Thus we have the exact sequence

$$1 \to S^o \xrightarrow{i} Pin(n) \xrightarrow{\rho} O(n) \to 1 .$$

Since S^o is in the center of $Pin(n)$, it follows that ρ is a central homomorphism. Finally, observe that

(a) $$\Phi \det \rho(a) = \deg a$$

where Φ denotes the isomorphism $S^o \to Z_2$.

11. O(n)-bundles

Let $P=(P,\pi,B,O(n)$ be an $O(n)$-bundle. Choose a simple covering of B and let $\gamma_{ij}: U_{ij} \to Pin(n)$ be liftings of the transition functions. Then the obstruction cocycle p is a 2-cocycle on B with values in S^o . Applying the isomorphism $\Phi: S^o \to Z_2$ we obtain a cocycle $\bar{P}$ with values in Z_2 . Its cohomology class will also be denoted by $\kappa(P)$.

Theorem I shows that P admits a Pin-structure if and only if $\kappa(P)=0$.

In sec. 12 it will be shown that $\kappa(P)$ coincides with the second Stiefel-Whitney class of P.

12. O(n+m)-bundles

Consider principal bundles P^n and P^m over the same base with groups $O(n)$ and $O(m)$, respectively. Then $P^n \times P^m$ is a bundle with group $O(n)\times O(m)$ (cf. sec. 6, §1). Let P^{n+m} denote the λ-extension of $P^n \times P^m$ where $\lambda: O(n)\times O(m) \to O(n+m)$ denotes the inclusion map.

Proposition III: The obstruction class of P^{n+m} is given by

$$\kappa(P^{n+m}) = \kappa(P^n) + \kappa(P^m) + \kappa_1(P^n) \cup \kappa_1(P^m) .$$

Proof: Identify $C(n+m)$ with $C(n) \hat{\otimes} C(m)$ 2).

Then it is easily checked that

$$\mathrm{ad}(a \hat{\otimes} b)(x \oplus y) = \mathrm{ad}(a)x \hat{\otimes} e + e \hat{\otimes} \mathrm{ad}(b)y$$

$$a \in C^+(n) , \quad b \in C^+(m) , \quad x \in R^n , \quad y \in R^m .$$

It follows from this relation that if $a \in Pin(n)$ and $b \in Pin(m)$ then $a \hat{\otimes} b \in Pin(n+m)$. Thus a map $\tilde{\lambda}: Pin(n) \times Pin(m) \to Pin(n+m)$ is defined by

$$\tilde{\lambda}(a \times b) = a \hat{\otimes} b (-1)^{\deg a \cdot \deg b}, \quad a \in Pin(n), \ b \in Pin(m) .$$

The map $\tilde{\lambda}$ has the following properties which are checked by straightforward computations:

(1) $\tilde{\lambda}[(a\times b)(a'\times b')] = (-1)^{\deg a\cdot\deg b'}\lambda(a\times b)\lambda(a'\times b')$

(2) The diagram

$$\begin{array}{ccc} Pin(n)\times Pin(m) & \xrightarrow{\tilde{\lambda}} & Pin(n+m) \\ \rho^n\times\rho^m \downarrow & & \downarrow \rho^{n+m} \\ O(n)\times O(m) & \xrightarrow{\lambda} & O(n+m) \end{array}$$

commutes. $\rho^n\times\rho^m$ and ρ^{n+m} are related by λ (compare section 7)

3) In particular, if $a\varepsilon S^0$ and $b\varepsilon S^0$, then

(*) $\lambda(a\times b) = a\cdot b$.

Thus we may apply Proposition II to obtain

(a) $q(i,j,k) = \theta(i,j,k)\tilde{\lambda}p(i,j,k)$

where p and q denote the obstruction cocycles for $P^n\times P^m$ and P^{n+m} .

To compute θ let a_{ij} and b_{ij} denote the transition maps for P^n and P^m , respectively, and let α_{ij} and β_{ij} be liftings of the a_{ij} and b_{ij} to $Pin(n)$ and $Pin(m)$. Then it is easily checked that

(b) $\theta(i,j,k) = (-1)^{\deg \alpha_{ij}\cdot \deg \beta_{jk}}$.

On the other hand, Proposition I yields

$$p(i,j,k) = p_1(i,j,k) \times p_2(i,j,k)$$

where p_1 and p_2 denote the cocycles for P^n and P^m.
Applying $\tilde{\lambda}$ and using formula (*)
we obtain

(c) $\tilde{\lambda} p_1(i,j,k) = p_1(i,j,k) \cdot p_2(i,j,k)$

Relations (a), (b) and (c) yield

$$q(i,j,k) = (-1)^{\deg \alpha_{ij} \deg \beta_{jk}} \cdot p_1(i,j,k) p_2(i,j,k) .$$

Now apply the isomorphism $\Phi \colon S^0 \to Z_2$ and set

$$\Phi(q) = Q \qquad \Phi(p_1) = \bar{P}_1 \qquad \Phi(p_2) = \bar{P}_2 .$$

Then we obtain

$$Q(i,j,k) = \deg \alpha_{ij} \cdot \deg \beta_{jk} + \bar{P}_1(i,j,k) + \bar{P}_2(i,j,k) .$$

Now

$$\deg \alpha_{jk} = \Phi \det \rho \alpha_{jk} = \Phi \det a_{jk} = A(j,k) ,$$

where A denotes the 1-cocycle for P^n defined in sec. 9.
Similarly,

$$\det \beta_{jk} = B_{jk} .$$

It follows that

$$Q(i,j,k) = A(i,j)B(j,k) + \bar{P}_1(i,j,k) + \bar{P}_2(i,j,k)$$

whence

$$Q = \bar{P}_1 + \bar{P}_2 + A \cup B$$

Finally, passing to cohomology we obtain

$$\kappa(P^{n+m}) = \kappa(P^n)+\kappa(P^m)+\kappa_1(P^n) \cup \kappa_1(P^m)$$

$$= \kappa(P^n)+\kappa(P^m)+\kappa_1(P^n) \cup \kappa_1(P^m) \ .$$

13. The relation to the Stiefel-Whitney classes

Following Hirzebruch[3] we define the Stiefel-Whitney classes of an $O(n)$-bundle to be the Stiefel-Whitney classes of the associated vectorbundle under the natural action of $O(n)$ on R^n.

Theorem IV: $\kappa_1(P)$ and $\kappa(P)$ coincide with the first and the second Stiefel-Whitney classes w_1 and w_2 of P.

Proof: Recall that $w_1(P)=\kappa(P)$ for $O(1)$-bundles[3].

To show that $\kappa(P)=\omega_2(P)$ observe first that $\omega_2(P')=0$ for every $O(1)$-bundle. On the other hand, since $Pin(1)\cong O(1)\times O(1)$, P' admits a Pin-structure (cf. sec. Hence Theorem I implies that $\omega_2(P')=0$.

Now assume by induction that

$$\kappa_1(P) = \omega_1(P) \quad \text{and} \quad \kappa(P) = \omega_2(P)$$

for every $O(n-1)$-bundle $(n\geq 2)$ and let P^n be an $O(n)$-bundle. By the splitting principle there is a principal bundle $(\hat{P},\hat{\pi},\hat{B},O(n))$ with the following properties[5]:

1) There is a bundle map

$$\begin{array}{ccc} \hat{P} & \xrightarrow{\phi} & P \\ \downarrow & & \downarrow \\ \hat{B} & \xrightarrow{\psi} & B \end{array}$$

which restricts to homeomorphisms on the fibres.

2) ψ induces a monomorphism in cohomology.

3) $\hat{P}$ is the λ-extension of an $O(n-1)$-bundle P and an $O(1)$-bundle P' under the inclusion map $\lambda: O(n-1)\times O(1)\to O(n)$.

By naturality,

$$\kappa_1(\hat{P}) = \psi^{\#}\kappa_1(P^n) \quad .$$

Next,

$$\kappa_1(\hat{P}) = \kappa_1(P^{n-1}) + \kappa_1(P')$$

and so by induction,

$$\kappa_1(\hat{P}) = \omega_1(P^{n-1}) + \omega_1(P') \quad .$$

But

$$\omega_1(P^{n-1}) + \omega_1(P') = \omega_1(\hat{P}) = \psi^{\#}\omega_1(P^n) \quad .$$

It follows that

$$\psi^{\#}\kappa_1(P^n) = \psi^{\#}\omega_1(P^n) \quad .$$

Since $\psi^{\#}$ is injective we obtain

$$\kappa_1(P^n) = \omega_1(P^n) \quad .$$

Similarly,

$$\kappa\ (\hat{P}) = \psi^{\#}\kappa(P^n) \quad .$$

By Proposition II,

$$\kappa(\hat{P}) = \kappa(P^{n-1}) + \kappa(P') + \kappa_1(P^{n-1}) \cup \kappa_1(P_1) \quad .$$

On the other hand

$$\omega_2(\hat{P}) = \omega_2(P^{n-1}) + \omega_2(P') + \omega_1(P^{n-1}) \cup \omega_1(P') .$$

Hence, by induction

$$\kappa(\hat{P}) = \omega_2(\hat{P})$$

But

$$\omega_2(\hat{P}) = \psi^{\#} \omega_2(P^n)$$

and so we obtain

$$\psi^{\#} \kappa(P^n) = \psi^{\#} \omega_2(P^n)$$

whence $\kappa(P^n) = \omega_2(P^n)$.

As an immediate consequence of Theorems I and IV we have

Theorem V: An O(n)-bundle admits a Pin-structure if and only if $\omega_2(P) = 0$.

SO(n)-bundles. Recall that Spin (n) is defined to be the 1-component of Pin (n). Thus the exact sequence $1 \to S_o \to P(n) \to O(n) \to 1$ restricts to an exact sequence

$$1 \to S^o \to \mathrm{Spin}\,(n) \to SO\,(n) \to 1 .$$

In particular, Spin (n) is a double covering group of SO(n). A Spin-structure on an SO(n)-bundle is a Γ-structure with respect to this covering.

Theorem VI: An SO(n)-bundle P admits a Spin-structure if and only if $\omega_2(P) = 0$.

Proof: Consider the commutative diagram

$$\begin{array}{ccc} \mathrm{Spin}(n) & \xrightarrow{\tilde{\lambda}} & \mathrm{Pin}(n) \\ \sigma \downarrow & & \rho \downarrow \\ SO(n) & \xrightarrow{\lambda} & O(n) \end{array}$$

where λ and $\tilde{\lambda}$ are the inclusion maps. Since $\tilde{\lambda}$ is a homomorphism, Proposition II implies that

$$\kappa(\hat{P}) = \tilde{\lambda}_* \kappa(P)$$

where P is an SO(n)-bundle and $\hat{P}$ is its λ-extension. Since $\ker \sigma = S^0$ and $\ker \rho = S^0$ it follows that $\tilde{\lambda}$ is the identity. Thus

$$\kappa(\hat{P}) = \kappa(P) \; .$$

But $\kappa(\hat{P}) = \omega_2(\hat{P})$ (cf. Theorem IV) and $\omega_2(\hat{P}) = \omega_2(P)$. It follows that

$$\kappa(P) = \omega_2(P) \; .$$

Now apply Theorem I.

14. U(n)-bundles

Recall that the universal covering group of U(n) is

$R\times SU(n)$. We shall denote it by $Sun(n)$. The covering projection is given by

$$\rho(t,\alpha) = \tau_n(t)\alpha$$

where

$$\tau_n(t) = \exp\frac{2\pi it}{n}, t\varepsilon R .$$

The kernel K of ρ consists of the elements $(k,\tau_n(k)e)$ with $k\varepsilon Z$. Thus the map

$$(k,\tau_n(k)e) \to k$$

defines an isomorphism $\Phi: K \overset{\cong}{\to} Z$.
Now let P be an $\kappa(n)$-bundle and set $\bar{P}=\Phi^{-1}(p)$ where p denotes the obstruction cocycle of P . Then $\bar{P}$ determines an element $\kappa(P)\varepsilon H^2(B,Z)$. By Theorem I, P admits a $Sun(n)$ structure, if and only if $\kappa(P) = 0$.
Now consider two bundles P^n and P^m with groups $U(n)$ and $U(m)$, respectively and let P^{n+m} denote the λ-extension of $P^n\times P^m$ where $\lambda: U(n)\times U(m) \to U(n+m)$ is the inclusion map.
Proposition IV: The obstruction class of P^{n+m} is given by

$$\kappa(P^{n+m}) = \kappa(P^n) + \kappa(P^m) .$$

Proof: Define a map $\tilde{\lambda}: SU(n)\times SU(m)\to SU(n+m)$ by

$$\tilde{\lambda}(s\times\sigma,t\times\tau) = (s+t) \times (\alpha(s,t)\sigma \times \beta(s,t)\tau)$$

where

$$\alpha(s,t) = \tau_{m+n}(s\cdot n/m-t)$$

and

$$\beta(s,t) = \tau_{m+n}(t\cdot m/n-s) \quad .$$

A straightforward computation shows that

1) $\tilde{\lambda}$ is a homomorphism .

2) The diagram

$$\begin{array}{ccc} \mathrm{Sun}(n) \times \quad (m) & \xrightarrow{\tilde{\lambda}} & \mathrm{Sun}(n+m) \\ \rho^n \times \rho^m \downarrow & & \downarrow \rho^{n+m} \\ U(n) \times U(m) & \xrightarrow{\lambda} & U(n+m) \end{array}$$

commutes.

Thus, by proposition II, the obstruction cocycles satisfy:

$$p^{n+m}(i,j,k) = \tilde{\lambda}(p^n(i,j,k) \times p^m(i,j,k))$$

Now apply the isomorphism $\Phi\colon K \to Z$ and observe that

$$\Phi\tilde{\lambda}(a \times b) = \Phi(a)+\Phi(b) \quad , a\varepsilon K, \ b\varepsilon K ,$$

as is easily checked. It follows that

$$\bar{p}^{n+m}(i,j,k) = \bar{p}^n(i,j,k) + \bar{p}^m(i,j,k) \quad .$$

Passing to cohomology we obtain the proposition.

Theorem VII : The class $\kappa(P)$ coincides with the first Chern class of P ,

$$\kappa(P) = c_1(P) \quad .$$

Proof: The theorem is known for U(1)-bundles[3].
Assume by induction that $\kappa(P^{n-1}) = c_1(P^{n-1})$ for $U(n-1)$-bundles and consider an $U(n)$-bundle P^n. By the splitting principle there is a bundle map, restricting to homeomorphisms on the fibres

$$\begin{array}{ccc} P' \times P^{n-1} & \xrightarrow{\phi} & P^n \\ \hat{\pi} \downarrow & & \downarrow \pi \\ \hat{B} & \xrightarrow{\psi} & B \end{array}$$

such that $\psi^{\#}$ induces a monomorphism in cohomology. Now imitate the argumente given in the proof of Theorem IV for O(n)-bundles.

As an immediate consequence of Theorems I and VII we have

Theorem VIII : An U(n)-bundle admits a Sun-structure if and only if its first Chern class vanishes.

References

1) N. Steenrod, The Topology of the fibre bundle (Princeton University Press, Princeton, N.J., 1951)

2) M.F. Atiyah, R. Bott and A. Shapiro, Topology 3, 3 (1964)

3) F. Hirzebruch, Topological Methods in Algebraic Geometry (Springer, N.Y. 1966)

4) A. Haefliger, C.R. Acad. Sc. Paris 243, 558 (1956)

5) D. Husemoller, Fibre bundles (Mac Graw Hill, N.Y., 1971)

ON THE NON-UNIQUENESS OF SPIN STRUCTURE IN SUPERCONDUCTIVITY

Herbert-Rainer Petry
Institut für Theoretische Kernphysik der Universität Bonn

Introduction: The definition of spinors

Let M denote an oriented, time-oriented Lorenz manifold of dimension four. The bundle ξ of oriented and time-oriented frames is a principal bundle with the special Lorenz group L_+ as structure group. Let ρ: Spin $\to L$ denote the universal covering of L_+ with kernel $K = \text{Kern}\,(\rho) = (e,-e)$.
A principal bundle $\tilde{\xi}$ over M with group Spin, such that

$$\begin{array}{ccc} \tilde{\xi} & \overset{\eta}{\rightarrow} & \xi \\ \tilde{\pi}\searrow & & \swarrow\pi \\ & M & \end{array}$$

commutes and such that

$$\eta(zg) = \eta(z)\rho(g) \qquad (z \in \tilde{\xi},\ g \in \text{Spin}) ,$$

holds, is called a Spin structure.
Let d be a representation of Spin in $Gl(C^m)$. A spinor field of type d is a section in the associated vector bundle $\tilde{\xi}\times_d C^m$. [1)]
It will be shown in this article that a spin structure is not uniquely defined. The physical consequences of this fact are investigated in a model for superconductivity.

Exotic Spinors: There is a natural definition of a covariant derivative in $\tilde{\xi}\times_d C^m$: let α denote the Riemannian connection form in ξ and let ρ_o denote the isomorphism of the Lie algebras of Spin and L_+ . Define a connection form $\tilde{\alpha}$ in $\tilde{\xi}$ by

$$\tilde{\alpha} = \rho_o^{-1}(\eta_+\alpha) .$$

The definition of a covariant derivative in $\tilde{\xi}\times_d C^m$ is then achieved by a standard procedure[1)]. According ref. 2) we know that $\tilde{\xi}$ exists if and only if the second Stiefel-Whitney class vanishes. Furthermore, if $\tilde{\xi}$ is a Spin structure, then one can construct a new inequivalent Spin structure $\tilde{\xi}(k)$ for each non-trivial $k \in H^1(M,K)$ with the properties listed below:

Let $\{U_\alpha\}$ be a simple covering of M and let k be represented by the cocycle $k_{\alpha\beta}: U_{\alpha\beta} \to K$.
There are local sections σ_α and σ_α^k (defined on U_α) for $\tilde{\xi}$ and $\tilde{\xi}(k)$ resp., and the corresponding transition functions $\phi_{\alpha\beta}, \phi_{\alpha\beta}^k$ satisfy

$$\phi_{\alpha\beta}^k = \phi_{\alpha\beta} \cdot k_{\alpha\beta} \ .$$

This yields a one-to-one correspondence between $H^1(M,K)$ and the set of inequivalent Spin structures[2]. If $H^1(M,K)$ is non-trivial, we call spinors exotic. Consider $\tilde{\xi}\times_d C^m$ and $\tilde{\xi}(K)\times_d C^m$ with $d(K) \simeq K$. Assume that we have functions

$$\lambda_\alpha: U_\alpha \to C \ , \quad |\lambda_\alpha| = 1$$

$$k_{\alpha\beta} = \lambda_\alpha / \lambda_\beta \ .$$

In particular,

$$\lambda_\alpha^2 = \lambda_\beta^2 \ \text{ in } \ U_{\alpha\beta} \ \text{ so } \ \lambda: M \to C$$

with $|\lambda|=1$ is uniquely defined by $\lambda = \lambda_\alpha^2$. k is said to be generated by local square roots of λ . As was said before, we have natural covariant derivatives ∇ and ∇^k for $\tilde{\xi}\times_d C^m$ and $\tilde{\xi}(K)\times_d C^m$.
It is proved in ref. 3) that there is a bundle isomorphism T , such that

$$\begin{array}{ccc} \tilde{\xi}(K)\times_d C^m & \xrightarrow{T} & \tilde{\xi}\times_d C^m \\ & \searrow \quad \swarrow & \\ & M & \end{array}$$

commutes and such that

$$T(\nabla_X^k \psi) = \nabla_X T\psi = \nabla_X T\psi - \frac{1}{2} \lambda^{-1} X(\lambda) \cdot T\psi$$

holds for all vector fields X and sections ψ .
This allows us to describe spinors ψ^k , arising from $\tilde{\xi}(k)$ in a different way, namely as sections in $\tilde{\xi}\times_d C^m$ but with another covariant derivative:

$$(1) \qquad \nabla_X^k \psi^k = \nabla_X \psi^k - \frac{1}{2} \lambda^{-1} X(\lambda) \psi^k \ .$$

∇ is the natural covariant derivative and k is generated by local square-roots of λ .
Example: Consider a full torus $T_3 \subset R^3$ centered around the z-axis. Choose $M \subset R^4$ as $R\times T^3$ ($R \simeq$ time), with the standard Minkowski metric. M is parallelizable and has a Spin structure $\tilde{\xi}$ namely $M \times \mathrm{Spin}$.

$H^1(M,K)$ is isomorphic to K and the only non-trivial element is generated by local square-roots of

$$\lambda = (x+y)(x^2+y^2)^{-1/2} .$$

Spinors arising from the trivial Spin-structure are, of course, represented by functions with values in C^m ; the covariant derivative is the ordinary one.

The second possible type of spinor is again such a function but with the covariant derivative given by (1). This example plays now a role in superconductivity.

A model for superconductivity

Let the full torus be made out of superconducting material below the critical temperature. The ions of the metal are fixed and produce an approximately constant potential which keeps the electrons inside the ring. The electron wave-function obeys a quasi-free equation. Single-particle states are Dirac spinors, but, as we have seen there are now two possible types with different covariant derivatives. (See formula (1)).

Let the electrons interact with the photons. This is most easily done by using field operators. But electrons may exist in two different types of spinor states, so let us describe them by two different Dirac field operators ψ,ψ' . With A denoting the photon field operator, the simplest system of quantum field equation reads as:

(1a) $$\gamma^\mu(i\partial_\mu + eA_\mu)\psi = m\psi$$

$$\gamma^\mu(i\partial_\mu + eA_\mu - \tfrac{i}{2}\lambda^{-1}\lambda_\mu)\psi' = m\psi'$$

(2) $$A_\mu = I_\mu$$

$$\operatorname{div} A = 0$$

(3a) $$I_\mu = \bar{\psi}\gamma_\mu\psi + \bar{\psi}'\gamma_\mu\psi'$$

Remark: Formula (1a) might suggest that the equation for ψ' is related to the equation for ψ by a simple gauge transformation. This would be true if we could find a well-defined square-root of λ . Setting $\psi'=\sqrt{\lambda}\tilde{\psi}$ and rewriting (1a) as an equation for $\tilde{\psi}$ indeed would yield an equation which is identical in form with the equation for ψ . We were then forced to regard ψ and ψ' only as different mathematical expressions of the same physical objects. But such a gauge transformation cannot exist since λ does <u>not</u> have a well-defined square-root. The square-root exists only <u>locally</u> and $\tilde{\psi}$ can, therefore, not be defined in a continuous way. It has to change sign at the discontinuity of the locally defined square--root of λ. Therefore, we have to regard ψ and ψ' as field--operators related to different electron states.

In addition I consider a model with a boson field of charge 2e replacing the spinors:

(1b) $$g^{\mu\nu}(i\partial_\mu + 2e\,A_\mu)(i\partial_\nu + 2e\,A_\nu)\Phi = m_\Phi^2\Phi$$

(3b) $$I_\mu = 2e\,\Phi^+(i\partial_\mu + 2e\,A_\mu)\Phi + h.c.$$

The second model is motivated by the pairing hypothesis which states that electrons form spinless bound states of charge 2e. For both types of models we derive (under idential assumptions) flux quantization and Josephson effect.

Observe that (1)-(2) are invariant under the substitutions

$$\Psi \to T(\Psi) = \Psi' \qquad \Psi' \to T(\Psi') = \Psi\cdot\lambda$$

$$T(\Phi) = \Phi\lambda \qquad T(A) = A - B$$

$$\text{with } B_\mu = \frac{i}{2e}\,\lambda^{-1}\lambda_{,\mu}$$

Furthermore, the equal-time commutation relations are invariant. This suggests that there is a unitary operator $\bar{T}$ which commutes with the hamiltonian, such that

$$\bar{T}\Psi\bar{T}^{-1} = T(\Psi) \text{ etc.}$$

For fixed charge consider now the subspace P of eigenstates with lowest possible energy. Let $\bar{P}$ denote the projection operator on P.

<u>Assumption I</u>:

$$\bar{P}A_\mu(x)(1-\bar{P}) = 0 .$$

Let us introduce the operator

$$\phi(t) = 2e \int_C \sum_{i=1}^{3} A_i(t)\,dx^i$$

where c is a path going once around inside the ring and t is any fixed time.

It follows that $\phi|\alpha\rangle \in P$ if $|\alpha\rangle \in P$.

<u>Assumption II</u>: For each $y \in R$ there is one and only one state $|y\rangle \in P$ such that $\phi|y\rangle = y|y\rangle$. These states span P and are normalized to

$$\langle y,y'\rangle = \delta(y-y') \ .$$

Since all states have the same energy

$$\phi(t_1)|y\rangle = \phi(t_2)|y) \ .$$

I and II imply that $|y\rangle$ is eigenstate of $A_\mu(x)$ and $I_\mu(x)$:

$$A_\mu(x)|y\rangle = A_\mu(x,y)|y\rangle \ ,$$

$$I_\mu(x)|y\rangle = I_\mu(x,y)|y\rangle \ ,$$

with time-independent functions $A_\mu(x,y)$ and $I_\mu(x,y)$.

Now we have

$$\bar{T}^{-1}A_\mu\bar{T} = A_\mu + B_\mu \ ,$$

$$\bar{T}^{-1}I_\mu\bar{T} = I_\mu \ .$$

It follows that

$$\bar{T}^{-1}\phi\bar{T} = \phi - 2\pi \ ,$$

$$\bar{T}|y\rangle = |y-2\pi\rangle \ .$$

Furthermore,

$$A_\mu(x,y) = A_\mu(x,y-2\pi) + B_\mu \ ,$$

$$I_\mu(x,y) = I_\mu(x,y-2\pi) \ .$$

Thus the eigenvalues of the current depend periodically on y .

<u>Assumption III</u> (Meissner-effect):

$$(A_{i,j} - A_{j,i})P = 0 \qquad (i,j = 1,2,3) \ .$$

This implies that the eigenvalues of $\phi(t)$ do not depend on the particular path which defines ϕ (Stokes' theorem). ϕ is recognized as the magnetic flux through a surface bounded by c . Since it is independent of c we can simply speak of the flux through the ring.

Flux-quantization

Up to now only the interior of the ring was discussed. We extend our solution to the exterior by requiring that in state $|y\rangle$ the exterior electrostatic potential vanishes (it is shielded by the ions) and that Biot-Savart's law holds for the exterior magnetic potential:

$$A_i^{out}(x,y) = \frac{1}{4\pi} \int_{T_3} d^3x' \, |x'-x|^{-1} I_i(x',y) \; .$$

The exterior magnetic field makes a contribution $\Delta E(y)$ to the total energy in state $|y\rangle$. The degenaracy in energy of these states is, therefore, somewhat (but not completely) removed: if ΔE has a minimum at $y=0$ (which is reasonable, since this corresponds to vanishing fields), then the minimum must also occur for $y=2\pi\cdot k$ (k integer), because ΔE is also periodic in y. Thus, in the states of minimal energy, the flux has the value

$$\pi k \; / \; e$$

in complete agreement with experiment.

Josephson effect

Consider the following experimental arrangement:

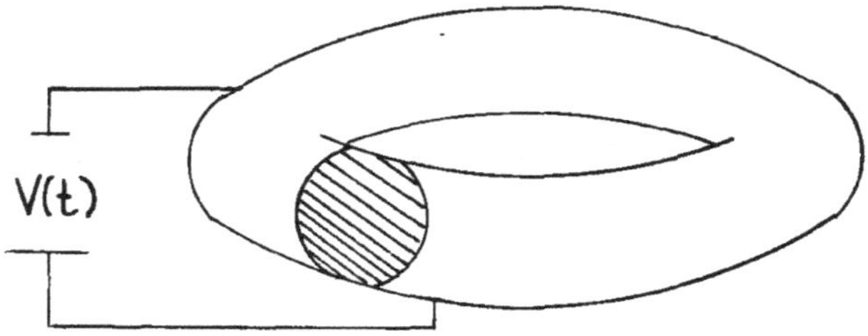

Voltage $V(t)$ is laid across a thin insulating layer which cuts the ring. It will transfer energy to the ring and, therefore, change the wave-function. We assume that this energy is instantaneously removed by cooling. The perturbation $V(t)$ will, therefore, create transitions between states $|y\rangle$ only. Assume now that the change in flux is dominated by the exterior field $V_o(t)$. Maxell's equation yield

$$\dot{\phi}(t) = 2e\, V(t) \ .$$

It follows that

$$\phi(t) = \phi(0) + 2e \int_0^t V(t)dt.$$

Thus the state $|y(t)>$ with $y(t) = y_o + 2e \int_0^t V((t)dt$

is a solution of the time-dependent problem. In these time-dependent states we measure the current $I(x,y(t))$ which is obviously a periodic function of

$$2e \int_0^t V(t)dt$$

This is the Josephson effect. (The fact that we do not get the usual sine-function should not disturb us, since only periodicity is verified experimentally)[4].

Conclusion

By making use of the two possible types of spinors, which arise from the non-trivial topology of the superconducting torus, we have been able to derive the correct results for the quantized magnetic flux and the Josephson-current. The results are the same, if one uses a charged scalar field of charge 2e which simulates an electron pair. This indicates that the assumptions which were made are not too unrealistic. Of course, both types of models do not represent an absolutely correct description of a superconducting torus; they are presented here merely to illustrate the effect of a non-trivial topology. The importance of the model using exotic spinors lies in the fact that it suggests that flux-quantization is an entirely topological effect as it was originally suggested by London. When it was discovered experimentally, there was a factor 1/2 missing in London's prediction[5]. This factor was supplied later on by the pairing hypothesis. By the discussion of exotic spinors presented here I wanted to show that this missing factor can also be supplied by a topological argument. The question now arises how the second type of spinor fits into the standard theory of superconductivity. As it is well-known, this theory is now based on the BCS-transformation[5]. It may, therefore, be conjectured that the second type of spinor is created in superconductivity as a quasiparticle state over a different vacuum solution. Several papers indicate such a possibility[6,7,8]. Let me conclude by mentioning that in this paper the torus was assumed to be highly symmetrical. In fact, we can allow it's shape to be deformed quite arbitrarily without disproving our results. More details can be found in ref. 3).

References

1 W. Greub et.al., Connections, Curvature and Cohomology (Academic Press, N.Y., 1973).

2 W. Greub and H.R. Petry, On the lifting of Structure Groups (contained in this volume).

3 H.R. Petry, Exotic spinors in Superconductivity (to be published in J. Math. Phys.).

4 I.R. Waldram, Rep. Progr. Phys. 39, 751 (1976)

5 I.R. Schrieffer, The Theory of Superconductivity (W.A. Benjamin, N.Y. 1964).

6 N. Byers and C.N. Yang, Phys. Rev. Lett. 7, 46 (1961).

7 A. Bohr and B.R. Mottelson, Phys. Rev. 125, 495 (1962).

8 D.A. Uhlenbrock and B. Zumino, Phys. Rev. 133, 350 (1964).

CONFORMAL INVARIANCE IN FIELD THEORY

V. De Alfaro
Istituto di Fisica Teorica - Università di Torino
Istituto Nazionale di Fisica Nucleare - Sezione di Torino

S. Fubini
CERN - Geneva

G. Furlan
Istituto di Fisica Teorica - Università di Trieste
Istituto Nazionale di Fisica Nucleare - Sezione di Trieste

1. - Introduction

In the recent years we have been interested in properties of field theories that contain no mass or dimensional parameter. Indeed a great deal of attention has been devoted by the community of theoretical physicists to such models as Yang - - Mills gauge theories, dual models, and some chiral models.

Apart from their relevance to physics, theories of this kind are of great mathematical interest, mainly because of the difficulties associated with infrared divergences which arise in these theories and because they show larger space time symmetries than the Poincare' group, as the conformal group or a reparametrization group in some variable. These groups are of great importance in the classification of classical solutions and, as we shall see, they can improve our understanding of massless theories.

A great development in the last decade or so has been the success of models with spontaneous breaking of invariance, and a justified hope could be that a similar pattern develops in massless models, so that lagrangian models without dimensional constants produce spectra with dimensional parameters because of the role of the "ground state" that contains a constant with non zero dimension. According to this view the dimensional constants appearing in hadrons (masses of resonances, slope of Regge trajectories) could be caused by a "vacuum" with less symmetry than the Lagrangian.

Of course there are two possible ways of looking at a massless theory; on one hand it may be taken as an exact theory and its consequences, as the spontaneous breaking, investigated thoroughly;on the other hand it may simply be considered as a high energy approximation to a theory with dimensional constants. In this last case the methods discussed here can be considered as just methods to be applied in the high energy region in order to avoid infrared divergences that should not be there.

In sect. 2 we shall discuss some general features of the conformal group in 4-dimensional space time. Section 3 is dedicated to a review of some interesting classical solutions of the Yang-Mills equations in which we have been interested recently. We shall also discuss a generally covariant model with reparametrization invariance for which we have found a class of solutions. We shall deal in section 4 with a very interesting simplified case of conformal invariance that exhibits the implications of conformal invariance. The model, formulated in just one dimension (i.e. time) is exactly solvable at the classical and quantum levels, and a number of results can be learnt that are useful in higher dimensions. In sect. 5 we develop very briefly a simple model in two dimensions that is solvable and shows the aims of the method we have been proposing. Indeed this model is interactionless in the usual sense, since the equations of motion are linear, and yet it gives non trivial results.

2. - The conformal group

The theories without dimensional constants that are renormalizable have a large group of symmetry, the conformal group (1). Among them the most widely known are the ϕ^4 theory and the Yang-Mills theory, whose Lagrangians are:

$$\mathcal{L} = \frac{1}{2} (\partial_\mu \phi)^2 - \frac{1}{4} g \phi^4$$

$$\mathcal{L} = - \frac{1}{4} F^\alpha_{\mu\nu} F^\alpha_{\mu\nu} \tag{1}$$

$$F^\alpha_{\mu\nu} = \partial_\mu A^\alpha_\nu - \partial_\nu A^\alpha_\mu + g' C_{\alpha\beta\gamma} A^\beta_\mu A^\gamma_\nu .$$

The canonical dimensions of both ϕ and A_μ is +1 in mass units, so that the constants g, g' are dimensionless. For these theories the classical action

$$A = \int \mathcal{L} \, d^4x \tag{2}$$

is invariant under the ordinay Poincare' group operations, and furthermore under dilatations and special conformal transformations,

$$x'_\mu = \frac{x_\mu + c_\mu x^2}{1 + 2c\cdot x + c^2 x^2} ,$$

hence a group with 15 parameters. Its algebra is given by

$$\begin{aligned}
&[L_{\mu\nu}, L_{\rho\sigma}] = i\{g_{\mu\rho} L_{\nu\sigma} + g_{\nu\sigma} L_{\mu\rho} - g_{\mu\sigma} L_{\nu\rho} - g_{\nu\rho} L_{\mu\sigma}\}, \\
&[L_{\mu\nu}, P_\alpha] = i\,(g_{\mu\alpha} P_\nu - g_{\nu\alpha} P_\mu), \\
&[L_{\mu\nu}, K_\alpha] = i\,(g_{\mu\alpha} K_\nu - g_{\nu\alpha} K_\mu), \\
&[L_{\mu\nu}, D] = 0, \\
&[P_\mu, D] = i P_\mu , \quad [K_\mu, D] = - i K_\mu , \\
&[P_\mu, K_\nu] = 2i\,(g_{\mu\nu} D + L_{\mu\nu}).
\end{aligned} \tag{3}$$

The conformal algebra is isomorphic to the algebra of the (pseudo) rotation group 0 (4, 2). This can be seen introducing a null-vector η_A, A=0, 1, 2, 3, 5, 6 such that

$$\eta_1^2 + \eta_2^2 + \eta_3^2 + \eta_5^2 = \eta_0^2 + \eta_6^2 . \tag{4}$$

0(4, 2) rotations of η_A correspond to conformal transformations or dilatations or plain Minkowski rotations of $\eta_\mu/(\eta_5 + \eta_6)$.
A customary choice of η_A is

$$\begin{aligned}
\eta_\mu &= 2a x_\mu , \\
\eta_5 &= a^2 - x^2 , \\
\eta_6 &= a^2 + x^2 .
\end{aligned}$$

We have of course $x_\mu = a\eta_\mu/(\eta_5 + \eta_6)$.
Denoting by L_{AB} the 15 generators of 0(4, 2) we have

$$D = -i L_{56}, \qquad (5)$$
$$P_\mu = \frac{1}{a}(L_{\mu 5} - i L_{\mu 6}),$$
$$K_\mu = a(L_{\mu 5} + i L_{\mu 6}).$$

We note the need to introduce the compactifying parameter a with physical dimensions of a length.

A tensor operator A with given Minkowski spin and canonical dilatation dimension d obeys the commutation relations

$$[A, P_\mu] = i\partial_\mu A,$$
$$[A, L_{\mu\nu}] = \{i(x_\mu \partial_\nu - x_\nu \partial_\mu) + \Sigma_{\mu\nu}\} A,$$
$$[A, D] = i(x\cdot\partial + d) A, \qquad (6)$$
$$[A, K_\mu] = -i\{x^2\partial_\mu + 2i x_\mu (x\cdot\partial + d) + 2i\Sigma_{\mu\nu} x_\nu\} A.$$

We have restrained ourselves to scalar representations in the "conformal spin" for simplicity (see ref. 1). These commutation relations can be formally written in 6-dimensional formalism if we introduce the operator

$$A^{(\eta)} = (\eta_5 + \eta_6)^{-d} A \qquad (7)$$

as a formal function of η_A with the homogeneity restriction

$$(\eta\cdot\partial_\eta + d) A^{(\eta)} = 0.$$

Then we have

$$[A^{(\eta)}, L_{AB}] = L_{AB}^{(\eta)} A^{(\eta)} \qquad (8)$$

where

$$L_{ab}^{(\eta)} = i\left(\eta_a \frac{\partial}{\partial \eta_b} - \eta_b \frac{\partial}{\partial \eta_a}\right), \quad a, b = 1, 2, 3, 5,$$

$$L_{06}^{(\eta)} = i\left(\eta_0 \frac{\partial}{\partial \eta_6} - \eta_6 \frac{\partial}{\partial \eta_0}\right),$$

$$L_{a6} = \eta_a \frac{\partial}{\partial \eta_6} + \eta_6 \frac{\partial}{\partial \eta_a}, \tag{9}$$

$$L_{a0} = \eta_a \frac{\partial}{\partial \eta_0} + \eta_0 \frac{\partial}{\partial \eta_a}.$$

In what follows we shall also be extremely interested in the largest compact subgroup of the conformal group, namely $0(4) \otimes 0(2)$. Notice that the group does not include Minkowski rotations 0(3, 1). It is generated by L_{i5}, L_{ij}, i, j=1, 2, 3 (generators of 0(4), and R_0, generator of 0(2). Appropriate variables to this subgroup can be defined by

$$z_a = \frac{\eta_a}{\eta_0 + i\eta_6}, \quad a = 1, 2, 3, 5. \tag{10}$$

On these variables the 0(4) group acts as a rotation group, while 0(2) rotations in $\eta_0 \cdot \eta_6$ by an angle τ generate the "rescaling" $z_\mu \to z_\mu e^{-i\tau}$ An equivalent set of coordinates for this symmetry has been used[(2-4, 5)]

$$z_a = \frac{\eta_a}{\eta_0^2 + \eta_6^2}, \quad a = 1, 2, 3, 5,$$

$$\rho_i = \frac{\eta_i}{\eta_0^2 + \eta_6^2}, \quad i = 0, 6. \tag{11}$$

Then 0(4) is a rotation leaving invariant the sphere $z^2 = 1$ and the 0(2) group leaves invariant the circle $\rho^2 = 1$. In terms of the Minkowski variables we have

$$z_i = 2x_i/K, \quad z_0 = (1 + t_+ t_-)/K,$$

$$\rho_0 = (t_+ + t_-)/K, \quad \rho_6 = (1 - t_+ t_-)/K, \tag{12}$$

$$K = (1 + t_+^2)^{1/2}(1 + t_-^2)^{1/2}, \quad t_\pm = t \pm |\vec{x}|.$$

We have also

$$\rho_0 = \cos\tau$$
$$\rho_6 = \sin\tau . \tag{13}$$

The appropriate representation of the manifold is given by a hypertorus obtained as the direct product of the threedimensional sphere S_3 times the circle S_1. The variable τ is connected to the Minkowski coordinates by

$$\tau = \operatorname{arctg} t_+ + \operatorname{arctg} t_- . \tag{14}$$

When $-\infty < t < \infty$ we have $-\pi < \tau < \pi$. In particular the generator $R_o = \frac{1}{2}(aH + \frac{1}{a}K_0)$ generates rotations in the $\rho^2 = 1$ circle and thus translations in the variable τ. The evolution of a system can be described by a canonical formalism in the τ variable describing movement from a surface τ = const to the next. It must be stressed that, because of the conformal invariance, there are many different (actually infinitely many) canonical formalisms which describe evolution[3, 7]. So R_o has been advocated by Fubini[3] as a most interesting operator since it generates displacements in the variable τ and the three-dimensional manifold left invariant by R_o is compact; so the movement itself can be represented as a rotation in 0(2). This has some nice consequences. The whole matter will be made clear by the discussion of the example in sect. 4.

There is a very convenient way to achieve correspondence between the x and the z formalism. One just performs a translation in x by 2a, an inversion, again a translation by -a, and then chooses a = (0, i) in Minkowski space. This procedure can be easily used in order to translate results from the Minkowski formalism to the less familiar $0(4) \otimes 0(2)$ one.

A very powerful help, when dealing with conformal invariant models, comes from investigating them is space time manifolds with different dimensions. For instance, if we pay our attention to a theory of a scalar field in D space time dimensions, the conformal invariant theory is described by a Lagrangian

$$\mathcal{L} = \frac{1}{2}(\partial_\mu \phi)^2 + g\,\phi^{\frac{2D}{D-2}} \tag{15}$$

with the equation of motion

$$\partial_\mu \partial_\mu \phi + g\, \phi^{\frac{D+2}{D-2}} = 0 . \qquad (16)$$

If we accept the criterion that theories with conformal invariance have common patterns for any dimension of space-time, then certainly we give up the idea that the interaction term is a polynomial in the field, since this happens for just a few exceptional values of D (D =4 is among them). According to such criterion the usual perturbation around $\phi = 0$ cannot be proposed in general, and most possibly in the case of 4 dimensions it deals badly with the group properties of the model. At the same time $\phi = 0$ is not a classical solution and the linear theory has nothing to do with the limit of the known classical solutions.

These considerations point to the interest in finding classical solutions around which to perform quantum expansion. In particular we shall be induced to consider regular solutions in the Minkowski manifold that have definite properties under the largest compact subgroup $0(4) \otimes 0(2)$. The reason for this will be apparent from the one-dimensional discussion of section 4.

3. - Classical solutions of the Yang-Mills equations and of a generally covariant gauge theory.

The classical solutions of conformal invariant field theories have been the subject of several recent investigations, both in the euclidean and in the Minkowski domain. The most important examples are the scalar field equation

$$\partial_\mu \partial_\mu \phi + g \phi^3 = 0 \qquad (1)$$

and the pure Yang-Mills equations for an SU(2) gauge field:

$$\partial_\rho F_{\mu\nu} = [F_{\mu\rho}, A_\rho] ,$$

$$F_{\mu\nu} = \partial_\mu A_\nu - \partial_\nu A_\mu + [A_\mu, A_\nu] , \qquad (2)$$

where $F_{\mu\nu}$ is a 2x2 matrix,

$$F_{\mu\nu} = \frac{\sigma^a}{2i} F_{\mu\nu}^a . \qquad (3)$$

There is a connection between solutions of (1) and solutions of (2) given by a theorem by Wilczek[8] and by Corrigan and Fairlie[9] Given a ϕ solution of (1) we may write a solution of eq. (2) under the form

$$A_\mu = i\,\sigma_{\mu\nu}\,\partial_\nu \ln\phi \tag{4}$$

where

$$\sigma_{\mu\nu} = \frac{1}{4i}\left(s_\mu \bar{s}_\nu - s_\nu \bar{s}_\mu\right),\quad \bar{\sigma}_{\mu\nu} = \frac{1}{4i}\left(\bar{s}_\mu s_\nu - \bar{s}_\nu s_\mu\right), \tag{5}$$

and

$$\begin{aligned} s_\mu &= (1, i\vec{\sigma}), \\ \bar{s}_\mu &= (1, -i\vec{\sigma}). \end{aligned} \tag{6}$$

A very well known solution of eq. (1) in the euclidean domain has been suggested recently[6]

$$\phi = \sqrt{2/g}\;\frac{2a}{x^2+a^2}\;. \tag{7}$$

This solution has a very simple group theoretical property, being invariant under the 0(5) subgroup of the 0(5,1) conformal group in 4-euclidean space. It is indeed easy to rewrite ϕ through the variables η_A previously introduced $(\eta_4 = i\eta_0)$:

$$\phi = (\eta\cdot\eta_a)^{-1} \tag{8}$$

with $\eta_a^2 = -a^2$. The solution (7) is regular everywhere in the euclidean domain and so has finite action; the energy-momentum tensor vanishes. This solution through eq. (4) gives the famous "instanton" solution directly found by Belavin, Poliakov, Schwarz and Tyupkin[10]:

$$\begin{aligned} A_\mu &= -2i\,\sigma_{\mu\alpha}\,\frac{x_\alpha}{x^2+a^2}\,, \\ F_{\mu\nu} &= 4i\,\frac{a^2\,\sigma_{\mu\nu}}{(a^2+x^2)^2}\,. \end{aligned} \tag{9}$$

This solution is regular everywhere in the euclidean domain; $F_{\mu\nu}$ has a fast convergence at large distances, while the potential A_μ converges to a gauge term:

$$A_\mu \sim g^{-1} \partial_\mu g \, , \tag{10}$$

$$g = \frac{s \cdot x}{\sqrt{x^2}} = \frac{x_4 - i \vec{x} \cdot \vec{\sigma}}{\sqrt{x^2}} \, .$$

A remarkable consequence is that the topological charge is 1:

$$q = \int D \, d^4x = \int_\infty j_\mu \, d\sigma_\mu = 1 \tag{11}$$

where

$$j_\mu = \frac{1}{16\pi^2} \varepsilon_{\mu\nu\alpha\beta} \, \mathrm{Tr} \left\{ A_\nu \partial_\alpha A_\beta + \frac{2}{3} A_\nu A_\alpha A_\beta \right\}$$

$$D = \partial_\mu j_\mu = \frac{1}{64\pi^2} \mathrm{Tr} \, \varepsilon_{\mu\nu\rho\sigma} F_{\mu\nu} F_{\rho\sigma} \, . \tag{12}$$

Let us spend a moment to understand the meaning of the topological quantum number. To be proper, if we have to define a solution of the Yang-Mills equation regular everywhere on the 4-dimensional euclidean space compactified by using the topology of the four-dimensional sphere S_4 we must give a set of A_μ, each one regular in a subdomain of S_4 and gauge equivalent in the sense that in the region where two solutions overlap they differ by a gauge transformation regular in that domain, which defines a non trivial bundle. In the case of the instanton we have two solutions for A_μ , one regular at infinity, the other at the origin, that differ by a gauge transformation:

$$A_\mu^{(\infty)} = -2i \sigma_{\mu\nu} \frac{x_\nu}{x^2 + a^2} \, , \tag{13}$$

$$A_\mu^{(0)} = -2i \sigma_{\mu\nu} \frac{x_\nu a^2}{x^2 (x^2 + a^2)} \, . \tag{14}$$

They are connected, in the intermediate region between $x^2=0$ and $x^2=\infty$, by the gauge transformation generated by $g = \frac{\bar{s} \cdot x}{\sqrt{x^2}}$ Consider the quantity

$$q = \int_R \left(j_\mu^{(\infty)} - j_\mu^{(0)} \right) d\sigma_\mu \tag{15}$$

on a 3-dimensional sphere S_3 centered around the origin, of radius R. Since $A_\mu^{(\infty)}$ and $A_\mu^{(0)}$ differ by a gauge transformation regular on S_3, q is only function of the gauge transformation between $A^{(\infty)}$ and $A^{(0)}$ and not of the individual gauges; it is a number that characterizes the bundle. Now we have

$$\int_R j_\mu^{(0)} d\sigma_\mu = -\int_{r>R} \partial_\mu j_\mu^{(0)} d^4x$$

because $A_\mu^{(0)}$ is regular at infinity, and

$$\int_R j_\mu^{(\infty)} d\sigma_\mu = \int_{r<R} \partial_\mu j_\mu^{(\infty)} d^4x$$

But $\partial_\mu j_\mu^{(\infty)} = \partial_\mu j_\mu^{(0)} = D(x)$, a gauge invariant quantity, and we finally have

$$q = \int d^4x \; D(x) \; . \tag{16}$$

Of course if $R \to \infty$ we have $q = \int_{R\to\infty} j_\mu^{(\infty)} d\sigma_\mu$

due to the regularity of $j_\mu^{(0)}$ at infinity, while for $R \longrightarrow 0$ we have

$$q = -\int_{R\to 0} j_\mu^{(0)} d\sigma_\mu \; .$$

Many instanton solutions with q=n have been found[11, 12] and we are not going to review this important development that has been discussed elsewhere in this conference. We rather proceed to discuss solutions with different symmetry that will be of interest especially in the Minkowski space.

Let us start from the euclidean space and look for a 0(4) invariant solution that possesses simple properties under dilatations.

$$\phi = \frac{1}{\sqrt{g}} \; \frac{1}{\sqrt{x^2}} \tag{17}$$

is a solution of eq. (1). It is singular in the euclidean space at $X^2=0, \infty$, which are the characteristic points of the 0(1, 1) dilatation group. However we can obtain other solutions simply by using the conformal invariance of the equations of motion. For instance we can displace the singular points to any positions u, v, obtaining the solution

$$\phi = \frac{1}{\sqrt{g}} \sqrt{\frac{(u-v)^2}{(x-u)^2(x-v)^2}} . \tag{18}$$

We can now choose $u=(1, \vec{0}) = -v$ and go to the physical Minkowski space by $x_4=ix_o$
The solution becomes

$$\phi = \frac{2}{\sqrt{g}} \frac{1}{(1+t_+^2)^{1/2}(1+t_-^2)^{1/2}} , \quad t_\pm = t \pm |\vec{x}| . \tag{19}$$

This solution is invariant under the largest compact subgroup $0(4) \otimes 0(2)$ of the con formal group in the Minkowski space time. This can be seen in many ways, the sim plest being to rewrite it in the η formalism:

$$\phi = \frac{1}{\sqrt{g}} \frac{1}{(\eta_0^2+\eta_6^2)^{1/2}} . \tag{20}$$

This shows that ϕ is invariant under separate rotations in the $\eta_1, \eta_2, \eta_3, \eta_5$ and in the η_0, η_6 spaces.

The solutions (17) and (19) are of no direct interest in the scalar Minkowski theory because the negative sign should be chosen for g in order to have a correct ϕ^4 behaviour of the potential term in the Lagrangian. They can be used, however, as superpotentials in order to obtain interesting Yang-Mills solutions. Indeed from the expression (4) we obtain from (17)[13]

$$A_\mu^m = \frac{1}{2} g^{-1} \partial_\mu g , \tag{21}$$

$$A_\mu^{\bar{m}} = \frac{1}{2} g \partial_\mu g^{-1} \tag{22}$$

where

$$g = \frac{\bar{s}\cdot x}{\sqrt{x^2}} , \quad g^{-1} = \frac{s \cdot x}{\sqrt{x^2}}$$

These two solutions are the original "meron"solutions established in a direct way in (13); they are related by a singular gauge transformation built through the function g itself:

$$A_\mu^{\bar{m}} = g\, A_\mu^{m} g^{-1} + g\, \partial_\mu g^{-1} ,$$
$$A_\mu^{m} = g^{-1} A_\mu^{\bar{m}}\, g + g^{-1} \partial_\mu g \tag{23}$$

The two solutions above are singular both at $x^2=0, \infty$ in the euclidean space. It is interesting to observe that the asymptotic behaviour of (21) for A_μ is the same as for the instanton solution (9) (apart an important factor 1/2; in (21) the asymptotic behaviour is not a pure gauge). It is easy to check explicitly that the pseudoscalar density D(x) vanishes in any regular domain in the finite euclidean space except at x=0. The surface integrals

$$q_m = \int_\infty j_\mu^{m}\, d\sigma_\mu = 1/2 ,$$
$$q_{\bar{m}} = \int_\infty j_\mu^{\bar{m}}\, d\sigma_\mu = -1/2 , \tag{24}$$

are halves of the corresponding integrals for the instanton (antiinstanton) solutions as seen by the asymptotic behaviour. The value of q is the same for any closed 3-dimensional surface enclosing the origin. We have

$$\partial_\mu j_\mu^{m} = D^{m} = \frac{1}{2}\, \delta^4(x) ,$$
$$\partial_\mu j_\mu^{\bar{m}} = D^{\bar{m}} = -\frac{1}{2}\, \delta^4(x) . \tag{25}$$

For these solutions the pseudoscalar density is concentrated in x=0 in the euclidean space. So the "meron" solution A_μ^{m} corresponds to pseudoscalar charge 1/2 at x=0 and -1/2 at $x \to \infty$ while the reverse holds for $A_\mu^{\bar{m}}$. For these reasons they have been named merons[14].

We use conformal invariance to obtain another solution with both singularities at finite. We perform a translation $x_\mu \to x_\mu - 2a_\mu$, an inversion $x_\mu \to x_\mu / x^2$ and again a translation $x_\mu \to x_\mu + a_\mu$ and we are led to the solutions[15]

$$A_\mu^{m\bar{m}} = \frac{1}{2} f^{-1} \partial_\mu f , \qquad f = \frac{s \cdot z}{\sqrt{z^2}} ,$$
$$A_\mu^{\bar{m}m} = \frac{1}{2} f\, \partial_\mu f^{-1} , \qquad z_\mu = \frac{a_\mu}{2a^2} - \frac{(x+a)_\mu}{(x+a)^2} . \tag{26}$$

These solutions have been obtained also with explicit $0(4) \otimes 0(2)$ formalism[2,3,4,5].
$A_\mu^{m\bar{m}}$ corresponds to a meron located at x=-a and an antimeron at x=a (the converse is true for $A_\mu^{\bar{m}m}$). Indeed we have[15]

$$\text{at } x=-a: \quad A_\mu^{m\bar{m}} \sim \frac{1}{2}(sy_+)\partial_\mu(\bar{s}y_+), \quad A_\mu^{\bar{m}m} \sim \frac{1}{2}(\bar{s}y_+)\partial_\mu(sy_+),$$

$$\text{at } x=+a: \quad A_\mu^{m\bar{m}} \sim \frac{1}{2}(\bar{s}y_-)\partial_\mu(sy_-), \quad A_\mu^{\bar{m}m} \sim \frac{1}{2}(sy_-)\partial_\mu(\bar{s}y_-), \tag{27}$$

$$y_\pm = x \pm a .$$

So for the $m\bar{m}$ and $\bar{m}m$ solutions respectively we have

$$D^{m\bar{m}} = \frac{1}{2}\left\{ \delta^4(x+a) - \delta^4(x-a) \right\},$$

$$D^{\bar{m}m} = -\frac{1}{2}\left\{ \delta^4(x+a) - \delta^4(x-a) \right\}, \tag{28}$$

in agreement with

$$\int_\infty j_\mu \, d\sigma_\mu = 0$$

due to the fast decrease of A_μ at infinity.

These solutions are gauge equivalent (through a singular gauge transformation) to the original dimeron solution. Indeed if we write

$$A_\mu^{m\bar{m}} = i \, \frac{\bar{s}\cdot y_+}{\sqrt{y_+^2}} \, \sigma_{\mu\nu} \left(\frac{y_{+\nu}}{y_+^2} - \frac{y_{-\nu}}{y_-^2} \right) \frac{s\cdot y_+}{\sqrt{y_+^2}} \tag{29}$$

and perform the gauge transformation

$$A_\mu \to h^{-1} A_\mu h + h^{-1} \partial_\mu h$$

with

$$h = \frac{\bar{s}\cdot y_+}{\sqrt{y_+^2}}$$

we obtain

$$A_\mu^{mm} = -i\,\sigma_{\mu\nu} \left(\frac{y_{+\nu}}{y_+^2} + \frac{y_{-\nu}}{y_-^2} \right). \tag{30}$$

This solution is not regular at infinity, its asymptotic behaviour being the same as for the instanton $-2i\sigma_{\mu\nu}\frac{x_\nu}{x^2+a^2}$ solution. It has two merons placed at $y=\pm a$:

$$D^{mm} = \frac{1}{2}\left\{\delta^4(x+a) + \delta^4(x-a)\right\} . \tag{31}$$

A remark, always in the euclidean space, provides further insight into the properties of the $A^{m\bar{m}}$, $A^{\bar{m}m}$ solutions in Minkowski space time[15]. The key is in the evaluation of the surface integral

$$I = \int_S j_\mu d\sigma_\mu \tag{32}$$

evaluated on the closed surface which encloses the $x=-a$ singular point and consists of the $x_4=0$ hyperplane and of the lower hemisphere at infinity. This integral has the value 1/2 for the $A^{m\bar{m}}$ solution. However, the asymptotic part vanishes and we are left with

$$I = \int_{x_4=0} j_4 d^3x = 1/2 \tag{33}$$

(for the $A^{\bar{m}m}$ solution we get of course - 1/2).
Note that the hyperplane $x_4=0$ is invariant under the Wick rotation $x_4 \to it$.

Let us now go to the Minkowski space. Clearly the $A^{m}, A^{\bar{m}}$ solutions are of little interest since they are singular on the light cone. However the conformally displaced solutions $A^{m\bar{m}}$, $A^{\bar{m}m}$ and their gauge equivalents are quite interesting because they are regular in the entire Minkowski domain including infinity. Let us concentrate our attention on the $A^{m\bar{m}}$ solution; the generating operator in Minkowski space is

$$f(\bar{x},t) = -\frac{(1-it_+\vec{\sigma}\cdot\vec{u})(1+it_-\vec{\sigma}\cdot\vec{u})}{(1+t_+^2)^{1/2}(1+t_-^2)^{1/2}} \tag{34}$$

where $\vec{u} = \vec{x}/|\vec{x}|$. Explicitly, we can write $A^{m\bar{m}}$ in the form[2,5]

$$A_0^{m\bar{m}} = \frac{it}{2K^2}\,\vec{\sigma}\cdot\vec{x}, \qquad (K^2 = (1+t_+^2)(1+t_-^2)),$$

$$A_i^{m\bar{m}} = \frac{-i}{2K^2}\left\{\varepsilon_{ijk}x_j\sigma_k + \frac{1+x^2}{2}\sigma_i + x_i\,\vec{\sigma}\cdot\vec{x}\right\} . \tag{35}$$

Under this form one sees at once that the gauge fields A_i^a are real (the reality of $F^a_{\mu\nu}$ had been checked earlier[13] from the A^{mm} gauge equivalent solution). This solution is regular everywhere in the compactified Minkowski space obtained through the introduction of the r_a, ρ_i variables of sect. 2, so that its topology is that of the $0(4) \otimes 0(2)$ hypertorus. Indeed, using an appropriate $0(4) \otimes 0(2)$ formalism for the field that separates out kinematical factors[2-5] and using the appropriate coordinates, the $A^{m\bar{m}}$ field becomes invariant on the torus. The symmetry $0(4) \otimes 0(2)$ is of course expected since the $A^{m\bar{m}}$ solution derives from the scalar solution (19) which is $0(4) \otimes 0(2)$ invariant. The canonical tensor has the same explicit form as for the scalar case:

$$\theta_{\mu\nu} = \frac{1}{2} w^2 (w^2 g_{\mu\nu} - 4 w_\mu w_\nu) \tag{36}$$

with

$$w_\mu = \frac{1}{2i} \frac{\partial}{\partial x_\mu} \ln\left(\frac{1+it_+}{1-it_+} \; \frac{1+it_-}{1-it_-}\right) = \frac{\partial}{\partial x_\mu} \tau . \tag{37}$$

In particular we have the following expressions for the Lagrangian, energy and action:

$$\mathcal{L} = -\frac{1}{8} \operatorname{Tr} F_{\mu\nu} F_{\mu\nu} = \frac{12}{(1+t_+^2)^2 (1+t_-^2)^2} ,$$

$$E = \frac{3}{2} \pi^2 , \tag{38}$$

$$A = \int \mathcal{L} d^4x = \frac{3\pi^3}{2} .$$

In the Minkowski domain the divergence of the "axial current" j_μ vanishes everywhere:

$$D(x) = 0 . \tag{39}$$

Following the usual prescription for the expression of conserved quantities, let us define

$$q = \int_s j_\mu \, d\sigma_\mu \tag{40}$$

where S represents a space-like surface. The vanishing of D and the fast decrease of $A^{m\bar{m}}$ for spacelike directions ensures that the integral (40) is the same for all spacelike surfaces. It can thus be computed on the t=o hyperplane, that is the same as we already computed in the euclidean case, therefore obtaining

$$q_{m\bar{m}} = 1/2 \ ,$$

$$q_{\bar{m}m} = -1/2 \ .$$

Hence the $0(4) \otimes 0(2)$ solutions of the Yang-Mills equations, and in particular the m-$\bar{m}$ solutions are the natural solutions in Minkowski space, regular everywhere on the $0(4) \otimes 0(2)$ torus and possessing a quantized axial charge. We shall not dwell here on the family of 0(4) invariant solutions for which we refer to the comprehensive review paper by Jackiw, Nohl and Rebbi[(4)].

It is interesting to note that both the original "meron" solution, eq. (22), and the generalized one eq. (26), are of the form

$$A_\mu = \frac{1}{2} \, G^{-1} \partial_\mu G \ . \tag{41}$$

The function G transforms as a scalar under conformal transformations. Indeed in eq. (22)

$$G = \frac{s \cdot x}{\sqrt{x^2}}$$

and under the conformal mapping $x \to z$ we have

$$G(x) \to G(z) = \frac{s \cdot z}{\sqrt{z^2}}$$

(see eq. (26).

Hence the quantity that has simple transformation properties under conformal transformations is indeed the matrix G of eq. (41).

A different way of writing G is of interest for further developments. Under the conformal transformation such that x=a, x=b correspond respectively to $z=0$, $z \to \infty$, i.e.

$$x_\mu \to z_\mu = \frac{(x-a)^2 (a-b)_\mu + (a-b)^2 (x-a)_\mu}{(x-b)^2} \tag{42}$$

the function G can be cast in the form

$$G = \frac{\bar{s}\cdot(x-a)}{|x-a|}\;\frac{s\cdot(x-b)}{|x-b|}\;\frac{\bar{s}\cdot(a-b)}{|a-b|} \tag{43}$$

from which it is easy to see that the solution corresponds to a meron at x=a and an antimeron at x=b. One could think that if we had a reparametrization invariance of the theory, $x_\mu \to z_\mu(x)$, larger than the conformal one, we could retrieve other solutions simply by having $G(x) \to G(z)$. Now the original Yang-Mills equations do not possess a space-time invariance group larger than the conformal one, and this is the reason that makes it so arduous to find multimeron solutions and determine their properties.

One is therefore tempted to reformulate gauge theories in an appropriate new framework, invariant under both gauge transformations and generalized coordinate transformations(19)

In this undertaking we shall follow the pioneering work by Wilczek(8), where it was show how a Riemann structure gives rise to a 0(4) gauge structure. Let us start from the so-called tetrad or vierbein formalism. These are vector fields $e^a_\mu(x)$ and are the simplest entities carrying internal symmetry indices "a" (a=1...4) and co-ordinate indices μ (we work Euclidean, i.e., μ = 1...4). Under co-ordinate transformations $x \to x' = z(x)$, the vierbeine transform covariantly, i.e.,

$$e'^a_\mu(x') = \frac{\partial x^\nu}{\partial x'^\mu}\, e^a_\nu(x) \tag{44}$$

while for local gauge transformations the rule is

$$\begin{aligned} e'^a_\mu(x) &= L^a{}_b\, e^b_\mu(x), \\ L^a_b(x)\, L^b{}_c(x) &= \delta_{ac}. \end{aligned} \tag{45}$$

The internal symmetry indices "a" are raised and lowered by the Kronecker symbol δ_{ab}, while for the co-ordinate indices "μ" we have to introduce a metric tensor $g_{\mu\nu}(x)$, according to the known rules of tensor analysis. In analogy with the gravitational case we define:

$$g_{\mu\nu}(x) = \delta_{ab}\, e^a_\mu(x)\, e^b_\nu(x), \tag{46}$$

according to this definition $g_{\mu\nu}(x)$ is a covariant tensor and a gauge scalar. For its inverse $g^{\mu\nu}(x)$ we have $g_{\mu\nu}(x)g^{\nu\lambda}(x) = \delta_{\mu\lambda}$ and so on.

The next step is the introduction of the 0(4) gauge potentials, whose definition is, as usual, related to the requirement of having gauge covariant derivatives The appropriate expression turns out to be:

$$A_{\mu ab}(x) = e_a^{\nu}(\partial_\mu e_{b\nu} - \partial_\nu e_{b\mu}) + \frac{1}{2} e_a^{\rho} e_b^{\sigma}(\partial_\sigma e_{c\rho} - \partial_\rho e_{c\sigma}) e_{c\mu} - (a \leftrightarrow b) . \tag{47}$$

The gauge fields $A_{\mu ab}$ are covariant vectors which behave in the familiar way under gauge transformations.

Let us finally define the covariant field tensor

$$F_{\mu\nu}^{ab} = \partial_\mu A_\nu^{ab} - \partial_\nu A_\mu^{ab} + A_\mu^{ac} A_{\nu c}{}^{b} - A_\nu^{ac} A_{\mu c}{}^{b} . \tag{48}$$

Next, for convenience, we shall introduce the operator notation

$$A_\mu = \frac{1}{2} \sigma^{ab} A_{\mu ab} \tag{49}$$

where σ^{ab} is the matrix generator of the gauge transformation. Then (48) reads

$$F_{\mu\nu} = \partial_\mu A_\nu - \partial_\nu A_\mu + [A_\mu, A_\nu] .$$

We now move to establish the equations of motion of the theory. The precise form of these equations depends on the invariance properties we want to impose on the theory. The simple requirements of Lorentz (and conformal) invariance plus gauge invariance leads to the familiar Yang-Mills expression. The departure from the conventional theory occurs if we further demand the covariance of the equations of motion under general reparametrizations: the usual prescriptions of tensor calculus then yield

$$\partial_\mu \left(\sqrt{g}\, F^{\mu\nu}(x)\right) = \sqrt{g}\, \left[F^{\mu\nu}(x), A_\mu(x)\right] , \tag{50}$$

where the metric tensor $g_{\mu\nu}$ has been used to build the controvariant quantity

$$F^{\mu\nu}(x) = g^{\mu\rho}(x)\, g^{\nu\sigma}(x)\, F_{\rho\sigma}(x) \tag{51}$$

and

$$g = \det g_{\mu\nu} .$$

Equation (50) defines our model we shall call a generally invariant gauge theory It is formally identical to a Yang-Mills theory in a curved space but the physical content is different. The quantities $g_{\mu\nu}(x)$ and $A_{\mu}(x)$ (or $F_{\mu\nu}$), one can consider as the effective fields, are not actually independent and according to Eqs. (47) and (48) they are entirely expressed in terms of the vierbeine $e^{a}(x)$, which are thus the fundamental dynamical variables of the theory.

It is clear that the gauge theory discussed here can be viewed from a pure general relativity point of view. It does then coincide with the alternative theory of gravitation proposed by C.N. Yang[20] and C.W. Kilminster[21]

The usual Yang-Mills theory corresponds then to a particularly simple case of these equations.

We will now concentrate on the determination of an extensive set of classical solutions of Eq. (50).

The symmetry properties of the above theoretical scheme allow us to derive a full class of solutions of Eq. (50), once an elementary one is known, just by applying general co-ordinate transformations. Since the simplest solution available is the single meron one (Eq. (22)), the first step in this construction is to verify that it does indeed satisfy the covariant equation (50).

This is easily done, noticing that a vierbein of the simple form

$$e_{\mu}^{a}(x) = (x^2)^{-1/2}\, \delta_{\mu a} \tag{52}$$

gives

$$g_{\mu\nu} = \frac{1}{x^2}\, \delta_{\mu\nu} \ , \quad A_{\mu} = -\, \sigma_{\mu\nu}\, \frac{x_\nu}{x^2} . \tag{53}$$

It is easy to verify that both the value of $g_{\mu\nu}$ given in the previous equation and the value $\overset{\circ}{g}_{\mu\nu} = \delta_{\mu\nu}$ lead to the same field equations for A , which the coincide with the familiar Yang-Mills equations(+). As a consequence, the single meron solution (22) is also a solution of the generalized gauge theory.

Next, it is immediate to see that under a general reparametrization $x_\mu \to z_\mu(x)$, $A_\mu \to \frac{1}{2} G^{-1} \partial_\mu G$ keeps its form unchanged, G(x) transforming as a scalar,

(+) This is true for any $g_{\mu\nu} = \rho(x)\,\delta_{\mu\nu}$, in particular for $\rho = (1+x^2)^{-1}$ leading to the elementary instanton solution.

i.e., $G(x) \rightarrow G(z(x))$. Therefore $A_\mu(x)$ as given by Eq. (41) is a solution of the generalized equation.

In order to have a physically meaningful characterization of a class of functions G(x) we will take into account the properties of the topological current. Since the general result is that for our solution D(x) = 0 in the domain of regularity, the non-vanishing contributions will be concentrated at the singularities of A (x) which derive from the zeros and the poles of G(x). Following the hint provided by eq. (43) we shall concentrate on fractional functions G(x). This class of functions corresponds of course to a given set of transformations $x \rightarrow z(x)$. On the other hand the nature of the problem is such that, to dispose of the Lorentz and gauge degrees of freedom, one has to resort to the matrix formulation based on the "quaternions" s_μ and $\bar{s}_\mu$. Thus the final expressions of $G(z(x))$ will be in terms of s'z, leading therefore to an effective correspondence $s \cdot x \rightarrow s \cdot z(x)$.

The general form of s'z we will consider is expressed in terms of zeros and poles, namely $\left((s \cdot a)^{-1} = \bar{s} \cdot a / a^2\right)$:

$$s \cdot z = s \cdot (x - a_1) \left[s \cdot (x - b_1) \right]^{-1} s \cdot (x - a_2) \ldots \tag{54}$$

Use of the simple properties of the s_μ, $\bar{s}_\mu$ matrices, like $s_\mu \bar{s}_\nu + s_\nu \bar{s}_\mu = 2\,\delta_{\mu\nu}$ and similar ones, will lead to the explicit function z(x). (We omit , for simplicity, terms which do not bear any x dependence and which can easily be introduced by appropriate limiting processes).

We can therefore conclude that the final form of G leading to a finite number of isolated singularities is:

$$G(x) = \frac{\bar{s} \cdot (x - a_1)}{|x - a_1|} \; \frac{s\,(x - b_1)}{|x - b_1|} \; \frac{\bar{s}\,(x - a_2)}{|x - a_2|} \; \ldots \tag{55}$$

It is now easy to ascertain the behaviour of $A_\mu(x)$ around the singular points a_i, b_j. One finds that, apart from a constant gauge transformation (different from point to point)

$$\text{for } x \sim a_i, \quad A_\mu \sim \frac{1}{2} \frac{s \cdot (x - a_i)}{|x - a_i|} \partial_\mu \frac{\bar{s}\,(x - a_i)}{|x - a_i|},$$

$$\text{for } x \sim b_i, \quad A_\mu \sim \frac{1}{2} \frac{\bar{s} \cdot (x - b_i)}{|x - b_i|} \partial_\mu \frac{s \cdot (x - b_i)}{|x - b_i|}.$$

which confirms that the configuration based on G(x) of Eq. (55) corresponds to having merons located at the points a_i and antimerons at the points b_j. This implies that

$$D(x) = \frac{1}{2}\left\{ \Sigma_i\ \delta^4(x-a_i) - \Sigma_j\ \delta^4(x-b_j)\right\}. \tag{56}$$

The non commutative nature of the matrices in the general solution (55) makes the properties of the general solution highly non trivial and deserving further accurate investigation. We think that many other aspects of this theory for which this class of solutions is known should be investigated.

4. - A one dimensional example.

When the dimension of space-time is one we have a single parameter t (time) and the theory is formally identical to quantum mechanics of a particle in a potential g/r^2.

The Lagrangian and equation of motion are[7]

$$L = \frac{1}{2}\left(\dot\phi^2 - g\phi^{-2}\right), \qquad \ddot\phi - g\phi^{-3} = 0. \tag{1}$$

The action of the theory is invariant under displacement $(t \to t+\varepsilon)$, dilatations $(t \to t+\varepsilon t)$ and conformal transformations $\left(t \to \frac{t}{1+at}\right)$. The set of these operations generates the projective group of transformations

$$t' = \frac{\alpha t+\beta}{\gamma t+\delta}, \qquad \alpha\delta - \beta\gamma = 1 \tag{2}$$

under which the action is invariant and the field ϕ transforms as

$$U^{-1}\phi(t)\,U = (\gamma t+\delta)\,\phi(t').$$

The three generators of the group are H, K and D, whose canonical expression is given by

$$\begin{aligned} H &= \frac{1}{2}\left(\dot\phi^2 + g\phi^{-2}\right), \\ K &= t^2 H - \frac{t}{2}\left(\phi\dot\phi + \dot\phi\phi\right) + \frac{1}{2}\phi^2, \\ D &= tH - \frac{1}{4}\left(\phi\dot\phi + \dot\phi\phi\right) \end{aligned} \tag{3}$$

and their algebra is

$$[H, D] = iH \qquad [H, K] = 2iD. \tag{4}$$
$$[K, D] = -iK$$

This algebra can be brought into an 0(2, 1) form by introducing

$$R = \frac{1}{2}\left(\frac{1}{a}K + aH\right), \tag{5}$$
$$L_1 = \frac{1}{2}\left(\frac{1}{a}K - aH\right), \quad L_2 = D,$$

(note the introduction of a constant a with dimension of time).
The algebra is then

$$[L_2, R] = iL_1, \qquad [L_1, L_2] = -iR.$$
$$[L_1, R] = -iL_2,$$

It is very important to notice that the eigenvalue of the Casimir operator is given in terms of the coupling constant g. Indeed we have

$$C \equiv \frac{1}{2}(HK + KH) - D^2 = R^2 - L_1^2 - L_2^2 \tag{6}$$
$$= \frac{g}{4} - \frac{3}{16}.$$

The representation is fixed by the coupling constant and the whole problem of motion is a group-theoretical investigation of a single representation. Note that in general there is no state which is invariant under the action of all the generators One recognizes that R is the generator of compact 0(2) rotations.
This 0(2) is the highest compact subgroup of 0(2, 1) and ts the reminder in one dimension of the largest compact subgroup $0(4) \otimes 0(2)$ in 4 dimensions. For comparison, the variables η_A are given by

$$\eta_0 = 2at \qquad z = \frac{1-it}{1+it}, \quad \tau = 2\,\mathrm{arctg}\, t/a. \tag{7}$$
$$\eta_5 = a^2 + t^2$$
$$\eta_6 = a^2 - t^2$$

An insight into the physical meaning of the group and its role can be obtained by some formal properties that are easy to deal with in this example. Consider a generator

$$G = uH + vD + wK \tag{8}$$

where u v w are real numbers. The sign of the quantity classifies the character of the generator G:

a) $\Delta < 0$: G is an elliptic operator (generates a compact transformation)

b) $\Delta > 0$: G is a hyperbolic operator (generates a non compact transformation)

c) $\Delta = 0$: G is a parabolic operator.

R belongs to class a, H and K to class c, D and L_1 to class b. All the different operators of class a can be brought into R by a time dilatation and a displacement.

This classification has of course immediate consequences on the spectrum of eigenvalues for each category, and also some other implications we are going to discuss now.

In principle motion, at least infinitesimally, can be generated by any operator G; the effect of G on the field and on a state function is given by

$$\begin{aligned} i[G,\phi] &= f_G \dot{\phi} - \tfrac{1}{2} \dot{f}_G \phi , \\ G|\psi(t)\rangle &= i f_G \frac{d}{dt} |\psi(t)\rangle , \end{aligned} \tag{9}$$

where

$$f_G = u + vt + wt^2 . \tag{10}$$

Now, for a given G we may find a "time" τ_G and a "rationalized field" φ_G such that G translates by a constant amount in τ_G :

$$d\tau_G = dt / f_G \quad , \tag{11}$$

$$\varphi_G = \phi / f_G^{-1/2} \quad . \tag{12}$$

Then

$$i\,[G, \varphi_G] = \frac{d\varphi_G}{d\tau_G}\ , \tag{13}$$

$$G\,|\psi(\tau_G)\rangle = i\,\frac{d}{d\tau_G}\,|\psi(\tau_G)\rangle \tag{14}$$

Not all generators G are good, however, in order to represent motion in t from $-\infty$ to $+\infty$. The connection between τ_G and t is given by

a) class a: $\Delta < 0$

$$\tau_G = \frac{2}{|\Delta|^{1/2}}\left\{\operatorname{arctg}\frac{2wt+v}{|\Delta|^{1/2}} - \operatorname{arctg}\frac{v}{|\Delta|^{1/2}}\right\}. \tag{15}$$

In this case τ_G remains finite for $-\infty < t < \infty$.

b) class b: $\Delta > 0$

$$\tau_G = \frac{1}{\Delta^{1/2}}\ \ln\frac{t-t_1}{t-t_2}\ \frac{t_2}{t_1} \tag{16}$$

where t_1, t_2 are the real roots of $f_G=0$. In particular D has $t=0$, ∞ as singular points.

c) class c: $\Delta = 0$:

$$\tau_G = \frac{1}{w}\left(\frac{1}{t-t_1} + \frac{1}{t_1}\right) \tag{17}$$

H and K belong to this class.

It is clear that if motion has to be generated for $-\infty < t < \infty$ only operators of class a can do it, besides from the hamiltonian that is of class c with the singular point at infinity. All these operators are (at least) as good as the hamiltonian in order to describe motion for all times. In particular this is true for R on which our interest will be next focused as the simplest representative of its class. R generates evolution in time through displacement in the variable

$$\tau = 2\ \operatorname{arctg} t/a \tag{18}$$

that is the one dimensional exact analogue of the $O(4)\otimes O(2)$ variable τ introduced in section 2. The rationalized field corresponding to R is

$$\varphi = \frac{\sqrt{2}\ \phi}{\sqrt{1+t^2}} \qquad (19)$$

In general the lagrangian can be recast as a function of φ_G and $\frac{d\varphi_G}{d\tau_G} \equiv \dot{\varphi}_G$ (a derivative with respect to τ_G, of no effect in the equations of motion, has been dropped from the lagrangian):

$$L_G = \frac{1}{2}\left\{\left(\frac{d\varphi_G}{d\tau_G}\right)^2 + \frac{\Delta}{4}\varphi_G^2 - \frac{g}{\varphi_G^2}\right\} . \qquad (20)$$

Then G is the formal "hamiltonian" of this lagrangian:

$$\begin{aligned} G &= \left(\frac{d\varphi_G}{d\tau_G}\right)^2 - L_G = \\ &= \frac{1}{2}\left(\frac{d\varphi_G}{d\tau_G}\right)^2 + W_G(\varphi_G) \end{aligned} \qquad (21)$$

where the "potential" W_G is

$$W_G = \frac{1}{2}\left(\frac{g}{\varphi_G^2} - \frac{\Delta}{4}\varphi_G^2\right) . \qquad (22)$$

We can see also from this point of view the relevance of the classification of generators according to $\Delta \gtreqless 0$.
Indeed, for $\Delta < 0$, W_G has the form

$$W_G = \frac{g}{\varphi_G^2} + \frac{|\Delta|}{4}\varphi_G^2 \qquad (23)$$

with a minimum at $\varphi_G^4 = 4g/|\Delta|$ and hence the operator G with $\Delta < 0$ has a discrete spectrum with a lower bound and normalizable states. This is in particular the case for R.
Classical constant solutions for this category can be easily found requiring that

$$\frac{d\varphi_G}{d\tau_G} = 0 \ , \qquad \frac{dV_G}{d\varphi_G} = 0 \qquad (24)$$

The constant solution for φ_G is

$$\varphi_G^{cl} = 2\, g^{1/4} / |\Delta|^{1/4} \tag{25}$$

from which the classical solution for ϕ is recovered

$$\phi^{cl} = \varphi_G^{cl}\, f_G^{1/2}(t) . \tag{26}$$

In particular for G=R we have

$$\varphi^{cl} = \sqrt{2}\, g^{1/4} , \tag{27}$$

$$\phi^{cl} = g^{1/4} \sqrt{1+t^2} . \tag{28}$$

Let us now turn our attention to eigenstates and eigenvalues of R; the algebraic approach gives them quite easily. One starts from the lowest eigenstate of R

$$R\,|0\rangle = \gamma\,|0\rangle \tag{29}$$

γ is connected to the eigenvalue of the Casimir operator by

$$\gamma(\gamma-1) = \frac{g}{4} - \frac{3}{16} .$$

From this state we can build all other eigenstates of R by repeated application of $L_+ = L_1 + iL_2$:

$$\begin{aligned} |m\rangle &= C_m\, L_+^m\, |0\rangle , \\ R\,|m\rangle &= (\gamma+m)\,|m\rangle . \end{aligned} \tag{30}$$

Of course inside the multiplet we can also find the eigenstates of the Hamiltonian:

$$H\,|E\rangle = E\,|E\rangle \tag{31}$$

The hamiltonian spectrum is continuous and bounded from below. In particular the lowest state is not normalizable.

It is possible to find the expression of $|E\rangle$ as a superposition of $|n\rangle$ and viceversa. To this aim it is more advisable to introduce a different way of labelling states inside the representation. We define a state $|t\rangle$ labelled by a continuous real parameter t with the following properties:

$$
\begin{aligned}
H|t\rangle &= i\frac{d}{dt}|t\rangle ,\\
D|t\rangle &= i\left(t\frac{d}{dt}+\gamma\right)|t\rangle ,\\
K|t\rangle &= i\left(t^2\frac{d}{dt}+2\gamma t\right)|t\rangle .
\end{aligned}
\tag{32}
$$

The states $|t\rangle$ can be analytically continued to $\mathrm{Im}\, t > 0$. From these states it is very easy to extract eigenstates of R or H or, for what matters, of any other operator. For instance in the case of R it is evident how to deal if one uses a different label $|\tau\rangle$ and defines

$$|\tau\rangle = (1+t^2)^{-\gamma}|t\rangle . \tag{33}$$

We have

$$R|\tau\rangle = i\frac{d}{d\tau}|\tau\rangle \tag{34}$$

So the eigenstates of R can be obtained by simply performing a harmonic analysis of $|\tau\rangle$. One obtains

$$
\begin{aligned}
|n\rangle &= \frac{1}{\sqrt{2\pi}}\int_{-\pi}^{\pi}|\tau\rangle e^{in\tau}\,d\tau ,\\
|t\rangle &= \sum_{m=0}^{\infty}\beta_m(t)\,|m\rangle ,\\
\beta_m(t) &= (-1)^n\left\{\frac{\Gamma(m+2\gamma)}{\Gamma(m+1)}\right\}^{1/2}\left(\frac{1+it}{1-it}\right)^{\gamma+m}(1+t^2)^{-\gamma} .
\end{aligned}
\tag{35}
$$

Equally simple is the expression for the eigenstates of the energy:

$$|E\rangle = \int dt\,|t\rangle\, e^{-iEt}\, 2^{\gamma} E^{\frac{1}{2}-\gamma} . \tag{36}$$

With the help of these states all scalar products of the kind e.g. $\langle m|E\rangle$ can be obtained immediately from $\langle t_1|t_2\rangle$; also the matrix elements of tensor operators (three point functions) can be obtained, from $\langle t_1|A(t)|t_2\rangle$.

In order to determine $\langle t_1|t_2\rangle$ we apply H, D and K and obtain

$$\left(\frac{d}{dt_1}+\frac{d}{dt_2}\right)\langle t_1|t_2\rangle=0$$

$$\left(t_1\frac{d}{dt_1}+t_2\frac{d}{dt_2}+2\gamma\right)\langle t_1|t_2\rangle=0 \tag{37}$$

$$\left(t_1^2\frac{d}{dt_1}+t_2^2\frac{d}{dt_2}+2\gamma t_1+2\gamma t_2\right)\langle t_1|t_2\rangle=0$$

and finally

$$\langle t_1|t_2\rangle = c\,(t_1-t_2)^{-2\gamma}. \tag{38}$$

In the case of $\langle t_1|A(t)|t_2\rangle$ where A(t) obeys the commutation relations that define its tensor character:

$$[H,A]=-i\frac{dA}{dt}$$

$$[D,A]=-i\left(t\frac{d}{dt}-a\right)A \tag{39}$$

$$[K,A]=-i\left(t^2\frac{d}{dt}-2at\right)A$$

one obtains with a procedure quite analogous to the previous one

$$\langle t_1|A(t)|t_2\rangle=A_0(t-t_1)^{-a}(t-t_2)^{-a}(t_1-t_2)^{a-2\gamma} \tag{40}$$

From this formula the matrix elements between, e.g., eigenstates of R can be recovered very simply. One result that immediately follows is that

$$\frac{\langle 0|A(0)|n\rangle}{\langle 0|A(0)|0\rangle}\sim\gamma^{-n} \tag{41}$$

for large γ (i.e. for large g). Hence in this limit the $0\longrightarrow 0$ amplitude is dominant over any $0\longrightarrow n$. This result could also be obtained algebraically.

The four point function is of interest. The formula holds (16)

$$\begin{aligned}\langle u|A(t_1)B(t_2)|v\rangle &= C_aC_b\,(v-u)^{-2\gamma-a-b}[(t_1-u)(t_1-v)]^a\cdot\\ &\cdot[(t_2-u)(t_2-v)]^b\,{}_1F_2(-a,-b;2\gamma;\xi)\end{aligned} \tag{42}$$

where

$$\xi=\frac{t_1-v}{t_1-u}\;\frac{t_2-u}{t_2-v} \tag{43}$$

It is interesting to see what happens for $t_1 \to t_2$. We have

$$\langle u | A(t_1) B(t_2) | v \rangle \simeq (v-u)^{-2\gamma-a-b} \cdot$$

$$\cdot \left\{ (t_1-u)(t_1-v) \right\}^{a+b} \left\{ C_1 + C_2 \left(\frac{u-v}{uv} \right)^{2\gamma+a+b} (t_1-t_2)^{a+b-2\gamma} \right\}. \tag{44}$$

We see that there is an "anomalous power" (remember that γ depends on g) whose presence is due to the fact that the multiplet of statesdepends on the coupling constant.

For reasons that will be clear later we are curious about the energy content of eigenstates of R. We have

$$\langle n | E \rangle \equiv C_n(E) = c_n E^{\gamma-1/2} e^{-aE} L_n^{2\gamma-1}(2aE) \tag{45}$$

(L_n is a Laguerre polynomial).
(we have inserted explicitly the constant a). For the picture of the relevant curves see ref. (7).
The position of the maximum in $C_0^{\ 2}$ is given by $(\gamma_0 - \frac{1}{2})\frac{1}{a}$.
The average value of the energy E_n in the large bump of C_n^2 behaves as $E_n \sim \frac{n}{a}$ for large n.

As a final point let us discuss the role of the classical solution with 0(2) invariance. Although this matter is not so important in this one dimensional case where the quantum problem is exactly solved, we shall discuss it because it is of great relevance for higher dim ensional theories, where the quantum problem cannot be solved exactly and this same method can be carried on.

There is a classical solution which is R-invariant and it has been obtained previously:

$$\phi^{cl} = g^{1/4} \sqrt{1+t^2}. \tag{46}$$

We have $\quad \mathcal{R}\, \phi^{cl} = 0$

where

$$\mathcal{R} = \frac{i}{2} \left\{ (1+t^2) \frac{d}{dt} - t \right\} \tag{47}$$

is the R-operator acting on functions with dimensionality 1/2.
We shall see that this classical solution has the same role as tha H-invariant classical solution (i.e. constant) for the case of the σ-model, namely as a value around which to perform a perturbative expansion. In our case the value is obviously no good because of the form of the Lagrangian. However, reminding that all compact operators can describe the motion, we may find substitutes using a classical solution that corresponds to a constant solution in τ, i.e. to a (constant) minimum of the "potential" $W = \frac{g}{\varphi^2} + \frac{1}{4}\varphi^2$ (of course any other solution corresponding to a class a operator would do the same).
We start the shifting procedure by using the R formalism and the rationalized field $\varphi = \frac{\sqrt{2}\,\phi}{\sqrt{1+t^2}}$. The action can be written as we have seen as

$$A = \frac{1}{2}\int d\tau \left\{\left(\frac{d\varphi}{d\tau}\right)^2 - W\right\} \tag{48}$$

and the stable equilibrium classical solution is

$$\varphi = \varphi^{cl} = \sqrt{2}\; g^{1/4}$$

as said.
We then proceed:

$$\varphi = \varphi^{cl} + \varphi' . \tag{49}$$

Expanding around this point we can linearize the equation for φ':

$$\frac{d^2\varphi'}{d\tau^2} + \varphi' = 0 \tag{50}$$

with the classical solutions $e^{\pm i\tau}$. φ' can be quantized:

$$\varphi' = \frac{1}{2}\left(a\, e^{i\tau} + a^+ e^{-i\tau}\right), \qquad [a, a^+] = 1 . \tag{51}$$

The space of states can be instantly built from a state $|0\rangle$ such that $a|0\rangle = 0$. We have of course $\langle 0|\varphi|0\rangle = \varphi^{cl}$. It is interesting to find the form of the operator R at this order:

$$R = \frac{1}{2} g^{1/2} + a^+ a . \tag{52}$$

So the states created by a^+ are eigenstates of R with eigenvalue $\frac{1}{2}g^{1/2}+n$; this is the large g approximation to the exact eigenvalues $\gamma+n$ where $\gamma=\frac{1}{2}\left(1+\sqrt{g+1/4}\right)$.

The shifting procedure and linear approximation correspond to approximating the potential W by a harmonic potential

$$W=\frac{1}{2}g^{1/2}+\left(\varphi-\sqrt{2}\,g^{1/4}\right)^2. \tag{53}$$

This approximation is easily seen, by comparison with the exact W, to be improving for large g.

It is interesting to compare these results with those derived from the equation of motion of φ switching off the interaction, g=0. We obtain $\ddot{\varphi}+\frac{1}{4}\varphi=0$. There is no constant contribution to φ and the fundamental frequency is 1/2. In quantum terms this means that diagonal matrix elements of φ are vanishing and that the eigenvalue difference between two successive levels is 1/2.

For $g\neq 0$ there is a constant contribution to φ increasing with g and the fundamental frequency is 1. In quantum terms this means that there is now a non vanishing diagonal matrix element of φ that becomes large when g increases; at the same time the level difference has now doubled.

Many of the general features of the exact solutions are already exhibited in the semiclassical approximation. The reason for this is not hard to understand, most features having an evident group theoretical origin which is not spoiled by the semiclassical approximation. Only those predictions depending in an essential way on the form of the Casimir operator are modified in the classical limit.

In conclusion, this simplest example of fully conformal invariant lagrangian presents a number of new physical features which could be a guidance towards more realistic conformal invariant theories.

In the first place the coupling constant g determines the Casimir eigenvalue and this is the only way g appears in all expressions. Secondly, there is no state which is invariant under the action of all conformal generators.

Another important feature that emerges is that the compact operators like $R=\frac{1}{2}(H+K)$ can describe the motion (at least) as well as the hamiltonian; R has a discrete spectrum and fully normalizable states, due to the presence of the time a acting as an infrared cutoff responsible for the good behaviour at large distance and discrete spectrum. As a consequence we get a strong indication that eigenstates of R rather than H should be used to solve the problem of motion. In the language of spontaneously broken symmetries we can say that invariance under transformations generated by e.g, H or D are spontaneously broken.

Of course in a field theory this approach is quite O.K. at the tree level and corresponds to a σ -model. We know however that in the usual approach the renormalization procedure introduces a breaking of the conformal invariance[17]. It is possible to speculate about the possibility of a regularization procedure different from the usual one in that R remains a constant of the motion. This would be a very interesting situation. Two different attitudes could be developed. The conservative one just allows for the use of R and the semiclassical expansion simply as a tool to get rid of infrared divergences through the introduction of a length a in the operator R to be eventually let go to infinity.

On the contrary a less moderate view could be that only average values of products of fields in $|0\rangle$ (lowest state of R) are observable. In this case one runs into trouble with energy-momentum conservation. Let us quote briefly the arguments[6] that have been advanced as a possible way out of this situation in a 4-dimensional theory.

In a conformal invariant theory only inclusive quantities can be observed, as usual with massless quantum theories. Consider an amplitude

$$M_{0n} = \int d^4x_1 \langle 0|A(x_1)|n\rangle\, e^{i\omega\cdot x_1} \tag{54}$$

The observable quantity will be

$$\begin{aligned}\sum_n |M_{0n}|^2 &= \int d^4x_1\, d^4x_2\, e^{i\omega\cdot(x_1-x_2)} \langle 0|A(x_1)A(x_2)|0\rangle = \\ &= \int d^4x\, e^{i\omega\cdot x} \int d^4X \langle 0|\, A(x+X)\, A(X-x)\,|0\rangle\end{aligned} \tag{55}$$

Now we have

$$\begin{aligned}\langle 0|\, A(X+x)\, A(X-x)\,|0\rangle &= \langle 0|e^{-iP\cdot X} A(x)\, A(-x)\, e^{iP\cdot X}|0\rangle = \\ &= \int d^4p_1\, d^4p_2 \langle p_1|A(x)\, A(-x)|p_2\rangle\, e^{i(p_2-p_1)\cdot X} \langle 0|p_1\rangle\langle p_2|0\rangle\end{aligned} \tag{56}$$

and

$$\sum_n |M_{0n}|^2 = \int d^4x\, e^{i\omega\cdot x} \int d^4p_1 \langle p_1|\, A(x)\, A(-x)|p_1\rangle\, |\langle 0|p_1\rangle|^2 . \tag{57}$$

This result shows that the recipe corresponds to averaging the matrix element $\langle p_1 | A(x) A(-x) | p_1 \rangle$ over the "energy-momentum content" of the lowest state, represented by $| \langle 0 | p_1 \rangle |^2$. Indeed the quantity $| \langle 0 | p_1 \rangle |^2$ is the energy-momentum spread of the ground state.

The recipe can also be read as a statistical average over the set of equivalent "vacua" that are obtained from $|0\rangle$ by a displacement:

$$|0_X\rangle = e^{iP.X} |0\rangle. \tag{58}$$

Indeed we can also write

$$\sum_m |M_{om}|^2 = \int d^4x \, e^{i\omega.x} \int d^4X \, \langle 0_X | A(x) A(-x) | 0_X \rangle. \tag{59}$$

In the one dimensional model that we have extensively examined, the energy spread of the ground state is given by

$$| \langle 0 | E \rangle |^2 = c \, E^{2\gamma-1} e^{-2aE} \tag{60}$$

and we may introduce the ground state statistical matrix

$$\rho = \int dE \, |E\rangle \langle E| \quad |\langle 0 | E \rangle|^2 . \tag{61}$$

It is amusing the note that this is approximately equivalent to a canonical distribution with temperature 1/2a.

5. - A simple model in two dimensions.

We want now to give an idea of the kind of results that can be obtained in a field theory with this approach. Reliable 4-dimensional theories are very difficult and we shall develop here a very simple model in two space-time dimensions. The model is a "free field" one and in two dimensions the dimensionality of the field is zero. This circumstance will make the model very non trivial since, in the absence of an interaction term, spontaneous breaking of conformal invariance will take place leading to Green's functions which will be substantially different from the usual free ones.

Consider a dimensionless field in two dimensions, obeying a free field equa tion of motion:

$$\Box \phi = 0 \tag{1}$$

Its usual handling is trivial. However, we shall use a non trivial representation for the field and shall see that the ensuing situation is of some interest. We shall suppose that the conformal invariance is broken.

For simplicity of notations we shall deal with euclidean x, so the conformal algebra will be 0(3, 1). The six conformal generators can be chosen as L, D, P , K , where L generates rotations in the x_1, x_4 plane while D generates dilatations. When in the end one wishes to use Minkowski variables the only thing one has to go through is considering the x variables as z ones and rename generators so that R will take the place of D.

We will now look for 0(2), that is L-invariant, solutions.
The only ones are given by 1, $\ln x^2$. So let us put

$$\phi = \phi_0 + \phi' \tag{2}$$

and

$$\phi_0 = q + \frac{p}{4\pi i} \ln x^2 . \tag{3}$$

This choice is consistent with the idea that ϕ_0 is the average value of ϕ in a state $|0\rangle$ that is an eigenstate of the 0(1, 1) generator D with non vanishing eigenvalue δ :

$$D|0\rangle = \delta |0\rangle \tag{4}$$

and $$\langle 0|\phi'|0\rangle = 0 . \tag{5}$$

Then

$$\langle 0|[D, \partial_\mu \phi]|0\rangle = 0 = i(x\cdot\partial+1)\partial_\mu \langle 0|\phi|0\rangle = \\ = i(x\cdot\partial+1)\partial_\mu \phi_0 \tag{6}$$

which is consistent with the form of ϕ_0 written above. Indeed $\partial_\mu \phi_0$ is a field with assigned dimensionality 1. On the contrary ϕ has an infinitesimal non zero dimensionality and, from (3), it is also not single-valued. However what we are doing is correct since the energy-momentum tensor (and hence all observable quantities) is expressed as a function of $\partial_\mu \phi$ only.

In order to quantize the field, let us look at the classical solutions. We have (in two dimensions)

$$x^2 \Box = -L^2 + (x \cdot \partial)^2 \tag{7}$$

where L is the rotation generator in the x_1 x_4 plane:

$$L^2 = -(x_1 \partial_4 - x_4 \partial_1)^2 = -\frac{\partial^2}{\partial \theta^2}$$

and

$$x \cdot \partial = r \frac{\partial}{\partial r} .$$

Thus we have

$$L^2 \phi = \left(r \frac{\partial}{\partial r}\right)^2 \phi$$

$$\left\{ \frac{\partial^2}{\partial \theta^2} + \left(r \frac{\partial}{\partial r}\right)^2 \right\} \phi = 0 \tag{8}$$

The solutions are

$$\phi_\ell = r^\ell e^{i\ell\theta}, \; r^{-\ell} e^{i\ell\theta} , \qquad \ell = \ldots -2, -1, 0, 1, 2, \ldots \tag{9}$$

So we put (hermitean field)

$$\begin{aligned} \phi = {} & q + \frac{p}{4\pi i} \ln x^2 + \\ & + \sum_{\ell=1}^{\infty} (4\pi\ell)^{-1/2} \left\{ a_\ell^+ x_+^\ell + a_\ell x_+^{-\ell} \right\} + \\ & + \sum_{\ell=1}^{\infty} (4\pi\ell)^{-1/2} \left\{ b_\ell^+ x_-^\ell + b_\ell x_-^{-\ell} \right\} \end{aligned} \tag{10}$$

where $x_\pm = r e^{\pm i\theta}$

and $[a_k, a_\ell^+] = [b_k, b_\ell^+] = \delta_{k\ell}$, $[q, p] = i$.

All other commutators vanish.

The canonical commutation relations at equal r follow:

$$\left[\phi(r, \theta_1), \, r \frac{d\phi(r, \theta_2)}{dr} \right] = \delta(\theta_1 - \theta_2). \tag{11}$$

A complete set of states can be very easily built starting from a state with the following properties:

$$\begin{aligned} a_\ell |0\rangle &= 0 , \\ b_\ell |0\rangle &= 0 , \\ q |0\rangle &= 0 , \\ p |0\rangle &= P_0 |0\rangle \neq 0 . \end{aligned} \tag{12}$$

The higher states can be obtained in the usual way by repeated application of a^+_k, b^+_k. With these conventions the $|0\rangle$ state is invariant under the $0(2)\otimes 0(1,1)$ subgroup generated by L and D. Before checking this, let us compute the Green function $\langle 0|\phi(x_2)\phi(x_1)|0\rangle$.

As usual when dealing with euclidean space, we can only define it for $r_2 > r_1$ [18]. We get

$$\langle 0|\phi(x_2)\phi(x_1)|0\rangle \,\vartheta(r_2 - r_1) = \tag{13}$$

$$= -\frac{1}{4\pi} \ln (x_1 - x_2)^2 + \frac{1}{4\pi} \ln r_2^2 +$$

$$+ \langle 0| \left\{ q^2 - \frac{p^2}{16\pi^2} \ln r_1^2 \ln r_2^2 + \frac{qp}{4\pi i} \ln r_1^2 + \frac{pq}{4\pi i} \ln r_2^2 \right\} |0\rangle .$$

From this expression one can obtain again the correct equal r commutator. Let us now find the form of the dynamical variables. The canonical tensor is (all products are meant to be normally ordered)

$$\theta_{\mu\nu} = \partial_\mu \phi \, \partial_\nu \phi - \frac{1}{2} \delta_{\mu\nu} \, \partial_\alpha \phi \, \partial_\alpha \phi . \tag{14}$$

Inserting $\phi = \phi_0 + \phi'$ we obtain

$$\theta_{\mu\nu} = -\frac{p^2}{4\pi x^2} \left(\frac{x_\mu x_\nu}{x^2} - \frac{1}{2} \delta_{\mu\nu} \right) + \tag{15}$$

$$+ \frac{p}{2\pi i} \left(x_\mu \partial_\nu \phi' + x_\nu \partial_\mu \phi' - \delta_{\mu\nu} \, x \cdot \partial \phi' \right) \frac{1}{x^2} +$$

$$+ \partial_\mu \phi' \partial_\nu \phi' - \frac{1}{2} \delta_{\mu\nu} \, \partial_\alpha \phi' \partial_\alpha \phi' .$$

The expressions for the 0(3, 1) generators are[18]

$$D = i\int_0^{2\pi} d\theta \, (x_\mu x_\nu \theta_{\mu\nu})_{r=1}$$

$$P_\mu = i\int_0^{2\pi} d\theta \, (x_\nu \theta_{\mu\nu})_{r=1}$$

$$K_\mu = i\int_0^{2\pi} d\theta \left\{ 2x_\mu x_\alpha x_\beta \theta_{\alpha\beta} - x^2 x_\alpha \theta_{\mu\alpha} \right\}_{r=1} \tag{16}$$

$$L = i\int_0^{2\pi} d\theta \left\{ x_1 x_\beta \theta_{4\beta} - x_4 x_\beta \theta_{1\beta} \right\}_{r=1} .$$

It is more advisable to use "light cone" variables:

$$x_\pm \, , \qquad \partial_\pm = \frac{\partial}{\partial x_\mp} \quad . \tag{17}$$

Then we have

$$\theta_{+-} = \theta_{-+} = 0$$

$$\theta_{++} \equiv \theta(x_-) = -\frac{p^2}{16\pi^2 x_-^2} + \frac{p}{2\pi i x_-} \partial_+ \phi' + (\partial_+ \phi')^2 \tag{18}$$

θ_{++} is just a function of x_-. Conversely

$$\theta_{--} = \theta(x_+)$$

Then we have the representation for the generators in terms of the field operators:

$$P_+ = \frac{p}{\sqrt{2\pi}} b_1^+ - i\sqrt{2}\sum_{l>1}^{\infty} \sqrt{l(l-1)} \, b_l^+ b_{l-1} \, ,$$

$$P_- = \frac{p}{\sqrt{2\pi}} a_1^+ - i\sqrt{2}\sum_{1}^{\infty} \sqrt{l(l-1)} \, a_l^+ a_{l-1} \, ,$$

$$K_+ = -\frac{p}{\sqrt{2\pi}} a_1 - i\sqrt{2}\sum_{1}^{\infty} \sqrt{l(l-1)} \, a_{l-1}^+ a_l \, , \tag{19}$$

$$K_- = -\frac{p}{\sqrt{2\pi}} b_1 - i\sqrt{2}\sum_{1}^{\infty} \sqrt{l(l-1)} \, b_{l-1}^+ b_l \, ,$$

(one can also easily recover the two algebras whose direct product is 0(3, 1)).

We have also

$$\Delta = iD = \frac{P^2}{4\pi^2} + \sum_1^\infty \ell \left(a_\ell^+ a_\ell + b_\ell^+ b_\ell \right),$$

$$L = i \sum_1^\infty \ell \left(a_\ell^+ a_\ell - b_\ell^+ b_- \right) \tag{20}$$

from which we see that states with definite occupation numbers are eigenstates of Δ We have

$$\Delta |0\rangle = \frac{P_0^2}{4\pi} |0\rangle . \tag{21}$$

The representation of the conformal group provided by applying the field operators to the "vacuum" $|0\rangle$ is of course highly reducible. We could build an irreducible representation by applying the raising operators $(P_+)^a$ or $(P_-)^a$ to $|0\rangle$. Another irreducible representation is provided by applying the raising operators to, e.g., $a_2^+ |0\rangle$. In this state Δ has the eigenvalue $\frac{P_0^2}{4\pi} + 2$.

We can compute vacuum expectation values of any operator. For instance, take the ordered product

$$4\pi^2 \langle 0 | \theta_{++}(x)\, \theta_{++}(y) | 0 \rangle\, \theta(\tau_x - \tau_y) =$$

$$= \frac{\ell_0^2}{x^2 y^2} + \frac{2\ell_0}{xy} \frac{1}{(x-y)^2} + \frac{1}{2} \frac{1}{(x-y)^4} \tag{22}$$

where $\ell_0 = P_0^2/8\pi$.

This formula shows that the model is non trivial in spite of the linearity of the equation of motion, because $\ell_0 \neq 0$, i.e. $\Delta|0\rangle = 2\ell_0 |0\rangle \neq 0$.

We see that the last term is translation invariant and is what one would have written for a conventional free field theory. The "classical" term $\propto \ell_0^2$ and the semiclassical $\propto \ell_0$ are due to the peculiar properties of the ground state.

An amusing way of considering the effect of the zero dimensionality of the field is to look at the "composite" scalar operator

$$S(x) = \partial_\mu \phi\, \partial_\mu \phi \tag{23}$$

which transforms under the conformal group as a scalar field of dimensionality +2.

The Green's function is

$$\langle 0|S(x)\,S(y)|0\rangle\,\Theta(x_4-y_4) = \frac{p^4}{16\pi^4}\,\frac{1}{x^2y^2} + \tag{24}$$
$$+\ \frac{p^2}{2\pi^3}\,\frac{1}{(x-y)^4}\left\{x\cdot y\left(\frac{1}{x^2}+\frac{1}{y^2}\right)-2\right\} + \frac{1}{\pi^2}\,\frac{1}{(x-y)^4}$$

which exhibits the customary p^4, p^2 terms related to the spontaneous breaking of conformal invariance.

REFERENCES

1) - G. Mack and Abdus Salam, Ann. Phys. 53, 174 (1969).

2) - D.M. Schechter, MIT preprint CTP 640.

3) - J. Cervero, L. Jacobs and C. Nohl.

4) - R. Jackiw, C. Nohl and C. Rebbi, MIT preprint CTP 675.

5) - M. Luscher, Hamburg University preprint.

6) - S. Fubini, N. Cimento 34A, 521 (1976).

7) - V. de Alfaro, S. Fubini, G. Furlan, N. Cimento 34A, 569 (1976).

8) - F. Wilczek in "Quark confinement and Field Theory," Stump and Weingarten eds. Wiley 1977.

9) - F. Corrigan and D. Fairlie, Phys. Lett. 67B, 69 (1977).

10)- A. Belavin, A. Polyakov, A. Schwartz and Y. Tyupkin, Phys. Lett. 59B, 85 (1975).

11)- G. 't Hooft, Coral Gables proceedings 1977.

12)- R. Jackiw, C. Nohl and C. Rebbi, Phys. Rev. D15, 1642 (1977).

13)- V. de Alfaro, S. Fubini, G. Furlan, Phys. Lett. 65B, 163 (1976).

14)- C. Callan, R. Dashen and D. Gross, Phys. Lett. 66B, 375 (1977).

15)- V. de Alfaro, S. Fubini, G. Furlan, CERN preprint TH 2397.

16)- V. de Alfaro, S. Fubini, G. Furlan, ICTP preprint 1976.

17)- S. Coleman, R. Jackiw, Ann. Phys. 67, 552 (1971).

18)- S. Fubini, A.J. Hanson and R. Jackiw, Phys. Rev. D7, 1732 (1973).

19)- V. de Alfaro, S. Fubini, G. Furlan, "Classical solutions of Generally Invariant Gauge Theories", CERN preprint TH 2438.

20) - C.N. Yang, Phys. Rev. Lett. 33, 445 (1974).

21) - C.W. Kilminster, in Perspectives in Geometry and Relativity, ed. Hoffman, Indiana University Press, Bloomington.

GEOMETRIC QUANTIZATION AND THE W K B APPROXIMATION

Nicholas Woodhouse
(The Mathematical Institute, 24-29 St. Giles, Oxford)

INTRODUCTION

Consider a classical system with phase space (M,ω) (a 2n-dimensional symplectic manifold) and Hamiltonian $h : M \to \mathbb{R}$. In [2], Blattner describes a general quantization construction in which the wave functions of the underlying quantum system are represented by half densities normal to a real polarization of M. He shows how the quantizations arising from two different polarizations can be related by pairing the corresponding wave functions and he explains why the pairing must be corrected by replacing half densities by objects called half forms.

In the first part of this paper, I shall explain, in the special case that M is the cotangent bundle of an n-dimensional configuration space Q, how Blattner's construction leads to a semiclassical description of the underlying quantum dynamics and I shall show why half forms are necessary to make this description consistent in the large. The ideas involved here are not new (see, in particular, Guillemin and Sternberg [4]), though the presentation may be unfamiliar. In the second part, I shall explain the connection with the older complex W K B method.

1.1 HALF DENSITY QUANTIZATION

First recall the half density quantization in the special case $M = T^*Q$ (see [2] and also [6]). The first step is to choose a real polarization F of M such that the space of leaves M/F is a Hausdorff manifold. An obvious possibility is to take the vertical polarization P, where the leaves are the individual cotangent spaces in T^*Q and $M/P = Q$; however, other choices must also be allowed for reasons which will emerge shortly.

Next, we introduce the wave function space W_F. If we let θ denote the canonical 1-form[1] on M, then this can be represented as the set of all half F-densities f which satisfy

$$i\hbar\, \mathcal{L}_X f - (X \lrcorner \theta) f = 0 \qquad 1.1$$

for every Hamiltonian vector field X tangent to F. (Informally, a half F-density is the square root of a volume element on M/F. More precisely, it is a complex valued function $f(m,X_a)$ depending on a point $m \in M$ and a complex basis $X_1, X_2, \ldots, X_n$ for the tangent space to F at m with the property that

$$f(m, X_a C_{ab}) = |\det C|^{-\frac{1}{2}} f(m, X_a) \qquad 1.2$$

for any $C \in GL(n,\mathbb{C})$; here, and below, the indices run over $1, 2, \ldots, n$ with a summation convention. The Lie derivative of a half F-density can be defined along any vector field X whose flow preserves F. See [2] or [6] for details.)

There is a natural inner product on W_F, given by an integration[2] over M/F, and any classical observable $k : M \to \mathbb{R}$ which generates canonical transformations leaving F invariant defines a symmetric operator K on W_F. This is given explicitly by

$$K(f) = i\hbar \mathcal{L}_{X_k} f + (k - X_k \lrcorner \theta) f \qquad 1.3$$

where X_k is the Hamiltonian vector field defined by

$$X_k \lrcorner \omega + dk = 0. \qquad 1.4$$

However, observables which do not leave F invariant cannot be dealt with so easily. Thus, if we are to obtain the quantum dynamics by this straightforward construction, then we must choose F so that it is invariant under the Hamiltonian flow generated by h. In particular, the vertical polarization will not do unless h is, at most, an inhomogeneous linear function of the momenta.

To make the calculations as simple as possible, we will assume that F is actually tangent to X_h. Locally, such polarizationscorrespond to the classical complete integrals of the Hamilton-Jacobi equation: for, suppose that F is transverse[3] to P in some neighbourhood $U \subset M$. Choose local coordinates $\{k_a\}$ on M/F and local coordinates $\{x^a\}$ on Q. Then, in U, the k_a's label the leaves of F and the x^a's label the individual points on each leaf. Together, therefore, the 2n functions $\{k_a, x^a\}$ can be used as local non-canonical coordinates on U (strictly, the coordinates are $x^a \circ pr_1$ and $k_a \circ pr_2$ where $pr_1 : M = T^*Q \to Q$ and $pr_2 : M \to M/F$ are the projections).

For simplicity, suppose that these coordinates take values in some open interval $V \times W \subset \mathbb{R}^n \times \mathbb{R}^n$. Then, for each $\mathbf{k} = (k_1, k_2, \ldots, k_n) \in V$, the corresponding leaf $\Lambda_{\mathbf{k}}$ of F defines a 1-form $\alpha_{\mathbf{k}}$ on $pr_1(U)$: the value of $\alpha_{\mathbf{k}}$ at $q \in pr_1(U)$ is the intersection point in U of $pr_1^{-1}(q)$ with $\Lambda_{\mathbf{k}}$. Since $\omega|\Lambda_{\mathbf{k}} = 0$ (from the definition of a polarization), $d\alpha_{\mathbf{k}} = 0$.

Thus, for each $\mathbf{k} \in V$, there is a function $S_{\mathbf{k}}$: $pr_1(U) \to \mathbb{R}$ (defined up to the addition of a constant) such that $\alpha_{\mathbf{k}} = dS_{\mathbf{k}}$. Clearly, the $S_{\mathbf{k}}$'s can be chosen so that $S(k_a, x^a) = S_{\mathbf{k}}(x^a)$ is a smooth function of the coordinates. The function S is called a *generating function* for F.

The condition that F should be tangent to X_h is simply the condition that h should be constant on the leaves of F. In other words, with[1] $h = h(p_a, q^a)$,

$$h(\partial_a S_{\mathbf{k}}, x^a) = \text{const.} \; ; \; \partial_a = \frac{\partial}{\partial x^a} \; ; \; \mathbf{k} \in V. \qquad 1.5$$

That is, the condition is that the generating function of F should be a complete integral of the Hamilton-Jacobi equation (eq. 1.5). Conversely, any complete integral locally determines a polarization tangent to X_h and transverse to P.

1.2 PAIRING AND WAVE FUNCTIONS IN CONFIGURATION SPACE

In the coordinates $\{k_a, x^a\}$

$$\theta = \frac{\partial S}{\partial x^a} dx^a \quad \text{and} \quad \omega = \frac{\partial^2 S}{\partial x^a \partial k_b} dk_b \wedge dx^a \qquad 1.6$$

and the F-wave functions have the form,

$$f = u(k_a) \exp\left[- iS(k_a, x^a)/\hbar\right] |d^n k|^{\frac{1}{2}} \qquad 1.7$$

where u is any complex valued function of the k_a's and $|d^n k|^{\frac{1}{2}}$ is the half F-density which takes the value 1 on the basis for F defined at each point of U by the Hamiltonian vector fields $X_{k_1}, X_{k_2}, \dots, X_{k_n}$.

Since F is tangent to X_h, the Hamiltonian operator H (which is defined as in eq. 1.3) acts on W_F by multiplication by h. Thus H has generalized eigenfunctions of the form

$$f = \delta(k_a - r_a) \exp(-iS/\hbar) |d^n k|^{\frac{1}{2}} \qquad 1.8$$

where $r = (r_1, r_2, \dots, r_n) \in V$ is fixed and δ is the Dirac distribution.

The next step is to pair the eigenfunctions of H with elements of the wave function space W_P. This enables us to identify the eigenfunctions with wave functions on Q, and hence, since the time evolution induced by the operator H on W_F can be found very easily, to obtain the quantum dynamics in configuration space. However, the result is independent of F only in the limit $\hbar \to 0$.

In detail, the construction is this: the elements of W_P have the form

$$g = v(x^a) |d^n x|^{\frac{1}{2}} \qquad 1.9$$

where v is a complex valued function which can be thought of as a wave function on Q and $|d^n x|^{\frac{1}{2}}$ is defined in a similar way to $|d^n k|^{\frac{1}{2}}$. At each $m \in U$, choose a symplectic basis $\{X_a, Y_a\}$ so that the X_a's span P and the Y_a's span F. Then for arbitrary $f \in W_F$ and $g \in W_P$, the product $f\bar{g}(m) = f(m, Y_a)\,\bar{g}(m, X_a)$ depends only on m and not on the particular choice made for X_a and Y_a (as a result of eq. 1.2). Thus, the quantity

$$(f,g)_{FP} = (2\pi\hbar)^{-\frac{1}{2}n} \int_M f\bar{g}\,\omega^n \qquad 1.10$$

is well defined: it is called the *pairing* of f and g. The idea is to try to associate with f a P-wave function $\hat{f}$ satisfying $(f,g)_{FP} = \langle \hat{f}, g\rangle$ for every $g \in W_P$ (here, $\langle .,. \rangle$ is the inner product in W_P).

If f and g are given locally by eqs. 1.8 and 1.9 then, written out explicitly in the coordinates $\{k_a, x^a\}$, eq. 1.10 is

$$(f,g)_{FP} = (2\pi\hbar)^{-\frac{1}{2}n} \int \bar{v}(x^a)\, \delta(k_a - r_a) \exp(-iS(x^a, k_a)/\hbar)\, \Delta^{\frac{1}{2}}\, d^n x\, d^n k$$

$$= (2\pi\hbar)^{-\frac{1}{2}n} \int \bar{v}(x^a) \exp(-iS_r(x^a)/\hbar)\, \Delta^{\frac{1}{2}}\, d^n x \qquad 1.11$$

where

$$\Delta = \det\left(\frac{\partial^2 S}{\partial k_a\, \partial x^b}\right) \qquad 1.12$$

and we have assumed that the k_a's have been chosen so that $\{k_a , x^a\}$ is a right-handed coordinate system in the natural orientation of M; that is, so that $\Delta > 0$. This suggests taking

$$\hat{f} = \Delta^{\frac{1}{2}} \exp(-iS_r / \hbar)\, |d^n x|^{\frac{1}{2}} \qquad 1.13$$

a choice which can be justified rigorously when F is transverse to P *globally*. However, trouble arises if, as will be true in general, there are singular points where P and F are not transverse. The integral curves of X_h in Λ_r project onto a congruence of classical trajectories in Q, and the singular points on Λ_r correspond to the caustics in this congruence. There are two possible methods of continuing the wave function on the right of eq. 1.13 across these caustics. The first is to reinterpret $\hat{f}$ as a distribution defined globally in Q. The second and older method is to go around the singularities in a complex analytic extension of Q. Both methods require the replacement of half densities by objects called *half forms*.

1.3 EIGENFUNCTIONS AS DISTRIBUTIONS

Suppose that the P-wave function in eq. 1.9 has an oscillating form, so that

$$v(x^a) = \rho(x^a) \exp(-i\alpha(x^a)/\hbar) \qquad 1.14$$

where α and ρ are smooth and real ρ is compactly supported in Q. Then as a function of $\hbar$, the quantity $(f , g)_{FP}$ has the asymptotic form

$$\begin{aligned}(f,g)_{FP} &= (2\pi\hbar)^{-\frac{1}{2}n} \int \rho \exp[i(\alpha - S_r)/\hbar]\, \Delta^{\frac{1}{2}}\, d^n x \\ &= \rho(q)\, \Delta^{\frac{1}{2}}\, |\det D|^{-\frac{1}{2}} \exp[i(\alpha(q) - S_r(q))/\hbar] \exp(-i\pi\, \mathrm{sign}\, D/4)(1 + O(\hbar)) \qquad 1.15\end{aligned}$$

by the method of stationary phase [3]. Here, q is the point where $\partial_a(\alpha - S_r) = 0$, D is the Hessian matrix

$$D_{ab} = [\partial_a \partial_b (S_r - \alpha)]_q \qquad 1.16$$

and sign D is the number of positive eigenvalues of D less the number of negative eigenvalues. It is assumed that α has been chosen so that D is nondegenerate; this ensures that that critical point q of the function $S_r - \alpha$ is isolated. (If there is more than one critical point in the support of ρ then the right hand side of eq. 1.15 must be replaced by a summation with a contribution from each of the critical points.)

The next step is to redefine the pairing $(f , g)_{FP}$ as a geometrical construction not involving a phase space integration. This new definition is equivalent to the old one only in the limit $\hbar \to 0$. However, it makes sense even at the singular points on Λ_r.

The phase function $\alpha(x^a)$ in eq. 1.14 itself generates a Lagrangian submanifold Λ_α of M, given by

$$\Lambda_\alpha = \{(q, d\alpha_q) | q \in Q\} \subset T^*Q = M; \qquad 1.17$$

Λ_α is necessarily transverse to P and the point q where $\partial_a(S_r - \alpha) = 0$ is the projection into Q of the intersection point m in U of Λ_α and Λ_r. The nondegeneracy of D is the condition for this intersection also to be transverse (for simplicity, we assume that m is unique).

Before proceeding further, it is necessary to consider in more detail the geometrical meaning of the distribution f appearing in eq. 1.8. If

$$f' = u'(k_a) |d^n k|^{\frac{1}{2}} \exp(-iS/\hbar)$$

is a half F-density, with u' an arbitrary complex test function with compact support on M/F, then $\langle f, f' \rangle$ is defined by identifying $f\bar{f}'$ with a density on M/F and integrating. Thus

$$\langle f, f' \rangle = \int \delta(k_a - r_a) \bar{u}'(k_a) d^n k = \bar{u}'(r_a) \qquad 1.18$$

Now, if the coordinates k_a are replaced by new coordinates $\tilde{k}_a$, then $u' |d^n k|^{\frac{1}{2}}$ becomes $|J|^{\frac{1}{2}} u' |d^n \tilde{k}|^{\frac{1}{2}}$ where

$$J = \det \left[\frac{\partial k_a}{\partial \tilde{k}_b} \right]. \qquad 1.19$$

Thus, if $\langle f, f' \rangle$ is to be independent of the choice of coordinates, then we must have

$$f = |J|^{-\frac{1}{2}} \exp(-iS/\hbar) \, \delta(\tilde{k}_a - \tilde{r}_a) |d^n \tilde{k}|^{\frac{1}{2}} \qquad 1.20$$

Hence, if we write $f = \nu_f \delta(k_a - r_a) |d^n k|^{\frac{1}{2}}$, then we can think of ν_f as a half density (in the usual sense[2]): ν_f is defined by the conditions:

(1) $\nu_f(m, X_{k_a}) = \exp(-iS/\hbar)$ in $\Lambda_r \cap U$ 1.21

(2) $i\hbar \mathcal{L}_X \nu_f - (X \lrcorner \theta) \nu_f = 0$ for every Hamiltonian vector field X tangent to Λ_r. 1.22

With this interpretation of ν_f, $\langle f, f' \rangle$ is defined by

$$\langle f, f' \rangle = \nu_f(z, X_a) \bar{f}'(z, X_a) \qquad 1.23$$

with the right hand side evaluated at any point $z \in \Lambda_r$ and on any complex basis $\{X_a\}$ for the tangent space to F at z. The result is independent of the choice made for z and $\{X_a\}$ and, as a means of defining f as a functional of f', it is equivalent to eq. 1.18 locally in U.

However, ν_f is defined globally and, unlike f, its definition does not depend on a particular choice of coordinates. (If the leaves of F are not simply connected, then ν_f will not be single valued unless the integral of θ over every 1-cycle in Λ_r is an integral multiple of $2\pi\hbar$. This is the *Bohr-Sommerfeld* condition, which restricts the allowed eigenvalues of H.)

Returning to the definition of the pairing $(f,g)_{FP}$, let L_m, P_m and F_m denote the tangent spaces to Λ_α, F and P at m. Choose a basis $\{X_a\}$ for P_m and let $\{U_a\}$ and $\{Y_a\}$ be the unique bases for L_m and F_m such that $\{U_a, X_a\}$ and $\{U_a, Y_a\}$ are symplectic frames at m; in other words, such that

$$\omega(U_a, X_b) = \omega(U_a, Y_b) = \delta_{ab}. \qquad 1.24$$

With ν_f interpreted as a half density on Λ_r, put

$$\langle f,g\rangle_{FP} = \nu_f(m,Y_a)\,\bar{g}(m,X_a). \qquad 1.25$$

Then, if $\{X_a\}$ is replaced by $\{X_a C_{ab}\}$, $C \in GL(n,\mathbb{C})$, $\{U_a\}$ is replaced by $\{U_a \hat{C}_{ab}\}$ and $\{Y_a\}$ by $\{Y_a C_{ab}\}$ and the right hand side of eq. 1.25 remains unchanged (here, $\hat{C}$ denotes the inverse transpose of C). Thus $\langle f,g\rangle_{FP}$ does not depend on the choice made for $\{X_a\}$.

Suppose, now, that $m \in U$. Then, we can take

$$X_a = \frac{\partial}{\partial p_a},\quad U_a = -\frac{\partial}{\partial q^a} - (\partial_a \partial_b \alpha)\frac{\partial}{\partial p_b}$$

and

$$Y_a = \hat{D}_{ab}\left(\frac{\partial}{\partial q^b} + (\partial_c \partial_b S_r)\frac{\partial}{\partial p_c}\right) \qquad 1.26$$

where $\{q^a, p_a\}$ is the canonical coordinate system on $M = T^*Q$ obtained by extending the coordinates[1] $\{x^a\}$ on Q, and $D_{ab} = \partial_a \partial_b(S_r - \alpha)$. Also,

$$X_{k_a} = \hat{B}_a{}^b\left(\frac{\partial}{\partial q^b} + \partial_b \partial_c S_r \frac{\partial}{\partial p_c}\right) = \hat{B}_a{}^b D_{bc} Y_c \qquad 1.27$$

where $B_a{}^b = \dfrac{\partial^2 S}{\partial x^a \partial k_b}$. Thus,

$$\nu_f(m, Y_a) = \Delta^{\frac{1}{2}}|\det D|^{-\frac{1}{2}} \exp(-iS/\hbar). \qquad 1.28$$

Hence,

$$\langle f,g\rangle_{FP} \exp(-i\pi \operatorname{sign} D/4) = (f,g)_{FP}\,(1 + O(\hbar)). \qquad 1.29)$$

Now, $\langle f,g\rangle_{FP}$ is well defined even if F_m and P_m are not transverse. Thus, to the highest order in $\hbar$ and modulo the factor $\exp(-i\pi \operatorname{sign} D/4)$, $\langle f,g\rangle_{FP}$ extends the definition of $(f,g)_{FP}$ to the singular points on Λ_r. With this extension, $\hat{f}$ is defined globally (in the limit $\hbar \to 0$) as a distribution dual to the oscillatory test functions of the type appearing in eq. 1.14. The only remaining task is to take care of the factor $\exp(-i\pi \operatorname{sign} D/4)$.

1.4 METAPLECTIC FRAMES

Let (V, ω) be a 2n-dimensional symplectic vector space and let $J: V \to V$ be a fixed positive complex structure on V. That is, J satisfies

$$J^2 = -I,\ \omega(JX, JY) = \omega(X, Y)\ \forall X,Y \in V \text{ and } \omega(X, JX) > 0\ \forall X \neq 0 \in V. \qquad 1.30$$

Some notation is needed: we denote by Lg(V) the set of all Lagrangian planes in V and by Lgf(V) the set of all Lagrangian frames. That is, each $L \in Lg(V)$ is an n-dimensional subspace of V on which ω vanishes and each $\{X_a\} \in Lgf(V)$ is an ordered set of n vectors in V with $\omega(X_a, X_b) = 0$ for each a,b. There is an obvious projection from Lgf(V) to Lg(V) obtained by taking linear spans. This will be represented by $\{X_a\} \to [X_a]$.

Similarly, we denote by $Lg(V_{\mathbb{C}})$ and $Lgf(V_{\mathbb{C}})$ the sets of Lagrangian planes and Lagrangian frames in the complexification $V_{\mathbb{C}}$ of V. Again, there is an obvious projection $Lgf(V_{\mathbb{C}}) \to Lg(V_{\mathbb{C}})$.

If $L \in Lg(V)$, then $Lgf(L)$ and $Lgf(L_{\mathbb{C}})$ denote the sets of Lagrangian frames whicl span L and $L_{\mathbb{C}}$.

Now let $P \in Lg(V)$ and choose an orientation for P; P and its orientation will be kept fixed throughout the following discussion.

Let $\{X_a\} \in Lgf(P)$ be a right handed basis for P such that $\omega(x_a, JX_b) = \delta_{ab}$ and define

$$\chi : Lgf(V_{\mathbb{C}}) \to \mathbb{C} \quad \text{by} \quad \chi(\{U_a\}) = \det(E) \tag{1.31}$$

where E is the matrix

$$E_{ab} = -\,\omega(iX_a + JX_a\,,\,U_b). \tag{1.32}$$

Since $\{X_a\}$ is a unique up to $X_a \longmapsto X_b H_{ba}$, $H \in SO(n)$, χ depends only on $\{U_a\}$ and not on the precise choice made for $\{X_a\}$.

A *metalinear* (ML) *Lagrangian frame* in $L \in Lg(V)$ is a pair (U_a, u) where $\{U_a\}$ is a complex basis for L and $u^2 = \chi(\{U_a\})$ (note that $\chi(\{U_a\}) \neq 0$ since L is real). We say that (U_a, u) is real whenever the U_a's are real and we denote the set of all real ML Lagrangian frames by Mgf(V): this is a double cover of Lgf(V) with the projection given by $(U_a, u) \to \{U_a\}$.

If $L, M \in Lg(V)$ are transverse (that is, if $L \cap M = \{0\}$), then, to each $\{U_a\} \in Lgf(L)$, there is a unique $\{V_a\} \in Lgf(M)$ such that $\omega(U_a, V_b) = \delta_{ab}$, and we indicate this relationship by writing $\{U_a\} \to \{V_a\}$. There is an analogous construction in Mgf(V): if $(U_a, u), (V_a, v) \in Mgf(V)$, then we write $(U_a, u) \to (V_a, v)$ whenever $\{U_a\} \to \{V_a\}$ and the path

$$t \longmapsto \chi(\{\cos t\, U_a + \sin t\, V_a\}) \; ; \quad t \in \left[0, \tfrac{\pi}{2}\right]$$

lifts to a path from u to v in $\mathbb{C} - \{0\}$. In this case, (U_a, u) and (V_a, v) are said to form a *metaplectic frame*.

Now let $L, F \in Lg(V)$ be such that L, P and F are mutually transverse. Let (U_a, u) be the real ML Lagrangian frame in L such that

$$(U_a, u) \to (X_a, 1) \tag{1.34}$$

and let (Y_a, y) and (W_a, w) be the real ML Lagrangian frames in F such that

$$(X_a, 1) \to (W_a, w) \quad \text{and} \quad (U_a, u) \to (Y_a, y). \tag{1.35}$$

Thus, $Y_b = W_a C_{ab}$ for some $C \in GL(n, \mathbb{R})$. Then,

1. <u>*Proposition*</u>: C is symmetric and $\arg(y/w) = -\frac{1}{4}\pi(\operatorname{sign} C - n)$;

<u>*Proof:*</u> Since $\{U_a\} \to \{X_a\}$ and $\{X_a\} \to \{W_a\}$,

$$W_b = -U_b + X_a A_{ab} \tag{1.36}$$

for some real symmetric A_{ab}. Thus

$$Y_b = (-U_a + X_c A_{ca}) C_{ab} \qquad 1.37$$

and therefore, since $\{U_a\} \to \{Y_a\}$, $A_{ca} C_{ab} = \delta_{cb}$. Hence $C = A^{-1}$ is symmetric.

Now assume that $F = J(P)$. Then $W_a = JX_a$, $w = \exp(-\frac{1}{4} n\pi i)$ and $Y_b = J X_a C_{ab}$.

Consider the one parameter family of Lagrangian frames

$$t \longmapsto V_b(t) = X_b - (1-t) U_a C_{ab} \;;\; t \in [0,1] \qquad 1.38$$

For each $t \in [0,1]$, $\{U_a\} \to \{V_a(t)\}$. Also, $V_a(0) = Y_a$ and $V_a(1) = X_a$. Thus, $\{V_a(t)\}$ lifts uniquely to a path $(V_a(t), v(t))$ in $Mgf(V)$ with

$$(U_a, u) \to (V_a(t), v(t)), \; \forall t \in [0,1] \qquad 1.39$$

and $(V_a(1), v(1)) = (X_a, 1)$ and $(V_a(0), v(0)) = (Y_a, y)$. Further

$$\begin{aligned} v(t)^2 &= \det(-\omega(Z_a, X_b - (1-t) U_c C_{cb}) \;;\; Z_a = JX_a + iX_a \\ &= \det(t\, \delta_{ab} - i(1-t) C_{ab}) \\ &= \prod_{a=1}^{n} (t - i(1-t)\lambda_a) \qquad 1.40 \end{aligned}$$

where the λ_a's are the eigenvalues of C_{ab}. Therefore, since $v(1) = 1$ and $v(t)$ is continuous, $\arg(v(0)) = -\frac{1}{4}\pi \operatorname{sign} C = \arg(y)$, and the statement follows.

The proof is completed by noting that, even if $F \neq JP$, $\arg(y/w)$ must still be an integral multiple of $\frac{1}{2}\pi$ and that C is independent of J. Thus the formula remains true as J is continuously deformed. Since it is certainly possible to find some positive complex structure such that $F = J(P)$, the proposition must hold in general. □

1.5 HALF FORMS AND THE MASLOV INDEX

Returning again to the pairing problem: let J be a positive almost complex structure on (M, ω). (That is, J is a smooth tensor field on M defining a positive complex structure in the tangent space at each point of M. For example, J might be defined by introducing a Riemannian metric into Q.) Suppose, further, that Q is oriented and that $\{x^a\}$ is a right handed coordinate system in $pr_1(U)$. The orientation of Q also induces an orientation in each of the tangent space to P at each point of M; in particular, the basis $\{\frac{\partial}{\partial p_a}\}$ is right handed at each point of U.

At each $m \in M$, choose a right handed basis $\{X_a\}$ in P_m such that

$$\omega(X_a, JX_b) = \delta_{ab} \;;$$

$\{X_a\}$ is unique up to $X_a \longmapsto X_a H_{ab}$, $H \in SO(n)$. Then, as in section 1.4, we can define metalinear frames in the tangent space at each point of M.

If G is a real polarization of M, then a half G-form is defined to be a complex valued function $g(m, U_a, u)$ depending on a point $m \in M$ and on a complex metalinear frame in the tangent space to G at m with the property

$$g(m, U_a, u) = vu^{-1} g(m, V_a, v). \qquad 1.41$$

Note that if $U_b = V_a E_{ab}$, $E \in GL(n, \mathbb{C})$, then $\det E = (u/v)^2$. Thus, this definition effectively removes the modulus sign the definition (eq. 1.2) of half G-densities.

If g and g' are half G-densities satisfying eq. 1.1 for every Hamiltonian vector field X tangent to G, then we can still identify $g\overline{g'}$ with a density on M/G; thus, the inner product can be defined as before.

Half P-forms can be canonically identified with half P-densities by evaluation on the ML Lagrangian frame $(X_a, 1)$ at each point of M.

Half forms on each leaf Λ_r of F are defined by a similar revision of the usual definition of half densities: thus, a half form on Λ_r is a function $\nu(m, U_a, u)$ depending on a point $m \in \Lambda_r$ and a complex ML Lagrangian frame (U_a, u) in the tangent space to Λ_r at m, which has the property that

$$\nu(m, U_a, u) = uv^{-1}\,\nu(m, V_a, v)\ . \qquad 1.42$$

Now, replace all half densities with half forms and, in the definitions of $(f,g)_{FP}$ and $\langle f,g\rangle_{FP}$, replace each Lagrangian frame by a corresponding metalinear Lagrangian frame, and each symplectic basis by a corresponding metaplectic basis. In particular, in place of eq. 1.26 take

$$(X_a, 1) = \left(\frac{\partial}{\partial p_a}, 1\right),\quad (U_a, u) = \left(-\frac{\partial}{\partial q^a} - \partial_a \partial_b \alpha \frac{\partial}{\partial p_b}, u\right) \qquad 1.43$$

and

$$(Y_a, y) = \left(\hat{D}_{ab}\left(\frac{\partial}{\partial q^b} + \partial_a \partial_b S_r \frac{\partial}{\partial p_c}\right), y\right) \qquad 1.44$$

at $m = \Lambda_\alpha \cap \Lambda_r$ and choose $(W_a, w) \in \mathrm{Mgf}(F_m)$ with u, y and w such that eqs. 1.34 and 1.35 are satisfied. Let ν_f be a half form on Λ_r such that, in $U \cap \Lambda_r$,

$$\nu_f(X_{k_a}, c) = \exp(-iS/\hbar)$$

where (X_{k_a}, c) is one of the two ML frames covering $\{X_{k_a}\}$, and extend ν_f to the whole of Λ_r in such a way that eq. 1.22 is satisfied[4]. As before, we can identify ν_f with the distributional half F-form defined in U by

$$f = \exp(-iS/\hbar)\,\delta(k_a - r_a)\,(d^n k)^{\frac{1}{2}}$$

where $(d^n k)^{\frac{1}{2}}$ is the half F-form taking the value 1 on the ML Lagrangian frame (X_{k_a}, c).

With these new definitions, eqs. 1.15 and 1.29 are replaced by

$$\begin{aligned}(f,g)_{FP} &= (2\pi\hbar)^{-\frac{1}{2}n} \int f(m, W_a, w)\,\bar{g}(m, X_a, 1)\,\Delta\, d^n x\, d^n k \\ &= cw^{-1}\rho(q)\Delta|\det D|^{-\frac{1}{2}} \exp\left[i(\alpha(q) - S_r(q)\right]/\hbar)\exp(-i\tfrac{\pi}{4}\operatorname{sign} D)(1 + O(\hbar))\end{aligned}$$

$$\langle f,g\rangle_{FP} = \nu_f(m, Y_a, y)\,\bar{g}(m, X_a, 1) = yc^{-1}\rho(q)\exp\left[i(\alpha(q) - S_r(q)\right]/\hbar).$$

Now, from eq. 1.22, $c^2 = \Delta^{-1}(\det D)y^2 = \Delta^{-1}w^2$. Therefore,

$$y/c = (c/w)(y/w)(w^2/c^2) = (c/w)\,\Delta|\det D|^{-\frac{1}{2}}\exp(i\arg(y/w)).$$

Also, from eq. 1.26, $C = \hat{D}$. Hence, using proposition 1, $\arg(y/w) = -\frac{1}{4}\pi(\operatorname{sign} D - n)$.

The factor $\exp(\frac{1}{4} in\pi)$ can be absorbed into the definition of $(f,g)_{FP}$, Then $(f,g)_{FP} = \langle f,g\rangle_{FP}\,(1 + O(\hbar))$ and it follows that, in the semi-classical limit $\hbar \to 0$, the pairing extends over the whole of M, and $\hat{f}$ is well defined globally as a distribution dual to the oscillatory test functions of the form eq. 1.14.

The practical consequence of all this is that the quantization condition must now be modified. In the notation of § 1.4, let $\gamma : [0,1] \to Lg(V)$ be a closed path. For each $t \in [0,1]$, choose a basis $\{Y_a\}$ for $\gamma(t)$ and put $\lambda(t) = \chi(\{Y_a\})^2$. Since the argument of $\lambda(t)$ depends only on $\gamma(t)$, the winding number $m(\gamma)$ of the path $t \longmapsto \lambda(t)$ about the origin in $\mathbb{C}$ is well defined: it is called the *Maslov index* of γ. Similarly if $\Gamma(t)$ is a closed path in some leaf Λ_r of F, choose a basis $\{Y_a\}$ for the tangent space to F at each $\Gamma(t)$ and define the Maslov index $m(\Gamma)$ to be the winding number of the path $t \longmapsto (\chi(\{Y_a\}))^2$ about the origin in $\mathbb{C}$.

If ν_f is to be single valued on Λ_r, then we now have that

$$\oint_\Gamma \theta + \frac{\pi}{2}\, m(\Gamma)$$

is an integral multiple of $2\pi\hbar$ for any closed path Γ in Λ_r. The term $\frac{\pi}{2}m(\Gamma)$ is the *Maslov correction* to the Bohr-Sommerfeld condition.

For example, for the one dimensional harmonic oscillator, the Bohr-Sommerfeld condition makes the energy levels $E_n = n\hbar$ while the Maslov correction leads to the correct $E_n = (n + \frac{1}{2})\hbar$.

Remarks

1. If the Y_a's are chosen so that $\omega(Y_a, JY_b) = \delta_{ab}$ then $Y_b = X_a A_{ab} - JX_a B_{ab}$ for some $A + iB \in U(n)$ and $\chi(Y_a) = \det(A + iB)$. Thus this definition of the Maslov index agrees with that given by Arnol'd [1].

2. In a more sophisticated language, what has been shown here is that, associated with any real polarization G of M, and any choice of orientation in M/G, there is a unique metaplectic structure on M. However, when M is a cotangent bundle, we obtain the correct semiclassical eigenvalues of M only if G is taken to be the vertical polarization: if, for example, we took the metaplectic structure associated with F and some orientation on M/F, we would obtain the uncorrected Bohr-Sommerfeld condition. Other choices of metaplectic structure (associated with other polarizations) would give even more absurd results.

2.1 COMPLEX LAGRANGIAN FRAMES

I will now turn to the problem of showing explicitly why the ancient technique of compex analytic continuation gives the same correction to the Bohr-Sommerfeld quantization.

In the notation of § 1.4, let Σ_p denote the set of all $\{U_a\} \in Lgf(V_{\mathbb{C}})$ such that $[U_a] \cap P_{\mathbb{C}} \neq \{0\}$. Then any $\{U_a\} \in Lgf(V_{\mathbb{C}}) - \Sigma_p$ is uniquely of the form

$$U_b = i(JX_a + X_c S_{ca}) G_{ab} \qquad 2.1$$

where $G_{ab} = -i\omega(X_a, U_b)$ is nonsingular and S_{ab} is a complex symmetric matrix.

Put $\mu(\{U_a\}) = \det G$ and note that μ is invariant under $X_a \longmapsto X_a H_{ab}$, $H \in SO(n)$.

If $\alpha : [0,1] \to Lgf(V_{\mathbb{C}}) - \Sigma_P : t \longmapsto \alpha(t) = \{U_a(t)\}$ is any path such that $U_b(0) = U_a(1)E_{ab}$ for some real E_{ab}, then the integer $Ind(\alpha)$ is defined to be the winding number of $(\mu \circ \alpha)^2$ about the origin in $\mathbb{C}$.

The relationship of Ind to the Maslov index is this: let $\gamma : [0,1] \to Lg(V)$ be a closed path in $Lg(V)$. For each $t \in [0,1]$, choose a real basis $\{Y_a(t)\}$ for $\gamma(t)$ such that $\omega(Y_a(t), JY_b(t)) = \delta_{ab}$. Then

$$Y_b(t) = X_a A_{ab}(t) - JX_a B_{ab}(t) \qquad 2.2$$

for some $(A(t) + iB(t)) \in U(n)$. For each t, let $\alpha(t)$ be the basis for[5] $V^{0,1}$ defined by

$$\alpha(t) = \{Y_b(t) + iJY_b(t)\} = \{(X_a + iJX_a)(A_{ab} + iB_{ab}(t))\} \qquad 2.3$$

Then $\mu(\alpha(t)) = \det(A(t) + iB(t)) = \chi(\{Y_a(t)\})$. Hence, the Maslov index of γ is equal $Ind(\alpha)$.

Now consider the map $\kappa : [0,1] \times [0,1] \to Lgf(V_{\mathbb{C}})$ defined by

$$\kappa(t,s) = \{Y_a(t) + isJY_a(t)\}\,; \quad (t,s) \in [0,1] \times [0,1]\,. \qquad 2.4$$

This satisfies:

(1) $\kappa(t,0) = \{Y_a(t)\}$ and $\kappa(t,1) = \{U_a(t)\}\ \forall t \in [0,1]$. 2.5

(2) $\kappa(t,s) \notin \Sigma_P\ \forall s \in (0,1]$ 2.6

It follows that we can compute the Maslov index of γ by computing Ind for the path $t \longmapsto \kappa(t,s)$ for any small positive value of s; that is, by deforming γ infinitesimally into the complex. The next task is to characterize in a more general way the direction in which γ should be deformed.

Let $s \longmapsto \rho(s)$, $s \in \mathbb{R}$, be a smooth path in $Lg(V_{\mathbb{C}})$ and let $U, V \in \rho(0)$. Choose a smooth one parameter family of vectors $V(s) \in \rho(s)$ such that $V(0) = V$ and put

$$R(U,V) = \omega(U, \dot{V}(0))\,; \quad \cdot = \frac{d}{ds}\,. \qquad 2.7$$

Then, R is a well defined (it depends only on U and V) symmetric complex bilinear form on $\rho(0)$; we can think of R as the tangent vector to ρ at $s = 0$.

2. *Proposition*: If $R/\rho(0) \cap P_{\mathbb{C}}$ is nondegenerate, then $\rho(s)$ is transverse to $P_{\mathbb{C}}$ for all small $s > 0$.

Proof: Assume $\rho(0) \cap P_{\mathbb{C}} \neq \{0\}$ (otherwise the statement is trivial). Suppose that, for each small $s \geq 0$, $\exists V(s) \in P_{\mathbb{C}} \cap \rho(s)$ with $V(s) \neq 0\ \forall s \geq 0$ and with V

depending smoothly on s. Then, for any $U \in P_{\mathbb{C}} \cap \rho(0)$,

$$\omega(U,V(s)) = 0 \qquad 2.8$$

since $P_{\mathbb{C}}$ is Lagrangian. But then $R(U,V(0)) = 0$ implying that $R|\rho(0) \cap P_{\mathbb{C}}$ is degenerate. It follows that such $V(s)$'s cannot be found, and thus that $\rho(s)$ is transverse to $P_{\mathbb{C}}$ for small $s > 0$. □

Now let $\sigma_0 : t \mapsto \{Y_a(t)\} \in Lgf(V)$ where the Y_a's are as above, and let $\sigma : (t,s) \mapsto \sigma(t,s) \in Lgf(V_{\mathbb{C}})$ be a smooth one parameter variation of σ_0 (that is, $\sigma(t,0) = \sigma_0(t)$ for each $t \in [0,1]$). Put $\tau(t,s) = [\sigma(t,s)]$. Then, at $s = 0$,

(a) differentiating with respect to t, we obtain a real symmetric bilinear form $R_1(t)$ on $\tau(t,0)$

(b) differentiating with respect to s, we obtain a complex symmetric bilinear form $R_2(t)$ on $\tau(t,0)$.

It follows from proposition 2 that $\tau(t,s)$ is transverse to $P_{\mathbb{C}}$ for all t and for all sufficiently small $s > 0$ if $xR_1 + yR_2$ is nondegenerate on $P \cap \tau(t,0)$ for all t and for all real x and y. This will certainly be true if $\mathrm{Im}(R_2)$ is positive definite on $P \cap \tau(t,0)$ for all t. In this case, we call σ a *positive variation* of σ_0. In particular, the variation κ defined by eq. 2.4 is positive.

We can write,

$$\sigma(t,s) = \{[Y_a(t) + JY_b(t)\, R_{ab}(t,s)]\, E_{ac}(t,s)\} \qquad 2.9$$

where, for small s, $E(t,s)$ is near the identity in $GL(n,\mathbb{C})$ and, to the first order in s, $R_{ab}(t,s)$ is the matrix of sR_2 in the basis $\{Y_a(t)\}$ for $\tau(t,0)$. It follows that if σ and $\tilde{\sigma}$ are two positive variations of σ_0, then, for sufficiently small s, we can deform σ into $\tilde{\sigma}$ through a sequence of positive variations by replacing $R_{ab}(t,s)$ by $(1-\lambda)R_{ab}(t,s) + \lambda\tilde{R}_{ab}(t,s)$ and $C_{ab}(t,s)$ by $(1-\lambda)\,C_{ab}(t,s) + \lambda\tilde{C}_{ab}(t,s)$, $\lambda \in [0,1]$.

Thus, if $\gamma : [0,1] \to Lg(V)$ is any closed path, then we can compute the Maslov index of γ by lifting to a (not necessarily closed) path in $Lgf(V)$ and computing Ind for any small positive variation into the complex. This is, in effect, what is done in the complex WKB method.

2.2 THE COMPLEX WKB METHOD

Returning, now, to the situation discussed in § 1.3, let Λ_r be a leaf of F and, as before, let ν_f be the half density on Λ_r which takes the value $\exp(-iS/\hbar)$ on the basis $\{X_{k_a}\}$ at each point of $\Lambda_r \cap U$. When Λ_r is transverse to P, we know how to associate to ν_f a half P-density $\hat{f}$ on Q. Indeed, if we think of $\hat{f}$ as a half density on Q, then it is simply the projection of ν_f into Q.

However, in general (dropping the subscript r), Λ is not transverse to P everywhere, and there will be some subset Σ of Λ where[6] $(T_m\Lambda) \cap P \neq \{0\}$. If Λ

is in 'general position', then Σ will consist of an $(n-1)$-dimensional hypersurface Σ_1 on which $(T_m\Lambda)\cap P_m$ is one dimensional, together with a boundary of lower dimension [1]. With an appropriate choice of coordinates near some $m_0\in\Sigma_1$, Λ is given by

$$q^1 = -\frac{\partial y}{\partial p_1} \quad \text{and} \quad p_\alpha = \frac{\partial y}{\partial p^\alpha} \quad (\alpha = 2, 3, \dots, n) \qquad 2.10$$

where y is a function of $p_1, q^2, \dots, q^n$.

From now on, it will be assumed that Q, F, M/F and y are real analytic. We will assume further that the critical points of $\frac{\partial y}{\partial p_1}$ (the points of Σ_1) as a function of p_1 are nondegenerate.

Allowing the coordinates to take on complex values, we obtain holomorphic coordinates z^a on the complexified configuration space $Q_{\mathbb{C}}$ and holomorphic canonical coordinates $\{w_a, z^a\}$ on $M_{\mathbb{C}}$ (in particular, on $M\subset M_{\mathbb{C}}$, $q^a = \mathrm{Re}(z^a)$ and $p_a = \mathrm{Re}(w_a)$).

By identifying the bases $\{\frac{\partial}{\partial q^a}, \frac{\partial}{\partial p_a}\}$ at each point $m\in M$ near m_0, we can map the tangent space T_mM to M onto a fixed symplectic vector space (V, ω). Similarly, identifying the bases $\{\frac{\partial}{\partial z^a}, \frac{\partial}{\partial w_a}\}$, we can map the holomorphic tangent space at each point of $M_{\mathbb{C}}$ onto the complexification $V_{\mathbb{C}}$ of V.

Then, for each $m\in\Lambda$ (near m_0) we have a Lagrangian subspace $L_m\subset V_{\mathbb{C}}$, identified with the holomorphic tangent space $T_m\Lambda_{\mathbb{C}}$; L_m is real whenever $m\in\Lambda$ and is given explicitly at these points as the linear span of the vectors

$$U(m) = \left[\frac{\partial}{\partial p_1} - \frac{\partial^2 y}{\partial p_1^2}\frac{\partial}{\partial q^1} + \frac{\partial^2 y}{\partial p_1 \partial q^\alpha}\frac{\partial}{\partial p_\alpha}\right]_m \qquad 2.11$$

and

$$V_\alpha(m) = \left[\frac{\partial}{\partial q^\alpha} + \frac{\partial^2 y}{\partial q^\alpha \partial q^\beta}\frac{\partial}{\partial p_\beta} - \frac{\partial^2 y}{\partial q^\alpha \partial p_1}\frac{\partial}{\partial q^1}\right]_m \quad (\alpha,\beta = 2, 3, \dots, n) \qquad 2.12$$

When $m\in\Lambda_{\mathbb{C}}$, we have the same expressions, but with the coordinates complexified. Using w_1 and z^α $(\alpha = 2, 3, \dots, n)$ as coordinates on $\Lambda_{\mathbb{C}}$, the analytic continuation of the generating function S is given on $\Lambda_{\mathbb{C}}$ by

$$dS = \theta = w_a\, dz^a = -w_1\left(\frac{\partial^2 y}{\partial w_1^2}dw_1 + \frac{\partial^2 y}{\partial w_1 \partial z^\alpha}dz^\alpha\right) + \frac{\partial y}{\partial z^\alpha}dz^\alpha . \qquad 2.13$$

Note that, if we move away from Σ_1 in $\Lambda_{\mathbb{C}}$ by giving w_1 a small imaginary value, then $\mathrm{Im}(S)$ and $\mathrm{Im}\,[\omega(U(m_0), U(m))]$ have opposite signs.

Now let m_1 and m_2 be two points on opposite sides of Σ_1 and let $\gamma: [-\varepsilon, \varepsilon]\to\Lambda$ be a path from m_1 to m_2 cutting Σ_1 at m_0. In $\Lambda_{\mathbb{C}}$, the complexification $(\Sigma_1)_{\mathbb{C}}$ of Σ_1 is a complex hypersurface so it is possible to deform γ into the complex so as to avoid $(\Sigma_1)_{\mathbb{C}}$. The question is: in which direction?

Near m_0, each fibre $pr_1^{-1}(q)$, $q\in Q$, of T^*Q cuts $\Lambda_{\mathbb{C}}$ either in two real points (in which case q is said to lie in the *classical region*), one real point or in two complex points (with which case q is said to lie in the *nonclassical region*. The answer given by the complex WKB method [5] is to deform γ so that the projection of the deformed curve passes over the nonclassical region of Q on the portion of $\Lambda_{\mathbb{C}}$ on which $\mathrm{Im}(S) < 0$. Then the (analytically extended) wave function is exponentially damped in the nonclassical region. It follows from the remarks above that this will be a positive variation in the sense that it will induce a positive variation of the path on Lg(V) obtained by taking the tangent space to Λ at each point of γ.

The function Δ appearing in eq. 1.13 is given by

$$\Delta^{-1} = (i)^n \det\left(-i\omega\left(\frac{\partial}{\partial P_a}, X_{k_b}\right)\right). \qquad 2.14$$

We can think of the half P-density $\hat{f}$ defined in U by eq. 1.13 as a half density on $\Lambda \cap U$ (that is, we identify $\hat{f}$ with ν_f in $\Lambda \cap U$), and we can then extend $\hat{f}$ to the whole $\Lambda - \Sigma$. However, instead of doing this by imposing eq. 1.22, we do the extension by

(1) defining $\Delta^{\frac{1}{2}}$ continuously, passing around Σ in the complex according to the above prescription,

(2) taking $S(\tilde{x}) = \int_x^{\tilde{x}} \theta$ with x some fixed point on Λ and with the integral taken along some path from x to $\tilde{x}$ in Λ.

It then follows (comparing eq. 2.14 with eq. 2.1) that the condition that $\hat{f}$ should be single valued on $\Lambda - \Sigma$ will give the same correction to the quantization condition as before. In effect, this is the complex WKB method.

Notes on notation

(1) A coordinate system $\{x^a\}$ on Q extends to a canonical coordinate system $\{p_a, q^a\}$ on T^*Q with the p_a's representing the components of covectors and the q^a's labelling the points of Q. The canonical 1- and 2-forms are given by

$$\theta = p_a\, dq^a \quad \text{and} \quad \omega = dp_a \wedge dq^a .$$

(2) An r-density on a manifold N is a function $\nu(x, X_a)$ depending on a point $x \in N$ and on a complex basis $\{X_a\}$ for the tangent space to N at x with the homogeneity property

$$\nu(x, X_a C_{ab}) = |\det C|^r \nu(x, X_a) \ ; \ C \in GL(n, \mathbb{C}), \quad n = \dim N .$$

The Lie derivative of ν along a vector field X is defined by regarding ν as a function on the complex frame bundle BN of N and lifting X to a vector field $\tilde{X}$ on BN by Lie propagation; ν is then differentiated along $\tilde{X}$.

When $r = 1$, ν is called a *density*: the integral of a density over an open set $U \subset N$ is defined by introducing coordinates $\{x^a\}$ and putting

$$\int_U \nu = \int_{x \in U} \nu\left(x, \frac{\partial}{\partial x^a}\right) d^n x .$$

If f and f′ are half F-densities satisfying eq.1.1 then $f\overline{f'}$ can be identified with a density on M/F. The inner product $\langle f, f' \rangle$ is defined by an integration over M/F. See [2] or [6] for details.

(3) That is, the tangent spaces to P and F span the tangent space to M at each point of U.

(4) The Lie derivative of a half form ν on Λ_r is defined by identifying ν locally with a half density.

(5) $V^{0,1} = \{X \in V_{\mathbb{C}} | JX = -iX\}$.

(6) $T_m\Lambda$ denotes the tangent space to Λ at m.

REFERENCES

[1] ARNOL'D, V.I., Funct. Anal. and its Appl., 1, 1 (1967).

[2] BLATTNER, R., Quantization and Representation Theory: In AMS Proc. of Symposia in Pure Math., 26 (1973).

[3] DUISTERMAAT, J.J., Commun. Pure and Appl. Math., 27, 207 (1974).

[4] GUILLEMIN, V. and STERNBERG, S., Geometric Asymptotics, AMS Surveys, 14 (1977).

[5] HEADING, J., Phase Integral Methods, (Methuen, London, 1962).

[6] SIMMS, D.J. and WOODHOUSE, N.M.J., Lectures on Geometric Quantization: Lecture Notes in Physics, 53 (Springer-Verlag, Berlin-Heidelberg-New York, 1976).

SOME PROPERTIES OF HALF-FORMS

J. H. Rawnsley
School of Theoretical Physics,
Dublin Institute for Advanced Studies,
Dublin 4, Ireland

Introduction

We are concerned with Geometric Quantization, and in particular, with the notion of half-forms. For background and terminology see [1,2,3,5,6,7,8]. The approach adopted here is that of [5]. The prequantization theory is not considered since all the difficulties reside with the polarizations and half-forms.

Let (X,ω) be a symplectic manifold and F a polarization. That is, X is a real manifold of even dimension, ω a closed 2-form of maximal rank and F a smooth sub-bundle of the complexified tangent bundle $TX^{\mathbb{C}}$, which is

(i) isotropic;

(ii) maximal with respect to (i);

(iii) integrable.

If $F^o \subset T^*X^{\mathbb{C}}$ denotes the annihilator of F then (i) and (ii) may be combined as: $\xi \mapsto \xi \lrcorner \omega$ maps F isomorphically onto F^o. $F \subset TX^{\mathbb{C}}$ is integrable if $F \cap \overline{F}$ has constant rank, F and $F+\overline{F}$ closed under Lie bracket of vector fields.

If dim X = 2n, $K^F = \Lambda^n F^o$ is a line bundle, the canonical bundle of F. Lie differentiation with respect to $\xi \in \Gamma F$ of sections of K^F defines a map $\nabla_\xi : \Gamma K^F \longrightarrow \Gamma K^F$ which is a flat F-connection, [4].

A polarization is positive if

$$-i\,\omega(\xi,\overline{\xi}) \geq 0$$

for all $\xi \in \Gamma F$. The Chern classes $c_i(F) \in H^{2i}(X,\mathbb{Z})$, i = 1,...,n are determined by ω if F is positive; they are the Chern classes $c_i(X,\omega)$ of a reduction to a U(n)-bundle of the symplectic frame bundle of (X,ω). $c_1(X,\omega)$ is the Chern class of K^F for any positive polarization F.

If F and G are both positive polarizations and $F \cap \overline{G} = 0$, then

$$\beta \otimes \overline{\gamma} \longmapsto \beta \wedge \overline{\gamma}$$

defines an isomorphism of $K^F \otimes \overline{K^G}$ with $\Lambda^{2n} T^*X^{\mathbb{C}}$. The latter is trivialized by the Liouville volume λ:

$$\lambda = (-1)^{n(n-1)/2} \omega^n / n! ,$$

and so we obtain a canonical isomorphism $K^F \xrightarrow{\sim} K^G$.

If $F \cap \overline{G} \neq 0$, F and G again positive polarizations, then K^F and K^G are still isomorphic, but not canonically so. If $F \cap \overline{G}$ has constant rank, it is the complexification of a real, integrable foliation D, and $K^F \otimes \overline{K^G}$ is isomorphic to $\mathcal{D}^{-2}(D)$, the bundle of densities on D of order -2. This arises as follows: Fix a frame $b = (\zeta_1,\ldots,\zeta_k)$ for D_x and extend it to a frame $b_1 = (\zeta_1,\ldots,\zeta_k,\xi_1,\ldots,\xi_{n-k})$ for F. Then any $\alpha \in K^F_x$

can be written

$$\alpha = a(\xi_1 \lrcorner \omega)_\wedge \cdots {}_\wedge(\xi_k \lrcorner \omega)_\wedge(\xi_1 \lrcorner \omega)_\wedge \cdots {}_\wedge(\xi_{n-k} \lrcorner \omega).$$

Let $D^\perp = \{\xi \in TX \mid \xi \lrcorner \omega | D = 0\}$, then $D \subset F \subset (D^\perp)^C$ and if $\pi: D^\perp \longrightarrow D^\perp/D$ is the projection (extended to be complex linear on the complexifications), $(\pi\xi_1, \ldots, \pi\xi_{n-k})$ is a frame for $(F/D^{\mathbb{C}})_x$. Then

$$\tilde{\alpha}_b = a(\pi\xi_1 \lrcorner \omega/D)_\wedge \cdots {}_\wedge(\pi\xi_{n-k} \lrcorner \omega/D)$$

is independent of the extension of b to a frame of F_x. Here ω/D is the symplectic structure induced by ω on $D^\perp/D$. If $g \in GL(k,\mathbb{R})$, b.g is again a frame for D_x and

$$\tilde{\alpha}_{b.g} = \mathrm{Det}[g^{-1}]\tilde{\alpha}_b.$$

Observe that $F/D^{\mathbb{C}}$ is a positive, maximally isotropic subbundle of $((D^\perp/D)^{\mathbb{C}}, \omega/D)$ and $\tilde{\alpha}_b \in (K^{F/D^{\mathbb{C}}})_x$. We can treat $\beta \in \Gamma K^G$ in the same fashion. Since $F/D^{\mathbb{C}} \cap \overline{G/D^{\mathbb{C}}} = 0$, exterior multiplication determines an isomorphism of $K^{F/D^{\mathbb{C}}} \otimes \overline{K^{G/D^{\mathbb{C}}}}$ with $\Lambda^{2(n-k)} D^\perp/D^{\mathbb{C}}$. The latter is again trivialized by the Liouville volume, so for $\alpha \in \Gamma K^F$, $\beta \in \Gamma K^G$ we may define

$$\langle \alpha, \beta \rangle (b) \ \lambda^{\omega/D} = \tilde{\alpha}_b \wedge \overline{\tilde{\beta}_b}$$

It is immediate that $\langle \alpha, \beta \rangle \in \Gamma \mathcal{D}^{-2}(D)$.

The Liouville volume on X defines an isomorphism $\mathcal{D}^{-2}(D) \xrightarrow{\sim} \mathcal{D}^2(TX/D)$ and so we may regard $\langle \alpha, \beta \rangle$ as a section of the latter. However if $\nabla_\xi \alpha = 0$, $\forall \xi \in \Gamma F$, $\nabla_\eta \beta = 0$ $\forall \eta \in \Gamma G$, it need not happen, in general, that $\langle \alpha, \beta \rangle$ is the pull-back of a density of order 2 on the space of leaves X/D. Blattner, [2], found an obstruction, the section Θ^D of D^* defined by

$$\Theta^D(v) = \sum_j \omega([v_j, w_j], v),$$

$v \in D$ where $v_1, \ldots, v_{n-k}, w_1, \ldots, w_{n-k}$ are vector fields on some open set U in X satisfying

$$\omega(v_i, v_j) = \omega(w_i, w_j) = 0, \quad \omega(v_i, w_j) = \delta_{ij}$$

and spanning a complement to D in $D^\perp$. Θ^D is independent of the choice of $v_1, \ldots, w_{n-k}$, and satisfies

$$\nabla_\xi \langle \alpha, \beta \rangle = \langle \nabla_\xi \alpha, \beta \rangle + \langle \alpha, \nabla_\xi \beta \rangle - \Theta^D(\xi) \langle \alpha, \beta \rangle,$$

with $\alpha \in \Gamma K^F$, $\beta \in \Gamma K^G$, $\xi \in \Gamma D$.

Half-forms

(X,ω) is called metaplectic if $c_1(X,\omega)$ is even. Then for each positive polarization F there is a square-root Q^F of K^F. ∇ in K^F induces a flat F-connection $\nabla^{1/2}$ in Q^F. The square-roots Q^F can be assigned so that $Q^F \otimes \overline{Q^G}$ is again trivial for all positive F and G. If $F \cap \overline{G} = D^{\mathbb{C}}$ there is a (unique up to a sign) pairing $Q^F \otimes \overline{Q^G} \longrightarrow \mathcal{D}^{-1}(D)$ such that

$$\langle \rho, \sigma \rangle^2 = \langle \rho \otimes \rho, \sigma \otimes \sigma \rangle$$

for $\rho \in \Gamma Q^F$, $\sigma \in \Gamma Q^G$. Q^F is the bundle of half-forms normal to F and $\langle \ , \ \rangle$ is the half-form pairing. If $\Theta^D = 0$, and $\nabla^{1/2}_\xi \rho = 0$, $\forall \xi \in \Gamma F$, $\nabla^{1/2}_\eta \sigma = 0$, $\forall \eta \in \Gamma G$ then $\langle \rho, \sigma \rangle$ is the pull-back of a density on X/D which can be integrated over X/D to give the pairing of Kostant, Blattner and Sternberg.

If ξ is an arbitrary vector field on X, there is, in general, no natural way by which ξ operates on ΓQ^F. If the (local) flow of ξ preserves F, then ξ operates by Lie differentiation.

The projection method

In [3] Kostant described, for F a real polarization, a projection method to make the Hamiltonian vector fields act on ΓQ^F. We use the methods described above to examine how we may extend Kostant's method to non-transverse, complex polarizations.

Let F and G be positive polarizations with $F\cap\overline{G} = D^{\mathbb{C}}$ of constant rank. The basic idea is to work with K^F and K^G instead of Q^F and Q^G, and use the isomorphism $(Q^F)^2 \cong K^F$ to deduce the result for Q^F. If ξ is a vector field on X and $\alpha\in\Gamma K^F$, then the Lie derivative $L_\xi\alpha$ of α with respect to ξ is not, in general, again in ΓK^F. If $\xi = \xi_\varphi$ is Hamiltonian, it turns out, we can still project $L_\xi\alpha$ into $\Lambda^{n-k}(D^\perp/D)^{\mathbb{C}}$ and so pair it to $\beta\in\Gamma K^G$. This needs, however, that $[\xi_\varphi, \Gamma D]\subset\Gamma D$.

If $b = (\xi_1,\ldots,\xi_k)$ is a local frame field for D and $[\xi_\varphi,\Gamma D]\subset\Gamma D$ then we determine a kxk matrix A_φ by

$$[\xi_\varphi, \xi_j] = \sum_I (A_\varphi)_{ij}\,\xi_i, \quad j = 1,\ldots,k.$$

Similarly, we obtain an $(n-k)\times(n-k)$ matrix B_φ by

$$[\xi_\varphi, \xi_j] = \sum_I (B_\varphi)_{ij}\,\xi_i \mod G$$

if $\xi_1,\ldots,\xi_{n-k}$ is an extension of b to a local frame field for F. Then any α in ΓK^F can be written

$$\alpha = f(\xi_1\lrcorner\omega)\wedge\cdots\wedge(\xi_k\lrcorner\omega)\wedge(\xi_1\lrcorner\omega)\wedge\cdots\wedge(\xi_{n-k}\lrcorner\omega)$$

with f a function, and

$$L_{\xi_\varphi}\alpha = (\xi_\varphi f + \mathrm{Tr}[A_\varphi]f + \mathrm{Tr}[B_\varphi]f)(\xi_1\lrcorner\omega)\wedge\cdots\wedge(\xi_k\lrcorner\omega)\wedge\{(\xi_1\lrcorner\omega)\wedge\cdots\wedge(\xi_{n-k}\lrcorner\omega)\bmod \Lambda^{n-k}G\}.$$

Thus $L_{\xi_\varphi}\alpha$ is projectable:

$$(L_{\xi_\varphi}\alpha)^\sim_b = (\xi_\varphi f + \mathrm{Tr}[A_\varphi]f + \mathrm{Tr}[B_\varphi]f)(\pi\xi_1\lrcorner\omega/D)\wedge\cdots\wedge(\pi\xi_{n-k}\lrcorner\omega/D)\bmod\Lambda G^0.$$

We can then define $\varphi.\alpha\in\Gamma K^F$ as the unique section satisfying

$$(\varphi.\alpha)^\sim_b\wedge\overline{\tilde\beta_b} = (L_{\xi_\varphi}\alpha)^\sim_b\wedge\overline{\tilde\beta_b}$$

for all $\beta\in\Gamma K^G$ (it can be checked that $(L_{\xi_\varphi}\alpha)^\sim_{b.g} = \mathrm{Det}[g^{-1}](L_{\xi_\varphi}\alpha)^\sim_b$, $\forall g\in GL(k,\mathbb{R})$, so this defines $\varphi.\alpha$ independently of b). Explicitly

$$\varphi.\alpha = (\xi_\varphi f + \mathrm{Tr}[A_\varphi]f + \mathrm{Tr}[B_\varphi]f)(\xi_1\lrcorner\omega)\wedge\cdots\wedge(\xi_k\lrcorner\omega)\wedge(\xi_1\lrcorner\omega)\wedge\cdots\wedge(\xi_{n-k}\lrcorner\omega).$$

From this formula it follows that

$$\varphi.(\psi\alpha) = \psi\,\varphi.\alpha + \xi_\varphi\psi\,\alpha, \tag{1}$$

for $\psi\in C(X)$.

If $\nabla_\xi\alpha = 0$ for $\xi\in\Gamma F$, the same will not, in general, be true of $\varphi.\alpha$. However, consider the expression

$$\nabla_\xi\,\varphi.\alpha - \varphi.\nabla_\xi\alpha - \nabla_{[\xi,\xi_\varphi]}\alpha.$$

An easy calculation using (1) shows this is C(X)-linear in $\xi\in\Gamma F$ and $\alpha\in\Gamma K^F$. It therefore has the form

$$\Theta_\varphi(\xi)\alpha$$

with $\Theta_\varphi\in\Gamma F^*$. Θ_φ depends on G as well as φ, F.

If we restrict ξ to be in D, then $\Theta_\varphi(\xi) = 0$. This may be seen by using also the following explicit formula for ∇_ξ : if

$$[\xi, \varsigma_j] = \sum_i (A_\xi)_{ij} \varsigma_i, \quad [\xi, \bar\xi_j] = \sum_i (B_\xi)_{ij} \bar\xi_i \bmod D,$$

then

$$\nabla_\xi \alpha = (\xi f + \mathrm{Tr}[A_\xi]f + \mathrm{Tr}[B_\xi]f)(\varsigma_1 \lrcorner \omega)_\wedge \cdots {}_\wedge(\varsigma_k \lrcorner \omega)_\wedge (\bar\xi_1 \lrcorner \omega)_\wedge \cdots {}_\wedge (\bar\xi_{n-k} \lrcorner \omega).$$

Thus

$$\Theta_\varphi(\xi) = \xi \mathrm{Tr}[A_\varphi] - \xi_\varphi \mathrm{Tr}[A_\xi] - \mathrm{Tr}[A_{[\xi,\xi_\varphi]}] + \xi \mathrm{Tr}[B_\varphi] - \xi_\varphi \mathrm{Tr}[B_\xi] - \mathrm{Tr}[B_{[\xi,\xi_\varphi]}]. \tag{2}$$

Using the Jacobi identity

$$[\xi, [\xi_\varphi, \varsigma_j]] = [[\xi, \xi_\varphi], \varsigma_j] + [\xi_\varphi, [\xi, \varsigma_j]],$$

we have

$$A_\xi A_\varphi + \xi A_\varphi = A_{[\xi,\xi_\varphi]} + A_\varphi A_\xi + \xi_\varphi A_\xi .$$

When we take traces this shows the A- terms cancel in (2). A similar argument is valid also for the B-terms.

Now suppose $\Theta^D = 0$ and $\nabla_\xi \alpha = 0, \forall \xi \in \Gamma F$, $\nabla_\eta \beta = 0$, $\forall \eta \in \Gamma G$ where $\alpha \in \Gamma K^F$, $\beta \in \Gamma K^G$. Let $\varphi \in C(X)$ preserve D, and $\xi \in \Gamma D$, then

$$\nabla_\xi \langle \varphi.\alpha, \beta\rangle = \langle \nabla_\xi \varphi.\alpha , \beta\rangle = \langle \varphi. \nabla_\xi \alpha + \nabla_{[\xi,\xi_\varphi]} \alpha , \beta\rangle = 0.$$

Thus $\langle \varphi.\alpha, \beta\rangle$ is projectable if α and β are. The action of φ on ΓQ^F is obtained as usual from:

$$\varphi.(\rho_1 \otimes \rho_2) = (\varphi.\rho_1) \otimes \rho_2 + \rho_1 \otimes (\varphi.\rho_2)$$

If $\sigma \in \Gamma_G Q^G$, $\rho \in \Gamma_F Q^F$ then $\langle \varphi.\rho, \sigma\rangle$ projects to a density on X/D which can be integrated.

In general one must now proceed formally. Although $\varphi.\rho$ is not in $\Gamma_F Q^F$, if the integrated pairing exists and defines an isomorphism $\Gamma_F Q^F \xrightarrow{\sim} \Gamma_G Q^G$, there will be $\varphi * \rho \in \Gamma_F Q^F$ such that

$$\langle \varphi * \rho, \sigma\rangle = \langle \varphi.\rho, \sigma\rangle$$

for all $\sigma \in \Gamma_G Q^G$. This makes φ act on $\Gamma_F Q^F$, generalizing the action obtained by Kostant in the real transverse case.

References

1. Blattner R. Quantization and representation theory. Proc. Symp. Pure Math. 26, A.M.S., Rhode Island, 1973.
2. Blattner R. The metalinear geometry of non-real polarizations. Lecture Notes in Mathematics 570. Springer-Verlag, Berlin-Heidelberg-New York, 1977.
3. Kostant B. On the definition of quantization. Colloque symplectique, CNRS, Aix-en-Provence, 1974.
4. Rawnsley J. On the cohomology groups of a polarization and diagonal quantization. Trans. Amer. Math. Soc. 230 (1977) 235-255.
5. Rawnsley J. On the pairing of polarizations. Comm. Math. Phys. (to appear, 1978).
6. Simms D. An outline of geometric quantization. Lecture Notes in Mathematics 570. Springer-Verlag, Berlin-Heidelberg-New York, 1977.
7. Simms D., Woodhouse N. Lectures on geometric quantization. Lecture Notes in Physics 53, Springer-Verlag, Berlin-Heidelberg-New York, 1976.
8. Sniatycki J. Geometric quantization and quantum mechanics, part I. Research paper 328, Department of Mathematics, University of Calgary, 1976.

ON SOME APPROACH TO GEOMETRIC QUANTIZATION

J. Czyz
Institute of Mathematics, Technical University of Warsaw

1. Introduction

The Kostant-Souriau (KS) geometric quantization is a beautiful and deep mathematical theory ([7], [13]). It gives us correct results in the case of Kepler problem (the hydrogen atom), see [11], but, alas, this theory does not agree with "classical" quantum mechanics in some cases. For instance the multi-odd--dimensional harmonic oscillator is unquantizable in the KS theory, see example in point 3.

The proposed modification allows us to quantize a bigger class of mechanical systems. In particular we obtain for the harmonic oscillator the same energy levels and multiplicities of degeneracy as we get it from the Schrödinger equation, see point 3.

The KS geometric quantization of symplectic manifold is roughly speaking a complex line bundle which is a tensor product of two terms: the first is a line bundle whose Chern class is a class of symplectic form, the second is a bundle of some half-forms. Each bundle is provided with a scalar product.

The paper presents a construction of geometric quantization which may be correct even if none of component-bundles exists but "their" tensor product can be defined. Namely we define the Chern class od bundle of quantization as a sum of class of the symplectic form and (plus or minus) half of first Chern class of a given complex structure on the symplectic manifold. Then we choose a harmonic form η in this class and define a Hermitian structure in this bundle to be invariant with respect to a connection whose curvative form is equal to η.

Such an approach allows us to observe a narrow connection between geometric quantization and the Maslov theory of a zeroth order term in the W.K.B. (Wentzel, Krammers, Brillouin) asymptotic series.

2. Quantization of a complex and almost complex symplectic manifold

2.1. Quantum bundle

All manifolds and mappings between them will be of C^∞ class. A point in a manifold and its image given by a fixed map will be identified. As a class of a complex line bundle we mean its first Chern class. Cohomology classes in sense of Cech and de Rham cohomology classes will be usually not distinguished. If $\varepsilon : H^k(P,\mathbb{Z}) \longrightarrow H^k(P,\mathbb{R})$ is the natural injection then $q \in H^k(P,\mathbb{Z})$ and $\varepsilon(q)$ will be not distinguished too.

Observe that to each complex manifold M^c (latter ^{c} means that we consider a complex structure) we can assign a real structure M such that to each coordinates system $z_1,\dots,z_n$ in M^c corresponds coordinates system $x_1,y_1,\dots,x_n,y_n$ in M such that $z_i = x_i + iy_i$, $i = 1,\dots,n$.

As a complex symplectic manifold $\langle M^c,\omega\rangle$ we mean a complex manifold M^c together with two-form ω which is a symplectic form in real structure M.

Definition

The complex line bundle Q over manifold M^c is called a quantum bundle of complex symplectic manifold $\langle M^c,\omega\rangle$ if the following axioms are satisfied:

q^1) ω is a form of type (1.1) on M^c.

q^2) $\operatorname{sgn} i\omega(X^h,\overline{X^h}) = \text{const.} := \tau$ where X^h is any non-zero holomorphic vector tangent to M^c.

q^3) $q := [\omega] + \tau/2\; c_1(M^c) \in H^2(M,\mathbb{Z})$ where q is the class of bundle Q and $c_1(\cdot)$ is a first Chern class of a complex manifold.

q^4) $\operatorname{sgn}[\omega]_{([a])} \neq -\operatorname{sgn} q_{([a])}$ for each $[a] \in H_2(M,\mathbb{R})$, $[a] \neq 0$.

If M has no 1 st. order torsion, e.i. $H^1(M,\mathbb{Z}) = \mathbb{Z}^k$, then the quantum bundle is determinated uniquely up to equivalence of bundles.

By virtue of axioms (q^1) and (q^3) any quantum bundle is a holomorphic line bundle over M^c, see [4].

If $\langle M^c,\omega\rangle$ admits a quantum bundle then M^c is a Kähler manifold with Kähler form $\omega_1 := -\tau\omega$. It is equivalent to (q^1) and (q^2).

Lemma 2.1.

1) If for a complex symplectic manifold $\langle M^c,\omega\rangle$ there exists a quantization bundle Ω then M^c possesses a Hodge structure given by form $\chi := 2\omega_1$. Moreover M^c may be considered as a complex submanifold of the complex projective space $P^r C$ for some $r \geq \dim_C M^c$ and χ is equal to the restriction of Hodge form γ on $P^r C$. The explicite formula for γ will be given in point 3.

2) Let $\langle M^c,\chi\rangle$ be a Hodge manifold and assume that there exists a number $l \in \{1/2,1\}$ such that

$$l\chi - c_1(M^c) \in H^2(M,\mathbb{Z})$$

Then there exists an infinite number of symplectic forms ω_j such that the complex symplectic manifolds $\langle M^c,\omega_j\rangle$ admit quantum bundles.

I do not know any Hodge manifold for which surely no symplectic form admitting any quantum bundle exists.

Theorem 2.1.

Let $\langle M^c,\omega\rangle$ be a complex symplectic manifold satisfying axioms (q^1) and (q^2). F denotes a polarization generated by all anti-holomorphic vectors if $\tau = 1$ (resp. holomorphic if $\tau = -1$). Assume that triplet $\langle M,\omega,F\rangle$ admits a quantization in sense of Kostant and Souriau with bundle $\Omega' := L \otimes D_F^{1/2}$ whose class is q' and $\mathrm{sgn}[\omega]_{([a])} \neq -\mathrm{sgn}\, q'([a])$ for each non-zero $[a] \in H_2(M,\mathbb{R})$. Then there exists a quantum bundle Ω of $\langle M^c,\omega\rangle$ and $q' = q$.

Remark 2.1.

Observe that a complex symplectic manifold may admit a quantum bundle even if the KS quantization does not exist. Then neither any prequantization nor any metalinear structure exist. Such a case appears for a multi-odd-dimensional harminic oscillator, see point 3.

The assumptions (q^2) and (q^4) are usually made in examples where the KS procedure is applied, see [8], [10].

2.2. Quantum bundle obtained by means of the Maslov quantization condition

We will show in this point that the Maslov quantization condition may be also used as a start point in a construction of a quantum bundle

Lemma 2.2.

Let us denote $\langle M', \omega' \rangle := \langle \mathbb{R}^{2n}, \sum_{i=1}^{n} dp_i \wedge dq_i \rangle$ and $M_E := h^{-1}(E)$ where $h : \mathbb{R}^{2n} \longrightarrow \mathbb{R}^1$ is a Hamiltonian. Assume that M_E is a compact, connected, simply connected, $(2n-1)$-dimensional submanifold of M'. Assume moreover that all orbits are closed, their set M is a manifold and $\pi : M_E \longrightarrow M$ is a projection. Then the following statements hold:

O1) M is simply connected.

O2) There exists a unique two-form $\omega \in \Lambda^2(M)$ which is symplectic and such that $\pi^* \omega = \omega'|M_E$

O3) $\int_{\mathcal{O}} \Theta$ = constant where $\Theta := \sum_{i=1}^{n} p_i dq_i$, $\mathcal{O}$ is an orbit in M_E.

O4) For each closed, orientable and simply connected surface a in M there exists a surface b in M_E such that

$$\int_a \omega = \int_b \omega' = \int_{\partial b} \Theta, \quad \pi(b) = a$$

and ∂b is either an orbit (perhaps with an integer multiplicity) or a void set.

Suppose that $\mathcal{L}$ is a Lagrangian surface in M_E ($\omega'|_{\mathcal{L}} \equiv 0$ and $\dim \mathcal{L} = n$). Let $\mathrm{ind}_{\mathcal{L}} \mathcal{O}_{\mathcal{L}}$, $\mathcal{O}_{\mathcal{L}}$ is an orbit in $\mathcal{L}$, be the Maslov index of $\mathcal{O}$ with respect to $\mathcal{L}$ (see [1]). One easy see that all orbits in belong to the same homology class in $H_1(\mathcal{L}, \mathbb{R})$ so that $\mathrm{ind}_{\mathcal{L}} \mathcal{O}_{\mathcal{L}}$ = const. for each orbit $\mathcal{O}_{\mathcal{L}} \subset \mathcal{L}$.

Let us put $\mathrm{ind}_{\mathcal{L}} \mathcal{O} := \mathrm{ind}_{\mathcal{L}} \mathcal{O}_{\mathcal{L}}$ for any orbit $\mathcal{O}$ in M_E.

Let us define

$$\tilde{q}(a) := \int_a \omega - \frac{1}{4} \mathrm{ind}_{\mathcal{L}} \partial b$$

(a and b are as in (O4).)

One easy verifies that $\tilde{q}$ is well defined on the space of homology classes $H_2(M,\mathbb{R})$. Thus the following theorem is clear.

Theorem 2.2.

If $n > 1$ and the Maslov quantization condition is satisfied, i.e. for some Lagrangian surface $\mathcal{L}$ in M_E there is

$$\int_c \Theta = k - \frac{1}{4}\,\mathrm{ind}_{\mathcal{L}}\, c\,, \quad k \in \mathbb{Z}$$

where c is any closed curve in $\mathcal{L}$ then $\tilde{q} \in H^2(M,\mathbb{Z})$.

Observe that a weaken condition, namely $\int_{\mathcal{O}} \Theta = k - \frac{1}{4}\,\mathrm{ind}_{\mathcal{L}}\mathcal{O}$ for some orbit $\mathcal{O}$ in M_E, implies $\tilde{q} \in H^2(M,\mathbb{Z})$ too.

One easy checks that $q = r[\omega]$, $r \in \mathbb{R}$. We had not such an additional condition in the case of the class of a quantum bundle.

The further considerations remain true for $\tilde{q}$ (instead of q) if we deal with a manifold of orbit M, there exists a Lagrangian surface $\mathcal{L}$ in M_E satisfying the (weaken) Maslov condition and axiom (q^4) holds. Then we have to introduce an almost complex structure such that (q^1) and (q^2) hold.

2.3. Hilbert space of quantum states.

To have any sensible model of quantum mechanics we must have a scalar product in a quantum bundle. We will define it using a Hermitian structure which exists by virtue of the following lemma.

Lemma 2.3. (B.Kostant, A.Weil)

Assume that $\mathcal{V} \in \Lambda^2(P)$. Then there exists a line bundle B over manifold P with connection ∇ such that $\mathrm{curv}\,(B,\nabla) = \mathcal{V}$ if and only if $[\mathcal{V}] \in H^2(P,\mathbb{Z})$. Then $[\mathcal{V}]$ is a first Chern class of B.

If $[\mathcal{V}] \in H^2(P,\mathbb{Z})$ then there exists connection ∇_1 and ∇_1-invariant Hermitian structure $(\cdot|\cdot)$ in B such that $\mathrm{curv}\,(B,\nabla_1) = \mathcal{V}$. If connection ∇_2 is equivalent to ∇_1 then the Hermitian structure $(\cdot|\cdot)$ is ∇_2-invariant too.

If manifold P is simply connected then there exists a unique class of such equivalent connections $[\nabla_1]$ and Hermitian structure $(\cdot|\cdot)$ is unique up to a positive multiplicative constant.

In general we can point out a bijective map between a set of all connections' classes and the group $(H^1(P,\mathbb{Z}))^*$; then a set of all such Hermitian structures is in one-to-one correspondence with $(H^1(P,\mathbb{Z}))^* \times \mathbb{R}^+$. ($G^*$ denotes here the group of all continuous homomorphisms from G into S^1).

The proof may be found in Ko stant's book [6].

To apply this lemma to construct a Hilbert space we must point out a concrete form in de Rham class $q \in H^2(M,\mathbb{Z})$. We choose a harmonic form. By virtue of the Hodge-Kodaira theorem each de Rham class of a Riemannian manifold P (so in particular a Kähler manifold $\langle M^c,\omega_1\rangle$) contains the only harmonic form. To justify this choice observe that form $r\omega_1$ $r \in \mathbb{R}$, on a complex symplectic manifold $\langle M^c,\omega_1\rangle$ is harmonic (see [4] in 15.8).

Definition

Let $\langle M^c,\omega\rangle$ be a compact complex symplectic manifold which admits a quantum bundle Ω with class q. Then the Hilbert space of quantum states $\mathcal{H}$ consists of all holomorphic (if $\tau = 1$) or antiholomorphic (if $\tau = -1$) sections of bundle Ω with the scalar product

$$(s_1,s_2) := \int_{M^c} (s_1,s_2)\ \omega_1^n$$

where $(\cdot|\cdot)$ is a Hermitian structure which is ∇-invariant and ∇ is a connection such that $\mathrm{curv}(\Omega,\nabla) = \eta$ where η is the harmonic form in de Rham class $q \in H^2(M,\mathbb{Z})$.

Form η will be called the quantum form.

If M is simply connected then Hilbert space $\mathcal{H}$ is defined uniquely up to a positive multiplicative constant. Dim $\mathcal{H}$ is always a finite number. It may be computed using the Riemann-Roch-Hirzebruch theorem (cf. [10]).

Considering holomorphic or anti-holomorphic sections allows us to change a sign before the symplectic form and for instance to distinguish the sign of potential energy.

2.4. Physical quantities.

Let $\langle M^c,\omega\rangle$ be henceforth a compact complex symplectic manifold which admits a quantum bundle Ω. Assume that $\tau = 1$. Let X_f^η be a vector field such that $df = X_f^\eta \lrcorner\, \omega$ (η is the quantum form). Consider the following set of functions on M

$$C_{ah}(M^c) := \{f \in C(M) \mid f \text{ is a real function and } X_f^\eta \text{ preserves the bundle of anti-holomorphic vectors on } M^c\}$$

Then the formula

$$A_f := \frac{1}{2\pi i} \nabla_{X_f^\eta} + f, \quad f \in C_{ah}(M^c)$$

determinates a Hermitian operator in $\mathcal{H}$.

If quantum form η is non-degenerated then operator A_f is defined uniquely and its domain is whole the space $\mathcal{H}$.

This definition differs to that of the KS theory (see [3]). The slighty difference will be shown in example 3.

The following definition allows us to extend quantization to a bigger class of functions. Let us put namely

$$(A_f s_1, s_2) := \int_{M^c} (s_1|s_2) f \omega_1^n + (s_1|s_2) \mathcal{L}_{(X_f^\eta)_2} \omega_1^n$$

where $(X)_2$ denotes a projection in $(TM)^C$ on the bundle of holomorphic vectors.

The above formula is in fact a quantization by means of repreoducing kernels, see [3].

The above methods hold for $\tau = -1$ too.

Observe that our quantization procedure of a complex symplectic manifold and physical quantities works also if we consider an almost complex structure on manifold M.

3. Example. Multidimensional harmonic oscillator

Let us consider the n-dimensional harmonic oscillator, $n \geq 2$, with diagonalized Hamiltonian. We may identify the space of orbits corresponding energy E with complex projective space $P^{n-1}C$. We note $U := \{[z] \in P^{n-1}C \mid z_n \neq 0\}$. Then we can write the following formula for symplectic form ω_E on $P^{n-1}C$

$$\omega_{E|U} = -\frac{iE}{h}\frac{\sum_{k=1}^{n-1} dw_k \wedge d\bar{w}_k}{(1+|w_1|^2+\ldots+|w_{n-1}|^2)^2} := -\frac{E}{\nu h}\gamma$$

where

$$w_1 = \frac{z_1}{z_n},\ldots,w_{n-1} = \frac{z_{n-1}}{z_n}$$

We have $\tau = 1$.
The Chern class of $P^{n-1}C$ is ng where g is a positive generator of group $H^2(P^{n-1}C,\mathbb{Z})$. There is $\gamma \in g$.
By virtue of (q^1) and (q^4) complex symplectic manifold $\langle P^{n-1}C,\omega_E\rangle$ possesses a quantum bundle if and only if $q := [\omega_E] + \frac{n}{2}g$ is a non-positive element of $H^2(P^{n-1}C,\mathbb{Z})$. Taking representants of classes we obtain $-\frac{E}{\nu h} + \frac{n}{2} = 0,-1,-2,\ldots$ thus

$$E = \nu h(N + \tfrac{n}{2}) := E_N, \quad N = 0,1,2,\ldots$$

Let us observe that if n is odd then E_N is not any integer multiple of νh. It implies that for odd-dimensional harmonic oscillator the KS condition of prequantization is not satisfied and the quantization in sense of Kostant and Souriau does not exist. Indeed, let us note $c := \{w \in U \mid w_2 = \ldots = w_{n-1} = 0\}$. Then it is

$$\int_c \omega_{E_N} = -(N+\tfrac{n}{2}) \notin \mathbb{Z}$$

Let us have a closer look at quantum bundle Ω_N corresponding to energy E_N. The quantum form η_N is

$$\eta_{N|U} := -\frac{iN}{2\pi}\frac{\sum_{k=1}^{n-1} dw_k \wedge d\bar{w}_k}{(1+|w|^2)^2}$$

where

$$|w|^2 = |w_1|^2 + \ldots + |w_{n-1}|^2$$

We can take connection ∇ which has connection form

$$\alpha_N\Big|_{\pi^{-1}(u)} := \frac{dz}{2\pi iz} + \frac{N}{2\pi i}\,\frac{\sum_{k=1}^{n-1} w_k d\bar{w}_k}{1+|w|^2}$$

The ∇-invariant Hermitian structure in Ω_N is $(\cdot|\cdot)_N$

$$|(w,\zeta)|_N^2 := |\zeta|^2(1+|w|^2)^N$$

If $s \in \Gamma(Q)$ is a holomorphic section of bundle Q_N then s restricted to U is

$$s|_U = \frac{p(w)}{(1+|w|^2)^N}$$

where $p(w)$ is a polynomial of variables $w_1,\ldots,w_{n-1}$ of degree $\leq N$.

The multiplicity of degeneracy of E_N is $\binom{n+N-1}{N}$. The same number we obtain in classical quantum mechanics solving the Schrödinger equation.

We may derive the following formula for a scalar product in Hilbert space of quantum states

$$(s_1,s_2) = (-1)^{\frac{(n-1)(n-2)}{2}}\, i^{n-1}\,\frac{(N+n/2)^{n-1}}{2\pi}\int_{C^{n-1}} \frac{p_1(w)p_2(w)\,dw\wedge d\bar{w}}{(1+|w|^2)^{N+2(n-1)}}$$

In the KS theory if n is even we have the following formula for elements of quantum Hilbert space (we take the anti-holomorphic polarization)

$$\sigma|_U = \frac{\tilde{p}(w)\sqrt{d\bar{w}}}{(1+|w|^2)^{N+n-1}}, \qquad \deg \tilde{p} \leq N$$

The formula for a scalar product as a function of coefficients of $\tilde{p}(\cdot)$ is the same as that in our quantization.

Algebra $C_{ah}(P^{n-1}C)$ consists of all functions

$$f([z_1,\dots,z_n]) = \frac{\sum_{i,j=1}^{n} a_{ij} z_i \bar{z}_j}{\sum_{i=1}^{n} |z_i|^2}$$

where matrix $[a_{ij}]$ is Hermitian.

Assume that $n = 2$.

Then $C_{ah}(P^{n-1}C)$ may be spanned by functions x_1, x_2, x_3 and a constant function where

$$[z_1, z_2] \xrightarrow{x_1} \frac{N}{2} \frac{z_1\bar{z}_2 + z_2\overline{z_1}}{|z_1|^2 + |z_2|^2}$$

$$[z_1, z_2] \xrightarrow{x_2} \frac{iN}{2} \frac{z_1\overline{z_2} - z_2\bar{z}_1}{|z_1|^2 + |z_2|^2}$$

$$[z_1, z_2] \xrightarrow{x_3} \frac{N}{2} \frac{|z_1|^2 - |z_2|^2}{|z_1|^2 + |z_2|^2}$$

Functions x_1, x_2, x_3 generate Lie subalgebra of $C_{ah}(P^{n-1}C)$ which will be denote by $C'_{ah}(P^{n-1}C)$. There is isomorphism $C'_{ah}(P^{n-1}C) = su(2)$. Identifying section s with sequence $\{a_j\}$ of coefficients of $p(\cdot)$ we obtain

$$A_{x_1}(\{a_j\}) = \{b_j\}$$

where

$$b_0 = \tfrac{1}{2}a_1, \quad b_j = [\tfrac{1}{2}(j+1)a_{j+1} + (N+1-j)a_{j-1}], \quad 1 \le j \le N-1$$

$$b_N = \tfrac{1}{2}a_{N-1}$$

$$A_{x_2}(\{a_j\}) = (\{c_j\})$$

where

$$c_0 = \frac{1}{2} i \cdot a_1 = ib_0, \quad c_j = ib_j, \quad 1 \leq j \leq N-1,$$

$$c_N = -\frac{i}{2} a_{N-1} = -ib_N$$

$$A_{X_3}(\{a_j\}) = \{(\frac{N}{2} - j)a_j\}$$

This gives us the standard Wigner representation of $su(2)$ for spin equal to $N/2$.

In the case of KS theory the same algebra is considered (cf. [3]). Then for each function $f \in C_{ah}(P^{n-1}C)$ one assigns the operator

$$A_f := \frac{1}{2\pi i} \nabla_{X_f} \otimes \mathcal{L}^{1/2}_{X_f} + f$$

where $\mathcal{L}^{1/2}_X$ denotes a Lie derivation in a half-forms bundle (see [3]). This quantization gives us the Wigner representation of $su(2)$ too but to get it we must take functions $\tilde{x}_1, \tilde{x}_2, \tilde{x}_3$ which have the real-number factor equal to $\frac{N+1}{2}$ instead of $\frac{N}{2}$.

In general we may represent as an operator in $\mathcal{H}$ any function f on P^1C such that the integral

$$(A_f s_1, s_2) := \frac{N+1}{2\pi} i \int_{C^1} \frac{f(w,\bar{w}) p_1(w) \overline{p_2(w)}}{(1+|w|^2)^N} \frac{dw \wedge d\bar{w}}{(1+|w|^2)^2} +$$

$$+ \frac{i}{4\pi} \mathcal{L}_{(X_f)_2} \frac{dw \wedge d\bar{w}}{(1+|w|^2)^2}$$

makes sense ($(X)_2$ denotes a projection on set of holomorphic vectors).

Let us apply to the multidimensional oscillator the WKB-Maslov method. The classical energy function is the Hamiltonian

$$h(m) := \frac{1}{2} \sum_{i=1}^{n} p_i^2 + \nu^2 q_i^2$$

Then the surface of constant energy is the $(2m-1)$-dimensional ellipsoid

$$\sum_{i=1}^{n} p_i^2 + \nu^2 q_i^2 = 2E$$

We take as Lagrangian surface $\mathcal{L}$ the surface given by equations

$$p_i^2 + \nu^2 q_i^2 = a_i^2, \quad i = 1,\dots,n, \quad a_i > 0, \quad \sum_{i=1}^{n} a_i^2 = 2E$$

The orbits on $\mathcal{L}$ are as follows

$$\mathcal{O} := \{(a_1 \cos t, \frac{a_1}{\nu}\sin t,\dots,a_n \cos t, \frac{a_n}{\nu}\sin t) \mid t \in \mathbb{R}\}$$

We can compute that $\operatorname{ind}_{\mathcal{L}} \mathcal{O} = -2n$. Thus the weaken Maslov condition is

$$\int_{\mathcal{O}} \frac{1}{h} \sum_{i=1}^{n} p_i dq_i + \frac{n}{2} \in \mathbb{Z}$$

for each orbit $\mathcal{O}$. Observe that surface $\mathcal{L}$ satisfies the Maslov quantization condition if and only if $a_i = N_i + 1/2$, $N_i = 0,1,\dots,N$.

Simply integration gives us the energy spectrum $E_N = h\nu(N + \frac{n}{2})$, $N = 0,1,\dots$. Taking $E = E_N$ we obtain $\tilde{q}_N = q_N$.

4. Quantization of symplectic manifold

The method of defining quantization bundle, which was carried out in the previous chapter for a symplectic manifold with a complex polarization, fails in the case of symplectic manifold $\langle M,\omega \rangle$ with real polarization F. Then the class of ω is an element of $H^2(M,\mathbb{R})$ and characteristic class of bundle of half-forms, i.e. the first Stiefel-Whitney class, is an element of $H^1(M,\mathbb{R})$. I do not see any natural way how to assign to these both elements an element of cohomology group $H^2(M,\mathbb{Z})$ which will be a class if a complex line bundle. Thus for a real polarization we will apply the K S quantization.

In the paper [3] we consider a general case of polarized symplectic manifold $\langle M,\omega,F \rangle$, i.e. polarization F may possess both a real and a complex part. If symplectic manifold $\langle M,\omega\rangle$ can be decomposed into a product of symplectic manifolds $\langle M_1,\omega_1\rangle$ and $\langle M_2,\omega_2\rangle$ and F induces a complex structure M_1^c on M_1 and real polarization on $\langle M_2,\omega_2\rangle$, then we can easily "product" the quantization procedure from the previous chapter and the KS-quantization.

In the paper [3] we deal with a more general case. Namely $M = M_1 \times M_2$ is a topological product, $\langle M_1,\omega|_{M_1}\rangle$ is a symplectic manifold and polarization F induces a complex structure on M_1 and a foliation of M_2, which is ω-orthogonal to M_1.

Acknowledgement

I would like to express my gratitude to dr N.M.J.Woodhouse from University of Oxford who suggested me these problems and sent me his probably unpublished manuscript. I used it in point 2.2. I am also deeply indebted to professor K.Maurin whose lectures made me clear the important theorems and to dr dr K. Gawędzki, J.Kijowski, W.Kondradzki, R.Rubinsztein, and A.Wawrzyńczyk for helpful and constructive talks.

Literature

1. Arnold, V.I., "A characteristic class related with quantiaztion conditions", Funct.Anal. and its Appl. 1(1967), 1-13, (in Russian).
2. Blattner, R., "Quantization and representation theory", Proc. Sympos. Pure Math., vol. 26, A.M.S., Providence R.L., 1974, 147.
3. Czyż, J., "On geometric quantization and its connections with the Marlov theory", to appear in Reports on Mathematical Physics.
4. Gawędzki, K., "Fourier-like kernels in geometric quantization", Dissertationes Mathematicae, CXXVIII, Warszawa 1976.
5. Hirzebruch, F., "Topological methods in algebraic geometry", New York 1966.
6. Ko stant, B., "On the definition of quantization", preprint MIT, 1975.
7. Ko stant, B., "Quantization and unitary representations", Lecture Notes in Mathematics, Vol. 170, Springer-Verlag, Berlin 1970, 87-208.
8. Newlander, A., L.Nirenberg, "Complex analytic coordinates in almost complex manifolds", Ann. of Math. 65(1957), 391-404.
9. Simms, D.J., "Geometric quantization of the harmaonic oscillator with diagonalised Hamiltonian", Proceedings of the Second International Colloquium on Group Theoretical Methods in Physics, Nijmegen 1973, 168.
10. ——, "Geometric quantization of symplectic amnifolds", Proc. Internat. Sympos. Math. Phys., Warsaw 1974.
11. ——, "Geometric quantization of energy levels in the Kepler problem", Conv. di Geom. Simp. e Fis. Math., INDAM, Rome 1973.
12. ——, N.M.J.Woodhouse, "Lecture on geometric quantization", Lecture Notes in Physics, Vol. 53, Springer-Verlag, Berlin 1976.
13. Souriau, J.-M., "Structure des systemes dynamiques", Dunod 1970.

REPRESENTATIONS ASSOCIATED TO MINIMAL CO-ADJOINT ORBITS

Joseph A. Wolf
Department of Mathematics
University of California
Berkeley, California 94720

ABSTRACT. The minimal dimensional co-adjoint orbits are determined for the real and complex classical Lie groups, and the representations associated to them by methods of geometric quantization are discussed. Some new computational methods are developed for the classical Lie algebras.

§1. Introduction

The unitary representations of reductive and semisimple Lie groups associated to maximal dimensional co-adjoint orbits now are well understood ([7], [12]). They correspond to regular semisimple elements in the Lie algebra, and the inner product on the representation space is defined by standard geometric techniques. The minimal dimensional orbits, however, present some new features. First, some of them are of physical interest, appearing for example in quantization of the harmonic oscillator ([8], [6], [5], [14]) and in certain completely integrable Hamiltonian systems such as the Calagero model ([4], [9]). Second, many of them correspond to nilpotent elements in the Lie algebra. Third, the definition of the inner product on the representation space remains problematic, and really has only been settled intrinsically in a few cases.

In addition to their physical interest, some of these representations are assuming important roles in algebraic geometry ([11], [3]), in the study of special functions [1], and in at least one area [15] closely related to much of the current work in algebraic number theory.

The construction of the inner product for the metaplectic representation (see [2]) is perhaps the major success of the Blattner-Kostant-Sternberg half form method. Within the framework of geometric quantization it associates the metaplectic representation to a minimal dimensional co-adjoint orbit of the real symplectic

---

Expanded version of part of a lecture at the conference "Differential Geometric Methods in Mathematical Physics," Bonn, July 13-16, 1977. Research partially supported by National Science Foundation Grant MPS-76-01692.

group. It turns out that, for many representations of semisimple or reductive groups, associated to minimal dimensional co-adjoint orbits, one can define the unitary structure of the representation space by realizing the representation within a metaplectic representation; see [10]. In that context one has a "moment map" [4] relating the metaplectic co-adjoint orbit to co-adjoint orbits of the group in question. In other cases one can define the inner product by various limit procedures ([5], [14]). And in a few of the remaining cases one has an *ad hoc* procedure.

In this paper, we determine all the minimal dimensional co-adjoint orbits of the real and complex classical (matrix) groups. Those are the linear, orthogonal, unitary, symplectic, and some other, groups. In many cases we associate representations to those orbits. This is done in §2 for real, complex and quaternionic linear groups; that seems to be new, but it is easy. Symplectic groups and metaplectic representations are done in §3, or rather just mentioned because references are available.

The real, complex, and quaternionic unitary groups appear in §4. There we find the minimal dimensional orbits in all three cases, and associate representations in the complex $(U(k, \ell))$ and quaternionic $(Sp(k, \ell))$ cases by means of the moment map and restriction of the metaplectic representation. The complex case was already known [10]. The inclusion $Sp(k, \ell) \subset U(2k, 2\ell)$ leads to some interesting relations between the representations of those two groups. The real case $(O(k, \ell))$ is treated later, in §6, by an extension of a limit method for quantization of the Kepler manifold that I introduced in [14] and reformulate in §5.

§7 deals with the complex orthogonal group, and the method is an adaptation of the technique used for the $O(k, \ell)$.

That leaves one series of classical groups, the real forms $SO^*(2m) = U(m, m) \cap O(2m; C)$ of the complex orthogonal groups $O(2m; C)$, with maximal compact subgroup $U(m)$. Their minimal co-adjoint orbits, and the associated representations, are determined in §8 by means of the moment map and restriction of the metaplectic representation.

Some comments on methodology are in order. In order to have a certain measure of uniformity for the various series of classical groups, we use quaternionic matrices for three of those series. In particular, $SO^*(2m)$ is taken as the automorphism group of Q^m with a skew-hermitian form. So one must remember to keep the scalars on the right and the linear transformations on the left. Also, then all but the general linear groups are defined by a form b that is symmetric or antisymmetric or hermitian or skew hermitian, and the Lie algebra is the real span of the

$$\xi_{u,v}: x \mapsto u \cdot b(v, x) \pm v \cdot b(u, x) .$$

Of course this already was known for the orthogonal groups $O(p, q)$ and $O(m; C)$, the unitary groups $U(p, q)$, and the symplectic groups $Sp(m; R)$ and $Sp(m; C)$. In the quaternionic cases it means that our hermitian and skew-hermitian forms are linear in the second variable and conjugate linear in the first. This explicit hold on the Lie algebra is especially useful for low-rank elements, and thus for the study of minimal co-adjoint orbits.

§2. Linear Groups

F will denote a real division algebra R (real numbers), C (complex numbers) or Q (quaternions). F^n is the right vector space of n-tuples from F. Scalars act on the right, so linear transformations act on the left and the elements of F^n are column vectors. The invertible F-linear transformations of F^n form a reductive real Lie group

$$\text{(2.1)} \qquad GL(n; F)\text{: general linear group of degree } n \text{ over } F .$$

View $GL(n; F)$ as the group of invertible $n \times n$ matrices over F; then its Lie algegra $\mathfrak{gl}(n; F) = F^{n\times n}$, space of all $n \times n$ matrices over F.

We identify $\mathfrak{gl}(n; F)$ with its real dual space $\mathfrak{gl}(n; F)^*$ by the $GL(n; F)$-invariant pairing $(\xi, \eta) = \mathrm{Re\ trace}(\xi\eta)$. Thus $\xi \in \mathfrak{gl}(n; F)$ is identified with the linear functional

$$\text{(2.2)} \qquad f_\xi: \mathfrak{gl}(n; F) \to R \quad \text{by} \quad f_\xi(\eta) = \mathrm{Re\ trace}(\xi\eta) .$$

This identifies co-adjoint orbits to adjoint orbits.

2.3. Proposition. *Every nontrivial co-adjoint orbit of* $GL(n; F)$ *has dimension* $\geqslant 2(\dim_R F)(n-1)$, *with equality just for the orbits* $\mathrm{Ad}^*(GL(n; F))\cdot f_\xi$ *where* $\xi \in \mathfrak{gl}(n; F)$ *is one of the* $\xi(\lambda, r)$, λ *and* r *central in* F (i.e. real in case $F = Q$), *given by*

$$\text{(2.4)} \quad \xi(\lambda, r) = \lambda I + \begin{pmatrix} r & & \\ & 0 & \\ & & \ddots & \\ & & & 0 \end{pmatrix} \textit{ for } r \neq 0 , \quad \xi(\lambda, 0) = \lambda I + \begin{pmatrix} 0 & \dots & 0 & 1 \\ & & & 0 \\ & & & \vdots \\ & & & 0 \end{pmatrix}$$

Indication of proof: We maximize the dimension of the centralizer $Z(\xi)$ of ξ in $GL(n; F)$, where ξ is not in the center of $\mathfrak{gl}(n; F)$. Arguing as in [10, §3], $\xi = \lambda I + \eta$ with λ central in F, η rank 1 and $Z(\eta)$ of maximal dimension. So ξ is conjugate to some $\xi(\lambda, r)$, r central in F, and $Z(\xi)$ has dimension $(\dim_R F)(n^2 - 2n + 2)$.

Proposition 2.3 shows that the minimal-dimensional co-adjoint orbits are the

$$\text{(2.5)} \qquad \mathcal{O}(\lambda, r) = \mathrm{Ad}^*(GL(n; F))\cdot f_{\xi(\lambda,r)} .$$

In all cases, $\mathcal{O}(\lambda, r)$ has invariant real polarization given by the maximal parabolic subalgebra of $\mathfrak{gl}(n; F)$,

$$(2.6)\qquad \mathcal{P} = \left\{ \left(\begin{array}{c|c} a & b \\ \hline 0 & A \end{array} \right) : a \in F\ ,\quad {}^t b \in F^{n-1}\ ,\quad A \in \mathfrak{gl}(n-1;\ F) \right\} .$$

The corresponding space $GL(n;\ F)/P$ is just the projective space $\mathbb{P}^{n-1}(F)$. Since

$$(2.7)\qquad f_{\xi(\lambda,r)}: \left\{ \begin{array}{c|c} a & b \\ \hline 0 & A \end{array} \right\} \to \operatorname{Re}\left\{ (\lambda + r)a + \lambda \cdot \operatorname{trace} A \right\} ,$$

$\sqrt{-1}\ f_{\xi(\lambda,r)}$ exponentiates to a unitary character $\chi(\lambda,\ r)$ on the identity component of P just when

$$(2.8)\qquad F \neq C:\ \text{no condition} \qquad F = C:\ \operatorname{Im} \lambda,\ \operatorname{Im} r \in \mathbb{Z} .$$

Now the unitary representations of $GL(n;\ F)$ associated to the co-adjoint orbits $\mathcal{O}(\lambda,\ r)$ are given as follows.

For $F = R$, let S consist of the four unitary characters $\begin{pmatrix} a & b \\ 0 & A \end{pmatrix} \to 1$, $a/|a|$, $\det A/|\det A|$, $a \cdot \det A/|a| \cdot |\det A|$ on P/P^0 . Each $\sigma \in S$ defines an extension $\chi(\lambda,\ r;\ \sigma)$ of $\chi(\lambda,\ r)$ to P . The resulting representations are the unitarily induced

$$\pi(\lambda,\ r;\ \sigma) = \operatorname{Ind}_{P \uparrow GL(n;R)} \chi(\lambda,\ r;\ \sigma) , \qquad \text{all } \lambda,\ r \in R .$$

For $F = C$, P is connected, and we have the unitarily induced representations $\pi(\lambda,\ r) = \operatorname{Ind}_{P \uparrow GL(n;C)} \chi(\lambda,\ r)$, all $\lambda,\ r \in R + \sqrt{-1}\,\mathbb{Z}$.

For $F = Q$, P is connected and we have the unitarily induced $\pi(\lambda,\ r) = \operatorname{Ind}_{P \uparrow GL(n;Q)} \chi(\lambda,\ r)$, all $\lambda,\ r \in R$.

If $r \neq 0$, we have another polarization by replacing $\mathcal{P}$ and P by their transposes. This gives the same set of representations.

The representations $\pi(0,\ r)$ of $GL(n,\ C)$ are implicit in some of the recent work ([4], [9]) on the Calagero model.

In all cases, the representations of $GL(n,\ F)$, just described, are on Hilbert spaces of half forms with values in bundles analogous to powers of the hyperplane bundle over $\mathbb{P}^{n-1}(F)$.

§3. Symplectic Groups

In this section F is R or C . Fix a nondegenerate antisymmetric bilinear form $\{\ ,\ \}$ on F^{2n} . Then the (real or complex) *symplectic group* is

$$(3.1)\qquad Sp(n;\ F) = \{g \in GL(2n;\ F) : \{gx,\ gy\} = \{x,\ y\} \ \text{ for all } x,\ y \in F^{2n}\} .$$

In a basis of F^{2n} relative to which $\{\ ,\ \}$ has matrix $\begin{pmatrix} 0 & I \\ -I & 0 \end{pmatrix}$, the Lie algebra is given by

(3.2) $$\mathcal{S}\mathcal{P}(n; F) = \left\{\begin{pmatrix} A & B \\ C & -{}^tA \end{pmatrix} : B = {}^tB \text{ and } C = {}^tC\right\}$$

in $n \times n$ blocks. Without basis, $\mathcal{S}\mathcal{P}(n; F)$ is spanned over R by the elements

(3.3) $$\xi_{u,v} : x \mapsto \{x, u\}v + \{x, v\}u \quad \text{where} \quad u, v \in F^{2n} .$$

Identify $\mathcal{S}\mathcal{P}(n; F)$ with its real linear dual space by $(\xi, \eta) = \text{Re trace}(\xi\eta)$. Then

(3.4) $$\xi_{u,v} \leftrightarrow f_{u,v} \quad \text{where} \quad f_{u,v}(\eta) = \{u, \eta(v)\} + \{v, \eta(u)\}$$

One can check (see [10, §3]) that the minimal dimensional co-adjoint orbits are

(3.5) $$\begin{cases} \text{for } R : \pm Ad^*(Sp(n; R))\cdot f_{u,u} \cong (R^{2n} - \{0\}, \pm\{\ ,\ \}) \\ \text{for } C : \ Ad^*(Sp(n; C))\cdot f_{u,u} \cong (\mathbb{C}^{2n} - \{0\}, \{\ ,\ \}) \end{cases}$$

for any nonzero $u \in F^{2n}$. These orbits do not have invariant polarization, but the metaplectic representations are associated to them by the Blattner-Kostant-Sternberg pairing of half forms; see [2]. The metaplectic representations also come out of a graded Lie algebra construction [10] which is convenient for restriction to subgroups $U(k, \ell)$. We will discuss that and other restrictions in the rest of this paper.

§4. Indefinite Unitary Groups

Let $k, \ell \geq 0$ be integers with $k + \ell = n$. We write $F^{k,\ell}$ for F^n together with the hermitian form

$$\langle x, y\rangle = \sum_{a=1}^{k} \bar{x}_a y_a - \sum_{b=k+1}^{n} \bar{x}_b y_b$$

which is conjugate-linear in the first variable and linear in the second. We follow this convention so that (4.3) below holds over Q as well as R and C . We still are led to the usual *unitary group* of $F^{k,\ell}$, which is

(4.1) $$U(k,\ell; F) = \{g \in GL(n; F): \langle gx, gy\rangle = \langle x, y\rangle \text{ on } F^{k,\ell}\} .$$

Here $U(k, \ell; R) = O(k, \ell)$, indefinite orthogonal group; $U(k, \ell; C) = U(k, \ell)$, ordinary indefinite unitary group; and $U(k, \ell; Q) = Sp(k, \ell)$, indefinite unitary-symplectic group.

If * denotes conjugate-transpose, the Lie algebra of $U(k, \ell; F)$ is

(4.2) $\mathcal{U}(k, \ell; F) = \left\{ \begin{pmatrix} A & B \\ B^* & D \end{pmatrix} : A = -A^* \in F^{k\times k} , \quad D = -D^* \in F^{\ell\times\ell} , \quad B \in F^{k\times\ell} \right\}$.

Without matrices: $\mathcal{U}(k, \ell; F)$ is spanned over R by the

(4.3) $$\xi_{u,v}: x \mapsto v\langle u, x\rangle - u\langle v, x\rangle \quad \text{where} \quad u, v \in F^{k,\ell} .$$

For these $\xi_{u,v}$ are F-linear, are in $\mathcal{U}(k, \ell; F)$ because $\langle \xi_{u,v}x, y\rangle + \langle x, \xi_{u,v}y\rangle = 0$ by calculation, and can be seen to span either by dimension count or because $Ad(g)\xi_{u,v} = \xi_{gu,gv}$ for $g \in U(k, \ell; F)$ so the $\xi_{u,v}$ span an ideal.

Identify $\mathcal{U}(k, \ell; F)$ with its real linear dual space $\mathcal{U}(k, \ell; F)^*$ under $(\xi, \eta) = \mathrm{Re\ trace}(\xi\eta)$. Then

(4.4) $$\xi_{u,v} \leftrightarrow f_{u,v} \quad \text{where} \quad f_{u,v}(\eta) = \mathrm{Re}\{\langle \eta v, u\rangle - \langle \eta u, v\rangle\} .$$

Also, if $F \neq R$ denote

(4.5) $$\begin{cases} \xi_u = \frac{1}{2}\xi_{u,ui}: x \mapsto ui\langle u, x\rangle & \text{for } u \in F^{k,\ell} ; \text{ and} \\ f_u = \frac{1}{2} f_{u,ui}: \eta \mapsto -\mathrm{Re}(i\langle \eta u, u\rangle) & \text{for } u \in F^{k,\ell} , \quad \eta \in \mathcal{U}(k, \ell; F) . \end{cases}$$

Finally, let $\{e_1, \ldots, e_{k+\ell}\}$ denote an orthonormal basis of $F^{k,\ell}$, i.e. one in which $\langle\ ,\ \rangle$ has matrix $\begin{pmatrix} I & 0 \\ 0 & -I \end{pmatrix}$.

<u>4.6. Proposition.</u> *The minimal dimensional co-adjoint orbits of* $Sp(k, \ell)$ *are the* $4(k + \ell - 1)$*-dimensional*

(4.7) $$\mathcal{O}(r) = Ad^*(Sp(k, \ell))\cdot f_u , \quad 0 \neq u \in Q^{k,\ell} , \quad \|u\|^2 = r \in R .$$

The minimal dimensional co-adjoint orbits of $U(k, \ell)$, $(k, \ell) \neq (1, 1)$, *are the* $2(k + \ell - 1)$*-dimensional*

(4.8) $$\begin{cases} \mathcal{O}_{\pm}(\lambda, r) = \pm Ad^*(U(k, \ell))\cdot(f_u + i\lambda\cdot \mathrm{trace}) \quad \textit{where} \\ \qquad \lambda \in R , \quad 0 \neq u \in C^{k,\ell} , \quad \|u\|^2 = r \in R . \end{cases}$$

The minimal dimensional co-adjoint orbits of $O(k, \ell)$ *are*

(4.9a) $$\begin{cases} \textit{if } \min(k, \ell) \geq 2\textit{: the } 2(k + \ell - 3)\textit{-dimensional orbit} \\ \mathcal{O} = Ad^*(O(k, \ell))\cdot f_{u,v} \quad \textit{where } u, v \textit{ span a totally isotropic plane,} \end{cases}$$

(4.9b) $$\begin{cases} \textit{if } (k, \ell) = (2, 4)\textit{: also the 6-dimensional orbits} \\ \mathcal{O}(r) = Ad^*(O(2, 4))\cdot r(f_{e_1,e_2} + f_{e_3,e_4} + f_{e_5,e_6}) , \quad r > 0 , \end{cases}$$

(4.9c) $$\begin{cases} \textit{if } (k, \ell) = (2, 2)\textit{: also the 2-dimensional orbits} \\ \mathcal{O}(r) = Ad^*(O(2, 2))\cdot r(f_{e_1,e_2} + f_{e_3,e_4}) , \quad r > 0 , \end{cases}$$

$$(4.10)\quad \begin{cases} \textit{if } \min(k,\ell)=1\textit{: the } 2(k+\ell-2)\textit{-dimensional orbits} \\ O(r,s) = \mathrm{Ad}^*(O(k,\ell))\cdot f_{u,v} \textit{ where } u,v \textit{ are linearly} \\ \textit{independent}, \ \langle u,v\rangle = 0, \ \|u\|^2 = r \leqslant s = \|v\|^2 \ ; \textit{ here} \\ \textit{we may take } (r,s) = (-1,0) \textit{ or } (0,1), \textit{ or } r=\pm 1 \textit{ with } s\neq 0, \textit{ or } s=\pm 1 \textit{ with } r\neq 0. \end{cases}$$

(4.11a) *if* $(k, \ell) = (4, 0)$: *the 2-dimensional* $O(r)$ *as in* (4.9c),

(4.11b) *if* $(k, \ell) = (6, 0)$: *the 6-dimensional* $O(r)$ *as in* (4.9b),

(4.11c) *if* $4 \neq k \neq 6$ *and* $\ell = 0$: *the* $2(k-2)$*-dimensional* $O(r, s)$ *as in* (4.10),

$$(4.11d)\quad \begin{cases} \textit{if } (k,\ell) = (8,0)\textit{: also the 12-dimensional} \\ O(r) = \mathrm{Ad}^*(O(8))\, r(\sum_1^4 f_{e_{2i-1},e_{2i}}) \ , \quad r > 0 \ . \end{cases}$$

Proof. Consider a minimal dimensional co-adjoint orbit $\mathrm{Ad}^*(O(k, \ell))\cdot f_\xi$, $f_\xi(\eta) = \mathrm{trace}(\xi\eta)$, $\xi \in \mathfrak{o}(k, \ell)$. In other words the $O(k, \ell)$-centralizer $Z(\xi)$ of ξ is maximal dimensional. So ξ is either semisimple or nilpotent. If $\min(k, \ell) \geqslant 1$ one sees that all such nilpotent elements in $\mathfrak{o}(k, \ell)$ are given by $\xi_{u,v}$ for (i) u, v span a totally isotropic plane, (ii) u, v span a plane on which $\langle\ ,\ \rangle$ has rank 1. In case (i), the $O(k, \ell)$-stabilizer of $f_{u,v}$ has codimension $2(k + \ell - 3)$, and in case (ii) it has codimension $2(k + \ell - 2)$. If $\min(k, \ell) = 1$ then only case (ii) occurs, and if $\min(k, \ell) = 0$ then $\mathfrak{o}(k, \ell)$ has no nonzero nilpotent element. In all cases, the maximal subgroups of $O(k, \ell)$ that are centralizers of semisimple elements have complexification of the form (iii) $O(2; \mathbb{C}) \times O(k + \ell - 2; \mathbb{C})$ or (iv) $GL(\frac{k+\ell}{2}; \mathbb{C})$. In case (iii) the corresponding co-adjoint orbits are those of the $f_{u,v}$ where u, v span a plane on which $\langle\ ,\ \rangle$ is nondegenerate; they have dimension $2(k + \ell - 2)$. In case (iv), the co-adjoint orbits are of the form $\mathrm{Ad}^*(O(k, \ell))\cdot rf \cong O(k, \ell)/U(\frac{1}{2}k, \frac{1}{2}\ell)$ where $k = 2k'$, $\ell = 2\ell'$, and $f = \sum_1^{k'} f_{e_{2i-1},e_{2i}} + \sum_{k'+1}^{k'+\ell'} f_{e_{2j-1},e_{2j}}$; those co-adjoint orbits have dimension $(k' + \ell')^2 - (k' + \ell')$. Now notice $(k' + \ell')^2 - (k' + \ell') - 2(k + \ell - 3) = (k' + \ell')^2 - 5(k' + \ell') + 6$ which is 0 for $k' + \ell' = 2, 3$ and is > 0 for $k' + \ell' > 3$. The statements on coadjoint $O(k, \ell)$-orbits follow.

The statement for $U(k, \ell)$ is [10, §3]. The same sort of argument shows that the minimal co-adjoint orbits of $Sp(k, \ell)$ are of the form $\mathrm{Ad}^*(Sp(k, \ell))\cdot f_u$, $0 \neq u \in Q^{k,\ell}$. There we may assume u real; then $f_u\colon \eta \mapsto \mathrm{Re}\langle\eta(u)i, u\rangle$ has stabilizer $\{g \in Sp(k, \ell)\colon gu = uc \text{ with } c \in Q,\ |c| = 1\}$. If $\|u\|^2 \neq 0$ this is the stabilizer of uQ, which has codimension $4(k + \ell - 1)$. If $\|u\|^2 = 0$, it has codimension 1 in the stabilizer of uQ, hence codimension $4(k + \ell - 1)$ in

$Sp(k, \ell)$. *q.e.d.*

View $U(k, \ell)$ as a subgroup of $Sp(k + \ell; R)$ as in [10]. Thus $Sp(k + \ell; R)$ consists of all R-linear transformations of $C^{k,\ell}$ that preserve the antisymmetric R-bilinear form $\{u, v\} = \mathrm{Im}\langle u, v\rangle$. Denote

(4.12) $$X\text{: the symplectic manifold } (C^{k,\ell} - \{0\}, \{\ ,\ \}) .$$

Then X is a Hamiltonian $Sp(k + \ell; R)$-space, and in fact

(4.13) $$\begin{cases} 0 \to R \to C^{\infty}(X) \to \mathrm{Hamiltonian}(X) \to 0 \\ \qquad\qquad \nwarrow^{\phi} \qquad \uparrow \\ \qquad\qquad\qquad \mathfrak{sp}(k + \ell; R) \end{cases}$$

The lift ϕ of $\mathfrak{sp}(k + \ell; R)$ from Hamiltonian vector fields to functions, is given by

(4.14) $$\phi(\xi)(v) = \frac{1}{2}\{\xi v, v\} ,$$

so the moment map is

(4.15) $$\Phi: X \to \mathfrak{sp}(k + \ell; R)^* \text{ by } \Phi(v)(\xi) = \frac{1}{2}\{\xi v, v\} ,$$

which, in terms of (3.4), is $v \mapsto -\frac{1}{4} f_{v,v}$ and gives the isomorphism of (3.5) for $F = R$.

Now view X as a Hamiltonian $U(k, \ell)$-space, using $\phi|_{\mathfrak{u}(k,\ell)}$. Then the moment map is, using (4.5),

(4.16) $$\Phi: X \to \mathfrak{u}(k, \ell)^* \text{ by } \Phi(u) = \frac{1}{2} f_u .$$

Thus Φ sends $\{u \in \mathbb{C}^{k,} : u \neq 0 \text{ and } 2\|u\|^2 = r\}$ to the orbit $\mathcal{O}_+(0, r)$ of (4.8). That orbit is integral just when r is an integer. On the level of prequantization - but leaving out the half forms - this decomposition

$$\Phi\text{: (Bohr-Sommerfeld set in } X\text{)/(null foliation)} \cong \bigcup_{d\in\mathbb{Z}} \mathcal{O}_+(0, d)$$

corresponds to the linearization of the restriction of the metaplectic representation to $U(k, \ell)$,

(4.17) $$\nu = \sum_{d\in\mathbb{Z}} \nu_d \quad \text{as in [10].}$$

More generally, $\mathcal{O}_{\pm}(\lambda, r)$ is integral just when λ and r are integers, and one expects the correspondence

(4.18) $$\mathcal{O}_+(\lambda, d) \leftrightarrow (\det)^{\lambda} \otimes \nu_d \quad\text{and}\quad \mathcal{O}_-(\lambda, d) \leftrightarrow (\det)^{-\lambda} \otimes \nu_d^* .$$

In fact, this has not yet been made precise within the framework of geometric quantization.

Given integers $k', \ell' \geq 0$, let $k = 2k'$ and $\ell = 2\ell'$. We inject $Sp(k', \ell') \subset Sp(k + \ell; R)$ and view X as a Hamiltonian $Sp(k', \ell')$-space. This can be done in two equivalent ways. First, we can view $Sp(k + \ell; R)$ as consisting of all R-linear transformations of $Q^{k',\ell'}$ that preserve the R-bilinear form $\{u, v\} = (\text{i-component of } \langle u, v\rangle) = -\mathrm{Re}(i\langle u, v\rangle)$. Second, we can view $Q^{k',\ell'} = C^{k',\ell'} + C^{k',\ell'}\cdot j \cong C^{k,\ell}$, which gives $Sp(k', \ell') \subset U(k, \ell) \subset Sp(k + \ell; R)$. At any rate, we then have $\phi|_{\mathfrak{sp}(k',\ell')}$, and using (4.5) the moment map is

$$(4.19) \qquad \Phi: X \to \mathfrak{sp}(k', \ell')^* \quad \text{by} \quad \Phi(u) = \frac{1}{2} f_u .$$

Thus Φ sends $\{u \in Q^{k',\ell'} : u \neq 0 \text{ and } 2\|u\|^2 = r\}$ to the orbit $\mathcal{O}(r)$ of (4.7). That orbit is integral just when r is an integer. On the prequantization level, leaving out half-forms, we have the decompostion

$$\Phi: (\text{Bohr-Sommerfeld set in } X)/(\text{null foliation}) \cong \bigcup_{d \in \mathbb{Z}} \mathcal{O}(d) ,$$

corresponding to the linearization of the metaplectic representation restricted from $Sp(k + \ell; R)$ to $Sp(k', \ell')$. It follows from a reproducing kernel argument [15] that this restriction is

$$(4.20) \quad \nu|_{Sp(k',\ell')} = \sum_{d \in \mathbb{Z}} \nu_d|_{Sp(k',\ell')} , \quad \text{each } \nu_d|_{Sp(k',\ell')} \text{ irreducible,}$$

where the $\nu_d \in U(k, \ell)^\wedge$ as in (4.17). Thus, as in (4.18), one expects

$$(4.21) \qquad \mathcal{O}(d) \leftrightarrow \nu_d|_{Sp(k',\ell')} \in Sp(k', \ell')^\wedge ,$$

and precision of (4.18) in the framework of geometric quantization will do the same for (4.21).

We defer consideration of representations associated to minimal $O(k, \ell)$-orbits to §§5 and 6.

§5. Quantization of the Kepler Manifold

The intimate relation between minimal dimensional co-adjoint orbits of $Sp(k', \ell')$ and of $U(2k', 2\ell')$, and their representations, has a special case related to quantization of the Kepler manifold. Consider

$$(5.1) \qquad Sp(1, 1) \subset SU(2, 2) \subset U(2, 2) \subset Sp(4; R) .$$

$Sp(1, 1)$ is isomorphic to the double cover $Spin(1, 4)$ of the identity component $SO(1, 4)$ of $O(1, 4)$, and $SU(2, 2)$ is isomorphic to the double cover $Spin(2, 4)$ of the (connected) conformal group $SO(2, 4)$. So we have

(5.2) $$SO(1, 4) \subset SO(2, 4) \subset Sp(4; R)/\{\pm I\} .$$

More generally consider $O(1, m) \subset O(2, m)$, $m \geqslant 2$, say by $O(1, m) = \{g \in O(2, m): ge_1 = e_1\}$ relative to the standard "orthonormal" basis of $R^{2,m}$. Then we have coadjoint orbits and projections

(5.3) $$\begin{cases} \mathcal{O} = \{f_{e_1+p,s(e_2+q)}\} \subset \mathfrak{o}(2, m)^* & \text{as in (4.9a)} \\ \alpha \downarrow & \\ \mathcal{O}(-1, 0) = \{f_{p,s(e_2+q)}\} \subset \mathfrak{o}(1, m)^* & \text{as in (4.10)} \\ \beta \downarrow & \\ T^+(S^{m-1}) = \{(p, sq)\} , & \text{the "Kepler manifold," consisting of nonzero cotangent vectors to the sphere } S^{m-1} , \end{cases}$$

where $s \neq 0$ and $p, q \in \mathrm{span}\{e_3, \ldots, e_{m+2}\}$ with $\|p\|^2 = \|q\|^2 = -1$ and $\langle p, q\rangle = 0$. In the Kepler manifold picture, p is a point on the sphere S^{m-1} and sq is a nonzero cotangent vector. Of course $f_{e_1+p,s(e_2+q)} \overset{\alpha}{\longmapsto} f_{p,s(e_2+q)} \overset{\beta}{\longmapsto} (p,sq)$. $\mathcal{O}(-1, 0)$ has an invariant real polarization, the usual fibration of $T^*(S^{m-1})$, which gives an irreducible unitary representation $\pi_{O(1,m)}$ of $O(1, m)$ on $L^2(S^{m-1})$. To complete the quantization of the Kepler manifold one wants to extend $\pi_{O(1,m)}$ to a unitary representation $\pi_{O(2,m)}$ of $O(2, m)$ on $L^2(S^{m-1})$. In the case of primary physical interest, the case $m = 4$, this is just (4.20) with $d = 0$. See [14], [5] and [16] for general m , and see §6 below for more general $O(k, \ell)$.

§6. Indefinite Orthogonal Groups

We will complete the discussion of the orthogonal groups $O(k, \ell)$ that was started in §4. In §7 and §8 we will discuss the other types of orthogonal group.

Suppose $\min(k, \ell) \geqslant 2$ and recall the $2(k + \ell - 3)$ dimensional orbit $\mathcal{O} \subset \mathfrak{o}(k, \ell)^*$ of (4.9a). We can obtain the associated representation $\pi_{\mathcal{O}}$ of $O(k, \ell)$ as a limit of principal series representations, using the method I described in [14, §6], as follows. Let $\{e_1, \ldots, e_{k+\ell}\}$ be an "orthonormal" basis of $R^{k,\ell}$ and denote

(6.1) $$\xi = \xi_{e_1+e_{k+1}, e_2+e_{k+2}} , \quad \eta = -\frac{1}{4}\xi_{e_1-e_{k+1}, e_2-e_{k+2}} , \quad \zeta = [\xi, \eta] .$$

These three elements of $\mathfrak{o}(k, \ell)$ are zero on $\mathrm{span}\{e_3, \ldots, e_k; e_{k+3}, \ldots, e_{k+\ell}\}$, and on $\mathrm{span}\{e_1, e_2; e_{k+1}, e_{k+2}\}$ their respective matrices are

$$\begin{pmatrix} J & -J \\ J & -J \end{pmatrix} , \quad \frac{1}{4}\begin{pmatrix} -J & -J \\ J & J \end{pmatrix} , \quad \begin{pmatrix} 0 & I \\ I & 0 \end{pmatrix}$$

where $J = \begin{pmatrix} 0 & -1 \\ 1 & 0 \end{pmatrix}$ and $I = \begin{pmatrix} 1 & 0 \\ 0 & 1 \end{pmatrix}$. Now ξ and η are nilpotent, ζ is semisimple with real eigenvalues, and $\{\xi, \eta, \zeta\}$ span an algebra that is isomorphic to $\mathfrak{sl}(2; R)$ under

$$\xi \to \begin{pmatrix} 0 & 1 \\ 0 & 0 \end{pmatrix}, \quad \eta \to \begin{pmatrix} 0 & 0 \\ 1 & 0 \end{pmatrix}, \quad \zeta \to \begin{pmatrix} 1 & 0 \\ 0 & -1 \end{pmatrix}.$$

From this,

$\xi + t\zeta$ is semisimple with real eigenvalues for $t \neq 0$.

We set

$$\mathcal{O}_t = \mathrm{Ad}^*(O(k, \ell))\cdot f_{\xi+t\zeta} \subset \mathfrak{o}(k, \ell)^* . \tag{6.2}$$

Let B be a minimal parabolic subgroup of $O(k, \ell)$ whose Lie algebra contains ξ and ζ. For $t \neq 0$, the representation π_t of $O(k, \ell)$ associated to $\mathcal{O}_t$ is a principal series representation on $L^2(O(k, \ell)/B)$. The representation $\pi_{\mathcal{O}}$ of $O(k, \ell)$ associated to $\mathcal{O} = \mathcal{O}_0$ is

$$\pi_{\mathcal{O}} = \lim_{t\to 0} \pi_t . \tag{6.3}$$

This limit can be made rigorous in two ways. One can just use the formulae for the π_t, or one can take the limit of the positive definite spherical functions ϕ_t associated to π_t for $t \neq 0$.

In the case $k = 2$, $\pi_{\mathcal{O}}$ can be realized on $L^2(S^{\ell-1})$, as described in §5. The analog for $2 \leqslant k \leqslant \ell$ goes as follows. First, we decompose

$$\mathcal{O} = \mathcal{O}_0 \cup \mathcal{O}_1 \cup \ldots \cup \mathcal{O}_{k-2} \quad \text{(disjoint)} \tag{6.4}$$

where $\mathcal{O}_j$ is a symplectic submanifold of codimension $2j$ in $\mathcal{O}$ that is isomorphic to the orbit $\mathcal{O}_j(-1, 0) = \mathcal{O}(-1, 0) \subset \mathfrak{o}(k - 1 - j, \ell)^*$ of (4.10). To do that, let $\{e_1, \ldots, e_{k+\ell}\}$ be an orthonormal basis of $R^{k,\ell}$ and view $O(k - a, \ell)$ $= \{g \in O(k, \ell) : ge_i = e_i$ for $1 \leqslant i \leqslant a\}$ for $a \leqslant k$. Let $f \in \mathcal{O}$, say $f = f_{u+p,s(v+q)}$ where $u, v \in \mathrm{span}\{e_1, \ldots, e_k\}$ (resp. $p, q \in \mathrm{span}\{e_{k+1}, \ldots, e_{k+\ell}\}$) are orthogonal unit vectors and $s \neq 0$. Split $\mathcal{O} = \mathcal{O}_0 \cup \mathcal{O}_0'$, disjoint, where $\mathcal{O}_0$ consists of all $f = f_{u+p,s(v+q)}$ in $\mathcal{O}$ with $\mathrm{span}\{u, v\} \not\perp e_1$. If $f \in \mathcal{O}_0$ we may assume $u = e_1$ and define

$$\alpha_0(f) = f_{p,s(v+q)} \in \mathcal{O}_0(-1, 0) \subset \mathfrak{o}(k - 1, \ell)^* .$$

So $\alpha_0 : \mathcal{O}_0 \cong \mathcal{O}_0(-1, 0)$. If $f \notin \mathcal{O}_0$, that is $f \in \mathcal{O}_0'$, then we define

$$\alpha_0(f) = f_{u+p,s(v+q)} \in \mathfrak{o}(k - 1, \ell)^*$$

and have $\alpha_0 : \mathcal{O}_0' \cong$ (the orbit in $\mathfrak{o}(k - 1, \ell)^*$ of (4.9a)). So $\mathcal{O}_0'$ is a replica of $\mathcal{O}$ -- but in $\mathfrak{o}(k-1, \ell)^*$. Repeating, $\mathcal{O}_0' = \mathcal{O}_1 \cup \mathcal{O}_1'$ with $\mathcal{O}_1 = \{f_{u+p,s(v+q)} \in \mathcal{O}_0' : \text{span}\{u, v\} \not\ni e_2\}$, and restriction maps

$$\alpha_1 : \mathcal{O}_1 \cong \mathcal{O}_1(-1, 0) \quad \text{and} \quad \alpha_1 : \mathcal{O}_1' \to (\text{orbit (4.9a) in } \mathfrak{o}(k-2, \ell)^*) .$$

Iterating, we obtain (6.4), the analog of the map α of (5.3).

The second step in obtaining the $2 \leqslant k \leqslant \ell$ analog of the representation of $O(2, m)$ on $L^2(S^{m-1})$, corresponds to the map β of (5.3). For that, view $R^{k-1,\ell} = \text{span}\{e_2, e_3, \ldots, e_{k+\ell}\} \subset R^{k,\ell}$ and let S denote the 2-sheeted covering of the projective light cone of $R^{k-1,\ell}$. In other words,

(6.5) $$S: \text{all light-like rays from } 0 \text{ in } R^{k-1,\ell} .$$

The cotangent bundle $T^*(S)$ consists of all $(r, v+q)$ where $v \in \text{span}\{e_2, \ldots, e_k\}$ and $q \in \text{span}\{e_{k+1}, \ldots, e_{k+\ell}\}$ are unit vectors and $r \in R^{k-1,\ell}$ with $r \perp v+q$. So $v+q$ represents the ray $\{a(v+q) : a > 0\} \in S$ and r represents a cotangent vector to S at that point. Let

$$T^{\pm}(S) = \{(r, v+q) \in T^*(S) : \pm\|r\|^2 < 0\} .$$

Then

(6.6a) $$\mathfrak{o}(k-1, \ell)^* \supset \mathcal{O}(-1, 0) \xrightarrow{\ \beta\ } T^+(S) \quad \text{by} \quad \beta(f_{p,s(v+q)}) = (sp, v+q)$$

and

(6.6b) $$\mathfrak{o}(k-1, \ell)^* \supset \mathcal{O}(0, 1) \xrightarrow{\ \beta\ } T^-(s) \quad \text{by} \quad \beta(f_{p,s(v+q)}) = (sp, v+q)$$

are symplectic manifold isomorphisms. The $O(k-1, \ell)$-invariant real polarizations of $T^{\pm}(S)$ that come from the usual fibration of $T^*(S)$, correspond to polarization of $\mathcal{O}(-1, 0)$ and $\mathcal{O}(0, 1)$ by the parabolic subalgebra of $\mathfrak{o}(k-1, \ell)$ that is the stabilizer of a null line. This gives us representations $\pi_{\pm}$ of $O(k-1, \ell)$ on $L^2(S)$; π_+ (resp. π_-) is associated to the co-adjoint orbit $\mathcal{O}(-1, 0)$ (resp. $\mathcal{O}(0, 1)$). If $k = 2$ then π_- and $\mathcal{O}(0, 1)$ don't occur, and π_+ is the $\pi_{\mathcal{O}(1,\ell)}$ mentioned at the end of §5.

Now combine (6.3), (6.4) and (6.6): this realizes the representation $\pi_{\mathcal{O}}$ of $O(k, \ell)$ on $L^2(S)$.

We now understand the nilpotent minimal dimensional co-adjoint orbits of the $O(k, \ell)$. To understand the semisimple ones, we embed $O(k, \ell) \subset Sp(k+\ell; R)$ and restrict the metaplectic representation.

One embeds $O(k, \ell)$ in $Sp(k+\ell; R)$ either through the inclusion $O(k, \ell) \subset U(k, \ell) \subset Sp(k+\ell; R)$ or, equivalently, by $R^{k,\ell} \otimes (R^2, \{\ ,\ \}) = (R^{2(k+\ell)}, \{\ ,\ \})$. View the manifold X of (4.12) as a Hamiltonian $O(k, \ell)$-space, using $\phi|_{\mathfrak{o}(k,\ell)}$.

To compute we identify $R^{2(k+\ell)} = C^{k,\ell} = R^{k,\ell} + (R^{k,\ell})i$; then for $\xi \in \mathfrak{o}(k, \ell)$ we have

$$\begin{aligned}
\phi(u + vi)(\xi) &= \tfrac{1}{2}\{\xi(u + vi), u + vi\} && \text{by } (4.14)\\
&= \tfrac{1}{2}\,\mathrm{Im}\,\langle \xi(u + vi), u + vi\rangle && \text{with } \langle\ ,\ \rangle \text{ of } C^{k,\ell}\\
&= \tfrac{1}{2}(\langle \xi(u), v\rangle - \langle \xi(v), u\rangle)\\
&= \tfrac{1}{2}\,\mathrm{Re}(\langle \xi u, v\rangle - \langle \xi v, u\rangle) && \langle\ ,\ \rangle \text{ real on } R^{k,\ell}\\
&= \tfrac{1}{2} f_{v,u}(\xi) && \text{by } (4.4)
\end{aligned}$$

So the moment map is

(6.7) $\Phi: X \to \mathfrak{o}(k, \ell)^*$ by $\Phi(u + vi) = f_{v,u}$.

Denote $D(u, v) = \begin{pmatrix} \langle u,u\rangle & \langle v,u\rangle \\ \langle u,v\rangle & \langle v,v\rangle \end{pmatrix}$ for $u, v \in R^{k,\ell}$. Suppose $f_{v,u} \neq 0$, so u, v span a plane E . If $D(u, v) = 0$ then E is totally isotropic, $\xi_{v,u}$ is nilpotent, $Ad^*(O(k, \ell))\cdot f_{v,u}$ is the $2(k + \ell - 3)$-dimensional orbit $\mathcal{O}$ of (4.9a), and that orbit is integral. If $D(u, v)$ has rank 1 then $\xi_{v,u}$ is nilpotent, and $Ad(O(k, \ell))\cdot f_{v,u}$ is one of the integral $2(k + \ell - 2)$-dimensional orbits $\mathcal{O}(-1, 0)$, $\mathcal{O}(0, 1)$ described in (4.10), although $\min(k, \ell)$ may be greater than 1 . If $\det D(u, v) < 0$ then $\xi_{v,u}$ is hyperbolic and $Ad^*(O(k, \ell))\cdot f_{v,u}$ is an (automatically integral) $2(k + \ell - 2)$-dimensional orbit $\mathcal{O}(-1, s)$, $s > 0$, as in (4.10). Now suppose $\det D(u, v) > 0$. We may assume $u = de_i$, $v = e_j$, $i \neq j$, in a standard "orthonormal" basis of $R^{k,\ell}$ with i, j both $\leqslant k$ or both $> k$. The orbit is integral just when $f_{v,u}(r\cdot\xi_{e_s,e_t}) \in 2\pi\mathbb{Z}$ whenever $\exp(r\cdot\xi_{e_s,e_t}) = 1$ in the universal covering group of $SO(k, \ell)$. That is automatic unless $s, t \leqslant k$ with $k > 2$ or $s, t > k$ with $\ell > 2$, in which cases it comes to the condition $\frac{1}{2} f_{v,u}(\xi_{e_s,e_t}) \in \mathbb{Z}$. But $f_{v,u}(\xi_{e_s,e_t}) = 2d(\delta_{jt}\delta_{si} - \delta_{js}\delta_{ti})$ and $d = \det D(u, v)$. In summary, now,

(6.8) *An orbit* $Ad^*(O(k, \ell))\cdot f_{v,u} \subset \mathfrak{o}(k, \ell)^*$ *is integral if and only if* (i) $D(u, v) \leqslant 0$, *or* (ii) $D(u, v)$ *is positive definite and* $k = 2$, *or* (iii) $D(u, v)$ *is negative definite and* $\ell = 2$, *or* (iv) $\det D(u, v) \in \mathbb{Z}$.

On the prequantization level, one expects, now, that the restriction to $O(k,\ell)$ of the metaplectic representation of $Mp(k + \ell; R)$ should be of the form

$$\int_{-\infty}^{0} \nu_{d,1}\, d\mu(\nu_{d,1}) + \sum_{\substack{d=-\infty \\ d\neq 0}}^{\infty} n_d \nu_d$$

where $\nu_{d,1}$ $(d < 0)$ is the principal series representation corresponding to the orbit $\mathcal{O}(d, 1)$ as in (4.10), where ν_d is a limit of holomorphic discrete series representations and corresponds to the orbit $\mathcal{O}(2d, -1)$ for $d < 0$, $\mathcal{O}(1, 2d)$ for $d > 0$, and where $\nu_{0,1}$ corresponds to the nilpotent orbits $\mathcal{O}(-1, 0)$, $\mathcal{O}(0, 1)$

and $\mathcal{O}$. This remains to be made precise.

§7. Complex Orthogonal Groups

Let $n \geqslant 1$ be an integer and let $(\, , \,)$ denote a nondegenerate symmetric bilinear form on C^n . Then we have the *complex orthogonal group*

$$(7.1) \qquad O(n; C) = \{g \in GL(n; C): (gx, gy) = (x, y) \text{ on } C^n\} .$$

Its Lie algebra

$$\mathfrak{o}(n; C) = \{\xi \in \mathfrak{gl}(n; C): (\xi x, y) + (x, \xi y) = 0 \text{ on } C^n\}$$

is spanned over R by the elements

$$(7.2) \qquad \xi_{u,v}: x \mapsto v(u, x) - u(v, x) \text{ where } u, v \in C^n .$$

Identify $\mathfrak{o}(n; C)$ with its real linear dual space by $(\xi, \eta) = \mathrm{Re\ trace}(\xi\eta)$; then

$$(7.3) \qquad \xi_{u,v} \leftrightarrow f_{u,v} \text{ where } f_{u,v}(\eta) = \mathrm{Re}\{(\eta v, u) - (\eta u, v)\} .$$

The analog of Proposition 4.6 is

7.4. Proposition. *If* $n \geqslant 4$ *then the minimal dimensional co-adjoint orbits of* $O(n; C)$ *are the orbits of real dimension* $4(n - 3)$ *given by*

$$(7.5a) \qquad \begin{cases} \mathcal{O}_C = \mathrm{Ad}^*(O(n, C))\cdot f_{u,v} & \textit{where } u, v \textit{ span a} \\ & \textit{complex totally isotropic plane ,} \end{cases}$$

$$(7.5b) \qquad \begin{cases} \textit{if } n = 6 : \textit{ also the 12-real-dimensional} \\ \mathcal{O}(r)_C = \mathrm{Ad}^*(O(6; C))\cdot r(f_{e_1,e_2} + f_{e_3,e_4} + f_{e_5,e_6}) , \quad r \neq 0 \end{cases}$$

$$(7.5c) \qquad \begin{cases} \textit{if } n = 4 : \textit{ also the 4-real-dimensional} \\ \mathcal{O}(r)_C = \mathrm{Ad}^*(O(4; C))\cdot r(f_{e_1,e_2} + f_{e_3,e_4}) , \quad r \neq 0 \end{cases}$$

The representation of $O(n; C)$ for the orbit $\mathcal{O}_C$ of (7.5a) is a limit of principal series representations obtained just as in (6.1), (6.2) and (6.3).

$O(6; C)$ has identity component isomorphic to $SL(4; C)/\{\pm I\}$, and the representations for the orbits $\mathcal{O}(r)_C$ are as in §2.

$O(4; C)$ has identity component isomorphic to $\{SL(2; C) \times SL(2; C)\}/\{\pm I\}$, and the representations for the orbits $\mathcal{O}(r)_C$ are obtained by tensor products of representations of §2.

§8. The Remaining Classical Group

The classical groups which we have not yet discussed are the groups $SO^*(2m)$, real form of $O(2m; C)$ with maximal compact subgroup $U(m)$. It is usual to realize

$$SO^*(2m) = \{g \in GL(2m; C): \langle gx, gy\rangle = \langle x, y\rangle \text{ and } (gx, gy) = (x, y) \text{ on } C^{2m}\}$$

where $\langle x, y\rangle = \sum_1^m x_a\bar{y}_a - \sum_{m+1}^{2m} x_a\bar{y}_a$ and $(x, y) = \sum_1^m (x^a y^{m+a} + x^{m+a}y^a)$. But we will find it more convenient to use

$$SO^*(2m) = \{g \in GL(m; Q): [gx, gy] = [x, y] \text{ on } Q^m\} \tag{8.1}$$

where $[\ ,\ ]$ is an arbitrary skew-hermitian form on Q^m, so there we may suppose

$$[x, y] = \sum_{a=1}^m \bar{x}_a i y_a \text{ on } Q^m . \tag{8.2}$$

See [13] for equivalence of (8.1) and the usual formulation. The advantage of (8.1) is that the Lie algebra $\mathfrak{so}^*(2m)$ is the real span of the

$$\xi_{u,v}: x \mapsto v[u, x] + u[v, x] , \quad u, v, x \in Q^m . \tag{8.3}$$

As usual, we identify $\mathfrak{so}^*(2m)$ with its real linear dual by $(\xi, \eta) = \mathrm{Re}\ \mathrm{trace}(\xi\eta)$, so, by calculation,

$$\xi_{u,v} \leftrightarrow f_{u,v} \text{ where } f_{u,v}(\eta) = \mathrm{Re}\{[\eta u, v] + [\eta v, u]\} . \tag{8.4}$$

For $u \in Q^m$ we now set

$$\xi_u = \tfrac{1}{2}\xi_{u,u}: x \mapsto u[u, x] \text{ and } f_u = \tfrac{1}{2} f_{u,u}: \eta \mapsto \mathrm{Re}[\eta u, u] . \tag{8.5}$$

Some examples are in order. Let $\{e_1, \ldots, e_m\}$ be the standard basis of Q^m. Then $u = e_1 + e_2 j$ satisfies $[u, u] = 0$. Now suppose $u, v \in Q^m$ with

$$[u, u] = [u, v] = [v, v] = 0 ,$$

that is

$$\sum_a \bar{u}_a i u_a = \sum_a \bar{u}_a i v = \sum_a \bar{v}_a i v_a = 0 .$$

Then

$$\xi^2_{u,v}(e_b) = \xi_{u,v}(v\bar{u}_b i + u\bar{v}_b i)$$

$$= \sum_a \{v(\bar{u}_a i v_a \bar{u}_b i + \bar{u}_a i u_a \bar{v}_b i) + u(\bar{v}_a i v_a \bar{u}_b i + \bar{v}_a i u_a \bar{v}_b i)\}$$

so $\xi^2_{u,v} = 0$. In particular if $[u, u] = 0$ then $\xi^2_u = 0$. More generally,

$\xi_u(x) = u[u, x]$ gives $\xi_u^{p+1}(x) = u[u, u]^p[u, x]$, so

$$\exp(t\xi_u)(x) = \begin{cases} x + u[u, x] & \text{if } [u, u] = 0 , \\ x + u\left(\frac{e^{[u,u]}-1}{[u,u]}\right) & \text{if } [u, u] \neq 0 . \end{cases}$$

Since $\overline{[u, u]} = -[u, u]$, i.e. $[u, u] \in \mathrm{Im}Q$, this says that ξ_u is nilpotent if $[u, u] = 0$, elliptic if $[u, u] \neq 0$.

Now we calculate the Lie algebra $\mathfrak{so}^*(2m)^{f_u}$ of $\{g \in SO^*(2m): Ad^*(g)\cdot f_u = f_u\}$. It is equal to the $\mathfrak{so}^*(2m)$-centralizer of ξ_u . If η is in that centralizer,

$$\begin{aligned} 0 = [\eta, \xi_u](x) &= \eta(u)[u, x] - u[u, \eta(x)] \\ &= \eta(u)[u, x] + u[\eta(u), x] \end{aligned}$$

for all $x \in Q^m$, so $\eta(u) = uc$ for some $c \in Q$ and

$$0 = uc[u, x] + u[uc, x] = u(c + \bar{c})[u, x]$$

shows $c \in \mathrm{Im}Q$. Furthermore,

$$0 = [\eta u, u] + [u, \eta u] = \bar{c}[u, u] + [u, u]c$$

shows that c commutes with $[u, u]$. Conversely, if $b \in Q$ one checks $f_{ub} = |b|^2 f_u$, which we can apply with $b = e^c$. In summary now,

(8.6) $\quad \mathfrak{so}^*(2m)^{f_u} = \{\eta \in \mathfrak{so}^*(2m): \eta u = uc , \ c \in \mathrm{Im}Q , \ c[u, u] = [u, u]c\}$.

In particular that gives us, for $0 \neq u \in Q^m$,

(8.7)
$$\dim Ad^*(SO^*(2m))\cdot f_u = \begin{cases} 4m - 6 & \text{if } [u, u] = 0 \\ 4m - 4 & \text{if } [u, u] \neq 0 \end{cases}$$

For that co-adjoint orbit is the quotient of the submanifold of real codimension 3 in Q^m given by $SO^*(2m)\cdot u = \{v \in Q^m: [v, v] = [u, u]\}$, by a circle group if $[u, u] \neq 0$, by an $Sp(1)$ if $[u, u] = 0$.

With the above calculations in mind, the method of the proof of the $O(k, \ell)$ part of Proposition 4.6 leads to

<u>8.8. Proposition</u>. *The minimal dimensional co-adjoint orbits of* $SO^*(2m)$ *are the* $2(2m - 3)$ *dimensional*

(8.9a) $\quad \mathcal{O}_{\pm} = \pm Ad^*(SO^*(2m))\cdot f_u$, $0 \neq u \in Q^m$, $[u, u] = 0$,

(8.9b)
$$\begin{cases} \textit{if } m = 3 : \textit{ also the 6-dimensional orbits} \\ \mathcal{O}(r) = Ad^*(SO^*(6))\cdot r(f_{e_1} + f_{e_2} + f_{e_3}) , \quad 0 \neq r \in \mathbb{R} , \end{cases}$$

(8.9c) $\begin{cases} \textit{if } m = 2 : \textit{ also the 2-dimensional orbits} \\ O(r) = Ad^*(SO^*(4))\cdot r(f_{e_1} + f_{e_2}) , \quad 0 \neq r \in R . \end{cases}$

Co-adjoint orbits of the form $Ad^*(SO(2m))\cdot f$, $f = r \sum_{a=1}^{m} f_{e_a}$, $0 \neq r \in R$, are the hermitian symmetric space $SO^*(2m)/U(m)$ with symplectic 2-form that is a real multiple of the imaginary part of the invariant kaehler metric. The associated representations occur on square integrable cohomology spaces for real powers of the canonical line bundle. See [12]. That gives the representations for the orbits (8.9b) and (8.9c).

Now we are going to realize the representations for the orbits $O_\pm$ of (8.9a) as components of the restriction of the metaplectic representations and its dual.

Embed $SO^*(2m) \subset Sp(2m; R)$ by $(Q^m, -Re[\ ,\ ]) = (R^{4m}, \{\ ,\ \})$. The centralizer of $SO^*(2m)$ in $Sp(2m; R)$ is

$$Sp(1) = \{u \mapsto uc \ \text{ on } \ Q^m \ \text{ where } \ c \in Q \ , \ |c| = 1\} .$$

The Lie algebra of $Sp(1)$ is $\mathfrak{sp}(1) = \{u \mapsto uc,\ c \in ImQ\} \cong ImQ$. The symplectic manifold $X = (R^{4m} \setminus \{0\}, \{\ ,\ \})$ of (4.12) is a Hamiltonian $SO^*(2m) \times Sp(1)$ space, where we use the restriction of the map ϕ of (4.14). Compare (4.14), (4.19) and (8.5) to see that the moment map is

$$\Phi: X \to \mathfrak{so}^*(2m) \oplus \mathfrak{sp}(1) \ \text{ by } \ \Phi(u) = \frac{1}{2}\left(-f_u \oplus \sum_{a=1}^{m} f'_{u_a}\right) \tag{8.10a}$$

where $u = \sum e_a u_a$ with $u_a \in Q$, $f_u \in \mathfrak{so}^*(2m)^*$ by $\eta \mapsto \{u, \eta u\}$, and $f'_b \in \mathfrak{sp}(1)^* = (ImQ)^*$ by $c \mapsto -|b|^2 Re(ic)$ where $b \in Q$. So

$$\Phi(u)(\eta \oplus c) = \frac{1}{2}\left[\{u, \eta u\} + \left(\sum_1^m |u_a|^2\right)(\text{i-component of } c)\right] . \tag{8.10b}$$

Note $\Phi(ut) = |t|^2\Phi(u)$ for $t \in Q$, so to test for integrality of $Ad^*(SO^*(2m) \times Sp(1))\cdot\Phi(u)$ we may assume

$$u = e_1 s \ , \quad s \in R \ , \quad \text{if } \ [u, u] \neq 0 \ ; \quad u = e_1 + e_2 j \ \text{ if } \ [u, u] = 0 .$$

Then one calculates

$$Ad^*(SO^*(2m) \times Sp(1))\cdot\Phi(u) \ \text{ integral} \Leftrightarrow \frac{1}{2}\,|[u, u]| \in \mathbb{Z} . \tag{8.11}$$

In view of [3] and [4], that gives us

<u>8.12. Proposition</u>. *The metaplectic representation* μ *of* $Mp(2m; R)$, *restricted to* $SO^*(2m) \times Sp(1)$ *and re-linearized, breaks up as*

$$\sum_{\ell=0}^{\infty} \beta_\ell \otimes \gamma_\ell \quad \textit{where} \quad \gamma_\ell \in Sp(1)^\wedge \ \textit{ has degree } \ n + 1$$

and where the $\beta_\ell \in SO^*(2n)^\wedge$ *are mutually inequivalent. The representation*

$\beta_\ell \otimes \gamma_\ell$ *corresponds to the integral co-adjoint orbit* $\mathrm{Ad}^*(SO^*(2m) \times Sp(1))\cdot\Phi(u)$ *where* $0 \neq u \in Q^m$ *and* $\frac{1}{2}\,|[u, u]| = \ell$. *In particular,*

$$\beta_\ell \text{ corresponds to } \mathrm{Ad}^*(SO(2m))\cdot f_u\,, \quad 0 \neq u \in Q^m\,, \quad \tfrac{1}{2}\,|[u, u]| = \ell\,,$$

and so β_0 *corresponds to the orbit* $\mathcal{O}_+$ *of* (8.9a).

In order to make Proposition 8.12 more explicit, we will recall the metaplectic representation space $\mathcal{H}$ and describe the decomposition $\mathcal{H} = \mathcal{H}_0 \oplus \mathcal{H}_1 \oplus \ldots$ such that $\mathcal{H}_\ell$ is the representation space of $\beta_\ell \otimes \gamma_\ell$.

In the identification $(Q^m, -\mathrm{Re}[\ ,\ ]) = (R^{4m}, \{\ ,\ \})$, we see that R^{4m} has real basis $\{p_1, \ldots, p_{2m}; q_1, \ldots, q_{2m}\}$ such that

$$\text{(8.13a)} \qquad \{p_a, q_b\} = \delta_{a,b}\,, \quad \{p_a, p_b\} = 0 = \{q_a, q_b\}$$

given by

$$\text{(8.13b)} \qquad p_{2a-1} = e_a\,, \quad p_{2a} = e_a j\,, \quad q_{2a-1} = e_a i\,, \quad q_{2a} = e_a k\,.$$

Recall (3.3); the Lie algebra $\mathfrak{sp}(2m; R)$ is spanned by the $\xi_{u,v}\colon x \mapsto \{x, u\}v + \{x, v\}u$ as u, v run over the basis (8.13b). Note that our $\xi_{u,v}$ is equal to -2 (the $\xi_{u,v}$ of [10, Eq. 4.2]).

The space $\mathcal{H}$ of the metaplectic representation μ of $Mp(2m; R)$ is the space of all holomorphic $f\colon C^{2m} \to C$ with $\int |f(z)|^2 e^{-|z|^2} d\lambda(z) < \infty$, where $|z|^2 = \sum_1^{2m} |z_j|^2$ and λ is Lebesgue measure on C^{2m}. It is a Hilbert space with inner product $\langle f, g\rangle = \pi^{-2m} \int f(z)\overline{g(z)}\, e^{-|z|^2} d\lambda(z)$. Given an integral multi-index $n = (n_1, \ldots, n_{2m})$, let $n! = \prod n_j!$ and $z^n = z_1^{n_1} \ldots z_{2m}^{n_{2m}}$ as usual. Then $\mathcal{H}$ has orthonormal basis consisting of the $z^n/\sqrt{n!}$ where the $0 \leq n_j \in \mathbb{Z}$. On the Lie algebra level, the metaplectic representation is given by

$$\text{(8.14)} \qquad \begin{cases} d\mu(\xi_{p_a,p_b}) = i(\partial_a\partial_b - z_a\partial_b - z_b\partial_a + z_a z_b - \delta_{ab}) \\ d\mu(\xi_{p_a,q_b}) = -(\partial_a\partial_b - z_a\partial_b + z_b\partial_a - z_a z_b) \\ d\mu(\xi_{q_a,q_b}) = -i(\partial_a\partial_b + z_a\partial_b + z_b\partial_a + z_a z_b + \delta_{ab}) \end{cases}$$

where $\partial_a = \partial/\partial z_a$. See [10, theorem 4.10] for this, taking account of the different normalization of $\xi_{u,v}$.

The Lie algebra $\mathfrak{sp}(1)$ consists of the operators on $R^{4m} = Q^m$ which are right multiplication by pure imaginary quaternions. Using (8.13b), now $\mathfrak{sp}(1)$ is the real span of

$$\begin{aligned} &\xi_i\colon p_{2a-1} \to q_{2a-1} \to -p_{2a-1}\,, \quad p_{2a} \to -q_{2a} \to -p_{2a} \\ &\xi_j\colon p_{2a-1} \to p_{2a} \to -p_{2a-1}\,, \quad q_{2a-1} \to q_{2a} \to -q_{2a-1} \\ &\xi_k\colon p_{2a-1} \to q_{2a} \to -p_{2a-1}\,, \quad q_{2a-1} \to -p_{2a} \to -q_{2a-1} \end{aligned}$$

From this, calculate

$$\xi_i = -\frac{1}{2}\sum_{a=1}^{m}(\xi_{p_{2a-1},p_{2a-1}} + \xi_{q_{2a-1},q_{2a-1}}) + \frac{1}{2}\sum_{a=1}^{m}(\xi_{p_{2a},p_{2a}} + \xi_{q_{2a},q_{2a}}) ,$$

$$\xi_j = \sum_{a=1}^{m}(\xi_{p_{2a-1},q_{2a}} - \xi_{p_{2a},q_{2a-1}}) , \quad \xi_k = -\sum_{a=1}^{m}(\xi_{p_{2a-1},p_{2a}} + \xi_{q_{2a-1},q_{2a}}) .$$

Using (8.14), now $d\mu(\mathfrak{sp}(1))$ is the real span of

(8.15)
$$\begin{cases} d\mu(\xi_i) = 2i\sum_{a=1}^{m}(z_{2a-1}\partial_{2a-1} - z_{2a}\partial_{2a}) , \\ d\mu(\xi_j) = 2\sum_{a=1}^{m}(z_{2a-1}\partial_{2a} + z_{2a}\partial_{2a-1}) , \\ d\mu(\xi_k) = 2i\sum_{a=1}^{m}(z_{2a-1}\partial_{2a} + z_{2a}\partial_{2a-1}) . \end{cases}$$

In particular, $d\mu(\mathfrak{sp}(1))_C$ is the complex span of the operators

(8.16a)
$$E = \sum_{a=1}^{m} z_{2a-1}\partial_{2a} , \quad F = \sum_{a=1}^{m} z_{2a}\partial_{2a-1}$$
$$\text{and} \quad H = \sum_{a=1}^{m}(z_{2a-1}\partial_{2a-1} - z_{2a}\partial_{2a})$$

which satisfy

(8.16b) $$[E, F] = H , \quad [H, E] = 2E , \quad \text{and} \quad [H, F] = -2F .$$

That is the standard form for generators of $\mathfrak{sl}(2; C)$, so the $d\mu$-image of the Casimir element of the universal enveloping algebra is

(8.16c) $$d\mu(\Omega) = H^2 + 2H + 4FE .$$

Recall that γ_ℓ denotes the irreducible representation of degree $\ell + 1$ of $Sp(1)$, so

(8.17a) $$\mathcal{H} = \sum_{\ell=0}^{\infty} \mathcal{H}_\ell \quad \text{where } \mathcal{H}_\ell \text{ is the } \gamma_\ell\text{-isotypic subspace.}$$

Using (8.16), we characterize

(8.17b) $\mathcal{H}_\ell$ is the $\ell^2 + 2\ell$ eigenspace of $H^2 + 2H + 4FE$.

One can write down a set of spanning polynomials for $\mathcal{H}_\ell$, though it is a bit of a mess, as follows. For each multi-index $t = (t_1, \ldots, t_m)$, t_a integers $\geqslant 0$, denote

$$\mathcal{H}(t)\text{: span of } \{z^n : n_{2a-1} + n_{2a} = t_a \text{ for } 1 \leqslant a \leqslant m\} .$$

The spaces $\mathcal{H}(t)$ are finite dimensional, and they are $d\mu(\otimes\mathfrak{p}(1))$-invariant by a glance at (8.15). Evidently Sp(1) acts on $\mathcal{H}(t)$ by $\gamma_{t_1} \otimes \gamma_{t_2} \otimes \ldots \otimes \gamma_{t_m}$. Given t , that representation is decomposed explicitly by writing down the H-eigenpolynomials in $\mathcal{H}(t)$ annihilated by E and iterating the action F on them. Of course for m = 2 this is just the implementation of $\gamma_r \otimes \gamma_s = \sum_{w=0}^{\min(r,s)} \gamma_{r+s-2w}$. One then has $\mathcal{H}(t) = \sum_{\ell} \mathcal{H}_{\ell}(t)$, finite sum where $\mathcal{H}_{\ell}(t)$ is the γ_{ℓ}-isotypic subspace, and thus has $\mathcal{H}_{\ell} = \sum_{t} \mathcal{H}_{\ell}(t)$.

References

[1] K.I. Gross and R.A. Kunze, *Bessel functions and representation theory*, I, J. Functional Analysis 22 (1976), 73-105; II, *ibid.* 25 (1977), 1-49.

[2] V. Guillemin and S. Sternberg, "Geometric Asymptotics," Math. Surveys No. 14, Amer. Math. Soc., Providence, 1977

[3] R. Howe, *Remarks on classical invariant theory*, Yale University preprint, 1976.

[4] D. Kazhdan, B. Kostant and S. Sternberg, *Hamiltonian group actions and dynamical systems of Calagero type*, Harvard University preprint, 1977.

[5] E. Onofri, *Dynamical Quantization of the Kepler Manifold*, Università Parma (Inst. di Fisica) preprint, 1975.

[6] J.H. Rawnsley, *On the cohomology groups of a polarization and diagonal quantisation*, Trans. Amer. Math. Soc. 230 (1977), 235-255.

[7] W. Schmid, *L^2-cohomology and the discrete series*, Ann. of Math. 103 (1976), 375-394.

[8] D.J. Simms, *Metalinear structures and the quantization of the harmonic oscillator*, Colloque Symplectique, Aix-en-Provence, 1974.

[9] S. Sternberg, article in this volume.

[10] S. Sternberg and J.A. Wolf, *Hermitian Lie algebras and metaplectic representations*, Trans. Amer. Math. Soc., to appear.

[11] A. Weil, *Sur certain groupes d'opérateurs unitaires*, Acta Math. 111 (1964), 143-311.

[12] J.A. Wolf, *The action of a real semisimple Lie group on a complex flag manifold, II: Unitary representations on partially holomorphic cohomology spaces*, Amer. Math. Soc, Memoir 138, 1974.

[13] J.A. Wolf, *Unitary representations of maximal parabolic subgroups of the classical groups*, Amer. Math. Soc. Memoir 180, 1976.

[14] J.A. Wolf, *Conformal group, quantization, and the Kepler problem,* in "Group Theoretical Methods in Physics," Fourth International Colloquium, Nijmegen 1975. Springer Lecture Notes in Physics 50 (1976), 217-222.

[15] J.A. Wolf, *Representations that remain irreducible on parabolic subgroups,* to appear.

[16] N. Woodhouse, *Twistor theory and geometric quantization,* in "Group Theoretical Methods in Physics," Fourth International Colloquium, Nijmegen 1975 Springer Lecture Notes in Physics 50 (1976), 149-163.

On the Schrödinger equation given by geometric quantisation.

D.J. Simms
School of Mathematics
Trinity College, Dublin.

Introduction

The fundamental work of Blattner, Kostant, and Sternberg in [1] has much enlarged the scope of the geometric quantisation techniques of Kostant [6] and Souriau [9]. In this talk I want to show in a quite explicit way how these techniques enable us to write down, in a formal way, a Schrödinger equation for the quantum mechanical time evolution of the wave functions on configuration space.

We deal with a time-dependent system with finitely many degrees of freedom. The constructions used are based on the differential geometry of the classical phase space, and are closely related to Hamilton - Jacobi theory. The essential ideas derive from [1] and are also treated in detail in [4] and, in a more conceptual way, in [5].

Let X be the configuration space of the system, which we suppose to be a manifold of finite dimension n. Then $X \times R$ is the space of events and we shall make the assumption that there is a unique classical path passing through any two points of $X \times R$ We assume further that X is a manifold on which square-roots of volume elements (half-forms) are defined. We then have an associated Hilbert space, denoted $L^2(X)$, which we shall call the space of wave functions. If ψ is a wave function on X then we shall construct another wave-function $T^{\tau,\tilde{\tau}}\psi$ which we shall call the quasi - classical evolution of ψ from time τ to time $\tilde{\tau}$.

From the quasi - classical evolution we obtain a Schrödinger equation by differentiation in the following way. Let $t \to \psi_t$ be the time - evolution in $L^2(x)$ of a quantum - mechanical wave function. Then

$$\frac{\partial}{\partial t}\psi_t = \lim_{t_1 \to t} \frac{T^{t,t_1}\psi_t - \psi_t}{t_1 - t}$$

For simple examples this gives the usual Schrödinger equation by a calculation similar to Feynman and Hibbs [3].

The transform $T^{\tau,\tilde{\tau}}$ has an explicit coordinate representation as follows. Let ψ have compact support contained in the domain V of coordinates $y^1, \ldots, y^n$ on X, and let

$$\psi = f(y^1, \ldots\ldots, y^n)\ (dy^1 \wedge \ldots \wedge dy^n)^{\frac{1}{2}}.$$

Then on the domain $\tilde{V}$ of coordinates $\tilde{y}^1, \ldots, \tilde{y}^n$ on X the transformed wave function $T^{\tau,\tilde{\tau}}\,\psi$ is equal to

$$g(\tilde{y}^1, \ldots, \tilde{y}^n)\ (d\tilde{y}^1 \wedge \ldots \wedge d\tilde{y}^n)^{\frac{1}{2}}$$

where the function g is given explicitly as follows. Let S, Hamilton's principal function, be the function of 2n + 2 variables such that

$$S(y^1(x), \ldots, y^n(x),\ \tilde{y}^1(\tilde{x}), \ldots, \tilde{y}^n(\tilde{x}),\ \tau, \tilde{\tau})$$

is the action of the classical path from (x,τ) to $(\tilde{x},\tilde{\tau})$ where $x \in V$ and $\tilde{x} \in \tilde{V}$. Let $\Delta\,(y^1, \ldots, y^n, \tilde{y}^1, \ldots, \tilde{y}^n)$ denote the determinant of the n x n matrix of partial derivatives

$$S_{,i,n+j}\quad (y^1, \ldots, y^n, \tilde{y}^1, \ldots, \tilde{y}^n, \tau, \tilde{\tau}).$$

Then $g(\tilde{y}^1, \ldots, \tilde{y}^n)$ equals

$$\left(\frac{i}{2\pi\hbar}\right)^{n/2} \int \left[\exp \frac{i}{\hbar}\, S(y^1, \ldots, y^n, \tilde{y}^1, \ldots, \tilde{y}^n, \tau, \tilde{\tau})\right]\ \Delta(y^1, \ldots, y^n, \tilde{y}^1, \ldots, \tilde{y}^n)^{\frac{1}{2}}\ f(y^1, \ldots, y^n)\, dy^1 \ldots dy^n$$

<u>Lagrangian foliations of the energy surface</u>

Let $T^*(X \times R)$ be the cotangent bundle of $X \times R$, let π be the projection onto $X \times R$, let β be the canonical 1 - form and $\omega = d\beta$ the symplectic form on the cotangent bundle. The dynamics is determined by a (2n + 1) - dimensional submanifold M, called the <u>energy surface</u> by Synge [10] and called the <u>espace d'evolution</u> by Souriau [9]. At each point <u>a</u> in M we denote by D_a the orthogonal complement of the tangent space $T_a(M)$ to M with respect to the symplectic form. Then $D_a \subset D_a^{\perp} = T_a(M)$, and D_a is the tangent space to the classical path through <u>a</u>.

Now fix a time τ and define maps $\sigma\colon M \to M$ and $\rho\colon M \to X$ so that <u>a</u> and $\sigma(a)$ lie on the same classical path and $\pi\sigma(a) = (\rho(a), \tau)$. The level sets of ρ give a partition F of M which we suppose to be a foliation of M by (n+1) - dimensional submanifolds. The symplectic form ω vanishes on each leaf of F. Thus F is a foliation of M by Lagrangian submanifolds of $T^*(X \times R)$. Let F_a denote the tangent space at <u>a</u> to the leaf of F. Let $(F/D)^o_a$ denote the annihilator of F_a/D_a in the dual of D_a/D_a.

Then the transpose of the derivative of ρ at $\underline{a}$ induces an isomorphism of $T^*(X)_{\rho(a)}$ onto $(F/D)^o_a$. Let $\Lambda^{\frac{1}{2}} T^*(X)$ denote a square root of the complex n^{th} exterior power of $T^*(X)$. Then the complex n^{th} exterior power of the vector bundle $(F/D)^o$ admits a square root denoted by $\Lambda^{\frac{1}{2}}(F/D)^o$ and we have a canonical isomorphism of $\Lambda^{\frac{1}{2}}T^*(X)_{\rho(a)}$ onto $\Lambda^{\frac{1}{2}}(F/D)^o_a$.

A section of $\Lambda^{\frac{1}{2}}(F/D)^o$ is said to be covariant constant along F if it is annihilated by the operator

$$L_\xi - \frac{i}{\hbar} < \beta, \xi>$$

for each vector field ξ on M tangent to F. Here L_ξ denotes the Lie derivative along ξ and $2\pi\hbar$ is Planck's constant. Thus we are using a covariant derivative whose curvature form is $\frac{-i}{\hbar}\omega$, as is customary in geometric quantisation [2,6,8]. We denote by Γ_F the space of sections of $\Lambda^{\frac{1}{2}}(F/D)^o$ which are covariant constant along F. Each element of Γ_F has its value at $\underline{a}$ determined uniquely by its value at $\sigma(a)$, and its values at $\sigma(a)$ and $\sigma(b)$ correspond to the same element of $\Lambda^{\frac{1}{2}}_{\rho(a)} T^*(X)$ if $\rho(a) = \rho(b)$. Thus we have a natural bijection

$$T^\tau: \Lambda^{\frac{1}{2}}X \rightarrow \Gamma_F$$

from the space $\Lambda^{\frac{1}{2}}X$ of sections of $\Lambda^{\frac{1}{2}}T^*(X)$.

Let now $\psi \in \Lambda^{\frac{1}{2}}X$ have support in the domain V of a coordinate system $y^1,\dots,y^n$ on X,

$$\psi = f(y^1,\dots,y^n)\ (dy^1\wedge\dots\wedge dy^n)^{\frac{1}{2}}$$

say. Let $p_1,\dots,p_n, q^1\dots,q^n$ be the canonical functions on $T^*(X \times R)$ associated with the coordinates $y^1,\dots,y^n$ on X. Let $P_1,\dots, P_n, Q^1,\dots,Q^n$ be the functions on $\rho^{-1}(V)$ given by $P_i(a) = p_i(\sigma(a))$ and $Q^i(a) = q^i(\sigma(a)) = y^i(\rho(a))$.

Let S(a) denote the integral of β over the classical path in M from $\sigma(a)$ to $\underline{a}$. Then

$$T^\tau\psi = \left[\exp \frac{i}{\hbar} S\right]\ f(Q^1,\dots,Q^n)\ (dQ^1 \tilde{\wedge}\dots\wedge dQ^n)^{\frac{1}{2}}.$$

The Pairing

Let $\tilde{\tau} \in R$. We define $T^{\tau,\tilde{\tau}}\psi$ to be the wave function whose scalar product with $\tilde{\psi}$ is equal to the pairing [2, 7] (described below) of $T^\tau\psi$ and $T^{\tilde{\tau}}\tilde{\psi}$ for any $\tilde{\psi}$ with compact support. To compute $T^{\tau,\tilde{\tau}}$ we choose $\tilde{\psi}$ with compact support in the domain $\tilde{V}$ of a coordinate system $\tilde{y}^1,\dots, \tilde{y}^n$ on X and suppose

$$\tilde{\psi} = \tilde{f}\ (\tilde{y}^1,\ldots,\tilde{y}^n)\ (d\tilde{y}^1 \wedge\ldots\wedge d\tilde{y}^n)^{\frac{1}{2}}$$

We then have

$$T^{\tilde{\tau}}\tilde{\psi} = \left[\exp \tfrac{i}{\hbar}\tilde{S}\right] \tilde{f}(\tilde{Q}^1,\ldots,\tilde{Q}^n)\ (d\tilde{Q}^1\wedge\ldots\wedge d\tilde{Q}^n)^{\frac{1}{2}}.$$

Also $S(a) - \tilde{S}(a)$ is equal to the integral of β over the classical path in M from $\sigma(a)$ to $\tilde{\sigma}(a)$ and therefore

$$S(a) - \tilde{S}(a) = \mathcal{S}(y^1(\rho(a)),\ldots,y^n(\rho(a)),\ \tilde{y}^1(\tilde{\rho}(a)),\ \ldots,\tilde{y}^n(\tilde{\rho}(a),\ \tau,\tilde{\tau}).$$

and thus

$$S - \tilde{S} = \mathcal{S}(Q^1,\ldots,Q^n,\ \tilde{Q}^1,\ldots,\tilde{Q}^n,\ \tau,\tilde{\tau}) \qquad (1)$$

Here $\tilde{\rho}$, $\tilde{\sigma}$, $\tilde{S}$, $\tilde{P}_i$, $\tilde{Q}^i$ are related to $\tilde{\tau}$ and $\tilde{y}^1,\ldots\tilde{y}^n$ in the same way as ρ, σ, S, P_i, Q^i are related to τ and $y^i,\ldots,y^n$.

The fundamental property of Hamilton's principal function $\mathcal{S}$ is that

$$d\mathcal{S}\ (Q^1,\ldots,Q^n,\ \tilde{Q}^1,\ldots,\tilde{Q}^n,\ \tau,\ \tilde{\tau}) = \Sigma P_i dQ^i - \Sigma\tilde{P}_i d\tilde{Q}^i .$$

Thus on M we have an invariantly defined 2 - form

$$\Omega = \Sigma\ dP_i \wedge dQ^i = \Sigma\ d\tilde{P}_i \wedge d\tilde{Q}^i$$

which is invariant under the classical flow. $\Omega^n = \Omega\wedge\ldots\wedge\Omega$ is a volume element transverse to the classical flow. Also

$$P_i = \mathcal{S},_i\ (Q^1,\ldots,Q^n,\ \tilde{Q}^1,\ldots\tilde{Q}^n,\ \tau\ ,\tilde{\tau}\ )$$

where $\mathcal{S},_i$ denotes the partial derivative of $\mathcal{S}$ with respect to its $_i$th variable. Thus

$$dP_i = \sum_{j=1}^{n}\left[\mathcal{S},_{i,j}\ (Q^1,\ldots,Q^n,\ \tilde{Q}^1,\ldots,\tilde{Q}^n,\tau,\ \tilde{\tau},)\ dQ^j + \mathcal{S},_{i,j+n}(Q^1,\ldots,Q^n,\tilde{Q}^1,\ldots,\tilde{Q}^n,\tau,\tilde{\tau})d\tilde{Q}^j\right]$$

and therefore

$$\Omega^n = \Delta(Q^1,\ldots,Q^n,\ \tilde{Q}^1,\ldots,\tilde{Q}^n)\ d\tilde{Q}^1\wedge\ldots\wedge d\tilde{Q}^n \wedge dQ^1\wedge\ldots\wedge\ dQ^n \qquad (2)$$

where Δ is the determinant of the n x n matrix of partial derivatives $(\mathcal{S},_{i,j+n})$ as in the introduction.

Since the exterior product of $dQ^1\wedge\ldots\wedge dQ^n$ with $d\tilde{Q}^1\wedge\ldots\wedge\ d\tilde{Q}^n$ equals

$$(-1)^n\ \Delta(Q^1,\ldots,Q^n,\tilde{Q}^1,\ldots,\tilde{Q}^n)^{-1}\quad \Omega^n$$

it follows that we should pair, up to sign, $(dQ^1\wedge\ldots\wedge\ dQ^n)^{\frac{1}{2}}$ and $(d\tilde{Q}^1\wedge\ldots\wedge d\tilde{Q}^n)^{\frac{1}{2}}$ to

$$i^n\ \Delta(Q^1,\ldots,Q^n,\ \tilde{Q}^1,\ldots\tilde{Q}^n)^{-\frac{1}{2}}$$

The sesquilinear pairing of $T^{\tau}\psi$ with $T^{\tilde{\tau}}\tilde{\psi}$ then given by

$$\frac{1}{(2\pi i\hbar)^{n/2}} \int \left[\exp \frac{i}{\hbar} S\right] f(Q^1,..,Q^n) \overline{\left[\exp \frac{i}{\hbar} \tilde{S}\right] \tilde{f}(\tilde{Q}^1,..,\tilde{Q}^n)}\; i^n \Delta(Q^1,..,Q^n,\tilde{Q}^1,...,\tilde{Q}^n)^{-\frac{1}{2}}\, \Omega^n$$

Using (1) and (2) and changing our integration variables to $y^1,..,y^n,\tilde{y}^1,...,\tilde{y}^n$ we obtain the formula for $T^{\tau,\tilde{\tau}}\psi$ given in the introduction.

I should like to thank Dr. John Rawnsley for helpful discussion on the pairing.

REFERENCES

1 R.J. Blattner. Quantization and representation theory, Proc. Sympos. Pure Math., vol 26, Amer. Math. Soc., Providence R.I. 1973, pp 147-165

2 R.J. Blattner. Pairing of half-form spaces. Geometrie symplectique et Physique Mathematique, Coll. Int. du C.N.R.S. No 237 Aix-en-Provence 1974, pp 175-186.

3 R.P. Feynman and A.R. Hibbs. Quantum mechanics and path integrals. McGraw-Hill, New York 1965.

4 V. Guillemin and S. Sternberg. Geometric asymptotics. Math surveys No 14, Am. Math. Soc., Providence R.I. 1977

5 J. Kijowski. Geometric structure of quantization. Differential geometric methods in mathematical physics p.p. 97 - 108, Lecture Notes in Mathematics 570, Springer, Berlin 1977

6 B. Kostant. Quantization and unitary representations. 1. Prequantisation Lecture Notes in Mathematics 170, Springer, Berlin 1970.

7 J. Rawnsley. On the pairing of polarizations, Comm. in Math. Physics (to appear).

8 D.J. Simms and N. Woodhouse. Lectures on geometric quantization. Lecture Notes in Physics No 53, Springer, Berlin 1976

9 J.M. Souriau. Structure des systemes dynamiques. Dunod, Paris 1970

10 J.L. Synge. Classical dynamics. Handbuch der Physik, Vol III /1, Principles of classical mechanics and field theory, (ed.) S. Flügge, Springer, Berlin 1960

APPLICATION OF GEOMETRIC QUANTIZATION

IN QUANTUM MECHANICS

by

Jędrzej Śniatycki

Department of Mathematics and Statistics

University of Calgary, Calgary, Alberta, Canada

1. Introduction

Geometric quantization is a well defined procedure leading from the classical to a quantum description of a physical system. The quantum theory obtained from a given classical one depends, in general, on the choice of the auxiliary structure used in quantization. Hence, in order to decide if it can serve as a model for the physical systems under consideration one has to compare the theoretical predictions with the experimental data.

In this lecture I would like to show how the generally accepted quantum mechanics of a relativistic charged particle in external electromagnetic and gravitational fields emerges as the result of the geometric quantization of the classical description of its dynamics. In the usual formulation of the classical dynamics of a charged particle the mass and the charge are treated asymmetrically, The mass is a dynamical variable while the charge is a fixed parameter in the theory. This asymmetry between mass and charge disappears in the five dimensional theory of Kaluza. The quantization of this theory in the representation in which the position and the charge operators are diagonal leads to the superselection rules for charge.

A relativistic formulation of the dynamics of a charged particle is given in Section 2. Geometric quantization of this dynamics in the representation in which the position operators are diagonal is given in Section 3. Section 4 contains a brief presentation of dynamics in the Kaluza theory and a derivation of the superselection rules for the charge operator.

It is assumed that the reader is familiar with the theory of geometric quantization. A comprehensive presentation of this theory is given in Ref. [1]. Details of the computations leading to the results presented here are contained in Ref. [2].

2. Relativistic dynamics of charged particles

Let Y denote an oriented 4-dimensional manifold representing the space-time. The gravitational field is represented in Y by a metric g with signature $(-,-,-,+)$, and the electromagnetic field is represented by a closed 2-form F. The phase space for a single particle in the space-time Y is the cotangent bundle space T^*Y of Y. Let $\pi : T^*Y \to Y$ denote the cotangent bundle projection, and θ_Y the canonical 1-form on T^*Y defined by

$$\theta_Y(u) = x(T\pi(u)) \tag{2.1}$$

for each $x \in T^*Y$ and each $u \in T_xT^*Y$. The Lagrange bracket, for a particle of charge e, is given by

$$\omega_e = d\theta_Y + e\pi^*F \tag{2.2}$$

where π^*F denotes the pull back of F to T^*Y, c.f. Ref. [3].

Classical dynamical variables are functions on the phase space. The position variables q are the pull-backs to T^*Y of smooth functions on Y;

$$q = \check{q} \circ \pi \tag{2.3}$$

where $\check{q}$ is a smooth function on Y. The momentum variables are associated to smooth vector fields on Y as follows. Given a vector field ζ on Y the associated momentum p_ζ is given by

$$(2.4) \qquad p_\zeta(x) = x(\zeta(\pi(x))),$$

for each $x \in T^*Y$. For each function f of the form

$$(2.5) \qquad f = \check{q}\circ\pi + p_\zeta ,$$

the hamiltonian vector field ξ_f^e of f, defined by

$$(2.6) \qquad \xi_f^e \lrcorner\, \omega_e = -df$$

preserves the fiber structure of T^*Y. The square of mass function M^2 is defined, in terms of the bilinear form $g : T^*Y \times_Y T^*Y \to \mathbb{R}$ induced by the metric g on Y, as follows. For each $x \in T^*Y$,

$$(2.7) \qquad M^2(x) = g(x,x).$$

For each $m > 0$, the integral curves of the hamiltonian vector field ξ_{M^2} of M^2 contained in $(M^2)^{-1}(m^2)$ yield, when projected to Y, the solutions of the equations of motion for a particle with mass m and charge e moving in the gravitational field g and the electromagnetic field F.

3. Quantization

The vertical distribution D in T^*Y, tangent to the fibres of $\pi : T^*Y \to Y$, is lagrangian with respect to the symplectic form ω_e given by Eq. (2.2). Hence, its complexification

$$(3.1) \qquad F = D^{\mathbb{C}}$$

is a polarization of (T^*Y, ω_e). The orientation in Y induces a metaplectic structure in T^*Y which, in turn, defines a metalinear structure of F. The associated bundle of half-forms normal to F has a covariant constant along F section ν_g such that the pairing of ν_g with itself yields the density $|\det g|^{\frac{1}{2}}$ on Y, c.f. Ref. [1] pp. 80-82.

The geometric quantization scheme requires the existence of a complex line bundle L_e over T^*Y, with a connection ∇ such that

$$\text{curvature } \nabla = -h^{-1}\omega_e, \tag{3.2}$$

and with a connection invariant hermitian metric. The condition (3.2) can be satisfied if and only if the de Rham cohomology class $[h^{-1}\omega_e]$ defined by the symplectic form $h^{-1}\omega_e$ is integral. In view of Eq. (3.1) this condition is equivalent to the Dirac condition (Ref. [4]) of integrality of the cohomology class $[h^{-1}eF] \in H^2(Y,\mathbb{R})$.

Let L_e be a line bundle with a connection ∇ satisfying the condition (3.2). There is an equivalence relation $\sim$ in L_e given by $\ell_1 \sim \ell_2$ if and only if there exists $y \in Y$ such that ℓ_1 can be joined to ℓ_2 by a horizontal curve in the restriction of L_e to T_y^*Y. The space $L_e^\sim$ of the $\sim$ equivalence classes has a structure of a complex line bundle over Y such that the canonical projection $\pi^\sim: L_e \to L_e^\sim$ is an isomorphism on each fibre and the diagram

$$\begin{array}{ccc} L_e & \xrightarrow{\pi^\sim} & L_e^\sim \\ \downarrow & & \downarrow \\ T^*Y & \xrightarrow{\pi} & Y \end{array} \tag{3.3}$$

in which the vertical arrows denote the line bundle projections, commutes. The connection ∇ in L_e induces a connection $\nabla^\sim$ in $L_e^\sim$ such that

$$\text{curvature } \nabla^\sim = -ieh^{-1}F . \tag{3.4}$$

Also, the connection invariant hermitian form $<\cdot,\cdot>$ in L_e induces a connection invariant hermitian form $<\cdot,\cdot>^{\sim}$ in $L_e^{\sim}$. There is a bijection between the space of the covariant constant along F sections of L_e and the space of sections of $L_e^{\sim}$ associating, to each covariant constant along F section λ of L_e, the section $\lambda^{\sim}$ of $L_e^{\sim}$ such that

$$\lambda^{\sim}(\pi(x)) = \pi^{\sim}(\lambda(x)) \tag{3.5}$$

for each $x \in T^*Y$. The space H_e of the quantum states of our system can be identified with the space of sections $\lambda^{\sim}$ of $L_e^{\sim}$ with the scalar product given by

$$\int_Y <\lambda_1^{\sim},\lambda_1^{\sim}>^{\sim} |\det g|^{\frac{1}{2}} \tag{3.6}$$

The geometric quantization associates to a function f on T^*Y a linear operator $Q_e f$ in H_e. For a function q constant along the fibres of $\pi : T^*Y \to Y$ given by

$$q = \check{q} \circ \pi \tag{3.7}$$

where $\check{q}$ is a smooth function on Y, we get

$$Q_e q \lambda^{\sim} = \check{q}\lambda^{\sim}, \tag{3.8}$$

for each $\lambda^{\sim} \in H_e$. Thus, we have a representation in which the "position' variables are diagonal. The quantization of the momentum function p_ζ associated to a smooth vector field ζ on Y, c.f. Eq. (2.3), yields

$$Q_e p_\zeta \lambda^{\sim} = -i\hbar(\nabla_\zeta^{\sim} + 1/2 \text{ Div } \zeta)\lambda^{\sim}, \tag{3.9}$$

where Div ζ denotes the covariant divergence of the vector field ζ.

In the absence of the electromagnetic field Eq. (3.9) agrees with the results of DeWitt, [6]. Since the hamiltonian vector field of the square of mass function M^2, given by Eq. (2.6), does not preserve the polarization, the quantization of M^2 requires the Blattner-Kostant-Stemberg kernels, c.f. Ref [1]. The resulting quantum operator $Q_e M^2$ is given by

$$(3.10) \qquad Q_e M^2 \lambda^{\sim} = -\hbar^2 (\Delta^{\sim} - 1/6 \; R)\lambda^{\sim}$$

where $\Delta^{\sim}$ denotes the Laplace-Beltrami operator defined in terms of the metric g in Y and the connection $\nabla^{\sim}$ in $L_e^{\sim}$, and R is the scalar curvature of the metric connection in Y. In terms of the local components g_{ij} of the metric g we have the following expression for R:

$$(3.11) \qquad R = g^{mn}(\Gamma^k_{mn,k} - \Gamma^k_{mk,n} + \Gamma^j_{mn}\Gamma^k_{jk} - \Gamma^j_{mk}\Gamma^k_{jn}) \;,$$

where comma denotes the differentiation with respect to the coordinate functions and Γ^k_{mn} denote the Christoffer symbols of the connection given by

$$(3.12) \qquad \Gamma^k_{mn} = 1/2 \; g^{kj}(g_{jm,n} + g_{jn,m} - g_{mn,j}) .$$

In Eqs. (3.11) and (3.12) the summation over the repeated indices is understood. In the absence of the dectromagnetic field Eq. (3.10) agrees with the results of DeWitt [8] and Cheng [9].

We have obtained quantization of the essential dynamical variables of a charged particle in the external electromagnetic field F without any reference to the potentials. This was possible only because we considered the presence of the electromagnetic field as modifying the symplectic form, Eq. (2.2), rather than the mass Eq. (2.6). Locally, one can find a 1-form A such that

$$F = dA \ . \tag{3.13}$$

Hence, there exists a local section $\lambda_0^{\sim}$ of $L_e^{\sim}$ such that

$$\nabla^{\sim}\lambda_0^{\sim} = -ie\hbar^{-1}A \otimes \lambda_0^{\sim} \ . \tag{3.14}$$

For a section $\psi\lambda_0^{\sim} \in H_e$, where ψ is a complex valued function on Y, we have

$$\nabla^{\sim}(\psi\lambda_0) = (d\psi - ie\hbar^{-1}\psi A) \otimes \lambda_0 \ . \tag{3.15}$$

Using Eq. (3.15) together with the local expressions for covariant differentiation in terms of a local coordinate system one can rewrite Eqs. (3.9) and (3.10) in the commonly used coordinate dependent form.

4. Superselection rules for charge

The charge appears as a dynamical variable in the 5-dimensional theory of Kaluza and its generalizations, c.f. Ref. [10]. We use here the version in which the 5-dimensional space Z is a $\mathbb{T}^1$ principal fibre bundle over the space-time manifold Y endowed with an invariant metric $\hat{g}$ of signature $(-,-,-,+,+)$ such that the fundamental vector field η_1 corresponding to the real number 1 in the Lie algebra of $\mathbb{T}^1$ has constant length 1. Here $\mathbb{T}^1$ is the multiplicative group of complex numbers of modulus 1 and its Lie algebra is identified with $\mathbb{R}$ by associating, to each $r \in \mathbb{R}$, the one parameter group $t \mapsto \exp(ie_0\hbar^{-1}rt)$, where e_0 is a parameter interpreted as the elementary charge. The distribution in TZ orthogonal to η_1 is the horizontal distribution of a connection in Z. The curvature of this connection is identified with the pull back to Z of the electromagnetic field F on Y. This means that the form $h^{-1}e_0F$ should define an integral cohomology class in $H^2(Y,\mathbb{R})$. The factor $h^{-1}e_0$ appears here because of our choice of identification of the Lie algebra of T^1 with $\mathbb{R}$. Similarly, the horizontal part of the metric $\hat{g}$ on Z is identified with the pull back to Z of the metric g describing the gravitational field in the space-time Y.

The dynamics of a single particle in the electromagnetic field F and the gravitational field g can be described as follows. The phase space is the cotangent bundle space T^*Z of Z and the Lagrange brachet is given by $d\theta_Z$, where θ_Z is the canonical 1-form of T^*Z defined in the same way as θ_Y, c.f. Eq. (3.1). There is a submersion $\sigma : T^*Z \to T^*Y$ such that, for each $w \in T_z^*Z$, the horizontal part of w coincides with the horizontal lift of $\sigma(w)$ to T_z^*Z. The dynamical variables q, p_ζ and M^2 defined in Sec. 2 can be pulled back by σ to functions on T^*Z, giving rise to the dynamical variables $q\circ\sigma$, $p_\zeta\circ\sigma$ and $M^2\circ\sigma$ in the 5-dimensional theory which have the same physical interpretation as the original ones. There is an additional dynamical variable Q, interpreted as the charge function, defined by

$$Q(w) = w(\eta_1(z)), \tag{4.1}$$

for each $z \in Z$ and each $w \in T_z^*Z$. The integral curves of the hamiltonian vector field of Q project to the fibres of $Z \to Y$. For each $m > 0$ and each $e \in \mathbb{R}$, the integral curves of the hamiltonian vector field of $M^2\circ\sigma$, which are contained in $(M^2\circ\sigma)^{-1}(m^2) \cap Q^{-1}(e)$, when projected to Y yield the solutions of the equations of motion for a particle of mass m and charge e moving in the gravitational field g and the electromagnetic field F.

Let $\hat{D}$ be the distribution on T^*Z spanned by the hamiltonian vector field of the charge function Q and the hamiltonian vector fields of the functions of the type $\check{q}\circ\pi\circ\sigma$, where $\check{q}$ is a smooth function on Y. Its complexification $\hat{F} = \hat{D}^{\mathbb{C}}$ is a polarization of $(T^*Z, d\theta_Z)$. The bundle of half-forms normal to $\hat{F}$ admits a nowhere zero covariant constant along F section.

Since the symplectic form $d\theta_Z$ is exact, the prequantization line bundle $\hat{L}$ for $(T^*Z, d\theta_Z)$ is trivial. Let $\hat{\nabla}$ be a connection in $\hat{L}$ and $\hat{\lambda}_0$ a section of $\hat{L}$ such that

$$\hat{\nabla}\hat{\lambda}_0 = -i\hbar^{-1}\theta_Z \otimes \hat{\lambda}_0 \quad . \tag{4.2}$$

The section $\hat{\lambda}_0$ is not covariant constant along $\hat{F}$. In fact, the covariant derivative of $\hat{\lambda}_0$ in the direction of the hamiltonian vector field of Q is given by

$$\hat{\nabla}_{\xi_Q} = i\hbar^{-1} Q\hat{\lambda}_0 \tag{4.3}$$

The representation space in this case can be identified with the space $\hat{H}$ of section of $\hat{L}$ which are covariant constant along $\hat{F}$. Every covariant constant along $\hat{F}$ section of $\hat{L}$ has its support in the Bohr-Sommerfeld set S given by

$$S = \{w \in T^*Z \mid e_0^{-1}Q(w) \in \mathbb{Z}\} \tag{4.4}$$

For each integer n let S_n be the component of S on which the charge function Q takes on the value ne_0 and let $\hat{H}_n$ denote the Hilbert space of the covariant constant along $\hat{F}|S_n$ sections of $\hat{L}|S_n$ (for the Hilbert space structure of the distribution sections see Ref. [11]). Then

$$\hat{H} = \oplus_n \hat{H}_n \tag{4.5}$$

and each subspace $\hat{H}_n$ of $\hat{H}$ is an eigensubspace of the charge operator $\hat{\mathcal{Q}}Q$ corresponding to the eigenvalue ne_0. For each one parameter group φ_t of symplectic diffeomorphisms of $(T^*Z, d\theta_Z)$, $\varphi_t(S_n) \cap S_{n'} = \emptyset$ if $n \neq n'$ and t is sufficiently close to zero. Hence, for each function f on T^*Z which can be quantized by means of the Blattner-Kostant-Sternberg kernels in the representation given by the polarization $\hat{F}$, the resulting operator $\hat{\mathcal{Q}}f$ in $\hat{H}$ commutes with the charge operator. Thus, we have obtained the super-selection rules for the charge.

Since, for each $n \in \mathbb{Z}$, $\hat{\mathcal{Q}}f(H_n) \subseteq H_n$, the operator $\hat{\mathcal{Q}}f$ is uniquely determined by the collection of its restrictions $\hat{\mathcal{Q}}_n f$ to the subspaces $\hat{H}_n$ of $\hat{H}$. Let $n \in \mathbb{Z}$ be fixed and $e = ne_0$. It follows from the definition of $\hat{H}_n$ that $\hat{\mathcal{Q}}_n Q$ is the operator of the multiplication by e. Moreover, there exists a unitary mapping $U_n \colon \hat{H}_n \to H_e$, where H_e is the representation space discussed in Sec. 3, such that

$$\hat{\mathcal{Q}}_n(\check{q} \circ \pi \circ \sigma) = U_n^{-1} \mathcal{Q}_e(q \circ \pi) U_n \tag{4.6}$$

and

$$\hat{\mathcal{Q}}_n(p_\zeta \circ \sigma) = U_n^{-1} \mathcal{Q}_e p_\zeta U_n \tag{4.7}$$

for each function $\check{q} : Y \to \mathbb{R}$ and each vector field ζ on Y. It is of interest to quantize $M^2 \circ \sigma$, which will involve the Blattner-Kostant-Sternberg kernels for non-transverse polarizations [12], and to verify if $\hat{\mathcal{Q}}_n(M^2 \circ \sigma)$ is equal to $U_n^{-1} \mathcal{Q}_e M^2 U_n$.

References

[1] D.J. Simms and N.M.J. Woodhouse, Lectures on Geometric Quantization, Lecture Notes in Physics, Vol. 53, Springer, Berlin, 1976.

[2] J. Śniatycki, Geometric Quantization and Quantum Mechanics, in preparation.

[3] J.-M. Souriau, Structure des Systèmes Dynamiques, Dunod, Paris, 1970.

[4] P.A.M. Dirac, Quantized Singularities in the Electromagnetic Field, Proc. Roy. Soc. A, Vol. 133, (1931), pp. 60-70.

[5] J. Śniatycki, Prequantization of Charge, J. Math. Phys. Vol. 15 (1974), pp. 619-620.

[6] B. DeWitt, Point Transformations in Quantum Mechanics, Phys. Rev. Vol. 85, (1952), pp. 653-661.

[7] R.J. Blattner, "Quantization and Representation Theory", in 'Harmonic Analysis on Homogeneous Spaces', Proceedings of Symposia in Pure Mathematics, Vol. 26, pp. 147-165, American Mathematical Society, R.I., 1973.

[8] B. DeWitt, Dynamical Theory in Curved Spaces, I, Rev. Mod. Phys. Vol. 29, (1957), pp. 377-397.

[9] K.S. Cheng, Quantization of a General Dynamical System by Feynman's Path Integration Formulation, J. Math. Phys., Vol. 13 (1972), pp. 1723-1726.

[10] P.G. Bergmann, Introduction to the Theory of Relativity, Prentice Hall, Englewood Cliffs, N.J., 1942.

[11] J. Śniatycki, Wave functions relative to a real polarization, Int. J. Theor. Phys., Vol. 14, (1975), pp. 277-288.

[12] R.J. Blattner, Pairing of half-form spaces, Lecture Notes in Mathematics, Vol. 570, pp. 11-45, Springer Verlag, Berlin, 1977.

THERMODYNAMIQUE ET GEOMETRIE

par

Jean-Marie SOURIAU *

RESUME : La mécanique classique ou relativiste peut se formuler en termes de géométrie symplectique ; cette formulation permet un énoncé rigoureux des principes de la mécanique statistique et de la thermodynamique.

Mais cette analyse met en évidence quelques difficultés fondamentales, qui restent cachées, dans le discours traditionnel, par une certaine ambiguïté.

Le "premier principe" de la thermodynamique peut échapper à cette ambiguïté à condition d'accepter un détour par le principe de relativité générale et par les équations d'Einstein de la gravitation. Les outils mathématiques utilisés sont la théorie des moments symplectiques, certaines formules cohomologiques et la notion de tenseur-distribution.

En ce qui concerne le second principe, nous nous contentons de montrer comment il est possible, en acceptant un certain statut géométrique pour la température et l'entropie, de construire un modèle relativiste de fluide dissipatif apte à décrire des situations expérimentales assez diverses (équilibres accélérés, transitions de phase, viscosité, conduction thermique). Dans l'approximation du fluide parfait, nous établissons certains résultats : extension relativiste des théorèmes d'Helmholtz et Ertel concernant les mouvements non isentropiques, étude géométrique des ondes de choc.

78/P.1008

* Université de Provence, et
Centre de Physique Théorique, CNRS Marseille

ADRESSE POSTALE : Centre de Physique Théorique
C.N.R.S.
31, chemin Joseph Aiguier
F-13274 MARSEILLE CEDEX 2 (France)

1. FORMULATION SYMPLECTIQUE DE LA DYNAMIQUE

Considérons d'abord un cas élémentaire de système dynamique : un point matériel newtonien, de masse m, de position $\vec{r}$, de vitesse $\vec{v}$, soumis à un champ de forces $(\vec{r},t) \mapsto \vec{F}$ ([1]) ; le triplet $y = (\vec{v},\vec{r},t)$ constitue une condition initiale d'un mouvement x ; y parcourt une variété V_7 (espace d'évolution) ; si on pose

$$(1.1) \qquad \sigma_V(dy)(\delta y) = \langle m\, d\vec{v} - \vec{F}\, dt, \delta\vec{r} - \vec{v}\,\delta t\rangle - \langle m\,\delta\vec{v} - \vec{F}\,\delta t, d\vec{r} - \vec{v}\, dt\rangle$$

d et δ étant deux variations arbitraires, les crochets $\langle\ ,\ \rangle$ désignant le produit scalaire dans $\mathbb{R}^3$, on définit sur V_7 une 2-forme σ_V de rang 6 ; les équations du mouvement s'écrivent $dy \in \ker(\sigma_V)$ ([2]) ; si $\vec{F}$ dérive d'un potentiel, σ_V est une forme fermée (sa dérivée extérieure est nulle) : σ_V est donc un invariant intégral absolu des équations du mouvement, découvert par E. Cartan, mais en fait déjà explicitement décrit par Lagrange. L'ensemble U des mouvements possibles possède alors une structure de variété symplectique (de dimension 6), muni de la 2-forme fermée et inversible σ_U dont l'image réciproque par la submersion $y \mapsto x$ coïncide avec σ_V (Figure 1).

Un tel schéma s'étend aux systèmes dynamiques généraux (systèmes de particules, particules à spin, mécanique relativiste, etc. ; voir [XI]) : dans tous ces cas, l'espace U des mouvements est une variété symplectique (donc de dimension paire) sur laquelle se projette l'espace d'évolution V ; chaque section $t = $ Cte de V est un "espace de phases" ; mais l'identification des espaces de phases correspondant à des dates différentes est une opération arbitraire, dépendant du choix d'un référentiel, qu'il vaut donc mieux éviter.

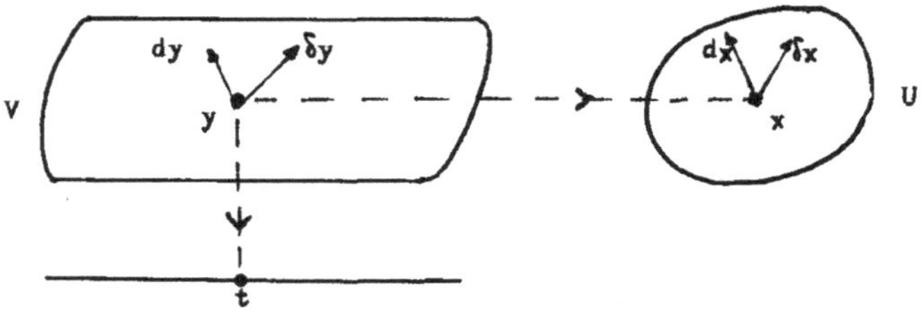

I

([1]) Sauf mention explicite, toutes les fonctions que nous considérerons dans ce travail seront supposées de classe C^∞ ; en particulier donc $(\vec{r},t) \mapsto \vec{F}$.

([2]) c'est-à-dire $\sigma_V(dy)(\delta y) = 0 \;\forall\, \delta y$: c'est le principe de d'Alembert généralisé.

2. FORMULATION SYMPLECTIQUE DE LA MECANIQUE STATISTIQUE

Dans cette représentation, un état statistique μ est simplement une loi de probabilité définie sur U (c'est-à-dire une mesure positive de masse 1) ; l'ensemble Prob(U) de ces lois de probabilité est un ensemble convexe, dont les points extrêmaux sont les mouvements classiques x (identifiés aux mesures de Dirac correspondantes). (Figure V).

Les états completement continus sont caractérisés par une densité de U , produit de la densité de Liouville ([1]) par un scalaire ρ qui s'identifie avec la fonction de répartition classique ([2]). On appelle entropie d'un état statistique μ la valeur moyenne S de $-\operatorname{Log}\rho$ dans cet état ; on peut définir une bonne classe d'états, les "états de Boltzmann" [XIV] , constituant un sous-convexe de Prob(U), pour lesquels l'intégrale de $-\operatorname{Log}\rho$ converge ; $\mu \mapsto S$ est une fonction concave sur ce convexe.

([1]) Une densité sur une variété est une fonction f définie sur les repères R et vérifiant $f(RM) = f(R)|\det(M)|$ pour toute matrice M ; sur une variété symplectique, il existe une densité f_0 -la densité de Liouville- telle que $f_0(R) = 1$ pour tout repère canonique. On peut définir l'intégrale d'une densité à support compact sur une variété independamment de tout système de coordonnée; ce qui permet d'identifier chaque champ de densités avec une mesure.

([2]) Par construction, ρ est une fonction définie sur U , qui se relève donc sur V par une intégrale première des équations du mouvement ; si on choisit une identification des divers espaces de phases, ceci implique que ρ est solution de l'équation de Liouville.

3. LES PRINCIPES DE LA THERMODYNAMIQUE

La mécanique statistique, telle qu'elle vient d'être décrite, est apte à décrire certains phénomènes réels -mais pas les phénomènes dissipatifs (frottement, conduction de chaleur, viscosité, etc.), qui sont l'objet de la thermodynamique. Les "deux principes" de la thermodynamique ne s'appliquent en fait qu'à une situation idéalisée : les transitions dissipatives, dans lesquelles un système est situé dans un état statistique μ_{in} avant les phénomènes dissipatifs, et aboutit à un nouvel état statistique μ_{out} après. Le second principe (Carnot-Clausius) s'écrit alors

$$S(\mu_{out}) \geqslant S(\mu_{in}) \tag{3.1}$$

alors que le premier principe exprime la conservation de la valeur moyenne de l'énergie E ; ce qui s'écrit

$$\mu_{out}(x \mapsto E) = \mu_{in}(x \mapsto E) \tag{3.2}$$

en considérant les mesures comme des fonctionnelles linéaires.

μ_{in} et μ_{out} appartiennent tous deux au convexe des états de Boltzmann donnant une valeur moyenne donnée Q à l'énergie. Sur ce convexe, il peut arriver que la fonction concave S soit bornée ; soit alors S_Q sa borne supérieure. On a évidemment

$$S(\mu_{in}) \leqslant S(\mu_{out}) \leqslant S_Q \tag{3.3}$$

ce qui majore la production d'entropie $S(\mu_{out}) - S(\mu_{in})$ par $S_Q - S(\mu_{in})$, connue en fonction de μ_{in} seulement. Il peut arriver en particulier que le maximum de S sur ce convexe soit atteint en un point unique μ_Q, appelé état de Gibbs ; si $\mu_{in} = \mu_Q$, la production d'entropie est nulle, et $\mu_{out} = \mu_{in}$: les états de Gibbs ne peuvent pas subir de phénomènes dissipatifs ; ils constituent ce qu'on appelle les équilibres thermodynamiques.

4. FORMULATION COVARIANTE DU PREMIER PRINCIPE

L'analyse précédente s'applique aux systèmes conservatifs ; la fonction $x \mapsto E$, définie sur la variété symplectique U , permet par le formalisme hamiltonien de définir un groupe à un paramètre de symplectomorphismes de U ([1]) ; le calcul montre que ce groupe se relève sur l'espace d'évolution V par le groupe des translations temporelles ; soit, dans le cas d'une particule

$$(4.1) \qquad \vec{v} \to \vec{v} \quad , \quad \vec{r} \to \vec{r} \quad , \quad t \to t + \text{Cte}$$

ce qu'on exprime assez incorrectement en disant que "le temps et l'énergie sont des variables conjugués" ([2]).

Il est clair que ces translations (4.1) sont liées à un référentiel particulier : le premier principe, tel qu'il est énoncé, ne respecte pas la covariance relativiste, même galiléenne ([3]) ; il doit donc exister un énoncé ne présentant pas cet inconvénient.

Une solution radicale consiste à remplacer le groupe (4.1) par le <u>groupe</u>

([1]) Un champ de vecteurs F défini sur une variété séparée U peut s'associer à l'équation différentielle $\frac{dx}{ds} = F(x)$; la solution de cette équation, prenant la valeur x_0 pour $s = 0$, se note $\exp(sF)(x_0)$; si elle existe pour tout $x_0 \in U$ et tout $s \in \mathbb{R}$, on dit que F est complet ; alors $s \mapsto \exp(sF)$ est un morphisme du groupe $(\mathbb{R},+)$ dans le groupe des difféomorphismes de U. Si U est symplectique, et si $x \mapsto u$ une fonction C^∞ sur U , on appelle <u>gradient symplectique</u> de la variable dynamique u le champ de vecteurs F défini par $\sigma(\delta x)(F(x)) = \delta u \quad \forall \delta$; l'équation associée est l'équation de Hamilton ; les $\exp(sF)$ préservent la forme symplectique σ, et s'appellent donc symplectomorphismes.

([2]) Avec les conventions de signe usuelles, il faut remplacer E par -E.

([3]) De façon précise, ces transformations (4.1) définissent un sous-groupe du groupe de Galilée qui <u>n'est pas</u> un sous-groupe invariant.

de Galilée ([1]) tout entier ; ou bien, si on veut faire de la mécanique relativiste, par le groupe de Poincaré ([2]).

L'action de ces groupes sur U par symplectomorphismes se définit naturellement si le système dynamique est isolé; sinon, on considère un système partiel auquel le "mécanisme" constitué par le système extérieur donné ne laisse que la symétrie correspondant à un sous-groupe du groupe de Galilée (resp.de Poincaré): par exemple une boîte immobile contenant un gaz - qui ne lui accorde que le sous-groupe (4.1) ; mais aussi une centrifugeuse, etc.

Soit donc G le groupe de symétries; nous cherchons un objet qui joue par rapport à G le même rôle que l'énergie par rapport au groupe (4.1). Il suffit pour cela de considérer tous les sous-groupes à un paramètre de G; chacun d'eux sera caractérisé par un élément Z de l'algèbre de Lie $\mathfrak{G}$ de G; il lui correspondra un hamiltonien que nous noterons M(Z). L'examen de la situation montre que l'on peut choisir la constante additive qui figure dans chaque hamiltonien de façon que la correspondance $Z \mapsto M(Z)$ soit linéaire; M devient donc une forme linéaire sur $\mathfrak{G}$,donc un élément de son dual $\mathfrak{G}^*$; il existe donc une application $x \mapsto M$ de U dans $\mathfrak{G}^*$; la variable M s'appellera moment du groupe; naturellement, nous remplacerons le premier principe (3.2) par l'énoncé [x]

(4.2) $$\mu_{out}(x \mapsto M) = \mu_{in}(x \mapsto M)$$

sans changer le second principe (3.1); les conclusions sont analogues. Sur le convexe des états de Boltzmann vérifiant $\mu(x \mapsto M) = Q$ ([3]), il peut exister un "état de Gibbs" μ_Q possédant la plus grande entropie S_Q ; comme précédemment, on obtient une majoration de la production d'entropie dans une transition dissipative, et on constate que les états de Gibbs ne sont plus susceptibles de phénomènes dissipatifs.

La fonction de distribution de ces états de Gibbs est l'exponentielle d'une fonction affine de M , ce qui s'écrit

---

([1]) C'est le groupe de Lie, de dimension 10, engendré par les isométries de l'espace $\mathbb{R}^3$, les translations temporelles et les transformations de Galilée $\vec{r} \to \vec{r} + \vec{a}t$, $\vec{v} \to \vec{v} + \vec{a}$.

([2]) Le groupe des isométries de l'espace de Minkowski. il est aussi de dimension 10.

([3]) Q est un élément de $\mathfrak{G}^*$ qui généralise la "chaleur" usuelle.

(4.3) $$\rho = e^{M\Theta - z}$$

Θ étant un élément de $\mathcal{G}$ (la "température géométrique"), z un nombre (le "potentiel thermodynamique de Planck" ; voir [VI]) que l'on obtient en fonction de Θ en écrivant que la masse de μ vaut 1 , ce qui donne

(4.4) $$z = \operatorname{Log} \int_U e^{M\Theta} \lambda(x)\, dx$$

λ étant la mesure de Liouville ; z est une fonction convexe de Θ , qui se trouve être la transformée de Legendre de $Q \mapsto -S_Q$:

(4.5) $$dQ\,\Theta = -dS \quad , \quad Q\, d\Theta = dz \quad , \quad Q\Theta = z - S \quad \forall d.$$

Toutes les formules classiques de la thermodynamique sont ainsi généralisées ; mais maintenant les variables sont pourvues d'un statut géométrique. Par exemple, la "température géométrique" Θ , élément de l'algèbre de Lie du groupe de Galilée ou de Poincaré, s'interprête comme champ de vecteurs de l'espace-temps ; dans la version relativiste, $\Theta(X)$ est un vecteur du genre temps ; son sens caractérise la "flèche du temps" ; sa direction est la quadri-vitesse du référentiel d'équilibre ; sa longueur (Minkowskienne) est $\beta = 1/kT$ (k : constante de Boltzmann, T : température absolue). Ce vecteur-température avait été proposé par Planck pour étudier la thermodynamique relativiste ; mais son correspondant galiléen est tout aussi pertinent.

Les formules auxquelles on est ainsi conduit s'appliquent correctement à un certain nombre de situations réelles : équilibre de particules à spin, centrifugeuses, rotation des corps célestes, etc.

De plus, il apparaît des relations d'un type nouveau, liées à la non-commutativité du groupe G , qui permettent certaines prédictions ; ainsi, sous des hypothèses faibles, on prévoit l'existence d'une température critique pour un système isolé, au-dessus de laquelle il n'existe plus aucun équilibre ; ce fait joue probablement son rôle en astrophysique (supernovae). Pour plus de détails, nous renvoyons à [XI] et [XIV] .

5. SUSCEPTIBILITE GRAVITATIONNELLE

La formulation covariante (4.2) du premier principe fait donc disparaître un paradoxe, tout en augmentant la valeur pratique de la thermodynamique. Mais elle laisse ouver un problème conceptuel.

Lors d'une transition dissipative, la mécanique statistique est nécessairement violée, puisque $\mu_{out} \neq \mu_{in}$; la variable dynamique énergie (ou plus généralement le moment) change de spectre pendant la transition [1]. Puisqu'on ne peut plus invoquer les lois de conservation de la mécanique classique ou statistique, il faut faire appel à d'autres lois de la nature pour comprendre comment celle-ci conserve la valeur moyenne de l'énergie, ou même simplement comment elle la mémorise.

De façon inattendue, la réponse est fournie par la relativité générale ; nous verrons comment au §7, après l'étude de notions préliminaires. Considérons un système dynamique qui évolue dans un champ de gravitation, champ qui est caractérisé en relativité générale par ses potentiels $g_{\mu\nu}$; l'espace des mouvements est toujours une variété symplectique U, dont la structure dépend du champ.

Choisissons maintenant un compact K de l'espace-temps E_4 (figure II), dans lequel nous perturbons les $g_{\mu\nu}$. Le nouvel espace des mouvements U' est encore une variété symplectique, que l'on peut rattacher à U par la technique de la diffusion ; technique que nous allons décrire dans le cas d'une particule sans spin, dont le mouvement est caractérisé par la ligne d'univers ; si celle-ci ne rencontre pas K, elle caractérise aussi bien un mouvement dans U qu'un mouvement dans U'.

Considérons maintenant un mouvement dans U, que noterons x_{in}, dont la ligne d'univers pénètre à un certain moment dans K ; avec les potentiels perturbés, elle sera déviée par rapport au mouvement initial (figuré en pointillé) ; lorsqu'elle ressort de K, elle emprunte un nouveau chemin qui s'identifie à un élément x_{out} de U ; la correspondance $x_{in} \mapsto x_{out}$, qui caractérise globale-

---

[1] En mécanique statistique non quantique, le spectre d'une variable dynamique u dans un état statistique μ est l'image par $x \mapsto u$ de la loi de probabilité μ ; c'est une loi de probabilité de $\mathbb{R}$ (ou de $\mathfrak{G}^*$ dans le cas du moment).

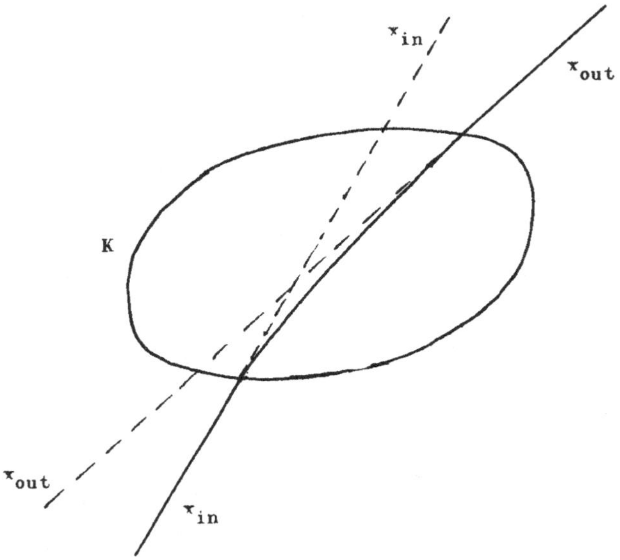

II

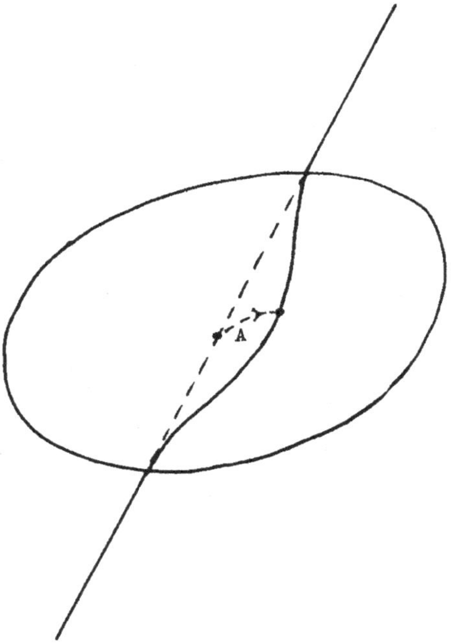

III

ment la diffusion gravitationnelle, est un symplectomorphisme local de U (parce que U et U' sont chacune symplectique, et que leur structure peut être obtenue en partant du même espace d'évolution). Un tel objet peut se caractériser au moyen d'une certaine variable dynamique, l'eikonal de diffusion.

Nous nous intéressons ici à la diffusion infinitésimale : si on donne aux $g_{\mu\nu}$ une variation $\delta g_{\mu\nu}$ nulle en dehors de K , le mouvement initial subira un déplacement $\delta x = F(x)$, qui dérive d'un certain hamiltonien φ (voir le §4) ; φ défini ainsi à une constant additive près, peut être déterminé complètement en le prenant nul sur les chemins qui ne rencontrent pas K ([1]).

Pour tout mouvement $x \in U$, notons $\mathcal{L}_x$ l'application qui fait correspondre ce nombre φ au champ de tenseurs $X \mapsto \delta g$ $(X \in E_4)$; $\mathcal{L}_x$ est une application linéaire, donc a priori une distribution ([2]) ; la connaissance de $\mathcal{L}_x$ permet de prévoir la réaction de la particule à toute "petite" modification du champ de gravitation : c'est pourquoi nous dirons que $\mathcal{L}_x$ est la susceptibilité gravitationnelle de la particule dans le mouvement x .

Utilisons maintenant le principe de relativité générale : il affirme qu'un difféomorphisme A de l'espace-temps E_4 , agissant simultanément sur les potentiels (selon les formules standard de la géométrie différentielle) et sur le mouvement (ici par image directe de la ligne d'univers) est inobservable (voir [XII]). Choisissons A tel qu'il laisse fixe les points extérieurs au compact K (figure III) ; il ne modifie les potentiels que dans K , et son action sur la particule laisse fixes les parties de la ligne d'univers qui sont extérieures à K ; la diffusion gravitationnelle correspondante est donc nulle.

Appliquons ce résultat au cas $A = \exp(sF)$ (voir le §4), $s \in \mathbb{R}$, F = champ de vecteurs nul en dehors de K . On voit que $\mathcal{L}_x(X \mapsto \delta g)$ est nul si δg est la dérivée par rapport à s , pour s = 0 , de l'image réciproque par A de $X \mapsto \delta g$; cette variation est, par définition, la dérivée de Lie de g associée au champ de vecteurs $\delta X = F(X)$; nous la noterons $\delta_L g$.

([1]) Une autre méthode, qui évite certaines difficultés topologiques, fait appel à l'algorithme de la préquantification (voir [XIV]).

([2]) La variable d'essai $X \mapsto \delta g$ étant un champ de tenseurs (covariants), on dit que $\mathcal{L}_x$ est un tenseur-distribution (contravariant).

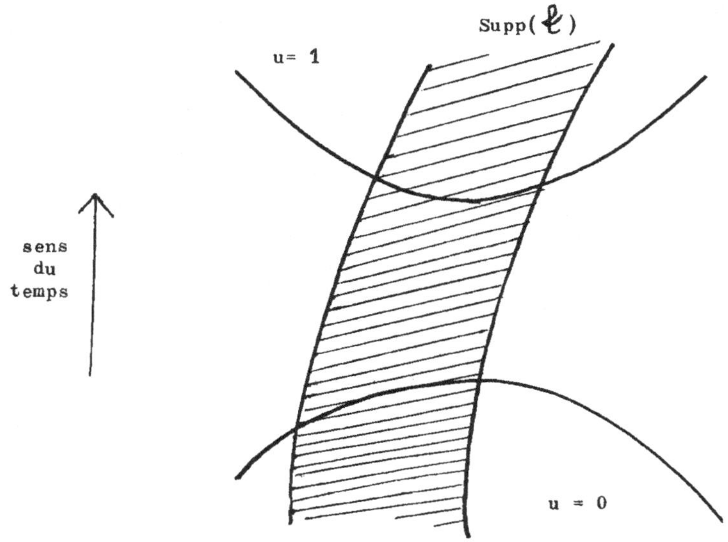
Supp($\mathcal{L}$)
u= 1
sens
du
temps
u = 0

IV

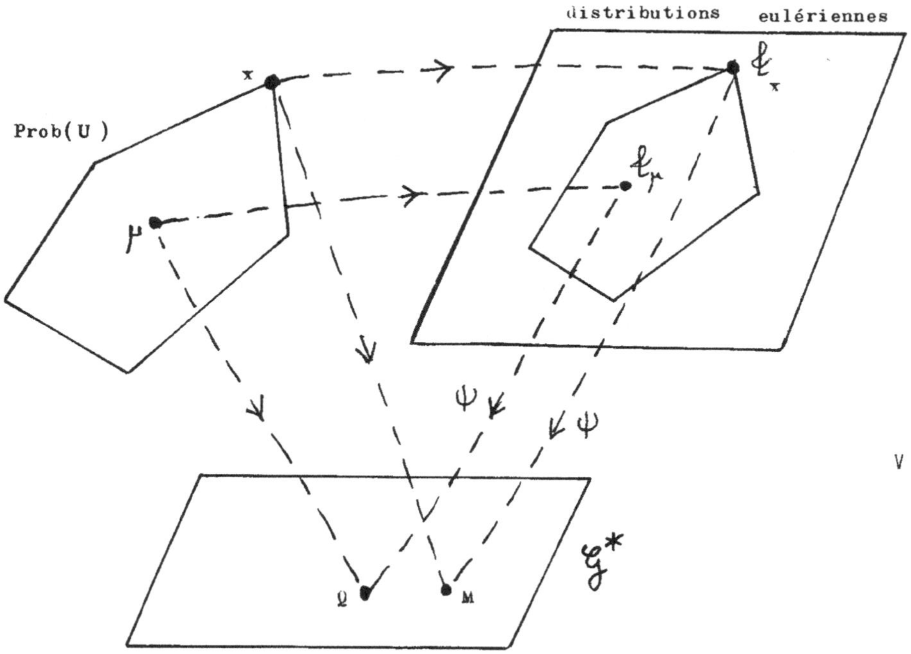
distributions eulériennes
x
$\mathcal{L}_x$
Prob(U)
μ
$\mathcal{L}_\mu$
Ψ
Ψ
$\mathcal{G}^*$
Ω
M

V

- Nous dirons, généralement, qu'une distribution $\mathcal{t}$ est <u>eulérienne</u> si elle vérifie la condition

(5.1) $\mathcal{t}(X \mapsto \delta_L g) = 0$ pour tout champ $X \mapsto \delta X$ à support compact

et nous savons donc que $x \mapsto \mathcal{t}_x$ est une application de l'espace des mouvements U dans l'espace vectoriel des distributions eulériennes de E_4 (figure V). Sous certaines hypothèses, une distribution eulérienne permet d'associer une grandeur conservée à tout vecteur de Killing Z de la métrique g ([1]) ; nous décrirons sommairement cette procédure dans le cas où E_4 est l'espace de Minkowski, et par conséquent Z un élément de l'algèbre de Lie du groupe de Poincaré ; la grandeur qui lui est associée est

(5.2) $I = \mathcal{t}(X \mapsto \delta_L g)$ avec $\delta X = u\, Z(X)$

u étant une fonction que l'on choisit nulle dans le passé, égale à 1 dans le futur (figure IV).

Contrairement à ce qu'on pourrait croire à l'examen de (5.1) et (5.2), I n'est pas nécessairement nulle, parce que $X \mapsto u\, Z(X)$ n'est pas un champ à support compact ; mais la condition eulérienne (5.1) permet de montrer que I ne dépend pas du choix de u , en faisant quelques hypothèses sur le comportement à l'infini de $\mathcal{t}$ ([2]).
On peut donc calculer I en faisant sauter u de 0 à 1 dans un petit voisinage d'une surface du genre espace ; l'indépendance du résultat par rapport au choix de cette surface exprime le caractère de "grandeur conservée" de I.

Il est clair que l'application $Z \mapsto I$ ainsi définie est linéaire ; elle associe donc à $\mathcal{t}$ un élément $\Psi(\mathcal{t})$ du dual $\mathcal{G}^*$ de l'algèbre de Lie $\mathcal{G}$ du groupe de Poincaré. Il est immédiat que l'application

---

([1]) Z est dit vecteur de Killing si exp(sZ) est une isométrie $\forall s$.

([2]) La plus simple est de supposer le support de $\mathcal{t}$ compact dans l'espace, c'est-à-dire que sont intersection avec toute tranche temporelle $t_0 \leqslant t \leqslant t_1$, t étant la date dans un référentiel de Lorentz arbitraire, est compacte. Cette condition est vérifiée pour les $\mathcal{t}_x$ que nous avons construit en considérant une particule (à condition que celle-ci ne soit pas un tachyon !).

(5.3) $$x \mapsto M = \Psi(\mathfrak{t}_x)$$

est un moment du groupe de Poincaré (§4) ; nous allons en trouver une autre propriété par des considérations d'équivariance.

Il est clair que le groupe de Poincaré G agit sur les champs de tenseurs à support compact, donc sur les tenseurs-distributions selon la formule

(5.4) $$a(\mathfrak{t})(a(X \mapsto \delta g)) = \mathfrak{t}(X \mapsto \delta g) \qquad \forall a \in G .$$

Il agit aussi sur les champs de vecteurs à support compact, et la dérivée de Lie de g est équivariante pour cette action, si bien que G agit sur les distributions eulériennes. Enfin G agit sur les vecteurs de Killing, et cette action coïncide avec la représentation adjointe de G sur $\mathcal{G}$.

Supposons maintenant que a soit un élément du sous-groupe orthochrone $G\uparrow$ ([1]) ; les constructions précédentes montrent facilement que :

(5.5) $$\Psi(a(\mathfrak{t}))(a(Z)) = \Psi(\mathfrak{t})(Z)$$

et, $\forall x \in U$, que

(5.6) $$a(\mathfrak{t}_x) = \mathfrak{t}_{a(x)}$$

d'où

(5.7) $$\Psi(\mathfrak{t}_{a(x)})(a(Z)) = \Psi(\mathfrak{t}_x)(Z) .$$

Cette formule possède une interprétation cohomologique : elle exprime la nullité d'un certain cocycle symplectique, et entraîne (Cf. [X])

([1]) La composante connexe de l'élément neutre d'un groupe de Lie est un sous-groupe invariant ; le quotient par ce sous-groupe est le groupe des composantes. Dans le cas du groupe de Poincaré, le groupe des composantes est le groupe de Klein à 4 éléments qui est abélien. Les éléments qui "respectent le sens du temps" constituent la réunion de deux composantes ; ils constituent donc un sous-groupe invariant, le groupe orthochrone.

(5.8) $$M\left[Z,Z'\right] = \sigma(Z(x))(Z'(x)) \qquad \forall Z,Z' \in \mathcal{G}$$

Cette formule permet de fixer la constante arbitraire qui figurait dans M (parce que l'algèbre de Lie du groupe de Poincaré est égale à son algèbre dérivée) ; elle montre en particulier, si Z est une translation temporelle infinitésimale, que l'intégrale I (5.2) est égale à l'énergie relativiste $E = mc^2$ du système dans le mouvement considéré.

Nous avons donc factorisé l'application "moment" $x \mapsto M$ par la composition de $x \mapsto \mathcal{E}_x$ et de $\mathcal{E}_x \mapsto M$ (figure V) ; c'est ce résultat qui va être essentiel pour la thermodynamique.

Indiquons les résultats détaillés dans le cas de la particule ; la forme symplectique (1.1) devient, en relativité générale

(5.9) $$\sigma_V(dy)(\delta y) = m\, g_{\mu\nu} \left[dX^\mu \hat{\delta} U^\nu - \delta X^\mu \hat{d} U^\nu\right]$$

ici une condition initiale y est un couple (X,U), X appartenant à la ligne d'Univers, U étant le vecteur unitaire tangent ; les chapeaux ^ désignent une dérivation covariante.

Le calcul de la susceptibilité gravitationnelle, assez technique, donne

(5.10) $$\mathcal{E}_x(X \mapsto \delta g) = \int \frac{1}{2}\, m\, \delta g_{\mu\nu}\, U^\mu \frac{dX^\nu}{ds}\, ds$$

$s \mapsto X$ étant une paramétrisation arbitraire de la ligne d'univers de la particule (du passé vers le futur). On voit que la distribution $\mathcal{E}_x$ est ici une mesure, ayant pour support cette ligne d'univers. Il se trouve que l'on connaît toutes les mesures eulériennes supportées par une courbe ; elles s'écrivent

(5.11) $$\mathcal{E}(X \mapsto \delta g) = \int \frac{1}{2}\, \delta g_{\mu\nu}\, P^\mu \frac{dX^\nu}{ds}\, ds$$

avec les conditions supplémentaires

(5.12) $$\frac{dX}{ds} \text{ parallèle à } P \quad , \quad \frac{\hat{d}P}{ds} = 0$$

(on trouvera la démonstration dans [XII]) ; ces conditions impliquent que la courbe est une géodésique -fait bien connu pour les particules, que l'on trouve aussi en utilisant le principe de d'Alembert $dy \in \ker(\sigma_V)$ à la forme (5.9). La quadri-impulsion $P = mU$ apparaît donc comme un élément de la susceptibilité gravitationnelle ; dans le cas de l'espace de Minkowski, la grandeur conservée associée à un élément Z de l'algèbre de Lie $\mathcal{G}$ du groupe de Poincaré est

(5.13) $$I = g_{\mu\nu} P^{\mu} Z(X)^{\nu}$$

X étant choisi arbitrairement sur la ligne d'univers ; en faisant varier Z dans $\mathcal{G}$, on met en évidence l'énergie, l'impulsion, le moment orbital, etc.

- Les structures que nous venons de mettre en évidence sur le cas le plus simple s'étendent à des circonstances très variées :

- On peut les transposer au cas de la mécanique classique ; on associe encore à chaque mouvement x une distribution $\mathcal{T}_x$; la condition eulérienne s'exprime, non plus par une dérivée de Lie, mais par une certaine connexion qui prend en compte le champ de gravitation sous sa forme newtonienne. On remarque que les grandeurs conservées associées au cas du champ nul font apparaître une algèbre de Lie de dimension 10 qui n'est pas celle du groupe de Galilée, mais assez paradoxalement celle du groupe de Caroll, contraction du groupe de Poincaré lorsque l'on fait tendre la vitesse c de la lumière vers 0. Ce phénomène peut être rapprochée de l'impossibilité, pour les moments du groupe de Galilée, d'être choisis de façon à vérifier la formule (5.8) : il apparaît une obstruction qui est une classe de cohomologie symplectique, et qui est mesurée par la masse totale du système -donc non nulle.

Physiquement, ces faits indiquent que, dans la formulation par distribution eulérienne de la mécanique classique, la masse cache l'énergie.

- On peut traiter par la même méthode les particules à spin, aussi bien en mécanique classique qu'en mécanique relativiste. La susceptibilité gravitationnelle met en jeu, à côté de la quadri-impulsion P^{μ}, le tenseur de spin antisymétrique $S^{\mu\nu}$.

- Cette méthode permet d'obtenir simplement les règles de collision ou de désintégration des particules : il suffit d'écrire que la somme des susceptibilités gravitationnelles supportées par les diverses lignes d'univers est encore une distribution eulérienne.

- Cette méthode peut s'étendre au cas de l'électro-dynamique : On calcule la susceptibilité du système pour une variation simultanée des potentiels de gravitation $g_{\mu\nu}$ et des potentiels électro-magnétiques A_{ρ} ; pour les particules cette susceptibilité introduit, à côté de la quadri-impulsion et du tenseur de spin, la charge électrique et le moment magnétique. Le principe de relativité générale, qui concernait le groupe des difféomorphismes d'espace-temps, se généralise à un groupe plus grand, produit semi-direct du précédent par le groupe des transformations de jauge électromagnétiques ; par conséquent la condition eulérienne devient

$$(5.14) \qquad \mathcal{E}(X \mapsto (\delta g, \delta A)) = 0 \text{ si } \delta g = \delta_L g\ ,\ \delta A = \delta_L A + \frac{\partial \alpha}{\partial X}$$

$X \mapsto \delta X$ et $X \mapsto \alpha$ étant respectivement un champ de vecteurs et un champ scalaire à supports compacts. On notera que ces structures sont particulièrement simples si on les écrit sur l'espace-temps à 5 dimensions de Kaluza.

6. LOCALISATION DES ETATS STATISTIQUES

Soit μ un état statistique d'un système dynamique ; c'est-à-dire, nous l'avons vu, une loi de probabilité de la variété U des mouvements (fig.V).

Si $X \mapsto \delta g$ est une variation à support compact des potentiels de gravitation , nous savons faire correspondre à chaque mouvement $x \in U$ un hamiltonien de diffusion $\varphi = \mathcal{E}_x(X \mapsto \delta g)$; φ est une fonction de x , c'est-à-dire une variable dynamique ; nous dirons que l'état μ est <u>localisable</u> si cette fonction est μ-intégrable pour tout $X \mapsto \delta g$; nous poserons alors

$$(6.1) \qquad \mathcal{E}_\mu(X \mapsto \delta g) = \int_U \mathcal{E}_x(X \mapsto \delta g)\, \mu(x)\, dx$$

cette quantité (6.1) est la <u>valeur moyenne</u> de φ dans l'état μ .

On vérifie immédiatement que :

(6.2) L'ensemble des états statistiques localisables est un <u>sous-convexe</u> de Prob(U), qui contient ses points extrêmaux ;

(6.3) Si μ est localisable, $\mathcal{E}_\mu$ est une <u>distribution eulérienne</u> ;

(6.4) Dans le cas de l'espace de Minkowski, l'élément $\Psi(\mathcal{E}_\mu)$ de $\mathcal{G}^*$ (procédure du §5) est égal à la valeur moyenne Q de M dans l'état μ (voir la figure V).

Il semble que les états localisables soient les seuls que l'on rencontre dans la nature ; notamment, les états de Gibbs sont localisables. Si l'état μ possède une fonction de distribution de classe C^{∞} (sur U) $\mathcal{E}_{\mu}$ sera une distribution complètement continue (sur E_4) ; ce qui s'écrira

$$(6.5) \qquad \mathcal{E}_{\mu}(X \mapsto \delta g) = \int_{E_4} \frac{1}{2} \delta g_{\nu\rho} \, \mathcal{E}^{\nu\rho}(X) \, dX$$

les $\mathcal{E}^{\nu\rho}$ étant des densités ($\mathcal{E}^{\nu\rho} = \mathcal{E}^{\rho\nu}$) ; ces $\mathcal{E}^{\nu\rho}$ sont les composantes d'une densité tensorielle, au sens de Brillouin ; on peut aussi les écrire $T^{\nu\rho} \lambda$, les $T^{\nu\rho}$ étant les composantes d'un tenseur symétrique et λ la densité riemannienne de l'espace-temps ; alors (6.5) devient

$$(6.6) \qquad \mathcal{E}_{\mu}(X \mapsto \delta g) = \int_{E_4} \frac{1}{2} T^{\nu\rho} \delta g_{\nu\rho} \lambda(X) \, dX$$

la condition eulérienne (5. 1) s'obtient par utilisation de la formule de Killing

$$(6.7) \qquad [\delta_L g]_{\alpha\beta} = g_{\alpha\gamma} \hat{\partial}_{\beta} \delta X^{\gamma} + g_{\gamma\beta} \hat{\partial}_{\alpha} \delta X^{\gamma}$$

on trouve facilement

$$(6.8) \qquad \hat{\partial}_{\nu} T^{\nu\rho} = 0$$

où l'on reconnaît l'écriture relativiste des équations d'Euler proposée par Einstein [II].

Par conséquent la localisation d'un état statistique permet d'interpréter celui-ci à l'aide d'un milieu continu, dont les $T^{\nu\rho}$ définis par (6.6) constituent le tenseur d'énergie, et sont automatiquement solutions des équations d'Euler. Cette interprétation peut se confirmer par un calcul détaillé ; ainsi, dans le cas d'une particuler, la composante T^{00}, qui s'interprète comme masse spécifique, est la valeur moyenne de la masse relativiste $m/\sqrt{1-v^2/c^2}$ dans un élément de volume au voisinage du point X considéré ; la pression, ou plus généralement le tenseur de contraintes s'interprète comme une mesure du caractère aléatoire des vitesses des mouvements qui passent par X ; etc.

- Traitons l'exemple d'un système de N particules relativistes sans spin, de masse m, constituant un équilibre de Gibbs dans une boîte de volume V , à la température T . Dans un référentiel lié à la boîte, le tenseur $T^{\nu\rho}$ est diagonal, et s'exprime au moyen d'une masse spécifique ρ et d'une pression p données par

(6.9) $\rho = \frac{Nm}{V}\,\frac{G''(x)}{-G'(x)}$ $\qquad$ $p = \frac{Nm}{V}\,\frac{G''(x) - G(x)}{-3\,G'(x)}$

où l'on a posé $x = m/kT$ et défini la fonction G par

(6.10) $$G(x) = \int_1^{\infty} e^{-xs}\sqrt{s^2-1}\,ds\,.$$

Il se trouve que $G(x) = \frac{K_1(x)}{x}$, K_1 étant la fonction de Bessel modifiée d'indice 1 ; G vérifie donc l'équation différentielle

(6.11) $$G''(x) + \frac{3}{x}\,G'(x) - G(x) = 0$$

qui entraîne

(6.12) $$pV = NkT\ ;$$

on retrouve donc exactement la loi classique des gaz parfaits (Boyle-Mariotte-Charles-Gay-Lussac-Avogrado-Kelvin-Boltzmann) ; les premiers termes du développement asymptotique de K_1 [XVI] fournissent la formule

(6.13) $$Q = \rho V = Nm + \frac{3}{2}NkT + \frac{Nm}{-2G'(m/kT)}\int_1^{\infty} e^{-ms/kT}(s-1)^{5/2}(s+1)^{1/2}\,ds.$$

On a fait $c = 1$; le premier terme est la masse au zéro absolu ; le second est la valeur moyenne classique de l'énergie (qui permet de calculer la capacité calorifique du gaz monoatomique) ; le troisième, positif, donne la correction relativiste.

On obtient simultanément le potentiel thermodynamique de Planck (4.4) :

(6.14) $$z = N\,\mathrm{Log}\big(-4\pi\,V\,m^3\,G'(m/kT)\big)$$

c'est effectivement une fonction convexe de $\beta = 1/kT$; la formule (4.5) donne ensuite l'entropie [1]

(6.15) $$S = kz + \frac{Q}{T}$$

(Cf. [XIV]).

---

[1] L'entropie usuelle est le produit par k de celle que nous avons utilisée ici. On peut toujours choisir une unité de température telle que $k = 1$.

- Il est remarquable que le tenseur $T^{\nu\rho}$, que nous avons construit comme caractéristique de la susceptibilité gravitationnelle, caractérise aussi l'action gravitationnelle de la matière définie par l'état statistique μ ; c'est en effet lui qui figure au second membre des équations d'Einstein de la gravitation

$$(6.16) \qquad R^{\nu\rho} - \frac{1}{2} R g^{\nu\rho} + \Lambda\, g^{\nu\rho} = 8\pi G\, T^{\nu\rho}$$

(Λ = constante cosmologique ; G = constante de Newton ; $c = 1$) ; en utilisant la définition (6.1), celles-ci s'écrivent

$$(6.17) \qquad \int_{E_4} \delta\mathcal{L}(X)\, dX = \int_U \mathcal{E}_x(X \mapsto \delta g)\, \mu(x)\, dx \qquad \forall \delta g$$

$\mathcal{L}$ désignant la densité lagrangienne du champ de gravitation

$$(6.18) \qquad \mathcal{L} = \frac{2\Lambda - g^{\nu\rho} R_{\nu\rho}}{8\pi G}\, \lambda \qquad (\lambda = \text{densité riemannienne})$$

sous cette forme, on remarque que la distribution définie par le premier membre est automatiquement eulérienne.

Toutes ces considérations s'étendent au cas électromagnétique ; les formules (6.6), (6.8) sont alors remplacées respectivement par

$$(6.19) \qquad \mathcal{E}_\mu(X \mapsto (\delta g, \delta A)) = \int_{E_4} \left[\frac{1}{2} T^{\nu\rho} \delta g_{\nu\rho} + J^\sigma \delta A_\sigma \right] \lambda(X)\, dX$$

et

$$(6.20) \qquad \hat{\partial}_\nu T^{\nu\rho} + F_\nu{}^\rho J^\nu = 0 \qquad\qquad \hat{\partial}_\sigma J^\sigma = 0$$

et les équations d'Einstein (6.16) par les équations couplées d'Einstein-Maxwell ; le quadri-vecteur J^σ s'interprète comme densité de courant-charge. L'application des formules précédentes aux états statistiques de particules à spin munies de moment magnétique permet de retrouver les principales caractéristiques du ferro-magnétisme (équivalence magnétique aimant-solénoïde ; en effet gyro-magnétique ; magnétostriction) [Voir [XII] et [XIV]].

7. INTERPRETATION GRAVITATIONNELLE DU PREMIER PRINCIPE

Plaçons-nous dans le cas d'une transition dissipative $\mu_{in} \to \mu_{out}$, et supposons qu'il existe une distribution eulérienne $\mathcal{E}$ qui coïncide avec μ_{in} avant les phénomènes dissipatifs, et avec μ_{out} après ; en d'autres mots, qui interpole μ_{in} et μ_{out}.

Nous saurons alors associer à $\mathcal{E}$ une grandeur conservée Q, qui prendra la même valeur que pour $\mathcal{E}_{\mu_{in}}$ ou pour $\mathcal{E}_{\mu_{out}}$ (puisqu'on peut la calculer à une date arbitraire) ; nous savons aussi que pour $\mathcal{E}_{\mu_{in}}$, Q est égale à la valeur moyenne du moment M de Poincaré ; de même pour $\mathcal{E}_{\mu_{out}}$; par conséquent le premier principe, sous sa forme covariante (4.2), sera assuré.

Or il suffit de poser (notation (6.17)) :

$$\mathcal{E}(X \mapsto \delta g) = \int_{E_4} \delta\mathcal{L}(X)\, dX \; ; \tag{7.1}$$

nous savons en effet que $\mathcal{E}$ est une distribution eulérienne, et qu'elle interpole $\mathcal{E}_{\mu_{in}}$ et $\mathcal{E}_{\mu_{out}}$, puisque les équations d'Einstein (6.16) sont valables avant et après les phénomènes dissipatifs ([1]).

Ce sont donc les potentiels de gravitation $g_{\nu\rho}$ qui mémorisent la valeur moyenne du moment M, et assurent la validité du premier principe (sous sa forme covariante (4.2)).

([1]) On remarquera que ce raisonnement utilise implicitement une approximation ; d'une part on se place en relativité restreinte pour construire les moments de Poincaré ; d'autre part on considère les $g_{\nu\rho}$ comme variables, puisqu'ils donnent par dérivation les $T^{\nu\rho}$ grâce aux équations d'Einstein. Ce niveau d'approximation revient à considérer la constante G comme petite, et à négliger la self-interaction gravitationnelle du système ; c'est l'usage en thermodynamique.

8. UN MODELE DE SYSTEME DISSIPATIF

Le problème d'une géométrisation analogue du second principe est encore loin d'être résolu ; une voie pour y parvenir serait la construction d'un quadrivecteur "flux d'entropie" S^μ, vérifiant

$$\hat{\partial}_\mu S^\mu \geqslant 0 \tag{8.1}$$

et qui, dans les états statistiques -ou au moins les états de Gibbs- vérifierait $\hat{\partial}_\mu S^\mu = 0$ et dont le flux -conservatif- à travers une hypersurface du genre espace coïnciderait avec l'entropie. Cette hypothèse a été envisagée par de nombreux auteurs (voir par exemple [V]).

Nous nous contenterons ici de construire un modèle phénoménologique de fluide dissipatif, comportant un tel vecteur, et de comparer son comportement à celui des fluides réels.

Le modèle est fondé sur les hypothèses suivantes [XIII] :

A) Existence, en tout point, d'un vecteur-température Θ ; nous poserons

$$\Theta = \beta\, U \tag{8.2}$$

U étant un quadrivecteur unitaire de genre futur (la "quadrivitesse du fluide"), $\beta = 1/kT$ la température réciproque.

Contrairement au cas des états de Gibbs (§4), Θ ne sera pas nécessairement un vecteur de Killing ; nous poserons

$$\gamma = \frac{1}{2}\, \delta_L g \qquad \text{pour} \qquad \delta X = \Theta \tag{8.3}$$

γ est un tenseur symétrique dont les composantes sont données par la formule de Killing (6.7).

B) Conservation du "nombre de particules" ; ceci étant matérialisé par un quadrivecteur N^μ, vérifiant

$$\hat{\partial}_\mu N^\mu = 0 \tag{8.4}$$

dont le flux est donc conservatif. Nous supposerons N parallèle à Θ en écrivant

(8.5) $$N^{\mu} = U^{\mu} n \qquad n \geqslant 0 .$$

Dans le "référentiel propre du fluide", défini par le vecteur U , n s'interprêtera comme valeur moyenne du nombre de particules par unité de volume.

C) Existence d'un "potentiel d'énergie" φ ; φ sera une fonction des variables

$$U , \beta , n , \gamma$$

telle que

(8.6) $$T^{\mu\nu} = \frac{\partial \varphi}{\partial \gamma_{\mu\nu}} \qquad (^1)$$

$T^{\mu\nu}$ étant le tenseur d'énergie du fluide -celui qui figure au second membre des équations d'Einstein ; il vérifie donc les équations

(8.7) $$\hat{\partial}_{\mu} T^{\mu\nu} = 0 .$$

D) Existence d'un "potentiel thermodynamique" ζ ; ζ est une fonction des variables

$$\beta , n$$

qui définit le vecteur flux d'entropie S^{μ} par la formule

(8.8) $$S^{\mu} = N^{\mu} \zeta + T^{\mu\nu} \Theta_{\nu}$$

- Il se trouve (voir [XIV]) que la condition (8.1)

(8.9) $$\hat{\partial}_{\mu} S^{\mu} \geqslant 0$$

est assurée si les potentiels φ et ζ vérifient les deux conditions

(8.10) $$\gamma = 0 \Rightarrow T^{\mu\nu} = \frac{n^2}{\beta} \frac{\partial \zeta}{\partial n} \left[g^{\mu\nu} - U^{\mu} U^{\nu} \right] - n \frac{\partial \zeta}{\partial \beta} U^{\mu} U^{\nu} ;$$

---

(1) Abus de notations signifiant :
$$T^{\mu\nu} = T^{\nu\mu} ; \quad \delta\varphi = T^{\mu\nu} \delta\gamma_{\mu\nu} \quad \text{si} \quad \delta U = 0, \ \delta\beta = 0, \ \delta n = 0.$$

(8.11) φ est une fonction convexe de γ [1]

qui seront donc prisescomme hypothèses.

Nous allons étudier trois types de mouvements du fluide.

I) <u>Mouvements non dissipatifs</u> (relativité restreinte)

Si φ est strictement convexe en γ, on peut montrer que les mouvements non dissipatifs ($\hat{\partial}_\mu S^\mu = 0$) n'existent que si $\gamma = 0$ -donc si le vecteur température est un vecteur de Killing. Dans l'hypothèse de la relativité restreinte, chacun de ces vecteurs est donné par un élément de l'algèbre de Lie du groupe de Poincaré ; ce qui définit en tout point X les variables U et β ; le mouvement sera donc complètement déterminé si on connaît la seule fonction $X \mapsto n$, pour laquelle on dispose des 5 équations (8.4), (8.7), $T^{\mu\nu}$ étant donné par (8.10).

Il se trouve -grâce à l'existence d'un facteur intégrant- que ces 5 équations sont compatibles, et s'intègrent en

(8.12) n constant sur les lignes de courant [2]

(8.13) $n\frac{\partial\zeta}{\partial n} + \zeta$ constant dans l'espace-temps.

Alors le vecteur flux d'entropie est donné par

(8.14) $$S^\mu = N^\mu s \quad , \text{ avec} \quad s = \zeta - \beta\frac{\partial\zeta}{\partial\beta} \ ;$$

la valeur de $T^{\mu\nu}$, donnée par (8.10), s'interprête dans le référentiel propre par <u>une masse spécifique</u> ρ et une <u>pression</u> p

(8.15) $$\rho = - n\frac{\partial\zeta}{\partial\beta} \qquad p = -\frac{n^2}{\beta}\frac{\partial\zeta}{\partial n} \ .$$

Les formules ci-dessus montrent que l'on peut interpréter ζ comme le <u>potentiel</u>

---

[1] Cette seconde condition n'est pas nécessaire pour tout γ ; elle l'est cependant au voisinage de $\gamma = 0$.

[2] Ce sont les courbes tangentes en tout point à U.

de Planck par particule ([1]). En particulier, si on fait le choix correspondant au modèle de gaz monoatomique (§6)

$$(8.16) \qquad = \mathrm{Log}\left(\frac{-G'(m\beta)}{n}\right) + \mathrm{Cte}$$

G étant la fonction définie par (6.10), on constate que l'équation (8.13) détermine n en tout point de E_4 à partir de sa valeur en un point, et que les valeurs de toutes les variables coïncident avec celles qui sont données par les états de Gibbs généralisés pour un système de particules sans spin de même masse m : les mouvements non dissipatifs décrits par le modèle sont donc conformes aux prédictions de la mécanique statistique.

- D'autres choix phénoménologiques sont possibles pour la fonction $\zeta(\beta,n)$; si ζ n'est pas concave par rapport au volume spécifique

$$(8.17) \qquad u = \frac{1}{n}$$

il peut exister des surfaces de discontinuité de n ; en traversant cette surface, les deux variables p et $n\frac{\partial\zeta}{\partial n}+\zeta$ sont continues (équation (8.7), prise au sens des distributions comme au §6, et équation (8.13)) ; on trouve les conditions classiques de transition de phase (équilibre liquide-vapeur).

II. Approximation du fluide parfait

On négligera les phénomènes dissipatifs en remplaçant la fonction par son développement limité au premier ordre (c'est une fonction affine, donc convexe). Alors :

- la production d'entropie $\hat{\partial}_\mu S^\mu$ est nulle ;
- le flux d'entropie conserve la valeur (8.14) ;
- le tenseur d'énergie $T^{\mu\nu}$ a la même valeur (8.10) que pour $\gamma = 0$.
- les équations (8.4) et (8.7) sont celles d'un fluide parfait (voir [IX]).

Si on introduit la 1-forme

---

[1] Cette possibilité de répartir le potentiel entre les particules suppose qu'il n'existe pas d'effet d'échelle -donc qu'on a négligé des phénomènes comme la capillarité. Nous sommes donc dans les conditions de la limite thermodynamique .

(8.18) $$H_\lambda = h\,U_\lambda \quad , \text{ avec } \quad h = \frac{p+\rho}{n}$$ (1)

et sa dérivée extérieure

(8.19) $$\Omega_{\lambda\mu} = \partial_\lambda H_\mu - \partial_\mu H_\lambda$$

on peut remplacer les équations du mouvement (8.4), (8.7) par

(8.20) $$\hat{\partial}_\lambda N^\lambda = 0 \quad , \quad \Omega_{\lambda\mu}\Theta^\mu + \partial_\lambda s = 0 \, .$$

Cette dernière équation montre que

(8.21) $$\delta s = 0 \quad \text{et} \quad \delta_L \Omega = 0 \quad \text{pour} \quad \delta X = \Theta$$

il en résulte que s est constant sur chaque ligne de courant, et que Ω est un invariant intégral du champ $X \mapsto \Theta$; son rang (4, 2 ou 0) est donc constant sur chaque ligne de courant, ainsi que le pseudo-scalaire

(8.22) $$\pi = \mathrm{pf}(\Omega)\,\frac{\rho}{n}$$

où $\mathrm{pf}(\Omega)$ désigne le pfaffien de la forme (2).

En général, $\pi \neq 0$, Ω est de rang 4 , le signe de π définit une orientation de l'espace. Il existe aussi des classes importantes de mouvements dans lesquelles $\pi = 0$:

- Les mouvements isentropiques (ceux où s prend la même valeur sur toutes les lignes de courant) ; alors (8.20) montre que $\Theta \in \ker(\Omega)$; en général le rang de Ω est 2 ; le noyau de Ω définit un feuilletage dont les feuilles, de dimension 2, s'interprêtent comme lignes de tourbillon emportées par le fluide. Ces mouvements sont barotropes : il existe une équation d'état, indexée par la va-

---

(1) A cause de l'équivalence relativiste entre masse spécifique et énergie spécifique (on a supposé $c = 1$) , h s'interprête comme enthalpie par particule.

(2) Ce pfaffien est défini par la relation $\frac{1}{2}\Omega \wedge \Omega = \mathrm{pf}(\Omega)\,\mathrm{vol}$, où vol désigne la 4-forme volume riemannien définie à l'aide d'une orientation de E_4 . π est l'équivalent relativiste du potentiel tourbillon de Ertel [III], [IV].

leur de s, obtenue par élimination de β et n entre p et ρ ; l'enthalpie particulaire h (8.18) coïncide avec l'indice défini par Lichnerowicz [IX].

- Les mouvements non isentropiques dans lesquels rang(Ω) = 2 (il suffit que ce soit vrai à une date arbitraire) ; ils constituent l'équivalent relativiste des mouvements <u>oligotropes</u> de Casal [I] ; les feuilles de Ω sont tracées sur les hypersurfaces s = Cte.

- les mouvements où $\Omega = 0$ (là aussi il suffit de le vérifier à une date arbitraire) ; (8.20) montre qu'ils sont isentropiques ; ce sont les mouvements <u>irrotationnels</u>, au sens de Lichnerowicz [IX].

- Il existe des solutions comportant des discontinuités sur une hypersurface Σ de E_4 (<u>ondes de choc</u>) ; les conditions que l'on obtient en écrivant les équations (8.4) et (8.5) au sens des distributions sont les suivantes [1]

$$(8.23)\qquad \begin{array}{lll} N'^{\lambda} - N^{\lambda} & \text{est tangent à } \Sigma \\ H'_{\lambda} - H_{\lambda} & \text{est normal à } \Sigma \\ h'^2 - h^2 & = [u'h' + uh][p'-p] \end{array}$$

si on ajoute que la discontinuité $s'-s$ est positive, on obtient les équations de choc de Rankine-Hugoniot, sous leur forme relativiste.

III. Mouvements faiblement dissipatifs.

On fait cette fois un développement au second ordre de $\gamma \mapsto \varphi$ au voisinage de $\gamma = 0$; on ajoute donc à la fonction affine du cas précédent une forme quadratique positive définie ; cette forme quadratique s'interprête comme <u>fonction de dissipation</u> (en un sens plus large que le sens usuel). Ce que nous écrirons

$$(8.24)\qquad \varphi = \varphi_0 + \overset{\circ}{T}{}^{\lambda\mu}\gamma_{\lambda\mu} + \frac{1}{2} C^{\lambda\mu,\nu\rho}\gamma_{\lambda\mu}\gamma_{\nu\rho}$$

Les $\overset{\circ}{T}{}^{\lambda\mu}$ ayant la valeur donnée en (8.10) ; d'où, grâce à (8.6)

$$(8.25)\qquad T^{\lambda\mu} = \overset{\circ}{T}{}^{\lambda\mu} + C^{\lambda\mu,\nu\rho}\gamma_{\nu\rho} .$$

---

[1] Les variables portant le signe ' sont prises après le choc.

La production d'entropie est

(8.26) $$\hat{\partial}_\lambda S^\lambda = C^{\lambda\mu,\nu\rho}\, \gamma_{\lambda\mu}\, \gamma_{\nu\rho}$$

donc le double de la fonction de dissipation ; la symétrie

(8.27) $$C^{\lambda\mu,\nu\rho} = C^{\nu\rho,\lambda\mu}$$

exprime la réciprocité d'Onsager (le "courant généralisé" $T - \mathring{T}$ est, grâce à (8.25), fonction linéaire de la "force généralisée" γ ; les $C^{\lambda\mu,\nu\rho}$ sont les coefficients de transport).

A priori, il existe 55 coefficients $C^{\lambda\mu,\nu\rho}$ indépendants ; mais la symétrie du fluide réduit ce nombre à 5 ([1]) ; on interprête le résultat (8.25) en se plaçant en relativité restreinte et en calculant les composantes $T^{\lambda\mu}$ dans le référentiel propre du fluide. On obtient ainsi, avec 5 coefficients A , B , C , E , F :

a) Le tenseur de contraintes

(8.28) $$\boxed{\tau_{jk} = -T_{jk} = \delta_{jk}\Big[-p + \Big[A - \frac{2E}{3}\Big]\partial_\ell v^\ell - B\frac{\partial\beta}{\partial t}\Big] + \beta E\big[\partial_j v_k + \partial_k v_j\big]}$$

(j , k, = 1, 2,3 ; v_j est la vitesse, nulle au point considéré).

b) Le flux de chaleur, dont les composantes sont les T^{jo} :

(8.29) $$\boxed{F\Big[\overrightarrow{\text{grad}}\,\beta - \beta\frac{\partial\vec{v}}{\partial t}\Big]}$$

c) La masse-énergie spécifique

(8.30) $$\boxed{T^{oo} = \rho + C\frac{\partial\beta}{\partial t} - B\beta\,\mathrm{div}\,\vec{v}}$$

([1]) On écrit que $\gamma \mapsto \varphi$ est invariant par la stabilisateur de U dans le groupe de Lorentz (c'est-à-dire le groupe des rotations dans le référentiel propre). Sans la réciprocité d'Onsager, il y aurait 6 coefficients indépendants.

A, B, C, E, F sont, à priori, des fonctions de β et n ; la convexité stricte de φ s'exprime par les inégalités

$$A > 0 \ , \quad C > 0 \ , \quad E > 0 \ , \quad F > 0 \ , \quad |B| < \sqrt{AC} \ . \tag{8.31}$$

L'interprétation est claire : on retrouve la description de Navier-Stokes de la viscosité, avec les deux coefficients

$$\lambda = \left[A - \frac{2E}{3}\right]\beta \qquad , \qquad \mu = E\beta \tag{8.32}$$

fonctions de la température et du volume spécifique, qui vérifient grâce à (8.31)

$$\mu > 0 \quad , \quad 3\lambda + 2\mu > 0 \ ; \tag{8.33}$$

la formule (8.29), où on rappelle que $\beta = 1/kT$, donne la conductivité thermique

$$\frac{F}{kT^2} \tag{8.34}$$

nécessairement positive ; le second terme de (8.29) est une correction relativiste [1].

Les coefficients C et B (ce dernier pouvant être nul) ont une interprétation plus délicate : ils modifient la conduction de la chaleur -en rendant elliptique le système des équations. C est peut-être mesurable expérimentalement.

---

[1] Ce terme montre qu'il peut exister des équilibres non isothermes : ce qui est évident à priori en mécanique statistique, parce que les vecteurs de Killing de Minkowski qui ne sont pas constants n'ont pas une longueur constante. Ce qui montre en particulier que la température propre, dans une centrifugeuse, est plus grande à la périphérie qu'au bord. Cet effet est probablement sensible dans le cas des pulsars, dans l'atmosphère desquels la vitesse d'entraînement peut approcher celle de la lumière.

CONCLUSION

Ce modèle est très schématique -mais cependant apte à décrire des comportements très divers des fluides réels. On peut donc penser que les méthodes géométriques peuvent contribuer aux progrès de la connaissance des phénomènes dissipatifs.

REFERENCES :

I	P.CASAL	Principes variationnels en fluide compressible et en magnétodynamique des fluides. Journ. de Mécanique, 5,2 (1966) p.149.
II	A.EINSTEIN	The Meaning of Relativity. Methuen, London (1922).
III	H.ERTEL	Meteorel.Ztschr.,59 (1942), p.277-385.
IV	B.L.GAVRILINE,M.M.ZASLAVSKII	Dauk.Ak.Nauk SSSR,192 , 1 (1970) p.48
V	W.ISRAEL	Nonstationary irreversible thermodynamics : a causal relativistic theory. O.A.P. 444, Caltech (1976).
VI	A.I.KHINCHIN	Mathematical Foundations of Statistical Mechanics , Dover , New York (1949).
VII	L.LANDAU,E.LIFCHITZ	Statistical Physics. Addison-Wesley (1969).
VIII	" "	Fluid Mechanics. Addison-Wesley (1958).
IX	A.LICHNEROWICZ	Théories relativistes de la gravitation et de l'électromagnétisme, Masson, Paris (1955).
X	J.M.SOURIAU	Définition covariante des équilibres thermodynamiques, Supp. Nuovo Cimento, 1, I , 4 (1966) p.203-216.
XI	"	Structure des systèmes dynamiques, Dunod, Paris (1969).
XII	"	Modèle de particule à spin dans le champ électromagnétique et gravitationnel, Ann.Inst.Henri Poincaré, XX,4 (1974)p.315-364
XIII	"	Thermodynamique relativiste des fluides. Rend.Sem.Mat., Univers.Politecn.Torino,35 (1976-77), p.21-34.
XIV	"	Structure of Dynamical Systems, North-Holland,Amsterdam (à paraître).
XV	J.L.SYNGE	The Relativistic Gas, Amsterdam (1957).
XVI	G.N.WATSON	A treatise of the Theory of essel Functions (1944).

SOME PRELIMINARY REMARKS ON THE FORMAL VARIATIONAL CALCULUS OF GEL'FAND AND DIKII

by

Shlomo Sternberg

This note consists of some comments on the variational calculus introduced by Gel'fand and Dikii in their study of the Korteweg-DeVries equation, [1]. In [1] an algebraic approach is used, introducing such concepts as differential rings and requiring the various functions to be polynomials. In this note we derive some of their formulas using a more geometrical approach, mainly a modification of the Hamilton Cartan formalism as developed in [2]. I obtained the results presented here about two years ago, but encountered some difficulties in extending them to the case of several independent variables, and put the problem aside. At the request of Prof. Gel'fand I am publishing these results here in their original form, and apologize to the reader for the fact that they are still incomplete. I want to thank Prof. Kazhdan for helpful conversations.

1. Notational set up. Let X be a differentiable manifold, for the time being of arbitrary dimension. (We shall, unfortunately, at the crucial juncture specialize to the case $\dim X = 1$.) We let $Y \to X$ denote a fibered manifold over X. We shall let $J \to X$ denote the bundle of infinite jets of sections of Y, regarded as an "inverse limit" of the manifolds J_k of k - jets of sections of Y. All functions, differential forms, etc. are pull backs of functions and forms defined on the finite jet bundles J_k. A vector field on J will be a derivation of the ring of functions and hence will have infinitely many "components" in a local coordinate system.

Any section s of Y over X gives rise to a section js of J over X. But not every section of J is the infinite jet of a section of Y. There is a simple differential condition on sections, u, of J which determines whether or not $u = js$ for some section s of Y. Indeed, cf. [2], there exists a "connection" ω, i.e. a section of $\mathrm{Hom}(TJ, TJ)$, defined on J such that

1) $\qquad\qquad \omega(\zeta) = \zeta$ if ζ is a vertical tangent vector

and

2) $\qquad\qquad u^*\omega = 0$ iff $u = js$.

We obtain an ideal, I, in the ring of differential forms on J consisting of all $\omega \overline{\wedge} \Omega$ where Ω is an arbitrary form. This ideal I satisfies $dI \subset I$ and so defines a "foliation" on J whose "leaves" are of $\dim X$. Of course there are no actual leaves but each vector field on X defines a unique vector field on J which is annihilated by ω. If $x^1, \cdots, x^n$ are coordinates on X so that $\frac{\partial}{\partial x^1}, \cdots, \frac{\partial}{\partial x^n}$ is a local basis for the vector fields on X, we shall denote the corresponding vector fields on J by $\frac{d}{dx^1}, \cdots, \frac{d}{dx^n}$. In particular

$$\left[\frac{d}{dx^i}, \frac{d}{dx^j}\right] = 0 \quad .$$

We let V denote the algebra of vertical vector fields, ξ, which preserve I. Thus $\xi \in V$ iff $D_\xi \Omega \in I$ for all $\Omega \in I$. This implies that

$$\left[\xi, \frac{d}{dx^j}\right] = a_1 \frac{d}{dx^1} + \cdots + a_n \frac{d}{dx^n} \quad .$$

If f is a function on X and ρ^*f denotes the corresponding function on J then $\frac{d}{dx^i}\rho^*f = \rho^*(\partial f/\partial x^i)$ while $\xi\rho^*f = 0$ for any vertical vector field ξ. Taking $f = x^k$ in the preceeding equation shows that $a_k = 0$. Thus

PROPOSITION 1. 1. <u>For any vertical vector field</u> ξ, <u>we have</u> $\xi \in V$ <u>if and only if</u>

$$\left[\xi, \frac{d}{dx^i}\right] = 0 \quad \text{for all } i \quad .$$

We now reduce to the case $n = \dim X = 1$.

We let x be a coordinate on X and u, v, etc. fiber coordinates. (We shall usually write the formulas as if we had only one fiber coordinate but this is irrelevant for much of what follows.) Coordinates on J will be x, u, v, u', v, u'', etc. The ideal I is generated by the forms

$$du - u'dx \ , \quad du' - u''dx \ , \quad \text{etc.}$$

PROPOSITION 1. 2. <u>The ideal</u> I <u>contains no non-trivial exact one forms; i. e. if</u> $df \in I$ <u>then</u> $df = 0$.

<u>Proof</u>: Write $df = a_0(du - u'dx) + \cdots + a_n(du^{(n)} - u^{(n+1)}dx)$ where a_n is the highest non-vanishing coefficient. Then $\partial f/\partial u^{(k)} = 0$ for $k > n$ so $\partial a_i/\partial u^{(n+1)} = \partial^2 f/\partial u^{(i)}\partial u^{(n+1)} = 0$ for all this. But then $\partial^2 f/\partial x\, \partial u^{(n+1)} = -a_n = 0$, contradicting the assumption that $a_n \neq 0$.

2. <u>The momenta and the variational operator.</u> We call a tangent vector, ζ, <u>very vertical</u> if $d\pi(\zeta) = 0$, where π denotes the projection of J onto Y. Thus $\partial/\partial u'$, $\partial/\partial u''$ etc. are very vertical vector fields.

THEOREM 2. 1. <u>For any one form</u> ω <u>there exists a unique one form</u> θ_ω <u>such that</u>

i) $$\theta_\omega \equiv \omega \mod I$$

and

ii) $\qquad \zeta \rfloor d\theta_\omega \equiv 0 \mod I$ for all very vertical vectors fields, ζ .

Proof: Uniqueness. Suppose that $\theta = a_0(du - u'dx) + \cdots + a_n(du^{(n)} - u^{(n+1)}dx)$. Then

$$(\partial/\partial u^{(n+1)}) \rfloor d\theta \equiv - a_n dx \mod I \quad .$$

Thus for the right-hand side to lie in I we must have $a_n = 0$. Proceeding recursively we see that $\zeta \rfloor d\theta \equiv 0 \mod I$ and $\theta \equiv 0 \mod I$ implies that $\theta =$

Existence: Let us write

$$\theta = Ldx + p_0(du - u'dx) + \cdots + p_n(du^{(n)} - u^{(n+1)}dx)$$

where

$$\omega \equiv Ldx \mod I$$

and the p_i are to be determined. Then

$$(\partial/\partial u^{(i)}) \rfloor d\theta \equiv (\partial L/\partial u^{(i)})dx - dp_i - p_{i-1}dx \quad .$$

For $i > 0$ we want the right-hand side to belong to I, which is the same as saying that the right-hand side is annihilated by the vector field $\frac{d}{dx}$. Applying this vector field we obtain the recursion relations

$$\frac{dp_i}{dx} + p_{i-1} = \partial L/\partial u^{(i)}$$

which have the explicit solution

$$p_i = \sum_{j=0}^{\infty} (-1)^j \left(\frac{d}{dx}\right)^j \frac{\partial L}{\partial u^{(i+j+1)}} \quad . \tag{2.1}$$

Compare with Section 4 in [1]. Since L depends on only finitely many of the variable $u^{(k)}$ the sum on the right is finite and vanishes for sufficiently large i. This proves the existence of θ_ω.

If $\omega = Ldx \mod I$ we define

$$\frac{\delta L}{\delta u} = \frac{d}{dx} \rfloor \frac{\partial}{\partial u} \rfloor d\theta_\omega \quad . \tag{2.2}$$

From the above formula for θ_ω we obtain

$$(\partial/\partial u) \rfloor d\theta_\omega \equiv (\partial L/\partial u)dx - dp_0$$

so

$$\frac{\delta L}{\delta u} = \sum_{j=0}^{\infty} (-1)^j \left(\frac{d}{dx}\right)^j \frac{\partial L}{\partial u^{(j)}} \quad , \tag{2.3}$$

which is the formula (10) given in [1].

It follows as an immediate corollary of the theorem that

$$\theta_{df} = df \quad \text{(by the uniqueness and } d(df) = 0)$$

and, conversely, $d\theta = 0$ iff $\theta = df$ (locally) since we may apply the Poincaré lemma on the finite dimensional manifold $J_k(Y)$. Thus

$$\theta_{\omega_1} = \theta_{\omega_2} \quad \text{iff} \quad \omega_1 \equiv \omega_2 \mod I$$

and

$$d\theta_{\omega_1} = d\theta_{\omega_2} \quad \text{iff} \quad \omega_1 \equiv \omega_2 + df \mod I \quad \text{for some } f \quad .$$

Notice that if s and r are sections of Y over X which agree outside some region of compact support, K, then

$$\int_D (jr)^* \omega_1 - \int_D (js)^* \omega_1 = \int_D (jr)^* \omega_2 - \int_D (js)^* \omega_2$$

if $\omega_1 \equiv \omega_2 + df \mod I$, and if $K \subset \operatorname{int} D$.

For each compact set K, we call s_t a K variation of s if s_t is a one parameter family of sections of Y over X such that $s_0 = s$ and $s_t = s$ outside K. A section s is called an extremal of ω if for any domain D and any $K \subset \operatorname{int} D$ we have

$$\frac{d}{dt} \int (js_t)^* \omega = 0 \quad \text{at} \quad t = 0$$

for all K variations. From the preceeding remarks we see that the set of extremals of ω depends only on $d\theta_\omega$.

Suppose that s_t is a K variation of s. Then $\left.\frac{ds_t}{dt}\right|_{t=0}$ is a vector field along s, i.e. a smooth map which assigns a tangent vector in $TY_{s(x)}$ to each $x \in X$, and this tangent vector is vertical, i.e. has zero projection onto X. We will denote this vector field by $\hat{\xi}$. Similarly, the one parameter family of sections js_t of J gives rise to a vector field along js which we denote by ξ. Thus $d\pi(\xi_{js(x)}) = \hat{\xi}(s(x))$ where π denotes the projection of J onto Y. Notice that $\xi_{s(x)} = 0$ if $x \notin K$. For any D containing K in its interior we have

$$\frac{d}{dt} \int_D (js_t)^* \omega = \frac{d}{dt} \int_D (js_t)^* \theta_\omega = \int_D \{(js)^*(\xi \rfloor d\theta_\omega) + d(js)^*(\xi \rfloor \theta_\omega)\}$$

$$= \int_D (js)^*(\xi \rfloor d\theta_\omega)$$

where all derivatives are evaluated at $t = 0$. In view of the defining properties of θ this last integral depends only on $\hat{\xi}$ and depends linearly on $\hat{\xi}$. Thus, for example, when $n = 1$ we can define the variational differential δL of any function L on J by

$$\delta L(\hat{\xi}) = \frac{d}{dx} \rfloor \xi \rfloor d\theta_{Ldx} \quad .$$

If V denotes the vertical tangent vectors to Y considered as a vector bundle over J, then δL is well defined as a section of V^* over J, and, if $\omega \equiv Ldx \mod I$ we can write the last integral as

$$\int_D (js)^* \, \delta L(\hat{\xi}) \, dx \quad .$$

Since $\hat{\xi}$ can be an arbitrary vertical vector field of compact support, we conclude that $\delta L(js) = 0$ along extremals.

3. Invariants and regular Lagrangians. Let F be a function on J such that

$$dF \equiv \xi \rfloor d\theta_\omega \mod I$$

for some vertical vector field ξ on J. If s is an extremal of ω then

$$d(js)^*F = (js)^*(\xi \rfloor d\theta_\omega) = 0 \quad .$$

Thus F is a constant. Thus each such function F defines a function, f, on the set of extremals. The collection of functions F satisfying the above equation is clearly an algebra and the restriction map $F \to f$ is a homomorphism into the algebra of functions on the set of extremals. In general this homomorphism need not be surjective onto the algebra of "smooth" functions on the set of extremals. For example, if $d\theta_\omega = 0$, then every section is an extremal, but the only functions F satisfying the above equation are the constants. However under suitable non-degeneracy assumptions, we can conclude that the set of extremals is a finite dimensional symplectic manifold whose symplectic structure is induced from $d\theta_\omega$ in a way that we shall make precise below, and that the above homomorphism is surjective.

For example, suppose that the fiber has dimension one with local coordinate u, so that $x, u, u', u'', \cdots$ are local coordinates on J. Suppose that $L = L(x, u, u', u'', \cdots, u^{(n)})$. Then in the formula for $\frac{\delta L}{\delta u}$ the term involving the highest derivative of u will come from

$$(-1)^n \left(\frac{d}{dx}\right)^n \frac{\partial L}{\partial u^{(n)}} \quad .$$

Now

$$\frac{d}{dx}\frac{\partial L}{\partial u^{(n)}} = \left(\frac{\partial}{\partial x} + u'\frac{\partial}{\partial u} + \cdots + u^{(n+1)}\frac{\partial}{\partial u^{(n)}} + \cdots\right)\frac{\partial L}{\partial u^{(n)}}$$

$$= u^{(n+1)}\frac{\partial^2 L}{(\partial u^{(n)})^2} + (\text{terms involving } u^{(k)} \text{ with } k < n+1) \quad .$$

Continuing to apply $\frac{d}{dx}$ we see that

$$\frac{\delta L}{\delta u} = (-1)^n u^{(2n)}\frac{\partial^2 L}{(\partial u^{(n)})^2} + (\text{terms involving } u^{(k)} \text{ with } k < 2n) \quad .$$

If we make the non-degeneracy assumption

$$\frac{\partial^2 L}{(\partial u^{(n)})^2} \neq 0 \quad ,$$

then the equation $\frac{\delta L}{\delta u} = 0$ is a non-singular $2n$-th order ordinary equation and we may use the "initial values" $u, u', \cdots, u^{(2n-1)}$ at some particular x to parametrize the space of extremals. Let us write

$$\theta_{Ldx} = Hdx + p_0 du + p_1 du' + \cdots + p_n du^{(n)}$$

where the p_i are given as above and where $H = L - p_0 u' - p_1 u'' - \cdots - p_n u^{(n+1)}$. We wish to show that the matrix whose i, j-th entry is

$$(\partial/\partial u^{(i)}) \rfloor (\partial/\partial u^{(j)}) \rfloor d\theta_{Ldx} \qquad i, j = 0, 1, \cdots, 2n-1$$

is non-singular. In evaluating this matrix we can ignore $dH \wedge dx$ and we should also observe that

$$(\partial/\partial u^{(2n-1)}) \rfloor d\theta_{Ldx} = \frac{\partial p_0}{\partial u^{(2n-1)}} du = (-1)^{n-1}\frac{\partial^2 L}{(\partial u^{(n)})^2} du$$

$$(\partial/\partial u^{(2n-2)}) \rfloor d\theta_{Ldx} = \frac{\partial p_0}{\partial u^{(2n-2)}} du + \frac{\partial p_1}{\partial u^{(2n-2)}} du'$$

etc. Thus the (antisymmetric) matrix has the form

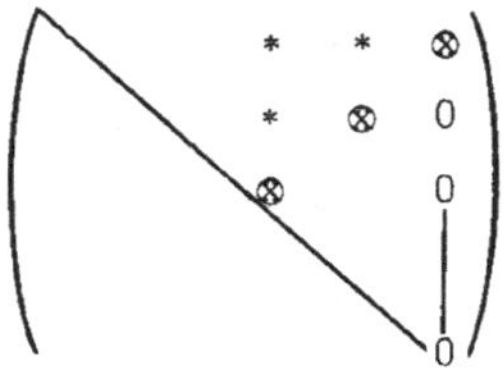

where the entry at the position marked ⊗ is $\pm\frac{\partial^2 L}{(\partial u^{(n)})^2}$ and so the matrix is non-singular.

It is clear that the same arguments work if the fiber dimension is greater than one and if we replace the non-degeneracy assumption by the assumption

$$\text{Hess } L \neq 0$$

where Hess L is defined by

$$\text{Hess } L = \left(\frac{\partial^2 L}{\partial u_i^{(n)} \partial u_j^{(n)}} \right)$$

where n is the smallest integer such that L is defined on $J_n Y$. We call n the order of L. We have proved the following

THEOREM 3.1. *Assume that* Hess $L \neq 0$, *where* L *is of order* n *and where the fiber dimension is* ℓ. *Then the set of all points of the form* js *where* s *is an extremal of* Ldx *is a submanifold of* J *of dimension* $2n\ell + 1$, *call it* S_{Ldx}. *The form* $d\theta_{Ldx}$ *when restricted to this manifold has rank* $2n\ell$ *and its null direction is spanned by the vector field* $\frac{d}{dx}$. *We let* E_{Ldx} *denote the quotient of this* $2n\ell + 1$ *dimensional manifold by the (one dimensional) foliation spanned by* $\frac{d}{dx}$. *(Locally* E_L *looks like a manifold of dimension* $2n\ell$ *and also globally, if we make some completeness assumptions about the ordinary differential equations* $\frac{\delta L}{\delta u} = 0$.) *If* f *is a function which is defined on* E_L *we can think of* f *as a function defined on* S *such that* $\frac{df}{dx} = 0$. *Then* $df = \zeta \rfloor d\theta_{Ldx}$ *on* S, *where* ζ *is a vector field tangent to* S *and determined up to a multiple of* $\frac{d}{dx}$.

$$\begin{array}{ccc} S & \longrightarrow & J \\ \big\downarrow & & \\ E & & \end{array}$$

We claim that we can choose the vector field ζ in the theorem so that $[\zeta, \frac{d}{dx}] = 0$. In fact on S we have, for any choice of ζ,

$$0 = \frac{df}{dx} = D_{\frac{d}{dx}} (\zeta \rfloor d\theta_{Ldx}) = \left[\frac{d}{dx}, \zeta\right] \rfloor d\theta_{Ldx} + \zeta \rfloor d\left(\frac{d}{dx} \rfloor d\theta_{Ldx}\right)$$

so

$$\left[\frac{d}{dx}, \zeta\right] = a \frac{d}{dx}$$

since $\frac{d}{dx}$ is the generator of the null line of $d\theta_{Ldx}$ on S. We now simply choose a function b on S such that

$$\frac{db}{dx} = a$$

and replace ζ by $\zeta - b\frac{d}{dx}$.

Now the set of vertical vector fields on J which preserve I can be identified with the set of sections of V(Y) over J. In particular, we can regard ζ as a section of V(Y) over S. Given any such section we can extend it to a section of V(Y) over J. We have thus proved the following

PROPOSITION 3.2. Suppose that $d\theta_{\omega}$ induces a symplectic structure on the space of extremals, for example suppose that Hess $L \neq 0$. Then for any Hamiltonian vector field, ζ on the space of extremals, we can find a vector field ξ in the algebra $\mathcal{V}$ such that ξ is tangent to the space of extremals and ξ coincides with ζ when restricted to this space.

Any vertical vector field, ξ, which commutes with $\frac{d}{dx}$ determines a section, $\hat{\xi}$ of V(Y), and, conversely, any section of V(Y) gives rise to a unique vector field commuting with $\frac{d}{dx}$. We can write the equation

$$\xi \rfloor d\theta_{Ldx} = df \quad \text{on} \quad S$$

as

$$\delta L(\hat{\xi})dx = df \quad \text{on} \quad S \quad ,$$

since the value of $\xi \rfloor d\theta$ mod I depends only on $\hat{\xi}$.

Suppose that we start with some vector field $\xi \in \mathcal{V}$ satisfying

$$\xi \rfloor d\theta_{Ldx} \equiv dF \quad \text{mod } I \quad .$$

Now $\theta_{Ldx} = Ldx + \alpha$ where $\alpha \in I$ and hence $D_{\xi}\theta_{Ldx} = D_{\xi}(Ldx) + D_{\xi}\alpha'$, where, for any form β we have $D_{\xi}\beta = \xi \rfloor d\beta + d(\xi \rfloor \beta)$. Since $\xi \in \mathcal{V}$, we know that $D_{\xi}\alpha \in I$ and hence that

$$D_{\xi}(Ldx) \equiv dF \quad \text{mod } I$$

which implies that ξ is tangent to the manifold S and induces a Hamiltonian vector field on S.

On the other hand, suppose that we start with a function f on S and assume that S is a manifold of dimension $2n\ell + 1$ and that $d\theta_{Ldx}$ has rank $2n\ell$. Let us also assume that there is some k so that the projection of J onto J_k induces an immersion on S, and such that $d\theta_{Ldx}$ is defined on J_k and has constant rank there. (For example, in the case of Lagrangian of order n with Hess $L \neq 0$, we can take $k = 2n$ and the projection of J onto J_{2n} induces a diffeomorphism of S with J_{2n}, while $d\theta_{Ldx}$ has exactly one singular direction at every point.) We obtain a foliation of J_k

consisting of the singular direction of $d\theta_{Ldx}$, and we can extend f to be a function, F, defined on all of J_k which is constant along the leaves of this foliation. As before, we can find $\hat{\xi}$ which induces a vector field ξ commuting with $\frac{d}{dx}$ and such that $\xi d\theta_{Ldx} = dF$. Thus the set of all vector fields satisfying $\xi d\theta \equiv dF \mod I$ maps surjectively onto the algebra of Hamiltonian vector fields on S.

REFERENCES

1] I. M. Gel'fand and L. A. Dikii, "Asymptotic behaviour of the resolvent of Sturm Liouville equations and the algebra of Korteweg-DeVries equations. Uspekhi Math Nauk 30:5 (1975) 67-100 or Russian Math Surveys 30:5 (1975) 77-113.

2] H. Goldschmidt and S. Sternberg, "The Hamilton Cartan formalism in the calculus of variations" Am. Inst. Four. 23 (1973) 203-267.

REDUCIBILITY OF THE SYMPLECTIC STRUCTURE OF MINIMAL INTERACTIONS

Pedro L. García

Antonio Pérez-Rendón

Universidad de Salamanca

Introduction

It is a well known fact that to every lagrangian system it is possible to associate a generalized symplectic structure (V,ω_2,Q) . V is the "manifold of solutions" of the Euler-Lagrange equations of the system, ω_2 is a "closed 2-form" on V , in general not irreducible, intimately related to the classical Poincaré-Cartán invariant of the Calculus of Variations, and Q is a "Poisson algebra" of functions on V to which belong all interesting dynamical quantities (energy, linear and angular moments etc.). Three recents references are [7] [13] and [15] . Following [7] , we shall call the 2-form ω_2 the "symplectic metric" of the system though in general it is not irreducible. In certain particular cases, V can be endowed with a differentiable manifold structure in such a way that ω_2 is a closed 2-form and Q is a subring of differentiable functions on V (finite-dimensional manifolds for systems with a finite number of degrees of freedom, and infinite-dimensional manifolds for classical fields). This "generalized symplectic structure" should be understood as an adequate geometrical lenguage unifying many notions and results in global Calculus of Variations (infinitesimal symmetries, Noether invariants, Jacobi fields along an extremal etc.). On the other hand, this language has been untill today the only leading line to define the differentiable structure on the manifold of solutions in all cases where this has been achieved. The case when ω_2 is irreducible is the best for a good dynamical theory. Nevertheless, there exist fundamental examples where ω_2 is not irreducible (generalized dynamical systems in the sense of Bergmann-Dirac, electromagnetic and Yang-Mills' fields, General Relativity etc.). In such cases it would be desirable to reduce the degeneration of ω_2 by trying to find a projection $\pi:V\longrightarrow\bar{V}$ such that ω_2 is projected on an irreducible metric $\bar{\omega}_2$, and Q on a subring $\bar{Q}$ of functions on $\bar{V}$. The new structure $(\bar{V},\bar{\omega}_2,\bar{Q})$, now a properly sumplectic one, would be the adequate one in order to develop the dynamical theory of the system. In General Relativity, where the degeneration of the metric is due to the fact that the system admits the group of space-time diffeomorphisms as symmetries (Gauge group), this type of reduction has been achieved by Marsden and Fischer in the realm of infinite-dimensional differentiable manifolds by using techniques of Global Analysis [2] [3] [4] . Yang-Mills' fields are another example of systems admitting a infinite-dimensional Lie group of symmetries (the automorphism group of a principal bundle), which it is also the origin of the degeneration of the corresponding symplectic metric. It seems now natural to trying to develop for these fields a study similar to the one done in General Relativity. The starting point for a such programme should be, we think, to stablishing of a precise way the relation between gauge-symmetries and degeneration of the metric. In [8] the first author announced in this sense two general results:

Theorem

Let (V,ω_2) be the symplectic manifold of solutions of a Yang-Mills' field with arbitrary lagrangian.

If D_σ is the tangent vector at $\sigma\varepsilon V$ defined by any infinitesimal gauge-symmetry, then one has:

$$i\, D_\sigma \cdot \omega_2 = 0$$

i.e. the gauge-symmetries of a Yang-Mills' field belong to the radical of its symplectic metric.

Theorem

Let $\mathfrak{h}$ be the idealizator of the Lie algebra of infinitesimal gauge-symmetries of a Yang-Mills' field in the Lie algebra of all its infinitesimal symmetries.

If ω_Y is the function on V defined by the Noether invariant corresponding to an infinitesimal symmetry $Y\varepsilon\mathfrak{h}$ and D is any gauge-symmetry, then for every point $\sigma\varepsilon V$ one has:

$$D_\sigma\, \omega_Y = 0$$

In particular, if Y is a gauge-symmetry, the function, ω_Y is identically zero.

The notion of Yang-Mills' field appearing in the above theorems was introduced by the first author in [9] [10] for an arbitrary principal bundle over an oriented manifold. Two important results of this formalism, on which the proofs of the above two theorems are based on, are: a geometric version of a classical theorem, due to Utiyama, about gauge-invariant lagrangians [19] , and the possibility of obtaining a Maxwell-type globalization of the field equations and of the Jacobi equations along an extremal. Other recent references on Geometry of Yang-Mills' fields are [1] [14] [20] and [21] .

Up to now, the fields mentioned are "free" in the sense that they are not "coupled" with any other field. Now, a natural problem would be to try to find out how the above theorems are affected by an interaction. In a more precise way, trying to find out whether if is possible to generalize, in some sense, the said theorems to the so-called minimal interactions, the only type of coupling between fields for which these is, up to now, an available symplectic formulation, as the papers [16] [17] [18] of the second author prove. This will be the aim of the present communication. The result could not be any simpler: starting from the Interaction Theory introduced by the second author, one proves that the above theorems hold for the symplectic structure of every minimal interaction.

As in the free field case, the main difficulty lies in having at hand good global field equations and Jacobi equations along an extremal. Being so, the main part of the exposition is centered on this point. The organization of the whole paper is presented in the following table of contents:

Contents

The authors wish to expres their acknowledgement to Professor J. Sancho Guimerá to whom they owe the central ideas, conjectures and orientations in the major part of the paper on Calculus of Variations published by them in the last years, and which have made possible the present one. They also thank to Professor K. Bleuler for his warm hospitality in Bonn during this Symposium.

1. Geometry of Minimal Interactions

The main novelty in the formalism introduced in [16], [17] and [18] by the second author, when he proposes a global definition of "Minimal Interaction" in the realm of Differential Geometry of fiber bundles, is that for the first time the celebrated "Yang-Mills' trick" does not appear at the basis of the doctrine. In this paragraph we shall give, without proofs, a more up-to-date version of this formalism, where we include some results on gauge algebras obtained by the first author in [9]. Proofs, tacit or explicit, can be found in the above referred four papers.

1.1. The Gauge Group and the Gauge Algebra of a Principal Bundle and their Natural Representations on the Affine Bundle of Connections and on Associated Vector Bundles. Yang-Mills' Fields: Utiyama's Theorem

Let $p:P \longrightarrow X$ be a principal bundle with structural group G with Lie Algebra L, and let A be the $\mathbb{R}$-algebra of differentiable functions on X.

To the bundle P there is an associated exact sequence of vector bundles on X:

$$(1.1) \qquad 0 \longrightarrow \mathrm{Ad}\,P \longrightarrow T_G(P) \xrightarrow{\ p\ } T(X) \longrightarrow 0$$

where $T_G(P)$ is the vector bundle of G-invariant vector fields on P, $\mathrm{Ad}\,P$ is the subbundle of $T_G(P)$ defined by the G-invariant vector fields tangent to the fibres of P, and $T(X)$ is the tangent bundle of X.

$\mathrm{Ad}\,P$ is a bundle of Lie algebras, where, if $D,D' \in (\mathrm{Ad}\,P)_x$, then $[D,D']$ is the ordinary Lie bracket of D and D'. On the other hand it is the bundle associated to P relative to the adjoint representation of G on L. It is called the adjoint bundle of P.

Definition 1.1

The group $G(P)$ of p-vertical automorphism of P is called the gauge group of P.

The infinitesimal generator of a uniparametric subgroup of $G(P)$ is p-vertical and G-invariant. This suggests the following definition for the Lie algebras of $G(P)$.

Definition 1.2

The Lie Algebra $\Gamma(\mathrm{Ad}\,P)$ of global sections of the adjoint bundle of P is called the gauge algebra of P.

This Lie algebra admits two natural representations by vector fields, which we now proceed to describe.

A connection on P can be defined as a splitting $\sigma:T(X) \longrightarrow T_G(P)$ of the exact sequence (1.1). In this way the connections on P can be identified with global sections of the affine bundle $\pi:E \longrightarrow X$ defined as follows: $x \in X$ being given, let E_x be the set of homomorphism $\sigma_x:T_x(X) \longrightarrow T_G(P)_x$ such that $p_x^* \cdot \sigma_x = 1$, let $E = \bigcup_{x \in X} E_x$, and let π be the natural projection of E onto X.

It can be proved that $\pi : E \longrightarrow X$ has a unique affine bundle structure such that, for every connection σ on P, the mapping $\sigma : \mathrm{Hom}(T(X), \mathrm{Ad}\,P) \longrightarrow E$ defined by $h_x \longmapsto \sigma(x) + h_x$ is an affine bundle isomorphism on X, it is the affine bundle corresponding to the vector bundle $\mathrm{Hom}(T(X), \mathrm{Ad}\,P)$ and it is called the bundle of connections on P.

The natural action of elements of the gauge group on connections on P induces a representation of the gauge algebra by π-vertical vector fields on E that can be directly described as follows:

Every element $s \in \Gamma(\mathrm{Ad}\,P)$ defines a uniparametric group τ_t of π-vertical automorphisms of E by the rule:

$$(1.2) \qquad \tau_t\, \sigma_x = \sigma_x + t\,[\sigma_x, s], \quad \sigma_x \in E$$

where $[\sigma_x, s] \in \mathrm{Hom}(T(X), \mathrm{Ad}\,P)$ is given by $[\sigma_x, s] D_x = [\sigma_x(D_x), s]$. If D_s is the infinitesimal generator of τ_t one has:

Proposition 1.1

The mapping $s \longmapsto D_s$ is a homomorphism from the Lie $\mathbb{R}$-algebra $\Gamma(\mathrm{Ad}\,P)$ into the Lie $\mathbb{R}$-algebra of π-vertical vector fields of E.

Now, let $\pi' : E' \longrightarrow X$ be a vector bundle associated to P with respect to a linear representation of G on a vector space V. The elements of the gauge group act in a natural way by π'-vertical automorphisms of E', giving thus rise to a representation of the gauge algebra by π'-vertical vector fields on E'. This can be described in the following way:

$s \in \Gamma(\mathrm{Ad}\,P)$ being given, let D'_s be the π'-vertical vector field of E' such that, for every function f of E', linear on the fibres, one has:

$$(1.3) \qquad (D'_s f)(e'_x) = f(s(x) \cdot e'_x), \quad e'_x \in E'_x$$

where, if $s(x)$ and e'_x are the equivalence classes $(\overline{\xi, A})$ and $(\overline{\xi, v})$, $\xi \in P$, $A \in L$ and $v \in V$, $s(x) \cdot e'_x$ is defined as the equivalence class $(\overline{\xi, Av})$, where Av is the action of A on v by the representation induced by G on its Lie algebra.

Proposition 1.2

The mapping $s \longmapsto D'_s$ is a homomorphism from the Lie A-algebra $\Gamma(\mathrm{Ad}\,P)$ into the Lie A-algebra of π'-vertical vector fields of E'.

The essential distinction between these two representations is that, while the first deals with Lie $\mathbb{R}$-algebras, the second deals with Lie A-algebras. In particular, $\{D_s\}$, $s \in \Gamma(\mathrm{Ad}\,P)$, is a Lie $\mathbb{R}$-algebra of π-vertical vector fields of E, and $\{D'_s\}$, $s \in \Gamma(\mathrm{Ad}\,P)$, is a Lie A-algebra of π'-vertical vector fields of E'.

The local case

If $P = X \times G$, the gauge group can be identified with the group of differentiable mappings from X into G, in such a way that, by taking constant maps, G can be considered as a subgruop of the gauge group. Analogously, $\mathrm{Ad}\,P = X \times L$ and then the gauge algebra can be identified with the tensor product $A\otimes_{\mathbb{R}} L$ endowed with the Lie product

$$[f\otimes e\,,\,f'\otimes e'] = (f\cdot f')\otimes[e\,,\,e']$$

$f,f'\varepsilon A$; $e,e'\varepsilon L$. Thus L can also be considered as a subalgebra of the gauge algebra in a natural way.

Let us suppose that we have a local coordinate system (x_i) on X and let (D_j) be the G-invariant vector fields defined on P by a basis (e_j) of L . Then the functions $(x_i\ A_{ij})$ on E , where

$$\sigma_x\left(\frac{\partial}{\partial x_i}\right) = \frac{\partial}{\partial x_i} + \sum_j A_{ij}(\sigma_x)D_j \quad , \quad \sigma_x \varepsilon E$$

define a local coordinate system on E .

With these new data, the gauge algebra is identified with the A-modulo of linear combinations

$$s = \sum_j f_j(x_i)D_j$$

endowed with the Lie product

$$[s,s'] = \sum_{i,j} f_i\, f'_j\, [D_i,D_j] = \sum_{i\,j\,k} f_i\, f'_j\, c^k_{ij}\, D_k$$

where c^k_{ij} are the structural constants of L .

The vector field D_s corresponding to $s=\sum_j f_j\, D_j$ is given by

$$(1.4) \qquad D_s = \sum_{ij} \left(\frac{\partial f_j}{\partial x_i} + \sum_{hk} c^j_{hk}\, A_{ih}\, f_k\right)\frac{\partial}{\partial A_{ij}}$$

Finally, let (v_k) be a basis of the typical fiber V of the associated bundle E' , let (a^i_{hk}) the matrix with respect to (v_k) of the element $e_j\varepsilon L$ in the representation of L on V induced by the given representation of G on V , and let (x_i,z_j) the system of local coordinates on $E' = X\times V$ where

$$z_j(x,v) = \lambda_j \quad , \quad v = \sum_j \lambda_j\, v_j$$

Relative to all these data, the vector field D'_s corresponding to $s=\sum_j f_j\, D_j$ is given by the formula

$$(1.5) \qquad D'_s = \sum_j \left(\sum_{\ell,m} f_\ell\, a^\ell_{jm}\, z_m\right)\frac{\partial}{\partial z_j}$$

Going back to the general case, let us suppose that the manifold X is orientable and endowed with a volume element η . Then we can define a <u>Yang-Mills' field</u> as a variational problem whose lagrangian density $\mathcal{L}\eta$ is defined on the bundle $j^1\pi : T^1E \longrightarrow X$ of the 1-jets of local sections of the bundle of connections on P .

<u>Definition 1.3</u>

We shall call a Yang-Mills field <u>gauge invariant</u> when its lagrangian density $\mathcal{L}\eta$ admits the representation $\{D_s\}$ of the gauge algebra $\Gamma(\mathrm{Ad}\,P)$ as infinitesimal symmetries, i.e.: $(j^1D_s)\mathcal{L} = 0$ for all $s\varepsilon\Gamma(\mathrm{Ad}\,P)$.

The lagrangians satisfying this condition can be characterised as follows: The curvature of a connection establishes a bundle epimorphism on X :

$$\begin{array}{ccc} J^1E & \xrightarrow{\quad K \quad} & \Lambda^2T^*(X)\otimes \mathrm{Ad}\,P \\ & \searrow \quad \swarrow & \\ & X & \end{array}$$

given by $K(j^1_x\sigma) = (K_\sigma)_x$, where K_σ is the curvature 2-form of a connection σ representative of the 1-jet $j^1_x\sigma$.

On the other hand, the gauge algebra $\Gamma(\mathrm{Ad}\,P)$ admits, as a Lie A-algebra, a new representation $s \longmapsto \hat{D}_s$ by vertical vector fields of $\Lambda^2T^*(X)\otimes\mathrm{Ad}\,P \longrightarrow X$ defined in the following way:

$s\varepsilon\Gamma(\mathrm{Ad}\,P)$ being given, $\hat{D}_s$ is the vertical vector field of $\Lambda^2T^*(X)\otimes\mathrm{Ad}\,P \longrightarrow X$ such that, for every point $(\omega_2)_x\varepsilon\Lambda^2T^*(X)\otimes\mathrm{Ad}\,P$ and every function f , linear on the fibres one has:

$$(1.6) \qquad (\hat{D}_s f)(\omega_2)_x = f([s(x)\ ,\ (\omega_2)_x])$$

where $[s(x)\ ,\ (\omega_2)_x]$ is the point of $\Lambda^2T^*(X)\otimes\mathrm{Ad}\,P$ given by:

$$[s(x)\ ,\ (\omega_2)_x](D,D' = [s(x)\ ,\ (\omega_2)_x(D,D')]$$

Now we have:

<u>Theorem 1.1.</u> (Utiyama)

A function $\mathcal{L} : J^1E \longrightarrow \mathbb{R}$ is gauge-invariant (i.e. is invariant by the Lie $\mathbb{R}$-algebra $\{j^1D_s | s\varepsilon\Gamma(\mathrm{Ad}\,P)\}$ if and only if:

$$\mathcal{L} = \bar{\mathcal{L}} \circ K$$

where $\bar{\mathcal{L}} : \Lambda^2T^*(X)\otimes\mathrm{Ad}\,P \longrightarrow \mathbb{R}$ is a function invariant by the Lie A-algebra $\{\hat{D}_s | s\varepsilon\Gamma(\mathrm{Ad}\,P)\}$ and K is the curvature epimorhism.

This version of Utiyama's theorem means that the condition of gauge-invariance of a function is equivalent, via the curvature, to this function being a first integral of an involutive distribution.

1.2. Classical Fields over an Associated Vector Bundle Endowed with a Linear Connection

The foundation of the classical field theory proposed by the first author in 1967, and developped in several papers [5] , [6] , [7] , [11] and [12] , can be summarized as follows.

A "classical field" is given by a variational problem defined by a fibered manifold $\pi : E \longrightarrow X$ over an orientable manifold X endowed with a volume element η , and by a real differentiable function $\mathcal{L}$ on the bundle $j^1\pi : J^1E \longrightarrow X$ of the 1-jets of local sections of E . The basic notion is that of the <u>structure 1-form</u> of the 1-jet bundle. Introduced from the notion of "vertical differential" of a section of

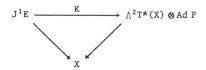

$\pi:E \longrightarrow X$, it is an 1-form on the manifold J^1E with values in the module M of sections of the induced vector bundle $q^*T^V(E)$, where T^VE is the vertical tangent bundle of E and q is the canonical projection of J^1E onto E . Schematically:

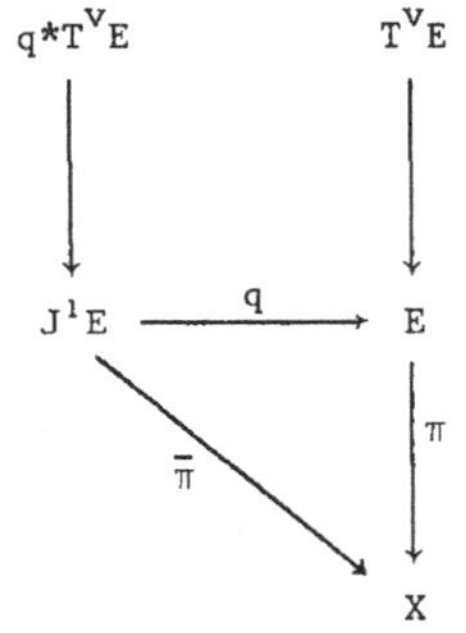

Its local expression with respect to a system of natural local coordinates $(x_i\, z_j\, p_{ij})$ on J^1E and a local basis $q^*\frac{\partial}{\partial z_j}$ for M , is:

(1.7) $$\theta = \sum_j (\, dz_j - \sum_i p_{ij}\, dx_i \,) \circ q^*\frac{\partial}{\partial z_j}$$

Now, it is posible introduce two fundamental differential forms: The <u>Legendre form</u> and the <u>Cartán form</u>, from them the whole of the theory can be developped.

The Legendre form Ω is an $(n-1)$-form on J^1E , $n = \dim X$, valued in the module M^* dual of M . It is introduced in a way like to the one employed in symplectic geometry to define the hamiltonian vector field corresponding to a function, as follows:

Ω is the unique M^*-valued $(n-1)$-form on J^1E such that:

(1.8) $$iF \cdot d\theta \equiv -d\mathcal{L}(\text{mod } Q^1_q) \,, \quad iF \cdot \eta = \Omega$$

where the differential of the M-valued 1-form θ is taken with respect to an arbitrary derivation law on M , F is an M^*-valued 1-contravariant tensor on J^1E , Q^1_q is the module of ordinary 1-forms on J^1E which are zero on the fibres of $q:J^1E \longrightarrow E$, and the interior products are taken with respect to the natural bilinear products between the corresponding modules.

We must make the following precision: the first equations defines F only locally and not in a unique way. Generally, this does not allow its globalization. But, when taking the interior product $iF\eta$ with any local solution of F , the local $(n-1)$-form thus obtained is unique, thus allowing us to globalize it in order to obtain Ω . Nevertheless , there are particular examples in which global solutions of F can be obtained. This is the case, e.g., of the lagrangians of electromagnetic and ordinary Yang-Mills' fields, where F es nothing but the contravariant 2-form defining the corresponding field intensities.

The local expression of Ω in a system of natural local coordinates $(x_i\, z_i\, p_{ij})$ on J^1E and the local basis q^*dz_j of M^* is:

$$(1.9) \qquad \Omega = \Phi \sum_{i,j} (-1)^{i-1} \frac{\partial \mathcal{L}}{\partial p_{ij}} dx_1 \wedge \ldots \wedge \widehat{dx_i} \wedge \ldots \wedge dx_n \circ q^* dz_j$$

where $\Phi = \eta(\frac{\partial}{\partial x_1} \ldots \frac{\partial}{\partial x_n})$.

The Cartan form Θ is the ordinary n-form on J^1E defined by:

$$(1.10) \qquad \Theta = \theta \wedge \Omega - \mathcal{L}\eta$$

where the exterior product is taken with respect to the bilinear product defined between M and M* by the duality notion.

The idea of the geometrization of the Calculus of Variations proposed in those papers consists in the formulation of all concepts and manipulations of the theory in terms of differential forms and operations of differential calculus on the manifold J^1E with values in the modules M , M* , Hom(M,M) etc. This same point of view has allowed to deal with several topics of greatest interest in a natural way providing simple solutions to a variety of problems (globalization of field equations, Cartan's formalism and symplectic structure, infinitesimal symmetries and Noether invariants etc.). At the same time this formalism is flexible enough so as to allows interesting variants and generalizations like the ones we now consider.

We start with an elementary remark. A connection on the fibered manifold $\pi: E \longrightarrow X$ being given, i.e. a splitting $\sigma: \pi^*T(X) \longrightarrow T(E)$ of the exact sequence of vector bundles on E :

$$0 \longrightarrow T^vE \longrightarrow T(E) \xrightarrow{\pi} \pi^*T(X) \longrightarrow 0$$

we can stablish an affine bundle isomorphism on E :

$$\begin{array}{ccc} J^1E & \xrightarrow{\quad\sigma\quad} & \mathrm{Hom}(\pi^*T(X), T^vE) \\ & \searrow \quad \swarrow & \\ & E & \end{array}$$

by the rule: $j^1_x s \longmapsto (ds)_x - \sigma_{s(x)}$, where s is a representative of the 1-jet $j^1_x s$. Now, we obtain a variational theory similar to the ordinary one by simply changing J^1E for $\mathrm{Hom}(\pi^*T(X),T^vE)$ and θ for the 1-form θ^σ obtained from θ by the above isomorphism. In this case, the lagrangian $\mathcal{L}$ is a real differentiable function on the manifold $\mathrm{Hom}(\pi^*T(X),T^vE)$ and the structure 1-form θ^σ is an M-valued 1-form on $\mathrm{Hom}(\pi^*T(X),T^vE)$, M being the module of sections of the induced vector bundle q^*T^vE , where q is now the canonical projection of $\mathrm{Hom}(\pi^*T(X),T^vE)$ onto E . In this case, we are able to have canonical field equations, for we have the connection σ as a data of the problem, and thus we can work the module-valued differential calculus with respect to this connection σ . Physically speaking, such a variational problem would describe a classical field with lagrangian $\mathcal{L}$ under the action of a fixed potencial given by the connection σ .

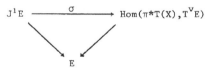

The interesting case for an "Interaction Theory" happens when $\pi:E\longrightarrow X$ is a vector bundle associated to a principal bundle $p:P\longrightarrow X$ and σ the connection defined on E by a connection on P. Following the notation introduced in §1.1 we shall write $\pi':E'\longrightarrow X$ such an associated fibration and we shall preserve $\pi:E\longrightarrow X$ for the bundle of connections on P. Now we have, for $T^vE'=\pi'^*E'$, that $\mathrm{Hom}(\pi'^*T(X),T^vE')=\mathrm{Hom}(\pi'^*T(X),\pi'^*E')=\pi'^*\mathrm{Hom}(T(X),E')$.

By using notations of §1.1 and by denoting with " ' " the objects related with the variational problem defined by E', if $(x_i\ z_j\ p_{ij})$ is a system of natural local coordinates on $\pi^*\mathrm{Hom}(T(X),E')$ and $q'^*\frac{\partial}{\partial z_j}$ is a local basis of M', we have the following local expression for the structure 1-form θ'^{σ}:

$$\theta'^{\sigma}=\sum_j\left[dz_j-\sum_i\left(p_{ij}+\sum_{\ell,m}a^{\ell}_{jm}A_{i\ell}(x_i)z_m\right)dx_i\right]\circ q'^*\frac{\partial}{\partial z_j}\tag{1.11}$$

The Legendre $(n-1)$-form Ω' is still having the same local expression.

On the other hand, the gauge-algebra $\Gamma(\mathrm{Ad}\,P)$ admits, as a Lie A-algebra, a representation $s\longmapsto\tilde{D}_s$ by vertical vector fields of $\pi'^*(\mathrm{Hom}(T(X),E')\longrightarrow X$ defined as follows:

$s\in\Gamma(\mathrm{Ad}\,P)$ being given, $\tilde{D}_s$ is the vertical vector field of $\pi'^*\mathrm{Hom}(T(X),E')\longrightarrow X$ such that, for every point $(e'_x,h_x)\in\pi'^*\mathrm{Hom}(T(X),E')$ and every function f linear on the fibres, one has:

$$(\tilde{D}_s f)(e'_x,h_x)=f(e'_x,s(x)\cdot h_x)\tag{1.12}$$

$(s(x)\cdot h_x)D_x=s(x)\cdot h_x(D_x)$ in the sense defined in §1.1 when the representation $s\longmapsto D'_s$ was introduced.

Locally, if $s=\sum_j f_j D_j$, $\tilde{D}_s$ is given by:

$$\tilde{D}_s=\sum_{\ell mj}f_\ell\, a^{\ell}_{jm}\left(z_m\frac{\partial}{\partial z_j}+\sum_i p_{im}\frac{\partial}{\partial p_{ij}}\right)\tag{1.13}$$

<u>Definition 1.4</u>

We shall say that the lagrangian $\mathcal{L}':\pi'^*\mathrm{Hom}(T(X),E')\longrightarrow\mathbb{R}$ is <u>gauge-invariant</u> when, for every $s\in\Gamma(\mathrm{Ad}\,P)$, $\tilde{D}_s\mathcal{L}=0$.

<u>Local meaning of the gauge-invariance condition for $\mathcal{L}'$</u>

Let $E'=X\times V$ be the vector bundle associated to $P=X\times G$ with respect to a given linear representation of G on V (we use hipothesis and notations of the local case in §1.1. We saw that in this case the Lie algebra L of G can be considered, in a natural way, like a subalgebra of the gauge algebra $A\otimes_{\mathbb{R}}L$. Now let us consider the variational problem defined on $\pi'^*\mathrm{Hom}(T(X),E')$ by the lagrangian $\mathcal{L}'$ and by the structure 1-form θ'^{σ_0} corresponding to the canonical flat connection σ_0 of $P=X\times G$. By (1.13), $\mathcal{L}'$ is gauge-invariant if and only if $\tilde{D}_s\mathcal{L}'=0$ for every $s\in L$. On the other hand, for the functions $A_{ij}(x_i)$ defined by σ_0 are null, then by (1.11), $\tilde{D}_s$, $s\in L$, are the infinitesimal contact transformations of

$\pi'^*\mathrm{Hom}(T(X),E')$ with respect to θ'^{σ_0} that are projected on D'_s , $s\in L$. Thus, the gauge-invariance condition for $\mathcal{L}'$ is equivalente to the real finite-dimensional Lie algebra $\{D'_s\}$, $s\in L$, being an algebra of infinitesimal symmetries of the variational problem under consideration. But, in general, those $\tilde{D}_s$, $s\notin L$, are not infinitesimal symmetries. This has lead us precisely to the starting-point of the ordinary Yang-Mills' theory: a variational problem with lagrangian $\mathcal{L}'$ which admits the Lie algebra of a Lie group G as infinitesimal symmetries, but not the infinite-dimensional Lie algebra $A\otimes_{\mathbb{R}} L$ in which L is embedded as a subalgebra. As is well known, the celebrated "Yang-Mills' trick" is a mechanism by means of which the given L-invariant field can be considered as a component of a larger field, which already admits $A\otimes_{\mathbb{R}} L$ as infinitesimal symmetries. It is precisely this last condition which, at least heuristically, allows one to introduce the new components of that larger field (Yang-Mills' field) in an essencially unique way.

We think that there is already enough motivation to consider the family of classical fields defined on $\pi'^*\mathrm{Hom}(T(X),E')$ by $\mathcal{L}'$ and θ'^{σ} , σ an arbitrary connection on P , as the kind of object that we wish to "minimally couple" with Yang-Mills' fields introduced at the end of §1.1.

1.3. The Definition of Minimal Interaction

The interaction theory we wish to show here is based upon the following idea:

It is possible to establish an epimorphism of fiber bundles on X between the fibered product $J^1E\times_X \pi'^*\mathrm{Hom}(T(X),E')$ and the 1-jet fiber bundle J^1E :

$$\begin{array}{ccc} J^1E\times \pi'^*\mathrm{Hom}(T(X),E') & \xrightarrow{\ \phi\ } & J^1E' \\ & \searrow \quad \swarrow & \\ & X & \end{array}$$

given by $(j^1_x\sigma,(e'_x,h_x))\longmapsto \sigma(e'_x,h_x)$, where σ is a connection representative of the 1-jet $j^1_x\sigma$ and $\sigma(e'_x,h_x)$ is the image in J^1E' of the point $(e'_x,h_x)\in\pi'^*\mathrm{Hom}(T(X),E')$ under the isomorphism $\pi'^*\mathrm{Hom}(T(X),E')\simeq J^1E'$ defined by the connection σ in §1.2.

Now, given a Yang Mills' field defined on J^1E by the lagrangian density $\mathcal{L}\eta$ and a field on $\pi'^*\mathrm{Hom}(T(X),E')$ with lagrangian density $\mathcal{L}'\eta$, we may think of developping a variational theory on the space $J^1E\times_X\pi'^*\mathrm{Hom}(T(X),E')$ for the lagrangian density sum $(\mathcal{L}+\mathcal{L}')\eta$ and with respect to the structure 1-form $\theta+\phi^*\theta'$, where θ and θ' are the structure 1-forms of J^1E and J^1E' . This is in fact possible with the important result that the Euler-Lagrange equations so obtained globalize the system of coupled equations for two classical fields in minimal interaction proposed by physics.

This suggests the following:

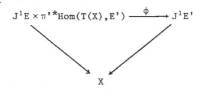

Definition 1.5.

We shall call Minimal Interaction of the Yang-Mills' field $\mathcal{L}\eta$ with the field $\mathcal{L}'\eta$, the variational problem defined on the space $J^1E \times_X \pi'^*\mathrm{Hom}(T(X),E')$ by the structure 1-form $\theta+\phi^*\theta'$ and by the lagrangian density $(\mathcal{L}+\mathcal{L}')\eta$.

In what follows we shall make some precisions and remarks on this new notion, as well as on the way the gauge-algebra $\Gamma(\mathrm{Ad}\,P)$ can be introduced in the theory.

First of all, by using the more simmetrical notation $J^1E = \bar{E}$ and $\pi'^*\mathrm{Hom}(T(E),E') = \bar{E}$, we have a scheme similar to one of an ordinary variational problem:

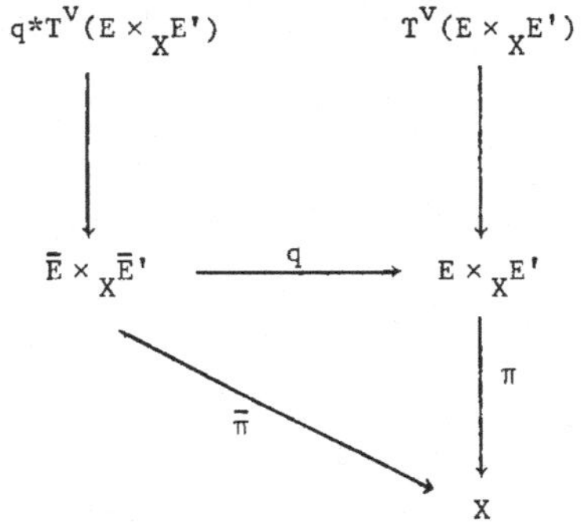

where q is given by $q(j^1_x\sigma,(e'_x,h_x)) = (\sigma(x),e'_x)$.

$T^V(E\times_X E)$ can be identified with the Whitney sum $T^V(E)\oplus_X T^V(E')$ and its natural projection on $E\times_X E'$. Correspondingly, $q^*T^V(E\times_X E')$ can be identified with $q^*T^V(E)\oplus_X q'^*T^V(E')$ and its natural projection on $\bar{E}\times_X\bar{E}'$. In this way θ can be considered as a 1-form on $\bar{E}\times_X\bar{E}'$ with values in $q^*T^V(E\times_X E')$, and analogously $\phi^*\theta'$. Thus, $\theta+\phi^*\theta'$ is a 1-form defined and valued as it is convenient for the Variational Calculus.

Now we observe that the descomposition $q^*T^V(E\times_X E') = q^*T^V(E)\oplus_X q'^*T^V(E')$ allows us to descompose forms as sums of two terms, taking its respective values in q^*T^VE and q'^*T^VE'. For instance, the structure 1-form $\theta+\phi^*\theta'$ splits into the sum of θ (with values in q^*T^VE) and $\phi^*\theta'$ (with values in q'^*T^VE'). Locally, if $(x_i\ A_{ij}\ B_{\ell ij}\ z_j\ p_{ij})$ is a natural system of local coordinates on $\bar{E}\times_X\bar{E}'$, and $(\frac{\partial}{\partial A_{ij}})$, $(\frac{\partial}{\partial z_j})$ are the corresponding local basis for the modules of sections of T^VE and T^VE', we have:

$$(1.14)\quad \theta=\sum_{h,k}(dA_{hk}-\sum_i B_{ihk}dx_i)\circ q^*\frac{\partial}{\partial A_{hk}}\ ,\quad \phi^*\theta'=\sum_j\Big[dz_j-\sum_i(p_{ij}+\sum_{\ell m}a^{\ell}_{jm}A_{i\ell}z_m)dx_i\Big]\circ q'^*\frac{\partial}{\partial z_j}$$

The Legendre form admits a similar descomposition $\Omega+\Omega'$. Locally:

$$(1.15)\quad \Omega=\Phi\sum_{ihk}(-1)^{i-1}\frac{\partial\mathcal{L}}{\partial B_{ihk}}dx_1\wedge\ldots\wedge\widehat{dx_i}\wedge\ldots\wedge dx_n\ ,\quad \Omega'=\Phi\sum_{ij}(-1)^{i-1}\frac{\partial\mathcal{L}'}{\partial p_{ij}}dx_1\wedge\ldots\wedge\widehat{dx_i}\wedge\ldots\wedge dx_n$$

The Cartan form splits as $\Theta + \Theta'$, where:

$$\Theta = \theta \wedge \Omega - \mathcal{L}\eta \;, \qquad \Theta' = \phi^*\theta' \wedge \Omega' - \mathcal{L}'\eta$$

Further on we shall see the corresponding descomposition for other forms. In particular, when dealing with the Euler-Lagrange equations we shall find the "two groups" of coupled equations defining two interacting field.

Two observations are here relevant. The first is that in the new definition of minimal interaction does not appear the "interaction lagrangian". In fact, the lagrangian we use is the sum $\mathcal{L} + \mathcal{L}'$ of the lagrangians of the interacting fields; the cause of the non trivial coupling between these fields is the special structure 1-form $\theta + \phi^*\theta'$ which we take, in which variables in $E \times_X \bar{E}'$ are not separated. We could, alternatively, have adopted a more traditional point of view by considering the epimorphism of fibre bundles on X :

$$\begin{array}{ccc} J^1E \times_X J^1E' & \xrightarrow{\ \rho\ } & \pi'^*\mathrm{Hom}(T(X), E') \\ & \searrow \quad \swarrow & \\ & X & \end{array}$$

defined in a way similar to ϕ . Defining now as structure form $\theta + \theta'$ and as lagrangian $\mathcal{L} + \rho^* \mathcal{L}'$, we obtain a theory in which we now have an interaction lagrangian (contained in $\rho^* \mathcal{L}'$) . Nevertheles, we believe that this alternative version of the formalism is less natural than the one we adopted, because it does not exhibit clearly that the real problem of the interaction theory is not in the lagrangian but in the geometry of the space on which it is defined.

The second observation concerns the group-theoretical aspect of the theory. It will be noticed that the interaction problem is not a problem of groups, as it is usually stated after the success of the Yang-Mills' trick. What it happens is that when $\mathcal{L}$ and $\mathcal{L}'$ are gauge-invariants (in the sense of Definitions 1.3 and 1.4), we can expect that this circumstance will produce "gauge-symmetries" of the corresponding minimal interaction. This is what actually happens, as is shown in the following new version of Utiyama's theorem:

<u>Theorem 1.2</u>

If the lagrangians $\mathcal{L}$ and $\mathcal{L}'$ of two fields in minimal interaction are gauge-invariants, then the Lie $\mathbb{R}$-algebra $\{D_s + D_s'\}$, $s \in \Gamma(\mathrm{Ad}\,P)$, of vertical vector fields on $E \times_X E'$ is of infinitesimal simmetries of the said minimal interaction.

From now on we shall suppose this gauge-invariance, thus getting in particular, the special form of the field equations we shall obtain, as well as the reducibility properties of the Poisson Algebra and the symplectic metric which are the main aims of the present paper.

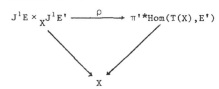

2. A Globalization of the Field Equations

In all variational problems mentioned above, the exterior differential of the Cartân form can be expressed by the basic formula:

$$(2.1) \qquad d\Theta = -\theta\Lambda(d\Omega + d \circ \eta)$$

where f is a section of the dual bundle of $q^*T^V(\ldots)$ (which is uniquely determinated by (2.1)), the exterior differential of Ω is taken with respect to a connection on $T^V(\ldots)$ with vertical torsion zero, and the exterior products are the naturals (Compare [7]) .

Once this formula is established, important results of the Calculus of Variations (first and second variation formulas, Euler-Lagrange equations, Cartan's equations, Noether invariants etc.) can be obtained in a automatic way. In particular, the condition for a section s of the corresponding fibered manifold to be extremal is that its 1-jet extension j^1s satisfies $(d\Omega + f\eta)_{j^1s} = 0$. Locally, this condition gives rise to the classical Euler-Lagrange equations of the variational problem under consideration. Thus, the differential system:

$$(2.2) \qquad (\theta, d\Omega + f \circ \eta)$$

is a globalization of these equations for every variational problem.

The system (2.2) depends essentially of the connection choosed on $T^V(\ldots)$, which afects the definition of f and the way to differentiate Ω . On the other hand, the equations thus obtained have a clear maxwellian structure: the formulas (1.8) defining the Legendre form correspond with the "first group" of Maxwell equations, while $d\Omega|_{j^1s} = -f|_{j^1s} \circ \eta$ are the generalization of the "second group".

We shall show in this section that this last remark is not a mere analogy for the type of variational problems defined by a Minimal Interaction. In fact, we shall see that $T^V(E \times_X E')$ has a canonical connection such that the corresponding system (2.2) for particular lagrangians, coincide with the global Maxwell equations and Yang-Mills equations (Compare e.g. [1]) . These last one, usually introduced by analogy with the ordinary Maxwell equations, and without showing their relation with the corresponding lagrangian theory will be obtained here as an application of our general theory of Minimal Interactions.In this way, we think the geometrical structure of these equations can be satisfactorily clarified.

2.1. Choosing a Connection on the Vertical Tangent Bundle $T^V(E \times_X E')$

The principal bundle, π^*P , induced of $p:P \longrightarrow X$ on its fibre bundle of connections $\pi:E \longrightarrow X$ by the projection π , has a canonical connection which can be described as follows.

The canonical morphism form π^*P onto P induces a vector bundles morphism $g:T_G(\pi^*P) \longrightarrow T_G(P)$, which in turn induces a morphism between the corresponding exact sequences (1.1) for π^*P and P :

$$\begin{array}{ccccccccc}
0 & \longrightarrow & Ad(\ \ ^*P) & \longrightarrow & T_G(\pi^*P) & \longrightarrow & T(E) & \longrightarrow & 0 \\
 & & \downarrow g & & \downarrow g & & \downarrow \pi^* & & \\
0 & \longrightarrow & Ad\ P & \longrightarrow & T_G(P) & \longrightarrow & T(X) & \longrightarrow & 0
\end{array}$$

As $Ad(\pi^*P) = \pi^*\, Ad\, P$, $g_{|Ad(\pi^*P)}$ is an isomorphism on each fibre. Then the exact sequence

$$0 \longrightarrow Ad(\pi^*P) \longrightarrow T_G(\pi^*P) \longrightarrow T(E) \longrightarrow 0$$

has a <u>canonical splitingg</u> $\rho: T_G(\pi^*P) \longrightarrow Ad(\pi^*P)$ defined by

$$\rho_{\sigma_x}(\bar{D}) = \rho_x(g\bar{D}) \ , \qquad \sigma_x \varepsilon E$$

where ρ_x is the projector $1-\sigma_x \cdot p^*_x$, and $\rho_x(g\bar{D})\varepsilon(Ad\, P)_x$ is considered as an element of the fibre $(Ad\,\pi^*P)_{\sigma_x}$ by the isomorphism $g:(Ad\,\pi^*P)_{\sigma_x} \longrightarrow (Ad\, P)_x$ which we mentioned before.

This splitting defines the connection we want on π^*P . In particular, we have now a canonical way of derivates covariantely sections of any bundle associated to π^*P (i e . $\pi^*Ad\, P$, π^*E' ...) .

Going back to our problem, as $\pi: E \longrightarrow X$ is the affine bundle corresponding to the vector bundle Hom(TX), Ad P) , the vertical tangent bundle T^VE can be identified with the induced vector bundle $\pi^*Hom(T(X), Ad\, P) = Hom(\pi^*T(X)\, ,\, \pi^*Ad\, P)$. Then, if we take the just introduced connection on π^*P and a <u>linear connection</u> on X , we can define in the usual way derivation law on the sections of $Hom(\pi^*T(X)\, , \pi^*Ad\, P)$. Thus a connection can be defined on the vertical tangent bundle T^VE . Locally we have:

$$\text{(2.3)} \qquad \frac{\partial^\nabla}{\partial x_i}\frac{\partial}{\partial A_{\ell j}} = \sum_{mk}\Big(\sum_h \delta_{\ell m}\, c^k_{hj}\, A_{ih} - \delta_{kj}\, \Gamma^\ell_{im}\Big)\frac{\partial}{\partial A_{mk}} \ , \ \frac{\partial^\nabla}{\partial A_{mh}}\frac{\partial}{\partial A_{\ell j}} = 0$$

where δ_{ij} is the Kronecker's symbol, and Γ^ℓ_{im} are the coefficients of the given linear connection on X .

On the other hand, as the vertical tangent bundle $T^V(E')$ of $\pi': E' \longrightarrow X$ can be identified with the induced vector bundle π'^*E' , the canonical connection of π^*P defines also a connection on the induced vector bundle of T^VE' on $E\times_X E'$ by the natural projection from $E\times_X E'$ onto E' . Locally we have:

$$\text{(2.4)} \qquad \frac{\partial^\nabla}{\partial x_i}\frac{\partial}{\partial z_j} = \sum_{hk} a^h_{kj}\, A_{ih}\frac{\partial}{\partial z_k} \ , \ \frac{\partial^\nabla}{\partial A_{mh}}\frac{\partial}{\partial z_j} = 0 \ , \ \frac{\partial^\nabla}{\partial z_j}\frac{\partial}{\partial z_k} = 0$$

By combining this two results a connection on the vertical tangent bundle $T^V(E\times_X E') = T^V(E)\oplus_X T^VE'$ can be finally established. Its local expression is given by formules (2.3) and (2.4) together with $\frac{\partial^\nabla}{\partial z_j}\frac{\partial}{\partial A_{\ell j}} = 0$. This is the connection with respect to which we are going to take the module-valued differential calculus on $\bar{E}\times_X \bar{E}'$ in what follows.

It can be observed that the only ambiguity of this connection on $T^V(E\times_X E')$ is the choosing of a linear connection on X . In particular, if X a pseudoriemannian manifold (as it is the case in classical field theory), one can take the Levi-Civitta connection corresponding to the metric, and so the ambiguity disappears . For reasons which appear below, we shall supose that the chosen linear connection has torsion zero.

Finally, another important remark is that the differential operations for forms on $\bar{E}\times_X\bar{E}'$ with values on $q^*T^V(E\times_X E') = q^*T^VE\oplus_X q'^*T^VE'$ with respect to the chosen connection, preserves the Whitney sum $q^*T^VE\oplus_X q'^*T^VE'$.

2.2. Current Tensor associated to a Minimal Interaction

By Theorem 1.2, the Cartan form $\Theta + \Theta'$ associated to the variational problem defined by a minimal interaction establishes a $\mathbb{R}$-linear mapping:

$$(2.5)\qquad s \longmapsto i(\overline{D_s + D_s'})(\Theta + \Theta')$$

between the gauge algebra $\Gamma(\mathrm{Ad}\,P)$ and the algebra of the corresponding Noether invariants.

As usual, $\overline{D_s + D_s'}$ will denote the 1-jet extension of $D_s + D_s'$ with respect to the structure 1-form $\theta + \phi^*\theta'$. From (1.16) one gets immediately:

$$(2.6)\qquad i(\overline{D_s + D_s'})(\Theta + \Theta') = i\bar{D}_s\Theta + i\bar{D}_s'\Theta'$$

The mapping (2.5) is not A-linear because of the term $i\bar{D}_s\Theta$, but, by not taking it under consideration, the resulting mapping:

$$(2.7)\qquad s \longmapsto i\,\bar{D}_s'\,\Theta'$$

is certainly A-linear.

This allows us the following:

Definition 2.1

We shall call Current Tensor associated to a Minimal Interaction, and shall denote it by J , the unique (n-1)-form on $\bar{E}\times_X\bar{E}'$ with values in the induced vector bundle $\bar{\pi}^*(\mathrm{Ad}\,P)^*$, which extends the mapping (2.7).

In terms of this tensor, the Noether invariant (2.6) can be so expressed:

$$(2.8)\qquad i\bar{D}_s\Theta + s\cdot J\ ,\quad s\in\Gamma(\mathrm{Ad}\,P)$$

where $s\cdot J$ is taken with respect to the bilinear product between sections of $\bar{\pi}^*(\mathrm{Ad}\,P)$ and $\bar{\pi}^*(\mathrm{Ad}\,P)^*$ defined by the notion of duality.

Thus, so long as the (n-1)-forms $i\bar{D}_s\Theta$, $s\in\Gamma(\mathrm{Ad}\,P)$, are identified with the Noether invariants corresponding to the gauge-simmetries of the (free) Yang-Mills' field defined on $\bar{E}$ by $\mathcal{L}\eta$, the tensor J express the difference between such invariants before and after a minimal interaction.

2.3. Field Equations: examples

Given a section $(\sigma,\sigma'):X\longrightarrow E\times_X E'$, defined by a connection σ on P and by a section σ' of E' , the restriction of the tensor J to the 1-jet extension

$\overline{\sigma,\sigma'}$) of (σ,σ') with respect to $\theta+\phi^*\theta'$ gives rise to on $(\mathrm{Ad}\,P)^*$-valued $(n-1)$-orm on X depending only of σ'. Its contraction with the contravariant volume lement η^* will then be an $(\mathrm{Ad}\,P)^*$-valued 1-contravariant tensor on X, which we hall denote by $J(\sigma')$. On the other hand, as $T^vE = q^*\pi^*\mathrm{Hom}(T(X),\mathrm{Ad}\,P)$, the corresonding restriction of the component Ω in the Legendre form defines a 1-contravaiant $(n-1)$-covariant hemisymmetric tensor on X with values in $(\mathrm{Ad}\,P)^*$ depending nly of σ, which we shall denote by $\Omega(\sigma)$.

Now we have:

Theorem 2.1

A section $(\sigma,\sigma'):X \longrightarrow E\times_X E'$ is an extremal of the variational problem defied by a minimal interaction if and only if:

$$d_\nabla \Omega(\sigma) = J(\sigma')\otimes\eta \tag{2.9}$$

where d_∇ is the covariant exterior differential for $(\mathrm{Ad}\,P)^*$-valued tensor on X efined by the connection σ and by the given linear connection on X), and the ection $\sigma':X\longrightarrow E'$ is an extremal of the variational problem defined on $\bar{E}'$ by he structure 1-form θ'^σ and the lagrangian density $\mathcal{L}'\eta$.

Proof

It is only a matter of computing the components $(d\Omega + f\circ\eta)_{\overline{(\sigma,\sigma')}} = 0$ and $d\Omega' + f'\circ\eta)_{\overline{(\sigma,\sigma')}} = 0$ of the field equations with the connection defined in §2.2. he first one will give rise to equation (2.9) while the second will be equivalent o rest of the assertion.

As we remarked at the end of §2.2, the differentials of the components coincide ith the components of the differentials, and then $(d\Omega)_{\overline{(\sigma,\sigma')}} = d\Omega_{\overline{(\sigma,\sigma')}}$. But this ast differential, when interpreted on the base manifold X, is $d_\nabla\Omega(\sigma)$ by definiion of covariant exterior differential. On the other hand, when computing f by he general method of [7] (Lemma 3.1), one has the following local expression:

$$f_{\ell m} = \frac{\partial\mathcal{L}}{\partial A_{\ell m}} - \sum_{ihk}\Big(\sum_\alpha \delta_{\ell h}\, c^k_{\alpha m} A_{i\alpha} - \delta_{km}\Gamma^\ell_{ih}\Big)\frac{\partial\mathcal{L}}{\partial B_{ihk}} - \sum_{j\beta} a^m_{j\beta} z_\beta \frac{\partial\mathcal{L}'}{\partial p_{\ell j}} =$$

$$= \underbrace{\frac{\partial\mathcal{L}}{\partial A_{\ell m}} - \sum_{ih\alpha} c^k_{\alpha m} A_{i\alpha}\frac{\partial\mathcal{L}}{\partial B_{i\ell k}}}_{(1)} + \underbrace{\sum_{ih}\Gamma^\ell_{ih}\frac{\partial\mathcal{L}}{\partial B_{ihm}}}_{(2)} - \sum_{j\beta} a^m_{j\beta} z_\beta\frac{\partial\mathcal{L}'}{\partial p_{\ell j}}$$

By Utiyama's theorem, $(1)=0$ and $\dfrac{\partial\mathcal{L}}{\partial B_{ihm}} + \dfrac{\partial\mathcal{L}}{\partial B_{him}} = 0$. Then, by the condition f torsion zero (locally $\Gamma^\ell_{ih} = \Gamma^\ell_{hi}$) imposed to the given linear connection on X, ne has $(2)=0$ also, and from here $f_{\ell m} = -\sum_{j\beta} a^m_{j\beta} z_\beta \dfrac{\partial\mathcal{L}'}{\partial p_{\ell j}}$, which are precisely he local components of the tensor $-J(\sigma')$. This proves (2.9).

Finally, the connection σ of P allows us to identify in an obvious way $\bar{E}'$ ith the submanifold $\{(j^1_x\sigma,(e'_x,h_x))\,|\,(e'_x,h_x)\in\bar{E}'\}$ of $\bar{E}\times_X\bar{E}'$ and θ'^σ with the restriction to it of $\phi^*\theta'$. By using this identification, the condition $(d\Omega' + f' \circ \eta)_{\overline{(\sigma,\sigma')}} = 0$ is nothing but what is said in the second part of the Theorem. q.e.d.

Remark

The hemisymmetry condition $\frac{\partial \mathcal{L}}{\partial B_{ihm}} + \frac{\partial \mathcal{L}}{\partial B_{him}} = 0$ together with the symmetry one $\Gamma^{\ell}_{ih} - \Gamma^{\ell}_{hi} = 0$ allows us to prove that $d_{\nabla}\Omega(\sigma)$ does not depend on the given linear connection on X . Thus, these field equations are canonically associated to a minimal interactions.

Equation (2.9) generalizes to an arbitrary minimal interaction the ordinary Maxwell equations and the global Yang-Mills equations recentily proposed by some authors, as we show with some examples.

Example 1.

In the electromagnetic field theory, $P = X \times U(1)$, where X is a pseudoriemannian orientable manifold and U(1) is the 1-dimensional unitary group. In this case, connections on P are identified with 1-forms on X (electromagnetic potencials). Their exterior differentials are the corresponding electromagnetic field tensors. The variational problem is defined by the (gauge-invariant) lagrangian density $\mathcal{L}\eta$, where η is the pseudoriemannian volume element and $\mathcal{L}$ is given by:

$$\mathcal{L}(j^1_x\omega) = 1/4\ \|F_2\|^2_x \ , \quad F_2 = d\omega$$

where $\|\ \|$ is the norm defined by the metric on 2-forms.

Equations (1.8), which define the Legendre form, give in this case $\Omega(\omega) = i\,F^2\eta$, where F^2 is the contravariant electromagnetic field and the interior products is taken with respect to the first indices in F^2 and η .

Acording to his we have:

$$d_{\nabla}\Omega(\omega) = d_{\nabla}(i\,F^2\eta) = \bullet\, d^{\nabla} F^2 \otimes \eta$$

where d^{∇} is the covariant differential with respect to the given linear connection on X and "•" indicates contraction of the first index of F^2 with the covariant index resulting from differentiation. As it is well known, as the linear connection is torsionfree and F^2 is hemisymmetric, $d^{\nabla}F^2$ is independent of the said linear connection. It is the divergence of the 2-contravariant hemisymmetric tensor F^2 .

Then (2.9) becomes finally the well known Maxwell equation:

$$\operatorname{div} F^2 = J(\sigma')$$

In this case, as $\operatorname{Ad} P = X \times \mathbb{R}$, $J(\sigma')$ is simply a vector field on X .

By identifying covariant and contravariant forms by means of the metric, the result can be expressed in terms of the codifferential $\delta = *^{-1}d*$ in the well known way:

$$F_2 = d\omega \quad , \quad \delta F^2 = J(\sigma')$$

Example 2

With the same hypothesis for the manifold X , let now P be a principal bundle with semisimple structural group. Then the Cartan-Killing metric induces a non-singular scalar product on $\Gamma(\mathrm{Ad}\,P)$, which, together with the pseudoriemannian metric in X , allows us to define a natural norm on Ad P-valued differential forms of X . Now the ordinary Yang-Mills' fields are defined by the variational problem whose lagrangian is given by:

$$\mathcal{L}(j^1_x\sigma) = 1/4\ \|F_2\|^2_x \quad , \quad F_2 = \mathrm{Curv}\,\sigma$$

By contravariating F_2 with the metric on X and with the Cartan-Killing one, we get on $(\mathrm{Ad}\,P)^*$-valued 2-contrariant hemisymmetric tensor, F^2 , in terms of which $\Omega(\sigma)$ is given in a way identical to the electromagnetic field case. Then (2.9) becomes:

$$\cdot\, d^\nabla F^2 = J(\sigma')$$

where d^∇ is the covariant differential with respect to the given linear connection on X and the connection σ . Once more, $d^\nabla F^2$ depends only of σ , and it is natural to call it <u>divergence</u> of the $(\mathrm{Ad}\,P)^*$-valued , 2-contravariant hemisymmetric tensor, F^2 , with respect to the connection σ .

As in the case of electromagnetic field, by identifying Ad P-valued covariant forms and $(\mathrm{Ad}\,P)^*$-valued contravariant forms on X by means the said metrics, the result can be expressed by:

$$F_2 = \mathrm{Curv}\,\sigma \quad , \quad \delta_\sigma F_2 = J(\sigma')$$

where $\delta_\sigma = *^{-1} d_\sigma *$, d_σ being the exterior differential for Ad P-valued forms on X with respect to the connection σ .

Remark

These last equations coincide with those proposed by M.E. Mayer in [1] and in his lecture at this Conference provided the source-field σ' is given a priori. We believe that the present paper sufficiently clarifies the geometric nature of the current term $J(\sigma')$ in the above equations (could this be a solution to the question mark in the so called Wu-Yang diccionary [20] ?). Of course, even though $J(\sigma')$ is related to Noether's invariants of σ' , it is no way a Noether invariant.

Example 3

The decomposition $\Omega(\sigma) = iF^2\eta$ in the preceding examples is always locally possible in virtue of equations (1.8) defining the Legendre form. When this descomposition can be globally made, the same argument proves that (2.9) becomes:

$$\cdot\, d^\nabla F^2 = J(\sigma')$$

In this way, what in examples 1 and 2 are $F_2 = d\omega$ and $F_2 = \mathrm{Curv}\,\sigma$ respectively (first group of Maxwell equations), is here the particular way of obtaining F^2 .

An inmediate consequence of this last example is the following general formula that we shall employ later on:

Corollary.

If the section $(\sigma,\sigma'):X \longrightarrow E \times_X E'$ is extremal, then, with the above notations, we have:

$$\cdot\, d^{\nabla}\Omega(\sigma) = iJ(\sigma')\eta \tag{2.10}$$

Proof.

It is enough to prove the equality locally. In this case, by example 3 we have $\Omega(\sigma) = iF^2\eta$, so (2.9) becomes $\cdot\, d^{\nabla}F^2 = J(\sigma')$. Then:

$$\cdot\, d^{\nabla}\Omega(\sigma) = \cdot\, d^{\nabla}(iF^2\eta) = i(\cdot\, d^{\nabla} F^2)\eta = iJ(\sigma')\eta$$

3. Jacobi Fields along an Extremal

The notions of hessiano and Jacobi field along an extemal, for the type of variational problems defined by a minimal interaction, can be stablished in a completely analogous way to the one in [7] for the ordinary case. In particular, Jacobi fields can be characterised as follows.

Given a π-vertical vector field Y on $E \times_X E'$ defined along a section (σ,σ') will be called the 1-jet extension of Y to the $\bar{\pi}$-vertical $\bar{Y}$ on $\bar{E} \times_X \bar{E}'$ defined along the 1-jet extension $\overline{(\sigma,\sigma')}$ of (σ,σ') such that:

$$L_{\bar{Y}}(\theta + \phi^*\theta')|_{\overline{(\sigma,\sigma')}} = 0 \tag{3.1}$$

Now, if (σ,σ') is an extremal, Y is a Jacobi field along it if and only if:

$$L_{\bar{Y}}\left[d(\Omega + \Omega') + (f + f')\circ\eta\right]|_{\overline{(\sigma,\sigma')}} = 0 \tag{3.2}$$

(3.1) and (3.2) are the version of the Jacobi equation for this particular type of variational problems.

As $T^V(E \times_X E') = T^V(E) \oplus_X T^V(E')$ one has $Y = Z + Z'$, where Z and Z' are, respectively, vertical vector fields on E and E' defined along σ and σ'. Then (3.1) and (3.2) should give rise to a system of coupled equations for Z and Z', which could be expressed in a way analogous to the one given for the field equations in Theorem 2.1. In fact it happens to be so, but we shall not deal with then here in order not to lengthen this paper too much. Nevertheless we shall try to give, for the $q^*T^V(E)$-component $L_{\bar{Y}}(d\Omega + f\eta)|_{\overline{(\sigma,\sigma')}} = 0$ in (3.2), an expression similar to the one for the field equation (2.9).

In this sense, if K is the curvature 2-form of the connection introduced in §2.1, one has:

$$L_{\bar{Y}}(d\Omega+f\eta)|_{\overline{(\sigma,\sigma')}} = L_{\bar{Y}}d\Omega|_{\overline{(\sigma,\sigma')}} + (\bar{Y}f)_{\overline{(\sigma,\sigma')}}\eta = d\, L_{\bar{Y}}\Omega|_{\overline{(\sigma,\sigma')}} + (i\bar{Y}K)\wedge\Omega|_{\overline{(\sigma,\sigma')}} + (\bar{Y}f)_{\overline{(\sigma,\sigma')}}\eta$$

Then, with $J(Y)$ denoting the contraction with the contravariant volume element of the (Ad P)*-valued (n-1)-form defined by $L_{\bar{Y}}J|_{\overline{(\sigma,\sigma')}}$, and analogously with $\Omega(Y)$ denoting the restriction $L_{\bar{Y}}\Omega|_{\overline{(\sigma,\sigma')}}$, a computation similar to the one in the proof of Theorem 2.1, gives the following equation on the base manifold, which has to satisface every Jacobi field along the extremal (σ,σ'):

(3.3) $$d_{\nabla}\Omega(Y) + (i\bar{Y}K)_{\overline{(\sigma,\sigma')}} \wedge \Omega(\sigma) = J(Y) \otimes \eta$$

It is an analogous equation to (2.9) except the second term which depends on the curvature of the choosen connection. If the structural group of the principal bundle P is abelian, $K = 0$, so this term does not appear and we have equations in all analogous to the field ones. This is, for instance, the case of electromagnetic field, where $G = U(1)$, and where one recovers the Maxwell equations.

Also, as in §2.3, one has the following formula analogous to (2.10):

(3.4) $$\bullet\, d^{\nabla}\Omega(Y) + \bullet\,(i\bar{Y}K)_{\overline{(\sigma,\sigma')}} \wedge \Omega(\sigma) = iJ(Y)\eta$$

where the "•" in the second tern denotes contraction of the contravariant index in $\Omega(\sigma)$ with the covariant one in $(i\bar{Y}K)_{\overline{(\sigma,\sigma')}}$.

4. Reducibility of the Poisson Algebra and the Symplectic Metric of a Minimal Interaction

Mimicking the ordinary case ([7], [13], [15]) we can associate to every minimal interaction a "generaliced symplectic structure" in the following way.

The set V of extremals of the variational problem defined by a minimal interaction can we thought of as a sort of "infinite-dimensional manifold", on which, we take as tangent space $T_{(\sigma,\sigma')}(V)$ in each point (σ,σ') , the vector space of Jacobi fields along the extremal (σ,σ') . This is the so called manifold of solutions of the variational problem under consideration. In this language, the notions of infinitesimal symmetry and Noether invariant acquire new interpretation. If D is an infinitesimal symmetry, then the vertical component $D_{(\sigma,\sigma')}$ of D with respect to an extremal (σ,σ') is a Jacobi field along it. Thus, infinitesimal symmetries can be considered as vector fields on the manifold of solutions. On the other hand, if ω is a Noether invariant, its restriction to the 1-jet extension $\overline{(\sigma,\sigma')}$ of an extremal (σ,σ') defines a closed (n-1)-form on the base manifold X (Noether's theorem); now, by taking the cohomology class in $H^{n-1}(X, \mathbb{R})$ defined by this form, ω can be interpreted as a function on V with values in $H^{n-1}(X, \mathbb{R})$. This suggests to take this cohomology space as "scalars" in place of real numbers. In particular, those Noether invariants defining an exact (n-1)-form should be interpreted as identically zero functions on the manifold of solutions. Consequently with this language, tensors on each point (σ,σ') are defined as $\mathbb{R}$-multilinear mappings from the tangent space $T_{(\sigma,\sigma')}(V)$ into $H^{n-1}(X, \mathbb{R})$. As in the ordinary case, two remarkable examples of this notion are the following.

If Y is a Jacobi field along an extremal (σ,σ') and ω is a Noether invariant, then $i\bar{Y}d\omega|_{\overline{(\sigma,\sigma')}}$ is closed, thus allowing us to define on $\mathbb{R}$-linear mapping

$(d\omega)_{(\sigma,\sigma')}:T_{(\sigma,\sigma')}(V) \longrightarrow H^{n-1}(X, \mathbb{R})$, which can be taken as definition of <u>differential</u> of the function ω at the point (σ,σ') . As usual, the scalar $(d\omega)_{(\sigma,\sigma')}(Y)$ should be called <u>derivative</u> of ω with respect to Y and be denoted by $Y\omega$. Analogously, if Y and $\bar{Y}$ are two Jacobi fields along the extremal (σ,σ') , $i\bar{Y}' \, i\bar{Y} \, d(\Theta+\Theta')|_{\overline{(\sigma,\sigma')}}$ is also closed, thus allowing us to define an $\mathbb{R}$-bilinear hemisymmetric mapping:

$$T_{(\sigma,\sigma')}(V) \times T_{(\sigma,\sigma')}(V) \xrightarrow{(\omega_2)_{(\sigma,\sigma')}} H^{n-1}(X,\mathbb{R})$$

which is the natural generalization to a minimal interaction of the symplectic 2-form of the ordinary Analytical Mechanics.

Once we are here, we can establish the main results of this paper.

<u>Theorem 4.1.</u> (Reducibility of the Poisson Algebra)

Let $\mathcal{D}$ be the idealizator of the Lie algebra $\{D_s + D_s'\}$ $s\varepsilon\Gamma(\mathrm{Ad}\,P)$ of gauge-symmetries of a minimal interaction in the Lie algebra of all its infinitesimal symmetries.

If ω_Y is the function on V defined by the Noether invariant corresponding to an infinitesimal symmetry $Y\varepsilon\mathcal{D}$ and $D_s + D_s'$ is any gauge-symmetry, then for every point (σ,σ') one has:

$$(4.1) \qquad (D_s + D_s')_{(\sigma,\sigma')}\omega_Y = 0$$

In particular, if Y is a gauge symmetry, the function ω_Y is identically zero.

<u>Proof</u>

As $\{D_s + D_s'\}$ $s\varepsilon\Gamma(\mathrm{Ad}\,P)$ is an ideal in $\mathcal{D}$, everything reduces to proving the last statement; i.e. the restriction of any Noether invariant $(\overline{D_s + D_s'})(\Theta+\Theta')$ to the 1-jet extension $\overline{(\sigma,\sigma')}$ of an extremal (σ,σ') is an exact (n-1)-form.

From (2.8) and the definitions of $\Omega(\sigma)$ and $J(\sigma')$, one has:

$$i(\overline{D_s + D_s'})(\Theta+\Theta')|_{\overline{(\sigma,\sigma')}} = i\bar{D}_s\Theta|_{\overline{(\sigma,\sigma')}} + s \cdot iJ(\sigma')\eta = \theta(\bar{D}_s)|_{\overline{(\sigma,\sigma')}}\Omega(\sigma) + s \cdot iJ(\sigma')\eta$$

where the upper " • " denotes contraction with respect to the bilinear product between sections of $\mathrm{Ad}\,P$ and $(\mathrm{Ad}\,P)^*$ defined by the notion of duality.

On the other hand, from (1.4) a simple local calculation proves that:

$$\theta(\bar{D}_s)|_{\overline{(\sigma,\sigma')}}\Omega(\sigma) = d^{\nabla}s : \Omega(\sigma)$$

where $d^{\nabla}s$ is the covariant differential of the section s with respect to the connection σ , the upper " • " is as before and the lower " • " denotes contraction of the contravariant index in $\Omega(\sigma)$ with the covariant one given by differentiation.

Thus, by the properties of d^{∇} , the computation of the Noether invariant can proceed in the following way:

$$d^{\nabla}s:\Omega(\sigma) + s \cdot iJ(\sigma')\eta = \cdot\, d^{\nabla}(s \cdot \Omega(\sigma)) - s\cdot(\cdot\, d^{\nabla}\Omega(\sigma)) + s \cdot iJ(\sigma')\eta =$$

$$= \cdot\, d^{\nabla}(s \cdot \Omega(\sigma)) - s \cdot \left[\cdot\, d^{\nabla}\Omega(\sigma) - iJ(\sigma')\eta\right]$$

The last term is zero by the equation (2.10), so everything reduces to proving that $\cdot\, d^{\nabla}(s \cdot \Omega(\sigma))$ is an exact (n-1)-form.

By proceeding as in the proof of (2.10) and using the same notations, $\cdot d^{\nabla}(s \cdot \Omega(\sigma))$ can be locally computed as follow:

$$\cdot\, d^{\nabla}(s \cdot \Omega(\sigma)) = \cdot d^{\nabla}(s \cdot iF^2\eta) = \cdot d^{\nabla}\left[i(s \cdot F^2)\eta\right] = i\left[\cdot d^{\nabla}(s \cdot F^2)\right]\eta = *\left[*^{-1} d * (s \cdot F^2)\right] =$$

$$= d * (s \cdot F^2)$$

But $*(s \cdot F^2)$ is the restriction to $\overline{(\sigma,\sigma')}$ of the (n-2)-form $*(s \cdot F)$ on $\bar{E} \times_X \bar{E}'$, where F is any local solution of the first equation (1.8) and s and $*$ are considered induced on $\bar{E} \times_X \bar{E}'$ by the projection $\bar{\pi}$. Then, although F is only locally and non-uniquely determined, the contraction $*(s \cdot F)$ is unique. Thus, the family of local (n-2)-forms $*(s \cdot F)$ defines a <u>global (n-2)-form</u> on $\bar{E} \times_X \bar{E}'$, and the exterior differential of its restriction to $\overline{(\sigma,\sigma')}$ coincides, by the above calculation, with $\cdot\, d^{\nabla}(s \cdot \Omega(\sigma))$. q.e.d.

<u>Theorem 4.2.</u> (Reducibility of the Symplectic Metric)

Let (V,ω_2) be the symplectic manifold of solutions associated to a minimal interaction.

If $(D_s + D_s')_{(\sigma,\sigma')}$ is the tangent vector at $(\sigma,\sigma')\in V$ defined by any gauge-symmetry, then one has:

$$i(D_s + D_s')_{(\sigma,\sigma')} \cdot \omega_2 = 0 \tag{4.2}$$

i.e. the gauge-symmetries of a minimal interaction belong to the radical of its symplectic metric.

<u>Proof</u>

We have to see that, if Y is an arbitrary Jacobi field along an extremal (σ,σ'), then $i\,\bar{Y}\, i(\overline{D_s + D_s'})d(\Theta + \Theta')_{\overline{(\sigma,\sigma')}}$ is an exact (n-1)-form on X.

As $D_s + D_s'$ is an infinitesimal symmetry then $L_{\overline{D_s + D_s'}}(\Theta + \Theta') = 0$, whence:

$$i\bar{Y}i(\overline{D_s + D_s'})d(\Theta + \Theta')|_{\overline{(\sigma,\sigma')}} = -\bar{Y}di(\overline{D_s + D_s'})(\Theta + \Theta')|_{\overline{(\sigma,\sigma')}} = -\bar{Y}d(i\bar{D}_s\Theta + i\bar{D}_s'\Theta')|_{\overline{(\sigma,\sigma')}} =$$

$$= -L_{\bar{Y}}(\theta(\bar{D}_s)\Omega)|_{\overline{(\sigma,\sigma')}} - L_{\bar{Y}}(\phi*\theta'(\bar{D}_s')\Omega')|_{\overline{(\sigma,\sigma')}} = -\bar{Y}\theta(\bar{D}_s)|_{\overline{(\sigma,\sigma')}}\Omega(\sigma) - \theta(\bar{D}_s)|_{\overline{(\sigma,\sigma')}}\Omega(Y) -$$

$$- s \cdot iJ(Y)\eta$$

A local calculus proves the following two formules:

$$-\bar{Y}\theta(\bar{D}_s)|_{\overline{(\sigma,\sigma')}}\Omega(\sigma) = s^{\bullet}\left|\bullet(i\bar{Y}K)_{\overline{(\sigma,\sigma')}} \wedge \Omega(\sigma)\right| \ , \ \theta(\bar{D}_s)|_{\overline{(\sigma,\sigma')}}\Omega(Y) = d^{\nabla}s : \Omega(Y)$$

so the above computation can proceed as follows:

$$s^{\bullet}\left[\bullet(i\bar{Y}K)_{\overline{(\sigma,\sigma')}} \wedge \Omega(\sigma)\right] - d^{\nabla}s{:}\Omega(Y) - s^{\bullet}\, iJ(Y)\eta = s^{\bullet}\left[\bullet(i\bar{Y}K)_{\overline{(\sigma,\sigma')}} \wedge \Omega(\sigma)\right] - \bullet d^{\nabla}(s^{\bullet}\Omega(Y)) +$$

$$+ s^{\bullet}(\bullet d^{\nabla}\Omega(Y)) - s^{\bullet}\, iJ(Y)\eta = -\bullet d^{\nabla}(s^{\bullet}\,\Omega(Y)) + s^{\bullet}\left[\bullet d^{\nabla}\Omega(Y) + \bullet(i\bar{Y}K)_{\overline{(\sigma,\sigma')}} \wedge \Omega(\sigma) - \right.$$

$$\left. - \, iJ(Y)\eta\right]$$

but $\bullet d^{\nabla}(s^{\bullet}\,\Omega(Y))$ is an exact (n-1)-form on X, by the same argument employed in the final part of the proof for Theorem 4.1, and the last term is zero by the equation (3.4), which has to satisface the Jacobi field Y along the extremal (σ,σ'). q.e.d.

References

1. W. Drechsler and M. Mayer, Fiber Bundle Techniques in Gauge Theories, Lecture Notes in Physics, vol. 67, Springer-Verlag (1977).
2. A. Fischer and J. Marsden, The Einstein equations of evolution, a Geometric approach, Journ. Math. Phys., 13 (1972), 546-568.
3. A. Fischer and J. Marsden, Linearization stability of the Einstein equations, Bull. Am. Math. Soc., 79 (1973), 997-1003.
4. A. Fischer and J. Marsden, General Relativity as a Hamiltonian System, Symposia Mathematica, vol. 14 (1974), 193-205.
5. P. García, Geometria Simplectica en la Teoría Clásica de Campos (Tesis), Universidad de Barcelona, 1967, Collect. Math., 19 (1968), 73-134.
6. P. García, Estructura Compleja en la Teoría Clásica de Campos, Collect. Math., 19 (1968), 155-175.
7. P. García, The Poincaré-Cartán Invariant in the Calculus of Variations, Symposia Mathematica, vol. 14 (1974), 219-246.
8. P. García, Reducibility of the Symplectic Structure of Classical Fields with Gauge-symmetry, Proc. Symp. on Diff. Geom. Meth. in Phys., Bonn 1975, Lecture Notes in Mathematics, vol. 570, Springer-Verlag (1977), 365-376.
9. P. García, Gauge Algebras, Curvature and Symplectic Structure, J. Differential Geometry, 12 (1977), 351-359.
10. P. García, Critical Principal Connections and Gauge-invariance, to appear in Reports on Math. Phys.
11. P. García and A. Pérez-Rendón, Symplectic Approach to the Theory of Quantized Fields I, Commun. Math. Phys., 13 (1969), 24-44.
12. P. García and A. Pérez-Rendón, Symplectic Approach to the Theory of Quantized Fields II, Arch. for Rat. Mech. and Anal, 43 (1971), 101-124.
13. H. Goldschmidt and S. Sternberg, The Hamilton-Cartán Formalism in the Calculus of Variations, Ann. Inst. Fourier, 23 (1973), 203-267.

13. H. Kerbrat-Lunc, Contribution à l'étude du champ de Yang-Mills, Ann. Inst. H. Poincaré 13 A (1970), 295-343.

15. J. Kijowski and W. Szczyrba, A Canonical Structure for Classical Field Theories, Commun. Math. Phys., 46 (1976), 183-206.

16. A. Pérez-Rendón, Principio de Mínima interaction y Variedad Funcional Asociada (Tesis) Universidad de Salamanca, 1973.

17. A. Pérez-Rendón, A Minimal Interaction Principle For Classical Fields, Symposia Mathematica, vol. 14 (1974), 293-321.

18. A. Pérez-Rendón, Yang-Mills Interactions: a problem not depending on the Gauge-invariance, Proc. of the 3rd. International Colloquium on Group Theoret. Meth. in Physics, Marseille, (1974).

19. R. Utiyama, Invariant theoretical interpretation of interaction, Phys. Rev. 101 (1956), 1597-1607.

20. T.T. Wu and C.N. Yang, Concept of Nonintegrable Phase Factors and Global Formulation of Gauge Fields, Phys. Rev. 12 D (1975), 3845-3857.

21. C.N. Yang, Integral Formalism for Gauge Fields, Phys. Rev. Lett., 33 (1974), 445-447.

AMBIGUITIES IN CANONICAL TRANSFORMATIONS OF CLASSICAL SYSTEMS AND THE SPECTRA OF QUANTUM OBSERVABLES

M. Moshinsky⁺ and T.H. Seligman
Instituto de Fisica, Universidad de México (UNAM)
México 20, D.F.

ABSTRACT. We discuss through the examples of the oscillator with centrifugal potential and the Coulomb problem in dilated form, the general procedure for finding the representation in quantum mechanics of non-linear and non-bijective canonical transformations. The ambiguity group associated with the canonical transformations and the irreducible representations of this group, which lead to the concept of ambiguity spin, suggest also that some features of the spectra of quantum operators are already present in the classical picture.

1. INTRODUCTION

The contribution presented by one of us (M.M.) at the 1977 Bonn Conference on Geometrical Quantization was based on two articles[1,2) that are already in press. Thus for the proceedings of this conference we wish to discuss some more recent developments of the subject.

In our previous papers[1,2,3) we were interested in a general procedure for finding the representation in quantum mechanics of non-linear and non-bijective canonical transformations. In this note we wish to indicate that the previous developments also suggest that the spectra of quantum observables may be already implicit in the corresponding classical problem.

We start by associating with any classical observable $f(q,p)$ (where q,p are coordinates and momenta and for simplicity we restrict this paper to problems of one degree of freedom) a canonical transformation, by considering that it defines a new generalized coordinate $\bar{q}$ i.e.,

---

⁺Member of the Instituto Nacional de Energia Nuclear and of El Colegio Nacional.

$$\bar{q} = f(q,p) \ . \tag{1.1}$$

As we shall see later we may have to modify (1.1) slightly if $f(q,p)$ does not take all real values between $-\infty$ and $+\infty$ which is the range of variation of $\bar{q}$. For the moment though let us consider what would be the momentum canonically conjugate to $\bar{q}$ which we designate by $\bar{p}$. If $f(q,p) = F(q)$ is a function of q only, $\bar{p} = [F'(q)]^{-1}p$ (where the prime indicates derivative with respect to q) as obviously then we get the value 1 for the Poisson bracket

$$\{\bar{q},\bar{p}\}_{q,p} \equiv \frac{\partial\bar{q}}{\partial q}\frac{\partial\bar{p}}{\partial p} - \frac{\partial\bar{q}}{\partial p}\frac{\partial\bar{p}}{\partial q} = 1 \ . \tag{1.2}$$

If $f(q,p)$ is a function of both q,p we can identify it with a Hamiltonian and make use of Hamiltons equation of motion to get the time as function of q,p . As the time is the canonically conjugate variable to the Hamiltonian[4] we have then a procedure for getting $\bar{p}$. More explicitly if $f(q,p)$ is taken as a Hamiltonian

$$\frac{dq}{dt} \equiv \dot{q} = \frac{\partial f(q,p)}{\partial p} \ , \quad \frac{dp}{dt} \equiv \dot{p} = -\frac{\partial f(q,p)}{\partial q} \ , \tag{1.3a,b}$$

and $f(q,p)$ is a constant of the motion which we designate by E . Thus from the implicit equation (1.3a), subject to well-known restrictions[5], we can get

$$p = g(q,\dot{q}) \ , \quad f[q,g(q,\dot{q})] \equiv h(q,\dot{q}) = E \ . \tag{1.4a,b}$$

From the implicit equation (1.4b) we in turn get[4]

$$\dot{q} = \Phi(q,E) \ , \quad t(q,E) = \int \Phi^{-1}(q,E)dq \ . \tag{1.5a,b}$$

Note that $t(q,E)$ is indetermined up to a function of E , thus leading to a family of canonically conjugate variables to q given by

$$\bar{p} = -t[q,f(q,p)] \ , \tag{1.6}$$

where the minus sign is due to the fact that $f(q,p)$ was taken as

a generalized coordinate $\bar{q}$ rather than the momentum $\bar{p}$ [4].
In reference 2 we considered the canonical transformation associated with

$$f(q,p) = \frac{1}{2}(p^2 \pm q^2) \ , \qquad (1.7)$$

i.e., the Hamiltonians, in appropriate units, of the attractive and repulsive oscillators. We found it to be of great importance to study the multivaluedness of the classical canonical transformation associated with these Hamiltonians.
We considered the groups of canonical transformations both on the original and on the new phase spaces that left unchanged the defining relations of the canonical transformations (1.1), (1.6) associated with the harmonic oscillator. We called these groups "ambiguity groups". For the repulsive oscillator we found the ambiguity group to be a simple inversion group in the original phase space, while for the attractive oscillator the ambiguity group was a semi-direct product $T \wedge I$ where T was a translation group by $2m\pi$ of the new momenta (i.e., $(\bar{p},\bar{q}) \rightarrow (\bar{p}+2m\pi,\bar{q})$, m integer) and I the inversion in the new phase space.
The two problems thus differ essentially in that in one case the ambiguity group is of finite order and its irreducible representations are characterized by a label taking a finite set of values, while in the other case the ambiguity group of denumerably infinite order leads to a representation label that ranges over a bounded but continuous set, as is well-known from Bloch functions, and to a row label that takes two values as the irreducible representations turn out to be two-dimensional.
If we now turn our attention to the quantum observables we note that the spectrum of $\bar{q}$ ranges once from $-\infty$ to $+\infty$; the one of the repulsive oscillator ranges twice over the same values and this degeneracy is implicit in the two irreducible representations of the ambiguity group of this problem. On the other hand the spectrum of the attractive oscillator is discrete and the continuous, but bounded, range of the representation label was found to be precisely adequate to smear this spectrum into a continuum.
The examples mentioned above and discussed in reference 2 are very special and the analysis of further examples will be important to obtain information about the general validity of the conclusions drawn. We shall, therefore, in this note retrace the basic argument

of reference 2 for other cases of interest.
To begin with we discuss in the next section the case of the oscillator both repulsive and attractive with a centrifugal force. This will provide a slightly more general problem than the one mentioned in the previous paragraph and at the same time establish the ground work for an analysis in section 3 of the Coulomb problem in the well-known dilated or pseudo-Coulomb form i.e., when we replace q by $(\pm 2E)^{-1/2}q$ where E is the energy of the state. Finally in the last section we discuss the conclusions we can derive from the present examples, as well as from the one related with the standard Coulomb problem, which we have also analyzed but, because of its length, reserve for another publication.

2. THE OSCILLATOR PROBLEM WITH A CENTRIFUGAL POTENTIAL.

The observables we want to discuss in this section are of the form

$$f_{\pm}(q,p) = \frac{1}{2}\,(p^2 + \Lambda^2 q^{-2} \pm q^2)\ , \tag{2.1}$$

where Λ is some positive real constant. They correspond, in appropriate units, to the Hamiltonians of attractive (+ sign) or repulsive (- sign) oscillators with a centrifugal potential of strength Λ^2. We shall first discuss both cases classically and then pass to the corresponding quantum mechanical problems.

a) Canonical Transformations Associated with the Repulsive Oscillator Problem.

We start by considering the case of the repulsive oscillator. As indicated in the previous section we can identify its Hamiltonian with the generalized coordinate i.e.

$$\bar{q} = \frac{1}{2}\,(p^2 + \Lambda^2 q^{-2} - q^2)\ , \tag{2.2}$$

where both the left and the right hand side of the equation can take all real values in the range $-\infty$ to $+\infty$. From the Hamiltonian equations of motion $p=\dot{q}$ and as the right hand side of (2.2) is a constant of motion, which we designate by E, we obtain for the time variable canonically conjugate[4] to the Hamiltonian[7]

$$t(q,E) = \int \frac{q\,dq}{(2Eq^2-\Lambda^2+q^4)^{1/2}} = \frac{1}{2}\ \ln\left[\frac{(q^4+2Eq^2-\Lambda^2)^{1/2}+(q^2+E)}{(E^2+\Lambda^2)^{1/2}}\right] \qquad (2.3)$$

where we easily check that the derivative of the right hand side with respect to q gives the integrand. The variable canonically conjugate to $\bar{q}$ i.e., $\bar{p}$ is then given by

$$\bar{p} = -t[q,f_-(q,p)] = -\frac{1}{2}\ \ln\left[\frac{qp+(p^2+\Lambda^2q^{-2}+q^2)/2}{[f_-^2(q,p)+\Lambda^2]^{1/2}}\right], \qquad (2.4)$$

for which we easily check that $\{\bar{q},\bar{p}\}_{q,p} = 1$. Note that equation (2.4) is always real as the argument of the logarithm is positive due to the fact that $(p^2+q^2+2qp) = (p+q)^2$.
The equations (2.2), (2.4) define the canonical transformation associated with the observable $f_-(q,p)$ and for later convenience we also write them in the implicit form

$$\bar{q} = \frac{1}{2}\ (p^2+\Lambda^2q^{-2}-q^2)\ . \qquad (2.5a)$$

$$(\bar{q}^2+\Lambda^2)^{1/2}\exp(-2\bar{p}) = qp+\frac{1}{2}(p^2 + \Lambda^2q^{-2} + q^2). \qquad (2.5b)$$

We note at this stage that there is no transformation in $(\bar{q},\bar{p})$ space that leaves the left hand side of the equations (2.5) invariant while on the other hand the transformation $(q,p) \to (-q,-p)$, leaves the right hand side invariant. Thus the ambiguity group mentioned in the introduction is restricted to inversions in the original phase space.
We note that in the case of the repulsive oscillator $\bar{q}$ and $f_-(q,p)$, when considered as quantum mechanical operators, have (up to a two fold degeneracy in $f_-(q,p)$) the same spectra taking all real values in the interval $(-\infty,+\infty)$. As we shall see this allows us to translate equation (2.5a) directily into operator form thanks to the fact that there is no ambiguity group in the $(\bar{q},\bar{p})$ phase space.

b) Canonical Transformations Associated with the Attractive Oscillator Problem.

The situation becomes quite different when we turn our attention to the attractive oscillator i.e., the observable $f_+(q,p)$ of (2.1). To begin with $f_+(q,p)$ is positive for all real values of q,p and besides the potential energy $\frac{1}{2}(\Lambda^2 q^{-2}+q^2)$ has a minimum at $q^2=\Lambda$ so that $f_+(q,p)>\Lambda$. Thus we cannot equate $\bar{q}$ with $f_+(q,p)$ as then for some real values of $\bar{q}$, (q,p) become complex. We can though write, using the absolute value of $\bar{q}$, the equation

$$|\bar{q}| + \Lambda = \frac{1}{2}(p^2 + \Lambda^2 q^{-2} + q^2) , \tag{2.6}$$

and have a relation valid for all real values of $\bar{q}$ and (q,p).

By the same reasoning that led to (2.3), the time variable canonically conjugate to the Hamiltonian $f_+(q,p)$ becomes now[7)]

$$t(q,E) = \int \frac{q\,dq}{(2Eq^2-q^4-\Lambda^2)^{1/2}} = \frac{1}{2} \text{ arc cos } \left[\frac{E-q^2}{(E^2-\Lambda^2)^{1/2}}\right]$$

$$= \frac{1}{2i} \ln\left[\frac{(E-q^2)+(-2Eq^2+q^4+\Lambda^2)^{1/2}}{(E^2-\Lambda^2)^{1/2}}\right] \tag{2.7}$$

To obtain the equation associated with canonically conjugate variables to $|\bar{q}|+\Lambda$ and $f_+(q,p)$, we see first that

$$\{|\bar{q}|+\Lambda , (\bar{q}\bar{p}/|\bar{q}|)\}_{\bar{q},\bar{p}} = 1 \tag{2.8}$$

and thus we can write

$$(\bar{q}\,\bar{p}/|\bar{q}|) = -t[q,f_+(q,p)]$$

$$= -\frac{1}{2i} \ln\left[\frac{(p^2+\Lambda^2 q^{-2}-q^2)/2 + iqp}{[f_+^2(q,p)-\Lambda^2]^{1/2}}\right]$$

$$= -\frac{1}{2} \text{ arc tan } [2qp/(p^2+\Lambda^2 q^{-2}-q^2)] , \tag{2.9}$$

where the last form indicates that equations (2.6), (2.9) define a <u>real</u> canonical transformation. We note incidentally that if we made the substitutions $q \to q \exp(-i\pi/4)$, $p \to p \exp(i\pi/4)$ then

$$qp \to qp \ , \ f_-(q,p) \to if_+(q,p) \ , \ \bar{q} \to i\bar{q}, \ \bar{p} \to -i\bar{p} \ , \qquad (2.10)$$

and (2.9) would follow from (2.4). This indicates that the undetermined function of E mentioned in the paragraph preceeding (1.6), was chosen in the same way for the repulsive and attractive oscillator. It is also interesting to note that the action variable[4)] $J = \oint p dq$ for the attractive harmonic oscillator with a centrifugal force is given by $J = \pi[f_+(q,p) - \Lambda]$ and from the considerations mentioned above will always be positive. Thus we can identify $|\bar{q}|$ with (J/π) and the canonical transformation defined by (2.6), (2.9) leads then essentially to the action-angle variables of the attractive oscillator problem.

Again it is convenient to write the canonical transformation associated with the observable $f_+(q,p)$ in an implicit fashion and from (2.6), (2.9) we get

$$|\bar{q}| + \Lambda = \frac{1}{2}(p^2 + \Lambda^2 q^{-2} + q^2) \ , \qquad (2.11a)$$

$$\left[|\bar{q}|(|\bar{q}| + 2\Lambda)\right]^{1/2} \exp(-2i\bar{q}\bar{p}/|\bar{q}|) = \frac{1}{2}(p^2 + \Lambda^2 q^{-2} - q^2) + iqp \quad (2.11b)$$

We see immediately that the right hand side of (2.11) remains invariant under the transformation $(q,p) \to (-q,-p)$, as was the case for (2.5) of the repulsive oscillator. We have also a group of canonical transformations in $(\bar{q},\bar{p})$ that leave the left hand side invariant i.e.,

$$\bar{q} \to \bar{q} \ , \ \bar{p} \to \bar{p} + m\pi \ , \ m \text{ integer, and } \ \bar{q} \to -\bar{q} \ , \ \bar{p} \to -\bar{p} \ . \qquad (2.12)$$

Thus the ambiguity group in the new phase space is the semidirect product $T \wedge I$, where T implies translations of the momentum variable by $m\pi$ and I is the inversion group associated with $(\bar{q},\bar{p}) \to (-\bar{q},-\bar{p})$.

When we look at equation (2.11a) and consider $|\bar{q}| + \Lambda$ and $f_+(q,p)$ as quantum mechanical operators we immediately notice that their spectra are very different. For $|\bar{q}| + \Lambda$ it takes all real values in the interval $(\Lambda,+\infty)$ while for $f_+(q,p)$ it is given by[6] $2n+\mu+1$ where $n=0,1,2...$ and $\mu=(\Lambda^2+\frac{1}{4})^{1/2}$, and both are doubly degenerate. As we shall see this discrepancy is related with the irreducible representations (irreps), of the ambiguity group $T\wedge I$ which we proceed to derive.

If we consider $x, -\infty\leqslant x\leqslant\infty$ as a variable subject to the transformation of the group $T\wedge I$, i.e. $x\to x+m\pi$, $x\to -x$, then the complete set of functions $(2\pi)^{-1/2}\exp(i\nu x)$ for $-\infty\leqslant\nu\leqslant\infty$ is a basis for a reducible representation of this group. To find the subsets of this set of functions which are basis for irreps we first notice that we can always write

$$\nu = 2\sigma(n+\lambda) \quad \text{where} \quad \sigma = \pm 1 \,,\ 0\leqslant\lambda<1 \quad \text{and} \quad n = 0,1,2... \tag{2.13}$$

If we consider then the two component function associated with $\sigma=\pm 1$ i.e.,

$$(2\pi)^{-1/2}\begin{bmatrix}\exp[i2(n+\lambda)x]\\ \exp[-i2(n+\lambda)x]\end{bmatrix} \tag{2.14}$$

the representations of translations and inversions are given respectively by

$$\begin{bmatrix}\exp(i2m\pi\lambda) & 0\\ & \\ 0 & \exp(-i2m\pi\lambda)\end{bmatrix},\quad \begin{bmatrix}0 & 1\\ & \\ 1 & 0\end{bmatrix}\,, \tag{2.15}$$

which are in irreducible form[2] for $0<\lambda<1$ and can be immediately reduced for $\lambda=0$. Thus the irreps of the ambiguity group $T\wedge I$ are characterized by λ where $0\leqslant\lambda<1$, and their row index by $\sigma=\pm 1$. In subsection d) we discuss the role of the irreps in the problem of the attractive oscillator.

c) Representation in Quantum Mechanics of the Canonical Transformations Associated with the Repulsive Oscillator Hamiltonian.

As in the previous subsections we consider first the repulsive oscillator whose Hamiltonian is $f_-(q,p)$ of (2.1) and the corresponding canonical transformation (2.5) associated with it. We wish now to obtain the representation in quantum mechanics of

(2.5) for which we follow the procedure already outlined in the Proceedings of the 1975 Bonn Conference on Geometrical Quantization[8] and in references 2 and 3.
To begin with, when we consider $\bar{q}$ and $f_-(q,p)$ as quantum mechanical operators their eigenvalues, which in both cases we denote by $\nu, -\infty \leq \nu \leq \infty$, and eigenfunctions satisfy

$$\bar{q}'\phi_\nu(\bar{q}') = \nu\phi_\nu(\bar{q}') , \qquad (2.16a)$$

$$\frac{1}{2}\left(-\frac{d^2}{dq'^2} + \frac{\Lambda^2}{q'^2} - q'^2\right)\psi^\mu_\nu(q') = \nu\psi^\mu_\nu(q') , \qquad (2.16b)$$

where we now employ Dirac's notation[9] with primes for the c--numbers associated with the operators $q, \bar{q}$ and we add an extra index $\mu = (\Lambda^2+1/4)^{1/2}$ to ψ to indicate its dependence on Λ. Up to a phase the normalized eigenfunctions of (2.16) are given by

$$\phi_\nu(\bar{q}') = \delta(\bar{q}'-\nu) , \qquad (2.17a)$$

$$\psi^\mu_\nu(q') = \frac{\exp(\pi\nu/4)\left[\Gamma(\frac{\mu}{2}+\frac{1}{2}+\frac{i\nu}{2})\Gamma(\frac{\mu}{2}+\frac{1}{2}-\frac{i\nu}{2})\right]^{1/2}}{(4\pi)^{1/2}\,\Gamma(\mu+1)}(q')^{-1/2}$$

$$M_{-\frac{i\nu}{2},\frac{\mu}{2}}(iq'^2) \qquad (2.17b)$$

where M is a Whittaker function[7] while Γ is a gamma function. The fact that $\psi^\mu_\nu(q')$ of (2.17b) satisfies (2.16b) can be seen immediately if we introduce the variable $z=iq'^2$ and make use of the equation[7] for the Whittaker function. The normalization constant can be derived in a fashion similar to how it was done for $\Lambda=0$ in Appendix B of reference 2.
Note that for $\Lambda \neq 0$, the expression (2.17b) of $\psi^\mu_\nu(q')$ is the only regular solution of the equation (2.16b) at the origin $q'=0$, taking there the value zero for itself and its derivative with respect to q'. Thus assuming the solution (2.17b) for $q'>0$ we can extend it and its derivative in a continuous fashion for $q'<0$ through the relations

$$\psi_\nu^\mu(-q') = \pm\, \psi_\nu^\mu(q') \ , \ q' \geqslant 0 \ . \tag{2.18}$$

The upper sign will give an even and the lower one an odd function of q' in the interval $-\infty \leqslant q' \leqslant \infty$ and we should add the index $\pm$ to the wave function when it is defined in this interval. Thus $\psi_\nu^{\mu\pm}(q')$ is even (+) or odd (-) under the inversion operation $q' \to -q'$ which corresponds in classical mechanics to $(q,p) \to (-q,-p)$.

The unitary representation in quantum mechanics of the canonical transformation (2.5) can then be written as[1,2,3,8]

$$\langle q'|U^\pm|\bar{q}'\rangle = \int_{-\infty}^{\infty} \exp\,[i\Phi(\nu)]\,\psi_\nu^{\mu\pm}(q')\,\delta(\bar{q}'-\nu)\,d\nu \ , \tag{2.19}$$

where $\Phi(\nu)$ is a phase determined by requiring that the equation (2.5b) holds for the corresponding quantum mechanical operators.

It is not at first sight obvious what is the operator form for the functions of (q,p), $(\bar{q},\bar{p})$ appearing on the right and left hand side of (2.5). Thus we shall rather follow an inverse procedure. We assume for the phase $\Phi(\nu)$ the value zero and then explicitly derive what is the operator in $(\bar{q},\bar{p})$ space corresponding to the hermitized one on the right hand side of (2.5b). Thus we start with the unitary representation

$$\langle q'|U^\pm|\bar{q}'\rangle = \psi_{\bar{q}'}^{\mu\pm}(q') \ , \tag{2.20}$$

and apply to it the operator

$$\left[\frac{1}{2i}\left(q'\frac{d}{dq'} + \frac{d}{dq'}q'\right) + q'^2 + \frac{1}{2}\left(-\frac{d^2}{dq'^2} + \frac{\Lambda^2}{q'^2} - q'^2\right)\right] \psi_{\bar{q}'}^{\mu\pm}(q')$$

$$= \left[(\bar{q}' - i)^2 + \mu^2\right]^{1/2} \psi_{\bar{q}'-2i}^{\mu\pm}(q') \ , \tag{2.21}$$

where the left hand side was obtained from the fact that $\psi_{\bar{q}'}^{\mu\pm}(q')$ satisfies (2.16b) as well as from well-known[7] properties of the Whittaker functions.
We now proceed to find an operator in the $\bar{q},\bar{p}$ canonically conjugate

variables whose action on $\psi^{\mu\pm}_{\bar{q}'}(q')$ gives the function on the right hand side of (2.21). From (2.5) and (2.21) it is immediately apparent that a possible hermitian operator with this property has the form

$$\exp(-\bar{p})(\bar{q}^2 + \mu^2)^{1/2} \exp(-\bar{p}) \ . \tag{2.22}$$

Taking into account that one has to apply the <u>conjugate</u>[2,3] of this operator to the unitary representation (2.20) we get the relation

$$\exp(-id/d\bar{q}')(\bar{q}'^2 + \mu^2)^{1/2} \exp(-id/d\bar{q}')\ \psi^{\mu\pm}_{\bar{q}'}(q')$$

$$= \left[(\bar{q}'-i)^2 + \mu^2\right]^{1/2} \psi^{\mu\pm}_{\bar{q}'-2i}(q') \ , \tag{2.23}$$

as the operator $\exp(-id/dq')$ acting on a function of $\bar{q}'$ only changes this variable to $\bar{q}'-i$.
As discussed in reference 2, the fact that the unitary representation (2.20) has two components $\pm$ associated with its even or odd character under the reflection $q' \rightarrow -q'$, requires that in the new Hilbert space we consider two component states, the first one associated with the even and the second with the odd states in the original Hilbert space. Thus to the operators in the original Hilbert space there correspond 2x2 <u>matrix</u> operators in the new Hilbert space. From (2.16), (2.21) and (2.23) we then see that the canonical transformation (2.5) is associated with the following relation between operators

$$\frac{1}{2}(p^2 + \Lambda^2 q^{-2} - q^2) \leftrightarrow \bar{q}\begin{bmatrix}1 & 0\\0 & 1\end{bmatrix}, \tag{2.24a}$$

$$\frac{1}{2}(qp + pq) + \frac{1}{2}(p^2 + \Lambda^2 q^{-2} + q^2) \leftrightarrow$$

$$\exp(-\bar{p})(\bar{q}^2 + \mu^2)^{1/2} \exp(-\bar{p}) \begin{bmatrix}1 & 0\\0 & 1\end{bmatrix} . \tag{2.24b}$$

The correspondance between operators cannot be deduced from trivial hermitization of the expressions in (2.5) as witnessed by the fact that the Λ^2 appearing in (2.5b) is replaced by $\mu^2 = \Lambda^2 + 1/4$ in (2.24b). Yet in the classical limit, when we disregard $1/4$ as compared to the values $\bar{q}^2$ takes in the present units[2], the quantum operator relations (2.24) reduce to the classical ones (2.5).

The analysis presented here has to be modified slightly when $\Lambda=0$ as then equation (2.14b) has regular solutions for $\mu=\pm\frac{1}{2}$. This was already done in reference 2 and as shown there the classical and quantum relations (2.5) and (2.24) continue to hold when $\Lambda=0$, $\mu^2=\frac{1}{4}$.

d) Representation in Quantum Mechanics of the Canonical Transformation Associated with the Attractive Oscillator Hamiltonian.

We now turn our attention to the attractive oscillator Hamiltonian $f_+(q,p)$ of (2.1). To begin with we consider the $|\bar{q}|+\Lambda$, $f_+(q,p)$ appearing in (2.11a) as quantum mechanical operators and their eigenvalues and eigenfunctions satisfy

$$(|\bar{q}'|+\Lambda)\phi_\nu(\bar{q}') = (|\nu|+\Lambda)\phi_\nu(\bar{q}'), \quad -\infty\leqslant\nu\leqslant\infty \tag{2.25a}$$

$$\frac{1}{2}\left(-\frac{d^2}{dq'^2}+\frac{\Lambda^2}{q'^2}-q'^2\right)\psi_n^\mu(q') = (2n+\mu+1)\psi_n^\mu(q'), \quad n = 0,1,2... \tag{2.25b}$$

where again we employ Dirac's notation with primes for the c-numbers associated with the operators $q,\bar{q}$. Up to a phase the normalized eigenfunctions of (2.25) are given by[6)]

$$\phi_\nu(\bar{q}') = \delta(\bar{q}'-\nu), \tag{2.26a}$$

$$\psi_n^\mu(q') = [2(n!)/\Gamma(n+\mu+1)]^{1/2}(q')^{\mu+1/2}\exp(-q'^2/2)L_n^\mu(q'^2) \tag{2.26b}$$

where L_n^μ is a Laguerre polynomial[7)].

Again we note that for $\Lambda\neq0$ the expression (2.26b) with $\mu=(\Lambda^2+1/4)^{1/2}$ is the only regular solution of the equation (2.25b) at the origin $q'=0$, taking there a zero value for itself and its derivative with respect to q'. Thus we can extend the solution from $q'\geqslant0$ to $q'<0$ through the relations

$$\psi_n^\mu(-q') = \pm\,\psi_n^\mu(q'), \quad q'>0. \tag{2.27}$$

As in the previous subsection we add an index $\tau=\pm$ to the wave function i.e., $\psi_n^{\mu\tau}(q')$ defined by its even $(\tau=+)$ or odd $(\tau=-)$ character in the full interval $-\infty\leqslant q'\leqslant\infty$.
The construction of the unitary representation in quantum mechanics of the canonical transformation (2.11) requires now more thought. To begin with we cannot simply correlate $|\bar{q}|+\Lambda$ with $f_+(q,p)$ when we consider them as quantum mechanical operators as from (2.25) we see that they have different spectra. At this stage the ambiguity group $T\wedge I$, that already makes its appearance in the classical problem, is crucial. From (2.13) we get irreps of $T\wedge I$ when ν , which we now identify with the eigenvalue of $\bar{q}$, is decomposed in the form $\nu=2\sigma(n+\lambda)$, $\sigma=\pm1$, $0\leqslant\lambda<1$, $n=0,1,2...$.
The states that are basis for these irreps are given by (2.14) where x can be interpreted as the momentum variable in view of its transformation properties under the $T\wedge I$ group. Thus when we pass to configuration space these states take the form

$$\delta\left[\bar{q}' - 2\sigma(n+\lambda)\right] . \tag{2.28}$$

We can now define a unitary transformation, for each irrep of $T\wedge I$ characterized by the (λ,σ), $0\leqslant\lambda<1$, $\sigma=\pm1$ appearing in (2.28) and $\tau=\pm$ appearing in $\psi_n^{\mu\tau}(q')$, through a relation similar to (2.19) i.e.,

$$\langle q'|U_{\lambda\sigma}^{\tau}|\bar{q}'\rangle = \sum_{n=0}^{\infty} e^{i\delta_n}\psi_n^{\mu\tau}(q')\delta\left[\bar{q}' - 2\sigma(n+\lambda)\right] . \tag{2.29a}$$

The phase δ_n can be determined by requiring that the unitary representation (2.29) transforms into each other the quantum mechanical operators associated with the classical observables from the right and left hand side of equation (2.11). As in the case of the repulsive oscillator we prefer to follow the opposite procedure i.e., we take

$$\delta_n = 0 \qquad \text{for all} \qquad n = 0,1,2,... \tag{2.29b}$$

and from the corresponding unitary representation (2.29a) derive the relations between operators in the original and new Hilbert spaces.

To achieve our objective we start by noting that when we apply $f_+(q,p)$ to (2.29) we get

$$\frac{1}{2}\left(-\frac{d^2}{dq'^2}+\frac{\Lambda^2}{q'^2}+q'^2\right)\langle q'|U^{\tau}_{\lambda\sigma}|\bar{q}'\rangle =$$

$$\sum_{n=0}^{\infty}(2n+\mu+1)\psi^{\mu\tau}_{n}(q')\delta\left[\bar{q}'-2\sigma(n+\lambda)\right] , \qquad (2.30a)$$

while when we apply $|\bar{q}|$ we obtain

$$|\bar{q}'|\langle q'|U^{\tau}_{\lambda\sigma}|\bar{q}'\rangle = \sum_{n=0}^{\infty} 2(n+\lambda)\psi^{\mu\tau}_{n}(q')\delta\left[\bar{q}'-2\sigma(n+\lambda)\right]$$

$$= \left[\frac{1}{2}\left(-\frac{d^2}{dq'^2}+\frac{\Lambda^2}{q'^2}+q'^2\right)+(2\lambda-\mu-1)\right]\langle q'|U^{\tau}_{\lambda\sigma}|\bar{q}'\rangle . \qquad (2.30b)$$

Thus passing $-(\mu+1)$ to the other side we obtain the following relation between quantum mechanical operators

$$\left[\frac{1}{2}(p^2+\Lambda^2 q^{-2}+q^2)+2\lambda'\right]\delta(\lambda'-\lambda'')\delta_{\sigma'\sigma''} \leftrightarrow (|\bar{q}|+\mu+1)\delta_{\tau'\tau''} \qquad (2.31)$$

where in the original Hilbert space we are dealing with <u>matrix</u> operators characterized by the continuous index $0\leq\lambda<1$ and the discrete one $\sigma=\pm1$, associated with irreps of $T\wedge I$, while in the new Hilbert space we have 2x2 matrix operators associated with the index $\tau=\pm$.

Turning now our attention to the classical observable

$$\frac{1}{2}(p^2+\Lambda^2 q^{-2}-q^2)+iqp , \qquad (2.32)$$

we note that the corresponding operator in quantum mechanics - in which qp is replaced by $\frac{1}{2}(qp+pq)$ - when acting on the unitary representation, gives[6)]

$$\left[\frac{1}{2}\left(-\frac{d^2}{dq'^2}+\frac{\Lambda^2}{q'^2}-q'^2\right)+\frac{1}{2}\left(q'\frac{d}{dq'}+\frac{d}{dq'}q'\right)\right]\langle q'|U^{\tau}_{\lambda\sigma}|\bar{q}'\rangle$$

$$= \sum_{n} 2\left[(n+\mu+1)(n+1)\right]^{1/2}\psi^{\mu\tau}_{n+1}(q')\delta\left[\bar{q}'-2\sigma(n+\lambda)\right] \qquad (2.33a)$$

The corresponding observable in the left hand side of (2.11b) in which, as in (2.5b), (2.22) in the repulsive oscillator case, we replace Λ by μ and symmetrize to get the operator

$$\exp\left[-i(\bar{q}/|\bar{q}|)\bar{p}\right]\left[|\bar{q}|(|\bar{q}|+2\mu)\right]^{1/2}\exp\left[-i(\bar{q}/|\bar{q}|)\bar{p}\right] , \qquad (2.34)$$

when applied in its hermitian conjugate form[2,3] to the unitary representation gives[2]

$$\exp\left[(\bar{q}'/|\bar{q}'|)d/d\bar{q}'\right]\left[|\bar{q}'|(|\bar{q}'|+2\mu)\right]^{1/2}\exp\left[(\bar{q}'/|\bar{q}'|)\partial/\partial\bar{q}'\right]$$

$$\langle q'|U^{\tau}_{\lambda\sigma}|\bar{q}'\rangle$$

$$= \sum_n 2\left[(n+\lambda-\tfrac{1}{2})(n+\lambda-\tfrac{1}{2}+\mu)\right]^{1/2}\psi^{\mu\tau}_{n}(q')\delta\left[\bar{q}'-2\sigma(n-1+\lambda)\right]$$

$$= \sum_n 2\left[(n+\lambda+\tfrac{1}{2})(n+\lambda+\tfrac{1}{2}+\mu)\right]^{1/2}\psi^{\mu\tau}_{n+1}(q')\delta\left[\bar{q}'-2\sigma(n+\lambda)\right] . \qquad (2.33b)$$

In this expression we made use of the fact that $\exp\left[(\bar{q}'/|\bar{q}'|)d/d\bar{q}'\right]$ replaces $\bar{q}'$ by $\bar{q}'+\sigma$ where $\sigma=\pm1$ has the same sign as $(\bar{q}'/|\bar{q}'|)$ and in the last term we replaced n by $n+1$.

Comparing then (2.33a) with (2.33b) we arrive at the following correspondance of operators

$$\left[\frac{1}{2}(p^2+\Lambda^2q^{-2}-q^2)+\frac{i}{2}(qp+pq)\right]\left\{\frac{[f_+(q,p)-\mu+2\lambda'][f_+(q,p)+\mu+2\lambda']}{[f_+(q,p)-\mu+1][f_+(q,p)+\mu+1]}\right\}^{1/2}$$

$$\delta(\lambda'-\lambda'')\delta_{\sigma'\sigma''}$$

$$\leftrightarrow \exp\left[-i(\bar{q}/|\bar{q}|)\bar{p}\right]\left[|\bar{q}|(|\bar{q}|+2\mu)\right]^{1/2}\exp\left[-i(\bar{q}/|\bar{q}|)\bar{p}\right]\delta_{\tau'\tau''} \qquad (2.35)$$

The relations (2.31), (2.35) are the quantum mechanical equivalent of the classical relations (2.11). We note that in the classical limit when the eigenvalues of $f_+(q,p)$ are large compared to 1 the expressions (2.31), (2.35), reduce to those in (2.11) if we suppress[2] the unit operators $\delta(\lambda'-\lambda'')\delta_{\sigma'\sigma''}$ and $\delta_{\tau'\tau''}$.
Again the case $\Lambda=0$ is special but as it was discussed in

reference 2 it requires no further analysis here.
We have thus completed our program for the oscillator problem and in the next section turn our attention to the Coulomb case.

3. THE COULOMB PROBLEM IN DILATED FORM

The observable we want to discuss in this section has the form

$$\Phi(Q,P) = \frac{1}{2}\left[P^2 + L^2Q^{-2} - |Q|^{-1}\right] , \tag{3.1}$$

where L is some real positive constant. It corresponds, in appropriate units, to the Hamiltonian of an attractive Coulomb problem with a centrifugal potential of strength L^2 . We designate the canonically conjugate variables by the capital letters Q,P as we want eventually to transform the observable $\Phi(Q,P)$ to $f_\pm(q,p)$ and thus we would like to reserve the lower case letters (q,p) for the latter problem.
We shall first carry out this transformation and then use the results of the previous section to arrive at some conclusion on the relavance of the ambiguity group of the Coulomb problem to its quantum mechanical spectrum.

a) <u>Transformation of the Coulomb Problem to the Harmonic Oscillator</u>.
Let us consider $\Phi(Q,P)$ as a quantum mechanical operator whose eigenstates $\chi_\rho^M(Q')$ satisfy

$$\frac{1}{2}\left(-\frac{d^2}{dQ'^2} + \frac{L^2}{Q'^2} - \frac{1}{|Q'|}\right)\chi_\rho^M(Q') = \pm\frac{1}{2\rho^2}\chi_\rho^M(Q') , \tag{3.2}$$

where again we designate by Q the c-number associated with Q and, for later convenience, denote the eigenvalues E by

$$E = \pm\ (2\rho^2)^{-1} , \tag{3.3}$$

with the sign depending on whether E is positive or negative.
The index M of the wave function is defined in analogy with $\mu=(\Lambda^2+1/4)^{1/2}$ for the oscillator problem i.e.,

$$M = (L^2 + \tfrac{1}{4})^{1/2} , \tag{3.4}$$

and is used to indicate the dependence of the wave function on L . Carrying out the dilatation transformation

$$Q' \to \rho Q' \tag{3.5}$$

the equation (3.2) becomes

$$|Q'|\left(-\frac{d^2}{dQ'^2}+\frac{L^2}{Q'^2}\mp 1\right)\psi_\rho^M(q') = 2\rho\psi_\rho^M(Q') , \tag{3.6}$$

where

$$\psi_\rho^M(Q') = \chi_\rho^M(\rho Q') . \tag{3.7}$$

The operator on the left hand side of (3.6) corresponds in classical mechanics to the observable

$$F_\mp(Q,P) = |Q|\left[P^2 + (L^2/Q^2) \mp 1\right] \tag{3.8}$$

associated with the Hamiltonian of a dilated Coulomb potential which in a previous publication[6] we referred to as the pseudo--Coulomb problem. Note that to the observable $\Phi(Q,P)$ there corresponds $F_-(Q,P)$ when the former is positive and $F_+(Q,P)$ when it is negative.

It is the pseudo-Coulomb problem we wish to analyze here, though in a later publication we will tackle the observable (3.1) directly. To begin with we note that if we carry out the point transformation

$$Q = \frac{1}{2}q^2 \qquad P = (p/q) , \tag{3.9}$$

which of course satisfies $\{Q,P\}_{q,p}=1$, then $F_\mp(Q,P)$ goes into $f_\mp(q,p)$ of (2.1) if

$$2L = \Lambda . \tag{3.10}$$

Thus the classical pseudo-Coulomb problem reduces to the harmonic oscillator discussed in the previous section.

It is important to notice though a slight change in the quantum mechanical problem (3.6) as compared with (2.16b , 2.25b) of the oscillator. When we make the transformation

$$Q' = \frac{1}{2} q'^2 , \tag{3.11}$$

the equation (3.6) becomes

$$\frac{1}{2} q'^2 \left(- \frac{1}{q'} \frac{d}{dq'} \frac{1}{q'} \frac{d}{dq'} + \frac{4L^2}{q'^4} \mp 1\right) \Psi_\rho^M(\tfrac{1}{2} q'^2)$$

$$= (q')^{1/2} \frac{1}{2} \left(- \frac{d^2}{dq'^2} + \frac{4L^2+3/4}{q'^2} \mp q'^2\right)(q')^{-1/2} \Psi_\rho^M(\tfrac{1}{2} q'^2)$$

$$= (q')^{1/2}\, 2\rho (q')^{-1/2} \Psi_\rho^M(\tfrac{1}{2} q'^2) \quad , \tag{3.12}$$

and thus we note that while the classical relation between the strengths of the centrifugal force in the pseudo Coulomb and oscillator problems is given by (3.10), the quantum mechanical relation is

$$2M = \mu . \tag{3.13}$$

Furthermore, from (2.16b) we see that for the - sign in (3.12)

$$\Psi_\rho^M (\tfrac{1}{2} q'^2) = (q')^{1/2} \psi_{2\rho}^{2M} (q') , \tag{3.14}$$

where the lower case ψ has the form (2.17b), while for the + sign in (3.12) we have from (2.25b) that

$$2\rho = 2n + 2M + 1 , \quad n = 0,1,2,\ldots \tag{3.15}$$

and thus

$$\Psi_\rho^M (\tfrac{1}{2} q'^2) = (q')^{1/2} \psi_n^{2M}(q') , \tag{3.16}$$

where the lower case ψ has the form (2.26b).

Finally, we wish to note that while the oscillator functions are orthonormal in a space of standard differential measure dq' we see from (3.14), (3.16) that the functions $\Psi_\rho^M(Q')$ are orthonormal in a space with the differential measure[6)] $dQ'/2Q'$. We note that the operator on the left hand side of (3.6) will be hermitian when we use this last measure.

b) Canonical Transformations Associated with the Pseudo Coulomb Problem and their Representations in Quantum Mechanics.

We are now in a position to translate all the results of section 2 to the pseudo-Coulomb problem by making use of the canonical transformation (3.9) and the relations (3.14), (3.16).

The canonical transformations associated with the pseudo-Coulomb problem corresponding to positive energies, i.e., for the - sign in (3.8), can be obtained from (2.5) leading to

$$\bar{Q} = |Q|\left[P^2 + (L^2/Q^2) - 1\right] \tag{3.17a}$$

$$(\bar{Q}^2 + 4L^2)^{1/2} \exp(-2\bar{P}) = 2QP + |Q|\left[P^2 + (L^2/Q^2) + 1\right] \tag{3.17b}$$

For the pseudo-Coulomb problem corresponding to negative energies, i.e., for the + sign in (3.8), we obtain from (2.11) that

$$|\bar{Q}| + 2L = |Q|\left[P^2 + (L^2/Q^2) + 1\right] , \tag{3.18a}$$

$$\left[|\bar{Q}|(|\bar{Q}|+4L)\right]^{1/2} \exp(-2i\bar{Q}\bar{P}/|\bar{Q}|) = |Q|\left[P^2+(L^2/Q^2)-1\right]+2iQP. \tag{3.18b}$$

For uniformity of notation we replace in (2.5), (2.11) $\bar{q},\bar{p}$ by $\bar{Q},\bar{P}$.

The unitary representation of the canonical transformation (3.17) is, from (2.20) given by

$$\langle Q'|U^{\pm}|\bar{Q}'\rangle = (2Q')^{-1/4}\, \psi^{M\pm}_{\bar{Q}'/2}(Q') , \tag{3.19}$$

while from (2.29) we see that for the canonical transformation (3.18) we have the unitary representation

$$\langle Q'|U^{\tau}_{\lambda\sigma}|\bar{Q}'\rangle = (2Q')^{-1/4} \sum_{n=0}^{\infty} \psi^{M\tau}_{n+M+1/2}(Q')\,\delta\left[\bar{Q}'-2\sigma(n+\lambda)\right] . \tag{3.20}$$

In (3.19), (3.20) the functions $\psi^{M}_{\rho}(Q')$ defined in (3.14), (3.16) in the interval $0\leqslant Q'\leqslant\infty$ are extended to the interval $-\infty\leqslant Q'\leqslant\infty$ through the relations

$$\psi^{M}_{\rho}(-Q') = \pm\, \psi^{M}_{\rho}(Q'),\ Q'\geqslant 0 , \tag{3.21}$$

and thus we added the upper index $\tau=\pm$ to the wave functions to indicate their even or odd character under reflections. The indices $0\leqslant\lambda<1, \sigma=\pm 1$ were discussed in subsection 2d and indicate that the ambiguity group for the pseudo Coulomb problem corresponding to negative energies continues to be $T \wedge I$.
From the discussion in (2.24) we see that to the canonical transformations (3.17) there correspond the operator relations

$$|Q|\left[P^2 + (L^2/Q^2) - 1\right] \leftrightarrow \bar{Q}\begin{pmatrix}1 & 0\\ 0 & 1\end{pmatrix} , \qquad (3.22a)$$

$$(QP+PQ)+|Q|\left[P^2+(L^2/Q^2)+1\right] \leftrightarrow \exp(-\bar{P})(\bar{Q}^2+4M^2)^{1/2} \exp(-\bar{P})\begin{pmatrix}1 & 0\\ 0 & 1\end{pmatrix} , \qquad (3.22b)$$

while from (2.31), (2.35) we obtain that to the canonical transformations (3.18) there correspond the operator relations

$$\{|Q|\left[P^2 + (L^2/Q^2) + 1\right] + 2\lambda'\}\delta(\lambda'-\lambda'')\delta_{\sigma'\sigma''}$$

$$\leftrightarrow (|\bar{Q}| + 2M + 1)\delta_{\tau'\tau''} , \qquad (3.23a)$$

$$\{|Q|\left[P^2+(L^2/Q^2)-1\right]+i(QP+PQ)\}\left\{\frac{[F_+(Q,P)-2M+2\lambda']\,[F_+(Q,P)+2M+2\lambda']}{[F_+(Q,P)-2M+1]\,[F_+(Q,P)+2M+1]}\right\}^{1/2} \delta(\lambda'-\lambda'')\delta_{\sigma'\sigma''}$$

$$\leftrightarrow \exp\left[-i(\bar{Q}/|\bar{Q}|)\bar{P}\right]\left[|\bar{Q}|(|\bar{Q}|+4M)\right]^{1/2} \exp\left[-i(\bar{Q}/|\bar{Q}|)\bar{P}\right]\delta_{\tau'\tau''} . \qquad (3.23b)$$

We have thus implemented fully our program for the pseudo-Coulomb problem and in the next section we analyze the conclusions that we can derive from it, as well as from the oscillator problem discussed in the previous section.

4. CONCLUSION

To put in perspective what we achieved in the present paper let us first apply the procedure we developed to the one canonical transformation whose representations is discussed in most quantum mechanical text books[9)] i.e.,

$$\bar{q} = p \qquad \bar{p} = -q \tag{4.1a,b}$$

The normalized eigenstates of the operators associated with the right and left hand side of (4.1a), again using Dirac's prime notation for the c-numbers, gives

$$\psi_\nu(q') = (2\pi)^{-1/2} \exp(i\nu q'), \quad \phi_\nu(\bar{q}') = \delta(\bar{q}'-\nu) \tag{4.2ab}$$

and thus assuming , as was done in (2.19), (2.20), the phase as zero, we get the unitary representation

$$\langle q'|U|\bar{q}'\rangle = \int_{-\infty}^{\infty} (2\pi)^{-1/2} \exp(i\nu q')\delta(\bar{q}'-\nu)d\nu = (2\pi)^{-1/2} \exp(iq'\bar{q}') \ , \tag{4.3}$$

which is the familiar Fourier transform kernel that takes us from configuration to momentum space. With its help we get the operator correspondance

$$\bar{q} \leftrightarrow p \qquad \bar{p} \leftrightarrow -q \ , \tag{4.4a,b}$$

which in this case is, of course, identical to the classical relations (4.1).

These results were known since the middle twenties and they are clearly summarized in Dirac's book[9]. For the repulsive oscillator and the dilated Coulomb problem corresponding to positive energy, we implemented essentially the same program, as only the ambiguity $=\pm$ appears in the eigenstates of the Hamiltonians when they are defined in the full interval $-\infty \leq q', Q' \leq \infty$. Note that here the operator relations (2.24), (3.22) are not identical to the canonical transformations (2.5), (3.17), but they do coincide in the classical limit in which the eigenvalues of the Hamiltonians are assumed large compared to unity.

For the attractive oscillator and the dilated Coulomb problem corresponding to negative energy the situation is quite different. In fact Dirac[9] gives no hint on how to approach this problem as he only considers canonical transformations in which all the observables have spectra from $-\infty$ to $+\infty$ when considered as quantum mechanical operators. In the present paper and the previous ones[1,2] of this series we managed to extend Dirac's program to cases in which some of the observables have discrete spectra, by noticing the non-bijective character of the canonical transformation involved, defining an ambiguity group related to the non--bijectiveness and finding the index characterizing the irreducible representations of this group i.e., the ambiguity spin[1,2].

Thus the Hamiltonians $f_+(q,p)$, $F_+(Q,P)$ in the classical mechanics picture become the <u>matrix</u> operators (2.31), (3.23a) in which the row and column indices are characterized by the continuous variable $0 \leq \lambda < 1$ and the discrete one $\sigma = \pm 1$.
The number of problems already discussed in this fashion[1,2], as well as the full analysis of the Coulomb problem that we will present in another publication, seems to indicate that we have a procedure for finding the unitary representation in quantum mechanics of an arbitrary canonical transformation.
But there is also another aspect to our problem. The indices $\lambda (0 \leq \lambda < 1)$, $\sigma = \pm 1$ in the ambiguity spin associated with the ambiguity group $T \wedge I$ already suggest the convenience of decomposing the set of values $-\infty \leq \nu \leq \infty$ that the classical observable $\bar{q}$ can take into subsets $\nu = 2\sigma(n+\lambda)$, $n = 0,1,2...$ of given λ, σ. Thus using the purely <u>classical</u> concept of ambiguity group, we in a sense prepare the problem so as to note that in <u>quantum mechanics</u> the operator $f_+(q,p)$ of (2.1) associated with $\bar{q}$ has the spectrum $2n+\mu+1$, $n=0,1,2...$, <u>up to a constant</u>.
Thus the relations between classical and quantum mechanics seem to flow in both directions and <u>not</u>, as is usually assumed[9], only in the direction that classical mechanics is an asymptotic form of quantum mechanics.

REFERENCES

1. P. Kramer, M. Moshinsky and T.H. Seligman, J. Math. Phys. 19 (1978) 683.

2. M. Moshinsky and T.H. Seligman, Ann. Phys. (N.Y.) (1978) (in press).

3. P.A. Mello and M. Moshinsky, J. Math. Phys. 16 (1975) 2017

4. H. Goldstein, "Classical Mechanics" (Addison-Wesley, Reading 1957)

5. R.S. Burington and C.C. Torrance, "Higher Mathematics", (Mc Graw-Hill Inc., New York 1939) p. 123

6. M. Moshinsky, T.H. Seligman and K.B. Wolf, J. Math. Phys. 13 (1972) 901

7. I.S. Gradstein and I.M. Ryzik, "Tables of integrals, series and products", (Academic Press, New York, 1965) p. 81; pp. 1057

8. M. Moshinsky, "Canonical transformations and their representation in quantum mechanics" in "Differential Geometric Methods in Mathematical Physics, Bonn Conference 1975, Lecture Notes in Mathematical Physics 570", Editors K. Bleuler and A. Reetz, (Springer, Heidelberg, 1977).

9. P.A.M. Dirac, "The principles of quantum mechanics", (Clarendon Press, Oxford, First Edition 1930, Third Edition 1947, pp. 103-107).

QUANTUM FIELD THEORY IN CURVED SPACE-TIMES
A GENERAL MATHEMATICAL FRAMEWORK

C.J. Isham
Blackett Laboratory
Imperial College
London SW7 2BZ.

§1. INTRODUCTION

There has been great interest over the last few years in the problem of quantizing fields propagating in a fixed (unquantized) curved space-time manifold . Understanding this subject seems a natural prerequisite for the study of quantum gravity proper and in any event it should play an important role as a semiclassical limit of such a theory. There is a fairly natural 'in/out' (S-matrix) quantization scheme for space-times possessing asymptotically stationary regions and results of fundamental physical importance have been obtained in such cases[1,2,3,4,]. For generic space-times possessing no such regions the problem is somewhat harder. Surprisingly little has been written about rigorous quantum field theory in such a situation and the present paper is an attempt to partially rectify this. Some of the ideas presented here are quite well known to workers interested in such matters and I have for example in places employed the results (written and verbal) of Ashtekar[5], Ashtekar-Magnon[6], Fulling, and especially, Kay[7,8,9]. However this is not a genuine review article since the presentation is unashamedly biased towards a particular viewpoint (the 'covariant' approach - see also Ref. 10). The treatment parallels as closely as possible the schemes that have been extensively developed to quantize a (flat space) linear field in the presence of an external potential[11-19].

I think that it is still not properly appreciated that many of the difficulties (both conceptual and technical) arising in a curved space-time possess analogues that have already been discussed in this earlier work. It is always important in any work related to quantum gravity to carefully isolate those problems which are genuinely associated with space-time structure and separate them from what are really fairly conventional quantum field theoretic difficulties.

Only linear fields are discussed in the present paper. This is partly because it is only for such systems that the existing external potential theory has been evolved and for which (in four space-time dimensions) one can reasonably expect a mathematically complete formalism. Superficially it seems unlikely that the introduction of interacting fields will lead to serious difficulties that are substantially different from those arising in Minkowski space. In particular questions of renormalizability, associated as they are with short distance behaviour, are unlikely to change. On the other hand the long range 'infrared' behaviour of an interacting theory could be modified by large scale space-time structure and of course if suitable asymptotic regions are not present there will be no conventional S-matrix interpretation. It is also interesting to note that Charap and Duff[20] have recently discovered instanton solutions to the coupled Einstein/Yang-Mills equations which possess no flat space analogue. This again suggests that <u>global</u> aspects of space-time structure may play an important role in interacting theories.

In the present paper only spin zero fields will be discussed. No major problem is anticipated in the spin half case once the well known restrictions on a manifold permitting it to admit a spin structure have been accomodated, although a fair amount of hopefully routine functional analysis will be necessary to achieve the same level of rigour currently enjoyed by the scalar field. Spin one and spin two fields raise partially unsolved problems of gauge invariance and/or removing spurious degrees of freedom and of course for spin 3/2, 5/2 and above there are the well known classical inconsistencies.[49]

The mathematical problems which arise in attempting to construct quantum fields on a curved space-time may be loosely divided into 'structural' and 'functional analytical'. The structural problems are concerned with the choice of general framework within which to work whereas the functional analytical problems refer to the need to prove specific technical results once such a framework has been selected. Within the context of linear systems a general analysis of the concept of quantization has been developed over the years by Segal[11,16-19] and many of his ideas are employed here. Broadly speaking one can distinguish at least four different (but needless to say interlinked) structural approaches:

1) The Covariant Method: Starting with classical field equations one attempts to find a quantum field, which is a function of space and time, satisfying the same linear equations. In addition covariant commutation relations are required (such unequal time commutation relations can be explicitly imposed for linear bose systems). Thus at least initially, only a single Hilbert space is employed.

2) The Canonical Method: A choice of time is made and the classical equations of motion are written in first order canonical form with, in general, a time dependent hamiltonian. Quantization is implemented by looking for representations of the equal time canonical commutation relations (CCR) and of the commutation relations

between the hamiltonian and the canonical variables. As is well known there are infinitely many unitarily inequivalent representations of the CCR and in the present context it is not unnatural to permit a different representation at each time. Thus in this approach one naturally uses infinitely many Hilbert spaces. In practice the representations are often chosen so that the instantaneous hamiltonians are realised as positive, self adjoint operators on their respective Hilbert spaces[5,7,8,9]. The resulting scheme is known as 'Hamiltonian Diagonalization' and has, in a rather heuristic form, been employed by various authors studying quantum field theory in a cosmological Robertson/Walker type of background. It should perhaps be emphasised that if the equal time fields are constructed in the covariant method they may also belong to inequivalent CCR representations even though the operators are all defined on the same Hilbert space. Indeed in an abstract algebraic sense (see below) the two schemes are identical although they do differ in the concrete choice of representation and hence to some extent in the physical interpretation of the underlying mathematics.

3) C*-algebra approach: The proliferation of representations of both the canonical and the covariant commutation relations leads rather naturally to an abstract algebraic approach. The C*-algebras can either be simply used as a very powerful tool in the representation theory of such commutation relations, or in perhaps a more fundamental way, they can be employed in a curved space-time version of the Haag-Kastler[21] theory of local observables.

4) Functional Integral methods: It is possible to drop all use of Hilbert spaces and start with a Feynman history integral[22]. This leads rather naturally to constructing fields and Schwinger functions on a Riemannian manifold and obtaining the desired pseudo-

Riemannian objects by a process of analytic continuation. Indeed it is quite easy to construct 'euclidean' (i.e. Riemannian) fields in the sense of Nelson although except in some very special cases there is no obvious way in which the analytical continuation to real space-times can be performed.

To some extent the choice of structural approach is dictated by the applications one has in mind. The first three methods are naturally employed in a 'conventional' treatment of quantum field theory in a curved space-time per se although if the subject is viewed as a first step in the quantum theory of gravity itself then the functional integral approach is currently an attractive one. In the present paper the main emphasis is on the covariant approach with a C*-algebra parent scheme. This has the virtue of being the one most closely related to the heuristic methods that have in practice been employed in, for example, black hole calculations. Whichever method is adopted however, great care needs to be taken over the physical interpretation of the mathematical constructs and in particular in the use of the language of particles or quanta of the field.

§ 2. THE COVARIANT APPROACH.

Let us consider a classical scalar field $\phi(x)$ linearly propagating in a C^{∞} spacetime manifold $\mathcal{M}$ with pseudo-Riemannian metric $g_{\mu\nu}(x)$ of signature (+1,-1,-1,-1). An appropriate Lagrangian density is

$$\mathcal{L} = \tfrac{1}{2}(-\det g)^{\frac{1}{2}} \, (g^{\mu\nu} \, \partial_{\mu} \phi \, \partial_{\nu} \phi - \mu^2 \phi^2) \tag{2.1}$$

In Minkowski space the constant μ would be regarded as the mass of the particles appearing in the quantum theory. Such an interpretation may however be untenable in a curved spacetime and it is best to simply regard μ as an arbitrary parameter in the theory. Indeed in most of what follows μ can even be a specified smooth function of space and time in which case the flat space external potential problem is a special case of the general situation described by eqn (2.1) . The conformally invariant scalar field can also be accommodated by choosing μ as 1/6 R.

The equations of motion are

$$L_x \, \phi(x) = 0 \tag{2.2}$$

where L_x is the linear differential operator

$$L_x \equiv g^{\mu\nu} \, \nabla_{\mu} \, \partial_{\nu} + \mu^2 \tag{2.3}$$

The requirements to be met by a covariant quantum field will now be listed with appropriate comments on each one. These are in a sense the Wightman-Gårding axioms[23] for the system although since only linear fields are being considered they barely merit such an illustrious title.

1) Field Existence

There should exist a densely defined operator $\hat{\phi}(\psi)$ defined on a Hilbert space $\mathcal{H}$ for all functions ψ defined in some test function space T. It is supposed that T is a topological vector space and then $\hat{\phi}$ is required to be an operator valued distribution in the usual sense that the map $T \rightarrow \mathbb{C}$

$$\psi \rightarrow \langle v, \hat{\phi}(\psi) w \rangle$$

is continuous and linear for all vectors v, w in the domain of $\hat{\phi}$.

Comments

(i) As usual one requires that $\hat{\phi}(\psi)$ possesses a common dense domain $D \subset \mathcal{H}$ for all $\psi \in T$ which is invariant in the sense that

$$\hat{\phi}(\psi) D \subset D \quad \forall \psi \in T$$

This enables polynomials to be formed in the smeared field operators $\hat{\phi}(\psi)$, $\psi \in T$

(ii) Classically the field ϕ in eqn. (2.2) is real. This would be reflected at the quantum level by the requirement that, at the very least, $\hat{\phi}(\psi)$ is a symmetric operator for all test functions ψ in T. One might in addition require that $\hat{\phi}(\psi)$ be essentially self-adjoint on the invariant dense domain D .

(iii) The precise choice of test function space plays an important role in conventional flat spacetime quantum field theory. Normally one employs the space $\mathcal{S}$ of C^∞ functions which vanish at infinity, together with all their derivatives, faster than any power of the euclidean distance[23]. This space has the important property that it is mapped isomorphically onto itself by Fourier transformation and elements of the dual of $\mathcal{S}$ can be conveniently identified as polynomial (distributional) derivatives of continuous functions. The choice of $\mathcal{S}$ is intimately linked with the known perturbative behaviour of renormalizable quantum field theories and with the experimentally observed behaviour of physical S-matrix elements. On a curved spacetime $\mathcal{M}$ there is in general no analogue of $\mathcal{S}$. On the other hand the space $\mathcal{D}(\mathcal{M})$ of C^∞ functions with compact support exists with a topological vector space structure which is of the same type as the usual flat space one[24]. In particular $\mathcal{D}(\mathcal{M})$ is a nuclear space - a prerequisite for T in order that the standard nuclear theorem on multilinear functionals be applicable. The nuclear structure also plays an important role in certain aspects of the representation theory of both canonical and covariant commutation relations and in general we shall identify T with $\mathcal{D}(\mathcal{M})$. If $\mathcal{M}$ is a C^k manifold, $k < \infty$, then an appropriate space of finitely differentiable test functions may be employed. This complicates things slightly at a formal level but poses no basic problems. We will for the sake of simplicity assume $k = \infty$ without any real loss in generality. If an interacting field theory were being considered one would need to think more carefully about the choice of T. The space $\mathcal{D}$ is, in the flat space case, appropriate for a <u>non</u> renormalizable theory and the question of the analogue of $\mathcal{S}$ for a formally renormalizable curved space interaction is an interesting (and unsolved) one.

Note finally that since ψ is a scalar the smeared field operator $\hat{\phi}(\psi)$ can be written heuristically as

$$\hat{\phi}(\psi) = \int d^4x\ (-\det g)^{\frac{1}{2}}\ \hat{\phi}(x)\psi(x);\ \psi \in \mathcal{D}(\mathcal{M}) \tag{2.4}$$

If desired the test function space T could be chosen as the space $\tilde{\mathcal{D}}(\mathcal{M})$ of C^∞, compact support, scalar densities in which case

$$\hat{\phi}(\psi) = \int d^4x\ \hat{\phi}(x)\psi(x);\quad \psi \in \tilde{\mathcal{D}}(\mathcal{M}) \tag{2.5}$$

<u>2) Microcausality</u>

Recall that the causal future $J^+(x)$ of a point $x \in \mathcal{M}$ may be defined as the set of all points in $\mathcal{M}$ which can be connected to x by a past directed (we are assuming that $\mathcal{M}$ is time orientable) timelike curve or null geodesic[25,26]. The causal past $J^-(x)$ is defined in a similar way.

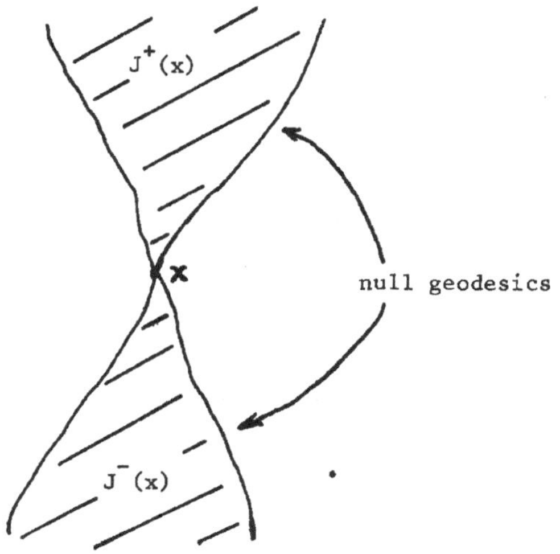

Heuristically we expect that if $y \in \mathcal{M}$ is not in $J^+(x) \cup J^-(x)$ then $[\hat{\phi}(x), \hat{\phi}(y)] = 0$, i.e. the observables corresponding to the fields evaluated at 'space-like separated' points are simultaneously measurable. Rigorously we require that if ψ_1, $\psi_2 \in T$ are test functions with compact support and if

$$\left[J^+ (\text{supp } \psi_1) \cup J^- (\text{supp } \psi_1)\right] \cap \text{supp } \psi_2 = \phi (\text{the empty set}) \tag{2.6}$$

then

$$[\hat{\phi}(\psi_1), \hat{\phi}(\psi_2)] = 0 \tag{2.7}$$

where $J^{\pm}$ (supp ψ) is defined in the obvious way. Schematically

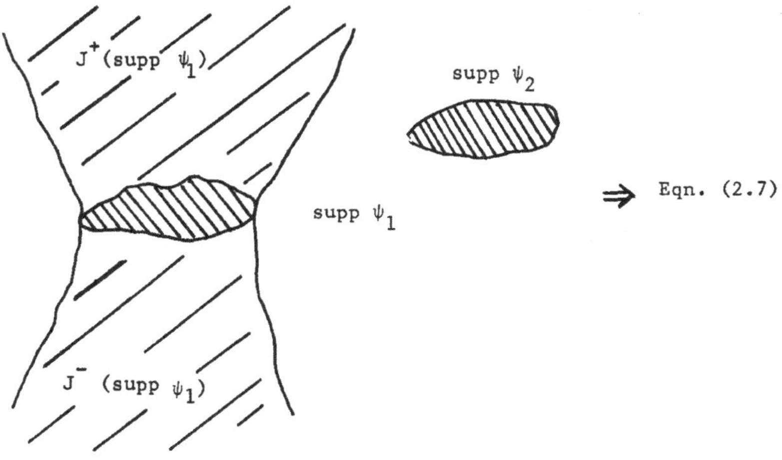

Comments

(i) Since $\hat{\phi}(\psi_{1,2})$ are unbounded operators (this will emerge shortly) eqn. (2.7) should only be regarded as being defined on the invariant domain $D \subset \mathcal{H}$. Alternatively we could write

$$\left[\hat{\phi}(\psi_1),\ \hat{\phi}(\psi_2)\right] \subset 0 \qquad (2.8)$$

(ii) Note that for this axiom to be non-empty there must exist at least some test functions in T which actually have compact support. In practice, in order to be able to get as 'close as one likes' to the non existent field-at-a-point $\hat{\phi}(x)$, it is desirable that T contains functions of arbitrarily small support. This requirement is of course met by the spaces $\mathcal{D}(\mathcal{M})$ or $\tilde{\mathcal{D}}(\mathcal{M})$.

(iii) If by some chance $\mathcal{M}$ contains closed timelike curves then the concept of microcausality becomes meaningless. One is tempted to rule out such pathological behaviour ab initio but manifolds of this type are not without interest. For example anti De Sitter spacetime arises naturally in certain supersymmetry contexts and in any event it would be difficult to prove or justify the statement that spacetime did not possess peculiar properties at say the Planck length scale. Of course one should really be considering quantum gravity proper in this situation but the very construction of such a theory might require an understanding of the quantum effects of pathological spacetimes. In the present case however the existence of closed time-like curves must be excluded for purely technical reasons. The classical Cauchy problem is not well posed in such manifolds and as a result the technique (§ 3) for actually constructing covariant quantum fields also breaks down. The reason for this will become clear later but for the moment we will simply assume that $\mathcal{M}$ admits a well defined classical Cauchy problem. This finds its formal expression in the requirement that $\mathcal{M}$ be globally hyperbolic.[25-29] We recall that $\mathcal{M}$ is said to be globally hyperbolic if

a) For all pairs of points $x, y \in \mathcal{M}$

$J^-(x) \cap J^+(y)$ is a compact subset of $\mathcal{M}$.

and b) $\mathcal{M}$ is strongly causal, that is there are no 'almost closed' timelike curves. More precisely if x is any point in $\mathcal{M}$ then every neighbourhood of x contains a neighbourhood of x which no non spacelike curve intersects more than once.

The physical significance of global hyperbolicity is not instantly obvious from this definition although property a) turns out to play a crucial role in the quantum field theory. The idea is best appreciated by considering the concept of a Cauchy surface. This is defined to be a spacelike hypersurface which every inextendable non spacelike curve intersects exactly once. Then $\mathcal{M}$ can be shown to be globally hyperbolic if and only if there exists a global Cauchy surface[30]. Furthermore $\mathcal{M}$ is diffeomorphic to $\mathbb{R} \times \Sigma$ where Σ is a three manifold and $\{r\} \times \Sigma$ is a Cauchy surface for all $r \in \mathbb{R}$[30]. If $\mathcal{M}$ is globally hyperbolic then any classical solution to the wave equation (2.2) is uniquely determined by its Cauchy data on any Cauchy surface.

(iv) A further property of globally hyperbolic manifolds is the existence of unique advanced and retarded Green's functions satisfying[27,29,3

$$L_x G^{adv}(x;y) = L_x G^{ret}(x;y) = -\delta(x,y) \qquad (2.9)$$

These may be constructed locally using parametrix methods[31] or globally using topological vector space techniques[27]. In fact they are two point distributions (elements of $\mathcal{D}'(\mathcal{M}) \times \mathcal{D}'(\mathcal{M})$) obeying (2.9) in the distributional sense. The words 'advanced' and 'retarded' arise because for fixed y the support in x of $G^{adv}(x;y)$ (resp. $G^{ret}(x;y)$) lies in $J^-(y)$ (resp. $J^+(y)$). More precisely it can be shown that for any test function $\psi \in \mathcal{D}(\mathcal{M})$ the distributions

$$G^{\substack{adv\\ret}}(x;\psi) = \int_{\mathcal{M}} G^{\substack{adv\\ret}}(x;y)\ \psi(y)\ (\det g)^{1/2} d^4y \qquad (2.10)$$

are actually C^∞ functions and

$$\operatorname*{supp}_x G^{adv}(x;\psi) \subset J^-(\operatorname{supp}\psi)\ ;\ \operatorname*{supp}_x G^{ret}(x;\psi) \subset J^+(\operatorname{supp}\psi) \qquad (2.11)$$

The microcausality condition can now be sharpened to the covariant commutation relations

$$\left[\hat{\phi}(\psi_1), \hat{\phi}(\psi_2)\right] = -i\hbar\, \tilde{G}(\psi_1;\psi_2) \text{ on } D\subset\mathcal{H},\ \forall\psi_1,\psi_2 \in \mathcal{D}(\mathcal{M}) \qquad (2.12)$$

where the 'commutator function' $\tilde{G}$ is defined as

$$\tilde{G}(\psi_1;\psi_2) := G^{adv}(\psi_1;\psi_2) - G^{ret}(\psi_1;\psi_2),\ \forall\psi_1,\psi_2 \in \mathcal{D}(\mathcal{M}) \qquad (2.13)$$

and $\hbar$ is Planck's constant. Eqn.(2.12) implies microcausality because

$$\operatorname*{supp}_x \tilde{G}(x;\psi) \subset J^+(\operatorname{supp}\psi) \cup J^-(\operatorname{supp}\psi) \qquad (2.14)$$

$$\operatorname*{supp}_y \tilde{G}(\psi;y) \subset J^+(\operatorname{supp}\psi) \cup J^-(\operatorname{supp}\psi) \qquad (2.15)$$

but the converse is clearly not true. Indeed since

$$L_x \tilde{G}(x;\psi) = L_y \tilde{G}(\psi;y) = 0 \qquad (2.16)$$

eqn.(2.12) is only valid for linear theories. It may be justified by noting that the Poisson bracket between the classical observables $\phi(\psi_1)$ and $\phi(\psi_2)$ is

$$\{\phi(\psi_1), \phi(\psi_2)\}_{P.B} = -\tilde{G}(\psi_1;\psi_2) \qquad (2.17)$$

From another point of view, if equal time fields can be constructed, eqn.(2.12) implies the canonical commutation relations and incorporates the feature that in a linear theory the commutator of two unequal time fields is a C-number.

Note that the test function space T has been assumed to be $\mathcal{D}(\mathcal{M})$ in the above. Leray's original work[27] led to the construction of Green's functions on significantly larger classes of functions. These include $\mathcal{D}^k(\mathcal{M})$ (for use in non C^∞ manifolds) and various local Sobolev spaces.

3) Irreducibility

The set of smeared field operators $\{ \hat{\phi}(\psi) \mid \psi \in T\}$ is irreducible.

Comments

(i) Irreducibility can, as usual, be defined in various ways[23]. For example, if $\hat{\phi}(\psi)$ is essentially self adjoint for all $\psi \in T$ then the Von Neumann algebra generated by the spectral projections of the associated self adjoint extensions is required to be the set of all bounded operators on $\mathcal{H}$.

(ii) In conventional Minkowski space quantum field theory[23], axiom 3 is usually replaced by the requirement that there exists a cyclic vacuum state. This, together with the assumption that the spectrum of the hamiltonian is positive, implies irreducibility. In the generic space-time case with no timelike Killing vectors there is no natural analogue of either of these requirements and irreducibility has to be postulated ab initio.

(iii) The physical meaning of axiom 3 is that any observable can be written as a function (in an appropriate sense) of the smeared field operators. Note that the algebra generated by fields smeared with test functions whose support lies in some subset of the space-time manifold $\mathcal{M}$ will not in general be irreducible. Indeed the thermal nature of the Hawking radiation[4] measured by an observer localised outside a black hole can be interpreted in this way. On the other hand in a conventional Minkowski space theory the algebra generated by such operators plus just the projection onto the vacuum is irreducible[23]. It would be an interesting problem to investigate the analogue of this Reeh-Schlieder theorem for a curved space-time with an event horizon (as in the black hole).

4) Field Equations

The quantum field is required to satisfy the covariant equations of motion (2.2) in the precise sense that:

$$(L_x\hat{\phi})\ (\psi) := \hat{\phi}\ (L_x\psi) = 0 \text{ on } D \subset \mathcal{H}\ ,\ \forall \psi \in T. \tag{2.18}$$

Comments

(i) This axiom only makes sense if $L_x\ \psi \in T$ for all $\psi \in T$. Since in the C^∞ manifold case L_x is a C^∞ differential operator this is trivially true if $T = \mathcal{D}(\mathcal{M})$. It can be easily adapted to the C^k, $k < \infty$, situation.

(ii) These field equations are almost implied by the covariant field commutation relations, equation (2.16) and irreducibility (axiom 3). One may in fact deduce from these that

$$(L_x\hat{\phi})\ (\psi) = \rho(\psi)\hat{\mathbb{1}} \quad \text{on } D \subset \mathcal{H} \tag{2.19}$$

where ρ is some distribution in $\mathcal{D}'\mathcal{M}$ and $\hat{\mathbb{1}}$ is the unit operator on $\mathcal{H}$. Such equations of motion would arise classically by adding an external source term to (2.1) of the form

$$\mathcal{L}(x) = \tfrac{1}{2}\ (-\det g(x))^{\frac{1}{2}}\ \ \rho\ (x)\ \phi(x) \tag{2.20}$$

and could if desired by treated as well.

5) Invariance Groups

If G is a group of isometries of $\mathcal{M}$ (μ-preserving if μ is a function) then $\mathcal{H}$ is required to carry a unitary representation of this group.

Comments

(i) Isometric μ preserving diffeomorphisms are regarded, in the passive sense as arising from, or corresponding to, physically equivalent descriptions of the same system. If $x \to g\,x$ denotes such a transformation, an induced action λ on T is defined by

$$(\lambda_g \psi)(x) := \psi(g^{-1} x) \quad \forall g \in G \tag{2.21}$$

and the quantum theory is rendered covariant by requiring the unitary operators $U(g)$, $g \in G$, in axiom 5 to satisfy

$$U(g)\,\hat{\phi}(\psi)\,U^{-1}(g) = \hat{\phi}(\lambda_{g^{-1}}\psi) \quad \forall \psi \in T \tag{2.22}$$

It is assumed in (2.22) that $U(g)\,D \subset D$, $\forall g \in G$.

(ii) Axiom 5 could be sharpened by requiring the existence of a cyclic vector $\Omega_o \in \mathcal{H}$ for the field operators which is G-invariant:

$$U(g)\,\Omega_o = \Omega_o \quad \forall g \in G \tag{2.23}$$

In the case of the Poincaré group acting on Minkowski space a further and crucial requirement is that the spectrum of the energy momentum four-vector (the generator of space-time translations) lies in the forward light cone. The nearest one is likely to come to this in a curved space-time is when a timelike Killing vector exists. It is then reasonable to look for an invariant 'vacuum' Ω_o with the self adjoint generator of the isometry group having a positive spectrum. Even then one needs to be careful if for example the Killing vector becomes null on an event horizon bounding the manifold (such as in the outside of an eternal black hole). In general caution should be employed when implementing any part of axiom 5.

This completes the list of mathematical requirements to be met by a covariant linear quantum field theory on the space-time manifold . The important questions now concern the existence and uniqueness of such a structure and its physical interpretation.

Existence will be discussed in the next section using an adaption of the existing schemes that were devised to deal with the Minkowski space external potential problem with lagrangian (J is a fixed function)

$$\mathcal{L} = \tfrac{1}{2} (\eta^{\mu\nu} \partial_\mu \phi \, \partial_\nu \phi - \mu^2 \phi^2 - J \, \phi^2); \; \eta_{\mu\nu} = \text{diag}\,(1, -1, -1, -1) \tag{2.24}$$

and equations of motion.

$$(\Box + \mu^2 + J\,(x))\,\phi\,(x) = 0 \tag{2.25}$$

We will discover that unfortunately the axioms listed above are rather flabby and, for a generic space-time, admit infinitely many realisations. Even for a free field in Minkowski space-time a unique quantization is obtained only after requiring the existence of a Poincaré group invariant cyclic vector and, most crucially, the positivity of the associated energy spectrum. It is the lack of such an isometry group in a general space-time $\mathcal{M}$ that leads to a proliferation of quantizations. Fortunately it transpires that to construct a physically well motivated unique quantum scheme it suffices that $\mathcal{M}$ possess asymptotic stationary regions - a requirement that is met by at least some space-times of interest. Note that precisely the same problem arises in the Minkowski external potential lagrangian of (2.24) and a unique quantization scheme can be obtained only if J(x) becomes asymptotically time independent. Even then there can be Klein paradox type problems if J(x) is not suitably restricted in form[32].

The physical interpretation of the mathematical structures raises difficult problems. The free field Minkowski space quantization mentioned above is realized on Fock space and as such possesses a natural particle interpretation which is ultimately linked back to the Poincaré group. Many of the curved space quantizations will be defined on a Hilbert space which is <u>a</u> Fock space but there is no reason why the interpretation of the resulting vector states as being associated with physical particles should be correct. There is a similar problem for the flat space external potential theory. If $J(\underline{x}, t)$ vanishes for $t < -T_1$ say ($T_1 > 0$) then the natural Fock quantization of

$$(\Box + \mu^2)\, \phi(x) = 0 \tag{2.26}$$

may be employed. The $J\phi^2$ interaction generates a time evolution of the early time canonical field variables which, when expressed as a Bogulubov transformation of the annihilation and creation operators, may be interpreted as producing pairs of particles of mass μ. More abstractly an automorphism is generated of the early time CCR C*-algebra which in turn induces a transformation on the states, converting an initial vacuum into a many particle state. (Note that since there is no Hamiltonian there is no Schrödinger picture in the usual sense and an algebraic view of the state evolution has to be adopted.) However even this interpretation is only really tenable if $J(\underline{x}, t)$ also vanishes for $t > T_2$ ($T_2 > 0$). Suppose for example that J varied smoothly in time between 0 for $t < -T$, and some constant $\delta > 0$ for $t > T_2$. Then it would be more meaningful to interpret the late time fields (or transformed states) in terms of the natural Fock quantization corresponding to a renormalized mass of $(\mu^2 + \delta)^{\frac{1}{2}}$. Thus two Hilbert spaces need to be employed and in fact even if J does vanish for $t > T_2$ an 'in/out' formalism is still a useful way of formulating the theory. It is important to note that there is no natural interpretation at all during periods when J $(\underline{x},t)$ is time-dependent although the 'hamiltonian diagonalization' method does provide a possible scenario for this and in any event correctly includes the mass renormalizations (this is one of the great virtues of that approach).

Translating the remarks above into a curved space-time context one anticipates that even for a space-time with asymptotically stationary regions two Hilbert spaces will be required - an 'in' space and an 'out' space. Thus if in the axioms above $\mathcal{H}$ is a Fock space associated with an 'in' region (this will all be explained properly later) it will be necessary to interpret the late time fields in terms of 'out' states and 'out' particle production.

This proliferation of Hilbert spaces and associated concrete realizations of the quantum field axioms ultimately enforces a C* algebraic approach. The different Hilbert spaces can then be associated, via the GNS construction, with different states on the underlying algebra. This point of view does not really assist much in solving the interpretational difficulties but does clarify the technical mathematical role played by the various spaces.

§3. IMPLEMENTATION OF THE COVARIANT SCHEME

As a motivation for the technique to be described let me briefly summarise the steps followed in quantizing the Minkowski space external potential system with classical field equations[12,13,14].

$$(\Box + \mu^2 + J(x))\ \phi\ (x) = 0, \qquad J \varepsilon\ C^\infty\ (\mathbb{R}^4) \tag{3.1}$$

1) Write (3.1) in two component hamiltonian form:

$$i \frac{\partial}{\partial t} \binom{\phi_t}{\pi_t} = H_t \binom{\phi_t}{\pi_t} \tag{3.2}$$

where H_t is a (time dependent) differential operator in the spatial variables and the Cauchy data

$$\phi_t\ (\underline{x}) := \phi\ (\underline{x}, t) \tag{3.3}$$

$$\pi_t\ (\underline{x}) := \frac{\partial}{\partial t}\ \phi\ (\underline{x}, t) \tag{3.4}$$

belong to some topological vector space - typically $(\phi_t, \pi_t)\ \varepsilon \mathcal{D}(\mathbb{R}^3) \oplus \mathcal{D}(\mathbb{R}^3)$.

2) The classical Cauchy problem can be solved in principle and the Cauchy data at time t related to that at some earlier time t_o by a linear operator $\mathcal{T}(t, t_o)$ on $\mathcal{D}(\mathbb{R}^3) \oplus \mathcal{D}(\mathbb{R}^3)$

$$\binom{\phi_t}{\pi_t} = \mathcal{T}(t, t_o) \binom{\phi_{t_o}}{\pi_{t_o}} \equiv \begin{pmatrix} T_{11} & T_{12} \\ T_{21} & T_{22} \end{pmatrix} \binom{\phi_{t_o}}{\pi_{t_o}} \tag{3.5}$$

If J(x) is suitably bounded (e.g. if $J\ \varepsilon \mathcal{S}(\mathbb{R}^4)$) then this evolution problem can be conveniently solved using an 'interaction picture' perturbation theory. Since $(\phi_t, \pi_t)\ \varepsilon\ \mathcal{D}(\mathbb{R}^3) \oplus \mathcal{D}(\mathbb{R}^3) \subset \mathcal{D}'(\mathbb{R}^3) \oplus \mathcal{D}'(\mathbb{R}^3)$ we can regard (3.5) as a distributional statement and smear it with $\mathcal{D}(\mathbb{R}^3) \oplus \mathcal{D}(\mathbb{R}^3)$ test functions (f, g) to get

$$\phi_t\ (f) = (T_{11}\ \phi_{t_o})\ (f) + (T_{12}\ \pi_{t_o})\ (f) = \phi_{t_o}\ (T^*_{11}\ f) + \pi_{t_o}\ (T^*_{12}\ f) \tag{3.6}$$

$$\pi_t(g) = (T_{21}\,\phi_{t_o})(g) + (T_{22}\,\pi_{t_o})(g) = \phi_{t_o}(T^*_{21}\,g) + \pi_{t_o}(T^*_{22}\,g) \qquad (3.7)$$

where * refers to the $L^2(\mathbb{R}^3)$ real Hilbert space adjoint. Using the canonical nature of these equations of motion it can be easily shown that

$$T_{11}\,T^*_{22} - T_{12}\,T^*_{21} = 1, \qquad (3.8)$$

$$T_{11}\,T^*_{12} - T_{12}\,T^*_{11} = 0,\; T_{21}\,T^*_{22} - T_{22}\,T^*_{21} = 0 \qquad (3.9)$$

3) Construct any representation of the canonical commutation relations for the $t = t_o$ fields with operators possessing a common domain D of essential self adjointness:

$$\left[\hat{\phi}_{t_o}(f), \hat{\pi}_{t_o}(g)\right] = i\hbar \int_{\mathbb{R}^3} f\,g \text{ on } D\;;\; \forall (f, g) \in \mathcal{D}(\mathbb{R}^3) \oplus \mathcal{D}(\mathbb{R}^3) \qquad (3.10)$$

$$\left[\hat{\phi}_{t_o}(f_1), \hat{\phi}_{t_o}(f_2)\right] = 0 \quad \text{on } D\;;\; \forall f_1, f_2 \in \mathcal{D}(\mathbb{R}^3) \qquad (3.11)$$

$$\left[\hat{\pi}_{t_o}(g_1), \hat{\pi}_{t_o}(g_2)\right] = 0 \quad \text{on } D\;;\; \forall g_1, g_2 \in \mathcal{D}(\mathbb{R}^3) \qquad (3.12)$$

4) <u>Define</u> the canonical fields at later times by mimicking (3.6), (3.7):

$$\hat{\phi}_t(f) := \hat{\phi}_{t_o}(T^*_{11}\,f) + \hat{\pi}_{t_o}(T^*_{12}\,f) \text{ on } D \qquad (3.13)$$

$$\hat{\pi}_t(g) := \hat{\phi}_{t_o}(T^*_{21}\,g) + \hat{\pi}_{t_o}(T^*_{22}\,g) \text{ on } D \qquad (3.14)$$

The CCR representation should be chosen so that the fields thus defined are densely defined with self adjoint closures. Then it can be shown that

a) $\hat{\phi}_t(f)$, $\hat{\pi}_t(g)$ satisfy the CCR for all t (by virtue of eqns. (3.8) and (3.9))

b) $\hat{\phi}(t, \underline{x}) := \hat{\phi}_t(\underline{x})$ satisfies the operator field equations (when suitably smeared).

c) $\hat{\phi}(t, \underline{x})$ satisfies covariant microcausal commutation relations.

One should note that

(i) Test function spaces other than $\mathcal{D}(\mathbb{R}^3) \oplus \mathcal{D}(\mathbb{R}^3)$ can be used. The main requirement is that this space should be mapped into itself under time evolution in order that (3.13) and (3.14) make sense.

(ii) If $J(\underline{x}, t) = 0$ for $t < -T_1$ $(T_1 > 0)$ then it is natural to use the usual free field Poincaré invariant Fock quantization for the initial CCR with t_o chosen to be less than $-T_1$.

(iii) A priori there will be no unitary operator $V(t, t_o)$ implementing the time evolution:

$$
\begin{aligned}
V(t,t_0)\, \hat{\phi}_{t_0}(f)\, V^{-1}(t,t_0) &= \hat{\phi}_t(f) \\
V(t,t_0)\, \hat{\pi}_{t_0}(f)\, V^{-1}(t,t_0) &= \hat{\pi}_t(f)
\end{aligned}
\tag{3.15}
$$

If a Fock representation is employed at $t = t_o$ then Shale's theorem[19] provides necessary and sufficient conditions for $V(t, t_o)$ to exist.

These four steps rather crudely summarise the approach to constructing linear quantum fields. The essential idea is really that, since the equations of motion are linear, the operator $U(t, t_o)$ which evolves the classical Cauchy data can also be employed to evolve quantum Cauchy data. In broad terms the same approach can be employed in the curved space-time case. However the following points should be noted:

(i) If the metric tensor can be sensibly decomposed around Minkowski space as $g_{\mu\nu} = \eta_{\mu\nu} + h_{\mu\nu}$ (h small in some sense) then the lagrangian (2.1) can be written as a sum of a 'free' part plus an interaction. This is in close analogue to the case above but is undesirable because:

a) $\mathcal{M}$ may not be topologically euclidean

b) The causal structure associated with $g_{\mu\nu}$ is different from that connected with $\eta_{\mu\nu}$. This would make it difficult to ensure that the quantum fields satisfied the correct microcausality condition.

c) The interaction term turns out to be an unbounded operator on the classical free field Hilbert space so the perturbative techniques used to describe the classical $J\,\phi^2$ time evolution are not directly applicable anyway.

d) Writing $g_{\mu\nu} = \eta_{\mu\nu} + h_{\mu\nu}$ flies in the face of the whole spirit of general relativity.

I will now outline in a series of steps a scheme which avoids these problems and provides concrete realizations of the axioms in §2.

1) Let Σ be a Cauchy surface in the globally hyperbolic manifold $\mathcal{M}$. If ϕ is a C^∞ real solution to the classical equations of motion (2.2) then, regarded as a distribution defined on the test function $\psi \in \mathcal{D}(\mathcal{M}) \equiv T$, it satisfies the relation

$$\phi(\psi) = \int_{\Sigma} \tilde{G}(\psi;y) \overleftrightarrow{\partial}_\mu \phi(y)\, d\sigma^\mu(y) \qquad (3.16)$$

where $\tilde{G}$ is the commutator function defined in eqn.(2.13) and $d\sigma^\mu$ is the normally directed volume element on Σ. Clearly (3.16) solves the classical Cauchy problem and expresses the solution ϕ in terms of its Cauchy data (field and normal derivative) on Σ. This can be regarded as a somewhat more covariant analogue of steps 1) and 2) above. It is convenient to introduce a local time coordinate so that Σ is the hypersurface $t = t_o$ (denoted Σ_{t_o}) and define the conjugate momentum by

$$\pi_{t_o} := \frac{\partial \mathcal{L}}{\partial(\partial_o \phi)} = (-\det g)^{\frac{1}{2}}\, g^{o\nu}\, \partial_\nu \phi \qquad (3.17)$$

in terms of which (3.16) becomes (I will assume throughout that Σ is orientable)

$$\phi(\psi) = \int_{\Sigma_{t_o}} \{ \tilde{G}(\psi;\underline{y}, t_o)\pi_{t_o}(\underline{y}) - (-\det g)^{\frac{1}{2}} \partial^o \tilde{G}(\psi;\underline{y}, y_o)\Big|_{y_o=t_o} \phi_{t_o}(\underline{y})\} d^3\underline{y} \tag{3.18}$$

where $\phi_{t_o}(\underline{y}) := \phi(\underline{y}, t_o)$. Since ϕ is a $C^\infty(\mathcal{M})$ solution, ϕ_{t_o} and π_{t_o} are respectively a $C^\infty(\Sigma)$ function and $C^\infty(\Sigma)$ scalar density (from (3.17)). Thus they can also be regarded as distributions:

$$\phi_{t_o} \in \tilde{\mathcal{D}}(\Sigma)', \; \phi_{t_o}(p) := \int_{\Sigma_{t_o}} \phi_{t_o}(\underline{y})\, p(\underline{y})\, d^3y \; ; \; p \in \tilde{\mathcal{D}}(\Sigma) \tag{3.19}$$

$$\pi_{t_o} \in \mathcal{D}(\Sigma)', \; \pi_{t_o}(q) := \int_{\Sigma_{t_o}} \pi_{t_o}(\underline{y})\, q(\underline{y})\, d^3y \; ; \; q \in \mathcal{D}(\Sigma) \tag{3.20}$$

Now, as mentioned in §2, supp $\tilde{G}(\psi; y) \subset J^+(\text{supp}\,\psi) \cup J^-(\text{supp}\,\psi)$. However if K is any compact subset of the globally hyperbolic manifold and Σ is any Cauchy surface it can be shown that $J^\pm(K) \cap \Sigma$ is a subset of Σ with compact closure. In particular the $C^\infty(\Sigma_{t_o})$ function $\tilde{G}(\psi; \underline{y}, t_o)$ has compact support as does $\partial^o \tilde{G}(\psi; \underline{y}, y_o)\Big|_{y_o=t_o}$. Thus regarding (ϕ_{t_o}, π_{t_o}) as an element of $\tilde{\mathcal{D}}(\Sigma)' \oplus \mathcal{D}(\Sigma)'$, eqn.(3.18) can be rewritten as

$$\phi(\psi) = \pi_{t_o}(\tilde{G}(\psi; (\underline{.}), t_o)) - \phi_{t_o}((-\det g)^{\frac{1}{2}} \partial^o \tilde{G}(\psi;(\underline{.})\, y_o)\Big|_{y_o=t_o}) \tag{3.19}$$

where $\tilde{G}(\psi; (\underline{.})\, t_o)$ denotes the $\mathcal{D}(\Sigma_{t_o})$ function $\underline{y} \rightsquigarrow \tilde{G}(\psi; \underline{y}, t_o)$. Eqn.(3.19) can be thought of as an analogue of eqns.(3.6) and (3.7).

2) Choose a representation of the CCR on Σ_{t_o}. Thus we look for operators $\hat{\phi}(p)$, $\hat{\pi}(q)$ defined on a Hilbert space $\mathcal{H}$ with a common domain of essential self adjointness D and satisfying

$$[\hat{\phi}(p), \hat{\pi}(q)] = i\hbar \int_{\Sigma_{t_o}} p\, q \text{ on } D \; ; \; \forall (p, q) \in \tilde{\mathcal{D}}(\Sigma_{t_o}) \oplus \mathcal{D}(\Sigma_{t_o}) \tag{3.20}$$

$$\left[\hat{\phi}(p_1), \hat{\phi}(p_2)\right] = 0 \text{ on } D \; ; \; \forall p_1, p_2 \in \tilde{\mathcal{D}}(\Sigma_{t_o}) \tag{3.21}$$

$$\left[\hat{\pi}(q_1), \hat{\pi}(q_2)\right] = 0 \text{ on } D \; ; \; \forall q_1, q_2 \in \mathcal{D}(\Sigma_{t_o}) \tag{3.22}$$

These are the analogue of eqns.(3.10) - (3.12) and step 3 above but notice that I am smearing $\hat{\phi}$ with scalar densities.

3) Similarly to step 4 above we now define the covariant field operators $\hat{\phi}(\psi)$ on D by mimicking the classical evolution equation (3.19) for all $\psi \in \mathcal{D}(\mathcal{M})$:

$$\hat{\phi}(\psi) := \hat{\pi}\left(\tilde{G}(\psi; (\underline{\cdot}), t_o\right) - \hat{\phi}\left((-\det g)^{\frac{1}{2}}\, \partial^o\, \tilde{G}(\psi; (\underline{\cdot}) y_o)\Big|_{y_o = t_o}\right) \tag{3.23}$$

The CCR representation should preferably be chosen so that $\hat{\phi}(\psi)$ thus defined is essentially self adjoint on D (it is clearly a symmetric operator). It possesses the following properties,

a) The following maps are linear and continuous

$$\begin{array}{ll} \mathcal{D}(\mathcal{M}) \to \mathcal{D}(\Sigma_{t_o}) , & \mathcal{D}(\mathcal{M}) \to \mathcal{D}(\Sigma_{t_o}) \\ \psi \rightsquigarrow \tilde{G}(\psi; (\underline{\cdot}), t_o) & \psi \rightsquigarrow \partial^o\, \tilde{G}(\psi; (\underline{\cdot}), y_o)\Big|_{y_o = t_o} \end{array} \tag{3.24}$$

Now suppose that the CCR representation is chosen so that the maps

$$\begin{array}{ll} \tilde{\mathcal{D}}(\Sigma_{t_o}) \to \mathbb{C} , & \mathcal{D}(\Sigma_{t_o}) \to \mathbb{C} \\ p \rightsquigarrow \langle v, \hat{\phi}(p)\, w \rangle & q \rightsquigarrow \langle v, \hat{\pi}(q)\, w \rangle \end{array} \tag{3.25}$$

are continuous for all vectors v, w in $D \subset \mathcal{H}$ (For example this is true for the Fock representations to be discussed later). Then the field existence axiom 1 in §2 is satisfied with the test function space T chosen as $\mathcal{D}(\mathcal{M})$ and the invariant domain D as above.

b) Since for $\psi \in \mathcal{D}(\mathcal{M})$ $\tilde{G}(y;\psi)$ is itself a $C^{\infty}(\mathcal{M})$ solution of the classical field equations it follows from (3.16) that

$$\tilde{G}(\psi_1;\psi_2) = \int_{\Sigma} \tilde{G}(\psi_1;y)\overleftrightarrow{\partial}_{\mu}\tilde{G}(y;\psi_2)\,d\sigma^{\mu}(y)\ ;\ \forall\psi_1,\psi_2 \in \mathcal{D}(\mathcal{M}) \tag{3.26}$$

from which it can be deduced that the operator $\hat{\phi}(\psi)$ constructed above satisfies microcausality (axiom 2) in the form of the covariant commutation relations in eqn.(2.12).

c) If an irreducible representation of the CCR is chosen then the irreducibility axiom 3 will be satisfied.

d) Since $\tilde{G}(L_x\psi\,;\,y) = 0\ \forall\psi \in \mathcal{D}(\mathcal{M})$ it follows that the operator field equations (axiom 4) are satisfied.

e) The group invariance requirement (axiom 5) will not be met by most choices of CCR representation although in practice there will be a suitable one.

We see from the above that the problem of constructing covariant fields is reduced to finding suitable representations of the CCR. The main problem is that there are just too many! The Fock representations (see §6) associated with any choice of initial hypersurface Σ_{t_o} satisfy the requirements but in general different hypersurfaces will lead to unitarily inequivalent covariant fields. For stationary or static space times however the work (§8) of Ashtekar, Magnon[5,6] and Kay[7-9] leads to a unique quantization which can in turn be applied in an 'in/out' formalism to space-times with asymptotically stationary regions.

The theory of CCR representations[19,33] is well developed and the fact that the canonical fields are defined on a three-manifold Σ does not lead to any significant new problems. Our treatment of this aspect of the subject can therefore be rather brief.

§4. USE OF SEGAL FIELDS

Canonical commutation relations can be discussed in a relatively covariant way by employing Segal fields. First we need some notation. Let $\mathcal{R}_c$ denote the vector space of real $C^\infty(\mathcal{M})$ solutions to the wave equation (2.2) whose Cauchy data on some Cauchy surface has compact support. The support properties of $\tilde{G}(x;y)$ imply that this feature then holds on any Cauchy surface. If Σ is a Cauchy surface a symplectic form σ_Σ can be defined on $\mathcal{R}_c$ by

$$\sigma_\Sigma(S_1,S_2) := \int_\Sigma S_1 \overleftrightarrow{\partial}_\mu S_2 \, d\sigma^\mu; \quad \forall S_1,S_2 \in \mathcal{R}_c \tag{4.1}$$

It follows at once from the equations of motion that σ_Σ does not depend on the choice of Σ and hence the Σ suffix may be dropped.

Now any real solution in $\mathcal{R}_c$ induces a unique set of Cauchy data on a Cauchy surface Σ. This will be indicated by the linear map:

$$\begin{array}{ccc} \mathcal{R}_c & \xrightarrow{i_\Sigma} & P_\Sigma := \mathfrak{D}(\Sigma) \oplus \tilde{\mathfrak{D}}(\Sigma) \\ S & \rightsquigarrow & (q_s, p_s)_\Sigma \end{array} \tag{4.2}$$

where q_s(resp. p_s) is the value of the solution $S \in \mathcal{R}_c$ (resp. its normal derivative) evaluated on Σ. We may regard P_Σ as the phase space associated with Σ. Conversely any pair of functions in P_Σ induces a unique solution as indicated by the linear map

$$\begin{array}{ccc} P_\Sigma & \longrightarrow & \mathcal{R}_c \\ (q,p)_\Sigma & \rightsquigarrow & S_{(q,p)_\Sigma} \end{array} \tag{4.3}$$

There is a natural symplectic form associated with each P_Σ defined by

$$\sigma((q_1,p_1)_\Sigma;(q_2,p_2)_\Sigma) := \int_\Sigma (q_1p_2 - q_2p_1) \tag{4.4}$$

which is precisely the same as eqn. (4.1) in the sense that

$$\sigma((q_1,p_1);(q_2,p_2)) := \sigma(S_{(q_1,p_1)_\Sigma}, S_{(q_2,p_2)_\Sigma}). \tag{4.5}$$

If Σ_1, Σ_2 are any two Cauchy surfaces the dynamical evolution may be described by the diagram

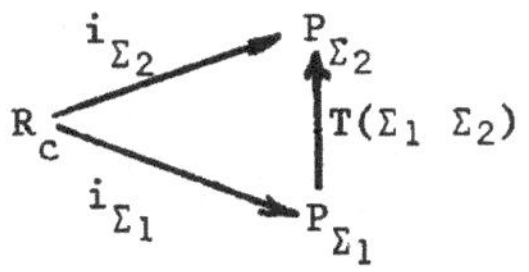

(4.6)

in which $\mathcal{T}(\Sigma_1,\Sigma_2)$, defined to make the diagram commutative, is the linear map connecting the Cauchy data on the two surfaces.

A <u>Segal Field</u> is defined to be a map $\hat{\Phi}$ from $\mathcal{R}_c$ into the unbounded operators on some Hilbert space $\mathcal{H}$ with the properties that:

i) $\{\hat{\Phi}(S) | S\varepsilon\mathcal{R}_c\}$ has a common, invariant, dense domain $D \subset \mathcal{H}$ of essential self adjointness.

ii) $S \rightsquigarrow \langle v, \hat{\Phi}(S)w\rangle$ is a real linear map for all $v, w \in D$. (4.7)

iii) $\left[\hat{\Phi}(S_1), \hat{\Phi}(S_2)\right] = i\hbar\,\sigma(S_1,S_2)$ on D (4.8)

Suppose that Σ is any Cauchy surface. Define

$$\hat{\phi}_\Sigma(p) := \hat{\Phi}(S_{(0,p)_\Sigma}) \qquad \forall\, p \in \tilde{\mathcal{D}}(\Sigma) \tag{4.9}$$

$$\hat{\pi}_\Sigma(q) := -\hat{\Phi}(S_{(q,0)\Sigma}) \qquad \forall\, q \in \mathcal{D}(\Sigma) \tag{4.10}$$

Then $$\left[\hat{\phi}_\Sigma(p), \hat{\pi}_\Sigma(q)\right] = i\hbar \int_\Sigma q\, p \qquad \text{on } D \subset \mathcal{H} \tag{4.11}$$

$$\left[\hat{\phi}_\Sigma(p_1), \hat{\phi}_\Sigma(p_2)\right] = 0 \qquad \text{on } D \subset \mathcal{H} \tag{4.12}$$

$$\left[\hat{\pi}_\Sigma(q_1), \hat{\pi}_\Sigma(q_2)\right] = 0 \qquad \text{on } D \subset \mathcal{H} \tag{4.13}$$

thus generating representations of the CCR on any Cauchy surface. In particular if we can construct Segal fields we will generate a CCR representation on the initial hypersurface Σ_{t_o} as required in the covariant method (eqns. (3.20) - (3.22)). Actually we also need the continuity conditions of eqn. (3.25) and these need to be checked in any specific case. Typically $\mathcal{R}_c$ will carry some topological vector space structure such that the map in (4.7) is continuous and such that a solution is a continuous function of its Cauchy data. The ubiquitous Fock representations described in §6 provide such a structure.

The Segal fields possess various interesting properties:

a) $\hat{\Phi}(S) = \hat{\phi}_\Sigma(p_s) - \hat{\pi}_\Sigma(q_s)$ on D $\quad$ (4.14)

This can be rewritten somewhat formally as (c.f. (4.4))

$$\hat{\Phi}(S) = \sigma((\hat{\phi}_\Sigma,\hat{\pi}_\Sigma);(q_s,p_s)_\Sigma) \tag{4.15}$$

or (c.f. (4.1))

$$\hat{\Phi}(S) = \sigma(\hat{\phi},S) \tag{4.16}$$

where $\hat{\phi}$ is the covariant field constructed by the methods of §3 in terms of initial quantum Cauchy data on the chosen initial surface Σ.

b) Classically observables are thought of as functions on phase space. In the present case $\mathcal{R}_c$ may be regarded as the phase space and for each $S \in \mathcal{R}_c$ a linear classical observable $\Phi(S)$ may be defined by[6,8]

$$\Phi(S) := \sigma(\cdot\,,S) \tag{4.17}$$

$$\text{i.e. } (\Phi\,(S))(S') := \sigma(S',S) \quad \forall S' \in \mathcal{R}_c \tag{4.18}$$

The Poisson bracket of such observables is

$$\{\Phi(S_1),\ \Phi(S_2)\}_{P.B} = \sigma(S_1,S_2) \tag{4.19}$$

Thus eqn. (4.8) can be viewed as a manifestation of the traditional Dirac quantization scheme in which quantum commutators of certain pairs of observables are set equal to i $\hbar$ times their classical Poisson bracket. It is in fact quite possible to start the axiomatization of quantum field theory on $\mathcal{M}$ from this point of view.

c) If $\psi \in \mathcal{D}(\mathcal{M})$ is any test function then

$$\psi_G(x) := \tilde{G}(x;\psi) \tag{4.20}$$

is an element of $\mathcal{R}_c$. Thus we can form $\hat{\Phi}(\psi_G)$ which turns out to be simply related to the covariant field $\hat{\phi}$ (constructed from $\hat{\Phi}$) by

$$\hat{\Phi}(\psi_G) = \hat{\phi}(\psi) \text{ on } D\subset\mathcal{H},\ \forall\psi \in \mathcal{D}(\mathcal{M}) \tag{4.21}$$

We could have actually defined $\hat{\phi}$ by this expression in the first place. The method of §3 however emphasises more clearly the relation with existing external potential work and the manner in which the classical evolu tion is used to construct the quantum fields, although (4.21) is admittedly superficially easier to handle.

d) It has been tacitly assumed in (4.21) that irrespective of which hypersurface Σ is used in (4.9), (4.10) the covariant fields obtained by the methods of §3 (evolving the quantum Cauchy data off Σ) are all the same. This is correct although it is not entirely obvious. In the language of §3 the question is the following:

"Suppose I construct a covariant field $\hat{\phi}$ in terms of a CCR representation on an initial hypersurface Σ_{t_o} using a Segal field as above. Then

i) Can $\hat{\phi}$ be used to construct equal time quantum canonical variable on a later time hypersurface Σ_{t_1}?

ii) How are these fields related to the canonical fields which $\hat{\Phi}$ induces on Σ_{t_1} via (4.14)?"

Question i) can only be answered affirmitively if $\hat{\phi}$ can be smeared with test functions which are Dirac δ-functions in time. Since such objects do not belong to $\mathcal{D}(\mathcal{M})$ it is by no means obviously possible. Looking back at eqn (3.23) (or 4.21), the question resolves into whether or not

$$\tilde{G}(p,t_1;\underline{y}\ t_o)\ ,\ p\in\mathcal{D}(\Sigma_{t_1})$$

is a smooth function of t_o and t_1 and is $\mathcal{D}(\Sigma_{t_o})$ in $\underline{y}$. Fortunately the nature of the singularity in the bi-distribution $\tilde{G}(x;y)$ is well understood[31]. For x and y sufficiently close it is a δ-function in the geodesic arc distance between these two points, i.e. the singularity is concentrated on the light cones emanating from the two points (as in the well known Minkowski space case.) This means that for t_1 sufficiently close to t_o, $\tilde{G}(p,t_1;\underline{y},t_o)$ has the desired properties and equal time fields can be constructed at $t = t_1$. A patching argument then gives them for all times. Question (ii) is answered by showing that these fields are in fact identical to those generated directly from $\hat{\Phi}$ so the formalism is in all respects consistent.

§5. USE OF C^*-ALGEBRAS

C^*-algebras can be used either as a mathematical tool in the concrete construction of Segal fields or in a superficially physically deeper way as the algebra of local canonical observables. The formalism will be developed here in a way which emphasises both these approaches.

One starts by observing as usual that eqn. (4.8) implies that $\hat{\Phi}(S)$ is necessarily an unbounded operator. The associated difficulties may be circumnavigated by looking instead for families of unitary operators $\hat{W}(S)$, $S\varepsilon\mathcal{R}_c$ on a Hilbert space $\mathcal{H}$ satisfying

$$\text{i) } \hat{W}(S_1)\hat{W}(S_2) = e^{-\frac{i\hbar}{2}\sigma(S_1,S_2)}\hat{W}(S_1+S_2), \quad \forall S_1,S_2\varepsilon\mathcal{R}_c \tag{5.1}$$

$$\text{ii) } \hat{W}(0) = 1 \tag{5.2}$$

$$\text{iii) } \mathbb{R}\ni t \rightsquigarrow \hat{W}(tS) \text{ is weakly continuous } \forall S\varepsilon\mathcal{R}_c. \tag{5.3}$$

Stone's theorem then implies the existence of a family of self adjoint operators $\hat{\Phi}(S)$ such that

$$\hat{W}(S) = e^{i\hat{\Phi}(S)}, \forall S\varepsilon\mathcal{R}_c \tag{5.4}$$

and for any finite set of elements $S_1 \ldots S_n \varepsilon\mathcal{R}_c$, Gårding's theorem provides a linear, dense, invariant subspace of $\mathcal{H}$ on which $\hat{\Phi}(S_1)\ldots\hat{\Phi}(S_n)$ are defined with the additive property $\hat{\Phi}(\lambda S_i + \mu S_j) = \lambda\hat{\Phi}(S_i) + \mu\hat{\Phi}(S_j)$, $(\lambda,\mu\varepsilon\mathbb{C})$ and on which $[\Phi(S_i), \Phi(S_j)] = i\hbar\sigma(S_i,S_j)$. Thus rather than directly finding representations of (4.8) one looks instead for Segal systems of unitary operators satisfying (5.1)...(5.3) (In practive one also trys to extend the Gårding domain to an invariant domain of essential self adjointness for all the $\hat{\Phi}(S)$, $S\varepsilon\mathcal{R}_c$).

The technique employed for constructing Segal systems is to associate a C^*-algebra with the pair $(\mathcal{R}_c,\sigma)$ in such a way that a representation of this algebra automatically generates a set of unitary operators satisfying (5.1) - (5.3). As this is essentially standard bookwork I will summarise the steps very briefly.

1) A suitable C^*- algebra $\mathcal{W}(\mathcal{R}_c,\sigma)$ can be constructed in various ways:

a) Segal's original construction[19] employed the fact that for finite dimensional quantum mechanical systems the Stone-Von Neumann theorem[33] shows that all non trivial Segal systems are unitarily equivalent. Thus one can pick any concrete representation and with it associate a C^*- algebra which it generates. The set of finite dimensional subspaces F of $\mathcal{R}_c$ (with σ_{1_F} non degenerate) form a directed set and $\mathcal{W}(\mathcal{R}_c,\sigma)$ is defined as the inductive limit of the associated algebras. Even at the finite dimensional level there are various C^*- algebras which can be associated with the commutation relations. This is all summarised clearly in ref. (19) where it is shown that when suitably defined $\mathcal{W}(\mathcal{R}_c,\sigma)$ has the property that all non trivial concrete representations are algebraically isomorphic (but not unitarily equivalent!) and that if α is any symplectic transformation on $(\mathcal{R}_c,\sigma)$, an automorphism of $\mathcal{W}(\mathcal{R}_c,\sigma)$ may be defined by its values on the generators $W(S)$:

$$\alpha(W(S)) := W(\alpha S) \tag{5.5}$$

b) A related 'minimal' algebra is that of Manuceau or Slawny[33-36]. This is simple and hence also has the desirable property that all non trivial representations are algebraically equivalent.

c) Ashtekar and Magnon[5,6] employ a free * algebra (no norm) over $\mathcal{R}_c$. This has the virtue of being associated directly with the observables $\Phi(S)$ but the lack of a norm reduces the amount of existing representation theory that can be used.

2) Recall that a state on any C^*- algebra $\mathfrak{A}$ is a linear map $w: \mathfrak{A} \longrightarrow \mathbb{C}$ such that $w(\mathbb{1}) = 1$ and $w(A^*A) \geq 0$, $A \in \mathfrak{A}$. This leads, via the GNS construction[33], to a representation ρ of $\mathfrak{A}$ by bounded operators on a Hilbert space $\mathcal{H}$ and a cyclic vector $\Omega \in \mathcal{H}$ with the

property that

$$w(A) = \langle\Omega, \rho(A)\Omega\rangle \quad \forall A \in \mathfrak{A} \tag{5.6}$$

Applying this technique to $\mathcal{W}(\mathcal{R}_c,\sigma)$ it can be shown[33] that concrete Segal systems (i.e. unitary operators satisfying (5.1)-(5.3)) are in one to one correspondence with maps $E: \mathcal{R}_c \to \mathbb{C}$ (known as generating functionals) such that

i) $E(0) = 1$ (5.7)

ii) For any $k < \infty$ and $\lambda_1 \dots \lambda_n \in \mathbb{C}$ $\sum_{i,j=1}^{k} E(S_i - S_j) e^{-\frac{i\hbar}{4}\sigma(S_i,S_j)} \bar{\lambda}_j \lambda_i \geq 0$ (5.8)

iii) $\mathbb{R} \ni t \rightsquigarrow E(tS)$ is continuous. (5.9)

The resulting representation possesses a cyclic vector Ω and $E(S) = \langle\Omega, \hat{W}(S)\Omega\rangle$.

As an alternative to the use of C^*-algebras one may exploit the nuclear structure on $\mathcal{R}_c$ identified as one of the phase spaces $P_\Sigma = \mathfrak{D}(\Sigma) \oplus \tilde{\mathfrak{D}}(\Sigma)$. The theory of measures on such spaces is well developed[37,38,39] and in particular if μ is a $\mathcal{R}_c$ quasi-invariant probability measure on $\mathcal{R}_c'$, then define on $\mathcal{L}^2(\mathcal{R}_c', d\mu)$,

$$(\hat{W}(S)\Psi)(\chi) = \left(\frac{d\mu(\chi+S)}{d\mu(\chi)}\right)^{\frac{1}{2}} e^{-i\hbar\sigma(\chi,S)} \Psi(\chi+S) \tag{5.10}$$

where $\frac{d\mu(\chi+S)}{d\mu(\chi)}$ is the Radon-Nikodym derivative associated with μ. The symplectic product $\sigma(\chi,S)$ is understood in the sense that $\chi = (\chi_1,\chi_2) \in \mathfrak{D}'(\Sigma) \oplus \tilde{\mathfrak{D}}'(\Sigma)$ and (c.f. (4.4))

$$\sigma(\chi,S) := \chi_1(q_s) - \chi_2(p_s) \tag{5.11}$$

where (q_s,p_s) are the Cauchy data induced by S on Σ. Clearly $\hat{W}$ defined in this way satisfies (5.1) and (5.2) and with some effort it can also be shown to satisfy[40] (5.3). The resulting representation is however reducible and irreducible subspaces need to be projected out (i.e. the system needs 'polarizing' in the language of geometric quantization).

A cyclic representation of the Weyl form of the CCR may also be obtained directly by picking a Cauchy surface Σ and a $\mathcal{D}(\Sigma)$ quasi-invariant probability measure on $\tilde{\mathcal{D}}(\Sigma)'$. Then define on $L^2(\tilde{\mathcal{D}}(\Sigma)', d\mu)$ the unitary operators:

$$(U(p)\Psi)(\chi) = e^{i\chi(p)}\,\Psi(\chi), \quad p \in \tilde{\mathcal{D}}(\Sigma) \tag{5.12}$$

$$(V(q)\Psi)(\chi) = \left(\frac{d\mu}{d\mu}\frac{(\chi+q)}{(\chi)}\right)^{\frac{1}{2}} \Psi(\chi+q), \quad q \in \mathcal{D}(\Sigma) \tag{5.13}$$

$$\text{Identifying } U(p) = e^{i\phi_\Sigma(p)}, \quad V(q) = e^{i\pi_\Sigma(q)} \tag{5.14}$$

we see that a Weyl representation of the CCR (cf. eqns (4.11) - (4.13)) is generated with cyclic vector $1 \in L^2(\tilde{\mathcal{D}}(\Sigma)', d\mu)$. Eqns. (5.12) and (5.13) rigorize the heuristic field analogue of the Schrödinger quantization rules:

$$\left.\begin{array}{l} \hat{x} \to x \\ \hat{p} \to -ih\dfrac{d}{dx} \end{array}\right\} \qquad \left.\begin{array}{l} \hat{\phi}(\underline{x}) \to \phi(\underline{x}) \\ \hat{\pi}(\underline{x}) \to -ih\dfrac{\delta}{\delta\phi(\underline{x})} \end{array}\right\} \tag{5.15}$$

and the state vectors have an immediate physical interpretation. Namely if the quantum system is in the state $\Psi \in L^2(\tilde{\mathcal{D}}(\Sigma)', d\mu)$ and a measurement is made of the classical field configuration, the probability P_B that the result lies in a borel set $B \subset \tilde{\mathcal{D}}'(\Sigma)$ is

$$P_B = \int_B |\Psi(\chi)|^2\, d\mu(\chi) \tag{5.16}$$

There has been a lot of discussion in the literature concerning the particle interpretation of a quantum field in a curved space-time. This concept is in general ambiguous and in any event can only be associated with Fock representations. The complementary field interpretation discussed above is however generally applicable and it might be worthwhile to reexamine some of the conceptual difficulties within this framework.

Apart from their role in CCR representation theory, C^*-algebras can also be used to formulate a more abstract approach to the quantum field dynamics. With each phase space P_Σ and symplectic form σ_Σ (eqn. 4.4), a C^*-algebra $\mathcal{W}(P_\Sigma,\sigma_\Sigma)$ can be associated by the methods mentioned above for $(\mathcal{R}_c,\sigma)$. The symplectic transformation $\mathcal{T}(\Sigma_1,\Sigma_2)$ then induces an isomorphism $\alpha(\Sigma_1,\Sigma_2)$ between $\mathcal{W}(P_{\Sigma_1}, \sigma_{\Sigma_1})$ and $\mathcal{W}(P_{\Sigma_2}, \sigma_{\Sigma_2})$. This can conveniently be described by the commutative map diagram.

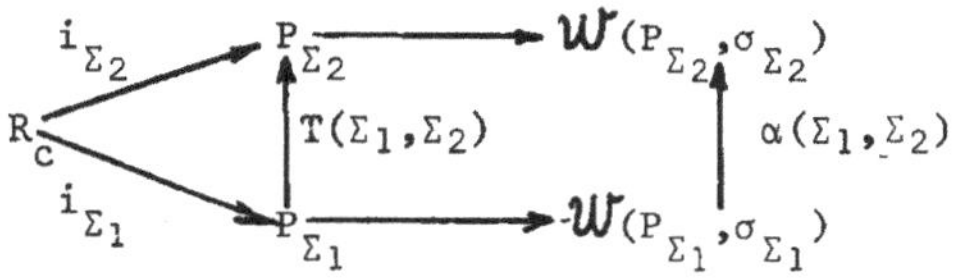

(5.17)

Both this particular use of C^* algebras and the associated diagrams have been introduced and exploited by Kay[8,9], and are very valuable for clarifying the relationships between various approaches to quantization. The isomorphism $\alpha(\Sigma_1, \Sigma_2)$ may be regarded as the abstract algebraic realisation of the quantum dynamics between the Cauchy surfaces Σ_1 and Σ_2. If concrete representations are chosen of the two C^* algebras, via functions E_{Σ_2} and E_{Σ_1} (eqns. (5.7) - (5.9), then the map diagram may be extended to

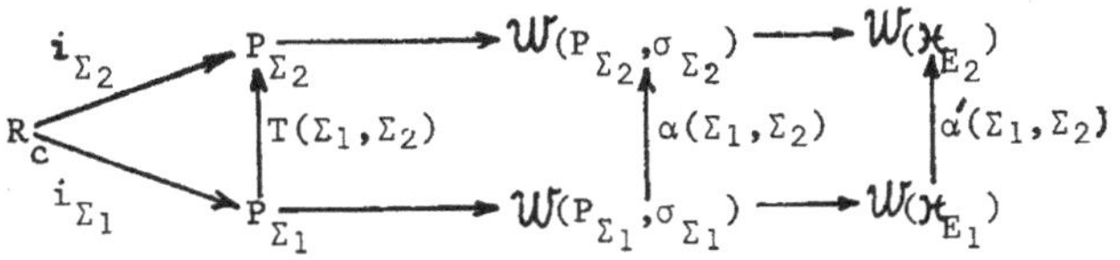

(5.18)

where $\mathcal{W}(\mathcal{H}_{E_i})$ i = 1,2 denotes the concrete C^*-algebra of bounded operators on the Hilbert space $\mathcal{H}_{E_i}$ constructed using the functional E_i. Quantum dynamics is now represented as the isomorphism $\alpha'(\Sigma_1, \Sigma_2)$ between these two concrete C^*-algebras. Since all non trivial representations produce isomorphic algebras the only virtue of choosing a concrete representation is if it s vector and density matrix states can in some way be naturally associated with physical observables. Fell's theorem[33] is of course applicable in this situation and guarantees that for any given Σ all representations of $\mathcal{W}(P_\Sigma, \sigma_\Sigma)$ are physically equivalent. Augmented with Haag's well known [41] remarks to the effect that in practice an experimentalist can only determine a weak * neighbourhood of a state, this means that by restricting one's attention to density matrices on any <u>single</u> representation no physical information is lost. Unfortunately however this helps us not one wit in relating real physical objects to the mathematical symbols.

The covariant approach described in § 2,3 can be usefully analysed in the present context. A representation of $\mathcal{W}(P_{\Sigma_o}, \sigma_{\Sigma_o})$ is selected and the quantum field evolved dynamically off this initial hypersurface. The representations of the C^*-algebras $\mathcal{W}(P_\Sigma, \sigma_\Sigma)$ for all other Cauchy surfaces are <u>uniquely fixed</u> by this concrete field operator. Thus the covariant method provides a natural concrete realisation of the structure depicted in (5.18). If there are asymptotically stationary regions in the far past and future then the remarks made earlier concerning an in/out situation may be interpreted in the framework of (5.18) by chosing Σ_1 (resp Σ_2) to lie in the deep past (resp. future) and imposing the natural Fock quantizations for these stationary cases (§ 6).

On the other hand in the Hamiltonian diagonalization scheme a different selection of CCR representations is made on each Cauchy surface with the aim mentioned previously of constructing positive, self-adjoint instantaneous Hamiltonians. It should be clear from all this discussion that in many ways (5.17) is the fundamental description of quantum dynamics. It sires various concrete realizations (5.18) of which the 'covariant method' and 'Hamiltonian diagonalization' are just two examples. Unfortunately the selection of the correct representation (if there be such a thing at all) requires a resolution of the still debated interpretational questions.

§6. FOCK SPACE

Let us now turn to the problem of constructing some concrete Segal fields for use either in the covariant method or within the framework of (5.18) which in principle permits a wider choice of CCR representation on different Cauchy surfaces. Since I am assuming that the reader is familiar with the Fock-Cook[42,43] construction the treatment is very terse.

Let V be any complex Hilbert space and let $\mathfrak{F}_n(V)$ denoted the symmetrized n-fold tensor product:

$$\mathfrak{F}_n(V) := S_n\,(\otimes^n V) \qquad n > 1 \tag{6.1}$$

$$\mathfrak{F}_o(V) := \mathbb{C} \tag{6.2}$$

Then the Fock space $\mathfrak{F}(V)$ over V is defined as the complete direct sum

$$\mathfrak{F}(V) = \overline{\bigoplus_{n=0}^{\infty} \mathfrak{F}_n(V)} \tag{6.3}$$

On the dense subspace $D := \bigoplus_{n=0}^{\infty} \mathfrak{F}_n(V)$, define the 'creation operator' $a^*(v)$ (* is just a notation not the adjoint!) by

$$(a^*(v)\,\Psi\,)_{n+1} := n^{\frac{1}{2}}\, S_n\,(v \otimes \Psi_n) \qquad v \in \mathbf{V}, \quad \Psi = (\Psi_o, \Psi_1, \Psi_2 \dots) \in D \tag{6.4}$$

Then both $a^*(v)$ and its adjoint $a^*(v)^*$ are closeable and $\Phi(v)$ defined by

$$\hat{\Phi}(v) := \frac{1}{i\sqrt{2}}\,(a^*(v) - a^*(v)^*) \tag{6.5}$$

is an essentially self adjoint operator on $D \subset \mathfrak{F}$ with the commutation relations

$$[\hat{\Phi}(v_1), \hat{\Phi}(v_2)] = i \;\mathrm{Im} < v_1, v_2 >_V \text{ on } D \tag{6.6}$$

In fact it can be shown that $\hat{W}(v) := e^{i\hat{\Phi}(v)}$ satisfies all three requirements (5.1) - (5.3) for a Segal system with the symplectic form chosen as $\hbar\,\sigma(v_1, v_2) \equiv \mathrm{Im}_{\hbar} < v_1, v_2 >$. Note however that the Fock-Cook construction sidesteps the general C^*- algebraic formalism by forming the Segal fields $\hat{\Phi}(v)$ directly.

It is clear from (6.5) that we will be able to use Fock space to obtain a concrete representation of the canonical commutation relations (4.8) provided that a complex Hilbert space structure can be imposed on R_c in such a way that the given symplectic form in (4.1) is precisely the imaginary part of the Hilbert space inner product. It is clearly sufficient to use a prehilbert space and this is formalised in the following definition of Segal:

A one-particle structure over (R_c, σ) is a real linear map K from R_c into a complex hilbert space V such that

i) $K(R_c)$ is dense in V

ii) $\hbar\,\sigma(S_1, S_2) = \mathrm{Im} < K\,S_1, K\,S_2 >_V \quad \forall S_1, S_2 \in R_c$.

(6.7)

Thus the problem of finding CCR representations is reduced to finding one-particle structures. Note that $\Omega := (1, 0, 0, 0....) \in D$ is a cyclic vector and that the Fock representations generating functional (§5) is

$$E_{\mathfrak{F}(V)}(S) = <\Omega, \hat{W}(S)\,\Omega>_{(V)} = e^{-{}^1/_4||S||^2_V} \qquad (6.8)$$

Fock space can also be written in the $L^2(\tilde{\Phi}(\Sigma)', d\mu)$ form discussed in §5 (c.f. eqns.(5.12) and (5.13)) in which μ is a gaussian measure. Thus the theory admits a field configuration interpretation but the usual one of course is in terms of particles. It is often assumed that vectors in V describes states of a single particle (in a first quantized sense) and that those in $\mathfrak{F}_n(V)$ correspond to n particles (bosons, because of the symmetrization). In the present situation for a generic space-time, such an interpretation is untenable for reasons discussed previously. In particular the lack of any timelike Killing vectors means that there is no 'one-particle' hamiltonian in V and hence no operator in $\mathfrak{F}(V)$ assigning additive energy to the 'n-particle' states.

As just remarked a given pair (R_c, σ) will generally admit numerous one-particle structures many of which will lead to unitarily inequivalent Fock representations. One way in practice of finding such Hilbert spaces is to look for complex structures on R_c with special properties. Such a complex structure is defined to be a real linear map $J : R_c \to R_c$ satisfying

$$\text{i)}\ J^2 = -1 \qquad (6.9)$$

$$\text{ii)}\ \sigma(J S_1, J S_2) = \sigma(S_1, S_2) \quad \forall S_1, S_2 \in R_c \qquad (6.10)$$

$$\text{iii)}\ \sigma(S, J S) > 0 \text{ unless } S = 0 \qquad (6.11)$$

A complex vector space structure may then be imposed on R_c in the usual way:

$$(a + i\, b)v := a\, v + b\, J\, v \qquad a, b \in \mathbb{R}, \quad v \in R_c \qquad (6.12)$$

and a complex inner product defined by

$$< S_1, S_2 >_{R_c J} := \sigma(S_1, J S_2) + i \quad \sigma(S_1, S_2) \qquad S_1, S_2 \in R_c \qquad (6.13)$$

where the Hilbert space completion of R_c with respect to this inner product is denoted $R_c^{\ J}$. Clearly the obvious injective map K embeds R_c as a dense linear subset of $R_c^{\ J}$ and the axioms for a one particle structure are met.

Viewing R_c as one of the phase spaces $\mathbf{P}_\Sigma$, a structure is

$$J : \mathfrak{D}(\Sigma) \oplus \tilde{\mathfrak{D}}(\Sigma) \to \mathfrak{D}(\Sigma) \oplus \tilde{\mathfrak{D}}(\Sigma)$$

$$(q, p) \quad \rightsquigarrow \left(\frac{-p}{(\det\gamma_\Sigma)^{\frac{1}{2}}}, \ (\det\gamma_\Sigma)^{\frac{1}{2}} q\right) \qquad (6.14)$$

where γ_Σ denotes the three metric induced by $g_{\mu\nu}$ on Σ. This proves, amongst other things, existence of covariant fields (via the construction in §3). However this particular complex structure is rather odd in the sense that even in Minkowski space-time (with Σ a constant time hypersurface) it does not lead to the correct Poincare invariant quantization of the free field. The correct result is obtained by using a complex structure involving terms like $(-\nabla^2 + \mu^2)^{\frac{1}{2}}$. This is a non local differential operator and as such cannot be sensibly defined as a map of R_c <u>into</u> R_c. Thus in practice it is useful to put some temporary Hilbert space structure on R_c (for example the Sobolev space $L^2(\Sigma) \oplus L^2(\Sigma)$ provided by (6.14)) enabling these non local operators to be properly defined. Alternatively of course one can forget about using complex structures and construct the required map K in (6.7) directly in some way.

§7. CONNECTION WITH THE HEURISTIC APPROACH

The covariant quantization of §3 can be related to the heuristic methods[1-4] that are used in practice to compute particle production in curved space. Indeed the possibility of making such a connection is one of the main virtues of the covariant approach although it should be remembered that since, by definition, the heuristic methods are not properly defined mathematically it is necessary to force them slightly into a certain mould before a sensible comparison can be made.

The heuristic quantization scheme can be paraphrased in the following steps:

1) On the complex vector space K_c of complex C^∞ solutions to the classical field equation (with compact support when restricted to any Cauchy surface) define the sesquilinear form

$$B(\alpha,\beta) = i \int_\Sigma \alpha^* \overleftrightarrow{\partial}_\mu \beta \; d\sigma^\mu \tag{7.1}$$

which, like σ in (4.1), is independent of the choice of Cauchy surface Σ. In Minkowski space

$$B(\alpha,\alpha) > 0 \text{ for positive frequency solutions} \tag{7.2}$$

$$B(\alpha,\alpha) < 0 \text{ for negative frequency solutions} \tag{7.3}$$

2) Choose some countable collection $\{f_i\}$ of complex solutions in K_c with the properties that

i) $B(f_i,f_j) = \delta_{ij} \quad (\Rightarrow B(f_i^*, f_j^*) = -\delta_{ij})$ (7.4)

ii) $B(f_i,f_j^*) = 0$ (7.5)

iii) $\sum_{i=1}^{\infty} \{f_i(x)f_i^*(y) - f_i^*(x)f_i(y)\} = -i\tilde{G}(x;y)$ (7.6)

Note that (7.4) and (7.5) mimic the behaviour of positive frequency solutions in Minkowski space and in this sense are said to form a 'basis for the positive frequency solutions in the space time manifold $\mathcal{M}$'. Since in a generic space time manifold (with no timelike Killing vectors) there is no intrinsic concept of positive frequency, this is merely a definition. Note also that the sense in which the sum in (7.6) is supposed to converge is usually unspecified but should at the very least be required to be true in a weak distributional sense;

$$\sum_{i=1}^{\infty} \{f_i(\psi_1)f_i^*(\psi_2) - f_i^*(\psi_1)f_i(\psi_2)\} = i\tilde{G}(\psi_1;\psi_2),\ \psi_1, \psi_2 \in \mathcal{Dm} \qquad (7.7)$$

where

$$f_i(\psi) := \int f_i(x)\psi(x)(-\det g)^{\frac{1}{2}}\, d^4x \qquad (7.8)$$

iv) It is further assumed that $\{f_i\}$ is chosen so that any real R_c solution ϕ can be expanded as

$$\phi(x) = \sum_{i=1}^{\infty} a_i f_i(x) + a_i^* f_i^*(x) \qquad (7.9)$$

for some set of complex coefficient a_i. Again a minimal requirement for (7.9) would be distributional convergence in thesense:

$$\phi(\psi) = \sum_{i=1}^{\infty} a_i f_i(\psi) + a_i^* f_i^*(\psi) \qquad \psi \in \mathcal{Dm} \qquad (7.10)$$

3) To heuristically quantize the system regard $\{a_i\}$ as operators satisfying

$$\left[\hat{a}_i, \hat{a}_j\right] = \left[\hat{a}_i^*, \hat{a}_j^*\right] = 0$$

$$\left[\hat{a}_i, \hat{a}_j^*\right] = \hbar\, \delta_{ij} \qquad (7.11)$$

then (7.6) formally implies that $\hat{\phi}(x)$ satisfies the covariant commutation relations

$$\left[\hat{\phi}(x), \hat{\phi}(y)\right] = -i\hbar\tilde{G}(x;y) \qquad (7.12)$$

The Fock representation is almost invariably chosen for (7.11) in an (implicit) infinite tensor product form associated naturally with the decomposition in (7.9) (with $\hat{}$ on the a_i) of the quantum field $\hat{\phi}$ into an infinite set of harmonic oscillators.

The field $\hat{\phi}$ is clearly defined on a single Hilbert space and the connection with the previously discussed rigorous covariant formalism is essentially the following (see also ref. 44). The set of all finite linear sums of the f_i (resp. f_i^*) is a vector subspace $W^{(+)}$ (resp. $W^{(-)}$) of K_c. On $W := W^{(+)} \oplus W^{(-)}$ define the function $||\alpha||$ by,

$$||\alpha||^2 := B(\alpha^{(+)}, \alpha^{(+)}) - B(\alpha^{(-)}, \alpha^{(-)}) \qquad \forall \alpha = \alpha^{(+)} + \alpha^{(-)} \in W \quad (7.13)$$

This is clearly a norm and since $||\alpha||^2 + ||\beta||^2 = \frac{1}{2} (||\alpha + \beta||^2 + ||\alpha - \beta||^2)$, it provides a scalar product on W. The Fock quantization (§6) over the complex Hilbert space $\overline{W}$ obtained by completing W with respect to this inner product then provides the desired rigorization of the heuristic quantum scheme. Note that (7.9) can now be understood as convergence in this particular Hilbert space. In general different choices of 'basis' $\{f_i\}$ is step 2 above lead to different Hilbert spaces and unitarily inequivalent Fock quantizations.

Frequently $\{f_i\}$ will have the property that $R_c \subset \overline{W}$. Under these circumstances let P_+ and P_- denote the projection operators onto $\overline{W}^{(+)}$ and $\overline{W}^{(-)}$ respectively and define

$$J = i\,(P_+ - P_-) \qquad (7.14)$$

Then this is a complex structure of the type previously discussed (§6), thus linking the heuristic method with this approach also.

There is a sort of converse of this latter statement. Within the rigorous framework of §6 let

$$K_J := R_c^J \oplus R_c^J \qquad (7.15)$$

equipped with the canonical complex structure. Thus K_J is a complex Hilbert space of complex solutions to the classical field equations. Now extend the map J to K_J by complex linearity and define

$$P_+ = (\mathbb{1} - iJ)/2, \quad P_- = (\mathbb{1} + iJ)/2 \tag{7.16}$$

Clearly

$$P_\pm^2 = P_\pm, \quad P_\pm^* = P_\pm, \quad \mathbb{1} = P_+ + P_- \tag{7.17}$$

and hence P_+ and P_- are projection operators on K_J. Decomposing K_J into the ranges of these operators as $K_J = K_J^{(+)} \oplus K_J^{(-)}$ and performing some algebra one finds

$$\text{a) If } \alpha \in K_J^{(+)} \quad \langle\alpha,\alpha\rangle_{K_J} = i \int_\Sigma \bar{\alpha} \overleftrightarrow{\partial}_\mu \alpha \, d\sigma^\mu \equiv B(\alpha,\alpha) \tag{7.18}$$

$$\text{b) If } \alpha \in K_J^{(-)} \quad \langle\alpha,\alpha\rangle_{K_J} = -i \int_\Sigma \bar{\alpha} \overleftrightarrow{\partial}_\mu \alpha \, d\sigma^\mu \equiv -B(\alpha,\alpha) \tag{7.19}$$

Since for any non trivial $\alpha \in K$ $\langle\alpha,\alpha\rangle \geq 0$ it follows that elements in $K_J^{(+)}$ and $K_J^{(-)}$, have the same properties of positive and negative B as those used in the heuristic formalism. One can paraphrase all this by saying that a choice of complex structure (and hence one-particle structure) on R_c provides a unique decomposition of complex classical solutions into positive and negative 'frequencies'.

§8. STATIONARY SPACE TIMES

A stationary spacetime is one possessing a global timelike Killing vector. There is then a natural choice of time t for which the components of the metric tensor are all time independent. We will restrict our attention to those space times for which all the equal t-time hypersurfaces are global Cauchy surfaces. The Cauchy data on any two such hypersurfaces can then be naturally related using the time lines and the classical time evolution described by a symplectic transformation $T(t)$ on the data on any arbitrarily chosen but fixed Cauchy surface Σ. In this case (4.6) becomes

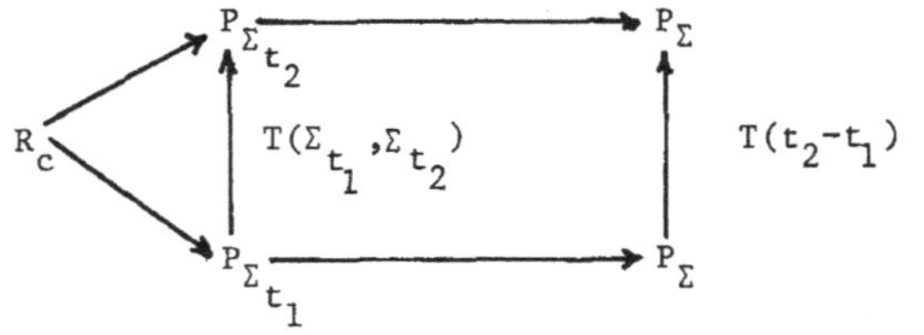

(8.1)

and the first order canonical equations of motion can be written as

$$\frac{\partial}{\partial t} \binom{q}{p} = - h \binom{q}{p} \tag{8.2}$$

where h is a time independent differential operator on $P_\Sigma \simeq R_c$. It is natural to quantize such a system so that the Hilbert space $\mathcal{H}$ on which the covariant field $\hat{\phi}(\psi)$ is defined carries a weakly continuous unitary representation $U(t)$ of the one parameter symplectic group $\{T(t) \mid t \in \mathbb{R}\}$. This will arise naturally via the GNS construction of §5 if a $T(t)$ invariant state can be found on the C^*–algebra $\mathcal{W}(P_\Sigma, \sigma)$. (Note that quantum dynamics is now abstractly defined as an automorphism of this single algebra). However if the self adjoint generator H of $U(t)$ is to physically correspond to energy then its spectrum should be positive. This will not be the case for a general invariant state. Fortunately Fock space comes to the rescue again. It is well known[42,43] that if $\hat{O}$ is any self adjoint operator on the complex Hilbert space then a 'second-quantized' form $d\Gamma(\hat{O})$ of $\hat{O}$ can be defined on $\mathfrak{F}(V)$ by

$$d\Gamma_n(\hat{O}): \mathfrak{F}_n(V) \to \mathfrak{F}_n(V),\ d\Gamma_n(\hat{O}) := \hat{O} \otimes \mathbb{1} \otimes .. \otimes \mathbb{1} + \mathbb{1} \otimes \hat{O} \otimes \mathbb{1} \otimes .. \otimes \mathbb{1} + \ldots + \mathbb{1} \otimes \mathbb{1} \otimes \ldots \otimes \hat{O} \qquad (8.3)$$

$$d\Gamma(\hat{O}) := \bigoplus_{n=0}^{\infty} d\Gamma_n(\hat{O}) \text{ on } \bigoplus_{n=0}^{\infty} \mathfrak{F}_n(V) \qquad (8.4)$$

If $D \subset V$ is any domain of essential self adjointness for $\hat{O}$ then $D' := \{\Psi \in \bigoplus_{n=0}^{\infty} \mathfrak{F}_n(V) \mid \Psi_n \in \otimes^n D\}$ is a domain of essential self adjointness for $d\Gamma(\hat{O})$. Furthermore if $\hat{O}$ is a positive operator then so is $d\Gamma(\hat{O})$. Thus it suffices to find a one particle structure on R_c such that $T(t)$ extends to a unitary group with strictly positive generator (I am assuming $\mu > 0$).

The use of Fock space to construct a stationary spacetime quantization has been extensively studied by Ashtekar, Magnon[5,6] and Kay[7-9] and I will simply state the key results with a few comments:

1) 'Given $(P_\Sigma, T(t))$ there is at most one-particle structure (K,V) such that $T(t)$ extends to a unitary group on V with a generator with strictly positive spectrum.' This result is essentially due to Weinless[45] but has been formulated in this way and applied in the present context by Kay[7-9]. It is a generalisation and functionally analytic cleaner proof of a more restricted earlier result by Ashtekar and Magnon[5,6] on the uniqueness of suitable complex structures on R_c.

2) 'If this (K,V) exists then the only quantization for which the map $R_c \ni S \rightsquigarrow \hat{W}(S)$ is weakly continuous is the Fock one.'

This is an old result of Segal[16-19] and explains why Fock space is so important - not only is it sufficient for the job it is almost necessary! The word 'almost' is inserted because strictly speaking we only require the weaker condition that $\mathbb{R} \ni t \rightsquigarrow \hat{W}(tS)$ is weakly continuous for all S in

R_c, however Weinless[45] has essentially closed this gap. Again there is an analogue in the work of Ashtekar/Magnon.

3) '(K,V) exists with the desired properties'[5,6,7-9]

The crucial steps in proving this vital theorem are

a) Equip P_Σ with the energy norm and complete to form a real Hilbert space P_E

b) Show that h (eqn. (8.2)) is a real skew-symmetric operator on $P_\Sigma \subset P_E$. Note that $-h^2$ is a positive symmetric operator on P_E and hence possesses a Friedrichs self adjoint extension $-\tilde{h}^2$. Then a complex structure on P_E with all the desired properties can be constructed as $J = (-\tilde{h}^2)^{\frac{1}{2}} h$ [5,6,46]. In principle such an extension is not unique but Kay[7-9] has shown using results of Chernoff[47] and Nelson[48] that h is actually essentially self adjoint on P_Σ and hence possesses a unique self adjoint extension $\bar{h}$. This leads to a unique J defined as $J = |\bar{h}|\bar{h}$.

It should be emphasized that Kay's rigorous results require two restrictions on the stationary spacetime:

i) The equal time hypersurfaces should be globally Cauchy (mentioned already).

ii) In the coordinates used, there should exist two positive finite real numbers ε_1 and ε_2 such that

$$\left.\begin{array}{l} 0 < \varepsilon_1 < g_{oo} g^{oo} \\ g^{oo} < \varepsilon_2 < \infty \end{array}\right\} \text{ on the Cauchy surface } \Sigma \text{ being employed.}$$

The results and techniques can be simplified for static manifolds (where the timelike Killing vector is hypersurface orthogonal) and the reader is referred to the original papers for details of this special case.

The stationary spacetime results are important in theirown right and as a tool for setting up an 'in/out' formalism on a general manifold whenever this is appropriate. Hamiltonian diagonalization [4,5,7-9] can now be easily described too. Make a specific choice of time and write the canonical equations of motion in the form in (8.2) but where h is now an explicit function of time h_t. Then at each time choose the Fock quantization above regarding h_t as frozen, i.e. taking on its value at the chosen moment in time. This thus provides a concrete example of the scheme in (5.8) with $\mathcal{W}(\mathcal{H}_{E_{1,2}})$ the two C^* algebras generated by the Fock quantizations for any <u>pair</u> of Cauchy surfaces. Kay[7-9] has pointed out that in fact for a complete formalism it is necessary to allow for a <u>different</u> choice of time on each Cauchy surface. More precisely he associates a Fock quantization with each Cauchy surface and each choice of lapse and shift function on that surface. The merits of the hamiltonian diagonalization scheme and its comparison with the covariant method have been discussed elsewhere in this paper. Perhaps one of its great advantages is that for any given manifold it does provide an unambigous quantization scheme which the covariant approach certainly does not.

§.9 CONCLUSION

We have seen that at both the structural and functional analytical levels the problem of constructing quantum fields on a curved, globally hyperbolic, spacetime is very similar to the Minkowski space external potential situation and possesses the same types of resolution. The major difficulty in both cases lies in the selection of some particular representation of the covariant or canonical commutation relations and in the associated physical interpretation within the interaction region. Stationary spacetimes and spacetimes with asymptotically stationary regions can be successfully handled but the generic case still poses a problem. In the case of an external potential it is often said that the difficulty disappears if the source J is itself quantized. The prime justification for this lies in the observation that Poincaré invariance is recovered in the ensuing closed system. Whether or not such a statement is valid for the curved spacetime case is however unclear. It seems a little hypothetical anyway in the absence of any remotely consistent quantum theory of gravity.

This particle interpretation problem has been discussed extensively by Hajicek in ref. (44) and also in the current volume. One often mooted approach is to adopt a Haag-Kastler local algebra of observables framework[21]. There are fairly obvious generalizations to a curved spacetime context of most of the Haag-Kastler axioms and the local algebras generated by the linear covariant fields constructed in §3 satisfy them. It is worth noting that in a globally hyperbolic manifold the class of all sets of the form $I^+(x) \cap I^-(y)$, ($I^+(x)$ and $I^-(y)$ respectively denote the chronological future and past of $x, y \in \mathcal{M}$) form a base for the manifold topology[25,26]. Thus these sets constitute an exact analogue of the much used 'diamond shaped regions' of Minkowski space algebraic quantum field theory. There is a fair literature on the particle interpretation of conventional field theories and the not unrelated problem of defining localized states.

Much of this has yet to be tried out in a curved spacetime setting although since it tends to use the Poincaré group and existence of a unique vacuum, the prognosis is not excessively bright.

One interesting question that can be asked within the context of the present paper is the precise role played by the global spacetime structure. The restriction to globally hyperbolic manifolds is very understandable but, as mentioned in the introduction, spacetimes lacking this property are not without interest. It would also be interesting to know whether the structure of the homotopy, homology, holonomy groups etc. associated with the spacetime could be woven into the quantum field theory in some explicit way.

Finally the study of higher spins certainly warrants further effort. Apart from the intrinsic interest in spin > 0 I strongly suspect that it is in these cases that one is most likely to see non-trivial effects of global topological structure.

ACKNOWLEDGEMENTS

I am most grateful to Bryce DeWitt who, over the years, has patiently weaned me away from my early strict canonical attitudes towards a more covariant approach to life. I have also enjoyed numerous discussions with Bernard Kay and many of the ideas thus mutually developed have found their way into the present text.

REFERENCES

The first three references are general review on heuristic quantum field methods in curved spacetimes.

1. B.S. DeWitt, Phys. Rep. 19C, 6, 297 (1975).
2. L. Parker 'The Production of Elementary Particles by Strong Gravitational Fields' In Proceedings of the Symposium on Asymptotic Properties of Space-Time. Plenum Publishing Company, New York 1977.
3. C.J. Isham 'Quantum Field Theory in Curved Space-Times - An Overview' In Proceedings of the Eighth Texas Symposium on Relativistic Astrophysics. To appear.
4. S.W. Hawking Comm. Math. Phys. 43, 199 (1975).
5. A. Ashtekar and A. Magnon Proc. Roy. Soc. A346, 375 (1975).
6. A. Magnon-Ashtekar 'Champs Quantique en espace-temps courbe'. Ph.D thesis, Universite de Clermont-Ferrand (1975).
7. B. Kay 'Structure of linear quantum fields in time dependent backgrounds' Imperial College preprint (1977).
8. B. Kay 'Quantum fields in time dependent backgrounds and in curved spacetimes' Ph.D thesis, London University (1977).
9. B. Kay, Trieste preprints - to appear.
10. M. Castagnino, Physics Department preprint Ciudad University, 1973.
11. P. Bongaarts 'Linear fields according to I.E. Segal'. In Mathematics of contemporary physics. Ed. R.F. Streater pp. 187-208, Academic Press, London (1972).
12. A. Capri, J. Math. Phys. 10, 575 (1969).
13. B. Schroer, R. Seiler and K. Swieca, Phys. Rev. D2, 2927 (1970).
14. R. Seiler, Comm. Math. Phys. 25, 127 (1972).
15. R. Seiler 'Remarks on the theory of particles with spin up to one in external fields'. In Troubles in the external field problem for invariant wave equations. Ed. A.S. Wightman pp. 23-43. Gordon and Breach.

16. I.E. Segal, Mat. Fys. Medd. Dansk Viel. Selsk 31, 12 1 (1959).

17. I.E. Segal, Can. J. Math. 13, 1 (1961).

18. I.E. Segal, Illinois J. Math. 6, 500 (1962).

19. I.E. Segal, 'Representations of canonical commutation relations'. In Applications of mathematics to problems in theoretical physics. Ed. F. Luciat pp. 107-170. Gordon and Breach, New York (1967).

20. J. Charap and M.J. Duff, Queen Mary College preprints, London (1977).

21. R. Haag and D. Kastler, Jour. Math. Phys. 5, 7, 848 (1964).

22. J.B. Hartle and S.W. Hawking, Phys. Rev. D13, 8, 2188 (1976).

23. See for example, R.F.S. Streater and A.S. Wightman, 'PCT, spin and statistics, and all that', Benjamin, New York (1964).

24. J. Dieudonne, 'Treatise on Analysis' Vol. III, Academic Press, New York (1972).

25. R. Penrose 'Techniques of differential topology in relativity'. SIAM Philadelphia (1973).

26. S.W. Hawking and G. Ellis, 'The large-scale structure of spacetime'. Cambridge University Press (1973)

27. J. Leray, 'Hyperbolic partial differential equations', Mimeographed notes, Princeton (1952).

28. A. Lichnerowicz, 'Propagateurs et commutateurs en relativite generale'. Publications I.H.E.S. No. 10 (1961).

29. Y. Choquet-Bruhat, 'Hyperbolic partial differential equations' In Battelle Rencontres 1967, Ed. C.M. DeWitt and J.A. Wheeler, Bhnjamin (1968).

30. R. Geroch, J. Math. Phys. 11 437 (1970).

31. F.G. Friedlander, 'The wave equation in a curved spacetime' Cambridge University Press (1975).

32. For a comprehensive discussion see S.A. Fulling, King's College preprint London (1975) and a short summary in Phys. Rev. D14, 8 1939 (1975).

33. G. Emch, 'Algebraic methods in statistical mechanics and quantum field theory', Wiley, New York (1972).

34. J. Manuceau, Ann. Inst. Henri Poincare 8 117 (1968).

35. J. Manuceau, Ann. Inst. Henri Poincare 8 139 (1968).

36. J. Slawny, Comm. Math. Phys. 24, 131 (1972).

37. N. Bourbaki, 'Elements de mathematique, Livre VI Integration, Chapitre IX' Hermann, Paris.

38. L. Schwartz 'Radon Measures', Oxford University Press (1973).

39. I. Gelfand and N. Vilenkin 'Generalized functions', Vol. 4, Academic Press, New York (1964).

40. H. Araki, Comm. Math. Phys. 20, 9 (1971).

41. For example see R. Haag in 'Mathematics of contemporary physics'. Ed. R.F. Streater, Academic Press, London (1972).

42. J. Cook, Trans. Amer. Math. Soc. 74, 222 (1953).

43. M. Reed and B. Simon, 'Methods of modern mathematical physics' Vol. I, II Academic Press (1972) and (1975).

44. P. Hajicek, Phys. Rev. D15 2757 (1977).

45. M. Weinless, Jour. Func. Analysis, 4 350 (1969).

46. P. Spindel, Bull.Acad. Relg. LXIII 51 (1977).

47. P.R. Chernoff, Jour. Func. Analysis 12, 401 (1973).

48. E. Nelson, Ann. Math. 70 572 (1959).

49. One of the remarkable spin offs from the supergravity programme is that in a supersymmetric framework there is a consistent formulation of spin 3/2 in an external gravitational field. See 'Broken supersymmetry and supergravity'. S. Deser and B. Zumino, preprint.

ON FUNCTIONAL INTEGRALS IN CURVED SPACETIME

G.W. Gibbons
Max-Planck-Institut für Physik und Astrophysik
Föhringer Ring 6, 8000 München 40, W-Germany

and

Department of Applied Mathematics and Theoretical Physics,
Silver Street, Cambridge

Abstract. We discuss the functional integral formulation of curved space quantum field theory for fields of spin 0, 1/2 and 1. We give a discussion of the Ghost problem and demonstrate gauge invariance of the formalism. We discuss the significance of the zero frequency modes, their relation to Black Hole, No Hair Theorems, and to the Topology of the space one works on. We use the formalism to evaluate the zero point energy of quantum fields in a static enclosure and the electromagnetic trace anomaly.

1. Introduction

The path integral formulation of quantum theory is intuitively perhaps the most direct and simplest of all of the various approaches in the literature. In particular it provides a compact and manifestly covariant means of handling delicate questions of gauge invariance (e.g. the theory of Fadeev-Popov ghosts) and should in principle be well suited to generalizations to curved space. The drawbacks are of course well known - the difficulty in giving a precise meaning to the various operations involved, functional sum, the inverse of differential operators and their "determinants" etc. Some, but not all of these problems may be alleviated by considering not the hyperbolic operators that appear naturally in the theory but rather related elliptic operators, obtained by analytic continuation. Under suitable circumstances these will possess unique inverses and further their determinants may be given a precise meaning. This is particularly gratifying since the determinants may be used to define a regularized stress tensor for the fields concerned. Indeed in some circumstances suitable determinants have a physical interpretation as partition functions of a canonical ensemble of particles. This enables one to

calculate a finite zeropoint energy for such an ensemble.
The purpose of this paper is to explore this formalism on a heuristic level. This is a necessary preliminary to any deeper, mathematically more rigorous treatment. Fields of spin 0, 1/2 and 1 are discussed (gravitons are discussed in [2]) and for the most part it is assumed that an analytic continuation is possible so that one may work on a Riemannian Space. This is not the general case but sufficiently interesting examples exist to make this worthwhile. It then turns out that one may use some of the considerable body of work on the spectral theory of elliptic operators acting on vector bundles over Riemannian Spaces. Especially interesting are the many relations to the topology of this Riemannian space. The use of some elementary Hodge de Rham theory enables one to show that in the case of electromagnetism physical quantities are gauge invariant. This is the main content of section 2. In section 3 we discuss the problem of finding the zeropoint energy of a quantum field confined to a finite region using this technique. In the case that g_{00} is independent of spatial variables we find that the evaluation of the low temperature partition function hinges on the value of a Hamiltonian zeta function. At high temperatures we recover the standard results for a thermal gas of bosons or fermions. In the special case of the Einstein Static Space these results coincide with those of other authors. In section 4 we discuss briefly the case of Black Hole backgrounds and in particular the significance of the zero frequency modes. Their number is determined by purely topological considerations, thus providing a topological approach to the No Hair Theorems. We find that the contribution of the zero frequency Maxwell fields which cannot be obtained from a vector potential must be included in the functional integral in order to obtain the correct total energy. Finally we comment on the applications to the Gravitational Instanton. The paper finishes with a short conclusion.

Conventions:	Signature:	$-+++$
	Riemann Tensor	$\nabla_{[\alpha}\nabla_{\beta]} K^{\sigma} = \frac{1}{2} R^{\sigma}{}_{\rho\alpha\beta} K^{\rho}$
	Ricci Tensor	$R_{\alpha\beta} = R^{\sigma}{}_{\alpha\sigma\beta}$
	units	$h=c=G=k=1$
	Greek letters	run from 0 to 3
	Latin letters	run from 1 to 3

. General Formalism

onsider a scalar field for simplicity. The expectation value of an perator O in an (unnormalized) mixed state ρ is postulated to be iven by

$$\mathrm{Tr}\ \rho\ O\ /\ \mathrm{Tr}\ \rho = \int_C \mathcal{D}[\phi] \exp iI[\phi] O[\phi] \quad \int_C \mathcal{D}[\phi] \exp iI(\phi). \quad (1)$$

r denotes the trace over a complete set of states. $\int \mathcal{D}[\phi]$ chematically denotes the functional integral. $I[\phi]$ is the lassical action functional and $O[\phi]$ denotes the classical funct-onal corresponding to the quantum operator O . C denotes the class f classical field histories to be summed over. The freedom to chose corresponds to the freedom to chose ρ . For "vacuum to vacuum" ransitions one conventionally states

$$\phi(|t| \to \infty) \quad : \quad \rho = |0-\rangle\langle 0+| \qquad (2)$$

or a Gibbs state

$$\phi(t+i\beta) = e^{\mu\beta}\phi(t) \qquad \rho = \exp-\beta(H-\mu N) \qquad (3)$$

being a Hamiltonian, β being the inverse temperature, μ the hemical potential. As usual we define a generating functional

$$Z[J] = \mathrm{Tr}\ \rho \rfloor \exp i \int J(x)\phi(x)\sqrt{-g}\, d^4x \qquad (4)$$

$\rfloor$ denotes Wick time ordering and propagators by

$$G_\rho(x^1,..x^n) = i^{\frac{n}{2}} \mathrm{Tr}\ \rho \rfloor \phi(x^1)\ldots\phi(x^n)/\mathrm{Tr}\ \rho \qquad (5)$$

$$= Z[J]^{-1} \frac{\delta^n}{\delta J(x^1)\ldots\delta J(x^n)} Z[J]_{|J=0} \qquad (6)$$

ote that

$$Z[0] = \mathrm{Tr}\ \rho \qquad (7)$$

f ρ is the Gibbs state $Z[0]$ is the partition function. Consider ow the "free field"

$$I(\phi) = -\frac{1}{2}\int \phi D\phi\sqrt{-g}\, d^4x \qquad (8)$$

$=-\nabla_\alpha\nabla^\alpha+m^2+\xi R$. If $\xi=\frac{1}{6}$; $m^2=0$ this gives rise to the conformally invariant wave equation. The functional integral is now a "Gaussian" and may be performed to yield

$$Z[J] = (\text{DET } D)^{-1/2} \exp i \iint J(x) G_\rho(x,y) J(y) \sqrt{g(x)} \sqrt{g(y)}\, d^4x\, d^4y \qquad (9)$$

where

$$D_x G_\rho(x,y) = \delta(x,y) . \qquad (10)$$

$\delta(x,y)$ is the Dirac delta function. For notational convenience we write (10) as

$$DG_\rho = I \qquad (11)$$

G_ρ is supposed to be the inverse of D subject to the boundary conditions denoted schematically be C. In order to specify these boundary conditions more precisely and to define DET D we perform an analytic continuation. That is we suppose the real spacetime $\{M, g_{\alpha\beta}\}$ with contravariant metric $g^{\alpha\beta} \frac{\partial}{\partial x^\alpha} \frac{\partial}{\partial x^\beta}$ to be embedded as a 4 real dimensional submanifold of a 4 complex dimensional manifold M_c, called the "complexification of M". M_c is supposed equipped with a complex contravariant tensor field of rank 2 and type (2,0) $g_c^{\alpha\beta}$ such that the restriction of $g_c^{\alpha\beta}$ (regarded as a quadratic form) to the (real) cotangent space of the submanifold is real. M is said to be a "real section" of M_c. Local coordinates for M_c are $x, \bar{x}$ and M is given by

$$x = \bar{x}$$

The original real slice and a metric of signature −+++ . A Riemannian real slice is one for which the signature is ++++. Not all spacetimes admit a complexification (they must be real analytic at least) and those that do may not admit a Riemannian real slice. We shall now make an even more restrictive assumption - that the real section is orientable and compact without boundary. This is mainly for pedagogic reasons. If the one has a compact region with boundary and one specifies some boundary conditions on this boundary some but not all of the results below would remain valid. The topological results would require modifications. In the absence of Riemannian sections one may be able to find "quasi-Riemannian sections" for which $g_c^{\alpha\beta}$ is non-vanishing for the real cotangent space. This condition ensures that second order differential operators with a leading term given by the metric will be elliptic. In what follows it is convenient to imagine the field ϕ as being in general vector valued (i.e. the field configurations are sections of a vector bundle upon which D acts mapping sections into sections). The vector space is supposed equipped with a Hermitian structure and this is extended to the sections by integration over the manifold with the standard volume measure. We suppose that D is self adjoint. The general form will be

$$D = -g^{\alpha\beta} \mathcal{D}_\alpha \mathcal{D}_\beta - \frac{1}{\sqrt{g}} (\sqrt{g}\ g)^{\alpha\beta}_{\ ,\beta} \mathcal{D}_\alpha \tag{12}$$

with

$$\mathcal{D}_\alpha = \partial_\alpha + W_\alpha \tag{13}$$

W_α is a set of matrix valued connection forms and E is some endomorphism of the fibres (i.e. a matrix acting on the vector space) W_α could include the effects of background electromagnetic or Yang-Mills fields as well as gravitational fields. W_α gives rise to curvature 2 forms

$$W_{\alpha\beta} = \partial_\alpha W_\beta - \partial_\beta W_\alpha + W_\alpha W_\beta - W_\beta W_\alpha \tag{14}$$

D is now a self adjoint operator on the space of sections and will possess a discrete spectrum $\{\lambda_n\}$ n=0,1... . If this spectrum does not include 0, D will possess a unique inverse G .
If D does have zero eigenvalues there must be projected out.
We shall see that they have interesting physical properties. We may define (via a spectral decomposition) various functions of D . Most interesting are $K = \exp -tD$ and D^{-s}, K satisfies the Heat equation

$$DK = -\frac{\partial K}{\partial t} \tag{15}$$

and

$$D^s = \frac{1}{2\pi i} \oint \frac{\lambda^s d\lambda}{D-\lambda} \tag{16}$$

the integration in (16) being taken along a curve in the complex λ plane encircling spec D . s may be any complex number. A fundamental analysis of Seeley [3] asserts that

$$\zeta_D(s) = \int \mathrm{Tr} D^{-s} \sqrt{g} d^4x = \sum_{n=0}^{\infty} \lambda_n^{-s} \tag{17}$$

will converge to an analytic function of s for Re s>2 . The analytic continuation to all s defines a meromorphic function of s with poles at 2, and 1. This provides the required definition of DET D as

$$\text{DET } D = (2\pi\mu^2)^{-\zeta_D(o)} \exp - \zeta_D'(0) \tag{18}$$

μ is the "renormalization mass".

$\zeta_D(0)$ may be found from Gilkey's general expression [4] for the asymptotic expansion of $K(x,y,t)$ for small t. He proves that

$$K(x,x,t) \simeq \sum_{m=0}^{\infty} E_m t^{m/2} (4\pi t)^{-2} \tag{19}$$

where E_m are endomorphisms of the fibres which vanish for odd m. He has computed E_m up to $m=6$. For $m=0,2,4$ we have

$$E_o = I \tag{20}$$

$$E_2 = E - 6^{-1}R \tag{21}$$

$$E_4 = (-30)^{-1}R^{;d}_{;d} + (72)^{-1}R^2 - 180)R_{\alpha\beta}R^{\alpha\beta}$$

$$-(180)^{-1}R_{\alpha\beta\gamma\delta}R^{\alpha\beta\gamma\delta} - 6^{-1}RE + 2^{-1}E^2 + 12^{-1}W_{\alpha\beta}W^{\alpha\beta}$$

$$+ 6^{-1}E^{;\alpha}_{;\alpha} \tag{22}$$

Defining $B_m = \text{Tr } E_m$ we have

$$\zeta_D(0) = \int B_4 (4\pi)^{-2} \sqrt{g} d^4x \tag{23}$$

The coefficients B_m are related to the deWitt coefficients in his expansion of an effective Lagrangian. If boundaries are present the E_m no longer vanish for odd m and are related to invariants of the boundary.

For scalars the application of these results is straight forward and is dealt with in [1]. For spinors the situation is slightly different. Formally one has

$$Z[0] = \int \mathcal{D}[\bar{\psi}] \mathcal{D}[\psi] \exp i \int \bar{\psi} P \psi \sqrt{g} d^4x \tag{24}$$

$^-$ denotes Dirac adjoint and P is the Dirac operator. It is usual to regard ψ as a classical anticommuting operator and the formal integration rules yield

$$Z[0] = \mathrm{DET}\ P \tag{25}$$

In addition the boundary conditions for the Gibbs ensemble are

$$\psi(t+i\beta) = -\psi(t)\exp \mu\beta \tag{26}$$

It is more convenient to introduce the adjoint of P, $\tilde{P}$ and write

$$P\tilde{P} = D_{1/2} \tag{27}$$

$$P = \gamma^{\alpha}\nabla_{\alpha} + m \tag{28}$$

$$D_{1/2} = -\nabla_{\alpha}\nabla^{\alpha} + \frac{R}{4} + m^2 \tag{29}$$

where the γ's obey

$$\gamma^{\alpha}\gamma^{\beta} + \gamma^{\beta}\gamma^{\alpha} = 2g^{\alpha\beta} \tag{30}$$

$$\nabla_{\alpha} = \partial_{\alpha} + \gamma_{\rho}\gamma^{\beta}\Gamma^{\ \rho}_{\beta\alpha} \tag{31}$$

(29) leads to Lichnerowic's Theorem: If $R>0$, there exist no zero eigenvalues. In terms of $D_{1/2}$ we have

$$Z = (\mathrm{DET}\ D_{1/2})^{1/2} \tag{32}$$

We remark that the "real section" must possess a spinor structure, if we are to be able to define spinor fields and evaluate the determinants. For more information on the zero frequency modes and axial anomalies see [5].
To describe spin 1 massive particles we use the Proca equations

$$\Delta A = -m^2 A \tag{33}$$

$$\delta A = 0 \tag{34}$$

A is a 1-form (covariant vector). If d is exterior differentiation and δ its formal adjoint, Δ is the de Rham Laplacian

$$\Delta = d\delta + \delta d\ . \tag{35}$$

(34) comes from the Lagrangian

$$L = -\frac{1}{2}(A_\mu \Delta A^\mu + m^2 A_\mu A^\mu) \tag{36}$$

but (34) does not. In effect one has to consider not all one forms but only the divergence free ones. It is possible to sum over all fields however provided one includes a scalar ghost. For massless spin 1 particles the situation is similar but in addition one must sum only over fields which are not pure gauge, i.e. such that $A \neq d\phi$ for some ϕ. This involves 2 ghost fields. Intuitively one thinks of A as having 4 degrees of freedom. A spin 1 particle should have 3 degrees of freedom if it has mass and 2 if it is massless. To make these ideas more precise we recall some facts from Hodge de Rham theory.

1) The space of p-forms which are square integrable splits into 3-disjoint-orthogonal components (Hodge decomposition)

the exact $A_p = dA_{p-1}$; $dA_p = 0$

the coexact $A_p = \delta A_{p-1}$; $\delta A_p = 0$

the harmonic $\Delta A_p = 0$; $dA_p = \delta A_p = 0$

The harmonic forms are finite in number, this number being a topological invariant, the p'th Betti number , b_p

2) $$\sum_{p=0}^{p=4} (-1)^p b_p = \chi$$

is the Euler Number of the manifold. The Gauss-Bonnet theorem states that

$$\chi = \frac{1}{128\pi^2} \int R_{\alpha\beta\gamma\delta} R_{\mu\nu\lambda\rho} \varepsilon^{\alpha\beta\mu\nu} \varepsilon^{\gamma\delta\lambda\rho} \sqrt{g}(x) d^4x$$

If M is Einstein and non flat $\chi > 0$ (Berger's Theorem)

3) $b_o = 1$; $b_p = b_{-p}$ (Poincaré duality)

4) $b_1 = 0$' if M is simply connected. $b_1 = 0$ if M is Ricci flat and non flat (Myer's Theorem).

5) If b_2^+ ; b_2^- are the numbers of self dual or antiself dual Harmonic 2-forms then

$$b_2^+ - b_2^- = \tau$$

s a topological invariant called the signature and given by the 'ollowing integral

$$\tau = \frac{1}{96\pi^2} \int R_{\alpha\beta\gamma\delta} R^{\alpha\beta}{}_{\lambda\rho} \varepsilon^{\gamma\delta\lambda\rho} \sqrt{g} d^4x$$

(Hirzebruch Signature Theorem) $3\tau = p$, the Pontryagin number of M. If M is Einstein then

$$\chi \geqslant \frac{1}{2} |p|$$

equality only being attained if the curvature is self or antiself dual (Hitchin's Theorem) [6].

6) The eigen forms with non zero eigenvalues are such that the coexact eigen p forms are in one-one correspondence with the exact eigen $p-1$ forms. The maps d and δ map them into one another. The coexact eigen p forms are in on-one correspondence with the exact eigen $4-p$ forms, the mapping being effected by the Hodge dual operator.

Consider now the Proca field. If one formed the determinant of Δ acting on 1 forms one would have included the exact eigen-values as well. These may be removed by dividing by the determinant of $\Delta + m^2$ acting on functions. Using Δ_p to denote the Laplacian on p forms we have for the Proca field

$$Z[0] = (\mathrm{DET}(\Delta_1 + m^2))^{-1/2} (\mathrm{DET}(\Delta_0 + m^2))^{+1/2} \qquad (36)$$

The second factor is often regarded as being due to fermions. However, this is not quite correct. If one uses (38) to evaluate a partition function one would have need to make the "fermions" obey boson statistics in order to get the correct results. It is better to regard the second factor just as a determinant and evaluate it directly.

For photons the situation is more complex. If one integrates only over fields satisfying the gauge condition

$$\nabla_\mu A^\mu = \omega \qquad (37)$$

where ω is a function the Fadeev-Popov prescription leads to the result

$$Z = \int \mathcal{D}[A] \exp - I[A] \delta(\int \nabla_\mu A^\mu - \omega) \mathrm{DET} \left| \frac{\partial \nabla_\mu A^\mu}{\partial \Lambda} \right| \tag{38}$$

$I[A]$ is the Maxwell action, $\delta(\quad)$ is a functional delta function obeying

$$\int \mathcal{D}[\omega] \delta(f) F[\omega] = F[f]$$

for any functional F. The interpretation of the functional determinant is as follows. Under an infinitesimal gauge transformation $\partial\Lambda$

$$A_\mu \to A_\mu - \partial_\mu \partial \Lambda \tag{39}$$

$$\nabla_\mu A^\mu \to \nabla_\mu A^\mu - \nabla_\mu \nabla^\mu \partial \Lambda \tag{40}$$

This suggests that

$$\mathrm{DET} \left| \frac{\partial \nabla_\mu A^\mu}{\partial \Lambda} \right| = \mathrm{DET}\ \Delta_0 \tag{41}$$

Δ_0 being the (minimally coupled) Laplacian on scalars.
Now in principle (41) should be independent of the gauge choice. Therefore, we multiply by a weighting function of ω and integrate over all possible ω using the relation

$$\int \exp - \int \frac{\alpha\omega^2}{2} \mathcal{D}[\omega] = (\mathrm{DET}\ \alpha I)^{-1/2} \tag{42}$$

Thus

$$Z = (\mathrm{DET}\ \alpha I)^{1/2} (\mathrm{DET}\ \Delta_0)$$

$$= \int \mathcal{D}[A] \exp - I[A] - \frac{\alpha}{2} \int (\Delta_\mu A^\mu)^2 \sqrt{g} d^4x \tag{43}$$

The argument of the exponential may be written (after integration by parts as)

$$- \int A^\mu D_\alpha A_\mu \sqrt{g} d^4x \tag{44}$$

where

$$\delta d + \alpha d \delta = D_\alpha \tag{45}$$

Thus

$$Z = (\mathrm{DET}\ D_\alpha)^{-1/2}(\mathrm{DET}\ \alpha I)^{1/2}(\mathrm{DET}\ \Delta_o) \qquad (46)$$

The new term in the action is called a gauge fixing term and should be independent of its particular choice, i.e. of α. Before checking this we must interpret DET α I since the definition above fails. We define it in such a way as to preserve the relation

$$\mathrm{DET}(\alpha\Delta_o) = (\mathrm{DET}\ \alpha I)\mathrm{DET}\ \Delta_o \qquad (47)$$

This may seem rather arbitrary but it is just what is needed to achieve independence of α. Now using the easily established result

$$\mathrm{DET}\ \alpha\Delta_o = \alpha^{\zeta_o(0)}\ \mathrm{DET}\ \Delta_o \qquad (48)$$

where $\zeta_p(0)$ is the zeta function for the Laplacian on p-forms we obtain

$$Z = (\mathrm{DET}\ D_\alpha)^{-1/2}\ \alpha^{1/2\ \zeta_o(0)}(\mathrm{DET}\ \Delta_o) \qquad (49)$$

To evaluate DET D_α we note that D_α has exact eigenforms with eigenvalues $\alpha\lambda_n^p$, where λ_n^p are the eigenvalues of the Laplacian on p-forms, and divergence free eigenforms with spectrum λ_n^τ. We have

$$\zeta_{D_\alpha}(s) = \zeta_o(s)\alpha^{-s} + \zeta_T(s) \qquad (50)$$

where $\zeta_T(s)$ is the zeta function corresponding to the divergence free modes. Using (53),(54) and (18) one obtains for photons:

$$Z = (2\pi\mu^2)^{(1/2)(\zeta_1(0)-2\ \zeta_o(0))} \exp \tfrac{1}{2}\ (\ \zeta_o'(0)-2\ \zeta_o(0)) \qquad (51)$$

This is independent of α as it should be. If $\zeta_1(0)-2\ \zeta_o(0)\neq 0$ the result will not be independent of the renormalization length. This violation of scale invariance results in a trace anomaly. The methods of [4] yield

$$T^\mu_\mu = \frac{1}{(4\pi)^2}\ (B_4^1 - 2B_4^o) \qquad (52)$$

where B_m^p are the Minakshisundarum-de Witt coefficients for the Laplacian on p-forms. Gilkey's results then yield

$$T^{\mu}_{\mu} = \frac{1}{(4\pi)^2}\frac{1}{180}\{-13\,R_{\alpha\beta\gamma\delta}R^{\alpha\beta\gamma\delta} + 62(R_{\alpha\beta}R^{\alpha\beta} - \tfrac{1}{3}R^2) - 18\,R^{;\alpha}_{\ ;\alpha}\} \tag{53}$$

This agrees with Dowker and Critchley [8] but not with Tsao [7], Brown and Cassidy [9] or Copper and Duff [10] who obtain +12 as the coefficient of $R^{;\alpha}_{\ ;\alpha}$. The massless Dirac spinor results obtained by the same method however agree with all other determinations. This shows that zeta function regularization and dimensional regularization are not quite equivalent. However, a $R^{;\alpha}_{\ ;\alpha}$ term is irrelevant on a compact manifold.

3. Zero Point Energies

We shall now apply these general methods to static spacetimes. Suppose that M is the product of the time line with coordinate t with a compact three dimensional manifold M_3 with metric h_{ij} . Let ${}_3d$ denote exterior differentiation with respect to the spatial variables and $*$ the Hodge dual operator on M_3 using the measure $(\det h_{ij})^{1/2}$. The riemannian section is evidently obtained by the replacement $t=i\tau$. If one considers the Gibbs state at temperature β one identifies τ modulo β . This ensures that all boson fields have the required periodicity. For fermion fields condition (26) is incomputable with this topology, one should stricting work on a manifold with period 2β , and consider only half the possible spinor functions. It is just as easy to work on the smaller interval bearing in mind that spinor fields are not really functions, they are double valued. The interesting operators take the form

bosons $$D = -|g_{oo}|^{-1}\left(\frac{\partial^2}{\partial\tau^2} - H^2\right) + m^2 \tag{54}$$

fermions $$\gamma^{\alpha}\nabla_{\alpha} + m = \gamma^{o}|g_{oo}|^{-1/2}\left(\frac{\partial}{\partial\tau} + H_{1/2}\right) + m \tag{55}$$

Here H generically denotes a Hamiltonian. Explicitly we have

scalars $$H^2 = -|g_{oo}|^{1/2}*{}_3d(|g_{oo}|^{1/2}*{}_3d) + \zeta R \tag{56}$$

ermions $$H_{1/2} = (\gamma^{o})^{-1}\gamma^{i}(\partial_i+\gamma_p\gamma^{q}\Gamma^{p}_{qi})|g_{oo}|^{1/2} \tag{57}$$

vectors $$H^2A_o = |g_{oo}|^{1/2}*_3d(|g_{oo}|^{1/2}*_3d)A_o \tag{58}$$

$$H^2A_i = |g_{oo}|^{1/2}*_3d(|g_{oo}|^{1/2}*_3d)A_i$$

$$+ |g_{oo}|^{1/2}*_3d(*|g_{oo}|^{1/2}{}_3d*)A_i$$

If E_n are the spectra of H, the standard expression for the partition functions are, in the massless case

$$\log Z^B = -\sum \frac{E_n\beta}{2} - \log(1-\exp-\beta\, E_n) \tag{59}$$

$$\log Z^F = +\sum \frac{E_n\beta}{2} + \log(1+\exp-\beta\, E_n) \tag{60}$$

B and F from now on denote boson and fermion respectively. The first terms represent the infinite zero point contributions. Using zeta functions we shall obtain finite results. This involves forming the following zeta functions

scalars $$\zeta^B(s) = \sum_{n=0}^{\infty}\sum_{r=-\infty}^{r=\infty} [E_n{}^2 + (2\pi r)^2\beta^{-2}]^{-s} \tag{61}$$

fermions $$\zeta^F(s) = \sum_{n=0}^{\infty}\sum_{r=1}^{\infty} [E_n{}^2 + \pi^2(2r-1)^2\beta^{-2}]^{-s} \tag{62}$$

This is because the fermion modes have eigenvalues of the form

$$E_n \pm i\pi\beta^{-1}(2r-1) \tag{63}$$

when allowance is made for the antiperiodicity. For photons we need

$$\zeta_1(s) - 2\,\zeta_o(s)$$

This amounts in the case that g_{oo} is independent of the spatial variable to a summation of the form (61) where the E_n are the energies of the 'transverse' modes.

We call modes with only an A_o component timelike those with no A_o component and such that $A_1=\partial_1\phi$ for some scalar ϕ 'longi tudinal'. Those which have $A_o=0$ and are divergence free in the sense that

$$\partial_i(|g_{oo}|^{1/2}A^i h^{1/2}) = 0 \tag{64}$$

we call 'transverse'. If one quantizes the electromagnetic field in Coulomb gauge, only the transverse components are quantized. Now the longitudinal modes have the same spectra as the scalar ghosts (a general result) so

$$\zeta_{photon} = \zeta_{transverse} + \zeta_{timelike} - \zeta_{ghost} \tag{65}$$

Now if g_{oo} is constant we may set it to unity with no loss of generality and easily see that the scalar ghosts and the timelike modes will have the same spectrum. This neat cancellation would not work if g_{oo} depended upon the spatial variables.
Now in [11] it was shown that if the n-summation in (61), (62) were finite one would be lead directly to (59), (60) together with zero point energies. If the n-summation is infinite, however, the interchange of differentiation and summation required is not permissable. One must adopt a different way of evaluating the answer. One approach is as follows. By performing a Mellin transformation one finds that

$$\zeta^B(s) = \frac{1}{\Gamma(s)} \int_o^\infty x^{s-1} Q(x)\,\theta_3\left(\frac{4\pi x}{\beta^2}\right) dx \tag{66}$$

$$\zeta^F(s) = \frac{1}{2\Gamma(s)} \int_o^\infty x^{s-1} Q(x)\left[\theta_3\left(\frac{\pi x}{\beta^2}\right) - \theta_3\left(\frac{4\pi x}{\beta^2}\right)\right] dx \tag{67}$$

where

$$Q(x) = \sum_{n=o}^{\infty} \exp - E_n^2 x \tag{68}$$

$$\theta_3(x) = \sum_{-\infty}^{+\infty} \exp - \pi n^2 x = x^{-1/2}\theta_3\left(\frac{1}{x}\right) \tag{69}$$

θ_3 is Jacobi's θ function. For small x one has the asymptotic relation

$$\theta_3(x) \simeq \frac{1}{x^{1/2}} \tag{70}$$

the error being exponentially small. $Q(x)$ falls off rapidly with large x and for small x spectral theory (in 3 dimensions) shows it to have the asymptotic form

$$Q(x) \simeq Vp(4\pi x)^{-3/2} \tag{71}$$

V is the volume of the spatial sections p=1 for scalars and p=2 for neutrinos or photons. For low temperatures we may use the asymptotic form of the θ_3 function to obtain the following asymptotic forms for the zeta functions

$$\zeta^B(s) \simeq \frac{\beta(4\pi)^{-1/2}\Gamma(s-\frac{1}{2})\zeta_H(s-\frac{1}{2})}{\Gamma(s)} \simeq 2\zeta^F(s) \tag{72}$$

where

$$\zeta_H(s) = \sum_{n=o}^{\infty} E_n^{-2s} \tag{73}$$

is the Hamiltonian zeta function. As usual we define it using (75) for sufficiently large real part of s and then by analytic continuation over the s-plane. Since H^2 is a second order elliptic operator in three dimensions general theory shows that $\zeta_H(s)$ may have poles at the half integers 3/2, 1/2, -1/2..... and vanishes at the negative integers.

Now using

$$E = -\frac{\partial \log Z}{\partial\beta} \tag{74}$$

we obtain

$$E^B = -\frac{1}{2}\frac{\partial}{\partial s}\left\{\frac{\Gamma(s-1/2)}{(4\pi)^{1/2}}\frac{1}{\Gamma(s)}\zeta_H(s-1/2)\right\}_{s=0} \tag{75}$$

$$-\log(2\pi\mu^2)\frac{1}{2}\left\{\frac{\Gamma(s-1/2)}{(4\pi)^{1/2}}\frac{1}{\Gamma(s)}\zeta_H(s-1/2)\right\}_{s=0}$$

$$E^F = +\frac{1}{2}\frac{\partial}{\partial s}\left\{\frac{\Gamma(s-1/2)}{(4\pi)^{1/2}}\frac{1}{\Gamma(s)}\zeta_H(s-1/2)\right\}_{s=0}$$

$$+\frac{1}{2}\log(2\pi\mu^2)\left\{\frac{\Gamma(s-1/2)}{(4\pi)^{1/2}}\frac{1}{\Gamma(s)}\zeta_H(s-1/2)\right\}_{s=0} \tag{76}$$

If it happens that $\zeta_H(s)$ is analytic at $s=1/2$ we have

$$E^B = \frac{1}{2}\zeta_H(-\frac{1}{2}) \tag{77}$$

$$E^F = -\frac{1}{2}\zeta_H(-\frac{1}{2}) \tag{78}$$

which is not unreasonable since formally

$$\zeta_H(-\frac{1}{2}) \text{ "=" } \sum_{n=0}^{\infty} E_n$$

For high temperatures instead of approximating $\theta_3(x)$ one uses its exact form and approximations Q by its small x behaviour to yield

$$\log Z^B \simeq \frac{Vp}{q_o}\frac{1}{\beta^3} \tag{79}$$

$$\log Z^F \simeq \frac{Vp}{q_o}\frac{1}{\beta^3}\frac{7}{8} \tag{80}$$

which are appropriate for a thermal gas of massless bosons or fermions respectively. (81) has already been obtained in ref. [1].

So far our results are general. If we now specialize M_3 to be the 3-sphere (Einstein Static Space) we may find the E_n exactly. They are, together with degeneracies [12] conformally invariant scalars:

	$E_n = (n+1)\frac{1}{R}$	$(n+1)(n+1)$	(81)
neutrinos	$E_n = (n+\frac{3}{2})\frac{1}{R}$	$2(n+1)(n+2)$	(82)
photons	$E_n = (n+2)\frac{1}{R}$	$2(n+1)(n+3)$	(83)

R is the radius of the 3-sphere. The resulting Hamiltonian zeta functions are

scalars $\qquad \zeta_H = R^{-2s}\zeta(2s-2) \qquad (84)$

neutrinos $\zeta_H = R^{-2s}2^{2s-1}\{(1-2^{2-2s})\zeta(2s-2)-(1-2^{-2s})\zeta(s)\}$ (85)

photons $\zeta_H = 2R^{-2s}\{\zeta(2s-2)-\zeta(2s)\}$ (86)

$\zeta(s)$ is Riemann's zeta function.
All three are analytic at $s=-1/2$. Thus the renormalization length does not enter the expression for the zero point energies (86), (87) and (88) yield the following energies

scalars 4/960R (87)

neutrinos 17/960R (88)

photons 88/960R (89)

These are in perfect agreement with other authors, [12], [13], [14]. The minimally coupled scalar field is more difficult, the degeneracies are of course the same but the energy levels are $(n(n+2))^{1/2}\frac{1}{R}$. The corresponding Hamiltonian zeta function is that for the Laplacian on the 3-sphere which has been discussed by Minakshisundaram [15]. He has shown that $\zeta_o(s)$ has poles at $s=3/2,\ 1/2,\ -1/2\ \ldots$ and vanishes for $s=-1,\ -2,\ \ldots$. This means that the zero point energy would depend on the renormalization length.

4. Zero Eigenvalues

It remains to discuss the zero eigenvalue modes. For manifolds which are the product of 2 compact manifolds M_1, M_2 , Kunneth's Theorem asserts

$$b_p(M) = \sum_{r=o}^{p} b_r(M_1)\,b_{p-r}(M_2)$$

For n-spheres $b_o=b_n$ and all others vanish. Thus for de Sitter space (which analytically continues to S^4) there are no zero frequency modes (other than the constants). For the Einstein static space $b_1=1$, $b_2=0$. The zero frequency mode is the closed but inexact one-form which has zero Maxwell field and contributes nothing to the functional integral.

A topologically more interesting example is the Riemannian--Schwarzschild solutions [16]. Since this is not compact the previous theorems do not go through completely. This space is homeomorphic to $R^2 x S^2$. It has $\chi=2$ and $\tau=0$. By gluing two such spaces together one could obtain a space homeomorphic to $S^2 x S^2$ which has $\chi=4$, $B_2=2$. This suggests that if one demands suitable fall off conditions there should exist just 2 harmonic 2-forms on Schwarzschild. This is in fact so. The 2 forms

$$\frac{1}{r}\, d\tau \Lambda ds \tag{90}$$

$$\sin\theta\; d\theta \Lambda d\phi \tag{91}$$

are 2 such 2-forms which correspond to adding electric or magnetic charge to the hole respectively. One is the dual of the other. By taking linear combinations one may construct a self or an antiself dual field with finite total action. The action of the fields (90) (91) is also finite and represents the electrostatic energy of the black hole. If it is not included in the total energy of the system as was done in [16] one would not obtain the correct energy. This shows that although they have no global vector potential associated with them (which is square integrable) these fields then contribute to the functional integral, i.e. Z should be written as

$$Z = (\mathrm{DET}\ \tilde{\Delta}_1)^{1/2} (\mathrm{DET}\ \tilde{\Delta}_0) \exp - \sum_i I_i \tag{92}$$

where $\tilde{\Delta}_1$, $\tilde{\Delta}_0$ are the Laplacians with the zero frequency modes projected out and I_i is the classical action of the i'th harmonic 2-form. The situation is analogous to the quantization of the electromagnetic field in the presence of sources. The coulomb fields are not quantized but nevertheless contribute to the total Hamiltonian of the system.

Note that if one wishes to couple these Maxwell fields to charged scalar a spinor fields one must impose the Dirac quantization condition that the total flux of magnetic field through any 2-sphere must be a half integral multiple of the inverse electron charge.

The finite number of these solutions is related to the No Hair Theorems. These are statements to the effect that time independent fields of a general sort do not exist on a (Lorentzian) black hole background. They are usually proven by means of a detailed examination of the relevant wave equations - scalar, Proca, Maxwell etc. Now such fields if analytically continued onto the Riemannian background would constitute zero eigenvalue solutions. For massive spin zero fields this is immediately ruled out because Δ_0+m^2 is positive. For spin 1 the situation is that Δ_1 is positive if the Ricci tensor vanishes. This rules out Proca Hair. In fact $S^2 x R^2$ is simply connected, so no harmonic 1-forms are allowed.

There are more complicated Riemannian metrics with 1 parameter groups of isometries (like that generated by time translations). The fix point sets of such isometries, providing they constitute smooth submanifolds, are related to the Euler number by a theorem of Hopf

$$\chi = \sum_i \chi_i \qquad (93)$$

The sum is over the connected fix point sets, the i'th one of which has Euler number χ_i . In the Schwarzschild case has one spherical fix point set, $\chi=2$, an example of a fix point set, $\chi=2$. An example of a fix point set of zero dimension is the Tau Nut Gravitational Instanton [5] with metric

$$ds^2 = \frac{1}{1+\frac{2M}{r}} (d\tau+4M\sin^2\frac{\theta}{2}\, d\phi)^2 + (1+\frac{2M}{r})\{dr^2+r^2(d\theta^2+\sin^2\theta d\phi^2)\} \qquad (94)$$

The reader is referred to [5] for more details about the coordinates. $\frac{\partial}{\partial\tau}$ is a Killing vector which generates a one-dimensional group of isometries which in fact have a fix point at $r=0$. Indeed the complete rotation group which acts on the r=constant surfaces (which are 3-spheres) has a fix point there. The topology of the Taub Nut Instanton is R^4 which has $\chi=1$ in accordance with formula (95). The more general Instanton described by Hawking [5] with metric

$$ds^2 = V^{-1}(d\tau+\underline{\omega}\underline{dx})^2 + V\underline{dx}\underline{dx} \qquad (95)$$

$$V = 1+2\sum_i M_i|r-r_i|^{-1} \qquad (96)$$

$$\text{curl } \omega = \text{grad } V \tag{97}$$

would possess in general only a one-parameter group of isometries corresponding to $\frac{\partial}{\partial\tau}$. These have N fix points and so $\chi = N$ and according to ref. [6] $p = 3\tau = 2N$. Since they have a self dual curvature they satisfy Hitchin's Relation. In the simpliest N=1 case since $\tau = 2/3$, the correct generalization of the Hirzebruch theorem must require some modification involving the boundary. Fermion Hair seems to play an important role in the axial current anomaly [5]. Further they can endow certain classical Yang-Mills solutions with fermionic properties. In the black hole case it is easy to verify that no such solutions can exist by a simple geometrical argument. The point is that any spinor field defined on the Riemannian Schwarzschild solution cannot have zero Lie derivative with respect to $\frac{\partial}{\partial\tau}$. This is because at $r=2M$, dragging along the orbits of $\frac{\partial}{\partial\tau}$ degenerates into a rotation through 2π keeping the point fixed. Under such change the spinor must change sign which contradicts its supposed constancy. Indeed one can see that any spinor field will satisfy (26) if one interprets as $\psi(\tau+\beta)$ dragging around the axis of rotation at $r=2M$.

Other solutions of the Einstein-cosmological constant equations which would be amenable to the sort of study described above would be

1) $S^2 x S^2$
2) PC^3
3) K3

1) $S^2 x S^2$ is simply the product of 2 spheres of radius $\Lambda^{-1/2}$. This has $\chi=4$ and $\tau=0$, $B_1=0$ and $B_2=2$. It may be obtained from the general (Schwarzschild-de Sitter family discussed in [18]

$$ds^2 = -Vd\tau^2 + \frac{dr^2}{V} + r^2(d\theta^2 + \sin^2\theta d\phi^2)$$

$$V = 1 - \frac{2M}{r} - \frac{\Lambda r^2}{3}$$

by setting $qM^2\Lambda=1$ whilst introducing a new coordinate just as one obtains the Robinson-Bertatti solutions as a limiting case of the Reisner-Nordstrom solutions when $Q^2=M^2$ [19].

2) PC^2 with its Fubini-Study metric has been discussed as a possible gravitational Instanton by Eguchi and Freund [20]. has $\chi=3$, $\tau=1$, $B_2=1$ and $B_1=0$. It has no Lorentzian and section, no spinor structure. It may be regarded as a special case of the Taub-Nut-De Sitter family. When the metric takes the form

$$ds^2 = \frac{dr^2}{(1+\frac{\Lambda r^2}{6})^2} + \frac{1}{4}\frac{r^2}{(1+\frac{\Lambda r^2}{6})^2}(d\psi+\cos\theta d\phi)^2 + \frac{r^2}{4(1+\frac{\Lambda r^2}{6})}(d\theta^2+\sin\theta d\phi^2) \tag{98}$$

ψ,θ,ϕ are euler angles $\frac{\partial}{\partial\psi}$ vanishes at $r=0$ (o-dimensional fixed point) and at $r=\infty$ (a spherical fixed point or horizon). It follows that $\chi=3$. PC^2 may be regarded as a compactification of R^4 by adding a sphere at infinity just as S^4 may be regarded as a compactification of R^4 by adding a point at infinity.

3) K3 is the only compact, self dual Ricci flat manifold. It has $\chi=24$, $\tau=16$; $B_2^+=36$ and $B_2^-=19$, $B_1=0$. It posseses spinor structure but no Lorentzian section. It also possesses no killing vector fields. Not surprisingly therefore no explicit form for K3 is known. Yau [21] however, has given a non-constructive existence proof.

A further study of these spaces is underway.

Conclusion

The main point of the paper is to show how the functional formulation of quantum field theory on curved backgrounds can be discussed in terms of mathematically well defined concepts in the spectral theory of elliptic operators, and that when one does so one encounters interesting relationships between the topology of a suitably complexified manifold and the answer to physical questions. Finally one may obtain in this way finite answers for such quantities as zero point energies.

The main drawback of this methods developed is that they are restricted to Riemannian sections. The generic case is likely to be that of quasi-Riemannian sections. How much of the present ideas can be carried over to that case would seem to be both a problem in mathematics as much as in physics since the theory does not seem to have been developed much yet.

Acknowledgements

I should like to thank J.S. Dowker, S.W. Hawking and C.J. Isham for valuable discussions, and M. Atiyah for telling me about formula (93).

References

[1] S.W. Hawking, Commun. Math. Phys. 55, 133 (1977)

[2] G.W. Gibbons, S.W. Hawking and M.J. Perry, in preparation.

[3] R.T. Seeley, Amer. Math. Soc. Proc. Symp. Pure Math. 10, 288-307 (1976)

[4] P. Gilkey, Journ. Diff. Geom. 10, 601 (1975)

[5] S.W. Hawking, Phys. Letts. A60, 81 (1977)

[6] N. Hitchin, Journ. Diff. Geom. 9, 435 (1974)

[7] H.S. Tsao, Phys. Letts. 68B, 79 (1977)

[8] J.S. Dowker and R. Critchley, "The Stress Tensor Conformal Anomaly of Scalar and Spinor Fields", Manchester preprint to appear in Phys. Rev. D 15

[9] L.S. Brown and J.P. Cassidy, Phys. Rev. D 15, 2810 (1977)

[10] D.M. Capper and M.J. Duff, Nuovo Cimento 23A, 175 (1974)

[11] G.W. Gibbons, Phys. Letts. 60A, 385 (1977)

[12] L.M. Ford, Phys. Rev. D 11, 3370 (1975)
L.M. Ford, Phys. Rev. D 14, 3304-3314 (1976)

[13] J.S. Dowker and Critchley, J. Phys. A9, 535 (1976)
Phys. Rev. D 15, 1484 (1976)

[14] J.S. Dowker and M.B. Altaie, "Spinor Fields in an Einstein Universe. The vacuum averaged stress-energy tensor", Manchester preprint

[15] S. Minakshisundaram, Journ. Ind. Math. Soc. 13, 41 (1949)

[16] G.W. Gibbons and S.W. Hawking, Phys. Rev. D 15, 2752 (1977)

[18] G.W. Gibbons and S.W. Hawking, Phys. Rev. D 15, 2738 (1977)

[19] B. Carter in "Black Holes" B.S. and C. de Witt ed. Gorden and Breach, London (1973)

[20] L.T. Eguchi and P. Freund, Phys. Rev. Letts. 37, 1251 (1976)

[21] S.T. Yau, Proc. Natl. Acad. Sci. USA 74, 1798 (1977)

OBSERVABLES FOR QUANTUM FIELDS ON CURVED BACKGROUND

Peter Hajicek
Institut für Theoretische Physik
Universität Bern
Sidlerstrasse 5, 3012 Bern, Schweiz

1. Introduction

The object of the present paper are linear quantum fields on curved background. That is to say, we restrict ourselves to field equations which are linear in the quantum field under consideration and in which the only allowed interaction is represented by a classical external gravitational field. Our motive is on the one hand, the basic importance of linear fields (not just free fields - recall quantization of kinks , solitons, etc. [1]) in quantum physics. As it is well known, the present day theory of non-linear fields is based on formal perturbation series starting with linear fields as zero approximation, and the interpretation of any calculations for the former is, therefore, dependent on our understanding of the latter. On the other hand, one encounters interpretation difficulties in quantum field theory on curved spacetimes already at the level of linear fields. The problem is, roughly speaking, to find a reasonable system of observables: in the flat spacetime, the most important observables are associated with spacetime symmetries and are of global character. This is an idealization which, in fact, does not correspond to real experiments. In curved spacetimes, there are no symmetries in general, and the measurements which men can do are always well localized.

An interesting proposal to solve the problem is to take a regularized energy-momentum operator as a basic local observable [2]. However, this operator is not an observable, even formally: it must be smeared. On the one hand, smearing a tensor field in a curved spacetime is a coordinate-dependent procedure. On the other hand, e.g. the energy of those field modes whose period is larger than the radius of the region over which the tensor has been smeared is not properly given by it: the corresponding total energy in this region of such a mode can even come out negative. It is, in fact, not clear at all, what measurements are represented by smeared energy-momentum operator.

We choose another way. We attempt to define localized particle-like observables, at least in some simple cases. Although such a notion can be felt as contradictio in adiecto, we show that it is not, if one chooses the basic observables properly.

The adequate mathematics seems to be the algebraic approach. In principle, as a basic space for a quantum theory, one can either take a space of states - e.g. a Fock space -, or a space of observables - a C*-algebra. In the Fock-space approach, one has to single out a particular state. This may be physically justified for simple systems in the flat spacetime, where there is a unique vacuum. In curved spacetimes, there need not exist anything similar to vacuum, and if, in some sense, it exists, then it need not be unique. Choosing a state in an appropriate but arbitrary manner is, of course, possible, but it leads to another difficulty. Due to thermal properties of quantum fields in curved backgrounds, we have to work with states which can never lie all in the same Fock space [3, 4].

None of these difficulties is present in C*-algebra approach. The algebras are uniquely and invariantly determined for any given spacetime (up to inclusion). The states are defined as certain functionals over the algebra and as such they form a set much larger than any Fock space. All states are dealt with on the same level. In such a way, the algebras are at least formally appealing.

However, there is another point which is less formal and which I consider crucial. Elements of the algebra are, roughly speaking, smeared fields. One can smear over arbitrarily small regions of space to obtain all generators of the algebra. There also are well defined sub-algebras localized to a given subregion of space. Thus, if one looks for some localized observables, the most direct way seems to go through the algebras. In fact, the generators of the algebras also are, at least in principle, observables [5]. However, one often performs measurements which are "complementary" to that of smeared fields and in which one sees the so-called particle-like properties. Our way, therefore, need not be too short.

The plan of the paper is as follows. In Sec. 2, we define the algebras and describe their most useful properties, going into details only where the difference between flat and curved spacetimes becomes important. We smear the fields only in space, not in time, so we work on Cauchy hypersurfaces. There are two constructions of C*-algebras from the canonical commutation relations of spatially smeared fields: that by Segal [6], and that of Manuceau [7] and Slawny [8]. The former contains the latter, but the latter seems sufficient for our purposes. We also analyze the dynamics of the field and show that it is determined only, if one prescribes what are "the same measurements at two different instants of time". Thus, even the dynamics is a problem of observable definition.

In Sec. 3, we modify the notion of complex structure as invented by Segal [6] to the so-called formal frequency splitting, which seems to be more adapted to curved spacetimes. We give a proof that the two notions are equivalent, at least in already complete one-particle Hilbert space.

In Sec. 4, generating functionals are introduced. The notion is slightly generalized to incorporate cases, in which the initial vacuum is different from the final one. From such functionals, some information can be read off very quickly, for instance the creation of particles in pairs, Cauchy data for n-point function and the nterrelationship between n-point functions.

Whereas the material of Sections 3 and 4 was rather formal, in Sec. 5 we return o our main problem of determining observables. We restrict ourselves to static lobally hyperbolic spacetimes. As it is well known, there is a non-trivial generlization of the flat spacetime field dynamics for these backgrounds, and one can ntroduce, at least formally, particle-like observables. Also contained in literaure, at least implicitly, is a gedankenexperiment describing a measurement of such bservables; it is based on the so-called Unruh's box [12]. We just collect these well nown things and fill some unimportant gaps. Then, in Sec. 6, simple two-dimensional spacetimes are introduced: Minkowski, Kruskal and de Sitter manifolds. In each of them, some proper sub-regions exist which are static and globally hyperbolic at the same time - the isometry group becomes spacelike elsewhere. As the quantum field on these backrounds, we choose massless hermitean scalar. Again, many results about the behaviour of this field in these spacetimes are scattered over the literature (for example, see 10, 12, 13). Our use of these simple exactly solvable examples is to prove that there can be well defined, measurable, particle-like observables confined to proper sub-regions of the whole spacetime. It seems even that only some special localized sub-algebras have physical representations, but there is in general no physical representation of the whole C*-algebra. This means, roughly speaking, that there is no global vacuum, for instance, but one has a unique Araki class of global states which may be considered as vacuum for some special sub-region. We use this fact to obtain interpretations of several well known results.

2. Algebraic Preliminaries

We consider a hermitean scalar field $\Phi(x)$ on a globally hyperbolic spacetime (M,g) (for the definition and properties of globally hyperbolic spacetimes, see Ref. 14). The restriction to globally hyperbolic spacetimes is necessary, for else there is no well defined dynamics even in classical sense.

In the formal approach, $\Phi(x)$ is an operator-valued distribution satisfying the Klein-Gordon equation

$$g^{kl} \nabla_k \nabla_l \Phi + m^2 \Phi = 0. \tag{1}$$

Let S be a Cauchy hypersurface in M given by the parametric relations

$$x^k = x^k(y), \tag{2}$$

where x^k are some coordinate on M, y^α on S. Let the y-components of the negative definite metric on S as induced by g_{kl} be denoted by $g_{\alpha\beta}$. The surface element dS of S is then given by

(3) $$dS = |\mathrm{Det}(g_{\alpha\beta})|^{1/2}\, d^3y.$$

Define

(4) $$\Phi(y) = \Phi(x)|_{x = x(y)} \quad , \quad \pi(y) = n^k(y)\, \partial_k\, \Phi(x)|_{x = x(y)} \quad ,$$

where $n^k(y)$ is the unit normal vector to S oriented to the future. Then,

(5) $$[\Phi(y)\ ,\ \Phi(y')] = [\pi(y)\ ,\ \pi(y')] = 0\ ,$$
$$[\Phi(y)\ ,\ \pi(y')]\, dS = i\ \delta(y - y')\ d^3y\ ,$$

where $\delta(y - y')\ d^3y$ is defined by

(6) $$\int_S f(y)\ \delta(y - y')\ d^3y = f(y')$$

for any C^∞ function $f(y)$ on S.

Instead of operator-valued distributions, one can work with <u>smeared field operators</u> defined as follows. Let $T(S)$ be the real linear space of all pairs $X = (f,h)$ of real C^∞ functions with compact support on S and let $B(X_1,X_2)$ be the Klein-Gordon bilinear form (also called symplectic form, if i is dropped) on $T(S)$:

(7) $$B(X_1,X_2) = -\, i \int_S (f_1(y)\ h_2(y) - f_2(y)\ h_1(y))\ dS.$$

The smeared field operator $\Phi(X)$ is formally defined by

(8) $$\Phi(X) = \int_S (\Phi(y)\ f(y) - \pi(y)\ h(y))\ dS$$

In terms of the smeared fields, the CCR read

(9) $$[\Phi(X_1)\ ,\ \Phi(X_2)] = B(X_1,X_2)$$

for all X_1, $X_2 \in T(S)$. The operators $\Phi(X)$ are essentially self-adjoint (for definition, see Ref. 16), but unbounded. Their exponentials $\exp(i\Phi(x))$ are then unitary and the CCR lead via Baker-Hausdorff identity to the so-called Weyl relation

(10) $$\exp(i\Phi(X_1)) \cdot \exp(i\Phi(X_2)) = \exp(-\tfrac{1}{2}B(X_1,X_2)) \cdot \exp(i\Phi(X_1+X_2))$$

for all X_1, $X_2 \in T(S)$.

In the algebraic approach, we start with the last relation. More precisely: we consider $T(S)$ as an Abelian group with a multiplier $\exp(\tfrac{1}{2}B(X_1,X_2))$ (see Ref. 16) and 'efine

Definition 1: Let H be a Hilbert space, $B(H)$ the algebra of all bounded linear operators in H, and $W: T(S) \to B(H)$ a map satisfying

(i) $W(X)$ is unitary for any $X \in T(S)$,

(ii) $W(X) \cdot W(Y) = \exp(-\frac{1}{2}B(X,Y)) \cdot W(X+Y)$
for any $X, Y \in T(S)$.

Then, the pair (H, W) is called a projective representation of the group $T(S)$ with multiplier $\exp(\frac{1}{2}B(X,Y))$.

The following important theorem has been shown in Ref. 7 and 8:

Theorem 1: Let (H_1, W_1) and (H_2, W_2) be two projective representations of the group $T(S)$ with the same multiplier $\exp(\frac{1}{2}(X,Y))$, and let the strongly closed algebras generated by $W_1(T(S))$ and $W_2(T(S))$ be denoted by A_1 and A_2, respectively. If the form $B(X,Y)$ is non-degenerate, then there is a unique *-isomorphism

$$\alpha: A_1 \to A_2$$

such that

$$\alpha\, W_1(X) = W_2(X)$$

for all $X \in T(S)$.

Strongly closed algebra A_i generated by $W_i(S(T))$ is defined as follows: form al finite linear combinations with complex coefficients of the form

$$\sum_{n=1}^{N} c_n\, W_i(X_n)\,.$$

Define a *-operation by

$$\left(\sum_{n=1}^{N} c_n\, W_i(X_n)\right)^* = \sum_{n=1}^{N} c_n^*\, W_1(-X_n)$$

and close the resulting *-algebra by the operator norm in H_i. Such algebras A_i are representations of one abstract C*-algebra, let us denote it by $A(S)$, because they are all *-isomorph according to Theorem 1. $A(S)$ is uniquely determined by $T(S)$ and $B(X,Y)$ and has well defined elements $W(X)$ for any $X \in T(S)$.

The construction by Segal [6] leads to a different C*-algebra, which contains the whole Manuceau-Slawny algebra together with many other observables. However, for our purposes - namely to define states and dynamics - the smaller algebra is sufficient (just the set of all unitary generators $W(T(S))$ would do).

We define next:

Definition 2: Let the projective representation (H,W) of $T(S)$ be such that $W(tX)$ is strongly continuous in the real argument t for any fixed $X \in T(S)$ (tX lies in $T(S)$) Then, (H, W) is called a Weyl system.

For a Weyl system, any $W(X)$ determines a self-adjoint operator $\Phi(X)$ by

$$W(X) = \exp(i\Phi(X))$$

(Stone's theorem), so we recover the smeared fields. It follows that only those representations of $A(S)$ which are generated by Weyl systems can have physical meaning.

An important feature of our construction is the dependence of the algebra $A(S)$ on the Cauchy hypersurface S. There are interesting maps of $A(S_1)$ onto $A(S_2)$ for given pairs of Cauchy hypersurfaces in M. The first one is constructed as follows.

If $X_1 \in T(S_1)$, $X_1 = (f_1, h_1)$, denote by $X(x)$ the unique classical solution of Klein-Gordon equation for the following Cauchy problem along S_1:

$$X(x)|_{S_1} = h_1 \quad , \quad n_1^k \partial_k X(x)|_{S_1} = f_1$$

Then, there is a unique $X_2 \in T(S_2)$, $X_2 = (f_2, h_2)$, defined by

$$h_2 = X(x)|_{S_2} \quad , \quad f_2 = n_2^k \partial_k X(x)|_{S_2} .$$

In order to perform this construction, we need the following

Assumption 1: The Cauchy problem has a unique global solution for any Cauchy hypersurface S and any initial data from $T(S)$.

Then, any $X_1 \in T(S_1)$ determines uniquely $X_2 \in T(S_2)$ and we have a map

$$\mathcal{O}_{12} \; : \; T(S_1) \rightarrow T(S_2)$$

given by

$$\mathcal{O}_{12} X_1 = X_2 \qquad \text{for all } X_1 \in T(S_1).$$

Assumption 1 implies that $\mathcal{O}_{12}$ is a bijection. Due to the linearity of the field equation, $\mathcal{O}_{12}$ is a linear map. In addition, $\mathcal{O}_{12}$ preserves the Klein-Gordon product $B(.,.)$ (see, e.g., Ref. 2 or 15):

$$B_2(\hat{\nu}_{12}X, \hat{\nu}_{12}Y) = B_1(X,Y)$$

for all X, Y ε $T(S_1)$, where $B_i(\cdot,\cdot)$ is the Klein-Gordon product in the linear space $T(S_i)$. Thus the structure of Abelian group with multiplier is preserved by $\hat{\nu}_{12}$, and Theorem 1 yields immediately a *-isomorphism ν_{12} : $A(S_1) \to A(S_2)$ with the property

$$\nu_{12}\, W(X) = W(\hat{\nu}_{12}X)$$

for all X ε $T(S_1)$. The *-isomorphism ν_{12} is, in such a way, uniquely and invariantly defined for any globally hyperbolic spacetime satisfying Assumption 1.

The meaning of ν_{12} is <u>not</u> what one usually calls "dynamical automorphism" (it is no automorphism at all). There seems to be no invariant definition of dynamics of a quantum field on a general globally hyperbolic manifold. Indeed, to observe a change of any system between two times t_1 and t_2, one must be able to perform at least two equal measurements, first at t_1, second at t_2, to see whether the results will be different. However, what is the precise meaning of "equality of measurements at different times" ? The generally used convention is based on symmetry: if our spacetime M allows a one-dimensional isometry group $\{g_t\}$ with timelike trajectories, then we can use the elements g_t of the group to translate our measuring device. For example, in the flat spacetime, the dynamics is defined by the time translation subgroup of Pincaré group (one can say <u>the</u> time translation subgroup, because they are all mutually conjugate). More precisely, let us formulate the following

<u>Assumption 2:</u> (M, g) admits a one-dimensional group $\{g_t\}$ of isometries such that there is a Cauchy hypersurface S in M satisfying

$$\{p \;\varepsilon\; g_t S \,|\, t \;\varepsilon\; R^1\} = M$$

Sometimes, one strengthens this by requiring the curves $g_t\, p$ to be timelike for any fixed p ε M.

Let Assumption 2 be satisfied. Then, we can construct another interesting *-isomorphism as follows.

Introduce the notation $S_t = g_t\, S$. $\{S_t\}$ is a one-parameter family of Cauchy hypersurfaces and t is a time function (see Ref. 14). We have a linear bijection

$$\hat{\gamma}_t \;:\; T(S_{t_1}) \to T(S_{t_1+t})$$

for all t_1 and t from R^1, which preserves the form $B(\cdot,\cdot)$ and which is defined as follows.

Let $X_1 \varepsilon T(S_{t_1})$, $X_1 = (f_1, h_1)$. Then, $\hat{\gamma}_t X_1 = X_2 \varepsilon T(S_{t_1+t})$, $X_2 = (f_2, h_2)$, where

$$f_2 = f_1 \circ g_t^{-1}, \quad h_2 = h_1 \circ g_t^{-1}.$$

The condition for g_t to be an isometry is important: for general maps, $B(\cdot,\cdot)$ will not be preserved. Again, by Theorem 1 there is a unique *-isomorphism γ_t: $A(S_{t_1}) \to$ $\to A(S_{t_1+t})$ satisfying

$$\gamma_t W(X) = W(\hat{\gamma}_t X)$$

for all $X \varepsilon T(S_{t_1})$. Let A be an observable of $A(S_{t_1})$, representing a measurement at the time t_1. Then $\gamma_t A$ is an observable in $A(S_{t_1+t_2})$ representing, per definition, the same measurement at the time $t_1 + t$.

Let us turn our attention to states. In the algebraic approach, roughly speaking, states are defined by the mean values of all observables:

Definition 3: A state, ϕ, over a given C*-algebra A is a normalized, positive linear functional (NPLF) on A:

(i) linear: $\phi(\xi A + \eta B) = \xi \phi(A) + \eta\phi(B)$ for all A, B ε A and $\xi, \eta \varepsilon C^1$.

(ii) normalized: $\phi(id) = 1$, where id is the identity element of A,

(iii) positive: $\phi(AA^*) \geq 0$ for all A ε A.

Applying this general definition to our case is not quite straightforward, because we have many algebras - one for any Cauchy hypersurface. Let us, therefore, consider families of NPLF's, $\{\phi_S\}$, consisting of states ϕ_S, each over another algebra $A(S)$. First, it is clear that not all such families are physically reasonable: measurements at different Cauchy hypersurfaces must be correlated. We define:

Definition 4: An allowed state family $\{\phi_S\}$ on the spacetime (M, g) is a collection of states ϕ_S over all $A(S)$ satisfying

$$\phi_{S_2}(\nu_{12} A) = \phi_{S_1}(A) \tag{11}$$

for any two Cauchy hypersurface S_1 and S_2 in M and for any A ε $A(S_1)$.

Thus, an allowed state family is determined by its values on just one of the algebras $A(S)$. Remark that the conditions (i) - (iii) of Definition 3 are satisfied by any state of the family, if they are satisfied at least by one, because ν_{12} is a *-isomorphism.

Let our spacetime M satisfy Assumptions 1 and 2. Then, we can define dynamics as follows.

Dynamics in Schrödinger's picture: given an allowed state family $\{\phi_S\}$, then its subfamily corresponding to the Cauchy hypersurfaces S_t can be denoted by $\{\phi_t\}$. The one-parameter set of states ϕ_t is the Schrödinger's time-dependent state. Observables are defined as equivalence classes by $\{\gamma_t\}$, that is $A \in A(S_{t_1})$ and $\gamma_t\, A \in A(S_{t_1+t})$ are in the same class, $\tilde{A}$, say. These classes form an abstract C^*-algebra $\tilde{A}$, say, which is independent on t. Each of the classes $\tilde{A}$ intersect any $A(S_t)$ in just one element, $\tilde{A}_t$, say:

$$\{\tilde{A}_t\} = \tilde{A} \cap A(S_t).$$

Thus, we can finally define

$$\phi_t(\tilde{A}) = \phi_t(\tilde{A}_t) \ .$$

Dynamics in Heisenberg's picture: We select one particular Cauchy hypersurface, S_0, say, from $\{S_t\}$. Any allowed state family determines a state ϕ_0 over $A(S_0)$. This is the time-independent Heisenberg state corresponding to $\{\phi_S\}$. The time development of observables in $A(S_0)$ is given by

(12) $$A_t = (\nu_{ot}^{-1} \circ \gamma_t)\, A_0$$

for any $A_0 \in A(S_0)$. From (11), it follows

(13) $$\phi_t\, (\tilde{A}) = \phi_0\, (A_t)$$

for any allowed family and $A_0 \in \tilde{A}$. The mapping

(14) $$\alpha_t = \nu_{ot}^{-1} \circ \gamma_t$$

is usually called the dynamical automorphism of $A(S_0)$.

Thus, the Assumptions 1 and 2 enable us to construct nice dynamics, at least formally (physical aspects of it remain obscure). However, only very special globally hyperbolic spacetimes satisfy Assumption 2. In fact, one can answer many physical questions without knowing the detailed dynamics: it is sufficient to know what are "the same measurements" just at the beginning and at the end. In such a case, we need no more than one isometry, g, say, that maps the "in-coming" Cauchy hypersurface onto the "out-going" one. Some sort of limiting case of such a situation could exist in asymptotically flat spacetimes.

We return to backgrounds satisfying the Assumption 2 even in the strong form later. Let us close this section by introducing some important sub-algebras.

Choose an arbitrary open sub-manifold S_1 of a given Cauchy hypersurface S in M. Then, $\mathcal{D}(S_1)$ is itself a globally hyperbolic spacetime. By $\mathcal{D}(S_1)$, we denote the so-called Cauchy development of S_1 in M (for definition see Ref. 14). Then, we can repeat everything which was said in this section replacing M by $\mathcal{D}(S_1)$. In particular, we obtain a unique C*-algebra $A(S_1)$.

There is an important *-injection of $A(S_1)$ into $A(S)$ defined as follows. Let the map

$$\hat{I} \; : \; T(S_1) \to T(S)$$

be determined by the relations:

if $X_1 \in T(S_1)$, $X_1 = (f_1, h_1)$, then $\hat{I}\, X_1 = (f, h)$,

where

$$f|_{S_1} = f_1 \; , \; h|_{S_1} = h_1 \; , \; f|_{S-S_1} = h|_{S-S_1} = 0.$$

$\hat{I}$ is a linear injection preserving the multiplier. Making use of Theorem 1, we obtain a unique *-injection $I : A(S_1) \to A(S)$ such that

$$I\, W(X_1) = W(\hat{I}X_1) \tag{15}$$

for all $X_1 \in T(S_1)$. The *-injection is uniquely and invariantly defined by S and S_1.

I determines a projection I' of states over $A(S)$ onto states over $A(S_1)$ (Araki classes of states which are equal on S_1, see Ref. 17) as follows. Let ϕ be a state over $A(S)$. Then $\phi_1 = I'\phi$ defined by

$$\phi_1\,(A) = \phi\,(IA) \tag{16}$$

for all $A \in A(S_1)$ is a state over $A(S_1)$ because I is a *-injection. In general, if ϕ is pure (external), ϕ_1 will not be such.

One can very roughly say that a great deal of Hawking's effect is produced by such an injection I.

3. Frequency Splitting

An important class of states is defined by the so-called frequency splitting.

Let S be a fixed Cauchy hypersurface in M. Denote by $T_c(S)$ the space of complex C^∞ function pairs with compact support on S. We regard $T(S)$ as a real linear sub-pace of $T_c(S)$ and $T_c(S)$ as the unique complex extension of $T(S)$: if $X \in T_c(S)$, hen $X = X_1 + iX_2$, where X_1, $X_2 \in T(S)$.

We can introduce two invariant operations in the complex linear space $T_c(S)$.

1) the anti-linear bijection $\hat{C} : T_c(S) \to T_c(S)$ defined by: If $X = (f, h) \in T_c(S)$, then

$$\hat{C}X = (f^*, h^*).$$

One can call this map <u>charge conjugation</u>, even if it has nothing to do with charge here, because the field $\Phi(x)$ is neutral. However, if we regard the elements of $T_c(S)$ as classical test solutions for a complex (charged) field $\Phi(x)$, then $\hat{C}$ coincides with charge conjugation. This holds for general field (e.g., real (Majorana) bispinor field, etc.). It can be easily verified, that $\hat{C}$ commutes with ν_{12}:

$$\hat{C}\,\nu_{12} = \nu_{12}\,\hat{C}\,,$$

because the charge conjugation is a symmetry of the field equation. In this sense, $\hat{C}$ is independent of Cauchy hypersurface.

2) The Klein-Gordon scalar product $B(X,Y)$ has a unique extension $B_c(X,Y)$ to $T_c(S)$ by antilinearity in the first and linearity in the second factor: let $X = X_1 + iX_2 = (f_1, h_1) \in T_c(S)$, $Y = Y_1 + iY_2 = (f_2, h_2) \in T_c(S)$; then, define

$$B_c(X,Y) = -i\,|B(X_2,Y_1) - B(X_1,Y_2)| + |B(X_1,Y_1) + B(X_2,Y_2)| = \tag{17}$$
$$= -i \int_S (f_1^* h_2 - h_1^* f_2)\, d S.$$

The form $B_c(\cdot,\cdot)$ is symmetric:

$$B_c(X,Y)^* = B_c(Y,X) \tag{18}$$

for all $X,Y \in T_c(S)$, non-degenerate:

$$B_c(X,Y) = 0 \quad \text{for all } X \in T_c(S) \text{ implies } Y = 0,$$

and satisfies

$$B_c(\hat{C}X,\hat{C}Y) = -B_c(X,Y)^* \tag{20}$$

for all $X,Y \in T_c(S)$.

Again, the form $B_c(X,Y)$ has a relation to charge: if we consider the elements $T_c(S)$ as Cauchy data for test solutions of complex (charged) field $\Phi(x)$, then $B_c(X,X)$

is a total charge carried by the solution corresponding to X. For $B_c(X,Y)$, defined at different Cauchy surfaces, we have

$$B_c(\nu_{12}X,\ \nu_{12}Y) = B_c(X,Y)$$

so the form is independent on Cauchy hypersurface.

After these preparatory remarks, we are ready to introduce

Definition 5: Frequency splitting is a form $\omega(\cdot,\cdot)$ on $T_c(S)$ with the properties

(i) $\omega(\cdot,\cdot)$ is symmetric, and real-valued on $T(S)$:

(21) $$\omega(X,Y) = \omega(Y,X)^* \ ,\ \omega(\hat{C}X,\hat{C}Y) = \omega(X,Y)^*$$

for all $X,Y \in T_c(S)$.

(ii) $\omega(\cdot,\cdot)$ is positive definite:

(22) $$\omega(X,X) \geq 0 \quad \text{for all } X \in T_c(S),$$
$$\omega(X,X) = 0 \quad \text{implies } X = 0.$$

(iii) Denote the Cauchy completion of $T_c(S)$ with respect to $\omega(\cdot,\cdot)$ by H. Then there is a self-adjoint operator N on H with the spectrum $\{-1, +1\}$ such that

(23) $$B_c(X,Y) = \omega(X,NY)$$

for all $X,Y \in T_c(S)$.

In a sense, our "frequency splitting" is equivalent to Segal's complex structure I (see Ref. 6). Our definition seems to be more appropriate for curved spacetimes. The reason is that I is not well defined on $T(S)$ in cases of physical interest, because it sends functions of compact support to functions with non-compact support. This is important as the extension to strongly decreasing functions is not of much use in curved spacetimes. In order to see the equivalence to Segal's complex structure, at least in the already complete space H, we can proceed as follows.

The operator N has two eigenspaces H^+ and H^- in H defined by

$$H^{\pm} = \{X \in H \mid NX = \pm X\}\ .$$

H^+ can, at least mathematically, be considered as a one-particle Hilbert space. Denote by the $P^{\pm}$ the projectors

$$P^{\pm}\, H = H^{\pm}.$$

We have the following

Theorem 2: Let $\overline{T(S)}$ be the closure of $T(S)$ in H. Then, $P^{\pm}$ restricted to $\overline{T(S)}$ is a linear bijection

$$P^{\pm} \; : \; \overline{T(S)} \to H^{\pm}$$

with the inverse

(25) $$(P^{+}|_{\overline{T(S)}})^{-1} = 2 \text{ Re} \; ,$$

where Re is the continuous extension of

$$\text{Re} \; : \; T_c(S) \to T(S)$$

to H.

With Theorem 2, it is straightforward to see that the map

$$I \; : \; \overline{T(S)} \to \overline{T(S)}$$

defined by

$$I = (P^{+}|_{\overline{T(S)}})^{-1} \; i \; P^{+} \; ,$$

where i is the usual imaginary unit, has all properties of Segal's complex structure We perform the proof in several steps.

Proof: 1) $\hat{C}$ is a bounded antilinear operator on $T_c(S)$ with respect to $\omega(\cdot,\cdot)$. This follows from the relations (21):

$$\omega(\hat{C}X,\hat{C}X) = \omega(X,X)^{*} = \omega(X,X).$$

In fact

(26) $$||\hat{C}||_{\omega} = 1 \; .$$

Thus, $\hat{C}$ can be extended to H. Define, for any $X \in H$,

(27) $$\text{Re } X = \frac{1}{2} (X + \hat{C}X) \; , \; \text{Im } X = - \frac{i}{2} (X - \hat{C}X).$$

The map Re and Im are also of ω-norm 1, and extend the maps $\text{Re} : T_c(S) \to T(S)$, and $\text{Im} : T_c(S) \to T(S)$ continuously to H. From the continuity of $\hat{C}$, it follows further that

(28) $$\{X \in H | \hat{C}X = X\} = \overline{T(S)}.$$

2) We proof that

$$(X \in H^{\pm}) \Leftrightarrow (B_c(X,X) = \pm\, \omega(X,X)). \tag{29}$$

From the left to the right, it follows directly from (23) and (24). To show it from the right to the left, set

$$X = X^+ + X^- \quad , \quad X^{\pm} = P^{\pm}X \in H^{\pm} \ , \tag{30}$$

use the ω-orthogonality of $H^{\pm}$ and again the relations (23) and (24). This leads to the relations

$$\begin{aligned} \omega(X,X) &= \omega(X^+,X^+) + \omega(X^-,X^-) \ , \\ B_c(X,X) &= \omega(X^+,X^+) - \omega(X^-,X^-). \end{aligned} \tag{31}$$

Thus, for the upper sign in (29), $\omega(X^-,X^-) = 0$, so $X^- = 0$, for the lower sign, $\omega(X^+,X^+) = 0$, so $X^+ = 0$.

3) We shall show that

$$(X \in H^{\pm}) \Leftrightarrow (CX \in H^{\mp}). \tag{32}$$

The left hand side of (32), together with (29), is equivalent to

$$B_c(X,X) = \pm\, \omega(X,X).$$

(20) and (21) yield

$$B_c(\hat{C}X,\hat{C}X) = -\, B_c(X,X) \ ,$$

$$\omega(\hat{C}X,\hat{C}X) = \omega(X,X) \ ,$$

so

$$B_c(\hat{C}X,\hat{C}X) = \mp\, \omega(\hat{C}X,\hat{C}X).$$

This, together with (29), is equivalent to the right hand side of (32). Thus, (32) is proved.

We have, therefore

$$P^{\pm}\hat{C} = \hat{C}P^{\mp} \ . \tag{33}$$

4) We prove: If $X \in \overline{T(S)}$, then $P^+X = 0$ implies $X = 0$. Similarly for P^-. That is $P^{\pm}|_{\overline{T(S)}}$ are linear injections.

$P^+X = 0$ implies $\hat{C}P^+X = 0$. Using (33) and (28), we obtain

$$P^-X = 0$$

Hence, $X = P^+X + P^-X = 0$, q.e.d.

5) We show that, if $X \in H^+$ and Re $X = 0$, then $X = 0$. Similarly for H^-. That is Re $|_{H^\pm}$ are linear injections.

The relations (27) imply

$$X = i \operatorname{Im} X ,$$

if Re $X = 0$. According to (29), we have

$$B_C (\operatorname{Im} X, \operatorname{Im} X) = \omega (\operatorname{Im} X, \operatorname{Im} X), \tag{34}$$

and from (17) it follows that

$$B_C(X,X) = 0 \quad \text{for all } X \in T(S) .$$

By continuity of $B_C(\cdot,\cdot)$, this must hold for all $X \in \overline{T(S)}$. However, Im $X \in \overline{T(S)}$, therefore

$$B_C (\operatorname{Im} X, \operatorname{Im} X) = 0$$

and (34) together with (22) yield the claim.

6) Finally, choose $X \in \overline{T(S)}$ and calculate

$$2 \operatorname{Re} P^\pm X = P^\pm X + \hat{C}P^\pm X = P^\pm X + P^\mp \hat{C}X = P^\pm X + P^\mp X = X ,$$

and this closes the proof of Theorem 2.

4. Generating Functionals

In the present section, we study some modifications of generating functionals which can be useful for curved spacetimes. We introduce well known notions only briefly; for more details see, e.g. Ref. 18. The formulas we obtain cannot be generalized to fermions. The way frequency splitting determines a generating functional and this, in turn, a state over $A(S)$ will be used in Sec. 5.

Choose a Cauchy hypersurface S. Let ϕ be a state over $A(S)$. Via Gel'fand-Naimark-Segal construction [19, 16], we obtain the representation $(\hat{\phi}, H)$ of $A(S)$ associated with ϕ. Here, H is a Hilbert space, $\hat{\phi}$ a *-injection

$$\hat{\phi} : \quad A(S) \to B(H)$$

and $B(H)$ the C*-algebra of all bounded operators in H. We call ϕ a regular state, if the projective representation

$$\hat{\phi} \circ W \ : \ T(S) \to B(H)$$

is a Weyl system. Clearly, if a state is regular over one of the algebras $A(S)$, then it is regular over all of them.

Regular states can be obtained from the so-called generating functionals:

Definition 6: Generating functional is a map $F : T(S) \to R^1$ with the properties

(i) $F(0) = 1$,

(ii) $F(tX)$ is continuous in t for any $X \in T(S)$,

(iii) for any N complex numbers $\lambda_1, \ldots, \lambda_N$ and N elements $X_1, \ldots, X_N$ of $T(S)$, it holds $\sum_{n,m=1}^{N} \exp(-\frac{1}{2} B_c(X_n, X_m))\, F(X_m - X_n)\, \lambda_n^* \lambda_m \geq 0$

If such a generating functional, F, is given, we can construct a state, ϕ, from F as follows. Let on the generators $W(X)$ of $A(S)$ ϕ have the values

$$\phi(W(X)) = F(X).$$

Then, ϕ can be extended by linearity to all finite linear combinations of $W(X)$'s. The resulting functional is linear by construction, normalized by condition (i) and positive by condition (iii) of Definition 6. Hence, it is continuous [19] and can be extended to the whole of $A(S)$. This extension is the desired state ϕ. The condition (ii) ensures that ϕ is regular.

We have the following

Theorem 3: Let $\omega(\cdot,\cdot)$ be a frequency splitting as defined in Sec. 3. Then, the map $F : T(S) \to R^1$ defined by

$$F(X) = \exp(-\frac{1}{2} B_c(P^+X, P^+X)) \tag{35}$$

for all $X \in T(S)$ is a generating functional.

The proof we give is not very straightforward, but its details will be useful later on. We construct a Fock representation of $A(S)$ with the help of $\omega(\cdot,\cdot)$ and show that the corresponding vacuum is the state defined by the generating functional $F(X)$. We do not claim any originality here; we just write well known thing over to ur notation (see, e.g. Ref. 20).

Proof: Denote by $S^p(H^+)$ the space of all symmetric tensors of rank p over H^+. The scalar product $B_c(\cdot\ \cdot) = \omega(\cdot,\cdot)$ on H^+ defines a scalar product $\langle\cdot|\cdot\rangle_p$ on $S^p(H^+)$, making it to a Hilbert space. Fock space $F(H^+)$ is the weak orthogonal sum of all these Hilbert spaces:

$$F(H^+) = C^1 \oplus_\perp S^1(H^+) \oplus_\perp \dots \oplus_\perp S^p(H^+) \oplus_\perp \dots .$$

Thus, the vectors

$$|\xi\rangle = |\xi_o, \dots, \xi_n, 0, \dots\rangle$$

with finite number of "components" $\xi_p \in S^p(H^+)$ form a dense linear subset of $F(H^+)$. On this subset, define the creation and annihilation operators $a^+(X)$ and $a(X)$ for any $X \in H^+$ as follows:

$$a^+(X)|\xi\rangle = |0, \xi_o X, \dots, \sqrt{n+1}\ \xi_n \otimes_S X, 0, \dots\rangle,$$

$$a(X)|\xi\rangle = |X \lrcorner \xi_1, \dots, \sqrt{n}\ X \lrcorner \xi_n, 0, \dots\rangle,$$

where $\lrcorner$ is the dual operation to the symmetrized tensor product $\otimes_S$:

$$\langle\xi_p | X \otimes_S \xi_{p-1}\rangle_p = \langle X \lrcorner \xi_p | \xi_{p-1}\rangle_{p-1}.$$

Hence:

$a^+(X)$, $a(X)$ are densely defined for any $X \in H^+$,

$a^+(X)$ is the adjoint operator to $a(X)$,

$a^+(X)$ is linear, $a(X)$ antilinear in the argument X.

Moreover, for any $X, Y \in H^+$, we have

(36)
$$[a(X), a(Y)] = [a^+(X), a^+(Y)] = 0,$$
$$\overline{[a(X), a^+(Y)]} = \omega(X,Y,) = B_c(X,Y),$$

where the bar denotes the closure of operators.

Define for any $X \in H$:

(37) $$\Phi(X) = a^+(P^+X) + a(\hat{C}P^-X) .$$

$\Phi(X)$ is densely defined, linear in its argument X and satisfies

$$\overline{[\Phi(x), \Phi(Y)]} = \omega(P^+\hat{C}X, P^+Y) - \omega(P^+\hat{C}Y, P^+X),$$

as direct computation using (36) and (37) gives. If we employ relations (21) and (31), the right-hand side can be reduced to

$$\overline{[\Phi(X),\ \Phi(Y)]} = B_c(\hat{C}X,\ Y)\ . \tag{38}$$

In particular, if $X \in T(S)$, then

$$\Phi(X) = a^+(P^+X) + a(P^+X) \tag{39}$$

is essentially self-adjoint (proof via Nelson theorem, see Ref. 16). Let us define its self-adjoint closure by the same symbol. Finally, define

$$W_1(X) = \overline{\exp(i\Phi(X))} \tag{40}$$

for all $X \in T(S)$. Each $W_1(X)$ is a unitary operator defined on the whole $F(H^+)$. The Baker-Hausdorff identity leads together with (28) and (38) to

$$W_1(X+Y) = W_1(X) \cdot W_1(Y) \cdot \exp(\tfrac{1}{2}\ B(X,Y\ )).$$

Hence, $(F(H^+),\ W_1)$ is a projective representation of $T(S)$ with multiplier $\exp(\frac{1}{2}\ B(X,Y))$. According to Theorem 1, the unitaries $W_1(X)$ generate a faithful representation of the abstract C*-algebra $A(S)$.

Introduce the notation

$$|\omega\rangle\ = |\ 1,\ 0,\ \ldots\ \rangle$$

for the vacuum in $F(H^+)$, and calculate

$$\omega(X_1(X)) = \langle\omega|W_1(X)|\omega\rangle$$

for any $X \in T(S)$. Using (36), (39), (40) and Baker-Hausdorff identity, we obtain

$$\langle\omega|\ W_1(X)\,|\omega\rangle = \langle\omega|\ \exp(ia^+(P^+X) + ia(P^+X))\,|\omega\rangle = \langle\omega|\ \exp(ia^+(P^+X))\ \cdot$$

$$\cdot\ \exp(ia(P^+X))\ \cdot\ \exp(-\ \tfrac{1}{2}\left[a(P^+X),\ a^+(P^+X)\right])\,|\omega\rangle = \exp(-\ \tfrac{1}{2}\ B_c(P^+X,\ P^+X)).$$

This shows that (35) is a generating functional and the state $\omega(\cdot) = \langle\omega|\ \cdot\ |\omega\rangle$ the corresponding state, q.e.d.

The notion of generating functional can be extended as follows. Let $\omega(\cdot,\cdot)$ and $\rho(\cdot,\cdot)$ be two frequency splittings such that

$$c\ \cdot\ \omega(X,X)\ <\ \rho(X,X)\ <\ C\ \cdot\ \omega(X,X) \tag{41}$$

for some constants c and C, $0 < c < C$, and for all $X \in T_c(S)$. We denote by index

(ρ) objects constructed by means of frequency splitting $\omega(\cdot,\cdot)$ ($\rho(\cdot,\cdot)$). Then, rom (41) we have

$$H_\omega = H_\rho ,$$

where the equality means equivalence of linear topological spaces.

If we consider $\omega(\cdot,\cdot)$ as defining positive frequencies for the incoming, $\rho(\cdot,\cdot)$ for the outgoing fields, then the condition (41) means, roughly speaking, the absence of infrared divergence [21].

Let us denote the linear topological space $H_\omega = H_\rho$ by H. Then, we have

Theorem 4: H is direct sum of linear spaces H_ω^+ and H_ρ^- :

(42) $$H = H_\omega^+ \oplus H_\rho^-$$

Proof: $H_\omega^\pm$ and $H_\rho^\pm$ are closed linear sub-spaces of H. Let us first show that

(43) $$H_\omega^+ \cap H_\rho^- = H_\omega^- \cap H_\rho^+ = \{0\},$$

where 0 means the null vector.

The relation (31) implies that $B_c(X,X) \geq 0$ for any $X \in H_\omega^+ \cup H_\rho^+$ and $B_c(X,X) \leq 0$ for any $X \in H_\omega^- \cup H_\rho^-$. Thus, $B_c(X,X) = 0$ for any $X \in H_\omega^+ \cap H_\rho^-$ or $X \in H_\omega^- \cap H_\rho^+$. However $B_c(\cdot,\cdot)$ restricted to H_ω^+ or to H_ρ^+ is positive definite ((31)). Hence, in both cases, $X = 0$.

It follows that H_1 defined by

(44) $$H_1 = H_\omega^+ \oplus H_\rho^-$$

is a closed linear sub-space of H. Let us look for a vector $X \in H$ satisfying

$$B_c(X,Y) = 0$$

for all $Y \in H_1$. (44) implies that X must be B_c-orthogonal on the whole of H_ω^+, and, therefore

$$X \in H_\omega^- .$$

Similarly, $X \in H_\rho^+$, but then, according to (43), $X = 0$.

Suppose that H_1 is a proper sub-space of H. Then, there is a non-zero vector, $Z \in H$, say, such that

$$\omega(Z,Y) = 0$$

for all $Y \in H_1$. Again, (44) implies that

$$Z \in H^-_\omega .$$

Hence, for any $Y \in H_1$, we obtain from (31)

$$B_c(Z,Y) = B_c(Z, P^-_\omega Y) = -\omega(Z, P^-_\omega Y) = -\omega(Z, P^-_\omega Y + P^+_\omega Y) = -\omega(Z,Y) = 0$$

This is a contradiction, so the theorem is shown.

From Theorem 4, it follows that any $X \in T(S)$ can be written uniquely as

(45) $$X = X_1 + X_2$$

where

(46) $$X_1 \in H^+_\omega \quad , \quad X_2 \in H^-_\rho .$$

X_1 and X_2 depend linearly on X. Warning:

$$X_1 \neq P^+_\omega X \quad , \quad X_2 \neq P^-_\rho X ,$$

unless $\rho(\cdot,\cdot) = \omega(\cdot,\cdot)$.

Let us now assume that the two states ρ and ω may be considered as lying in the same Fock space, say, $|\rho\rangle \in F(H^+_\omega)$. This is a stronger assumption than (41), see Ref. 21. Then, the matrix element

$$\langle\omega|\exp(i\Phi(x))|\rho\rangle$$

is well defined for all $X \in T(S)$, and we can calculate it using (45) and the linearity of $\Phi(X)$:

$$\langle\omega|\exp(i\Phi(X))|\rho\rangle = \langle\omega|\exp(i\Phi(X_1) + i\Phi(X_2))|\rho\rangle = \langle\omega|\exp(i\Phi(X_1) \cdot$$
$$\cdot \exp(i\Phi(X_2)|\rho\rangle \cdot \exp(\tfrac{1}{2}[\Phi(X_1), \Phi(X_2)]).$$

Relations (46), (37) and (38) imply

(48) $$\langle\omega|W_1(X)|\rho\rangle = \langle\omega|\rho\rangle \cdot \exp(\tfrac{1}{2} B_c(\hat{C}X_1,X_2)).$$

Thus, defining

(49) $$F_{\omega\rho}(X) = \langle\omega|\rho\rangle \cdot \exp(\tfrac{1}{2} B_c(\hat{C}X_1,X_2)),$$

we have a generalization of the formula (35), which is obtained from (49) if we set

$$F(X) = F_{\omega\omega}(X).$$

Of course, $F_{\omega\rho}(X)$ does not satisfy the conditions of Definition 6, and no states can be constructed from it. However, it contains information about matrix elements of field products between $\langle\omega|$ and $|\rho\rangle$. For example

$$\langle\omega|\Phi(X)|\rho\rangle = -\, i\left(\frac{\partial}{\partial t} F_{\omega\rho}(tX)\right)_{t=0} = 0$$

for any $X \in T(S)$. Thus, $|\omega\rangle$ contains only pairs of ρ-particles. Similarly,

$$\langle\omega|\Phi(X)\Phi(Y)|\rho\rangle = \frac{1}{2}\langle\omega|\rho\rangle\, B(X,Y) - \left[\frac{\partial^2 F(t_1X+t_2Y)}{\partial t_1 \partial t_2}\right]_{t_1=t_2=0}$$

etc.

The extended generating functional contains as much relevant information, as it is contained in the Bogoliubov transformation corresponding to the two frequency splittings $\omega(\cdot,\cdot)$ and $\rho(\cdot,\cdot)$.

5. Static Globally Hyperbolic Spacetimes

A spacetime (M,g) is said to be static, if it admits a one-dimensional Lie group G of isometries, whose trajectories are everywhere timelike and hypersurface orthogonal. If such a spacetime is globally hyperbolic at the same time, then spacelike hypersurfaces which are orthogonal to the group trajectories are Cauchy hypersurfaces [22].

If the condition of hypersurface orthogonality is dropped, we have the so-called stationary spacetimes. Most stationary spacetimes satisfying Einstein's equations will not be globally hyperbolic (cf. Ref. 23). However, we could replace the condition of stationarity by a weaker one of pseudo-stationarity [22, 23] and use our Assumption 2 in its general form. Thus, a more general treatment would be possible. In the present paper, we restrict ourselves to the static case, because it is sufficient for our purposes: to find examples of physical representations of localized field algebras.

In many cases, the full isometry group of (M,g) has more than one dimension, and there are more than one sub-group G of it making (M,g) static. It should be emphasized that one always means a definite one-dimensional group of isometries, if one speaks of (M,g) being static; a static spacetime is, in fact, a triple (M,g,G). Two such triples, (M_1,g_1,G_1) and (M_2,g_2,G_2), are equivalent, if there is an isometry $\iota\colon M_1 \to M_2$ such that

$$G_2 = \iota \circ G_1 \circ \iota^{-1}.$$

Hence, the triples (M,g,G_1) and (M,g,G_2) can be considered as the same, if G_1 and G_2 are conjugated sub-groups of the full isometry group.

It is interesting to observe that any static globally hyperbolic spacetime (M,g,G) is minimal in the following sense: there is no static globally hyperbolic spacetime (M_1,g_1,G_1) such that $M_1 \subset M$, $g_1 = g|_{M_1}$, $G_1 = G|_{M_1}$, unless $M_1 = M$. It follows that (M,g,G) is maximal in the same sense.

Let (M,g,G) be a static globally hyperbolic spacetime, and let S be a Cauchy hypersurface in M orthogonal to the trajectories of G. We can construct the algebra $A(S)$ and its dynamical automorphism α_t corresponding to $G = \{g_t\}$ as described in Sec. 2. Prof. Isham has shown in his contribution to this volume that then there is a unique Segal's complex structure satisfying the so-called ground state condition. This result can be reformulated for our frequency splittings as follows.

Theorem 4: Let (M,g,G) be a static globally hyperbolic spacetime, S, $A(S)$ and α_t be defined as above. Then, there is a unique frequency splitting $\omega(\cdot,\cdot)$ on $T_c(S)$ satisfying the conditions:

(i) the state ω corresponding to $\omega(\cdot,\cdot)$ is invariant:

$\omega(\alpha_t A) = \omega(A)$ for all $A \in A(S)$,

(ii) the spectrum condition is satisfied. That is, on the Fock space $F(H^+)$, α_t is implemented by unitary operator U_t for any t such that all U_t form a strongly continuous group. The generator of this group, Hamiltonian H defined by

$$H = \frac{\partial U_t}{\partial t}\Big|_{t=0}$$

is a self-adjoint operator with a non-negative spectrum.

If (M,g) is the flat spacetime and G a time-translation subgroup of Poincaré-group, one calls the representation of $A(S)$ given by Theorem 4 the physical representation. Let us generalize this notion to any static globally hyperbolic spacetime (M,g,G). Justification of calling this representation "physical" is not the formal analogy, but the existence of a reasonable gedankenexperiment, in which the particles defined by the structure of the Fock space $F(H^+)$ as well as their energy represented by operator H can be detected.

We propose a particular detection method based on Unruh box [12] following the group trajectories. (In fact, already in Unruh's original paper, the box is put on some group trajectories). According to our opinion, this is even the only sensible use of Unruh box; measurements with the box following, e.g. arbitrary geodesics must e of a rather chaotic character impossible to interpret. Let us describe the method n more details.

We place a box on a trajectory $g_t\, p$ of G through a point $p \in S$. The box is so mall that the "gravitational field" (acceleration) can be regarded as homogeneous nside it. It also is time independent. We put a non-relativistic Schrödinger particle $\Psi(x)$ into the box. The box walls are unpenetrable for $\Psi(x)$, so the particle has a well defined energy spectrum $\{E_n\}$ depending on the uniform acceleration on the curve $g_t\, p$. $\Psi(x)$ is weakly coupled to $\Phi(x)$, and the walls of the box are completely transparent to $\Phi(x)$. Thus, if Ψ is in its ground state $|E_o\rangle$ and $\Phi(x)$ is in a given state $|\Phi\rangle$ one can calculate the probability that $\Psi(x)$ jumps over into some higher level E_n pro unit time. (Of course, the box itself is not a classical device and it must be supplemented by some classical device able to read E_n).

In order to calculate the transition probabilities as Unruh has done in Ref. 12, the field $\Phi(x)$ has to be expanded in orthonormal modes which are periodic in group time t. Such modes are described in general case by Fulling in Ref. 10. Moreover, it must be shown that these modes diagonalize $H|_{H^+}$ as defined by Theorem 4.

Let us choose a Killing vector field ξ^k of G, and coordinates x^o, x^α, $\alpha = 1,2,3$, such that relation (2) reads

(52) $$x^o = 0, \quad x^\alpha = y^\alpha \quad , \quad \alpha = 1,2,3 \quad ,$$

and in addition,

(53) $$\xi^k = (1,0,0,0) \; .$$

Let

(54 $$\xi = \sqrt{g_{kl}\xi^k\xi^l} > 0$$

be the norm of ξ^k and let

$$\gamma = \mathrm{Det}(g_{\alpha\beta}) \quad , \quad g = \mathrm{Det}(g_{kl})$$

Then,

(55) $$n^k = \xi^{-1}\xi^k$$

and

(56) $$g = \xi^2\,\gamma \, .$$

The Klein-Gordon equation (1) is

(57) $|g|^{-1/2}\ \partial_k\ (|g|^{1/2}\ g^{kl}\ \partial_l\Phi) + m^2\ \Phi = 0$

This equation can be separated in the coordinates (x^0, x^α). If we put

(58) $\Phi(x^0, x^\alpha) = \psi(x^\alpha)\ \exp\ (\pm\ i\ E\ x^0)$

in (57), then an eigenvalue equation on S results of the form

$$K\,\psi = E^2\ \psi,$$

where

(60) $K = -\ \xi|\gamma|^{-1/2}\ \partial_\alpha\ (\xi|\gamma|^{1/2}\ \gamma^{\alpha\beta}\ \partial_\beta) + \xi^2\ m^2$,

K is symmetric differential operator on the space $C_0^\infty(S)$ of all C^∞ complex functions with compact support on S, if we introduce the auxiliary scalar product on $C_0^\infty(S)$ by

(61) $(\phi,\psi) = \int_S \xi^{-1}\ \phi^*\ \psi\ d\,S.$

The form $(\phi, K\,\psi)$ is positive [10].

Conjecture: On the Hilbert space resulting from $C_0^\infty(S)$ by Cauchy completion with respect to the auxiliary from (61), K is essentially self-adjoint.

It is the global hyperbolicity of (M,g) which makes this conjecture plausible. (According to a priviate communication by C. Isham, the conjecture is already proved by B. Kay.) If the conjecture is right, then we have a unique non-negative spectrum, $E^2(k)$, for K and a complete orthonormal system of generalized eigenfunctions $\psi(k, x^\alpha)$. Here, k is an abbreviation for a set of some indices, some discrete, some continuous. The space of all k is denoted by $\hat{S}$ and the summation over $\hat{S}$ by

$$\int_{\hat{S}} \mu(k)\ .$$

All these objects are well defined, according to the well known spectral theorem due to von Neumann (see, e.g. Ref. 24). We have

(62) $K\,\psi(k, x^\alpha) = E^2(k)\ \psi(k, x^\alpha)$

for all $k \in \hat{S}$, where E(k) is a non-negative function on $\hat{S}$, and

(63) $\int_S \xi^{-1}\ \psi^*(k', x^\alpha)\ \psi(k, x^\alpha)\ d\,S = \delta(k'-k)$,

where $\delta(k'-k)$ is an abbreviation for a product of Kronecker and Dirac δ's corresponding to $\hat{S}$. Moreover, from (62) and reality of K it follows that $\psi(k, x^\alpha)$ can be chosen real:

$$\psi^*(k, x^\alpha) = \psi(k, \alpha^\alpha) .$$

Any solution $\psi(x^0, x^\alpha)$ of (57) whose Cauchy data along S lies in $T_c(S)$ can be written as follows

$$\Phi(x^0, x^\alpha) = \int_{\hat{S}} \mu(k)\, \psi(k, x^\alpha)\, (A^{(+)}(k) \exp(iE(k)x^0) + A^{(-)}(k) \exp(-iE(k)x^0))$$

with uniquely given functions $A^{(+)}(k)$ and $A^{(-)}(k)$ on $\hat{S}$. The Cauchy data of $\phi(x^0, x^\alpha)$ on S are given by

$$h(y) = \Phi|_S \quad , \quad f(y) = n^k(y)\, \partial_k \Phi(x)|_S \quad ,$$

where n^k is defined by (55). Using (64), we obtain

(65)
$$h(y^\alpha) = \int_{\hat{S}} \mu(k)\, \psi(k, y^\alpha)\, (A^{(+)}(k) + A^{(-)}(k)) ,$$
$$f(y^\alpha) = i \int_{\hat{S}} \mu(k)\, \psi(k, y^\alpha)\, \xi^{-1}(y)\, E(k)\, (A^{(+)}(k) - A^{(-)}(k)) .$$

The inverse relations following from (63) are

(66)
$$A^{(+)}(k) = \frac{1}{2} \int_{S} dS\, \psi(k, y^\alpha)\, (\xi^{-1}(y^\alpha)\, h(y^\alpha) - i\, E^{-1}(k)\, f(y^\alpha)) ,$$
$$A^{(-)}(k) = \frac{1}{2} \int_{S} dS\, \psi(k, y^\alpha)\, (\xi^{-1}(y^\alpha)\, h(y^\alpha) + i\, E^{-1}(k)\, f(y^\alpha)) .$$

Let us denote by F the integral transformation defined by the right hand sides of (66). F is well defined on the whole of $T_c(S)$ and maps any element X of $T_c(S)$ on a pair, $\hat{X} = (A^{(+)}(k), A^{(-)}(k))$ of functions on S. Denote by $\hat{T}_c(S)$ the image of $T_c(S)$ by F. The form $B_c(\cdot,\cdot)$ on $T_c(S)$ defines a form $\hat{B}_c(\cdot,\cdot)$ on $\hat{T}_c(S)$ as follows

$$\hat{B}_c(\hat{X},\hat{Y}) = B_c(F^{-1}\hat{X}\, F^{-1}\hat{Y})$$

where F^{-1} is given by (65). Simple calculation yields

$$\hat{B}_c(\hat{X}_1,\hat{X}_2) = 2 \int_{\hat{S}} \mu(k)\, E(k)\, (A_1^{(+)*}(k)\, A_2^{(+)}(k) - A_1^{(-)*}(k)\, A_2^{(-)}(k)) .$$

Define the from $\hat{\omega}(\cdot,\cdot)$ on $\hat{T}_c(S)$ by

$$\hat{\omega}(\hat{X}_1, \hat{X}_2) = 2 \int_{\hat{S}} \mu(k)\, E(k)\, (A_1^{(+)*}(k)\, A_2^{(+)}(k) + A_1^{(-)*}(k)\, A_2^{(-)}(k)).$$

The formulas (67) - (69) imply immediately that $\omega(\cdot,\cdot)$ defined by

$$\omega(X,Y) = \hat{\omega}(FX, FY)$$

is a frequency splitting. We have to show next that $\omega(\cdot,\cdot)$ satisfies the conditions (i) and (ii) of Theorem 4, in order to justify the use of the modes $\psi(k, x^{\alpha}) \cdot \exp(iE(k)x^{0})$ in calculation of transition probability for Unruh's box. However, this is not difficult, if we use the following well known [25]

Theorem 5: A given state ω satisfies the conditions (i) and (ii) of Theorem 4, iff for any $A_1, A_2 \in A(S)$ there is a function $\lambda(z)$, analytic for Im $z < 0$, uniformly bounded for Im $z \leq 0$, continuous on the real axis and such that

$$\lambda(z) = \omega((\alpha_z A_1)A_2)$$

for all real z.

It is sufficient to show that the premises of Theorem 5 are satisfied for all A_1 and A_2 chosen just under the unitary generators of the algebra $A(S)$. To this end, we need the explicit form of α_t on $W(X)$ for any $X \in T(S)$.

Let $X \in T(S)$ and $FX = (A^{(+)}(k), A^{(-)}(k))$. As $\hat{C}X = X$, we have $A^{(-)}(k) = A^{(+)*}(k)$. Then it can easily be shown that

$$\alpha_t W(X) = W(F^{-1}(A^{(+)}(k) \exp(-iE(k)t), A^{(-)}(k) \exp(iE(k)t))$$

Utilizing the last equation, the Weyl relation ((ii) of Definition 1) to calculate

$$(\alpha_t W(X_1)) W(X_2) ,$$

the linearity of the NPLF ω, the values of ω on the generators $W(X)$ given by the generating functional $F(X)$, which is according to (35) defined by

$$F(X) = \exp\left(-\frac{1}{2}\omega(P^+X, P^+X)\right) ,$$

and, finally, the relations

$$P^+F^{-1}(A^{(+)}(k), A^{(-)}(k)) = F^{-1}(A^{(+)}(k), 0) ,$$

$$P^-F^{-1}(A^{(+)}(k), A^{(-)}(k)) = F^{-1}(0, A^{(-)}(k)) ,$$

one can prove the desired claim. We shall not give the rather lengthy details of this simple proof.

. Examples of Localized Field Algebras Having Physical Representations

As it is well known [26], vacuum Einstein's equations have no complete static solution except for the flat spacetime. Thus, either there are Killing horizons on singularities or both; we find many examples of one-dimensional isometry groups whose trajectories are timelike only in some proper, globally hyperbolic, sub-regions of the spacetimes. In such cases, we also can apply the results of the foregoing section: the field algebras localized to these sub-regions have physical representations. It seems that nothing similar exists, at least in general, for the global algebra.

Let us illustrate these claims on very simple examples where the field equation is explicitly solvable. We consider the massless scalar field in two-dimensional spacetimes. Penrose diagrams [27] of three particular cases are given in Figures 1, 2 and 3.

Example I shows that even in the Minkowski spacetime, G can be chosen so that its trajectories are timelike only in a sub-region: Let us take one of the mutually conjugated boost sub-groups of the Poincaré-group. The group G has a fixed point, p, and the two null geodesics, K_1 and K_2, through p form the two Killing horizons. There are two globally hyperbolic and static regions, $\mathcal{D}(S_1)$ and $\mathcal{D}(S_2)$. $S = S_1 \cup \{p\} \cup S_2$ is one of the global Cauchy hypersurfaces orthogonal to the group trajectories.

Example II is the u-v-surface in the Kruskal manifold. The full isometry group is one-dimensional in this case. The Killing horizons K_1 and K_2 coincide with the bifurcate hypersurface $r = 2\,M$ (Schwarzschild radius), where M is the mass parameter of the Schwarzschild family.

Example III is the two-dimensional de Sitter spacetime. The full isometry group is three-dimensional and all its one-dimensional sub-groups are mutually conjugated, so we choose just one of them for G. G has two fixed points, p_1 and p_2, which are "antipodal" to each other, and four Killing horizons given by the four null geodesics K_1, K_2, K_3, K_4 through p_1 and p_2. A global Cauchy hypersurface $S = S_1 \cup \{p_1\} \cup S_2 \cup \{p_2\}$ is orthogonal to the group trajectories; S is compact, homeomorph to S^2.

In all these cases, there is an invariant way to choose the group parameter t uniquely. The action of G on K_i is given in general as follows (see Ref. 28). Let λ_i be an affine parameter along K_i increasing in the future direction and such that $\lambda_i(p) = 0$ ($\lambda_i(p_1) = \lambda_j(p_2) = 0$ in de Sitter case. Then, for any group parameter t, there is a constant c_i such that

$$\lambda_i(g_t q) = \exp((-1)^{i+1} c_i t) \cdot \lambda_i(q)$$

for all $q \in K_i$. There is exactly one t that makes all $c_i = 1$. We denote the corresponding Killing vector by ξ^k; its orientation is indicated by arrows in the Figures 1, 2 and 3.

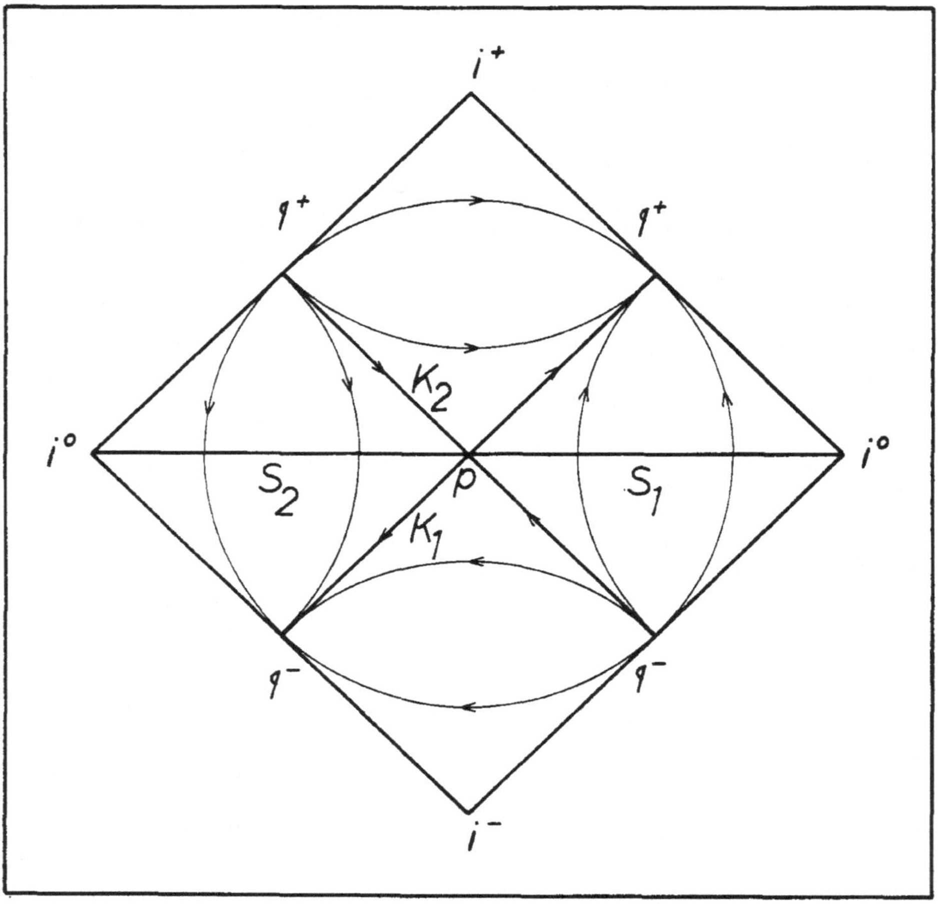

Fig. 1: Two-dimensional Minkowski spacetime with boosts about p as the group G. The fixed point p, the Killing horizons K_1, K_2, and some trajectories of G are shown, as well as a Cauchy hypersurface orthogonal to the trajectories.

Given a static observer O through a point $q \in \mathcal{D}(S_1)$, we have in all three cases:

$$\tau(g_t q) - \tau(q) = \xi(q) \cdot t ,$$

where ξ is defined by (54) and τ is the proper time of O. In Examples I and II, the asymptotic observers are distinguished by zero acceleration; denoting the corresponding ξ's by $\xi_g^{I,II}$, we have

$$\xi_g^{I} = \infty \quad , \quad \xi_g^{II} = 4\,M .$$

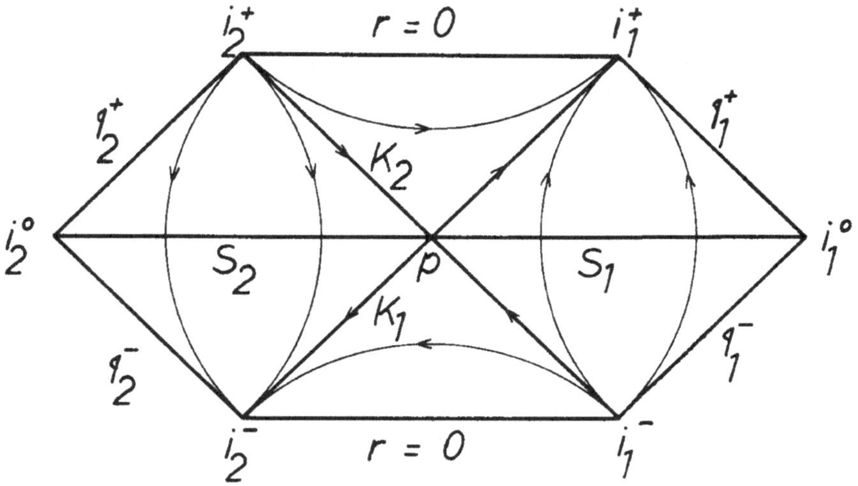

Fig 2: Two-dimensional Kruskal manifold. The fixed point p, the Killing horizons K_1 and K_2, and some trajectories of the isometry group G are shown, as well as a Cauchy hypersurface orthogonal to the group trajectories.

In Example III, there even exists a geodesic observer in $\mathcal{D}(S_1)$ and we have for his ξ,

$$\xi_g^{III} = R ,$$

where R is the radius of de Sitter spacetime.

In the regions $\mathcal{D}(S_1)$ and $\mathcal{D}(S_2)$, we can apply the constructions of the preceding sections to obtain the algebras $A(S_1)$, $A(S_2)$ and the physical representations for them. We have then well-defined vacui, ω_1 and ω_2, and detectable n-particle states with positive energy represented by Hamiltonians H_1 and H_2, all in regions $\mathcal{D}(S_1)$ and $\mathcal{D}(S_2)$.

Given a state, ϕ, on $A(S)$ (global state), we use the injections $I_1 : A(S_1) \to A(S)$ and $I_2 : A(S_2) \to A(S)$ to define the projections,

$$\phi_1 = I_1' \phi, \quad \phi_2 = I_2' \phi ,$$

of ϕ to $A(S_1)$ and $A(S_2)$. Measurable properties of these projections can be studied with the help of our observables in $\mathcal{D}(S_1)$ and $\mathcal{D}(S_2)$.

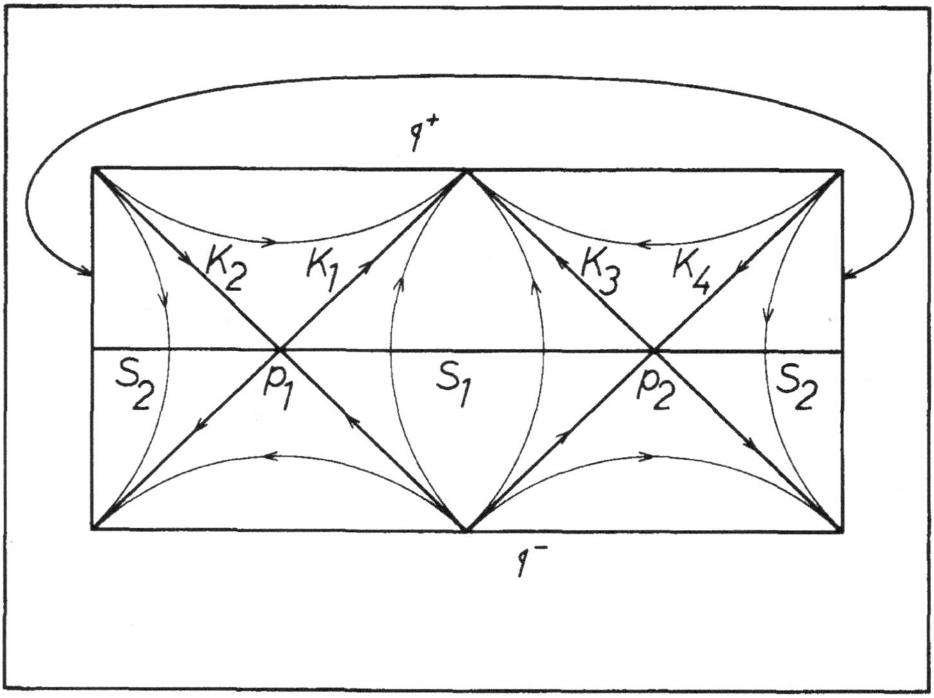

Fig. 3; Two-dimensional de Sitter anifold. The extreme left and right vertical lines should be identified. In this case, two fixed points, p_1, and p_2 and four Killing horizons, $K_1, \dots, K_4$, are present.

In Example I, the total algebra $A(S)$ has physical representation, based on the usual flat spacetime vacuum ω. Very interesting global states have been constructed for Example II by Hartle and Hawking [29], and for Example III by Gibbons and Hawking [13]. Their construction has been generalized by Israel [30] to any spacetime with bifurcate Killing horizon for which the two regions, $\mathcal{D}(S_1)$ and $\mathcal{D}(S_2)$, are isometric. It turns out that Israel's construction leads to ω in case of Example I. Hence, the Israel states are in some formal aspect analogous to flat spacetime vacuum. However, in general case like Examples I and II, there is no well defined global positive Hamiltonian, which would be minimized by the Israel states. In addition, it is not clear how such Israel states can be prepared: in both Examples, this would require organized action of observers in two causally disconnected regions.

Nevertheless, given such a state, our localized observables can be measured, at least in principle. These gedankenexperiments give remarkable results: for any static observer O, the Israel states look as thermal equilibriums of boson gas with temperature T_0 given by

$$T_0 = \frac{1}{2\pi\xi(0)} \quad ,$$

n our algebraic language, the projection of any Israel state satisfies the Kubo-artin-Schwinger conditions [25] with $\beta = 2\pi$. The Israel state is pure (extremal) and ts projection is mixed just in the very peculiar way to give Gibbs' distribution.

References

1. R. Rajaraman: Phys.Letters C21, 227 (1975).
2. B. DeWitt; Phys.Letters C19, 295 (1975)
3. R.M. Wald: Commun.Math.Phys. 45, 9 (1975).
4. P. Hajicek: Acta Phys.Austriaca, Suppl. XVIII, 835 (1977).
5. N. Bohr, L. Rosenfeld: Det.Kgl.Dan.Vid.Selskab.Math.-fys.Med. 12, No 8 (1933).
6. I.E. Segal: Det.Kgl.Dan.Vid.Selskab.Mat.-fys.Med. 31, No 12 (1959), Can.J.Math. 13, 1 (1961).
7. J. Manuceau: Ann.Inst.H.Poincaré 8, 139 (1968).
8. J. Slawny: Commun.Math.Phys. 24, 151 (1972).
9. G. Gibbons: Commun.Math.Phys. 44, 245 (1975).
10. S.A. Fulling: Phys.Rev. D7, 2850 (1973).
11. C. Isham: private communication.
12. W.G. Unruh: Phys.Rev. D14, 870 (1976).
13. G.W. Gibbons, S.W. Hawking: Phys.Rev. D15, 2738 (1977).
14. S.W. Hawking, G.E.R. Ellis: The Large Scale Structure of Space-Time. University Press, Cambridge, 1973.
15. S.A. Fulling: Thesis, Princeton University.
16. B. Simon: in "Mathematics of Contemporary Physics". Ed. by R.F. Streater. Academic Press, London, 1972.
17. A.L. Licht: J.Math.Phys. 7, 1956 (1966).
18. P.J.M. Bongaarts: in "Mathematics of Contemporary Physics". Ed. by R.F. Streater Academic Press, London, 1972.
19. M.A. Neumark: "Normierte Algebren". VEB, Berlin, 1959.
20. J.M. Cook: Trans.Amer.Math.Soc. 74, 222 (1953).
21. P. Hajicek: Phys.Rev. D15, 2757 (1977).
22. B. Carter: GRG J., to be published.
23. B. Carter: J.Math.Phys. 10, 70 (1969).
24. K. Maurin: "Methods of Hilbert Spaces". Polish Scientific Publishers, Warszawa, 1972.
25. N.M. Hugenholtz: in "Mathematics of Contemporary Physics". Ed. by R.F. Streater. Academic Press, London, 1972.
26. A. Lichnerowicz: "Théories Relativistes de la Gravitation et de l'Electromagnétisme". Masson, Paris, 1955.

27. R. Penrose: in "Relativity, Groups and Topology", Ed. by C. DeWitt and B. DeWitt Gordon and Breach, New York, 1964.

28. R.H. Boyer: Proc.Roy.Soc. A311, 245 (1969).

29. J.B. Hartle, S.W. Hawking: Phys.Rev. D13, 2188 (1976).

30. W. Israel: Phys.Letters 57A, 107 (1976).

QUANTIZATION OF FIELDS ON A CURVED BACKGROUND

by P. Spindel
Mons State University, 7000 MONS - BELGIUM

In 1960, Professor Lichnérowicz wrote a procedure to quantize the fields generalizing very naturally the methods used in the Minkowski space-time framework [1]. His principle has been discussed during this meeting by Professor Isham, so we will limit ourselves by illustrating it in a heuristic way and by proposing an extention whose physical implications seem interesting to us.

The quantum fields $\hat{\phi}$ are defined as operator valued distributions acting on a linear manifold, dense in a Hilbert space. The dynamics of the fields is deduced from classical theory by asking that they be distribution solutions of the motion equations supposed of linear hyperbolic type. The works of Choquet-Bruhat, Leray and Lichnérowicz [2,3,4] show that on a globally hyperbolic manifold such equations always admit one and only one elementary solution $\Delta(x,x')$ whose support is contained in and on the x' light cone and furnish - in a neighbourhood of any Cauchy surface Σ - the distribution solution of a Cauchy problem on the surface by a composition with the initial data according to the tensorial nature of the fields [4].
More particulary, let us consider a scalar field whose motion equations are given by a self-adjoint differential operator L :

$$L(\phi) \equiv (\Box + m^2)\phi = 0 \tag{1}$$

In this case the elementary solution Δ - difference between the retarded and advanced solutions of the inhomogenous equation associated to (1) - is skewsymmetrical. The solution of the Cauchy problem on Σ is given by :

$$\phi(x) = \int_\Sigma \left\{ \phi(x') \nabla_\lambda \Delta(x,x') - \Delta(x,x') \nabla_\lambda \phi(x') \right\} d\sigma^\lambda_{x'} \equiv \phi * \Delta \tag{2}$$

and the quantization postulate :

$$[\hat{\phi}(x), \hat{\phi}(x')] = i\hbar \Delta(x,x') \tag{3}$$

has still a meaning.
Nevertheless the knoweldge of Δ is not sufficient to quantize. Moreover it is necessary to define the space of states and the action of the fields on it; or equivalently to give a rule of decomposition of the fields in negative and positive frequency parts.
In this goal Lichnérowicz [1] postulates the existence of a real symmetrical distribution kernel $\Delta_1(x,x')$, solution of equation (1) and filling the following composition relation :

$$\Delta(x,x') = \int_\Sigma \left\{ \Delta_1(x,y) \nabla_\lambda \Delta_1(x',y) - \Delta_1(x',y) \nabla_\lambda \Delta_1(x,y) \right\} d\sigma^\lambda_y \equiv \Delta_1 * \Delta_1 \tag{4}$$

which allows him to build two projectors :

$$\Delta^{\oplus} = \frac{1}{2}\left(\Delta - i\Delta_1\right) \qquad \text{and} \qquad \Delta^{\ominus} = \frac{1}{2}\left(\Delta + i\Delta_1\right) \tag{5}$$

This leads to the frequency decomposition by :

$$\phi^{\oplus} = \phi * \Delta^{\oplus} \qquad \text{and} \qquad \phi^{\ominus} = \phi * \Delta^{\ominus} \tag{6}$$

and to the construction of the Fock space if the composition :

$$\phi * \phi_1 = \| \phi \|^2 \qquad \text{with} \qquad \phi_1 = \phi * \Delta_1 \tag{7}$$

defines a norm on the space of solutions of (1).

These distribution kernels Δ and Δ_1 can be locally build from a complete system of solutions $\{u_A, u_A^*\}$ such as :

$$u_A * u_B = -i\,\delta_{AB} = u_B^* * u_A^* \quad ; \quad u_A^* * u_B = 0 \tag{8}$$

One can then build two kernels :

$$\Delta_u(x,x') = i\, S_A \left\{ u_A(x)\, u_A^*(x') - u_A(x')\, u_A^*(x) \right\} \tag{9}$$

$$\Delta_{1u}(x,x') = S_A \left\{ u_A(x)\, u_A^*(x') + u_A(x')\, u_A^*(x) \right\} \tag{10}$$

The first is to be identified with the kernel Δ while the second one fulfills all the Δ_1 properties. However if one changes the system of solutions this last one is not invariant. So if one sets :

$$v_A = S_B \left\{ c(A,B)\, u_B + d(A,B)\, u_B^* \right\} \tag{11}$$

the kernels built with this new system - supposed to fulfil analogous relations to (8) - are such that:

$$\Delta_v = \Delta_u \qquad \text{but} \qquad \Delta_{1v} = \Delta_{1u} \qquad \text{iff} \quad d(A,B) \equiv 0 \tag{12}$$

Of course we have no a priori criterium to select a choice of modes more than another. Some authors even claim that this choice is unessential and matter of convenience [5]. Yet if the space admits a timelike Killing vector field ξ - if the space is stationary - then it is possible to fix univoquely the kernel Δ_1 by asking its invariance for translations along the ξ integral curves or in other terms by asking that :

$$(\mathcal{L}_\xi \phi) * \Delta_{1u} = \mathcal{L}_\xi (\phi * \Delta_{1u}) \tag{13}$$

where $\mathcal{L}_\xi$ is the Lie derivative along ξ .

This prescribes to the modes u_A to satisfy :

$$\mathcal{L}_\xi u_A = -i\,\kappa(A)\, u_A \quad ; \quad \kappa(A) > 0 \tag{14}$$

in the basis where Δ_1 is diagonal [6].

It is the condition (13) which is the foundation of the frequency decomposition on the Minkowski space that allowed Chevalier and Moréno [6,7] to build - in a rigorous way - a kernel Δ_1 , unique, on a stationary space-time.

f space-time is not stationary we can try to generalize the method as following. Let us chooce a congruence of timelike curves - with ξ as tangent vector - and isochronous sections realised with a family of Cauchy hypersurfaces $\{\Sigma_t\}$.

If ϕ is a solution of the Klein-Gordon equation (1) it is entirelly determined by its Cauchy data on a hypersurface $\Sigma \in \{\Sigma_t\}$. On the space $\mathcal{D}_\Sigma$ of C^∞ Cauchy data with compact support on Σ - with sufficient regularity conditions for Σ, ξ and the metric - we can define a sesquilinear form starting from the conserved current associated to (1) :

$$\psi * \phi = \int_\Sigma \left(\psi^* \phi_{;\alpha} - \psi^*_{;\alpha} \phi \right) d\sigma^\alpha \tag{15}$$

and define a scalar product starting from the energy-momentum tensor :

$$\langle \phi, \psi \rangle = \int_\Sigma \left\{ \psi^*_{;\alpha} \phi_{;\beta} + \phi_{;\alpha} \psi^*_{;\beta} - g_{\alpha\beta} \left[\psi^*_{;\rho} \phi^{;\rho} - m^2 \psi^* \phi \right] \right\} \xi^\alpha d\sigma^\beta \tag{16}$$

In this way $\mathcal{D}_\Sigma$ receives a prehilbertian structure. By completion we obtain a Hilbert space H_Σ admitting $\mathcal{D}_\Sigma$ as dense linear submanifold.

By writing (15) and (16) we misuse terms, identifying the fields to their Cauchy data :

$$\phi = \begin{pmatrix} \varphi \\ \pi \end{pmatrix} \tag{17}$$

where φ and π represent the restriction of ϕ to Σ and its normal derivative on Σ. In addition to the regularity conditions on Σ, ξ and the metric, we assume the existence of two strictly positive numbers a and b such that : $a < \xi . n < b$

So the convergence for the topology of H_Σ implies the convergence in the topological space $L^2(\Sigma) \oplus L^2(\Sigma)$ and so the continuity of the composition (16).

Let us write ${}_\Sigma\mathcal{L}_\xi$ the linear operator defined on the domain $\mathcal{D}_\Sigma \subset H_\Sigma$ by :

$${}_\Sigma\mathcal{L}_\xi \phi = \begin{pmatrix} \mathcal{L}_\xi \phi \\ \frac{n_\alpha}{\sqrt{-g}} \mathcal{L}_\xi \sqrt{-g}\, \phi^{;\alpha} \end{pmatrix}_\Sigma \quad ; \left(n_\alpha \div \partial_\alpha \Sigma \; ; n_\alpha n^\alpha = 1 \right) \tag{18}$$

One verifies, integrating by parts that :

$${}_\Sigma\mathcal{L}_\xi \phi * \psi = \langle \psi, \phi \rangle \qquad \forall \phi \in \mathcal{D}_\Sigma \tag{19}$$

In fact this operator is the natural generalization of the Lie derivative acting on the solutions of (1) when ξ is a Killing vector, in which case (16) and (18) become independant of the choice of Σ.

<u>Theorem I.</u>

${}_\Sigma\mathcal{L}_\xi$ is inversible on $\mathcal{D}_\Sigma$.

Proof :

${}_\Sigma\mathcal{L}_\xi$ is injective on $\mathcal{D}_\Sigma$ because it maps $\mathcal{D}_\Sigma$ in $\mathcal{D}_\Sigma$ and if ${}_\Sigma\mathcal{L}_\xi \phi = {}_\Sigma\mathcal{L}_\xi \psi$ we deduce

$$0 = {}_\Sigma\mathcal{L}_\xi (\phi - \psi) * (\phi - \psi) = \| \phi - \psi \|^2 \implies \phi = \psi$$

${}_\Sigma\mathcal{L}_\xi$ is surjective on $\mathcal{D}_\Sigma$ because the equation ${}_\Sigma\mathcal{L}_\xi \Omega = \phi$ can be writen explicitly, with $\Omega = \begin{pmatrix} \omega \\ \sigma \end{pmatrix}$:

$$\sigma = \frac{1}{\xi . n} \left(\psi - \xi_\alpha h^{\alpha\beta} \omega_{;\beta} \right) \tag{A}$$

$$\xi . n \left[h^{\alpha\beta} (h_\alpha{}^\gamma \omega_{;\gamma})_{;\beta} + \frac{\xi^\alpha}{\xi . n} h_\alpha{}^\beta \left(\frac{\xi^\gamma}{\xi . n} h_\gamma{}^\delta \omega_{;\delta} \right)_{;\beta} + m^2 \omega \right]$$

$$+ \left[\xi^\alpha h_\alpha{}^\beta n^\gamma{}_{;\beta} h^\delta{}_\gamma + \frac{1}{\xi . n} \left(\xi^\alpha{}_{;\alpha} - \xi . n \, h^{\alpha\beta} n_{\alpha;\beta} - n^\alpha n^\beta \xi_{\alpha;\beta} \right) \xi^\gamma h_\gamma{}^\delta + \right.$$

$$\left. + \; n^\alpha \xi_{\alpha;\beta} h^{\beta\delta} \right] \omega_{;\delta} = \frac{1}{\xi . n} \left(\xi^\alpha{}_{;\alpha} - \xi . n \, h^{\alpha\beta} n_{\alpha;\beta} - n^\alpha n^\beta \xi_{\alpha;\beta} \right) \psi +$$

$$+ \; \xi^\alpha h_\alpha{}^\beta \left(\frac{\psi}{\xi . n} \right)_{;\beta} - \pi \tag{B}$$

with : $h_{\alpha\beta} = g_{\alpha\beta} - n_\alpha n_\beta$ the induced metric on Σ .
The equation (B) is an inhomogenous elliptic partial derivative equation whose quadratical form is negative definite and whose coefficient of the function ω is positive. This equation admits one solution in $\mathcal{D}_\Sigma$ which gives σ by (A). This defines Ω .

Theorem II.

${}_\Sigma\mathcal{L}_\xi$, $[{}_\Sigma\mathcal{L}_\xi^{-1}]$, is skewsymmetric.

Proof :

$$< \phi, {}_\Sigma\mathcal{L}_\xi \psi > = < {}_\Sigma\mathcal{L}_\xi \psi, \phi >^* = ({}_\Sigma\mathcal{L}_\xi \phi * {}_\Sigma\mathcal{L}_\xi \psi)^* = - ({}_\Sigma\mathcal{L}_\xi \psi * {}_\Sigma\mathcal{L}_\xi \phi) = - < {}_\Sigma\mathcal{L}_\xi \phi, \psi >$$

Theorem III.

$i_\Sigma\mathcal{L}_\xi$, $[-i_\Sigma\mathcal{L}_\xi^{-1}]$, is essentially self-adjoint.

Proof :

Let ${}_\Sigma\mathcal{L}_\xi^*$ be the adjoint of ${}_\Sigma\mathcal{L}_\xi$. It exists because $\mathcal{D}_\Sigma$ is dense. If ψ is a solution of one of the equations : ${}_\Sigma\mathcal{L}_\xi^* \psi = \pm \psi$ there exists a sequence of elements ϕ_n belonging to $\mathcal{D}_\Sigma$ (dense in H_Σ) converging to ψ such that:

$$\forall \varepsilon > 0 \quad \exists \; N(\varepsilon) \in \mathbb{N} \; \S \; \forall n > N(\varepsilon) : \; | \psi * \phi_n | < \varepsilon \; , \; \| \psi - \phi_n \| < \varepsilon$$

In this case we deduce :

$$| \phi_n * {}_\Sigma\mathcal{L}_\xi^* \psi | = | < {}_\Sigma\mathcal{L}_\xi^* \psi , {}_\Sigma\mathcal{L}_\xi^{-1} \phi_n > | = | < \psi, \phi_n > | < \varepsilon$$

hereby $\| \psi \| = 0$, which means there is no other solution than the trivial one.
By consequence the operator $-{}_\Sigma\mathcal{L}_\xi^{-1} {}_\Sigma\mathcal{L}_\xi^{-1}$ is hermitian and positive. It can be extended univoquely to a positive self-adjoint operator which admits a unique positive square (fourth) root.
The operator :

$$ {}_\Sigma\Delta^1_\xi = - \left(- {}_\Sigma\mathcal{L}_\xi^{-1} {}_\Sigma\mathcal{L}_\xi^{-1} \right)^{\frac{1}{2}} {}_\Sigma\mathcal{L}_\xi \tag{20}$$

as the following properties :

) ${}_{z}\Delta^{1}_{\xi}$ commutes which ${}_{z}\mathcal{L}_{\xi}$

i) $({}_{z}\Delta^{1}_{\xi})^{2} = -\mathbb{1} = -{}_{z}\Delta$

ii) $\phi * {}_{z}\Delta^{1}_{\xi}\,\phi = \phi * \left[-(-{}_{z}\mathcal{L}^{-1}_{\xi}\,{}_{z}\mathcal{L}^{-1}_{\xi})^{\frac{1}{4}}(-{}_{z}\mathcal{L}^{-1}_{\xi}\,{}_{z}\mathcal{L}^{-1}_{\xi})^{\frac{1}{4}}\,{}_{z}\mathcal{L}_{\xi}\,\phi\right] = \|(-{}_{z}\mathcal{L}^{-1}_{\xi}\,{}_{z}\mathcal{L}^{-1}_{\xi})^{\frac{1}{4}}\phi\|^{2} \geq 0$

<u>Theorem IV.</u>

${}_{z}\Delta^{1}_{\xi}$ is univoquely determined by these properties.

Proof [6] :

If ${}_{z}\Lambda^{1}_{\xi}$ is an operator fulfilling these properties, ${}_{z}\Lambda^{1}_{\xi}$ commutes with ${}_{z}\Delta^{1}_{\xi}$. Let us set $\psi = ({}_{z}\Lambda^{1}_{\xi} - {}_{z}\Delta^{1}_{\xi})\phi$ then :

$$\psi * ({}_{z}\Lambda^{1}_{\xi} + {}_{z}\Delta^{1}_{\xi})\psi = 0 = \psi * {}_{z}\Lambda^{1}_{\xi}\,\psi + \psi * {}_{z}\Delta^{1}_{\xi}\,\psi$$

By consequence ψ is nul because each term in the last sum of the equalities is positive definite. Hereby ${}_{z}\Lambda^{1}_{\xi}\phi = {}_{z}\Delta^{1}_{\xi}\,\phi$.

The first quantization method mentioned above is arbitrary in the choice of the kernel Δ_1 . On the other side as this kernel is a solution of motion equations the quantum states built by it are stable in the acceptation that the existence of a gravitational field even not stationary does not lead to the spontaneous creation of particles. Eventually the existence of an isometry group on the manifold allows to reduce the ambiguity in the definition of Δ_1, but if a timelike Killing vector field does not exist we cannot conclude definitively [8]. The second method is equivalent to the hamiltonian diagonalization technique [9,10,11]. It specifies a class of observers by the choice of $\{\Sigma_t\}$ and ξ and reduces itself to the first one when ξ is a Killing vector. Here the theory predicts creations of pairs, but generally in infinite number which does not allow to build the Fock space anymore. The own definition of particles depends of the observers. So, even on the Minkowski space, an accelerated observer detects particles where an inertial one does not see any [12], the link between the different choices is realised by a Bogoliubov transform [13].

So in the framework of spaces admitting a 3-parameters abelian group of spatial isometries, by choosing adapted coordinates such that the metric can be written :

$$g = dx^{0} \otimes dx^{0} - g_{ij}(x^{0})\, dx^{i} \otimes dx^{j}$$

the Cauchy surfaces be : $\{x^{0} = t\}$ and the vector field be : $\xi = n = \frac{\partial}{\partial x^{0}}$

the Fourier transform of ${}_{t}\Delta^{1}_{\xi}$ is given by :

$$ {}_{t}\tilde{\Delta}^{1}_{\xi} = \begin{pmatrix} 0 & -\omega^{-1}_{\vec{k}}(t) \\ \omega_{\vec{k}}(t) & 0 \end{pmatrix}$$

with : $\omega_{\vec{k}}(t) = \left(g^{ij}(t)\, k_i k_j + m^2 \right)^{1/2}$.

We can deduce from it a family of kernels ${}_t\Delta^1_\xi$ depending of the surfaces $x^0 = t$, solution of (1) for the Cauchy problem :

$$ {}_t\Delta^1_\xi (x,x')\big|_{x^0 = x'^0 = t} = -(2\pi^2)^{-1} \frac{\partial}{\partial r} K_0(mr) \; ; \quad \frac{\partial}{\partial x^0}\, {}_t\Delta^1_\xi (x,x')\big|_{x^0 = x'^0 = t} = 0 $$

where : $r^2 = g_{ij}(t)\, (x^i - x'^i)(x^j - x'^j)$.

To make an end let us draw that the method can be extended to the spinor case by inverting the roles played by current and energy [13].

REFERENCES.

1. A. Lichnérowicz, Propagateurs et quantification en relativité générale, Conf. int. sur les théories de la gravitation, (1962), Warszawa.
2. Y. Bruhat, Solutions élémentaires d'équations du second ordre de type quelconque, Coll. int. C.N.R.S., (1956), Nancy.
3. J. Leray, Hyperbolic differential equations, (1952), Princeton.
4. A. Lichnérowicz, Propagateurs et commutateurs en relativité générale, Publication I.H.E.S., 10 (1961), Paris.
5. L. parker, S. Fulling, Quantized matter fields and the avoidance of singularities in general relativity, Phys. Rev. D, 7, 2357, (1973).
6. M. Chevalier, Opérateurs de création et annihilation sur un espace-temps stationnaire, J. Math. Pures et Appliquées, L. 53, (1974).
7. C. Moréno, Thèse d'Etat, (1975).
8. C. Schomblond, P. Spindel, Conditions d'unicité pour le propagateur Δ_1 du champ scalaire dans l' univers de de Sitter, Ann. I.H.P., XXV 1, 67, (1976), Propagateurs des champs spinoriels et vectoriels dans l' univers de de Sitter, Bull. Classe des Sciences, 5, LXII, (1976).
9. A. Magnon, Thèse, (1976).
10. A. Astekar, A. Magnon, Quantum fields in curved space-times, Proc. Roy. Soc. London, A 346, (1975).
11. A. Grib, S. Mamayev, V. Mostepanenko, Particle creation from vacuum in homogenous isotropic models of the universe, G.R.G. 7, 6, (1976).
12. W. Troost, H. Van Dam, Thermal effects for an accelerating observer, Preprint Rijksuniversiteit Utrecht, Netherlands.
13. P. Spindel, Quantification du champ scalaire dans un champ de gravitation extérieur non stationnaire, Bull. Classe des Sciences, 5, LXIII, (1977).
14. This idea seems to originate in the works of L. Parker, see for example Phys. Rev. 183, 1057, (1969).

SUPERGRAVITY*

S. Deser
Department of Physics
Brandeis University
Waltham, Massachusetts 02154

January 1978

ABSTRACT

Supergravity, a new fermionic gauge theory unifying gravitation with spin, is described and some of its achievements and problems assessed. The theory is governed by a vector field whose values lie in a graded Lie algebra rather than the usual Poincaré algebra, and corresponds to a Dirac square root of general relativity.

Based on a lecture delivered at the conference on Differential Geometric Methods in Mathematical Physics, Bonn 1977.

*Supported in part by NSF Grant PHY 76 07299 A01

INTRODUCTION

Supergravity [1, 2] is not much more than a year old, but already possesses a vast literature. We cannot attempt any complete survey here, but only describe some of the features which make it an exciting new candidate among the gauge fields that seem to play so fundamental a role in current theoretical physics (and perhaps also in nature). At its least ambitious, this theory is a unification of Einstein's theory of gravitation with matter, particularly of half-integer spin (fermions), totally different from any Einstein ever dreamt of. At its most extreme, it seeks nothing less than to incorporate into a single supermultiplet all fundamental particles of spin two and lower, including both the traditional vector gauge fields (e.g. photons) and the scalars and spinors which are not gauge fields at all. We shall discuss some of the developments it has taken and the problems it faces, but concentrate primarily on those of its formal properties and underlying ideas which are well understood, such as its relation to the graded Lie algebras which have been discussed in Kac's beautiful lecture at this conference. We shall also sketch another very interesting way to understand the theory, namely as the natural Dirac square root of general relativity, in a very precise sense [3].

The unification of gravity with matter means that the traditional "no-go" theorems forbidding unification of internal symmetries with space-time properties have been circumvented. This is due precisely to gauging the graded extensions rather than the Poincaré algebra itself. There are deep conceptual differences involved here, because the new (fermionic) gauge invariance of the unified theory which mixes spacetime with matter implies that the former is no longer the invariant (albeit dynamical) arena of Einstein's theory, but that its structure is affected by choice of gauge. This new invariance is, however, essential to the theory since we shall see that it guarantees the consistency of the gravity-matter coupling, which had hitherto been unattainable for the spin 3/2 field companion of the graviton.

Work on supergravity models has proceeded in a number of directions, of which we will mention the following. One, closely related to the original motivation underlying the theory, has been its ultraviolet behavior (renormalizability) where traditional Einstein theory appears to fail very badly. Another has sought to find various generalizations of the original theory to include more particles into the gauge supermultiplet as well as coupling of the latter to [supersymmetric] matter. [For an interesting class of models involving an even more radical fusion of spacetime and spinorial variables, the so-called supergauge theories, see [4]]. Also, as always in physics, the original degree of symmetry of the basic models is too high to be realistic and symmetry breaking has to be considered. Understanding of the dynamical content of a theory, and therefore of its quantum properties always requires canonical Hamiltonian analysis; since supergravity is both coordinate and fermionic gauge invariant, the constraint structure is more involved than for gravity and eluidation of the proper Cauchy data is also closely related to the Dirac square root spect mentioned earlier. The Hamiltonian treatment has had one remarkable conse-uence for the long-standing problem of positive energy in gravitation. It will urn out that not only is a physically transparent proof of positive energy possible or full second quantized supergravity, but that this can then be used to establish ositivity for classical relativity, which has been impossible hitherto within its wn framework. This result will therefore stand whatever the ultimate physical value f supergravity. Finally, we shall list some of the many open problems in the theory, ncluding those of possible mathematical and classical relativistic interest.

LOBAL SUPERSYMMETRY

This ancestor of our local gauge theory is by now all of 3-4 years old, and therefore needs little introduction (for a very recent review, see [5]). We shall only sketch those of its general aspects which will be needed for later development. Global supersymmetry deals with realizations of graded extensions of the Poincaré algebra, the simplest of which adjoins to the latter,

$$i[P_a,P_b] = 0 , \quad i[J_{ab},P_c] = \eta_{ac}P_b-\eta_{bc}P_a$$

$$i[J_{ab},J_{cd}] = \eta_{ac}J_{bd} + \eta_{bd}J_{ac} - \eta_{bc}J_{ad} - \eta_{ad}J_{bc} \tag{1a}$$

the following relations

$$i[P_a,Q_\alpha] = 0 , \quad i[J_{ab},Q_\alpha] = \tfrac{1}{2}(\sigma_{ab})_\alpha{}^\beta Q_\beta$$

$$\{Q_\alpha,\bar{Q}_\beta\} = -2\,(\gamma^a)_{\alpha\beta}P_a . \tag{1b}$$

Here the supercharge Q_α is a Majorana (real) spinor, and we use the Majorana representation of the Dirac algebra with $\sigma_{ab} \equiv \tfrac{1}{4}[\gamma_a,\gamma_b]$. The commutators in (1b) tell us that Q_α is conserved and transforms as a spinor under Lorentz rotations. As fermionic operators, the components Q_α must obey anticommutation relations among themselves, and these link to the translation operators, a fact which will be of great importance in the positive energy application. If we introduce anticommuting spinor parameters α_i, the anticommutator may be rewritten as

$$[\bar{\alpha}_1 Q,\bar{\alpha}_2 Q] = -2\,\bar{\alpha}_1\gamma^a\alpha_2 P_a . \tag{1c}$$

In this way, one may exponentiate the algebra to obtain a "group" involving some parameters lying in a Grassmann algebra. A somewhat more general from of (1c) exists when the Q_α also have an internal symmetry index, in which case the right hand side may acquire additional terms proportional to $(1,\gamma_5)$ whose coefficients are called central charges and commute with the other generators. The words simple and extended are often used to distinguish the cases without and with internal labels.

Before going on to treat supergravity as the gauging of the algebra (1) in terms of a vector field

$$e^a_\mu P_a + \frac{1}{2}\omega^{ab}_\mu J_{ab} + \psi^\alpha_\mu Q_\alpha \tag{2}$$

with values in the algebra and coefficients $(e;\omega;\psi)$ corresponding respectively to ordinary gravity and to a new spinor-vector component ψ^α_μ, let us first discuss possible realizations of the global graded algebra. All irreducible one-particle state representations are known [6]; in particular, for massless systems one may classify the supermultiplets in terms of their helicity content, using the fact that the Q^i_α (i = 1..N is the internal index) are essentially ladder operators for helicity. Thus, starting from a singlet state of helicity $|\lambda|$, one may reach helicity $|\lambda|-N/2$, the multiplicity at each step $|\lambda| - \frac{1}{2}K$ being given by $\frac{N!}{K!(N-K)!}$, with transformation properties corresponding to a rank-K tensor of SO(N). If one adds to this the set starting with the CPT conjugate of helicity $-|\lambda|$ and raising, then if $-|\lambda|=\lambda-N/2$, i.e., $|\lambda|=N/4$, one has a self-conjugate system, which permits one to look for (but does not in itself guarantee existence of) local field models. For a given top helicity singlet, the maximal internal symmetry O(N) is thereby determined. For a vector theory (λ=1), this is O(4), while for the case of present interest (λ=2), the internal group may be as large as O(8). Some parenthetical remarks: First, a given realization may have a higher symmetry than O(N); however, these new symmetries are purely global. Such additional invariances have in fact been of great usefulness in constructing the details of the complicated extended models of supergravity; the most recent example is the N=4 case [7]. Second, the maximal models seem to have particularly interesting properties. This is the case for the O(4) vector supermultiplet [8], which also displays surprisingly anomaly-free behavior in that its β function vanishes through two loops at least [9]. We also note that the particle contents mentioned here are within the framework of standard relativistic particle theory with $p^2_\mu=0$. Third, since N=8 is too small to span all the currently popular strong, electromagnetic and weak symmetries, higher order actions are being considered (see for example [10]), involving fourth order field equations, in the hope of obtaining a larger particle spectrum, at the possible cost of additional non-physical "ghost" states.

Let us now consider some examples of realizations of global supersymmetry by simple models of (free) local fields, in order to exhibit concrete forms for the spinor charges and the transformations they generate. The simplest cases involve massless systems of spin $(\frac{1}{2},0)$ or $(1,\frac{1}{2})$. In the former, we have a scalar (A) and a pseudoscalar (B) field, together with a real Majorana spinor λ. Using the Majorana representation of the Dirac algebra, the action is

$$I = -\frac{1}{2}\int (dx)\ [(\partial_\mu A)^2 + (\partial_\mu B)^2 + i\bar{\lambda}\not{\partial}\lambda] \tag{3}$$

t is invariant under

$$\delta\lambda = \partial_\mu(A + \gamma_5 B)\gamma^\mu\alpha \quad , \quad \delta A = i\bar{\alpha}\lambda \ , \quad \delta B = i\bar{\alpha}\gamma_5\lambda \tag{4}$$

ith α a constant Grassmann parameter, preserving the fermionic or bosonic character of the field variations; the associated conserved Noether current is

$$J^\mu = \partial_\nu(A+\gamma_5 B)\gamma_5^\nu\gamma^\mu\lambda \tag{5}$$

and its conservation follows from the free field equations implied by (3). The total supercharge, $Q = \int(d^3x)J^0$, generates the transformations (4), namely $\delta X=[\bar{\alpha}Q,X]$ for all three fields (A,B,λ). To check the above assertions involves use of Dirac algebra identities (clever use of the latter and of Fierz transformations is an essential requirement in this subject).

The same pattern is followed by the $(1,\frac{1}{2})$ system,

$$I = -\frac{1}{2}\int (dx)[\frac{1}{2} F_{\mu\nu}^2 + i\bar{\lambda}\not{\partial}\lambda] \tag{6}$$

with

$$\delta\lambda = -F_{ab}\sigma^{ab}\alpha \quad , \quad \delta A_\mu = i\bar{\alpha}\gamma_\mu\lambda \tag{7}$$

(of course, $\delta F_{\mu\nu} = \partial_\mu\delta A_\nu - \partial_\nu\delta A_\mu$) and

$$J^\mu = 2F\cdot\sigma\gamma^\mu\lambda \tag{8}$$

The existence of a global gauge invariance immediately prompts the question of a possible generalization to a local one, with a space-time dependent parameter $\alpha(x)$. Clearly it will be necessary to introduce a massless spinor-vector gauge field ψ_μ to couple to the current J^μ, just as we need the vector potential A_μ to transform gauge invariance of the first kind to local invariance by coupling to the conserved electromagnetic current j^μ. But ψ_μ will itself require a massless companion of spin 1 or 2 to preserve the supersymmetry. Since gravity (spin 2) is in any case necessarily coupled to the stress tensor of all matter, it is natural to unify the two into a single massless (2, 3/2) multiplet. This can already be envisaged at the linearized level, i.e., one may consider the free linear spin 2 field (linearized gravity) together with a free spin 3/2 field as another realization of global supersymmetry. [The possibility of linking gravity with a spin 5/2 field may also be envisaged, but there are several arguments against it: 1) the supercurrent is a spin 3/2 object; 2) even the free spin 5/2 action does not possess sufficient gauge invariance to exclude lower helicities, and it seems unlikely that it can be consistently linked to gravity anyhow; 3) in any extended version with internal symmetry, these would have to be more than one graviton]. The spin 3/2 action was given long ago (in slightly different form) by Rarita and Schwinger [11]. The total action reads

$$I_{SG}^{LIN} = I_2^{LIN} + I_{3/2}^{LIN} = \int (dx) \; [(h^{\mu\nu} - \tfrac{1}{2}\eta^{\mu\nu}h^{\alpha}_{\alpha})(\partial_\nu\omega_\mu - \partial_\lambda\omega_{\nu\mu}{}^{\lambda}) + \tfrac{1}{2}(\omega_{\mu\lambda\nu}\omega^{\nu\lambda\mu} - \omega_\lambda^2)] - \tfrac{i}{2}\int (dx)\,\epsilon^{\mu\nu\alpha\beta}\bar{\psi}_\mu\gamma_5\gamma_\nu\partial_\alpha\psi_\beta \tag{9}$$

where $\omega_\mu \equiv \omega_{\lambda\mu}{}^{\lambda}$ and $\omega_{\mu\nu\lambda} = -\omega_{\mu\lambda\nu}$ are the linearized Ricci rotation coefficients, and $h^{\mu\nu}$ the linearized deviation from the Minkowski metric $\eta^{\mu\nu}$. For future comparison, we expressed the spin 2 action in first order form with the vierbein and Ricci rotations as independent variables. Note that the theory (9) possesses two abelian gauge invariances, under the usual linearized coordinate transformations and under

$$\delta\psi_\mu = \partial_\mu\alpha \ . \tag{10}$$

We will not insist on the global invariance details, since they can be recovered as the linearized limit of the full supergravity theory, just as the separate local abelian gauge and global rotation invariances of a set of N Maxwell fields, $L = -\tfrac{1}{4}\sum_{i=1}^{N}(F_{\mu\nu}{}^{i})^2$, under $\delta A^i_\mu = \partial_\mu\lambda^i(x) + C^{ijk}\rho^j A^k_\mu$ can be obtained from the corresponding coupled Yang-Mills model's invariance under the local invariance where (infinitesimally) $\rho^j \to \rho^j(x) = \lambda^j(x)$. Indeed, one could construct supergravity in the same way as the Yang-Mills or ordinary gravity actions are obtainable [12] by requiring self-coupling of the initially linear system (9) to its own conserved "currents", $T_{\mu\nu}$ and J^μ, the stress-tensor and supercurrent. This is one of several roads which lead to supergravity, in addition to the historical developments we will be following. The above path simply makes use of the fact that we have here a doubly gauge invariant theory, which once coupled at all, must bootstrap itself completely with its own sources, in order that the coupling be consistent (if one exists at all). All roads (including superspace methods, recently surveyed in [13]), do lead to the same supergravity.

SPIN 3/2 FIELD

Because the properties of the spin 3/2 field may be unfamiliar, we give a brief introduction to its Hamiltonian form, both massless and massive, to exhibit its degrees of freedom and the essentially Dirac-like character of their propagation.

lthough our discussion is mostly of the free field, much of the content carries ver qualitatively to the case where gravitational coupling is included, since e concentrate essentially on kinematical properties. A treatment of the linearized hoery may also be found in [14], while the Hamiltonian dynamics of full supergrav-ty are given in [15, 16]. Other useful formal properties of the system are given n [17]. The basic physical quantity is not the potential ψ_μ but the gauge-invari-nt field strength,

$$f_{\mu\nu} \equiv \partial_\mu\psi_\nu - \partial_\nu\psi_\mu \ , \qquad {}^*f^{\mu\nu} \equiv \frac{1}{2}\,\varepsilon^{\mu\nu\alpha\beta}f_{\alpha\beta} \tag{11}$$

The action is just

$$I^{LIN}_{3/2} = -\frac{i}{2}\int (dx)\,\bar{\psi}_\mu\gamma_5\gamma_\nu{}^*f^{\mu\nu} \tag{12}$$

which is manifestly invariant under the abelian gauge transformation (10), owing to the Bianchi identity $\partial_\mu{}^*f^{\mu\nu} \equiv 0$. The field equations may be written in many forms, including one,

$$f_{\mu\nu} + \gamma_5{}^*f_{\mu\nu} = 0 \tag{13}$$

which exhibits the invariance under duality ($f\to{}^*f$) and chirality ($f\to\gamma_5 f$) transformations, and their close relation.

The existence of a local invariance implies, as always the presence of a constraint. As is clearest from (9), the coefficient of ψ_0 is independent of time derivatives. When the action is decomposed into noncovariant form and the content of the constraints fully utilized, only the gauge-invariant quantities ψ_i^{TT} ($i=1,2,3$), which satisfy

$$\gamma^i\psi_i^{TT} = 0 = \partial^i\psi_i^{TT} \tag{14}$$

appear in the action. The two conditions leave only one spinor out of the three ψ_i, corresponding to the two helicity $\pm$ 3/2 degrees of freedom of a single Majorana fermion. Indeed, the action simply reads

$$I^{LIN}_{3/2} = -\frac{i}{2}\int (dx)\ \bar{\psi}_i^{TT}\not\partial\psi_i^{TT} \tag{15}$$

In the massive case, there are also helicity 1/2 excitations present, and the action becomes

$$I^{LIN}_{3/2}(m) = -\frac{i}{2}\int (dx)\ [\bar{\psi}_i^{TT}(\not\partial+m)\psi_i^{TT} + \bar{\chi}(\not\partial+m)\chi] \tag{16}$$

where $\chi \equiv (3/2)^{\frac{1}{2}} \gamma_5 \gamma^i \psi_i^T$, with $\partial^i \psi_i^T \equiv 0$, is the lower helicity component.

The Hamiltonian analysis of full supergravity leads to an action which is formally quite complicated, as might be expected from the many nonlinear constraints present, but whose generic form has the same degree of freedom count and the structure to be expected from that of general relativity in vierbein form [18] on the one hand, and from that of higher spin fields interacting with gravity [19], on the other

The count of degrees of freedom may also be performed in a covariant way as for other spin systems [20]: The Rarita-Schwinger equation implies that $p^\mu \psi_\mu = 0$, as well as $p^2 \psi_\mu = 0$. The general ψ_μ can therefore be represented as a combination of three spinors with polarization vector $e^{\pm}_\mu(p)$ and p_μ coefficients (since $p \cdot e^{\pm} = 0$). Two of these three spinors may be removed by appropriate gauge choice, since as always, gauge freedom removes components in pairs. However, we mention that the usual count of ghost compensation of unphysical degrees of freedom in the quantum formulation is somewhat altered when gauge fixing is performed and that the Feynman rules for supergravity differ from those in other gauge theories, requiring quartic terms in the ghost fields [20, 21]. These modifications are of course consistent with the Hamiltonian formulation.

SIMPLE SUPERGRAVITY

Having seen some of the background of and motivation for supergravity, we shall simply write down the correct O(1) supergravity action and verify the properties from which it was originally derived: local gauge invariance [1] and consistency [2], respectively. The action is simply the generally covariant (and locally Lorentz covariant) form of (9), namely (in units of $\kappa=1$)

$$I_{SG} = I_2(e,\omega) + I_{3/2}(\psi; e,\omega) \tag{17a}$$

with

$$I_2 = \frac{1}{2} \int e e^{\mu a} e^{\nu b} R_{\mu\nu ab}(\omega) = \frac{1}{2} \int e_{\mu a} e_{\nu b} {}^{**}R^{\mu\nu ab}(\omega) \tag{17b}$$

and

$$I_{3/2} = -\frac{i}{2} \int \varepsilon^{\mu\nu\alpha\beta} \bar{\psi}_\mu \gamma_5 e_{\nu a} \gamma^a D_\alpha(\omega) \psi_\beta \tag{17c}$$

Here $R_{\mu\nu ab}(\omega)$ is the curvature tensor expressed in terms of the Ricci coefficients, $e_{\mu a}$ and $e^{\mu a}$ are the co- and contra-variant vierbein components (e is the determinant) and ${}^{**}R$ the double dual of the curvature, (${}^{**}R = \frac{1}{4}\varepsilon\varepsilon R$). Note that only the curl enters in (17c), which permits us to restrict D_α to be to the covariant derivative corresponding to spin 1/2, namely

$$D_\mu \equiv \partial_\mu - \frac{1}{2}\omega_\mu{}^{ab}\sigma_{ab} \quad ; \quad [D_\mu, D_\nu] = -\frac{1}{2}R_{\mu\nu}{}^{ab}\sigma_{ab} \ . \tag{18}$$

his is in complete accord with that coupling of a vector field to gravity which espects its gauge invariance, namely $(\partial_\mu A_\nu - \partial_\nu A_\mu)$ rather than $(D_\mu A_\nu - D_\nu A_\mu)$ which iffers from it, for an asymmetric affinity (torsion) by the non-gauge invariant art $(\Gamma^\alpha_{\mu\nu} - \Gamma^\alpha_{\nu\mu})A_\alpha$, and is the most minimal possible coupling in first-order form where $e,\omega)$ are to be varied independently. The equations of motion are

$$\delta\psi_\mu: \ R^\mu \equiv \varepsilon^{\mu\nu\alpha\beta}\gamma_5\gamma_a(e_{\nu a}D_\alpha - \frac{1}{4}C^a_{\nu\alpha})\psi_\beta = 0 \tag{19a}$$

$$\delta\omega: \ C^a_{\nu\alpha} \equiv D_\nu e^a_\alpha - D_\alpha e^a_\nu = \frac{i}{2}\bar{\psi}_\nu\gamma^a\psi_\alpha \tag{19b}$$

$$\delta e: \ G^{a\mu} \equiv R^{a\mu} - \frac{1}{2}e^{a\mu}R = \frac{i}{2}e^{-1}\varepsilon^{\nu\mu\alpha\beta}\bar{\psi}_\nu\gamma_5\gamma^a D_\alpha\psi_\beta \equiv T^{a\mu} \tag{19c}$$

The non-vanishing of $C^a_{\nu\alpha}$ implies that space-time has torsion, while the Einstein tensor in (19c) is non-symmetric (and $R_{\mu a} \equiv R_{\mu\nu ab}e^{\nu b}$). So far, the system does not seem particularly different from, say, the usual Einstein-Dirac coupling with torsion. However, there are two (related) profound differences: the first is that the matter equation (19a) has an open index and therefore the question of its consistency with the others immediately arises -- does $D_\mu R^\mu$ vanish by virtue of (19) themselves? This is not an idle question, because precisely this problem of consistency has stymied all previous attempts to couple higher spin fields to gauge theories such as gravity or electromagnetism, and we shall see how delicate is the matching of equations. Second is the fact that the free ψ_μ field possessed the local invariance (10), which is now lost, being replaced by a non-manifest invariance under the covariant version of (10), with (e,ω) transforming simultaneously.

What makes consistency at all possible is that when we compute $D_\mu R^\mu$, the commutation of the two covariant derivatives reduces to the Einstein, rather than full Riemann tensor, through the interplay of Dirac algebra and Ricci identities. The result is the identity

$$D_\mu R^\mu \equiv -\frac{1}{2}G^{a\mu}\gamma_a\psi_\mu + \frac{1}{4}\varepsilon^{\mu\nu\alpha\beta}C^a_{\mu\nu}\gamma_5\gamma_a D_\alpha\psi_\beta \ . \tag{20}$$

But there is another identity, this time derivable from Fierz rearrangements (for anticommuting spinors) which tells us that

$$\frac{1}{2}T^{a\mu}\gamma_a\psi_\mu - \frac{1}{4}\varepsilon^{\mu\nu\alpha\beta}\bar{\psi}_\mu\gamma^a\psi_\nu\ \gamma_5\gamma_a D_\alpha\psi_\beta \equiv 0 \tag{21}$$

Adding (20) and (21) provides the desired relation among the field equations (19), which we may write as

$$-2D_\mu \delta I/_{\delta\psi_\mu} + i\gamma^a \psi_\mu \delta I/_{\delta e^a_\mu} + W^{ab}_\mu \delta I/_{\delta\omega_\mu}{}_{ab} \equiv 0 \qquad (22)$$

Multiplying this spinorial identity by an arbitrary spinor function $\alpha(x)$ and integrating over spacetime exhibits the new gauge invariance: the total variation of I_{SG} vanishes under the set

$$\delta e^a_\mu = i\bar{\alpha}\gamma^a \psi_\mu , \qquad \delta\psi_\mu = 2D_\mu \alpha$$
$$\delta\omega_{\mu ab} = B_{\mu ab} - \tfrac{1}{2}(e_{\mu b} B_{\nu ac} e^{\nu c} - e_{\mu a} B_{\nu bc} e^{\nu c}) \qquad (23)$$
$$B_{\mu ab} \equiv i\epsilon_{ab}{}^{\alpha\beta} \bar{\alpha}\gamma_5 \gamma_\mu D_\alpha \psi_\beta \; . \quad W_{\mu ab} \equiv "\delta\omega_{\mu ab}/\bar{\alpha}".$$

Note that as usual, interdependence of field equations was related to gauge invariance (just as the contracted Bianchi identities for any geometrical action, $D_\mu(\delta I(g)/\delta g_{\mu\nu}) \equiv 0$ are equivalent to coordinate invariance because $\int \delta_c g_{\mu\nu} \delta I/\delta g_{\mu\nu} = 0$ with $\delta_c g_{\mu\nu} \equiv D_\mu \rho_\nu + D_\nu \rho_\mu$). One could also have begun with the demand that a local invariance of the form (23) exist and try to reach an action which possesses it. This procedure was performed [1] in second order form, where appropriate spin-spin contact terms replace the torsion conditions (since the affinity is an auxiliary field, obeying the non-dynamical equations (19b), this is always possible). Note that the correct coupling is quite rigid: anything but minimal first order coupling (equivalently, very special non-minimal second-order coupling) would be inconsistent (with the exception of the cosmological version below). The action (17) represents the gauge field (e,ω,ψ) in the same sense as the Einstein action does (e,ω) or the Maxwell action does for A_μ. The three invariances, coordinate, Lorentz and supersymmetry are not, however, independent and indeed the latter is really the fundamental one, since the commutator of two independent supersymmetry transformations characterized by (α_1,α_2) yields a general coordinate transformation characterized by the displacement $2i\bar{\alpha}_1\gamma^\mu\alpha_2$, together with field-dependent Lorentz and supersymmetry transformations. This is another symptom of the "Dirac square root" character of supergravity.

As stated earlier, higher spin (including 3/2) had never been successfully coupled to gravity, consistency of the field equations being but one form of the problem. Others include anomalous characteristic surfaces and resultant unphysical propagation properties. Here those problems are also avoided and there is perfectly compatible propagation of the Cauchy data. Since spacetime is no longer an invariant object (we may vary $\delta g_\mu = i\bar{\alpha}(\gamma_\mu\psi_2 + \gamma_2\psi_\mu)$ which is not a coordinate transformation) we should really say, for example, that the ψ-field propagates on the local null cone in any α-gauge.

The masslessness and (real) Majorana rather than (complex) Dirac form of the 3/2 field were both essential for consistency. Had we attempted the construction with a complex field, the Fierz identities would have failed, as is to be expected from our earlier global symmetry considerations. A complex field is really an O(2) object and as such requires an additional vector field by the count made there; this corresponds, in fact, to the extension of supergravity [22] to include O(2) internal symmetry. As for the mass term, there is surprisingly enough a version of the theory which does ostensibly permit a mass term, provided it is accompanied by a cosmological term of appropriate magnitude and sign [23]: one may add to (17) the following combination

$$- \frac{i}{2} m \, \varepsilon^{\mu\nu\alpha\beta} \bar{\psi}_\mu \gamma_5 \sigma_{\nu\alpha} \psi_\alpha + 3m^2 \kappa^{-2} e \qquad (24)$$

with a slightly modified variation of ψ and ω. This is perplexing at first sight, because we know that a cosmological term does not correspond to a graviton mass (since it preserves general coordinate invariance and the number of graviton degrees of freedom). How then can the ψ-field acquire mass and thereby lead to a supermultiplet of different masses? The answer is that the "mass" term does not imply that there is a ψ-mass, as we shall see later. [The possibility of truly massive gravity coupled to massive spin 3/2 must await the construction of a physically acceptable massive graviton, which is not yet available [24]].

Note finally that in this local gauge theory the total "currents" are identically conserved corresponding to the Bianchi identities. However, if we take the weak field limits, we regain the global symmetry free linear (2, 3/2) doublet discussed previously.

COUPLING TO MATTER AND EXTENDED SUPERGRAVITY

One can extend the simple supergravity model of the previous section in two ways: couple it to global supermatter multiplets as discussed earlier, and look for extensions with internal symmetry. The coupling procedure is, in principle, straight forward. One takes the global action, makes it generally covariant in the usual way (to take care of the normal coupling to gravity) and adds a term $\sim \psi_\mu J^\mu$ to take care of the supercurrent, together with whatever contact ("seagull") terms may be needed to complete the gauge invariance. This is also necessary in electromagnetism coupled to a charged scalar field, say, where one must include the usual $A_\mu{}^2 |\phi|^2$ term in addition to $A_\mu j^\mu_{em}$. Unfortunately, the analog of the seagull is not obvious here and requires an enormous amount of detailed calculation of a "non-algorithmic" kind, though the final result is just what one would arrive at by a naive copying of the electromagnetic procedure. We will not give details here, since they are algebraically complicated, but refer to the literature (see for example [25]) for the various known couplings, including to the (1/2, 0) and (1, 1/2) cases in simple supergravity and also in extended theory. These couplings can also be recovered as limiting cases of extended source-free supergravity models in which some of the component fields are suppressed. An important technical advance would be one which gave a uniform coupling procedure akin to the replacement $\partial_\mu \to D_\mu$ in other gauge cases; although the notion of a "supercovariant derivative" has been introduced, it is not sufficient so far for this purpose. One point worth noting concerning matter coupling is that the transformation laws for the ψ_μ (but not the vierbein) field are altered, and depend on the matter axial currents in a non-obvious way. The technical difficulties in checking consistent couplings may be due to the fact that auxiliary (non-dynamical) fields play an important role in uniformizing globally symmetric procedures and we have not yet found a geometrically clear set of them here. Certainly the affinities ω (which are auxiliary) led to an enormous simplification compared to the use of second order form. A related fact is that the algebra of the transformations only closes when the field equations are used; the commutator of two supersymmetry transformations on a field variable is just a combination of coordinate, Lorentz and supersymmetry transformations only when the field equations are used. To remove the on shell requirement requires very complicated auxiliary structure [26].

In addition to the matter couplings just mentioned, there are some amusing applications to lower dimensional models. The first is to the spinning string of dual models [27], from which supersymmetry was in fact discovered [28]. A previously unsolved problem there was the incorporation of the constraints into the string action. The latter was simply the sum of a free scalar and spinor field evolving in the (1+1) dimensions representing the intrinsic geometry of the string, subject to the vanishing of the stress tensor and supercurrent. If one now couples this system to two-dimensional supergravity [29], the (e,ω,ψ) fields are essentially Lagrange multipliers because both the Einstein and spin 3/2 actions are empty [R is a total divergence and no gauge invariant action of the form $\bar{\psi}\gamma\partial\psi$ can be written]. The Einstein and Rarita-Schwinger equations will therefore simply read

$$0 = T_{\mu\nu} \qquad 0 = J_\alpha \tag{25}$$

which are precisely the desired constraints. It is also necessary (and possible) to verify that there exists a gauge for (e,ω,ψ) in which the coupled action does reduce to that of the original string. Even more degenerate is the (0+1) dimensional system describing a spinning particle, where the constraints $(p^2+m^2) = 0$ and $(\not{p}+m) = 0$ are again recovered by coupling it to the trivial gravity fields (e,ψ); this exercise [30] provides an interesting insight into the significance of the Dirac equation as the vanishing of the supercurrent and the $(p^2+m^2)=0$ condition as a consequence of the graded algebra of the constraints.

We turn now to extended supergravity, where explicit models have been constructed for N=2,3,4 [22,31]. The field contents follow the binomial rule. In addition to the singlet graviton, O(2) includes two (real) spin 3/2 and a singlet vector. In O(3) there are three (3/2, 1) fields and a singlet spin 1/2. It is significant in that for the first time a non-gauge particle (a neutrino) is incorporated into a supermultiplet with higher spin gauge fields. For O(4), the content is (1,4,6,4,1+1) where the "1+1" refers to a scalar plus pseudoscalar field. Here, non-polynomial xpressions in the scalar field begin to appear. Very recently [32], the O(4) sys-em has been expressed in particularly simple form, and it has been noted that it ontains additional hidden symmetries which raise O(4) to global SU(4) and also in-lude a new global SU(1,1) non-compact invariance. Beyond N=4, only the maximal (8) case has been considered, although only partial results are available due to he algebraic complication involved [33]. Various possible matter couplings, e.g. f O(4) supergravity to the O(4) vector supermatter multiplet could also be consid-red. Coupling to matter has two drawbacks: the first is that it is less elegant han to include everything in the gauge multiplet. However, to the extent that even)(8) does not include all observed symmetries in nature, inclusion of matter may rovide an additional spectrum. [The additional hidden symmetries referred to above re global and therefore of no direct help in constructing empirically large enough ultiplets.] The second drawback, as we shall see, is that, at least for O(1) or)(2), renormalizability is already lost at one loop order, which is less attractive than the pure gauge models.

SYMMETRY BREAKING

Any realistic supersymmetric model must allow for the observed mass differences in nature, where there are very few zero mass particles or equal masses among particles of different spins. This can be approached either by extending the scheme to include massive supermatter (but this will then also require symmetry breaking) or by spontaneous symmetry breaking within the original scheme. The problem has recently been discussed [34], with encouraging results, and we shall summarize the findings of that work. We begin with the apparently massive version of simple supergravity mentioned earlier. The action is the sum of (17) and (24)

$$I = I_{SG} - \frac{i}{2} m \, \varepsilon^{\mu\nu\alpha\beta} \bar{\psi}_{\mu} \gamma_5 \sigma_{\nu\alpha} \psi_{\beta} + 3m^2 e \; . \qquad (26)$$

[The sign of m is of course a convention for spinors; however the sign of the cosmological term is uniquely fixed.] Note first that the gravitational part is simply Einstein theory plus a cosmological term, which does not represent massive gravitons. Instead, the latter forbids Minkowski space as a solution of the Einstein equations, but rather it is DeSitter space which is the "flattest" background solution. [The necessary sign of the cosmological term leads to O(3,2) rather than to the anti-DeSitter O(4, 1), which is comforting since the latter suffers from causality difficulties.] Thus, we can only analyze the meaning of the mass term in DeSitter and not Minkowski terms (where it would of course represent a true mass). Now in De-Sitter space, mass ($p_\mu p^\mu$) is no longer a Casimir operator, the consequences of which were noted long ago for "massive" spin 1/2 systems [35]. In particular, one must investigate the propagation character of the ψ-field in DeSitter space to see if it is along local null cones. The field equation reads (in DeSitter space without torsion),

$$R^{\mu} \equiv \epsilon^{\mu\nu\alpha\beta}\gamma_5\gamma_\nu \mathcal{D}_\alpha\psi_\beta = 0 , \quad \mathcal{D}_\alpha \equiv D_\alpha + \tfrac{1}{2}m\gamma_\alpha \tag{27}$$

with

$$[\mathcal{D}_\alpha, \mathcal{D}_\beta]^\alpha = 0 \tag{28}$$

on any spinor $\alpha(x)$. One can deduce from this and the field equations that the field strengths $\psi_{\mu\nu} \equiv \mathcal{D}_\mu\psi_\nu - \mathcal{D}_\nu\psi_\mu$ (and its dual) propagate according to the Maxwell-like equations $D^\mu\psi_{\mu\nu} = 0 = D^\mu{*\psi_{\mu\nu}}$, corresponding to massless behavior. This ceases to be true if the exact relation between cosmological and mass terms is violated (in particular, if there is no cosmological term, or even if there is no mass term in the presence of a cosmological term), and there are then four (massive) degrees of freedom. The above conclusions hold also when gravity is dynamical.

Let us now come to the question of symmetry breaking and the Goldstone-Higgs mechanism. Suppose there is spontaneous supersymmetry breaking which results in generation of a Goldstone fermion λ. Does the latter, when coupled to supergravity, get "swallowed" by the spin 3/2 field and make it massive as in the corresponding bosonic situation where the Goldstone scalar gives mass to the vector gauge field through a Higgs mechanism? The answer is yes, although we do not yet have a complete description of Goldstone fermion couplings to supergravity in four dimensions. A generic action for the Goldstone particle in flat space is furnished by the nonlinear realization of global supersymmetry [36], with action

$$I_\lambda = -(2a^2)^{-1} \det\left[\delta^a_\mu + ia^2\bar{\lambda}\gamma^a\partial_\mu\lambda\right] \tag{29}$$

invariant under

$$\delta\lambda = a^{-1}\alpha + ia(\bar{\alpha}\gamma^\mu\lambda)\partial_\mu\lambda \ . \tag{30}$$

When coupled to supergravity, the system's invariance will involve the function $\alpha(x)$, which will permit one to gauge away λ altogether, leaving behind only the cosmological term $-1/2a^2\, e$, of opposite sign to that in supergravity (26). But as we have just seen, if the relation between mass and cosmological coefficients is altered (including the case where neither was present), in particular if m is chosen so as to cancel $-1/2a^2$, the ψ field does acquire a mass and the additional helicity $\pm$ 1/2 degrees of freedom associated with it, which were originally present in the fermion action (29). Since the typical particle mass scale will be reflected in the size of a^2, it is thereby also possible to avoid the embarrassingly large cosmological term (radius of the universe comparable to particle Compton wavelength) hitherto present, while endowing the spin 3/2 with a very small mass. It has been very recently uggested [37] that these helicity states might still be related to weak interaction effects, as was already discussed in the global case for Goldstone fermions. In any case, the above considerations present a coherent picture of possible symmetry breaking and its effects within supergravity which are encouraging. The above conclusions about the Goldstone-Higgs effect are reinforced by detailed calculations in (1+1) and (2+1) dimensions [38], where the coupling can be obtained in detail and the transfer of degrees of freedom to the massive ψ_μ occurs as expected. The action (29) couples to the gauge fields according to

$$I_\lambda(\lambda;e,\omega) = -(2a^2)^{-1} \int (d^3x) \det [e^a_\mu + ia^2\bar{\lambda}\gamma^a(D_\mu\lambda - a^{-1}\kappa\psi_\mu)] \qquad (31)$$

with the invariances of $I_\lambda + I_{SG}$ given by

$$\delta\lambda = a^{-1}\alpha(x) \;, \quad \delta\psi_\mu = 2\kappa^{-1}D_\mu\alpha \;, \quad \delta\omega^{ab}_\mu = 0 \qquad (32)$$

$$\delta e^a_\mu = i\kappa\bar{\alpha}\gamma^a\psi_\mu - D_\mu(ia\bar{\alpha}\gamma^a\lambda)$$

in (2+1) dimensions. Note that we have "linearized" the λ transformations by transferring (for convenience) the nonlinear part to a field-dependent coordinate transformation on the vierbein. Here, it is clear that λ can be removed, leaving the $-1/2a^2$ e remainder, whose effect on the ψ-field is precisely as stated above for four dimensions.

A rather different, but instructive consequence of these ideas lies in their ability to solve an old problem: to give a consistent theory of charged spin 3/2 systems, with non-vanishing electromagnetic coupling. This is not possible in flat space (with or without mass), the obvious symptom being that

$$D_\mu R^\mu \simeq {}^*F^{\mu\nu}\gamma_5\gamma_\mu\psi_\nu \qquad (33)$$

where the D_μ are the electromagnetic covariant derivatives. Unlike the gravitational analog ($G^{\mu\nu}$), $^*F^{\mu\nu}$ is not part of a field equation. However, when gravity is also included, without a det e term, one has a non-supersymmetric theory which is nevertheless consistent provided the gravitational constant, mass, charge and anomalous magnetic moment of the ψ-field are suitably related [34]. This is an important lesson, that consistency may require full use of all the couplings to which a field is subject, as well as relations among the parameters involved. The causal propagation requirements are also satisfied, of course.

THE SQUARE ROOT OF GRAVITY

To gain further insight into the properties of supergravity, we shall show here that it may be understood as the Dirac square root of general relativity in a very precise way [3]. Let us first discuss this at the level of the constraints which determine the character of any gauge theory, (and it is of course the $p^2+m^2=0$ constraint whose square root is the Dirac constraint on a spinning particle). In addition to the usual Einstein constraints ($G_{0\mu}=T_{0\mu}$), the spin 3/2 field has its own constraint, as we saw in the Hamiltonian section, namely the coefficient of ψ_0 in the action. The total Hamiltonian density of the theory is itself a constraint (by coordinate and supersymmetry invariance) and may be written in terms of its tangent space projections as

$$H(x) = e_o{}^a G_a + \frac{1}{2}\,\omega_{oab}J^{ab} + \bar{\psi}_o S \tag{34}$$

with a,b tangent spacetime indices. Here C_a are essentially the Einstein constraints (generating local translations); J^{ab} the generators of tangent space Lorentz rotations, while S is the spin 3/2 constraint. In this representation, the algebra of the constraints takes on a particularly transparent form. In addition to the gravitational part,

$$[J_{ab}(x),\ J_{cd}(x')] = \delta^3(x,x')[\eta_{ac}J_{bd} + \eta_{bd}J_{ac} - \eta_{bc}J_{ad} - \eta_{ad}J_{bc}]$$

$$[C_c(x),\ J_{ab}(x')] = \delta^3(x,x')(\eta_{cb}C_a - \eta_{ca}C_b)$$

$$[C_a(x),\ C_b(x')] = \delta^3(x,x')(\tfrac{1}{2}\,\Omega_{abcd}J^{cd} + \bar{f}_{ab}S) \tag{35a}$$

we now have

$$[S(x),\ J_{ab}(x')] = -\,\delta^3\,\tfrac{1}{2}\,\sigma_{ab}S\ ,\quad [S,C_a^{\prime}] = \delta^3\,\tfrac{1}{2}\,\Sigma_{cab}J^{ab} \tag{35b}$$

$$\{S(x),\bar{S}(x')\} = -2\delta^3\gamma^a C_a$$

with $\Sigma_{cab} = \gamma_5(\gamma_c{}^*f_{ab}+\tfrac{1}{2}\delta_{ac}\gamma\cdot{}^*f_b - \tfrac{1}{2}\delta_{bc}\gamma\cdot{}^*f_c)$, $\Omega_{abcd} \equiv R_{abcd}-\bar{\psi}_a\Sigma_{bcd}+\bar{\psi}_b\,\Sigma_{acd}$.
S is thus a local spinor whose anticommutator transforms into the translation constraints, while it transforms into the spin 3/2 curvature under local translations, just as the latter commute to the spacetime curvature (augmented by f_{ab} contributions) locally. In particular, if the curvatures vanish, we simply have copies at those points of the global graded algebra. Note that if a state obeys the 3/2 constraints it will, by the anticommutation relation in (35b), automatically obey the Einstein ones.

The derivation of the S constraint itself comes, not surprisingly in view of the above, by taking the square root of the basic (G_{00}) Einstein constraint [39]

$$\pi^{ij}\pi_{ij} - \frac{1}{2}\,\pi^i_i\pi^j_j - \sqrt{{}^3g}\ {}^3R = 0 \tag{36}$$

to linearize it in terms of the gravitational momenta π^{ij}. Despite the non-polynomial dependence of ${}^3R$ on the metric, this can be accomplished by the ansatz

$$S = \gamma_i \psi_j \pi^{ij} + 4\sigma^{ij} D_i \psi_j \qquad (37a)$$

ogether with the fundamental identity (where D_i is the metric covariant derivative)

$$\{\sigma_{ij} D_i \psi_j(x), \sigma_{em} D_e \psi_m(x')\} = 4^{-3} \delta^3(x,x')\, {}^3R \qquad (37b)$$

'or ψ_i obeying fundamental anticommutation relations

$$\{\psi_{i\alpha}(x), \psi_{j\beta}(x')\} = -\frac{1}{8}(\gamma_j \gamma_i)_{\alpha\beta} \delta^3(x,x') \qquad (37c)$$

appropriate to spin 3/2 (of course the square root introduces ψ field sources in the resulting Einstein constraints). We refer to the original work [3] for details. Once the constraints are derived, the supergravity Lagrangian may be constructed from them just as Einstein's action follows from <u>its</u> constraint structure.

POSITIVE ENERGY

Once the constraint structure of supergravity and its Hamiltonian form are understood, one may add to the usual concepts of energy-momentum and angular momentum as surface (flux) integrals at spatial infinity in asymptotically flat spacetimes [39] (which exist in the presence of any bounded matter source of gravity) by including supercharge. These quantities do not a priori have any relation to the flat space global supersymmetry quantities bearing the same names, for they have an entirely different origin. The total supercharge Q will be well defined if the ψ_i-fields fall off sufficiently fast at spatial infinity for total energy to be well-defined (essentially $\psi_i \sim 1/r^2$). Then, just as P^0 is (in suitably cartesian coordinates) the flux integral

$$P^0 = -\oint d\underline{S} \cdot \underline{\nabla} g^T \qquad (38a)$$

where g^T is essentially the Newtonian component of the spatial metric, the spinor charge may be written as

$$Q = 2\oint dS_i \sigma^{ij} \psi_i \qquad (38b)$$

It is then possible to show [3] that these flux integral charges do obey the global graded algebra (1) appropriate to the asymptotic Minkkowski space. For the most general case of O(N) supergravity coupled to supermatter, the basic anticommutator is given by

$$\{\bar{Q}_i, Q_j\} = -2\,\delta_{ij}\gamma^a P_a (+ \text{ central charges}) \quad . \tag{39}$$

This immediately implies, upon tracing (39) with γ^0, that

$$P^o = \frac{1}{4N} \sum_{\alpha=1}^{4} \sum_{L=1}^{N} (Q_{\alpha i})^2 > 0 \tag{40}$$

since any central charges are proportional to 1 or γ_5 and vanish when the trace is performed. Thus, we have formally established (without worrying about delicate ordering problems) that for full second-quantized supergravity models, with supermatter sources and internal symmetry, energy is necessarily positive [40]. No details of the constraints were needed, but only that supergravity was a "square root". This is to be contrasted with the fact that even the existing partial proofs (see e.g. [41]) of the corresponding conjecture in classical Einstein theory required detailed calculations involving the constraints' structure. Amazingly enough, it turns out to be possible to use the above quantum supergravity result to establish positivity of classical gravity as well [42]. The idea is simply to consider the energy in states devoid of spin 3/2 particles, and then to note that since it is positive for all values of $\hbar$, in particular for $\hbar=0$, this removes all closed loops (where virtual spin 3/2 particles are present) and implies that the source-free classical theory's energy is positive. Thus, even if supergravity is not the successor to general relativity, the very possibility of its existence may be exploited to prove results in the latter. [We mention parenthetically that, similarly, the [Q,J] relation may be used to derive properties of helicity amplitudes in tree level graviton-graviton scattering in general relativity [43] which are also not manifest within the latter alone]. As is not unknown in other domains, extension from the real to the complex has enabled one to establish hitherto elusive properties of the real variables.

RENORMALIZABILITY

We complete our survey with a discussion of one of the chief motivations for supergravity, that it might overcome the ultraviolet problems of ordinary gravity. To assess how far this hope has been realized, we first review the situation in quantum gravity. As a classical, tree, theory general relativity is essentially the unique model (at low frequencies) compatible with the qualitative observed properties of gravitation and the requirements of special relativistic particle theory, as can be shown in full generality [44]. The equivalence principle, which is the basis of the theory and its classical virtues, becomes its downfall when closed loops are included. For the universal coupling of all stress-tensors involves the imensional gravitational coupling constant κ which accompanies the rising of the ormer as a function of energy (linear or quadratic for fermions or bosons). As a esult, simple power counting already implies that the counterterms at each loop rder (i.e. each extra κ^2) will be different from those at previous order and from he tree action. Standard renormalization is thus impossible, unless there are ither "accidental" cancellations or additional symmetries to forbid all counter-erms. The only permitted ones are those which vanish on mass shell, since they an then be absorbed in (somewhat nonstandard) field renormalizations [45]. Further-ore, since gravity is universally coupled, its nonrenormalizability would (at least n principle) destroy the renormalizability enjoyed by flat space theories such as electrodynamics, and thereby also many of its predictive powers.

These general considerations have, in recent years, been confirmed by explicit calculations at one-loop level involving source-free Einstein theory [45], as well as its coupling to spin 0, 1/2 and 1 fields, including Yang-Mills, [46]. For source-free gravity, one can immediately write down the possible one-loop counterterms by covariance arguments alone. The general divergent ΔL will be proportional to the three possible invariants of dimension four of the Cartan classification, namely

$$\Delta L = \alpha_1 R^2_{\mu\nu\alpha\beta} + \alpha_2 R^2_{\mu\nu} + \alpha_3 R^2 \tag{41}$$

All three α_i turn out to be non-vanishing, although the last two terms are of course harmless on mass shell ($R_{\mu\nu}=0=R$) for the external gravitons emerging from the loop. However, the first term does not vanish a priori, but does so for an "accidental" reason, namely the Gauss-Bonnet identity, which states that

$$\int (d^4x) R^2_{\mu\nu\alpha\beta} = \int (d^4x)(4R^2_{\mu\nu} - R^2) \ . \tag{42}$$

As a result, the one-loop correction to pure gravity happens to be finite (for S-matrix elements, rather than Green's functions), although this situation is no longer expected to hold for higher loops where terms $\sim \kappa^{2(n-1)} (R_{\mu\nu\alpha\beta})^{n+1}$ could be present, without benefit of further identities like (42). [There is some possibility that two loops may also be "safe" because of duality invariance of the theory, but this is still under investigation.] However, as soon as matter is coupled to gravity (and this includes spin 0, 1/2 or 1 or combinations such as electrodynamics [47]), then the explicit one-loop claculations show that ΔL no longer vanishes on the combined Einstein-matter mass shell.

Since supergravity may be viewed as a particular matter system interacting with gravity, with the usual dimensional coupling constant, one would expect it to fail as well. However, we must remember that it possesses an additional invariance which must be respected by the counterterms, at least in a regularization procedure compatible with supersymmetry, and in the absence of anomalies. Explicit calculations at one loop do indeed indicate that pure supergravity is finite there, but that this is no longer the case when it is coupled to matter [48], just as for pure gravity itself. Just as one can write down (41) without explicit calculation and thereby conclude one-loop finiteness of gravity, the corresponding features of supergravity can also be understood without calculation by examining [50] the possible supersymmetric extensions of (41). There turns out to be a "super" Gauss-Bonnet analogy of (42) which guarantees that pure simple supergravity is one-loop finite, but when supermatter is also present, it is possible to construct a non-vanishing term whose globally symmetric image is of the form

$$\int \Delta L_{IM} \sim \int (T^2_{\mu\nu} + i\bar{J}^{\mu}\partial\!\!\!/ J_{\mu} - \tfrac{3}{2} C^{\mu}\Box C_{\mu}) \tag{43}$$

where (T,J and C) are the stress-tensor, supercurrent and axial current of the matter (1, 1/2) or (1/2, 0) supermultiplet. Its invariance can be checked from the generic transformation properties of these quantities under the action of supersymmetry [49], and it does not vanish on shell. The supergravity sector itself is safe because $T_{\mu\nu}$, for example, is not even invariant under the local abelian transformations of the free spin 2 or 3/2 fields. More generally, any of the O(N) extended supergravities will be one-loop finite, but this is no longer the case in the presence of N=1, 2 or 3 supermatter (at least). Similar arguments based on helicity conservation can also be given at the one- and two-loop levels [48]. At two loops, the methods of [50] also show that pure supergravity is still safe, i.e., that there is no acceptable supersymmetric generalization of $\kappa^2(R_{\mu\nu\alpha\beta})^3$. Unfortunately, this same approach uncovers possible invariants at three loops and beyond, where an explicit invariant generalizing (43) may be given. Its globally symmetric part is proportional to [50]

$$\int \Delta L_{3SG} \sim \int \{T^2_{\mu\nu\alpha\beta} + i\bar{J}^{\mu\alpha\beta}\partial\!\!\!/ J_{\mu\alpha\beta} - \tfrac{3}{2} C^{\mu\alpha\beta}\Box C_{\mu\alpha\beta}\} \tag{44}$$

where

$$-T_{\mu\nu\alpha\beta} \equiv (R^{\lambda\rho}_{\alpha\mu}R_{\lambda\beta\rho\nu} + {}^{*}R^{\lambda\rho}_{\alpha\mu}{}^{*}R_{\lambda\beta\rho\nu}) + \tfrac{i}{2}\bar{f}^{\lambda}{}_{\alpha}(\gamma_{\mu}\partial_{\nu} + \gamma_{\nu}\partial_{\mu})f_{\beta\lambda} \tag{45a}$$

$$J_{\mu\alpha\beta} \equiv R^{\lambda}_{\alpha ab}\sigma^{ab}\gamma_{\mu}f_{\lambda\beta} \tag{45b}$$

$$C_{\mu\alpha\beta} \equiv -\tfrac{i}{2}\bar{f}{}_{\alpha}\gamma_5\gamma_{\mu}f_{\beta\lambda} \tag{45c}$$

The first quantity is the "Bel Robinson tensor" in our supergravity (which generalizes the usual stress tensor), and the other two generalize the super- and axial currents. Of course, it is conceivable that there is some further symmetry (or an accidental cancellation) forbidding ΔL_3; the prohibitive nature of even two loop calculations makes it unlikely that we can settle this question explicitly in the near future, ut at present there is no known principle forbidding (44) and its higher loop uccessors. Do the additional (global) internal symmetries which the extended odels possess help? All we know so far is that, at least for the O(2) case [51], nd perhaps also for O(3), it is possible to extend (44) in a way which maintains lobal supersymmetry and the internal O(2) but does not vanish on shell. This is ot to imply that all O(N) models will fail; the actual arguments used so far were pecific to the O(2) model and need not carry over to O(4) or O(8). There is still ope, therefore, both for pure and matter coupled higher-N extensions of supergravity o be finite. In any case, one must not underestimate the triumphs of supergravity n the ultraviolet domain: it provides a matter coupling to gravity which is safe hrough two loops (instead of failing at one). This is particularly impressive since)(2) supergravity, for example, includes as a special case the one-loop nonrenormalizable Einstein-Maxwell theory, while O(3) includes the equally unacceptable Einstein-Maxwell-Dirac system. So the existence of extra symmetry always improves loop behavior, although there are still a number of open questions concerning the properties of cosmological supergravity, and its spontaneously broken variants, because it is difficult to do meaningful quantum perturbation theory in a cosmological background. Also, there may be problems due to anomalies, which have recently been discovered [52] in supersymmetric systems and in supergravity itself.

SUMMARY

The new gauge theory of supergravity is an elegant formal generalization of Einstein's theory, uniting spacetime with spinning matter and internal symmetries. Based on the graded extensions of the Poincaré algebra, it is the natural Dirac square root of relativity. Its observationally predictable deviations from the latter are negligible in the macroscopic domain because exchange of single spin 3/2 fermions has no macroscopic consequences, while multiple fermion exchange is necessarily short range. At present, therefore, its attraction is conceptual and at the microscopic level rather than empirical, although there may well be important consequences to the singularity theorems and to cosmological predictions. The new gauge invariance it embodies incorporates into single multiplets hitherto separate particle types, including non-gauge lower spin systems. It has solved and clarified many old problems of higher spin field couplings to gravity and vector gauge fields. Its very possibility has led to results in ordinary gravity not previously obtained, especially the positive energy theorem.

Particularly spectacular has been the significant improvement in ultraviolet behavior due to the constraints imposed by the new invariance. Since gravity couples to all matter, its nonrenormalizability would, in principle, destroy the predictive power of our basic renormalizable flat space theories (because of gravitational radiative corrections). These improvements are therefore highly significant physically, irrespective of whether supergravity models rich enough to encompass all the other known symmetries can be constructed. It may be that wider generalizations, involving higher derivatives (or higher spins) will be required to achieve these ends, although there appear to be a number of immediate difficulties with such actions.

Within the short span of its existence, the subject has expanded enormously and raised a great many new questions. In this respect, there is a great deal that mathematics can contribute to its development. First, we need a better formalism to handle the technical construction of the various models, whether in terms of superspace or by other means. For example, a generalization of Cartan's original simple classification of Riemannian invariants in terms of curvatures and their covariant derivatives would be an extremely powerful tool, as would an understanding of whether internal symmetries can be extended in terms of wider algebras. Even on the classical front, we have as yet almost no insights into the effect on singularity theorems, collapse and classical solutions. Topological properties of solutions involving "super" Gauss-Bonnet and Pontrjiagin invariants would also be very enlightening. Given what has already been obtained from the simplest notions of graded algebras, any mathematical progress is bound to have fruitful physical application.

ACKNOWLEDGMENTS

It is a pleasure for the author to thank the Max-Planck-Institut für Physik und Astrophysik for its hospitality, and several of its members for enlightening discussions, while this work was begun.

REFERENCES

1. D.Z. Freedman, P. van Nieuwenhuizen and S. Ferrara, Phys. Rev. D13, 3214 (1976).
2. S. Deser and B. Zumino, Phys. Lett. 62B, 335 (1976).
3. C. Teitelboim, Phys. Rev. Lett. 38, 1106 (1977); Phys. Lett. 69B, 240 (1977). R. Tabensky and C. Teitelboim, Phys. Lett. 69B, 453 (1977).
4. R. Arnowitt and P. Nath, Nucl. Phys. B122, 301 (1977); Phys. Rev. Lett. 36, 1526 (1976); Phys. Lett. 65B, 73 (1976), and earlier papers.
5. P. Fayet and S. Ferrara, Physics Reports 32C, 249 (1977).
6. R. Haag, J.T. Lopuszanski and M. Sohnius, Nucl. Phys. B88, 257 (1975). W. Nahm, CERN preprint TH 2341 (1977).
7. E. Cremmer, J. Scherk and S. Ferrara, LPTENS preprint (1977).

8. L. Brink, J.H. Schwarz and J. Scherk, Nucl. Phys. B121, 77 (1977).
F. Gliozzi, J. Scherk and E. Olive, Nucl. Phys. B122, 253 (1977).
M. Gell-Mann, P. Ramond and R. Slansky, Rev. Mod. Phys. (to be published).

9. E. Poggio, H.N. Pendleton; D.R.T. Jones, Phys. Lett. B (both in press). See also L.F. Abbott, M.T. Grisaru & H.J. Schnitzer, Phys. Lett. 71B, 161 (1977).

10. S. Ferrara, M. Kaku, P. van Nieuwenhuizen and P.K. Townsend, Nucl. Phys. B (in press); A. Ferber and P.G.O. Freund, Nucl. Phys. B122, 170 (1977); S. Ferrara and B. Zumino, CERN preprint TH 2418 (1977).

11. W. Rarita and J. Schwinger, Phys. Rev. 60, 61 (1941).

12. S. Deser, J. GRG 1, 9 (1970).

13. J. Wess and B. Zumino, Phys. Lett. 66B, 361 (1977).
J. Wess, Lectures at GIFT Seminar (1977).

14. A. Das and D.Z. Freedman, Nucl. Phys. B114, 271 (1977); G. Senjanovic, Phys. Rev. D16, 307 (1977).

15. S. Deser, J.H. Kay and K.S. Stelle, Phys. Rev. D16, 2448 (1977).

16. M. Pilati, Nucl. Phys. B (in press).

17. D.Z. Freedman, P. van Nieuwenhuizen and S. Ferrara, Phys. Rev. D14, 912 (1976)

18. S. Deser and C.J. Isham, Phys. Rev. D14, 2505 (1976).

19. K. Kuchar, J. Math. Phys. 17, 792; 801 (1976); 18, 1589 (1977).

20. G. Sterman, P. van Nieuwenhuizen and P.K. Townsend, ITP-SB-77-65 preprint (1977).

21. E.S. Fradkin and M.A. Vasiliev, Phys. Lett. 72B, 70 (1977).
R.E. Kallosh, JETP Letters 26, 575 (1977).

22. S. Ferrara and P. van Nieuwenhuizen, Phys. Rev. Letters 37, 1669 (1976).
D.Z. Freedman and A. Das, Nucl. Phys. B120, 221 (1977); P.K. Townsend Phys. Rev. D (in press).

23. S.W. MacDowell and F. Mansouri, Phys. Rev. Lett. 38, 739, 1376(E) (1977).
P.K. Townsend and P. van Nieuwenhuizen, Phys. Lett. B (in press).

24. D. Boulware and S. Deser, Phys. Rev. D6, 3368 (1972).

25. S. Ferrara, F. Gliozzi, J. Scherk and P. van Nieuwenhuizen, Nucl. Phys. B117, 333 (1976), and with D.Z. Freedman and P. Breitenlohner, Phys. Rev. D15, 1013 (1977).
A. Das, M. Fishler and M. Rocek, preprints ITP-SP-77-15 / 38 (1977).
E. Cremmer and J. Scherk DAMTP preprint 77/7 (1977).
S. Ferrara, J. Scherk and P. van Nieuwenhuizen, Phys. Rev. Letters 37, 1037 (1976); D.Z. Freedman and J.H. Schwarz, Phys. Rev. D15, 1007 (1977); D.Z. Freedman, Phys. Rev. D (in press).

26. P. Breitenlohner, Nucl. Phys. B124, 500 (1977); Phys. Lett 67B, 49 (1977).

27. A. Neveu and J.H. Schwarz, Nucl. Phys. B31, 86 (1971); P. Ramond, Phys. Rev. D3, 2415 (1971); Y. Aharonov, A. Casher and L. Susskind, Phys. Letters 35B, 512 (1971); J.L. Gervais and B. Sakita, Nucl. Phys. B34, 633 (1971).

28. J. Wess and B. Zumino, Nucl. Phys. B70, 39 (1974).
29. S. Deser and B. Zumino, Phys. Lett. 65B, 369 (1976).
30. L. Brink, S. Deser, B. Zumino, P. Di Vecchia and P. Howe, Phys. Lett. 64B, 435 (1976).
31. S. Ferrara, J. Scherk and B. Zumino, Phys. Letters 66B, 35 (1977).
D.Z. Freedman, Phys. Rev. Letters 38, 105 (1976).
A. Das preprint ITP SB 77-4 (1977).
E. Cremmer and J. Scherk (in press).
32. E. Cremmer and J. Scherk LPTENS 77/19 preprint (1977).
33. B. De Wit and D.Z. Freedman, Nucl. Phys. B130, 105 (1977).
34. S. Deser and B. Zumino, Phys. Rev. Letters 38B, 1433 (1977).
35. P.A.M. Dirac, Ann. Math. 36, 657 (1935); F. Gursey and T.D. Lee, Proc. Nat. Acad. Sci. 49, 179 (1963); L. Halpern J. GRG 8, 623 (1977).
36. D.V. Volkov and V.P. Akulov, Phys. Lett. 46B, 109 (1973).
37. P. Fayet, Ecole Polytechnique preprint (1977).
38. T. Dereli and S. Deser, J. Phys. A10, L149 (1977) and J. Phys. A (in press).
39. R. Arnowitt, S. Deser and C.W. Misner, Phys. Rev. 117, 1595 (1960); 118, 1100 (1960); 122, 997 (1961).
40. S. Deser and C. Teitelboim, Phys. Rev. Letters 39, 249 (1977).
41. D.R. Brill and S. Deser, Ann. Phys. 50, 548 (1968); with L.D. Faddeev, Phys. Lett. 26A, 538 (1968); P.S. Jang and R. Wald, J. Math. Phys. 18, 41 (1977).
42. M.T. Grisaru, Phys. Lett. B (in press).
43. M.T. Grisaru and H.W. Pendleton, Nucl. Phys. B124, 81 (1977).
44. D. Boulware and S. Deser, Ann. Phys. 89, 173 (1975).
45. G. 't Hooft and M. Veltman, Ann. Inst. H. Poincaré 20, 69 (1974).
46. S. Deser and P. van Nieuwenhuizen, Phys. Rev. D10, 401, 411 (1974); D10, 3337 (1974), with H.-S. Tsao.
47. P. van Nieuwenhuizen and J.A.M. Vermaseren, Phys. Rev. D (in press).
48. M.T. Grisaru, P. van Nieuwenhuizen and J.A.M. Vermaseren, Phys. Rev. Lett. 37, 1662 (1976).
M.T. Grisaru, Phys. Lett. 66B, 75 (1977); E. Tomboulis Phys. Lett. 67B, 417 (1977).
49. S. Ferrara and B. Zumino, Nucl. Phys. B87, 207 (1975).
50. S. Deser, J.H. Kay and K.S. Stelle, Phys. Rev. Letters 38, 527 (1977).
51. S. Deser and J.H. Kay (to be published).
52. L.F. Abbott, M.T. Grisaru and H.J. Schnitzer, Phys. Lett. B (to be published).

REPRESENTATIONS OF CLASSICAL LIE SUPERALGEBRAS

V. Kac
Department of Mathematics
Massachusetts Institute of Technology, Cambridge, MA 02139

These notes contain an exposition of the results of [1] and [2] concerning the finite-dimensional representations of basic classical Lie superalgebras. Among the new results we note a) the explicit construction of typical representations for all superalgebras (modules $V(\Lambda)$), b) some new characterizations of typical representations (see Theorem 1), c) formula for the multiplicities of the restriction of a representation to the even part (formula (2.8)) and d) structure theory of modules $\tilde{V}(\Lambda)$ (Theorem 3).

The list of complex finite-dimensional simple Lie superalgebras consists of two essentially different parts – classical and Cartan superalgebras [1]. Classical superalgebras are those for which the even subalgebra is reductive. The list of classical Lie superalgebras is in turn divided into two parts – basic and strange superalgebras. The basic classical Lie superalgebras are from many points of view the closest to the ordinary simple Lie algebras. For instance they can be constructed in terms of Cartan matrix and canonical generators. In [1] these Lie superalgebras are called contragredient superalgebras.

First we describe the structure of basic classical Lie superalgebras . Then we construct irreducible representations with highest weight of these algebras and classify all the finite-dimensional irreducible representatons . The central part of the notes is the computation of the characters and supercharacters for typical finite-dimensional irreducible representations of basic classical Lie superalgebras. We call an irreducible representation <u>typical</u>, if it splits in any finite-dimensional representation. An equivalent definition requires that the infinitesimal character determines the representation uniquely among the irreducible finite-dimensional representations.

We then find necessary and sufficient conditions for a representation to be typical in terms of highest weight (these conditions are expressed in a finite system of inequalities). A crucial point in the proof is the invariants restriction theorem for Lie superalgebras (Theorem 2), which is a generalization of the classical Chevalley theorem for semisimple Lie algebras. An essential circumstance here is that Chevalley theorem only holds in a weakened form. This is related to the presence of "degenerate" (non-typical) representations, for which the problem of computing the characters remains unsolved (cf.

formula (2.1)).

The ground field is the field $\mathbb{C}$ of complex numbers.

§1. Basic classical Lie superalgebras

1. Some examples [1]. Let $V = V_{\bar{0}} \oplus V_{\bar{1}}$ be a Z_2-graded space, $\dim V_{\bar{0}} = m$, $\dim V_{\bar{1}} = n$. The associative algebra End V becomes an associative superalgebra if we let

$$\mathrm{End}_i V = \{a \in \mathrm{End}\, V \,|\, a\, V_s \subseteq V_{i+s}\}\ , \quad i,s \in \mathbb{Z}_2 \ .$$

The bracket $[a,b] = ab - (-1)^{(\deg a)(\deg b)} ba$ makes End V into a Lie superalgebra, denoted by $\ell(V)$ or $\ell(m,n)$. If we regard the same decomposition $V = V_0 \oplus V_1$ as a $\mathbb{Z}$-gradation of V, then the same construction gives a consistent $\mathbb{Z}$-gradation: $\ell(V) = G_{-1} \oplus \ell(V)_{\bar{0}} \oplus G_1$. On $\ell(V)$ there is defined the supertrace, a linear function $\mathrm{str} : \ell(V) \to k$, which is uniquely defined by the properties $\mathrm{str}([a,b]) = 0$, $a,b \in \ell(V)$, and $\mathrm{str}\ \mathrm{id}_V = m - n$.

The subspace

$$s\ell(m,n) = \{a \in \ell(m,n) \,|\, \mathrm{str}\ a = 0\}$$

is an ideal in $\ell(m,n)$ of codimension 1. In some (homogeneous) basis of V, $\ell(m,n)$ consists of matrices of the form $\begin{pmatrix}\alpha & \beta\\ \gamma & \delta\end{pmatrix}$ and $\mathrm{str}\begin{pmatrix}\alpha & \beta\\ \gamma & \delta\end{pmatrix} = \mathrm{tr}\ \alpha - \mathrm{tr}\ \delta$. The $\mathbb{Z}$-gradation $s\ell(m,n) = G_{-1} \oplus s\ell(m,n)_{\bar{0}} \oplus G_1$ looks as follows: $s\ell(m,n)_{\bar{0}}$ is the set of matrices of the form $\begin{pmatrix}\alpha & 0\\ 0 & \delta\end{pmatrix}$, where $\mathrm{tr}\ \alpha = \mathrm{tr}\ \delta$, G_1 is the set of matrices of the form $\begin{pmatrix}0 & \beta\\ 0 & 0\end{pmatrix}$ and G_{-1} of the form $\begin{pmatrix}0 & 0\\ \gamma & 0\end{pmatrix}$ (where α is an $(m\times m)$-, δ an $(n\times n)$-, β an $(m\times n)$-, and γ an $(n\times m)$-matrix).

$s\ell(n,n)$ contains the one-dimensional ideal consisting of the scalar matrices $\lambda 1_{2n}$. We set

$$\mathbb{A}(m,n) = s\ell(m+1,n+1) \quad \text{for} \quad m \neq n, \quad m,n \geq 0 \ ,$$

$$\mathbb{A}(n,n) = s\ell(n+1,n+1)/\{\lambda 1_{2n+2}\}\ , \qquad n > 0 \ .$$

Let F now be a non-degenerate bilinear form on V, such that $V_{\bar{0}}$ and $V_{\bar{1}}$ are orthogonal and the restriction of F to $V_{\bar{0}}$ is a symmetric and to $V_{\bar{1}}$ a skew-symmetric form (in particular, $n = 2r$ is even). We define in $\ell(m,n)$ the subalgebra

sp(m,n) = $osp(m,n)_{\bar 0} \oplus osp(m,n)_{\bar 1}$ by setting
$sp(m,n)_s = \{a \in \ell(m,n)_s \mid F(a(z),y) = -(-1)^{s \deg x} F(x,a(y))\}, \; s \in \mathbb{Z}_2$.
he algebra $osp(m,n)$ is called ortho-symplectic Lie superalgebra.

Let us find the explicit matrix form of the elements of $osp(m,n)$. 'e put

$$\mathcal{J}_m = \begin{bmatrix} & & 1 \\ & \cdot^{\cdot^{\cdot}} & \\ 1 & & \end{bmatrix}, \quad B = \left[\begin{array}{c|cc} \mathcal{J}_m & 0 & 0 \\ \hline 0 & 0 & \mathcal{J}_r \\ 0 & -\mathcal{J}_r & 0 \end{array}\right]$$

and let d^r mean the transposition of a matrix d with respect to the side diagonal. In some homogeneous basis of V the matrix of the form ' is B and $osp(m,n)$ consists in this basis of the matrices of the form:

$$\left[\begin{array}{c|cc} \alpha & \xi & \eta \\ \hline -\eta^r & \beta & \gamma \\ \xi^r & \delta & -\beta^r \end{array}\right]$$

where $\alpha = -\alpha^r$, $\gamma = \gamma^r$, $\delta = \delta^r$, β is any $(r\times r)$-matrix, ξ and η are any $(m\times r)$-matrices.

By analogy with Cartan's notation we set:

$$\begin{aligned} &\underset{\sim}{B}(m,n) = osp(2m+1,2n), && m \geq 0, && n > 0; \\ &\underset{\sim}{D}(m,n) = osp(2m,2n), && m \geq 2, && n > 0; \\ &\underset{\sim}{C}(n) = osp(2,2n-2), && n \geq 2. && \end{aligned}$$

We now examine the Lie superalgebra $\underset{\sim}{C}(n)$. Subalgebra of the form $\underset{\sim}{C}(n)_{\bar 0}$ consists of the matrices

$$\left[\begin{array}{cc|cc} \alpha & 0 & & \\ 0 & -\alpha & & \\ \hline & & \beta & \gamma \\ & & \delta & -\beta^r \end{array}\right]$$

where β, γ, and δ are $(n-1\times n-1)$-matrices, $\gamma^r = \gamma$, $\delta^r = \delta$, and $\alpha \in \underset{\sim}{C}$. $\underset{\sim}{C}(n)$ has the consistent $\underset{\sim}{Z}$-gradation:
$\underset{\sim}{C}(n) = G_{-1} \oplus \underset{\sim}{C}(n)_{\bar 0} \oplus G_1$, where G_{-1} and G_1 consist of matrices of the form (respectively):

$$\left[\begin{array}{cc|cc} & & 0 & 0 \\ & & \xi_2 & \eta_2 \\ \hline -\eta_2^r & 0 & & \\ \xi_2^r & 0 & & \end{array}\right] \qquad \left[\begin{array}{cc|cc} & & \xi_1 & \eta_1 \\ & & 0 & 0 \\ \hline 0 & -\eta_1^r & & \\ 0 & \xi_1^r & & \end{array}\right]$$

2. Classification. We call a bilinear form F on Lie superalgebra $G = G_{\bar{0}} + G_{\bar{1}}$ invariant if it satisfies the following three conditions:

a) $F(a,b) = (-1)^{\alpha\beta} F(b,a)$, $a \in G_\alpha$, $b \in G_\beta$;

b) $F(a,b) = 0$, $a \in G_{\bar{0}}$, $b \in G_{\bar{1}}$;

c) $F([a,b],c) = F(a,[b,c])$.

We call a Lie superalgebra $G = G_{\bar{0}} \oplus G_{\bar{1}}$ basic classical if a) G is simple, b) Lie algebra G_0 is a reductive subalgebra and c) there exists a non-degenerate invariant bilinear form on G.

From [1] one derives the following statement.

Proposition 1.1. a) A complete list of basic classical Lie superalgebras is as follows:

1) simple Lie algebras,

2) $A(m,n)$, $B(m,n)$, $C(n)$, $D(m,n)$, $D(2,1;\alpha)$, $F(4)$, $G(3)$.

b) For the Lie superalgebras $B(m,n)$, $D(m,n)$, $D(2,1;\alpha)$, $F(4)$, $G(3)$, the $G_{\bar{0}}$-module $G_{\bar{1}}$ is irreducible and isomorphic to the modules in the following list:

G	$G_{\bar{0}}$	$G_{\bar{0}} \mid G_{\bar{1}}$	G	$G_{\bar{0}}$	$G_{\bar{0}} \mid G_{\bar{1}}$
$B(m,n)$	$B_m \oplus C_n$	$so_{2m+1} \otimes sp_{2n}$	$F(4)$	$B_3 \oplus A_1$	$spin_7 \otimes s\ell_2$
$D(m,n)$	$D_m \oplus C_n$	$so_{2m} \otimes sp_{2n}$	$G(3)$	$G_2 \oplus A_1$	$G_2 \otimes s\ell_2$
$D(2,1;\alpha)$	$A_1 \oplus A_1 \oplus A_1$	$s\ell_2 \otimes s\ell_2 \otimes s\ell_2$			

c) The Lie superalgebras $A(m,n)$ and $C(n)$ admit a unique consistent Z-gradation of the form $G_{-1} \oplus G_0 \oplus G_1$, where the G_0-modules G_1 and G_{-1} are irreducible and contragredient; they are isomorphic to the modules in the following list:

G	G_0	$G_0 \mid G_{-1}$
$A(m,n)$	$A_m \oplus A_n \oplus C$	$s\ell_{m+1} \otimes s\ell_{n+1} \otimes C$
$A(n,n)$	$A_n \oplus A_n$	$s\ell_{n+1} \otimes s\ell_{n+1}$
$C(n)$	$C_{n-1} \oplus C$	csp_{2n-2}

3. Borel subalgebras. Let $G = G_{\bar 0} \oplus G_{\bar 1}$ be a Lie superalgebra and $B_{\bar 0}$ be a Borel subalgebra (maximal solvable subalgebera of $G_{\bar 0}$). We call Borel subalgebra of Lie superalgebra G any maximal solvable subalgebra B of G, which contains $B_{\bar 0}$. All Borel subalgebras in a Lie algebra are conjugate. Lie superalgebra in general contains several classes of conjugacy of Borel subalgebras but of course any class of conjugacy contains a Borel subalgebra B with given $B_{\bar 0}$. From this remark we easily obtain the following proposition (see also [1]).

Proposition 1.2. a) There is a one-to-one correspondence between classes of conjugacy of Borel subalgebras in $s\ell(m,n)$ and the orderings $\sigma: e_{i_1}, \dots, e_{i_{m+n}}$ of a fixed homogeneous basis of V, such that the sequence of indices of even elements and respectively of odd ones increases. This correspondence is given by:

$$\sigma \to B_\sigma = \{a \in s\ell(m,n) \mid a \langle e_{i_1}, \dots, e_{i_s} \rangle \subset \langle e_{i_1}, \dots, e_{i_s} \rangle, \quad s = 1,2,\dots,m+n\}.$$

b) There is a one-to-one correspondence between classes of conjugacy of Borel subalgebras in $osp(m,n)$ and orderings $\sigma: e_{i_1}, \dots, e_{i_{[m+n/2]}}$ of a fixed homogeneous basis of a fixed maximal isotropic subspace of V with the same property as in a). The correspondence is given by:

$$\sigma \to B_\sigma = \{a \in osp(m,n) \mid a \langle e_{i_1}, \dots, e_{i_s} \rangle \subset \langle e_{i_1}, \dots, e_{i_s} \rangle, \quad s = 1,2,\dots,[(m+n)/2]\}.$$

c) The numbers of conjugacy classes of Borel subalgebras in $D(2,1;\alpha)$, $F(4)$ and $G(3)$ is 4, 4 and 1 respectively (the corresponding systems of simple roots are described in [1]).

4. Root system [1]. Let $G = G_{\bar{0}} \oplus G_{\bar{1}}$ be a basic classical Lie superalgebra and let H be a Cartan subalgebra of $G_{\bar{0}}$. For $\alpha \in H^*$, $\alpha \neq 0$, we set

$$G_\alpha = \{a \in G \mid [h,a] = \alpha(h)a, \quad h \in H\}.$$

We call α a root if $G_\alpha \neq 0$. A root α is called even (respectively odd) if $G_\alpha \cap G_{\bar{0}} \neq 0$ (respectively $G_\alpha \cap G_{\bar{1}} \neq 0$).

We denote by Δ, Δ_0 and Δ_1 the sets of all roots, even roots and odd roots respectively. We introduce also the following two sets of roots:

$$\bar{\Delta}_0 = \{\alpha \in \Delta_0 \mid \alpha/2 \notin \Delta_1\}$$
$$\bar{\Delta}_1 = \{\alpha \in \Delta_1 \mid 2\alpha \notin \Delta_0\}.$$

Below we describe the systems of even and odd roots for all basic classical Lie superalgebras.

In all the examples the Cartan subalgebra H is a subspace of the space D of diagonal matrices; the roots are expressed in terms of the standard basis ε_i of D^* (more precisely the restrictions of the ε_i to H).

$\underset{\sim}{A}(m,n)$. The roots are expressed in terms of linear functions $\varepsilon_1,\ldots,\varepsilon_{m+1}, \delta_1 = \varepsilon_{m+2},\ldots,\delta_{n+1} = \varepsilon_{m+n+2}$.

$$\Delta_0 = \{\varepsilon_i - \varepsilon_j;\ \delta_i - \delta_j\};\ \Delta_1 = \{\pm(\varepsilon_i - \delta_j)\}, \quad i \neq j.$$

$\underset{\sim}{B}(m,n)$. The roots are expressed in terms of linear functions $\varepsilon_1,\ldots,\varepsilon_m, \delta_1 = \varepsilon_{2m+1},\ldots,\delta_n = \varepsilon_{2m+n}$.

$$\Delta_0 = \{\pm\varepsilon_i\pm\varepsilon_j;\pm2\delta_i;\pm\varepsilon_i;\pm\delta_i\pm\delta_j\}; \quad \Delta_1 = \{\pm\delta_i;\pm\varepsilon_i\pm\delta_j\}, \quad i \neq j.$$

$\underset{\sim}{C}(n)$. The roots are expressed in terms of linear functions $\varepsilon_1, \delta_1 = \varepsilon_3,\ldots,\delta_{n-1} = \varepsilon_{n+1}$.

$$\Delta_0 = \{\pm2\delta_i;\pm\delta_i\pm\delta_j\}; \quad \Delta_1 = \{\pm\varepsilon_1\pm\delta_i\}.$$

$\underset{\sim}{D}(m,n)$. The roots are expressed in terms of linear functions $\varepsilon_1,\ldots,\varepsilon_m, \delta_1 = \varepsilon_{2m+1},\ldots,\delta_n = \varepsilon_{2m+n}$.

$$\Delta_0 = \{\pm\varepsilon_i\pm\varepsilon_j;\pm 2\delta_i;\pm\delta_i\pm\delta_j\},\quad \Delta_1 = \{\pm\varepsilon_i\pm\delta_j\},\quad i \neq j.$$

$D(2,1;\alpha)$. The roots are expressed in terms of linear functions ε_1, ε_2 and ε_3.

$$\Delta_0 = \{\pm 2\varepsilon_i\};\quad \Delta_1 = \{\pm\varepsilon_1\pm\varepsilon_2\pm\varepsilon_3\}.$$

$F(4)$. The roots are expressed in terms of linear functions ε_1, ε_2, ε_3, corresponding to B_3, and δ, corresponding to A_1.

$$\Delta_0 = \{\pm\varepsilon_i\pm\varepsilon_j;\pm\varepsilon_i;\pm\delta\},\quad i \neq j;\quad \Delta_1 = \{\tfrac{1}{2}(\pm\varepsilon_1\pm\varepsilon_2\pm\varepsilon_3\pm\delta)\}.$$

$G(3)$. The roots are expressed in terms of linear functions ε_1, ε_2, ε_3, corresponding to G_2, $\varepsilon_1 + \varepsilon_2 + \varepsilon_3 = 0$, and δ, corresponding to A_1.

$$\Delta_0 = \{\varepsilon_i - \varepsilon_j;\pm\varepsilon_i;\pm 2\delta\};\quad \Delta_1 = \{\pm\varepsilon_i\pm\delta;\pm\delta\}.$$

We fix an invariant non-degenerate bilinear form $(\ ,\ )$ on G.

Proposition 1.3. A basic classical Lie superalgebra G satisfies the following properties:

a) $G = H \oplus \left(\bigoplus_{\alpha\in\Delta} G_\alpha\right)$,

b) $\dim G_\alpha = 1$, $\alpha \in \Delta$, except for $A(1,1)$.

c) $[G_\alpha, G_\beta] \neq 0$ if and only if $\alpha, \beta, \alpha+\beta \in \Delta$.

d) If $\alpha \in \Delta$ (or Δ_0, or Δ_1, or $\bar{\Delta}_0$, or $\bar{\Delta}_1$), then so is $-\alpha$.

e) $(G_\alpha, G_\beta) = 0$ for $\alpha \neq -\beta$, the form $(\ ,\ )$ determines a non-degenerate pairing of G_α and $G_{-\alpha}$ and the restriction of $(\ ,\ )$ on H is non-degenerate.

f) $[e_\alpha, e_{-\alpha}] = (e_\alpha, e_{-\alpha})h_\alpha$, where h_α is a non-zero vector determined by $(h_\alpha, h) = \alpha(h)$, $h \in H$.

g) The bilinear form on H^* defined by $(\alpha,\beta) = (h_\alpha, h_\beta)$ is a non-degenerate W-invariant form.

h) $k\alpha \in \Delta$ for $\alpha \neq 0$, $k \neq \pm 1$, if and only if $\alpha \in \Delta_1$, and $(\alpha,\alpha) \neq 0$; in this case $k = \pm 2$.

In particular, root α belongs to $\bar{\Delta}_1$ if and only if $(\alpha,\alpha) = 0$.

Let now $B_{\bar{0}}$ be a Borel subalgebra of $G_{\bar{0}}$, containing H. We fix a Borel subalgebra $B = B_{\bar{0}} \oplus B_{\bar{1}}$ of G. Since the adjoint representation of H in G is diagonalizable we obtain the following decomposition of G:

$$G = N^- \oplus H \oplus N^+, \quad B = H \oplus N^+,$$

where N^- and N^+ are subalgebras and $[H,N^+] \subset N^+$, $[H,N^-] \subset N^-$.

A root α is called positive (respectively negative) if $G_\alpha \cap N^+ \neq 0$ (respectively $G_\alpha \cap N^- \neq 0$. We denote by Δ^+, Δ_0^+, Δ_1^+, $\bar{\Delta}_0^+$, $\bar{\Delta}_1^+$ the subsets of positive roots in the sets Δ, Δ_0 etc. respectively.

We let ρ_0 (respectively ρ_1) denote the half-sum of all of the even (respectively odd) positive roots. We set $\rho = \rho_0 - \rho_1$. We let also $\bar{\rho}_0$ denote the half-sum of all the roots from $\bar{\Delta}_0^+$.

A positive root α is called simple if it cannot be decomposed into a sum of two positive roots. Let $\Pi = \{\alpha_1,\ldots,\alpha_r\}$ be the set of all simple roots.

Proposition 1.4. *For a basic classical Lie superalgebra* G *the following properties hold.*

a) *All the subspaces* $G_\alpha \cap N^\pm$ *are one-dimensional.*

b) *One can choose non-zero elements* $e_i \in G_{\alpha_i} \cap N^+$, $f_i \in G_{-\alpha_i} \cap N^-$ *and* $h_i \in H$, $i = 1,\ldots,r$, *such that* e_i, f_i, h_i, $i = 1,\ldots,r$, *is the system of generators of* G, *satisfying the following relations:*

$$(1.1)\quad [e_i,f_j] = \delta_{ij}h_i,\ [h_i,h_j] = 0,\ [h_i,e_j] = a_{ij}e_j,\ [h_i,f_j] = -a_{ij}f_j$$

c) $(\rho,\alpha_i) = \frac{1}{2}(\alpha_i,\alpha_i)$; *in particular*, $(\rho,\alpha_i) = 0$ *if* $\alpha_i \in \bar{\Delta}_1$.

We can (and will) assume that the matrix $A = (a_{ij})$ satisfies the following normalizing conditions:

$$(1.2)\begin{cases} a_{ii} = 2 \quad \text{or} \quad 0, \quad i = 1,\ldots,r; \\ \text{if } a_{ii} = 0, \text{ then the first non-zero element among } a_{ii+k}, \\ \qquad k = 1,2,\ldots, \text{ is } 1. \end{cases}$$

The matrix A is called the Cartan matrix of Lie superalgebra G. We emphasize that A depends on the choice of B.

Let τ be a subset of $\{1,\ldots,r\}$, consisting of those i for which α_i is an odd root.

It is convenient to describe the pair (A,τ) by a Dynkin diagram. t consists of r nodes, of the form O, ⊗ and ●, which are called hite, grey and black respectively. The i-th node is white if $i \in \tau$ nd grey or black if $i \in \tau$ and $a_{ii} = 0$ or 2 respectively. The -th and j-th nodes are joined by $|a_{ij}a_{ji}|$ lines unless the case $(2,1;\alpha)$. We note that always $a_{ij}a_{ji} = 0$ implies $a_{ij} = a_{ji} = 0$ nd that in the i-th row of A all the entries are non-positive ntegers if $a_{ii} = 2$.

roposition 1.5. *Let G be a basic classical Lie super algebra, let* ʜ *be a Cartan subalgebra and let B be a Borel subalgebra in G,* ontaining H.

a) *The elements e_i, f_i, h_i, $i = 1,\dots,r$, generate G.*

b) *The elements $h_1,\dots,h_r$ span H. They are linearly independent for all G except for $G = A(n,n)$. For $A(n,n)$ there is a unique linear dependence:* $(h_1+h_{2n+1}) + 2(h_2+h_{2n}) + \cdots + (n-1)+(h_{n-1} + h_{n+1}) + nh_n = 0$.

c) *Let $\bar{G} \to G$ be the universal central extension of G. Then* $\bar{G} = G$ *if* $G \neq A(n,n)$ *and* $\bar{G} = s\ell(n+1,n+1)$ *if* $G = A(n,n)$.

d) G *is uniquely determined up to isomorphism by the pair* (A,τ).

Basic classical Lie superalgebras admit a Borel subalgebra B, *for which the corresponding Dynkin diagram has the form represented in Table 1. The labels c_i in the Table are the coefficients of the decomposition of the highest root with respect to simple roots,* s *being the number of the exceptional nonwhite node,* r *being the total number of nodes.*

We call this B a *distinguished* Borel subalgebra. The pair (A,τ) is uniquely determined by the Dynkin diagram except for the case $D(2,n)$ and $D(2,1;\alpha)$. The Cartan matrix of $D(2,1;\alpha)$ is

$$D_\alpha = \begin{pmatrix} 0 & 1 & \alpha \\ -1 & 2 & 0 \\ -1 & 0 & 2 \end{pmatrix}$$

and (3×3)-submatrix corresponding to the last 3 nodes of Dynkin diagram of $D(2,n)$ is D_1.

Table 1

G	Diagram	s	r
A(m,n)	1 1 1 1 1 O—O— ··· —O—⊗— ···—O	m + 1	m + n + 1
B(m,n), m > 0	2 2 2 2 2 2 O—O— ···—O—⊗— ··· —O=>O	n	m + n
B(0,n)	2 2 2 2 2 O—O— ··· —O—O=>●	n	n
C(n), n > 2	1 2 2 2 1 ⊗—O— ··· —O—O<=O	1	n
D(m,n)	2 2 2 2 2 O—O— ···—O—⊗— ··· —O< (O1, O1)	n	m + n
F(4)	2 3 2 1 ⊗—O<=O—O	1	4
G(3)	2 4 2 ⊗—O<≡O	1	3
D(2,1;α)	2⊗< (→O1, →O1)	1	3

Let B be the distinguished Borel subalgebra of a basic classical Lie superalgebra G. Let e_i, f_i, h_i, $i = 1,\dots,r$, be the corresponding generators. We introduce a distinguished $\mathbb{Z}$-gradation $G = \bigoplus_{i\in\mathbb{Z}} G_i$ of G by putting:

$$\deg h_i = 0, \quad i = 1,\dots,r; \quad \deg e_i = \deg f_i = 0, \quad i \neq s;$$

$$\deg e_s = -\deg f_s = 1.$$

We call Lie superalgebras A(m,n) and C(n) superalgebras of I type and all the other basic classical Lie superalgebras - superalgebras of II type.

roposition 1.6. The distinguished $\mathbb{Z}$-gradation $G = \bigoplus_{i\in\mathbb{Z}} G_i$ of a basic lassical Lie superalgebra G satisfies the following properties.

a) $G_i = 0$ for $|i| > 1$ for superalgebras of I type and $G_i = 0$ for $|i| > 2$ for superalgebras of II type.

b) G_0 is a direct sum of a one-dimensional center C (except for $A(n,n)$, when $C = 0$) and a semisimple Lie algebra $[G_0, G_0]$, whose Dynkin diagram is a subdiagram of the Dynkin diagram of Lie superalgebra G with a non-white (s-th) node removed.

c) The representations of G_0 and G_i and G_{-i} are dual. The lowest weight of G_0-module G_1 is α_s. The lowest weight of G_0-module G_2 in case of superalgebras of II type is $\delta = \sum_{i=s}^{r} c_i\alpha_i$, where c_i are the labels in Table 1.

d) The system $\Pi' = \{\alpha_i, i \neq s\}$ for superalgebras of I type and $\Pi' = \{\alpha_i, i \neq s; \delta\}$ for superalgebras of II type is a system of simple roots for $G_{\bar{0}}$.

5. Weyl group and functions L and K[2]. Let W be the Weyl group of the even part $G_{\bar{0}}$ of a basic classical Lie superalgebra $G = G_{\bar{0}} \oplus G_{\bar{1}}$. Every element $w \in W$ can be written as a product of a certain number of reflections with respect to the even roots. We introduce two homomorphisms ε and ε': $W \to \{\pm 1\}$ as follows. Put $\varepsilon(w) = 1$ if the number of reflections in the expression of w is even and $\varepsilon(w) = -1$ if the number is odd; put $\varepsilon'(m) = 1$ if the number of reflections with respect to the roots of $\bar{\Delta}_0^+$ in this expression is even and $\varepsilon'(w) = -1$ if the number is odd. ε and ε' are well-defined. It follows from Proposition 1.7 below.

For $\Lambda \in H^*$ let $D(\Lambda)$ be the set of linear functions on H^* of the form: $\Lambda - \sum_{\alpha\in\Delta^+} n_\alpha \alpha$, $n_\alpha \in \mathbb{Z}_+$.

For $\lambda = \Lambda - \Sigma n_\alpha \alpha \in D(\Lambda)$ we set $\omega(\lambda) = (-1)^{\Sigma n_\alpha \deg \alpha}$.

Let E be the algebra of functions on H^* that vanish outside of the union of finitely many sets of the form $D(\Lambda)$, with the convolution operation. $e^\lambda \in E$ is defined by $e^\lambda(\lambda) = 1$, $e^\lambda(\mu) = 0$ for $\mu \neq \lambda$. Clearly, $e^\lambda \cdot e^\mu = e^{\lambda+\mu}$. We set:

$$L = \prod_{\alpha\in\Delta_0^+} (e^{\alpha/2} - e^{-\alpha/2}) / \prod_{\alpha\in\Delta_1^+} (e^{\alpha/2} + e^{-\alpha/2})$$

$$L' = \prod_{\alpha\in\Delta_0^+} (e^{\alpha/2} - e^{-\alpha/2}) / \prod_{\alpha\in\Delta_1^+} (e^{\alpha/2} - e^{-\alpha/2}) .$$

We define the Kostant function K on H^* by setting $K(\lambda)$ equal to the number of sets $\{n_\alpha\}$, $n_\alpha \in \mathbb{Z}_+$ for $\alpha \in \Delta_0^+$ and $n_\alpha = 0,1$ for

$\alpha \in \Delta_1^+$ such that $\sum_{\alpha\in\Delta^+} n_\alpha \alpha = -\lambda$. We set $K'(\lambda) = \omega(\lambda)K(\lambda)$.

Proposition 1.7. a) The sets Δ, Δ_0, Δ_1, $\bar{\Delta}_0$, $\bar{\Delta}_1$ are W-invariant.

b) W acts transitively on $\bar{\Delta}_1$ and the function $\prod_{\alpha\in\bar{\Delta}_1^+} \alpha$ is W-invariant.

c) For superalgebras of I type the set Δ_1^+ is W-invariant; In particular, $w(\rho_1) = \rho_1$ for $w \in W$.

d) Let T be a subgroup in W generated by reflections with respect to the roots from $\Delta_0^+ - \bar{\Delta}_0^+$.

  (i) W is a semidirect product of the normal subgroup T and a subgroup W_0 such that $\varepsilon'(w_0 t) = \varepsilon(w_0)$ for $w_0 \in W_0$, $t \in T$.

  (ii) $\prod_{\alpha\in\bar{\Delta}_0^+} (e^{\alpha/2} - e^{-\alpha/2})$ is T-invariant;

  (iii) the order of T is given by the following formula:

  $\#T = 1$ if $G \neq B(m,n)$ or $G(3)$,

  $\#T = 2^n$ for $B(m,n)$,

  $\#T = 2$ for $G(3)$.

e) $w(L) = \varepsilon(w)L$, $w(L') = \varepsilon'(w)L'$ for $w \in W$.

f) $L\cdot K = L'\cdot K' = e^\rho$.

Proof. a) is evident, b) and c) can be checked case by case verification. Also from the list of the root systems Δ_0 it is clear that $\Delta_0 \neq \bar{\Delta}_0$ only for $G = B(m,n)$ or $G(3)$, i.e., $(W:W_0) \neq 1$ only for these superalgebras. For these two cases d) follows from the structure of Weyl group of Lie algebra C_n. For the proof of e) and f) see [4].

§2. Finite-dimensional representations of basic classical Lie superalgebras.

1. Universal enveloping superalgebra and induced representations ([1], [3]). For associative Z_2-graded algebra A a bracket is defined by:

$$[a,b] = ab - (1)^{(\deg a)(\deg b)} ba .$$

This bracket defines a Lie superalgebra A_L.

A pair $(U(G),i)$ where $U(G)$ is an associative Z_2-graded algebra and $i : G \to U(G)_L$ is a homomorphism of Lie superalgebras, is called a universal enveloping superalgebra of G if for any other pair (U',i') there is a unique homomorphism $\theta : U \to U'$ for which $i' = \theta \circ i$.

The universal enveloping superalgebra of $G = G_{\bar{0}} \oplus G_{\bar{1}}$ is con-tructed as follows. Let $T(G)$ be the tensor superalgebra over the pace G with the induced $\mathbb{Z}_2$-gradation, and R the ideal of $T(G)$ enerated by the elements of the form:

$$[a,b] - a \otimes b + (-1)^{(\deg a)(\deg b)} b \otimes a .$$

e set $\mathfrak{U}(G) = T(G)/R$. The natural map $G \longrightarrow \mathfrak{U}(G)$ evidently induces homomorphism $i : G \to \mathfrak{U}(G)_L$, and the pair $(\mathfrak{U}(G), i)$ is the re-uired enveloping superalgebra.

In Lie superalgebra case the Poincaré-Birkhoff-Witt theorem has he following form.

The Poincaré-Birkhoff-Witt Theorem. Let $G = G_{\bar{0}} \oplus G_{\bar{1}}$ be a Lie super-algebra, $a_1,\dots,a_m$ a basis of $G_{\bar{0}}$, $b_1,\dots,b_n$ a basis of $G_{\bar{1}}$. Then he elements of the form

$$a_1^{k_1} \dots a_m^{k_m} b_{i_1} \dots b_{i_s}, \quad \text{where } k_i \geq 0 \text{ and } 1 \leq i < \dots < i_s \leq n ,$$

form a basis of $\mathfrak{U}(G)$.

Let $V = V_{\bar{0}} \oplus V_{\bar{1}}$ be a $\mathbb{Z}_2$-graded linear space. A linear representation π of a Lie superalgebra $G = G_{\bar{0}} \oplus G_{\bar{1}}$ in V is a homomorphism $\pi : G \to \ell(V)$.

For brevity we often say in this case that V is a G-module, and instead of $\pi(g)(v)$ we write $g(v)$, $g \in G$, $v \in V$. Note that, by definition, $G_i(V_j) \subseteq V_{i+j}$, $i,j \in \mathbb{Z}_2$, and $[g_1,g_2](\ ) = g_1(g_2(v)) - (-1)^{(\deg g_1)(\deg g_2)} g_2(g_1(v))$. Note also that the map $\mathrm{ad} : G \to \ell(G)$ for which $(\mathrm{ad}\, g)(a) = [g,a]$ is a linear representation of G. It is called the adjoint representation.

A submodule of a G-module V is assumed to be $\mathbb{Z}_2$-graded; a G-module V is said to be irreducible if it has no non-trivial submodules. By a homomorphism of G-modules $\Phi : V \to V'$ we mean one that preserves the $\mathbb{Z}_2$-gradation in the sense that $\Phi(V_i) = V'_{\phi(i)}$, where ϕ is a bijection $\mathbb{Z}_2 \to \mathbb{Z}_2$.

Schur's lemma in Lie superalgebra case has the following form.

Schur's Lemma. Let $V = V_{\bar{0}} \oplus V_{\bar{1}}$, $\mathfrak{M}$ an irreducible family of operators from $\ell(V)$, and $C(\mathfrak{M}) = \{a \in \ell(V) \mid [a,m] = 0,\ m \in \mathfrak{M}\}$. Then either $C(\mathfrak{M}) = \langle 1 \rangle$ or $\dim V_{\bar{0}} = \dim V_{\bar{1}}$ and $C(\mathfrak{M}) = \langle 1, A \rangle$, where A is a non-degenerate operator in V permuting $V_{\bar{0}}$ and $V_{\bar{1}}$, and $A^2 = 1$.

By the symmetric superalgebra over $\mathbb{Z}_2$-graded space $V = V_{\bar 0} \oplus V_{\bar 1}$ we mean an algebra $S(V) = \bigoplus_{k \geq 0} S^k(V)$, where $S^k(V) = \bigoplus_{r=0}^{k} (S^r(V_{\bar 0}) \otimes \Lambda^{k-r}(V_{\bar 1}))$ is a $\mathbb{Z}_2$-graded space with a natural gradation induced from V. Analogously the exterior superalgebra over V is an algebra $\Lambda(V) = \bigoplus_{k \geq 0} \Lambda^k(V)$, where $\Lambda^k(V) = \bigoplus_{r=0}^{k} (\Lambda^r(V_{\bar 0}) \otimes S^{k-r}(V_{\bar 1}))$ is a $\mathbb{Z}_2$-graded space.

Any representation of Lie superalgebra G in V produces representations of G in $S^k(V)$ and $\Lambda^k(V)$; they are called symmetric and exterior powers of G-module V. The definition of these representations in "super" case differs from the usual case only by a minus sign which we put each time there is a permutation of two neighbouring odd elements.

Let G be a Lie superalgebra, $\mathfrak{U}(G)$ its universal enveloping superalgebra, H a subalgebra of G, and V an H-module. V can be extended to a $\mathfrak{U}(H)$-module. We consider the $\mathbb{Z}_2$-graded space $\mathfrak{U}(G) \otimes_{\mathfrak{U}(H)} V$ (this is the factor space of $U(G) \otimes V$ by the linear span of the elements of the form $gh \otimes v - g \otimes h(v)$, $g \in \mathfrak{U}(G)$, $h \in U(H)$, with an obvious $\mathbb{Z}_2$-gradation). This space can be endowed with the structure of a G-module as follows: $g(u \otimes v) = gu \otimes v$, $g \in G$, $u \in U(G)$, v V. The so constructed G-module is said to be induced from the H-module V and is denoted by $\mathrm{Ind}_H^G V$.

We list some of the simplest properties of induced modules, which follow from the Poincaré-Birkhoff-Witt theorem.

Proposition 2.1. a) Let G be a Lie superalgebra, H a subalgebra, V a simple G-module, and W an H-submodule of V considered as an H-module. Then V is a factor module of the G-module $\mathrm{Ind}_H^G W$.

b) If $H_2 \subset H_1 \subset G$ are subalgebras of G and W is an H_2-module, then $\mathrm{Ind}_{H_1}^G (\mathrm{Ind}_{H_2}^{H_1} W) \simeq \mathrm{Ind}_{H_2}^G W$.

c) Let $H \subset G$ be a subalgebra of G containing $G_{\bar 0}$, and $g_1, \ldots, g_t$ be odd elements of G whose projections onto G/H form a basis. Let W be an H-module. Then $\mathrm{Ind}_H^G W = \bigoplus_{1 \leq i_1 < \cdots < i_s \leq t} g_{i_1}, \ldots, g_{i_s} W$ is a direct sum of subspaces; in particular, $\dim \mathrm{Ind}_H^G W = 2^t \dim W$.

2. Representations with the highest weight [1]. From now on $G = G_{\bar 0} \oplus G_{\bar 1}$ is one of the basic classical Lie superalgebras $A(m,n)$, $m \neq n$, $B(m,n)$, $C(n)$, $D(m,n)$, $D(2,1;\alpha)$, $F(4)$, $G(3)$. We exclude the exceptional case of $A(n,n)$ (see Proposition 1.5c)) for simplicity of exposition. All the statements below hold for the central extension $s\ell(n+1,n+1)$ of $A(n,n)$.

et H be a Cartan subalgebra of $G_{\bar{0}}$. From now on we fix a Borel sub- lgebra G, containing H; then $B = H \oplus N^{+}$.

Let $\Lambda \in H^*$ be a linear function on H. We define a one-dimen- ional even B-module $\langle v_{\Lambda} \rangle$ by:

$$h(v_{\Lambda}) = \Lambda(h)v_{\Lambda}, \quad h \in H; \quad N^{+}(v_{\Lambda}) = 0; \quad \deg v_{\Lambda} = \bar{0} .$$

'e set $\tilde{V}(\Lambda) = \mathrm{Ind}_B^G \langle v_{\Lambda} \rangle$. This G-module contains a unique maximal sub- odule $I(\Lambda)$; we set $V(\Lambda) = \tilde{V}(\Lambda)/I(\Lambda)$. G-module $V(\Lambda)$ is called an rreducible representation with highest weight.

roposition 2.2. a) v_{Λ} is a unique vector in $V(\Lambda)$ up to a constant actor, for which $N^{+}(v_{\Lambda}) = 0$. In particular, for G-module $V(\Lambda)$ the irst case of Schur's lemma always takes place.

b) The G-modules $V(\Lambda_1)$ and $V(\Lambda_2)$ are isomorphic if and only if $\Lambda_1 = \Lambda_2$.

c) If V is a quotient module of the G-module $\tilde{V}(\Lambda)$ then the weight decomposition takes place:

$$V = \bigoplus_{\lambda \in D(\Lambda)} V_{\lambda}, \quad \text{where} \quad V_{\lambda} = \{v \in V \mid h(v) = \lambda(h)v,\ h \in H\},$$

all V_{λ} being finite-dimensional.

d) Any finite-dimensional irreducible representation of G is one of $V(\Lambda)$.

Proposition 2.3. Let B be a distinguished Borel subalgebra of G (the corresponding Dynkin diagram is contained in Table 1). Let e_i, f_i, h_i, $i = 1,\dots,r$, be the generators of G described in sec. 2.1. Let $\Lambda \in H^*$; we set $a_i = \Lambda(h_i)$, $i = 1,\dots,r$. The representation $V(\Lambda)$ is finite-dimensional if and only if the following conditions are satisfied:

1) $a_i \in \mathbb{Z}_+$ for $i \neq s$;

2) for superalgebras of II type $k \in \mathbb{Z}_+$, where k is given by the following table:

Table 2

G	k	b
$B(0,n)$	$\frac{1}{2}a_n$	0
$B(m,n)$, $m>0$	$a_n - a_{n+1} - \cdots - a_{m+n-1} - \frac{1}{2}a_{m+n}$	m
$D(m,n)$	$a_n - a_{n+1} - \cdots - a_{m+n-2} - \frac{1}{2}(a_{m+n-1}+a_{m+n})$	m
$D(2,1;\alpha)$	$(1+\alpha)^{-1}(2a_1 - a_2 - \alpha a_3)$	2
$F(4)$	$1/3(2a_1 - 3a_2 - 4a_3 - 2a_4)$	4
$G(3)$	$\frac{1}{2}(a_1 - 2a_2 - 3a_3)$	3

3) <u>for</u> $k < b$ <u>(in table 2) there are the supplementary conditions:</u>

$B(m,n)$: $a_{n+k+1} = \cdots = a_{m+n} = 0.$

$D(m,n)$: $a_{n+k+1} = \cdots = a_{m+n} = 0,\ k \leq m-2;\ a_{m+n-1} = a_{m+n},\ k = m-1.$

$D(2,1;\alpha)$: <u>all</u> $a_i = 0$ <u>if</u> $k=0$; $(a_3+1)\alpha = \pm(a_2+1)$ <u>if</u> $k=1$.

$F(4)$: <u>all</u> $a_i = 0$ <u>if</u> $k=0$; $k \neq 1$; $a_2 = a_4 = 0$ <u>if</u> $k=2$; $a_2 = 2a_4 + 1$ <u>if</u> $k=3$.

$G(3)$: <u>all</u> $a_i = 0$ <u>if</u> $k=0$; $k \neq 1$; $a_2 = 0$ <u>if</u> $k=2$.

<u>Remarks</u>. a) Properties 1) and 2) (in proposition 2.3) are equivalent to the property of Λ being dominant for Lie algebra $G_{\bar{0}}$.

b) <u>For Lie superalgebras of I type</u> a_s is an arbitrary complex number, i.e. there exist parameter families of finite-dimensional representations. For Lie superalgebras of II type there is no such family. Note however, that for both cases the set of Λ for which $\dim V(\Lambda) < \infty$ is a Zariski dense set in H^*.

Now we give a more explicit construction of the finite-dimensional G-modules $V(\Lambda)$. Let $G = \bigoplus_{i\in\mathbb{Z}} G_i$ be the distinguished $\mathbb{Z}$-gradation of G. We set $P = \bigoplus_{i\geq 0} G_i$, $P_+ = \bigoplus_{i>0} G_i$. Let $\Lambda \in H^*$ be a linear form, which satisfies the conditions 1) - 3) of Proposition 2.3.

Let $V^0(\lambda)$ denote a simple G_0-module with highest weight λ.

y condition 1) G_0-module $V^0(\Lambda)$ is finite dimensional. We extend $V^0(\Lambda)$ to a P-module by putting $P_+(V^0(\Lambda)) = 0$. We take the induced module $\mathrm{Ind}_P^G V^0(\Lambda)$. When G is of II type this module contains a submodule,

$$M = U(G) G_{-\delta}^{k+1} v_\Lambda .$$

Here $G_{-\delta}$ is a (1-dimensional) root space, which corresponds to the even root $\delta = \sum_{i=s}^{r} c_i \alpha_i$ (see Proposition 1.6 c)) $k = 2(\Lambda,\delta)/(\delta,\delta)$ is an integer given by Table 2 and v_Λ is the highest weight vector of $V^0(\Lambda)$ (see the proof of theorem 8 in [1]). We set

$$\bar{V}(\Lambda) = \mathrm{Ind}_P^G V^0(\Lambda) \quad \text{if } G \text{ is of I type,}$$

$$\bar{V}(\Lambda) = \mathrm{Ind}_P^G V^0(\Lambda)/M \quad \text{if } G \text{ is of II type.}$$

From the proof of theorem 8 in [1] we obtain,

Proposition 2.4. Suppose that $\Lambda \in H^*$ satisfies conditions 1) - 3) of Proposition 2.3. Then G-module $\bar{V}(\Lambda)$ is finite-dimensional; $V(\Lambda)$ contains a unique maximal submodule $\bar{I}(\Lambda)$, and $V(\Lambda) = \bar{V}(\Lambda)/\bar{I}(\Lambda)$.

3. Characters, supercharacters, and infinitesimal characters. Let V be a quotient module of G-module $\tilde{V}(\Lambda)$ and let $V = \bigoplus_{\lambda \in D(\Lambda)} V_\lambda$ be the weight decomposition of V. We set

$$\mathrm{ch}\, V = \sum_{\lambda \in D(\Lambda)} (\dim V_\lambda) e^\lambda , \quad \mathrm{sch}\, V = \sum_{\lambda \in D(\Lambda)} \omega(\lambda)(\dim V_\lambda) e^\lambda .$$

The functions ch V and sch V are called the character and supercharacter, respectively, of G-module V.

Let Z(G) be the center of the enveloping superalgebra $U(G)$. An element $z \in Z(G)$ can be uniquely written in the form:

$$z = u_z + \sum_i u_i^- u_i^0 u_i^+ , \quad \text{where } u_z, u_i^0 \in U(H),\ u_i^\pm \in N^\pm U(N^\pm).$$

The map $z \to u_z^0$ gives a monomorphism (Harish-Chandra homomorphism):

$$\beta : Z(G) \to S(H) \ (= \mathbb{C}[H^*]).$$

We define an automorphism $\alpha:\underset{\sim}{C}[H^*] \to \underset{\sim}{C}[H^*]$ by setting $\alpha(P(\lambda)) = P(\lambda-\rho)$ and set $\lambda = \alpha\circ\beta$: $Z(G) \to S(H)$.

We note that if v_λ is an eigenvector of the Borel subalgebra B, then

$$z(v_\lambda) = u_z(\lambda)v_\lambda \ .$$

Therefore all the elements of $Z(G)$ act as scalar operators $(u_z(\Lambda)I)$ in $\tilde{V}(\Lambda)$ and any quotent module V of $\tilde{V}(\Lambda)$. We have:

$$z(v) = \chi_\Lambda(z)v \ , \quad z \in Z(G), \quad v \in V,$$

where $\chi_\Lambda : Z(G) \to \underset{\sim}{C}$ is a homomorphism. χ_Λ is called the <u>infinitesimal character</u> of G-module V (it depends only on Λ).
We set $B(\Lambda) = D(\Lambda) \cap \{\lambda \in H^* | \chi_\lambda = \chi_\Lambda\}$.

<u>Proposition 2.5.</u> <u>Let</u> V <u>be a quotient module of</u> G-<u>module</u> $\tilde{V}(\Lambda)$, $\Lambda \in H^*$.

a) <u>For</u> $V = \tilde{V}(\Lambda)$ <u>the following formulas hold</u>

$$\mathrm{ch}\ \tilde{V}(\Lambda) = \sum_{\lambda\in D(\Lambda)} K(\lambda-\Lambda)e^\lambda, \quad \mathrm{sch}\ \tilde{V}(\Lambda) = \sum_{\lambda\in D(\Lambda)} K'(\lambda-\Lambda)e^\lambda.$$

$$L\cdot\mathrm{ch}\ \tilde{V}(\Lambda) = L'\cdot\mathrm{sch}\ \tilde{V}(\Lambda) = e^{\Lambda+\rho} \ .$$

b) $$L\cdot\mathrm{ch}\,V = \sum_{\lambda\in B(\Lambda)} c_\lambda e^{\lambda+\rho}, \quad L'\cdot\mathrm{sch}\,V = \sum_{\lambda\in B(\Lambda)} c'_\lambda e^{\lambda+\rho},$$ <u>where</u> $c_\Lambda = c'_\Lambda = 1$.

c) $(\lambda+\rho,\lambda+\rho) = (\Lambda+\rho,\Lambda+\rho)$ <u>for</u> $\lambda \in B(\Lambda)$.

d) <u>We define the Laplace operator</u> Δ <u>on</u> E <u>by the formula</u> $\Delta(e^\lambda) = (\lambda,\lambda)e^\lambda$. <u>Then</u> $L\cdot\mathrm{ch}\,V$ <u>and</u> $L'\cdot\mathrm{sch}\,V$ <u>are eigenfunctions of</u> Δ <u>with eigenvalue</u> $(\Lambda+\rho,\Lambda+\rho)$. <u>In particular</u>:

$$\Delta(L) = (\rho,\rho)L, \quad \Delta(L') = (\rho,\rho)L' \ .$$

Let $\delta: S(G) \to U(G)$ be the supersymmetrization map (it generalizes the symmetrization map in Lie algebra case; the difference is that each permutation of two neighbouring odd elements produces a minus sign). Let $S(G)^{\natural}$ denote a subalgebra in $S(G)$ of elements which are annihilated by the action of G. Then $\delta : S(G)^{\natural} \to Z(G)$ is an isomorphism. Finally let $i: S(G) \to S(H)$ be the homomorphism induced by the orthogonal projection of G onto H. It induces a homomorphism $i: S(G)^{\natural} \to S(H)^W$.

Proposition 2.6. a) $\gamma(Z(G)) \subset S(H)^W$.

b) $i = \gamma\circ\delta$ *is a monomorphism* $S(G)^{\natural} \to S(H)^W$. *The fields of fractions of* $i(S(G)^{\natural})$ *and* $S(H)^W$ *coinside.*

c) $\chi_{w(\Lambda+\rho)} = \chi_{\Lambda+\rho}$ *for* $w \in W$, $\Lambda \in H^*$.

d) *If* $\Lambda \in H^*$ *and* $(\Lambda+\rho,\alpha) = 0$, $\alpha \in \bar{\Delta}_1$, *then* $\chi_\Lambda = \chi_{\Lambda-t\alpha}$ *for any* $t \in \mathbb{C}$.

Proof. For a), b) and c) see [2]. To prove d) we first note that d) is true if the root $\alpha = \alpha_s$ is simple. Indeed, let v_Λ be the highest vector of $\tilde{V}(\Lambda)$ and $(\Lambda+\rho,\alpha_s) = (\Lambda,\alpha_s) = 0$. Then $e_{-\alpha_s}v_\Lambda$ is obviously an eigenvector for B. Therefore $z(e_{-\alpha_s}v_\Lambda) = u_z(\Lambda-\alpha_s)v_\Lambda = \chi_{\Lambda-\alpha_s}(z)e_{-\alpha_s}v_\Lambda$, $z \in Z(G)$, which proves that $\chi_\Lambda = \chi_{\Lambda-\alpha_s}$.

Now $w(\alpha) = \alpha_s$ for some $w \in W$. We have:

$$\chi_{\Lambda-\alpha} = \chi_{w(\Lambda-\alpha+\rho)-\rho} = \chi_{(w(\Lambda+\rho)-\rho)-\alpha_s} .$$

We set $M = w(\Lambda+\rho)-\rho$; then $(M+\rho,\alpha_s) = (w(\Lambda+\rho),\alpha_s) = (\Lambda+\rho,\alpha) = 0$. Therefore $\chi_{M-\alpha_s} = \chi_M$ and using again c) we obtain: $\chi_{\Lambda-\alpha} = \chi_\Lambda$.

Since $(\alpha,\alpha) = 0$ we obtain by iteration that $\chi_{\Lambda-t\alpha} = \chi_\Lambda$, $t \in \mathbb{Z}_+$, and therefore $\chi_{\Lambda-t\alpha} = \chi_\Lambda$ for any $t \in \mathbb{C}$.

4. Typical representations.

Our first main result is the following

Theorem 1. Let Λ be the highest weight of a finite-dimensional irreducible module $V(\Lambda)$ of a basic classical Lie superalgebra G. Then the following statements are equivalent.

a) If $\chi_\Lambda = \chi_\lambda$ for some $\lambda \in H^*$, then $\lambda+\rho = w(\Lambda+\rho)$ for some $w \in W$.

b) If all the polynomials of $i(S(G)^{\natural})$ have the same value at Λ and at some $\lambda \in H^*$, then $\lambda+\rho = w(\Lambda+\rho)$ for some $w \in W$.

c) $V(\Lambda)$ is uniquely determined up to an isomorphism among the irreducible finite-dimensional G-modules by its infinitesimal character χ_Λ.

d) $V(\Lambda)$ splits in any finite-dimensional representation (i.e. if $V(\Lambda)$ is a submodule or a factor-module of a finite-dimensional G-module, then $V(\Lambda)$ is a direct summand).

e) $V(\Lambda) = \bar{V}(\Lambda)$.

f) $\bar{V}(\Lambda)$ is an irreducible G-module.

g) $V(\Lambda) = \tilde{V}(\Lambda)/I(\Lambda)$, where $I(\Lambda) = \sum_{i\neq s} U(G) f_i^{a_i+1} v_\Lambda$ for G of I type and $I(\Lambda) = \sum_{i\neq s} U(G) f_i^{a_i+1} v_\Lambda + U(G) G_{-\delta}^{k+1} v_\Lambda$ for G of II type ($a_i = \Lambda(h_i)$, $k = 2(\Lambda,\delta)/(\delta,\delta)$).

h) $\mathrm{ch}\, V(\Lambda) = L^{-1} \sum_{w\in W} \varepsilon(w) e^{w(\Lambda+\rho)}$.

i) $\mathrm{ch}\, V(\Lambda)(\lambda) = \sum_{w\in W} K(\lambda+\rho-w(\Lambda+\rho))$.

j) $\mathrm{sch}\, V(\Lambda) = L'^{-1} \sum_{w\in W} \varepsilon'(w) e^{w(\Lambda+\rho)}$.

k) $\mathrm{sch}\, V(\Lambda)(\lambda) = \sum_{w\in W} K'(\lambda+\rho-w(\Lambda+\rho))$.

ℓ) $(\Lambda+\rho,\alpha) \neq 0$ for any $\alpha \in \bar{\Delta}_1^+$.

<u>Definition</u>. A finite-dimensional irreducible G-module $V(\Lambda)$ with highest weight Λ is called <u>typical</u> if one of the equivalent conditions a) - ℓ) of Theorem 1 is satisfied.

An essential part of Theorem 1 is contained in [2]. In these notes this theorem follows from Propositions 2.6, 2.7, 2.8, 2.12 and Theorem 2.

<u>Proposition 2.7. Let V be a finite-dimensional G-module with highest weight Λ. We set $B_+(\Lambda) = \{\lambda \in B(\Lambda) \mid \lambda \neq \Lambda$ and $\lambda+\rho$ is dominant for $G_{\bar{0}}\}$. Then the following formulas take place:</u>

$$(2.1) \qquad L\,\mathrm{ch}\, V = \sum_{w\in W} \varepsilon(w) e^{w(\Lambda+\rho)} + \sum_{\lambda\in B^+(\Lambda)} c_\lambda(V) \sum_{w\in W} \varepsilon(w) e^{w(\lambda+\rho)}$$

$$(2.1') \qquad L'\mathrm{sch}\, V = \sum_{w\in W} \varepsilon'(w) e^{w(\Lambda+\rho)} + \sum_{\lambda\in B^+(\Lambda)} c'_\lambda(V) \sum_{w\in W} \varepsilon'(w) e^{w(\lambda+\rho)}.$$

<u>Proof</u>. The action of $G_{\bar{0}}$ on V can be extended to the action of the corresponding connected simply connected Lie group G_0, and in particular to the normalizer of H in G_0. It proves that $\mathrm{ch}\, V$ and $\mathrm{sch}\, V$ are W-invariant functions. Since any $\mu \in H^*$ can be transformed by W into a dominant for $G_{\bar{0}}$ function, by Proposition 2.5 b) and 1.7 e) we obtain (2.1) and 2.1').

roposition 2.8. Let V be a finite-dimensional G-module with highst weight Λ, which satisfies property a) of Theorem 1. Then

2.2) $$\mathrm{ch}\, V = L^{-1} \sum_{w\in W} \varepsilon(w) e^{w(\Lambda+\rho)}$$

2.2') $$\mathrm{sch}\, V = L'^{-1} \sum_{w\in W} \varepsilon'(w) e^{w(\Lambda+\rho)}$$

2.3) $$\dim V_\lambda = \sum_{w\in W} K(\lambda+\rho-w(\Lambda+\rho))$$

2.3') $$\omega(\lambda)\dim V_\lambda = \sum_{w\in W} K'(\lambda+\rho-w(\Lambda+\rho)).$$

Proof. Since $B_+(\Lambda) = \phi$ if a) is satisfied, (2.2) and (2.2') follow from (2.1) and (2.1'). Formulas (2.3) and (2.3') are equivalent forms of (2.2) and (2.2') (cf. Proposition 1.7 f).

The following implications in Theorem 1 are either evident or follow from Propositions 2.6, 2.7, 2.8:

$$\begin{array}{l} \ell) \longleftarrow a) \longleftrightarrow b) \longrightarrow c) \longrightarrow d) \longrightarrow e) \longleftrightarrow f) \longleftrightarrow g) \\ \qquad\quad \downarrow \\ \qquad\quad h) \longleftrightarrow i) \longleftrightarrow j) \longleftrightarrow k) \end{array}$$

We also show how to prove that f) <—> b) for Lie superalgebras G of I type. Together with preceding remarks and Theorem 2 below for G [2] it proves Theorem 1 for superalgebras of I type.

Proposition 2.9. Let G be a Lie superalgebra of I type. Then G-module $\bar{V}(\Lambda)$ is irreducible if and only if $(\Lambda+\rho,\alpha) \neq 0$, for $\alpha \in \bar{\Delta}_1^+$.

Proof. We note that $\Delta_1^+ = \bar{\Delta}_1^+$ for Lie superalgebras of I type, and that $[G_\alpha, G_\beta] = 0$, if $\alpha,\beta \in \Delta_1^+$. We set: $T_- = \prod_{\alpha\in\Delta_1^+} e_{-\alpha}$, $T_+ = \prod_{\alpha\in\Delta_1^+} e_\alpha$. Clearly any submodule of $\bar{V}(\Lambda)$ contains the element $T_- v_\Lambda$ and therefore $\bar{V}(\Lambda)$ is irreducible if and only if $T_+T_-v_\Lambda \neq 0$. We consider G-module $\tilde{V}(\Lambda)$ for any $\Lambda \in H^*$. We have the following relation: $T_+T_-v_\Lambda = P(\Lambda)v_\Lambda$, where $P(\Lambda)$ is a polynomial of degree equals to $\#\Delta_1^+$. From Proposition 2.6 we obtain that $P(\Lambda) \neq 0$ and from the proof of this proposition that $P(\Lambda)$ is divisible by the linear form $\phi(\Lambda) = (\Lambda,\alpha_s) = (\Lambda+\rho,\alpha_s)$ where α_s is the unique odd simple root.

Suppose that $\Lambda(h_i) \subset \mathbb{Z}_+$ for $i \neq s$. Since evidently $f_i, i \neq s$, commutes with T_+T_-, we obtain that $P(\Lambda) = P(\Lambda-(\Lambda(h_i)+1)\alpha_i)$, $i \neq s$, and therefore that $P(\Lambda) = P(w(\Lambda+\rho)-\rho)$, $w \in W$. Since the last equality holds for Zariski dense set in H^*, it holds for any $\Lambda \in H^*$.

Since W acts transitively on Δ_1^+ we obtain now that $P(\Lambda)$ is divisible by $\prod_{\alpha\in\Delta_1^+}(\Lambda+\rho,\alpha)$. Since $P(\Lambda)$ is a polynomial of degree $\#\Delta_1^+$, we obtain that $P(\Lambda) = c\prod_{\alpha\in\Delta_1^+}(\Lambda+\rho,\alpha)$, $c \in \mathbb{C}^*$, which proves the proposition.

The following theorem, which is a generalization of Chevalley invariants restriction theorem to Lie superalgebra case, is a crucial point in the proof of Theorem 1.

Theorem 2. Any element of $S(H)^W$ can be represented in the form P/Q^t, where $P \in i(S(G)^{\mathfrak{h}})$, $Q = \prod_{\alpha\in\bar{\Delta}_1^+} h_\alpha \in i(S(G)^{\mathfrak{h}})$, $t \in \mathbb{Z}_+$.

The proof of a slightly weakened form of Theorem 2 can be found in [2]. We note that Theorem 2 implies that if the property b) of Theorem 1 holds then the W-orbit of Λ can be separated from any other W-orbit by polynomials from $i(S(G)^{\mathfrak{h}})$. Therefore, by Proposition 2.6 we obtain the implication ℓ) —> a) of Theorem 1.

Remark. It follows from the results above that all the finite-dimensional representations of G are typical if and only if $G = B(0,n)$ (since $\bar{\Delta}_1 = \phi \iff G = B(0,n)$). This result is consistent with Djokovic-Hochshild Theorem [6] that any finite-dimensional representation of a Lie superalgebra G is completely reducible if and only if G is a direct sum of a semisimple Lie algebra and several copies of superalgebras of type $B(0,n)$.

Now we will compute the dimensions of typical representations and their even and odd parts [2]. We introduce a homomorphism $f_\mu: E \to \mathbb{C}[[t]]$, where t is an indeterminate, $\mu \in H^*$, by setting

$$f_\mu(e^\lambda) = e^{t(\lambda,\mu)} .$$

Two cases of f_μ are especially important:

$$f = f_{\rho_0} \quad \text{and} \quad \bar{f} = f_{\bar{\rho}_0} .$$

roposition 2.10. Let G be a basic classical Lie superalgebra. We et $d = \#\Delta_1^+$, $\bar{d} = \#\bar{\Delta}_1^+$. Let V by a typical G-module with highest eight Λ. Then

$$\text{(2.4)}\qquad \dim V = 2^d \prod_{\alpha\in\Delta_0^+} \frac{(\Lambda+\rho,\alpha)}{(\rho_0,\alpha)} .$$

$$\text{(2.5)*}\qquad \bar{f}(\mathrm{sch}\, V) = t^{\bar{d}} \prod_{\alpha\in\bar{\Delta}_0^+} \frac{(\Lambda+\rho,\alpha)}{(\bar{\rho}_0,\alpha)} \prod_{\alpha\in\bar{\Delta}_1^+} (\bar{\rho}_0,\alpha) + o(t^{\bar{d}}) .$$

$$\text{(2.6)}\qquad \begin{cases} \dim V_{\bar{0}} - \dim V_{\bar{1}} = 0 \quad \underline{\text{if}} \quad G \neq B(0,n), \\ \dim V_{\bar{0}} - \dim V_{\bar{1}} = \prod_{\alpha\in\bar{\Delta}_0^+} \frac{(\Lambda+\rho,\alpha)}{(\bar{\rho}_0,\alpha)} \quad \underline{\text{if}} \quad G = B(0,n). \end{cases}$$

roof. Clearly, $\dim V = \lim_{t\to 0} f(\mathrm{ch}\, V)$. Let $c = \#\Delta_0^+$. We have:

$f\left(\sum_{w\in W} \varepsilon(w) e^{w(\Lambda+\rho)}\right) = \sum_{w\in W} \varepsilon(w) e^{(w(\Lambda+\rho),\rho_0)t} = \sum_{w\in W} \varepsilon(w) e^{(\Lambda+\rho,w(\rho_0))t} =$

$= f_{\Lambda+\rho}\left(\sum_{w\in W} \varepsilon(w) e^{w(\rho_0)}\right) =$ (by Weyl denominator formula) $=$

$= f_{\Lambda+\rho} \prod_{\alpha\in\Delta_0^+} (e^{\alpha/2} - e^{-\alpha/2}) = t^c\left[\prod_{\alpha\in\Delta_0^+} (\Lambda+\rho,\alpha) + o(1)\right]$. On the other hand $f(L) = t^c\left[2^{-d} \prod_{\alpha\in\Delta_0^+} (\rho_0,\alpha) + o(1)\right]$. Hence $f(\mathrm{ch}\, V) = 2^d \prod_{\alpha\in\Delta_0^+} (\Lambda+\rho,\alpha)/(\rho_0,\alpha) + o(1)$, which proves (2.4).

For the proof of formula (2.5) we use the Weyl denominator formula for the root system $\bar{\Delta}_0^+$:

$$\sum_{w\in W} \varepsilon(w) e^{w(\bar{\rho}_0)} = \prod_{\alpha\in\bar{\Delta}_0^+} (e^{\alpha/2} - e^{-\alpha/2}) .$$

From Proposition 1.7 d) we obtain the following form of this identity:

$$\text{(2.7)}\qquad \sum_{w\in W} \varepsilon'(w) e^{w(\bar{\rho}_0)} = \#T \prod_{\alpha\in\bar{\Delta}_0^+} (e^{\alpha/2} - e^{-\alpha/2}) .$$

Now we apply $\bar{f}$ to both sides of formula (2.2'). The same argument as before together with (2.7) gives (2.5).

Since $\dim V_0 - \dim V_1 = \lim_{t\to 0} \mathrm{sch}\, V$ and $\bar{d} = 0$ if and only if $G = B(0,n)$, formula (2.5) gives (2.6).

---

* There is a misprint in the formula for $\bar{f}(\mathrm{sch}\, V)$ in the article [2] (in the case $B(m,n)$).

Remark. If $G = B(0,n)$, then clearly $(\rho,\alpha) = (\bar{\rho}_0,\alpha)$ for $\alpha \in \bar{\Delta}_0^+$ and we can rewrite (2.6) as follows (see [4]):

$$\dim V_{\bar{0}} - \dim V_{\bar{1}} = \prod_{\alpha \in \bar{\Delta}_0^+} \frac{(\Lambda+\rho,\alpha)}{(\rho,\alpha)} .$$

Examples. 1. $A(m,n)$. A finite-dimensional representation $V(\Lambda)$ is typical if and only if*

$$a_{m+1} \neq \sum_{t=m+2}^{j} a_t - \sum_{t=1}^{m} a_t - 2m - 2 + i + j ,$$

for $1 \leq i \leq m+1 \leq j \leq m+n+1$. Under these conditions

$$\dim V(\Lambda) = 2^{(m+1)(n+1)} \prod_{j \leq i \leq j \leq m} \frac{a_i + a_{i+1} + \cdots + a_j + j - i + 1}{j-i+1} \prod_{m+2 \leq i \leq j \leq m+n+1} \frac{a_i + \cdots + a_j + j - i + 1}{j-i+1}$$

For a non-typical finite-dimensional representation $\dim V(\Lambda) <$ right hand-side.

2. $C(n)$. A finite-dimensional representation $V(\Lambda)$ is typical if and only if

$$a_1 \neq \sum_{t=2}^{i} a_t + i - 1$$

$$a_1 \neq \sum_{t=2}^{i} a_t + 2 \sum_{t=i+1}^{n} a_t + 2n - i - 1,$$

for $1 \leq i \leq n-1$. Under these conditions

$$\dim V(\Lambda) = 2^{2n-2} \prod_{2 \leq i \leq j \leq n-1} \frac{a_j + \cdots + a_j + j - i + 1}{j-i+1} \prod_{2 \leq i \leq j \leq n} \frac{a_i + \cdots + a_{j-1} + 2a_j + \cdots + 2a_n}{2n-i-j+2} .$$

For a non-typical finite-dimensional representation $\dim V(\Lambda) <$ right hand-side.

3. $B(m,n)$. A finite-dimensional representation $V(\Lambda)$ is typical if and only if

$$\sum_{t=i}^{n} a_t - \sum_{t=n+1}^{j} a_t + 2n - i - j \neq 0$$

---

* We assume that $\sum_{t=i}^{j} = 0$ if $j < i$.

$$\sum_{t=i}^{n} a_t - \sum_{t=n+1}^{j} a_t - 2\sum_{t=j+1}^{m+n-1} a_t - a_{m+n} - i + j - 2m + 1 \neq 0$$

for $1 \leq i \leq n \leq j \leq m+n-1$. Under these conditions

$$\dim V(\Lambda) = 2^{(2m+1)n} \prod_{1\leq i<j\leq n-1} \frac{a_i+\cdots+a_j+j-i+1}{j-i+1} \prod_{n+1\leq i<j\leq m+n-1} \frac{a_i+\cdots+a_j+j-i+1}{j-i+1} \cdot$$

$$\prod_{1\leq i<j\leq n} \frac{(a_i+\cdots+a_{j-1}) + 2(a_j+\cdots+a_n-a_{n+1}-\cdots-a_{m+n-1}) - a_{m+n} + 2n-2m+1-i-j}{2n+2-i-j} \cdot$$

$$\prod_{n+1\leq i<j\leq m+n-1} \frac{(a_i+\cdots+a_{j-1}) + 2(a_j+\cdots+a_{m+n-1}) + a_{m+n} + 2m-i-j+1}{2m-i-j+1} .$$

$\mathbf{B}(0,n)$. All the finite-dimensional representations $V(\Lambda)$ are typical.

$$\dim V_{\bar{0}}(\Lambda) - \dim V_{\bar{1}}(\Lambda) = \prod_{1\leq i<j\leq n} \frac{(a_i+\cdots+a_j) + 2(a_{j+1}+\cdots+a_{n-1})+a_n+2n-i-j}{2n-i-j}$$

$$\dim V(\Lambda) = (\dim V_{\bar{0}}(\Lambda) - \dim_{\bar{1}}(\Lambda)) \prod_{1\leq i\leq n} \frac{2(a_i+\cdots+a_{n-1}) + a_n + 2n-2i+1}{2n-2i+1}$$

4. $\mathbf{D}(m,n)$. A finite-dimensional representation $V(\Lambda)$ is typical if and only if

$$\sum_{t=i}^{n} a_t - \sum_{t=n+1}^{j} a_t + 2n - i - j \neq 0$$

for $1 \leq i \leq n \leq j \leq m+n-1$,

$$\sum_{t=i}^{n} a_t - \sum_{t=n+1}^{m+n-2} a_t - a_{m+n} + n-m-i+1 \neq 0$$

for $1 \leq i \leq n$, and

$$\sum_{t=i}^{n} a_t - \sum_{t=n+1}^{j} a_t - 2\sum_{t=j+1}^{m+n-2} a_t - a_{m+n-1} - a_{m+n} - i+j-2m \neq 0$$

for $1 \leq i \leq n \leq j \leq m+n-2$.

The formula for $\dim V(\Lambda)$ is similar to one for $\underset{\sim}{B}(m,n)$.

5. $\underset{\sim}{D}(2,1;\alpha)$. A finite-dimensional representation $V(\Lambda)$ is typical if and only if $a_1 \neq 0$, $a_1 \neq a_2 + 1$, $a_1 \neq \alpha a_3 + \alpha$, $a_1 \neq a_2 + \alpha a_3 + 1 + \alpha$. Under these conditions

$$\dim V(\Lambda) = 16(a_2 + 1)(a_3 + 1)\left((1+\alpha)^{-1}(2a_1 - a_2 - \alpha a_3)-1\right).$$

6. $\underset{\sim}{G}(3)$. A finite-dimensional representation $V(\Lambda)$ is typical if and only if $a_1 \neq 0$, $a_1 \neq a_2 + 1$, $a_1 \neq a_2 + 3a_3 + 4$, $a_1 \neq 3a_2 + 3a_3 + 6$, $a_1 \neq 3a_2 + 6a_3 + 9$, $a_1 \neq 4a_2 + 6a_3 + 10$. Under these conditions

$$\dim V(\Lambda) = \frac{8}{15}(a_2+1)(a_3+1)(a_2+a_3+2)(a_2+3a_3+4)(a_2+2a_3+3) \times$$

$$(2a_2+3a_3+5)(a_1-2a_2-3a_3-5).$$

7. $\underset{\sim}{F}(4)$. A finite-dimensional representation $V(\Lambda)$ is typical if and only if

$$a_1 \neq 0, \quad a_1 \neq a_2+1, \quad a_1 \neq a_2+2a_3+3, \quad a_1 \neq 2a_2+2a_3+4$$

$$a_1 \neq a_2+2a_3+2a_4+5, \quad a_1 \neq 2a_2+2a_3+2a_4+6,$$

$$a_1 \neq 2a_2+4a_3+2a_4+8, \quad a_1 \neq 3a_2+4a_3+2a_4+9.$$

Under these conditions

$$\dim V(\Lambda) = \frac{32}{45}(a_2+1)(a_3+1)(a_4+1)(a_2+a_3+2)(a_3+a_4+2)(a_2+2a_3+3) \times$$

$$(a_2+a_3+a_4+3)(a_2+2a_3+2a_4+5)(a_2+2a_3+a_4+4)(2a_1-3a_2-4a_3-2a_4-9)$$

<u>Proposition 2.11.</u> <u>Let</u> V <u>be a typical</u> G-module with highest weight Λ. <u>Let</u> $m_\Lambda(\mu)$ <u>denote the multiplicity of an irreducible</u> $G_{\bar{0}}$-module <u>with (dominant) highest weight</u> μ <u>in</u> $G_{\bar{0}}$-module V, <u>and let</u> $K_1(\lambda)$ <u>denote the number of partitions of</u> $\lambda \in H^*$ <u>into a sum of distinct</u> <u>roots from</u> $\Delta_{\bar{1}}^+$. <u>Then the following formula takes place:</u>

$$(2.8) \qquad m_\Lambda(\mu) = \sum_{w \in W} \varepsilon(w)K_1(w(\Lambda+\rho) - (\mu+\rho)).$$

roof. Let $V^0(\mu)$ denote an irreducible $G_{\bar 0}$-module with highest eight μ. By formula (2.2) we have:

$$\mathrm{ch}\, V(\Lambda) = \prod_{\alpha\in\Delta_1^+}(1+e^{-\alpha})\prod_{\alpha\in\Delta_0^+}(1-e^{-\alpha})^{-1}\sum_{w\in W}\varepsilon(w)e^{w(\Lambda+\rho)-\rho} =$$

$$= \sum_{\nu} m_\Lambda(\nu)\prod_{\alpha\in\Delta_0^+}(1-e^{-\alpha})^{-1}\sum_{w\in W}e^{w(\Lambda+\rho_0)-\rho_0} =$$

$$= \left(\prod_{\alpha\in\Delta_0^+}(1-e^{-\alpha})\right)^{-1}\sum_{w\in W}\varepsilon(w)\sum_{\mu}m_\Lambda\left(w^{-1}(\mu+\rho_0)-\rho_0\right)e^{\mu} =$$

$$= \left(\prod_{\alpha\in\Delta_0^+}(1-e^{-\alpha})\right)^{-1}\sum_{w\in W}\varepsilon(w)\sum_{\mu}m_\Lambda(\mu)e^{\mu},$$

because $m_\Lambda(w^{-1}(\mu+\rho_0)-\rho_0) = 0$ for $w \neq e$, since $w^{-1}(\mu+\rho_0)-\rho_0$ is not a dominant function for $w \neq e$. Therefore we have:

$$\sum_{\mu} m_\Lambda(\mu)e^{\mu} = \prod_{\alpha\in\Delta_1^+}(1+e^{-\alpha})\sum_{w}\varepsilon(w)e^{w(\Lambda+\rho)-\rho},$$

which is equivalent to (2.8).

Remark. The proof of Proposition 2.11 is the same as Kostant's proof of his formula of multiplicities with respect to a regular subalgebra of a semi-simple Lie algebra. There is an obvious generalization of both formula (2.8) and Kostant's formula for the multiplicities with respect to a regular subalgebra of $G_{\bar 0}$.

Example. Let $G = A(1,0)$ and let $V = \bar V(\Lambda)$, where $\Lambda(h_1) = k \in \mathbb{Z}_+$, $\Lambda(h_2) \in \mathbb{C}$, be a G-module. Formula (2.8) gives the following decomposition of the G-module V with respect to $G_{\bar 0} = g\ell_2$:

$$V_{\bar 0} = V^0(\Lambda) \oplus V^0(\Lambda-\alpha_1-2\alpha_2)$$

$$V_{\bar 1} = V^0(\Lambda-\alpha_2) \oplus V^0(\Lambda-\alpha_1-\alpha_2) \quad \text{for } k \geq 1$$

and $$V_{\bar 1} = V^0(\Lambda-\alpha_2) \quad \text{for } k = 1.$$

Here $V^0(\lambda)$ denotes a $G_{\bar 0}$-module with highest weight λ. Their dimen sions are respectively:

$\dim V^0(\Lambda) = \dim V^0(\Lambda-\alpha_1-2\alpha_2) = k + 1,$

$\dim V^0(\Lambda-\alpha_2) = k + 2, \quad \dim V^0(\Lambda-\alpha_1-\alpha_2) = k.$

A non-typical G-module $S^k s\ell(2,1)$ is isomorphic to $\bar{V}(\Lambda)/\bar{I}(\Lambda)$, where $\Lambda(h_1) = k$, $\Lambda(h_2) = 0$ and where $\bar{I}(\Lambda)$ is an irreducible G-module with highest weight $\Lambda - \sigma_2$.

5. <u>Bilinear form on $U(G)$ and the structure of modules $\tilde{V}(\Lambda)$.</u> In the same way as in [7] we introduce a bilinear form $A: U(G) \otimes_{\mathbb{C}} U(G) \to U(H)$. For that we denote by σ an involutive anti-automorphism of $U(G)$, which is identical on H and for which $\sigma(N^+) = N^-$. We have decomposition $U(G) = U(H) \oplus (N_- U(G) + U(G)N_+)$. Denote by β the projection $\beta: U(G) \to U(H)$. We set:

$$A(x,y) = \beta(\sigma(x)y), \qquad x,y \in U(G).$$

We have a weight decomposition of $U(N_-)$ with respect to H: $U(N_-) = \bigoplus_{\eta \in H^*} U(N_-)^{-\eta}$. Let A_η be the restriction of A to $U(N_-)^{-\eta}$, $\eta = \Sigma k_i \alpha_i$, $k_i \in \mathbb{Z}_+$. The same proof as in [8] gives the following result (cf. [5] and [7]).

<u>Theorem 3.</u> a) Up to a non-zero constant factor

$$\det A_\eta = \prod_{n>0} \prod_{\alpha \in \Delta_0^+} \left(h_\alpha + (\rho,\alpha) - n\frac{(\alpha,\alpha)}{2}\right)^{K(\eta - n\alpha)} \prod_{\beta \in \Delta_1^+} \left(h_\beta + (\rho,\beta) - \frac{(\beta,\beta)}{2}\right)^{K(\eta-\beta)}.$$

b) If G-module $\tilde{V}(\Lambda)$ has a subquotient isomorphic to $V(\mu)$, then there exists a chain $\Lambda = \mu_0, \mu_1, \ldots, \mu_k = \mu$ such that $\mu_{i+1} = \mu_i - n_i \beta_i$, where $\beta_i \in \Delta^+$, n_i is a positive integer, and $2(\mu_i + \rho, \beta_i) = n_i(\beta_i, \beta_i)$.

c) If there exists a chain $\Lambda = \mu_0, \mu_1, \ldots, \mu_k = \mu$ with properties described in b), and in addition $\beta_i \neq \beta_{i+1}$, $n_i = 1$ for $\beta_i \in \Delta_1^+$, then there exists a homomorphism of G-modules: $\tilde{V}(\mu) \to \tilde{V}(\Lambda)$.

Theorem 3 has the following consequence for finite-dimensional representations.

<u>Proposition 2.12. The statements f), h), and ℓ) of Theorem 1 are equivalent.</u>

. Some remarks and open questions.

a) For superalgebras G of I type the expression for ch $\bar{V}(\Lambda)$ s given by the right hand-side of (2.2) (it follows easily from the act that $w(\rho_1) = \rho_1$, $w \in W$, in this case). Therefore, ch $\bar{V}(\Lambda)$ is iven by right hand-side of (2.4) and for a non-typical G-module (Λ) we have (cf. (2.4) and (2.8)):

$$\dim V(\Lambda) < 2^d \prod_{\alpha \in \Delta_0^+} \frac{(\Lambda+\rho,\alpha)}{(\rho_0,\alpha)} \tag{2.9}$$

$$m_\Lambda(\mu) \leq \sum_{w \in W} \varepsilon(w) K_1(w(\Lambda+\rho) - (\mu+\rho)) . \tag{2.10}$$

)ne can show that (2.9) and (2.10) take place for superalgebras of II ype if $\frac{2(\Lambda+\rho,\delta)}{(\delta,\delta)}$ is a positive integer. It is interesting to find he expression for ch $\bar{V}(\Lambda)$ for superalgebras of II type and corres- onding analogues of (2.9) and (2.10) in general case.

b) It is interesting to compute $c_\lambda(V)$ in formula (2.1) for any finite-dimensional G-module V; I do not know the answer even for the one-dimensional V. By Theorem 3 $c_\lambda(V(\Lambda)) = 0$ if Λ and λ can not be connected by a chain, described in Theorem 3.

c) Find the irreducible subquotients of G-modules $\bar{V}(\Lambda)$ (in the non-typical case) (cf. Theorem 3).

d) Find the formulas of $\dim V(\Lambda)$ and $\dim V_0(\Lambda) - \dim V_1(\Lambda)$ for non-typical representations.

e) In the simplest case of $G = A(1,0)$ a complete list of non-typical representations is $S^k s\ell(2,1)$ and its dual, $k \in \mathbb{Z}_+$. It indicates that all non-typical representations can be constructed from the standard representation by some tensor operations. We note that one-dimensional G-module for $G \neq B(0,n)$, standard representations $s\ell(m,n)$ and $osp(m,n)$, $m \neq 0$, and the adjoint representations of $G \neq A(1,0)$ or $B(0,n)$ are non-typical (because $\dim G_{\bar{0}} \neq \dim G_{\bar{1}}$).

f) The supertrace form $(a,b)_V = \operatorname{str} ab$ of a G-module $V = V(\Lambda)$ is nondegenerate if and only if $(\dim V_{\bar{0}} - \dim V_{\bar{1}})(\Lambda,\Lambda+2\rho) \neq 0$, except for the case $G = A(1,0)$ [1]; in the latter case $(a,b)_V$ is non-degenerate for all V. Proposition 2.8 shows that for $G \neq A(1,0)$ and typical G-module V the form $(a,b)_V$ is 0. Find the conditions in terms of Λ for $(a,b)_{V(\Lambda)}$ to be non-degenerate.

g) Conjecture (cf. Proposition 2.6). $\chi_\lambda = \chi_\mu$ if and only if $w(\mu+\rho) = \lambda+\rho+\beta$ for certain $w \in W$, where $\beta = \Sigma c_i \beta_i$, $\beta_i \in \bar{\Delta}_1$, $c_i \in \mathbb{C}$ and $(\lambda + \rho + c_1\beta_1 + \ldots + c_i\beta_i, \beta_{i+1}) = 0$, $i = 0,1,\ldots$.

h) It follows from Theorem 1, that $H^1(G,V(\Lambda)) = 0$ for a typical G-module $V(\Lambda)$. One can be shown that $\dim H^1(A(1,0),s\ell(2,1)) = 1$ and $H^1(\underset{\sim}{A}(1,0),V) = 0$ for all the other modules $V = V(\Lambda)$. Compute $H^1(G,V(\Lambda))$ in general case.

i) Classify all the finite-dimensional indecomposable G-modules. All the irreducible quotients of any such reducible module are evidently non-typical. Note that the modules $\bar{V}(\Lambda)$ are indecomposable.

References

[1] V. G. Kac, Lie superalgebras, Advances in Math., 26, no. 1 (1977), 8-96.

[2] V. G. Kac, Characters of typical representations of classical Lie superalgebras, Communications in Algebra, 5(8) (1977), 889-897.

[3] L. Corwin, Y. Ne'emen, S. Sternberg, Graded Lie algebras in mathematics and physics, Rev. Mod. Phys. 47 (1975), 573-604.

[4] V. G. Kac, Infinite-dimensional algebras, Dedekind's η-function, classical Möbius function and the very strange formula, Advances in Math., to appear.

[5] I. N. Bernstein, I. M. Gelfand, S. I. Gelfand, Structure of representations generated by vectors of highest weight, Funk. Anal. Appl., 5 (1971), 1-9.

[6] D. Ž. Djoković and G. Hochschild, Semi-simplicity of $\mathbb{Z}_2$-graded Lie algebras II, Illinois J. Math. 20 (1976), 134-143.

[7] N. N. Shapovalov, On a bilinear form on the universal enveloping algebra of a complex semisimple Lie algebra, Funk. Anal. Appl. 6 (1972), 307-312.

[8] V. G. Kac, D. A. Kazhdan, Representations with highest weight of infinite-dimensional Lie algebras, to appear.

21: G. Cherlin, Model Theoretic Algebra – Selected Topics. 4 pages. 1976.

22: C. O. Bloom and N. D. Kazarinoff, Short Wave Radiation ems in Inhomogeneous Media: Asymptotic Solutions. V. 104 . 1976.

23: S. A. Albeverio and R. J. Høegh-Krohn, Mathematical y of Feynman Path Integrals. IV, 139 pages. 1976.

24: Séminaire Pierre Lelong (Analyse) Année 1974/75. Edité Lelong. V, 222 pages. 1976.

525: Structural Stability, the Theory of Catastrophes, and cations in the Sciences. Proceedings 1975. Edited by P. Hilton. 8 pages. 1976.

26: Probability in Banach Spaces. Proceedings 1975. Edited Beck. VI, 290 pages. 1976.

527: M. Denker, Ch. Grillenberger, and K. Sigmund, Ergodic ry on Compact Spaces. IV, 360 pages. 1976.

528: J. E. Humphreys, Ordinary and Modular Representations evalley Groups. III, 127 pages. 1976.

529: J. Grandell, Doubly Stochastic Poisson Processes. X, pages. 1976.

530: S. S. Gelbart, Weil's Representation and the Spectrum e Metaplectic Group. VII, 140 pages. 1976.

531: Y.-C. Wong, The Topology of Uniform Convergence on r-Bounded Sets. VI, 163 pages. 1976.

532: Théorie Ergodique. Proceedings 1973/1974. Edité par Conze and M. S. Keane. VIII, 227 pages. 1976.

533: F. R. Cohen, T. J. Lada, and J. P. May, The Homology of ted Loop Spaces. IX, 490 pages. 1976.

534: C. Preston, Random Fields. V, 200 pages. 1976.

535: Singularités d'Applications Differentiables. Plans-sur-Bex. 5. Edité par O. Burlet et F. Ronga. V, 253 pages. 1976.

536: W. M. Schmidt, Equations over Finite Fields. An Elementary roach. IX, 267 pages. 1976.

537: Set Theory and Hierarchy Theory. Bierutowice, Poland 5. A Memorial Tribute to Andrzej Mostowski. Edited by W. Marek, Srebrny and A. Zarach. XIII, 345 pages. 1976.

538: G. Fischer, Complex Analytic Geometry. VII, 201 pages. 6.

539: A. Badrikian, J. F. C. Kingman et J. Kuelbs, Ecole d'Eté de babilités de Saint Flour V-1975. Edité par P.-L. Hennequin. IX, pages. 1976.

. 540: Categorical Topology, Proceedings 1975. Edited by E. Binz H. Herrlich. XV, 719 pages. 1976.

. 541: Measure Theory, Oberwolfach 1975. Proceedings. Edited A. Bellow and D. Kölzow. XIV, 430 pages. 1976.

. 542: D. A. Edwards and H. M. Hastings, Čech and Steenrod motopy Theories with Applications to Geometric Topology. VII, 6 pages. 1976.

l. 543: Nonlinear Operators and the Calculus of Variations, uxelles 1975. Edited by J. P. Gossez, E. J. Lami Dozo, J. Mawhin, d L. Waelbroeck, VII, 237 pages. 1976.

l. 544: Robert P. Langlands, On the Functional Equations Satisd by Eisenstein Series. VII, 337 pages. 1976.

l. 545: Noncommutative Ring Theory. Kent State 1975. Edited by H. Cozzens and F. L. Sandomierski. V, 212 pages. 1976.

ol. 546: K. Mahler, Lectures on Transcendental Numbers. Edited nd Completed by B. Diviš and W. J. Le Veque. XXI, 254 pages. 976.

ol. 547: A. Mukherjea and N. A. Tserpes, Measures on Topological emigroups: Convolution Products and Random Walks. V, 197 ages. 1976.

ol. 548: D. A. Hejhal, The Selberg Trace Formula for PSL (2, $\mathbb{R}$). olume I. VI, 516 pages. 1976.

ol. 549: Brauer Groups, Evanston 1975. Proceedings. Edited by). Zelinsky. V, 187 pages. 1976.

ol. 550: Proceedings of the Third Japan – USSR Symposium on robability Theory. Edited by G. Maruyama and J. V. Prokhorov. VI, 22 pages. 1976.

Vol. 551: Algebraic K-Theory, Evanston 1976. Proceedings. Edited by M. R. Stein. XI, 409 pages. 1976.

Vol. 552: C. G. Gibson, K. Wirthmüller, A. A. du Plessis and E. J. N. Looijenga. Topological Stability of Smooth Mappings. V, 155 pages. 1976.

Vol. 553: M. Petrich, Categories of Algebraic Systems. Vector and Projective Spaces, Semigroups, Rings and Lattices. VIII, 217 pages. 1976.

Vol. 554: J. D. H. Smith, Mal'cev Varieties. VIII, 158 pages. 1976.

Vol. 555: M. Ishida, The Genus Fields of Algebraic Number Fields. VII, 116 pages. 1976.

Vol. 556: Approximation Theory. Bonn 1976. Proceedings. Edited by R. Schaback and K. Scherer. VII, 466 pages. 1976.

Vol. 557: W. Iberkleid and T. Petrie, Smooth S^1 Manifolds. III, 163 pages. 1976.

Vol. 558: B. Weisfeiler, On Construction and Identification of Graphs. XIV, 237 pages. 1976.

Vol. 559: J.-P. Caubet, Le Mouvement Brownien Relativiste. IX, 212 pages. 1976.

Vol. 560: Combinatorial Mathematics, IV, Proceedings 1975. Edited by L. R. A. Casse and W. D. Wallis. VII, 249 pages. 1976.

Vol. 561: Function Theoretic Methods for Partial Differential Equations. Darmstadt 1976. Proceedings. Edited by V. E. Meister, N. Weck and W. L. Wendland. XVIII, 520 pages. 1976.

Vol. 562: R. W. Goodman, Nilpotent Lie Groups: Structure and Applications to Analysis. X, 210 pages. 1976.

Vol. 563: Séminaire de Théorie du Potentiel. Paris, No. 2. Proceedings 1975–1976. Edited by F. Hirsch and G. Mokobodzki. VI, 292 pages. 1976.

Vol. 564: Ordinary and Partial Differential Equations, Dundee 1976. Proceedings. Edited by W. N. Everitt and B. D. Sleeman. XVIII, 551 pages. 1976.

Vol. 565: Turbulence and Navier Stokes Equations. Proceedings 1975. Edited by R. Temam. IX, 194 pages. 1976.

Vol. 566: Empirical Distributions and Processes. Oberwolfach 1976. Proceedings. Edited by P. Gaenssler and P. Révész. VII, 146 pages. 1976.

Vol. 567: Séminaire Bourbaki vol. 1975/76. Exposés 471–488. IV, 303 pages. 1977.

Vol. 568: R. E. Gaines and J. L. Mawhin, Coincidence Degree, and Nonlinear Differential Equations. V, 262 pages. 1977.

Vol. 569: Cohomologie Etale SGA $4^{1}/_{2}$. Séminaire de Géométrie Algébrique du Bois-Marie. Edité par P. Deligne. V, 312 pages. 1977.

Vol. 570: Differential Geometrical Methods in Mathematical Physics, Bonn 1975. Proceedings. Edited by K. Bleuler and A. Reetz. VIII, 576 pages. 1977.

Vol. 571: Constructive Theory of Functions of Several Variables, Oberwolfach 1976. Proceedings. Edited by W. Schempp and K. Zeller. VI, 290 pages. 1977

Vol. 572: Sparse Matrix Techniques, Copenhagen 1976. Edited by V. A. Barker. V, 184 pages. 1977.

Vol. 573: Group Theory, Canberra 1975. Proceedings. Edited by R. A. Bryce, J. Cossey and M. F. Newman. VII, 146 pages. 1977.

Vol. 574: J. Moldestad, Computations in Higher Types. IV, 203 pages. 1977.

Vol. 575: K-Theory and Operator Algebras, Athens, Georgia 1975. Edited by B. B. Morrel and I. M. Singer. VI, 191 pages. 1977.

Vol. 576: V. S. Varadarajan, Harmonic Analysis on Real Reductive Groups. VI, 521 pages. 1977.

Vol. 577: J. P. May, E_∞ Ring Spaces and E_∞ Ring Spectra. IV, 268 pages. 1977.

Vol. 578: Séminaire Pierre Lelong (Analyse) Année 1975/76. Edité par P. Lelong. VI, 327 pages. 1977.

Vol. 579: Combinatoire et Représentation du Groupe Symétrique, Strasbourg 1976. Proceedings 1976. Edité par D. Foata. IV, 339 pages. 1977.

Vol. 580: C. Castaing and M. Valadier, Convex Analysis and Measurable Multifunctions. VIII, 278 pages. 1977.

Vol. 581: Séminaire de Probabilités XI, Université de Strasbourg. Proceedings 1975/1976. Edité par C. Dellacherie, P. A. Meyer et M. Weil. VI, 574 pages. 1977.

Vol. 582: J. M. G. Fell, Induced Representations and Banach *-Algebraic Bundles. IV, 349 pages. 1977.

Vol. 583: W. Hirsch, C. C. Pugh and M. Shub, Invariant Manifolds. IV, 149 pages. 1977.

Vol. 584: C. Brezinski, Accélération de la Convergence en Analyse Numérique. IV, 313 pages. 1977.

Vol. 585: T. A. Springer, Invariant Theory. VI, 112 pages. 1977.

Vol. 586: Séminaire d'Algèbre Paul Dubreil, Paris 1975–1976 (29ème Année). Edited by M. P. Malliavin. VI, 188 pages. 1977.

Vol. 587: Non-Commutative Harmonic Analysis. Proceedings 1976. Edited by J. Carmona and M. Vergne. IV, 240 pages. 1977.

Vol. 588: P. Molino, Théorie des G-Structures: Le Problème d'Equivalence. VI, 163 pages. 1977.

Vol. 589: Cohomologie l-adique et Fonctions L. Séminaire de Géométrie Algébrique du Bois-Marie 1965–66, SGA 5. Edité par L. Illusie. XII, 484 pages. 1977.

Vol. 590: H. Matsumoto, Analyse Harmonique dans les Systèmes de Tits Bornologiques de Type Affine. IV, 219 pages. 1977.

Vol. 591: G. A. Anderson, Surgery with Coefficients. VIII, 157 pages. 1977.

Vol. 592: D. Voigt, Induzierte Darstellungen in der Theorie der endlichen, algebraischen Gruppen. V, 413 Seiten. 1977.

Vol. 593: K. Barbey and H. König, Abstract Analytic Function Theory and Hardy Algebras. VIII, 260 pages. 1977.

Vol. 594: Singular Perturbations and Boundary Layer Theory, Lyon 1976. Edited by C. M. Brauner, B. Gay, and J. Mathieu. VIII, 539 pages. 1977.

Vol. 595: W. Hazod, Stetige Faltungshalbgruppen von Wahrscheinlichkeitsmaßen und erzeugende Distributionen. XIII, 157 Seiten. 1977.

Vol. 596: K. Deimling, Ordinary Differential Equations in Banach Spaces. VI, 137 pages. 1977.

Vol. 597: Geometry and Topology, Rio de Janeiro, July 1976. Proceedings. Edited by J. Palis and M. do Carmo. VI, 866 pages. 1977.

Vol. 598: J. Hoffmann-Jørgensen, T. M. Liggett et J. Neveu, Ecole d'Eté de Probabilités de Saint-Flour VI – 1976. Edité par P.-L. Hennequin. XII, 447 pages. 1977.

Vol. 599: Complex Analysis, Kentucky 1976. Proceedings. Edited by J. D. Buckholtz and T. J. Suffridge. X, 159 pages. 1977.

Vol. 600: W. Stoll, Value Distribution on Parabolic Spaces. VIII, 216 pages. 1977.

Vol. 601: Modular Functions of one Variable V, Bonn 1976. Proceedings. Edited by J.-P. Serre and D. B. Zagier. VI, 294 pages. 1977.

Vol. 602: J. P. Brezin, Harmonic Analysis on Compact Solvmanifolds. VIII, 179 pages. 1977.

Vol. 603: B. Moishezon, Complex Surfaces and Connected Sums of Complex Projective Planes. IV, 234 pages. 1977.

Vol. 604: Banach Spaces of Analytic Functions, Kent, Ohio 1976. Proceedings. Edited by J. Baker, C. Cleaver and Joseph Diestel. VI, 141 pages. 1977.

Vol. 605: Sario et al., Classification Theory of Riemannian Manifolds. XX, 498 pages. 1977.

Vol. 606: Mathematical Aspects of Finite Element Methods. Proceedings 1975. Edited by I. Galligani and E. Magenes. VI, 362 pages. 1977.

Vol. 607: M. Métivier, Reelle und Vektorwertige Quasimartingale und die Theorie der Stochastischen Integration. X, 310 Seiten. 1977.

Vol. 608: Bigard et al., Groupes et Anneaux Réticulés. XIV, 334 pages. 1977.

Vol. 609: General Topology and Its Relations to Modern Analy and Algebra IV. Proceedings 1976. Edited by J. Novák. XVIII, 2 pages. 1977.

Vol. 610: G. Jensen, Higher Order Contact of Submanifolds Homogeneous Spaces. XII, 154 pages. 1977.

Vol. 611: M. Makkai and G. E. Reyes, First Order Categorical Log VIII, 301 pages. 1977.

Vol. 612: E. M. Kleinberg, Infinitary Combinatorics and the Axiom Determinateness. VIII, 150 pages. 1977.

Vol. 613: E. Behrends et al., L^p-Structure in Real Banach Spac X, 108 pages. 1977.

Vol. 614: H. Yanagihara, Theory of Hopf Algebras Attached to Gro Schemes. VIII, 308 pages. 1977.

Vol. 615: Turbulence Seminar, Proceedings 1976/77. Edited P. Bernard and T. Ratiu. VI, 155 pages. 1977.

Vol. 616: Abelian Group Theory, 2nd New Mexico State Universi Conference, 1976. Proceedings. Edited by D. Arnold, R. Hunter ar E. Walker. X, 423 pages. 1977.

Vol. 617: K. J. Devlin, The Axiom of Constructibility: A Guide for th Mathematician. VIII, 96 pages. 1977.

Vol. 618: I. I. Hirschman, Jr. and D. E. Hughes, Extreme Eigen Value of Toeplitz Operators. VI, 145 pages. 1977.

Vol. 619: Set Theory and Hierarchy Theory V, Bierutowice 197 Edited by A. Lachlan, M. Srebrny, and A. Zarach. VIII, 358 page 1977.

Vol. 620: H. Popp, Moduli Theory and Classification Theory c Algebraic Varieties. VIII, 189 pages. 1977.

Vol. 621: Kauffman et al., The Deficiency Index Problem. VI, 112 pages 1977.

Vol. 622: Combinatorial Mathematics V, Melbourne 1976. Proceedings. Edited by C. Little. VIII, 213 pages. 1977.

Vol. 623: I. Erdelyi and R. Lange, Spectral Decompositions or Banach Spaces. VIII, 122 pages. 1977.

Vol. 624: Y. Guivarc'h et al., Marches Aléatoires sur les Groupes de Lie. VIII, 292 pages. 1977.

Vol. 625: J. P. Alexander et al., Odd Order Group Actions and Witt Classification of Innerproducts. IV, 202 pages. 1977.

Vol. 626: Number Theory Day, New York 1976. Proceedings. Edited by M. B. Nathanson. VI, 241 pages. 1977.

Vol. 627: Modular Functions of One Variable VI, Bonn 1976. Proceedings. Edited by J.-P. Serre and D. B. Zagier. VI, 339 pages. 1977.

Vol. 628: H. J. Baues, Obstruction Theory on the Homotopy Classification of Maps. XII, 387 pages. 1977.

Vol. 629: W. A. Coppel, Dichotomies in Stability Theory. VI, 98 pages. 1978.

Vol. 630: Numerical Analysis, Proceedings, Biennial Conference, Dundee 1977. Edited by G. A. Watson. XII, 199 pages. 1978.

Vol. 631: Numerical Treatment of Differential Equations. Proceedings 1976. Edited by R. Bulirsch, R. D. Grigorieff, and J. Schröder. X, 219 pages. 1978.

Vol. 632: J.-F. Boutot, Schéma de Picard Local. X, 165 pages. 1978.

Vol. 633: N. R. Coleff and M. E. Herrera, Les Courants Résiduels Associés à une Forme Méromorphe. X, 211 pages. 1978.

Vol. 634: H. Kurke et al., Die Approximationseigenschaft lokaler Ringe. IV, 204 Seiten. 1978.

Vol. 635: T. Y. Lam, Serre's Conjecture. XVI, 227 pages. 1978.

Vol. 636: Journées de Statistique des Processus Stochastiques, Grenoble 1977, Proceedings. Edité par Didier Dacunha-Castelle et Bernard Van Cutsem. VII, 202 pages. 1978.

Vol. 637: W. B. Jurkat, Meromorphe Differentialgleichungen. VII, 194 Seiten. 1978.

Vol. 638: P. Shanahan, The Atiyah-Singer Index Theorem, An Introduction. V, 224 pages. 1978.

Vol. 639: N. Adasch et al., Topological Vector Spaces. V, 125 pages. 1978.

ʒ40: J. L. Dupont, Curvature and Characteristic Classes. X, ›ages. 1978.

641: Séminaire d'Algèbre Paul Dubreil, Proceedings Paris -1977. Edité par M. P. Malliavin. IV, 367 pages. 1978.

ʒ42: Theory and Applications of Graphs, Proceedings, Michigan . Edited by Y. Alavi and D. R. Lick. XIV, 635 pages. 1978.

643: M. Davis, Multiaxial Actions on Manifolds. VI, 141 pages. .

644: Vector Space Measures and Applications I, Proceedings . Edited by R. M. Aron and S. Dineen. VIII, 451 pages. 1978.

645: Vector Space Measures and Applications II, Proceedings . Edited by R. M. Aron and S. Dineen. VIII, 218 pages. 1978.

646: O. Tammi, Extremum Problems for Bounded Univalent ;tions. VIII, 313 pages. 1978.

647: L. J. Ratliff, Jr., Chain Conjectures in Ring Theory. VIII, 133 ɜs. 1978.

648: Nonlinear Partial Differential Equations and Applications, ;eedings, Indiana 1976-1977. Edited by J. M. Chadam. VI, 206 ɜs. 1978.

649: Séminaire de Probabilités XII, Proceedings, Strasbourg, ɜ-1977. Edité par C. Dellacherie, P. A. Meyer et M. Weil. VIII, pages. 1978.

650: C*-Algebras and Applications to Physics. Proceedings 7. Edited by R. V. Kadison. V, 192 pages. 1978.

651: P. W. Michor, Functors and Categories of Banach Spaces. 9 pages. 1978.

652: Differential Topology, Foliations and Gelfand-Fuks-Coho- ogy, Proceedings 1976. Edited by P. A. Schweitzer. XIV, 252 es. 1978.

653: Locally Interacting Systems and Their Application in ogy. Proceedings, 1976. Edited by R. L. Dobrushin, V. I. Kryukov A. L. Toom. XI, 202 pages. 1978.

654: J. P. Buhler, Icosahedral Golois Representations. III, 143 es. 1978.

. 655: R. Baeza, Quadratic Forms Over Semilocal Rings. VI, pages. 1978.

. 656: Probability Theory on Vector Spaces. Proceedings, 1977. ted by A. Weron. VIII, 274 pages. 1978.

. 657: Geometric Applications of Homotopy Theory I, Proceedings 77. Edited by M. G. Barratt and M. E. Mahowald. VIII, 459 pages. 78.

. 658: Geometric Applications of Homotopy Theory II, Proceedings 77. Edited by M. G. Barratt and M. E. Mahowald. VIII, 487 pages. 78.

l. 659: Bruckner, Differentiation of Real Functions. X, 247 pages. 78.

l. 660: Equations aux Dérivée Partielles. Proceedings, 1977. Edité r Pham The Lai. VI, 216 pages. 1978.

l. 661: P. T. Johnstone, R. Paré, R. D. Rosebrugh, D. Schumacher, J. Wood, and G. C. Wraith, Indexed Categories and Their Applica- ns. VII, 260 pages. 1978.

ɔl. 662: Akin, The Metric Theory of Banach Manifolds. XIX, 306 ages. 1978.

ol. 663: J. F. Berglund, H. D. Junghenn, P. Milnes, Compact Right opological Semigroups and Generalizations of Almost Periodicity. ., 243 pages. 1978.

ol. 664: Algebraic and Geometric Topology, Proceedings, 1977. :dited by K. C. Millett. XI, 240 pages. 1978.

Vol. 665: Journées d'Analyse Non Linéaire. Proceedings, 1977. :dité par P. Bénilan et J. Robert. VIII, 256 pages. 1978.

Vol. 666: B. Beauzamy, Espaces d'Interpolation Réels: Topologie et Géometrie. X, 104 pages. 1978.

Vol. 667: J. Gilewicz, Approximants de Padé. XIV, 511 pages. 1978.

Vol. 668: The Structure of Attractors in Dynamical Systems. Proceedings, 1977. Edited by J. C. Martin, N. G. Markley and W. Perrizo. VI, 264 pages. 1978.

Vol. 669: Higher Set Theory. Proceedings, 1977. Edited by G. H. Müller and D. S. Scott. XII, 476 pages. 1978.

Vol. 670: Fonctions de Plusieurs Variables Complexes III, Proceedings, 1977. Edité par F. Norguet. XII, 394 pages. 1978.

Vol. 671: R. T. Smythe and J. C. Wierman, First-Passage Perculation on the Square Lattice. VIII, 196 pages. 1978.

Vol. 672: R. L. Taylor, Stochastic Convergence of Weighted Sums of Random Elements in Linear Spaces. VII, 216 pages. 1978.

Vol. 673: Algebraic Topology, Proceedings 1977. Edited by P. Hoffman, R. Piccinini and D. Sjerve. VI, 278 pages. 1978.

Vol. 674: Z. Fiedorowicz and S. Priddy, Homology of Classical Groups Over Finite Fields and Their Associated Infinite Loop Spaces. VI, 434 pages. 1978.

Vol. 675: J. Galambos and S. Kotz, Characterizations of Probability Distributions. VIII, 169 pages. 1978.

Vol. 676: Differential Geometrical Methods in Mathematical Physics II, Proceedings, 1977. Edited by K. Bleuler, H. R. Petry and A. Reetz. VI, 626 pages. 1978.